Aruba Certified Associate - Campus Access
OFFICIAL CERTIFICATION STUDY GUIDE
(EXAM HPE6-A85)

First Edition

Richard Deal, Peter Debruyne, and Jakub Stelmaszczyk

HPE Press
660 4th Street, #802
San Francisco, CA 94107

Aruba Certified Associate - Campus Access
Official Certification Study Guide (Exam HPE6-A85)
Richard Deal, Peter Debruyne, and Jakub Stelmaszczyk

Published by:

Hewlett Packard Enterprise Press
660 4th Street, #802
San Francisco, CA 94107

ISBN: 978-1-960058-05-8

Printed in Mexico

WARNING AND DISCLAIMER

This book provides information about the topics covered in the Aruba Certified Associate (ACA) - Campus Access certification exam (HPE6-A85). Every effort has been made to make this book as complete and as accurate as possible, but no warranty or fitness is implied.

The information is provided on an "as is" basis. The author, and Hewlett Packard Enterprise Press, shall have neither liability nor responsibility to any person or entity with respect to any loss or damages arising from the information contained in this book or from the use of the discs or programs that may accompany it.

The opinions expressed in this book belong to the author and are not necessarily those of Hewlett Packard Enterprise Press.

FEEDBACK INFORMATION

At HPE Press, our goal is to create in-depth reference books of the best quality and value. Each book is crafted with care and precision, undergoing rigorous development that involves the expertise of members from the professional technical community.

Readers' feedback is a continuation of the process. If you have any comments regarding how we could improve the quality of this book, or otherwise alter it to better suit your needs, you can contact us through email at hpepress@epac.com. Please make sure to include the book title and ISBN in your message.

We appreciate your feedback.

Publisher: Hewlett Packard Enterprise Press

HPE Press Program Manager: Michael Bishop

About the Authors

Richard Deal is an independent consultant who designs wireless and wired networks and provides network management services for small companies. He has created and developed many networking courses and instructs IT professionals on the best ways to design and implement Aruba's wired and wireless network solutions.

Peter Debruyne was co-author for the ACA-Campus Access course. He is an Aruba Certified Instructor, Aruba Certified Edge Expert: ACCX, ACMX, ACSX, and ACDX certified.

Jakub Stelmaszczyk is an Aruba Certified Instructor, Aruba Certified Edge Expert: ACCX, ACMX, and ACSX certified.

Introduction

This book teaches you the fundamental knowledge and practical skills required to configure and manage modern, open standards-based networks for small-to-medium enterprises. You will learn how to implement and validate a wide variety of essential networking solutions using Aruba's wired, wireless, security and management technologies. To help you learn as you go, Learning Check questions are included in each chapter. In addition, Chapter 15 has a 30-question sample test that simulates the kinds and styles of questions that you can expect to meet in the certification exam (HPE6-A85). Answers to the Learning Checks and sample test questions can be found in Appendix A.

Certification and Learning

Hewlett Packard Enterprise (HPE) Certification and Learning provides end-to-end continuous learning programs and professional certifications that can help you open doors and accelerate your career.

We provide:

- **Professional sales and technical training and certifications** to give you the critical skills needed to design, manage and implement the most sought-after IT disciplines; and
- **Continuous learning activities and job-role based learning plans** to help you keep pace with the demands of the dynamic, fast paced IT industry
- **Advanced training** to help you navigate and seize opportunities within the top IT transformation areas that enable business advantage today.

As a Certification and Learning certified member, your skills, knowledge, and real-world experience are recognized and valued in the marketplace. To continue your professional and career growth, you have access to our large HPE community of world-class IT professionals, trend-makers and decision-makers. Share ideas, best practices, business insights, and challenges as you gain professional connections globally.

To learn more about HPE Certification and Learning certifications and continuous learning programs, please visit: http://certification-learning.hpe.com.

Audience

This book is designed for individuals who will implement wireless, wired, and security solutions for small campus networking deployments. The audience includes network technicians, network engineers, and network architects.

Assumed Knowledge

It is recommended that candidates have foundational networking experience or attend Aruba's Essentials eLearning series to glean knowledge on Aruba's Campus Access design solution. Also, the course is based on Aruba's *ESP Campus Design* Validated Solution Guide (VSG); it is highly recommended that you read that document, which can be found here:
https://www.arubanetworks.com/techdocs/VSG/docs/010-campus-design/esp-campus-design-000/.

Minimum Qualifications

The ideal candidate has 1+ years of experience with networking, vendor agnostic understanding of basic network protocols. Under the direction of a Professional or Expert, they are able to apply the configuration and verify the status of a campus network.

Relevant Certifications

After you pass this exam, your achievement may be applicable toward more than one certification. To determine which certifications can be credited with this achievement, log in to The Learning Center and view the certifications listed on the exam's More Details tab. You might be on your way to achieving additional certifications.

Preparing for Exam HPE6-A85

This self-study guide does not guarantee that you will have all the knowledge you need to pass the exam. It is expected that you will also draw on real-world experience and would benefit from completing the hands-on lab activities provided in the instructor-led training. However, the labs taught in the official course are provided for you here in this book, assuming you have the correct hardware and software to perform the labs. The book covers the same material as the instructor-led and eLearning course. Therefore, to pass the certification exam, you should be intimately familiar with the material presented here. Aruba recommends reading the book and doing the lab exercises three times before taking the exam, as well as gaining practical experience in the field on the topics covered in this book. Please note that successful completion of this book, corresponding course, or additional study material does not ensure you will pass the HPE6-A85 exam.

Recommended HPE Training

Recommended training to prepare for each exam is accessible from the exam's page in The Learning Center. See the exam attachment, "Supporting courses," to view and register for the courses. For the Aruba Campus Access Fundamentals course, you can find more information here: https://certification-learning.hpe.com/tr/datacard/Course/00011199937.

Obtain Hands-on Experience

You are not required to take the recommended, supported courses, and completion of training does not guarantee that you will pass the exams. HPE strongly recommends a combination of training, thorough review of courseware and additional study references, and sufficient on-the-job experience prior to taking an exam.

Exam Registration

To register for an exam, go to: https://certification-learning.hpe.com/tr/learn_more_about_exams.html

CONTENTS

1 Networking Fundamentals

OBJECTIVES

After completing this chapter, you should be able to:

- ✓ Define networking, LAN, WAN, and their components.
- ✓ Explain the OSI model and encapsulation.
- ✓ Discuss different types of physical media.
- ✓ Compare unicast, multicast, and broadcast.
- ✓ Explain the TCP/IP stack.
- ✓ Discuss different types of networking devices.

Overview

Basic network concepts provide a foundation for deeper understanding. After completing this chapter, you will be able to define networking, LAN, WAN, and networking components. You will be able to explain the OSI model and encapsulation, and discuss different types of physical media. You will be able to compare unicast, multicast, and broadcast, explain the TCP/IP protocol stack, and discuss different types of networking devices.

Basic network concepts

This chapter begins with an introduction to basic networking concepts.

What is a computing network?

Computing communication is the process of sending or receiving data from other entities. Computing network engineering makes this happen. It is all resources and infrastructure – both hardware and software – working to create bridges, paths, and highways used to transport data between devices.

Taking pictures from your smartphone is great, but it's worthless if you cannot share them. Moving data out of your smartphone is an example of computing communication (see Figure 1-1).

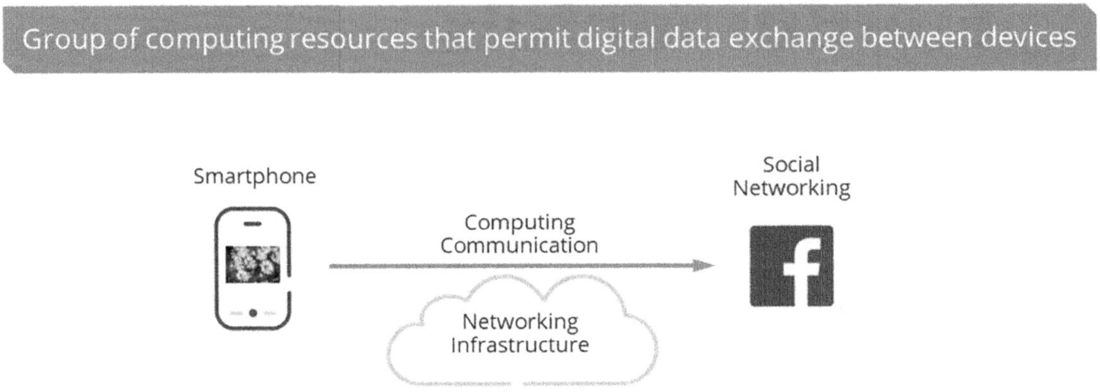

Figure 1-1: What is a computing network?

Network classifications

Based on the geographical coverage, a computing network can be categorized as a local area network (LAN) or a wide area network (WAN) (see Figure 1-2).

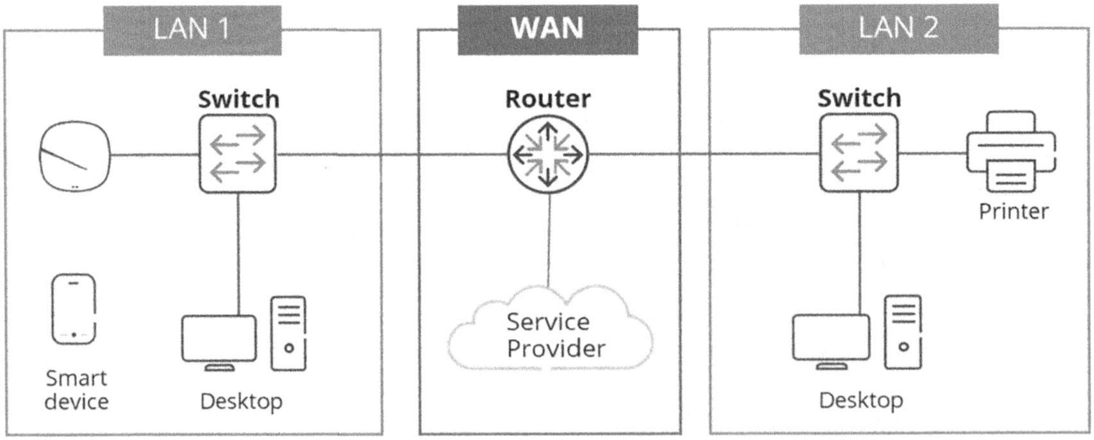

Figure 1-2: Network classifications

A LAN is a group of computer devices that are geographically in the same place. For example, a group of devices within a building can be considered a LAN.

LANs are used in several settings:

- Small office/home office (SOHO)
- Office LANs
- Building LANs
- Campus LANs

A WAN on the other hand is a group of computer resources that can communicate over large geographical distances – typically a few kilometers or miles, and perhaps thousands of miles, such as the Internet. The Internet is considered a WAN since it permits communication across countries and continents.

Typically, WANs are deployed by Internet service providers (ISPs) since those companies have the economic resources to interconnect long distances. Examples of WAN technologies include the following:

- Internet
- Multi-Protocol Label Switching (MPLS)
- Asynchronous Transfer Mode (ATM) (obsolete)
- Frame relay (largely obsolete)
- Dark fiber

What is a protocol?

Communication is the main purpose of a computing network, and this communication is enabled using protocols. A protocol is a group of rules or instructions that computer devices follow in order to properly establish and maintain communications (see Figure 1-3).

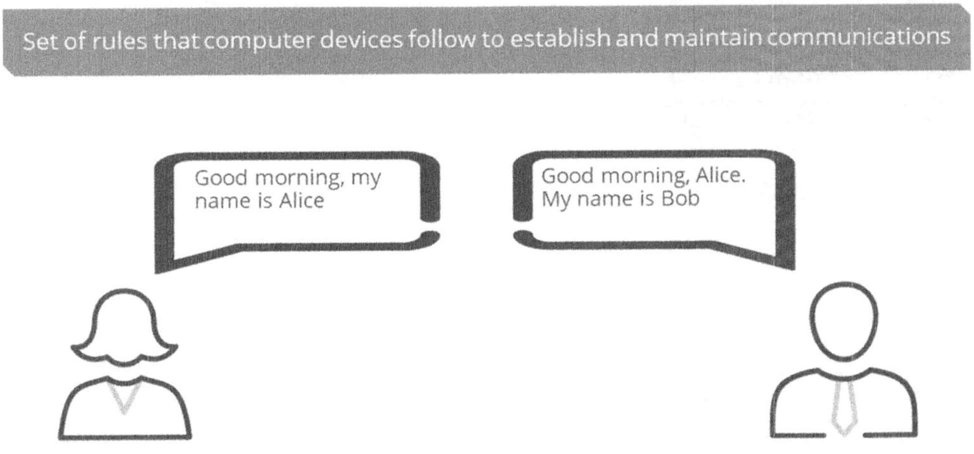

Figure 1-3: What is a protocol?

Consider this analogy:

- Alice meets Bob for the first time, and starts the conversation by saying "Good morning, my name is Alice".
- Bob replies, "Good Morning Alice, my name is Bob."

This brief conversation is actually a procedure. Notice that Alice starts the communication with a greeting, and then she identifies herself. Bob's reply is also a procedure. He acknowledges Alice, and then he identifies himself by name. The implicit rules in this conversation help to establish and maintain a conversation. Likewise, computing devices exchange messages in a specific order, following specific rules.

OSI model and encapsulation

Let's move on to the OSI reference model and encapsulation.

OSI reference model

In the mid-1980s during the fast evolution of computing, every vendor wanted to implement their own, proprietary communication protocol. These proprietary protocols created interoperability challenges. The International Organization for Standardization (ISO) solved the problem by presenting a standard communication model for computing devices: The Open Systems Interconnection (OSI) model.

The OSI model organizes computing communication in 7 layers (see Figure 1-4). Each layer defines a phase of message processing.

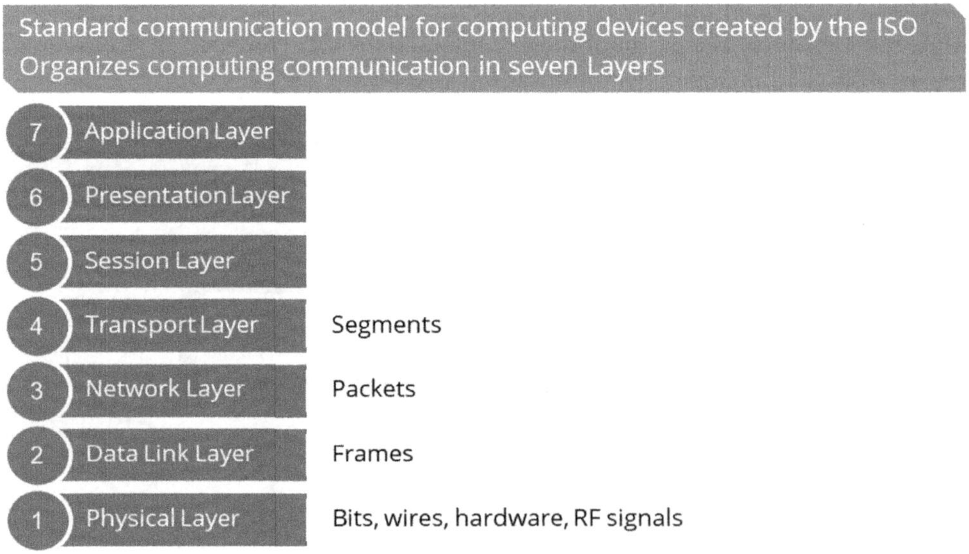

Figure 1-4: OSI reference model

The OSI layers are shown in the figure, and described below:

- **Layer 7**: Application layer
- **Layer 6**: Presentation layer
- **Layer 5**: Session layer
- **Layer 4**: Transport layer, which organizes data into segments
- **Layer 3**: Network layer, which organizes data into packets
- **Layer 2**: Data link layer, which organizes data into frames
- **Layer 1**: Physical layer, which organizes the data into bits, and transmits those bits using physical hardware over wires, fiber optic cable, or RF signals

Layer 1: Physical layer

The physical layer dictates the physical aspects of how signals are transmitted and received across some media (see Figure 1-5). Computing devices convert logical data bits into the correct physical principle depending on the media in use; this process is known as modulation. The inverse process, converting signals into logical data bits, is known as demodulation.

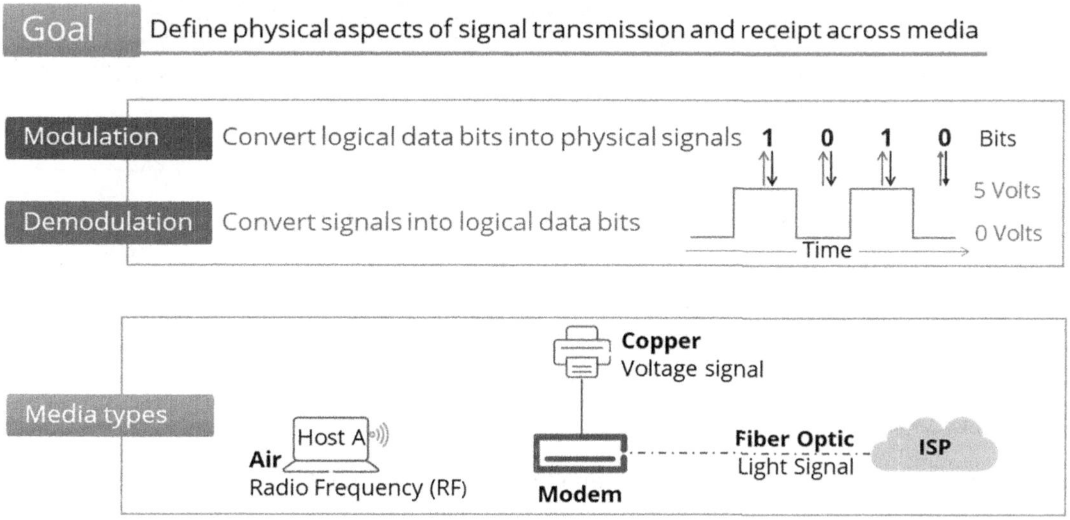

Figure 1-5: Layer 1—Physical layer

This layer also defines the material characteristics and the components to use to achieve the correct transmission of the messages. For example, consider the modem that is used in a home setting; your tablet connects to the modem using radio frequency (RF) signals that travel across the air. This modem connects to the ISP network using fiber optic cables. Thus, the router converts data into optical signals. Also, the printer connects directly to the modem via a twisted pair copper patch cable. This means that print jobs are converted into electrical voltage signals before being sent to the printer.

Layer 2: Data link layer

The data link provides three main functions (see Figure 1-6):

- Media access control
- Link layer addressing
- Error detection

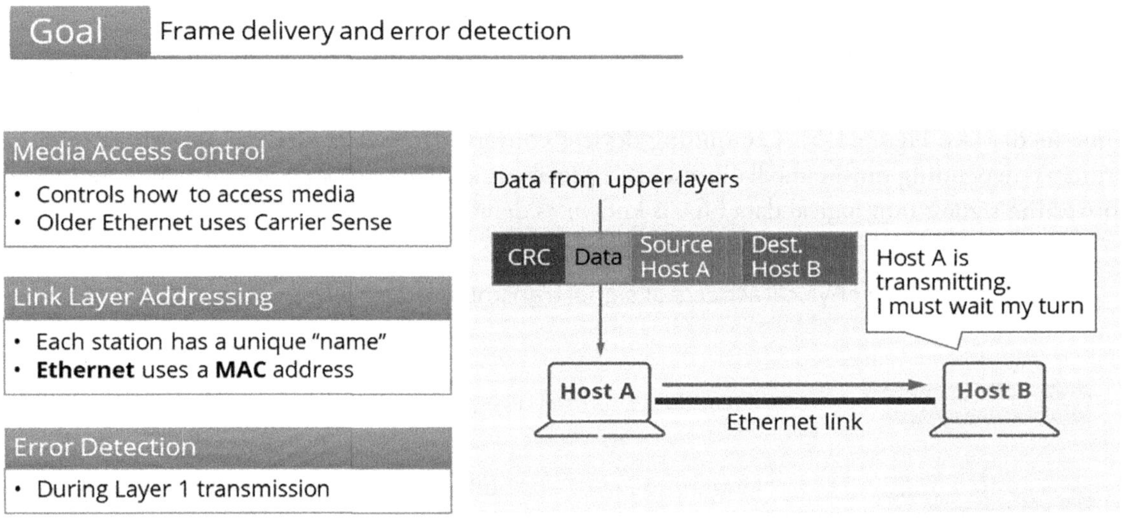

Figure 1-6: Layer 2—Data link layer

Media access control

In polite human conversation, only one person talks at a time. While Bob talks, Alice politely listens. She detects when Bob stops talking and knows that it is her turn to speak. Similarly, for most media, only one device may successfully transmit at a time, so access to the media must be controlled. Likewise, many media access control techniques leverage a *carrier sense* mechanism: before a station may transmit, it must sense the state of the carrier or media. If a transmission signal is detected, the station must wait to transmit, until the currently transmitting station is complete.

Link layer addressing

In a crowded room, Bob may call out, “Hey, Alice can I buy you an ice cream?” Everyone hears these sound waves, which travel over a shared media – the air in the room. However, only Alice responds – she was identified as the intended recipient. Likewise, each station on a LAN has a unique “name”. Instead of an alphanumeric name like Alice or Bob, each station is identified by a 6-byte hexadecimal number called a MAC address. All stations on a shared media receive the message, but only the device identified as the intended recipient processes and responds. Like humans, the other stations realize, “This message is not for me” and simply ignore the message. This information is added to the data from the upper layers as so-called *header information*, about which you will soon learn.

Error detection

For the receiver side, layer 2 helps to detect errors that could occur during layer 1 transmission. This avoids unnecessary processing of corrupted or incomplete messages. This is accomplished by adding a checksum of the data to the *trailer*. This is also called a cyclic redundancy check (CRC).

Layer 3: Network layer

The main goal of the network layer is to establish device communications across multiple LANs or WANs, using the best available path (see Figure 1-7). This is achieved using two fundamental techniques:

- **Logical addressing**: A unique layer 3 identifier for the source and destination is maintained across the path.
- **Path discovery and selection**: The network layer runs algorithms and protocols to find all possible paths, and then choose the best path. Later in this book, you will learn more about this, and things like the Routing Information Protocol (RIP) and Open Shortest Path First (OSPF).

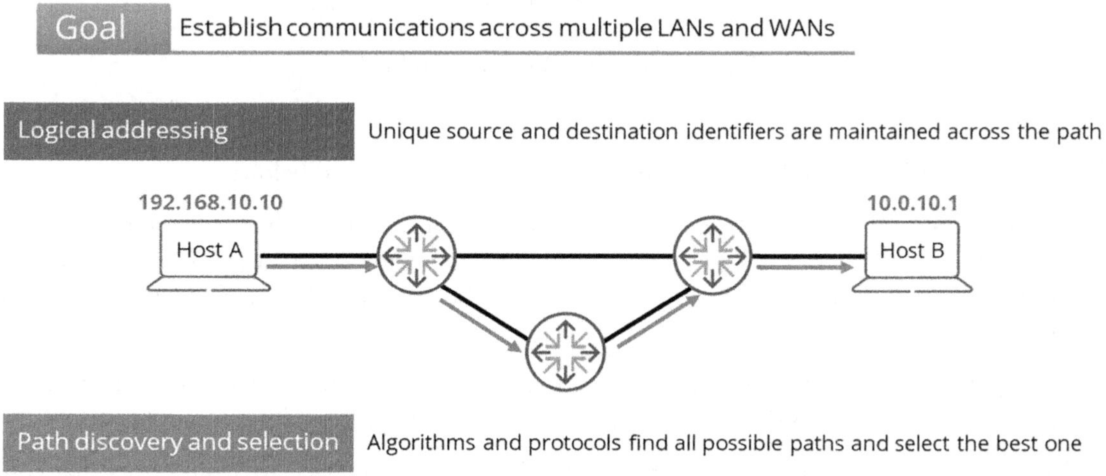

Figure 1-7: Layer 3—Network layer

The communication between two computing devices can take a specific path but not necessary the same one will be used in the future. Protocols and algorithms used in this layer could update the path any time depending on multiple factors. You will learn more about it in this book.

Layer 4: Transport layer

The transport layer controls the reliability of a given link through segmentation, de-segmentation, and error control (see Figure 1-8). In this layer some protocols, like the Transmission Control Protocol (TCP), are connection-oriented. This means that the transport layer can keep track of the messages and retransmit those that fail. Other protocols, like the User Datagram Protocol (UDP), are stateless or connectionless. This means the transport layer does not keep track of the messages. The advantage of this is that processing these connections is relatively fast and easy to compute.

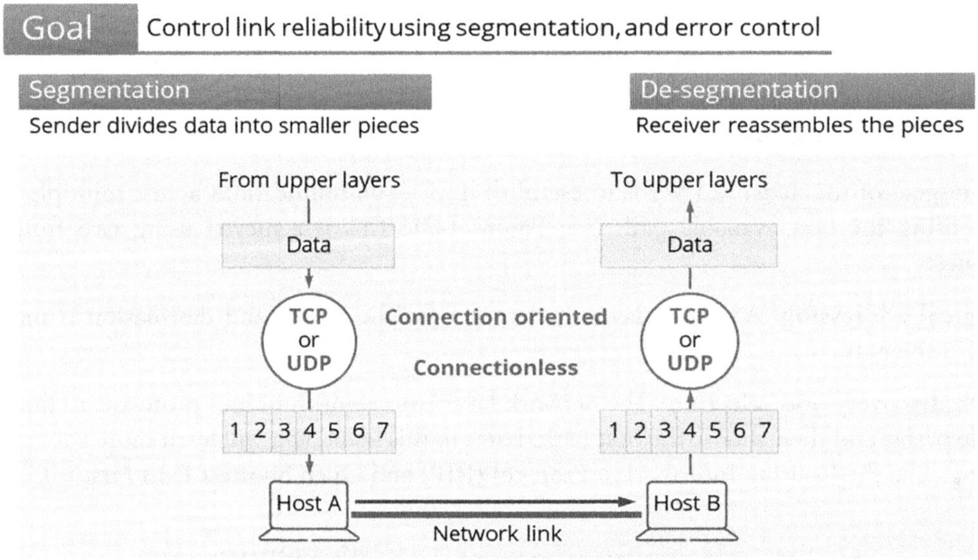

Figure 1-8: Layer 4—Transport layer

There are three main responsibilities of the transport layer, as described below:

- **Segmentation**. The sender's TCP or UDP process accepts files from the application and divides them into smaller pieces (typically 1,500 bytes) called segments. Each piece is passed down to the lower layers and transmitted individually – over an Ethernet link in the example shown in the figure.
- **De-segmentation**. The receiver accepts each segment, puts them back in the correct order if need be, and reassembles the information. This then can be processed by the application.
- **Error control**. Refers to the verification of the information received to avoid errors that could occur at the lower layers (1-3).

Note

Error detection is a process that happens in different layers: 2, 4, and sometimes in 7 of the OSI model.

Layer 5: Session layer

Layer 5 is responsible for setup, maintenance, and tear down of sessions between two computing devices. A session is a particular conversation between two computer devices (see Figure 1-9).

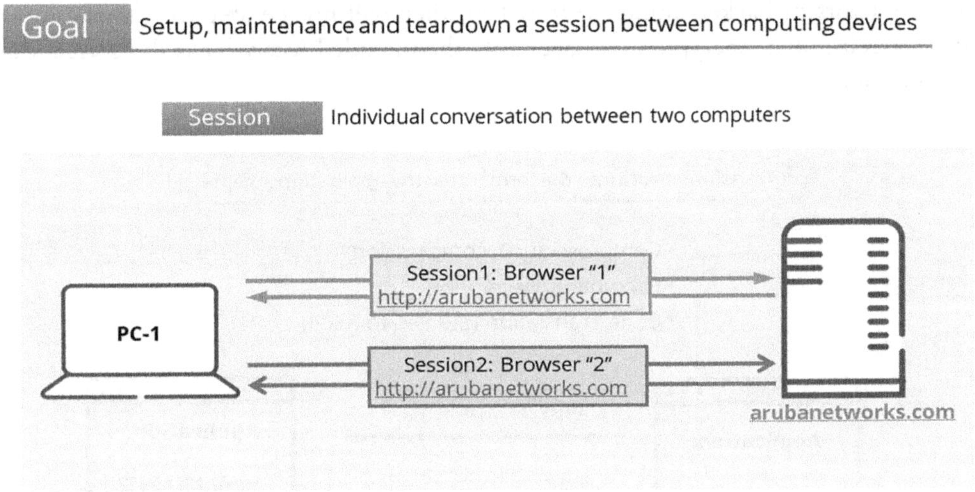

Figure 1-9: Layer 5—Session layer

Suppose that some user opens a browser and connects to a web page, like http://arubanetworks.com. A session is created. Then the same user opens a different browser to the same destination, since the application is different, a new conversion or session is created. Thus, two separate sessions are maintained.

Then the user might establish a connection to a different host for purposes of file transfer, and seven more sessions to remotely configure seven Aruba switches. A typical computer could generate thousands of sessions.

Layer 6: Presentation Layer

The presentation layer transforms data into the formats that the application accepts. Typical processes include:

- Compression/decompression
- Encryption/decryption
- Code translation (EBCDIC to ASCII)

For example, Figure 1-10 shows how an application passes the clear-text message "Hello" to the presentation layer process, which encrypts this message before transmission. This provides confidentiality. If any bad actors or hackers intercept this data, they will not be able to read the message. Of course, upon receipt of an encrypted message, only the intended receiver has the correct digital key to decrypt the data.

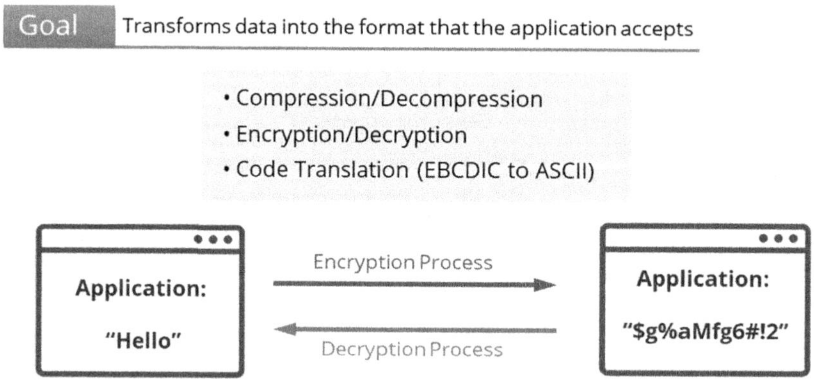

Figure 1-10: Layer 6—Presentation layer

Layer 7: Application layer

The application layer is the closest to the end user (see Figure 1-11), which means that both the OSI application layer and the user interact directly with the software application.

Goal End user interacts with this layer via an application

Functions
- Identify communication partners
- Provide network resources to user applications

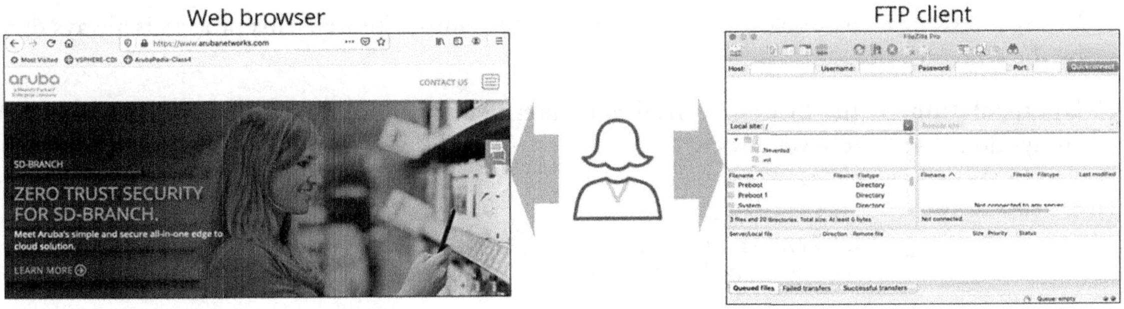

Figure 1-11: Layer 6—Application layer

Application layer functions include:

- Identifying communication partners. The application layer determines the identity and availability of communication partners for an application with data to transmit.
- Provide network resources. This layer provides network services to user applications, such as file transfer, email, video conferencing, and many others.

Some examples of application layer include:

- Hypertext Transfer Protocol (HTTP), which relies on TCP for transport
- Hypertext Transfer Protocol Secure (HTTPS), which relies on TLS for the presentation and TCP for transport
- File Transfer Protocol (FTP), which also relies on TCP
- Domain Name System (DNS), which typically uses UDP, but sometimes also uses TCP
- Trivial File Transfer Protocol (TFTP), which uses UDP as its transport mechanism

Layer headers and encapsulation

In the OSI model, each layer has a specific responsibility during network communications. In the computing world, devices use headers to establish a communication exchange and to control information on a particular layer, or the layers above it.

> **Note**
> A header that is generated on a specific layer by the sender can only be read at the same level on the receiver side.

See Figure 1-12 for an introduction to the encapsulation and decapsulation process.

- **Encapsulation.** Is the process where each OSI Layer adds a header. This process is always done by the sender device.
- **Decapsulation.** Is the process to read and interpret the header information. This process is always done by the receiver device.

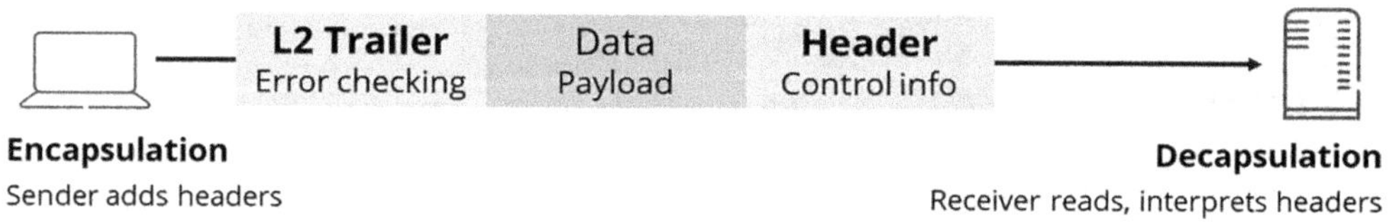

Figure 1-12: Layer headers and encapsulation

Protocol data units (PDUs)

The OSI model introduces the concept of a protocol data unit (PDU). This is simply a structure that considers the header and payload or data for each layer. Figure 1-13 summarizes the PDUs from Layer 7 to Layer 2.

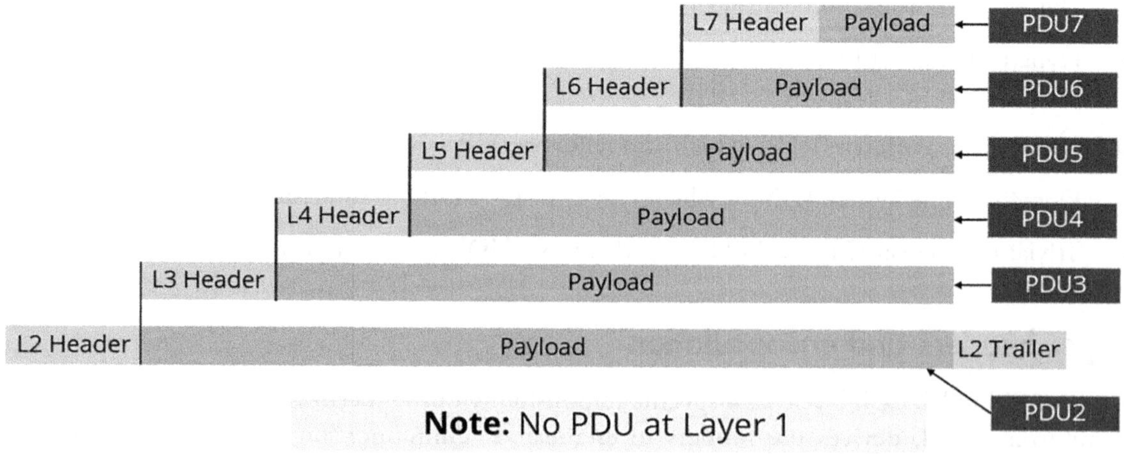

Figure 1-13: Protocol Data Units (PDUs)

There are three key terms related to PDUs that you should know:

- **Segment**. Refers to the encapsulation that is done in layer 4. A segment is equivalent to PDU4. So, you might hear networking people speak about a TCP segment or a UDP segment.
- **Packet**. Refers to the encapsulation that is done in layer 3. A packet is equivalent to PDU3. You might talk about layer 3 IP packets.
- **Frame**. Refers to the encapsulation that is done in layer 2. A frame is equivalent to PDU2. You might talk about Ethernet frames or Wi-Fi frames.

You might notice that PDU2 – a layer 2 frame – not only includes a header but also a trailer that is appended after data, as shown in Figure 1-13.

The trailer is typically used to detect errors during the transmission of the message. You recently learned about this during the discussion of layer 2 of the OSI model. Layer 2 protocols like Ethernet and Wi-Fi add a trailer, often labeled CRC or perhaps as a *frame check sequence* (FCS).

Also, notice that there is a maximum amount of data that can be transmitted over certain media – the so-called maximum transmission unit (MTU) size.

Maximum transmission unit (MTU)

The MTU is the maximum transmission size an interface can handle (see Figure 1-14), not including the Ethernet header of 18 or 20 bytes. It could be different for different interfaces, and for different media types. Upper layer protocols (ULP) break up data into appropriate size to fit within the interface maximum size.

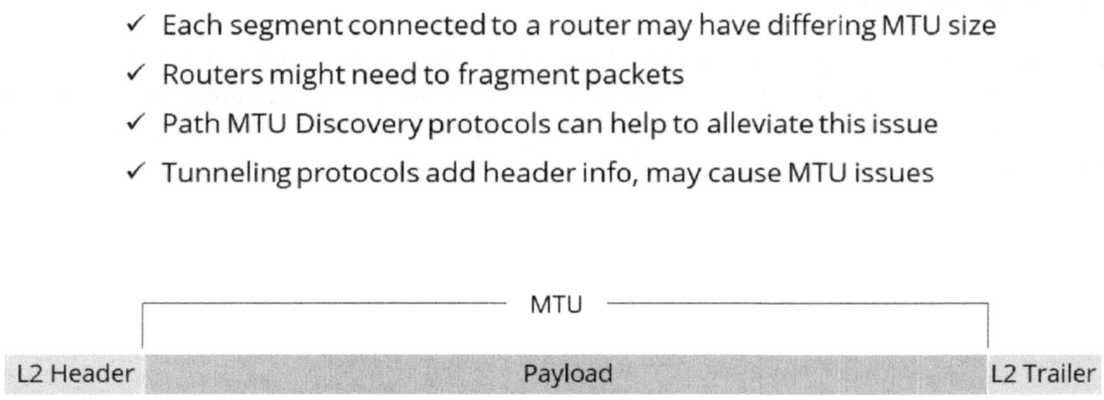

Figure 1-14: Maximum Transmission Unit (MTU)

Ethernet networks have classically used an MTU of 1,500 bytes. However, as data rates increased, this small MTU size limited performance. Higher speed Ethernet interfaces may use MTUs much larger—4 KB, 8 KB, or larger–and are referred to as *jumbo frames*.

You just learned that all the ULP headers are considered part of the layer 2 payload. Thus, this detracts from actual raw data that can be sent.

Also, a router might connect several networks with different MTU sizes. When a router receives a frame from a network that uses 8 KB MTU size, it may have to break it up into smaller, 1,500-byte chunks. However, if the IP header sets the *Do Not Fragment* (DF) flag, the router must discard the packet and send ICMP message to the originator of the packet.

The need for routers to handle differing MTU sizes has been recognized as a performance inhibitor, and so it might be best to ensure that MTUs are consistent throughout most of your network. Newer *Path MTU Discovery* protocols might help you by auto-negotiating an end-to-end MTU. This alleviates routers of the need to break up packets, so they can focus on their main job: moving packets along the best path toward their destination.

Some people run into MTU issues when doing tunneling of packets, especially when using IPsec, PPPoE, GRE, and etcetera. This is because tunneling protocols add additional header overhead, thus increasing the MTU size. This may happen without knowledge of the end systems sending packets. Packets get dropped, and you must be aware of this possibility and look for dropped packets as a part of your troubleshooting process.

Physical media

Airwaves for Wi-Fi and copper or fiber optics for Ethernet: let's talk about physical media in this section.

Wired connection using copper

The most popular cable type is unshielded twisted pair (UTP) (see Figure 1-15). It is a cheap solution to interconnect devices in a LAN. The main consideration is the cable's maximum length, 300 feet or 100 meters. This connection type uses electrical signals to communicate information with data rates up to 40 gigabit per second.

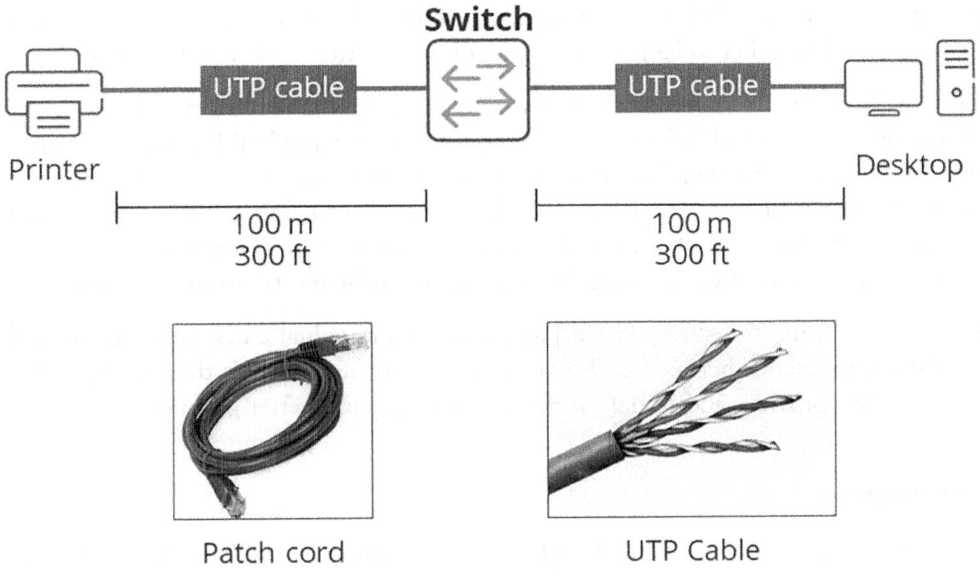

Figure 1-15: Wired connection using copper

UTP cabling contains 8 color-coded wires that group into 4 pairs. The connector UTP uses is an RJ-45 connector. UTP cables have categories that represent cable quality and the signal frequency band that can be carried by the cable. For example, UTP category (or cat for short) 6 cabling supports 10 Gbps speeds for up to 55 meters in distance; in contrast UTP cat 6E cable allows 10 Gbps can span a distance of up to 100 meters.

STP versus UTP

Copper cables used in networking come in several variants. Three common variants are shown in Figure 1-16.

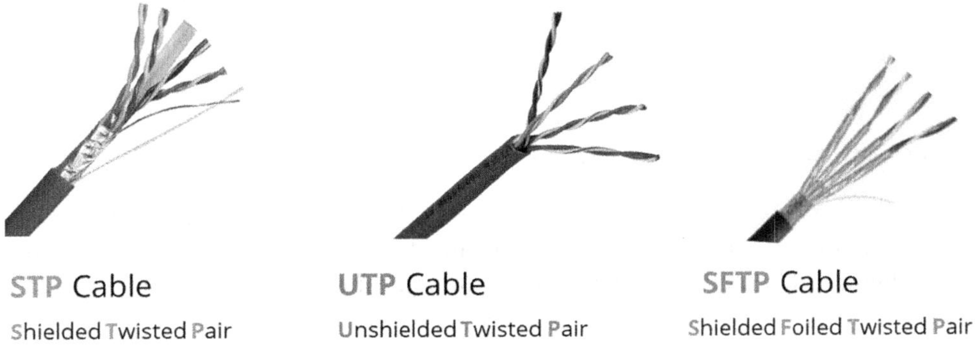

Figure 1-16: STP versus UTP

UTP cable has no shielding. This is the most used and most basic type of cable. The cable contains pairs of wires twisted together to help reduce and prevent electromagnetic interference.

STP cables are like UTP cables, where the wires are twisted together and then wrapped with a shielding or screening material which consists of foil wrapping or a copper braid jacket. Shielding the copper cables can have a tremendous effect on the stability, performance, speed, and longevity of the cable and ultimately your network. Because STP cables have shielding they are more resistant to electromagnetic interference, like fluorescent lighting, nearby motors, or any other electro-mechanical equipment. This provides better signal quality and allows for higher transmission speeds.

STP and UTP cables use the same twisted pairs, but the individual pairs are wrapped in foil. This helps to reduce interference between each pair. Plus, the entire bundle is then wrapped in a strong braided screen. This provides additional mechanical strength and better grounding.

Cable category

The *Cat* in Cat5e, Cat6, and etcetera, is short for *Category*. Network cables are divided into categories based mainly on bandwidth (measured in MHz), maximum data rate (measured in megabits per second), and shielding. Figure 1-17 summarizes commonly used categories.

Category1	Cable Type	Maximum Transmission Speed	Signaling rate
Cat. 3	UTP	10 Mbps	16 MHz
Cat. 5		100Mbps	100 MHz
Cat. 5e		1Gbps	100 MHz
Cat. 6 (6e)		1Gbps	250 MHz
Cat. 6a	STP	10Gbps	500 MHz
Cat. 7		10Gbps	600 MHz
Cat. 8		40Gbps	2000 MHz

Figure 1-17: Cable category

Wired connection using fiber optics

Use fiber optic cables to connect devices in your LAN that are more than 300 feet or 100 meters apart. This technology uses light impulses to transmit data using fiber. Maximum length can vary widely depending on cable and transceiver type, from a few meters, to around 500 meters, perhaps up to 40 kilometers.

Optic fiber achieves faster data rates. A less expensive fiber transfers 10 gigabits per second while a higher-quality fiber provides 100 gigabits per second data rates.

Fiber optic cables are organized into two main groups (see Figure 1-18): Multimode (MM) is less expensive for shorter distances. Single mode (SM) is more expensive but supports longer distances.

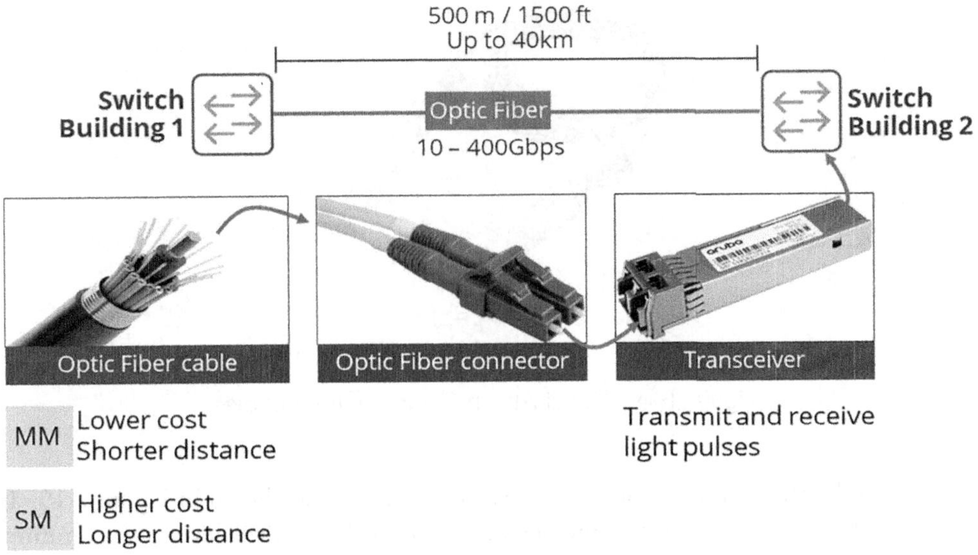

Figure 1-18: Wired connection using fiber optics

There are several standard fiber connectors, but you typically use a Lucent Connector (LC), as shown in Figure 1-18. You connect each end of the cable into a transceiver, which you have installed into a network device or server. Fiber optic transceivers are the active components that transmit and receive data, in the form of light pulses.

Direct attach copper (DAC) cable

Direct attached copper (DAC) cables are twinax cables that are factory-terminated to a transceiver (see Figure 1-19). You use them for relatively short connections, often from a switch in one equipment rack to another switch, server, or other device in the same rack.

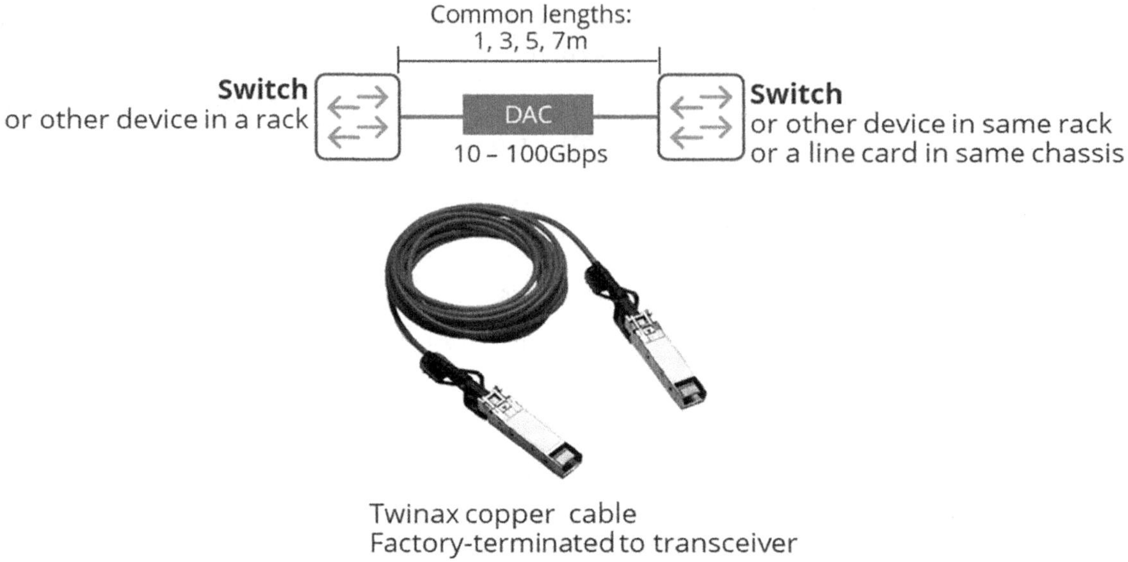

Figure 1-19: Direct Attach Copper (DAC) cable

There are DAC cables that connect from one line card to another, as shown in Figure 1-19. There are also breakout DACs that connect from one device and up to four other devices. A common option from Aruba is a DAC cable with SFP+ transceivers. For more information, search for "ArubaOS-Switch and ArubaOS-CX Transceiver Guide" currently at this URL: https://www.arubanetworks.com/techdocs/Switches/xcvrs/5200-3362/index.html#GUID-D7147C7F-2016-0901-0A24-0000000004A4.html.

Wireless connection using radio frequency

Most endpoints only use a wireless connection. This connection type has a typical range of 30 to 45 feet, or 10 to 15 meters (see Figure 1-20). The range depends on variables like the access point (AP) position, the endpoint type, and the furniture in the area. Objects affect the radio frequency (RF) signal by absorbing or reflecting it. Optimal distance between APs (and thus APs and endpoints) is best determined by understanding client need, and then performing a site survey to accommodate that need. These topics are beyond the scope of this book.

The data rate depends on the Wi-Fi technology in use. The latest Wi-Fi technology is IEEE 802.11ax, with data rates up to 4.8 gigabits per second.

To transmit data, Wi-Fi devices modify aspects of an electromagnetic signal, like amplitude, phase, or frequency. These modulated radio waves transmit through the air. This wireless transmission technology frees users from wired connections and allows them to roam.

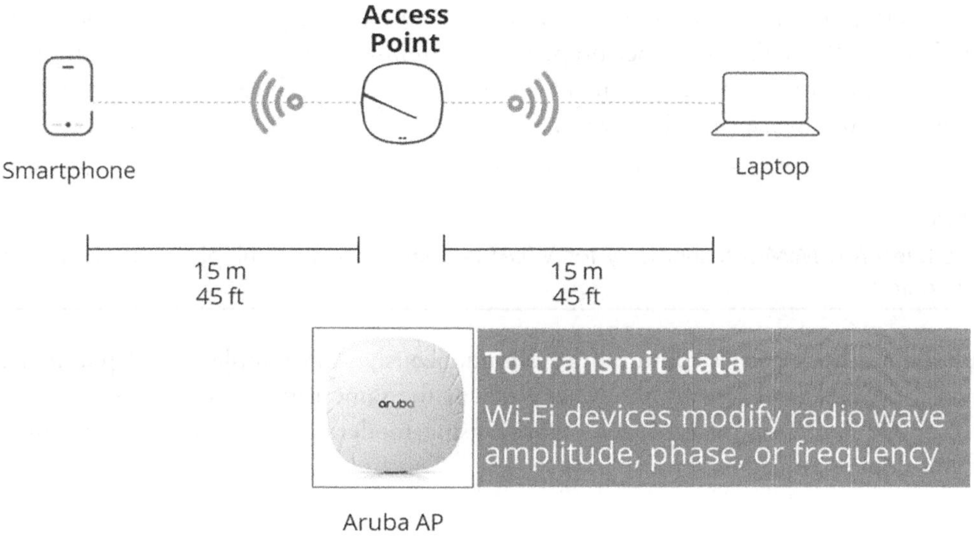

Figure 1-20: Wireless connection using radio frequency

Duplex

A duplex communication system has two or more connected parties that communicate in both directions (see Figure 1-21). There are two duplex communication system types: Half duplex (HDX) and full duplex (FDX).

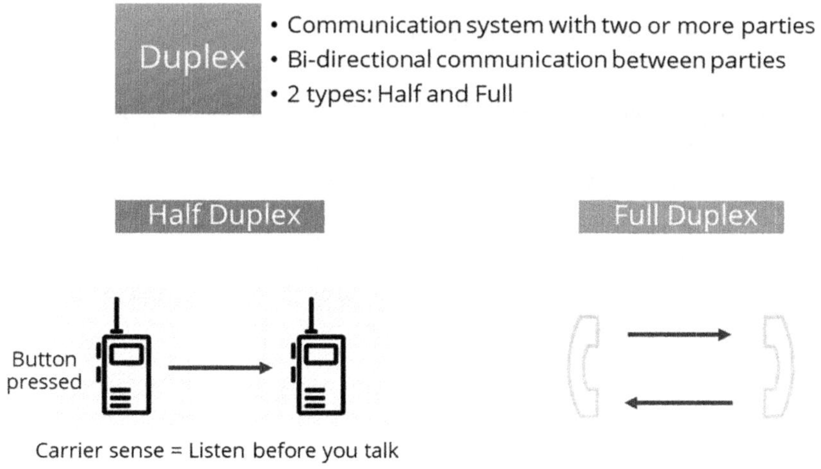

Figure 1-21: Duplex

With half duplex, both parties do not communicate simultaneously. It is one direction at the time. An example is a walkie-talkie. Each person presses a *push-to-talk* button. When the button is pressed, the user does not hear the remote person. To listen, they release the button. A more modern use case for half duplex is the typical Wi-Fi network.

Note

The latest MU-MIMO technology for WLANs does allow for multiple transmissions at the same time.

With full duplex, both parties communicate simultaneously. An example is a telephone. Parties on both ends speak and are heard by the other party at the same time. Again, a more data network-specific example is the typical Ethernet network using modern switches and devices: all of which support full-duplex connectivity.

Types of traffic: Unicast, multicast, and broadcast

Computing communications have three types (see Figure 1-22):

- Unicast
- Multicast
- Broadcast

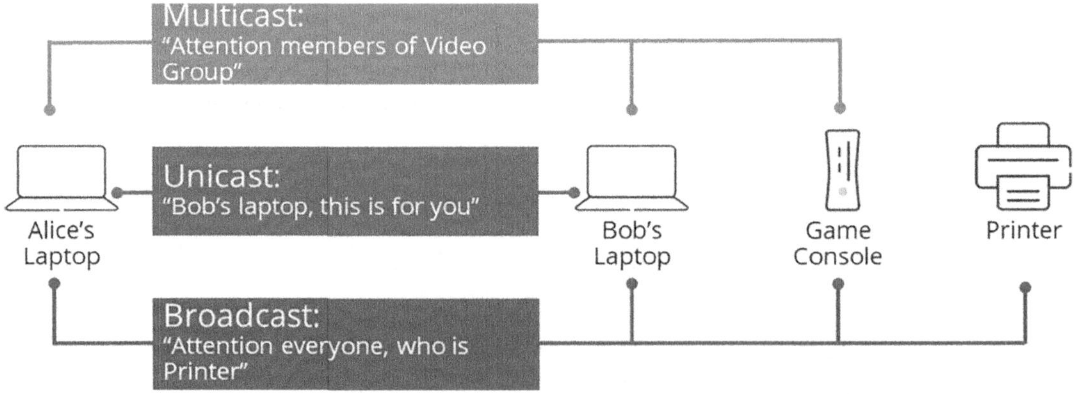

Figure 1-22: Types of traffic—unicast, multicast, and broadcast

Unicast communications

Unicast traffic refers to one-to-one communication: a transmitter and a receiver. Imagine several learners in a classroom. Bob is the instructor. He calls out, "Alice, I have a message for you from the front desk." This message came from one source, Bob, and is destined to a single destination, Alice. Similarly, when a PC needs to transfer a file to a server, the two devices use unicast communications.

Multicast communications

Multicast traffic refers to one-to-many communication: a transmitter and multiple receivers. In our classroom analogy, suppose lunch is brought in. Bob calls out, "All vegetarian meals are on the green table". The message came from one source, Bob. It is destined for several people, the vegetarians. A common example is video streaming. A video source sends a video stream to capable devices interested in receiving that information: on their PC, tablet, or smartphone.

Note
Multicast traffic includes many devices, but typically not all network devices.

Broadcast communications

Broadcast traffic communication is one-to-all. In our classroom, Bob calls out, "Attention, everyone! It is break time, and there are free doughnuts in the lobby." The word "everyone" means all people in the classroom are intended to receive this message.

Message types are more than just concepts. This is real traffic that every network-enabled device generates. Devices choose the type based on data they want to exchange and the protocol generating it. Later, you learn how traffic's destination addresses define whether it is unicast, multicast, or broadcast.

TCP/IP stack

By far, the most widely-used protocol stack is TCP/IP. This section provides an introduction to the TCP/IP protocol stack.

TCP/IP stack comparison

The Internet Protocol (IP) suite is a conceptual model and set of communication protocols used in Internet and computing networks. It is commonly known as the TCP/IP stack because the foundational protocols in the suite are Transmission Control Protocol (TCP) and IP. This model provides end-to-end data communication. It specifies how data should be packetized, addressed, transmitted, routed, and received. This functionality is organized into four abstraction layers, which can be seen in Figure 1-23.

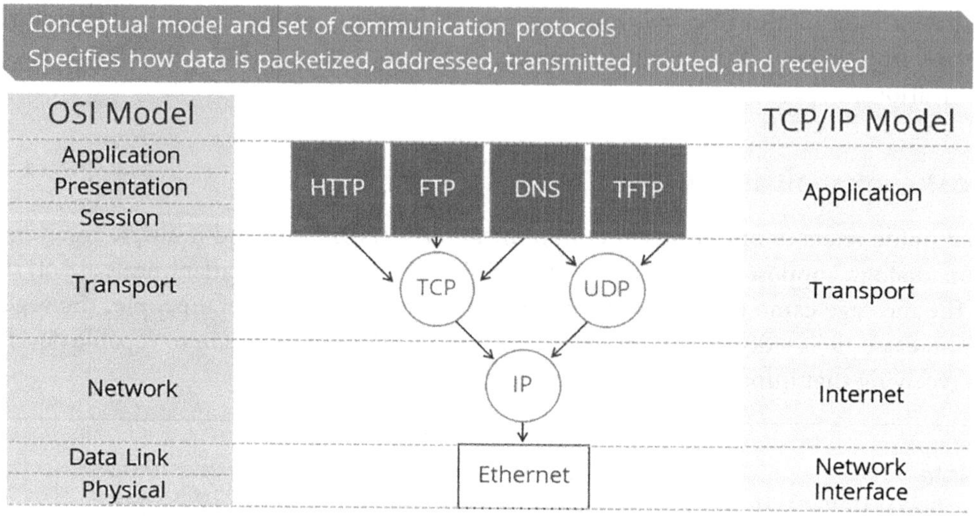

Figure 1-23: TCP/IP stack comparison

The TCP/IP model is often compared to the OSI model, as shown in Figure 1-23. You see that the OSI model has seven layers, which map to the four layers of the TCP/IP model. As you can see the functionality of the OSI model's application, presentation, and session layers all map to a single TCP/IP Application layer.

This means that TCP/IP-based applications are responsible for interfacing with users and creating the data (application layer), putting it in the proper format (presentation layer), and managing sessions (session layer).

The transport layers of each model directly map. Recall that this layer is responsible for flow and error control. This layer establishes an end-to-end connection of two devices whose logical connection traverses a series of networks. The two choices used in a TCP/IP network are UDP and TCP.

The OSI Network Layer maps to the TCP/IP Internet layer. Both perform the same function - routing data across logical network paths by defining a packet format and an addressing format. IP version 4 (IPv4) and IP version 6 (IPv6) are the protocols that work at this layer.

The OSI models data link and physical functions are all defined in a single layer in the TCP/IP model: the network interface layer. As you learned in the section about the OSI model, these layers contain protocols relating to the physical communication medium (copper wires, fiber optics, and radio waves). Recall that physical media access control also occurs here. Ethernet and Wi-Fi are very common protocols that work at this layer.

Ethernet

Ethernet is a family of computer networking technologies that are used at layer 1 and layer 2 of the OSI model. At layer 2, this standard provides for the assignment of 6-byte (48-bit) media access control (MAC) addresses to each device (see Figure 1-24). Each manufacturer assigns a globally unique hexadecimal MAC address to each Network Interface Card (NIC). This address is stored in permanent read-only memory (ROM) on the NIC. In other words, every Ethernet NIC in the world has a unique MAC address. How is this possible?

The most significant first 3 bytes of each MAC address are referred to as the manufacturer's Organizationally Unique Identifier (OUI). A standards body assigns each manufacturer in the world a unique 3-byte number. All NICs from that vendor therefore have the same number in the first 3 bytes.

The last 3 bytes of the MAC address comprise the NIC serial number. Recall that with hexadecimal a single byte is comprised of two digits. Thus, each MAC address is represented with 12 hexadecimal characters; each byte or octet is typically separated by either a colon or dash.

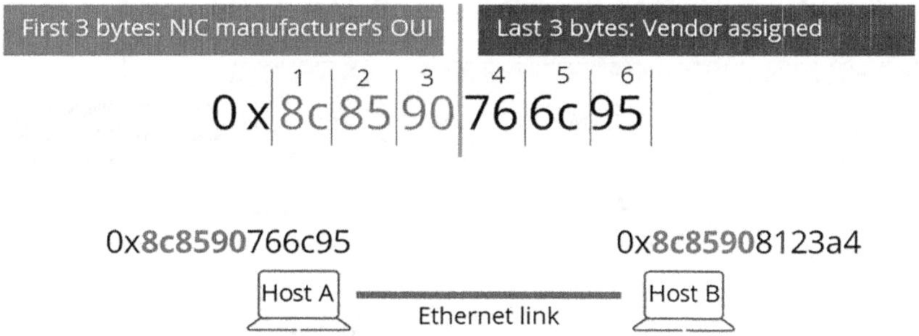

Figure 1-24: Ethernet

For example, for the MAC address 8c:85:90:76:6c:95, the OUI is 8c:85 :90, while the NIC serial number is 76:6c:95. As you progress through this guide, you will often see fictitious, 2-byte MAC addresses such as 8c::95. This just makes it easier for you to focus on fundamental concepts, as opposed to always trying to read out large, 6-byte hexadecimal numbers.

To verify the vendor of a NIC by its MAC address, access a web site like https://macvendors.com, which allows you to perform OUI lookups. If you were to lookup the address in this example, you would determine that the manufacturer is Apple.

Note

Private IP address ranges are defined in RFC 1918. These are only to be used within an organization and never on the Internet. Likewise, there is a set of layer 2 MAC address ranges—any address where the most significant octet begins with is x2, x4, xA, or xE—that are private MAC addresses. Other than these private addresses, MAC addresses are globally unique. But understand that when a router receives a layer 3 IP packet, wrapped in a layer 2 frame, it strips off that frame, adds a new frame, and transmits it toward its destination. Thus, the requirement is that MAC addresses only need to be unique on the same layer 2 network (subnet/VLAN).

Also, newer Apple, Windows, and Linux platforms can *pseudo-randomize* their MAC address to improve privacy. These systems use the private MAC address ranges for this purpose.

Ethernet frame

Suppose that Host B, in Figure 1-25, receives data from the upper layers of the OSI model. Think of Ethernet as an employee of the upper layers. They have given Ethernet this payload, and asked Ethernet to get this over to Host A. To do this, the Ethernet process must wrap this payload with a header and trailer, as shown in Figure 1-25.

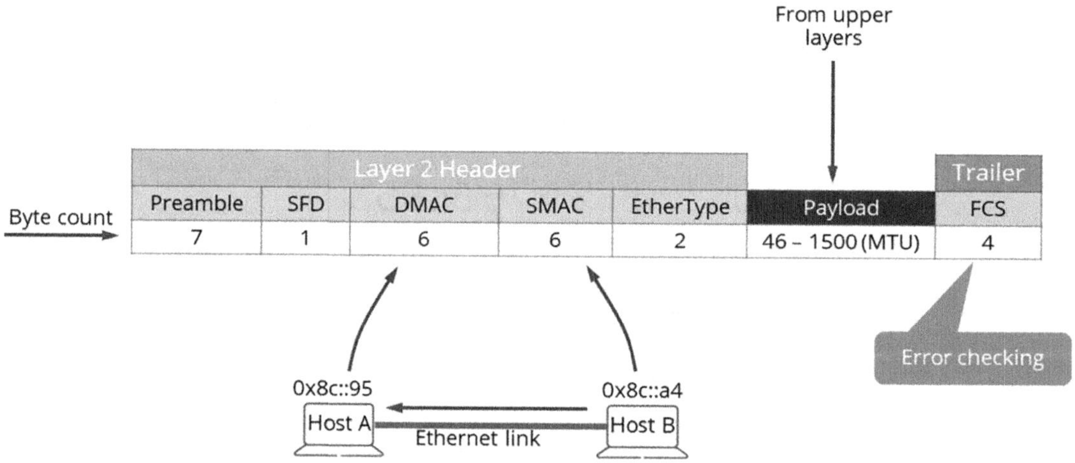

Figure 1-25: Ethernet frame

Each field in the header and trailer is described below:

- **Preamble**. Informs the receiving system that a frame is starting and enables synchronization.
- **SFD (Start Frame Delimiter)**. Indicates that the destination MAC address field begins with the next byte. There might be several hosts on this Ethernet segment. The preamble and SFD are seen by every host as a kind of signal, "Hey, here comes a frame, and it might be for you!" How does the receiving station know whether the frame is for it, or some other host? By reading the next field: the destination MAC address.
- **Destination MAC (DMAC)**. Identifies the NIC of the receiving computing system. In this example Host A reads the MAC address and knows, "Hey, this is my MAC address – I should accept this frame". All other stations read the MAC address, see that the message is not for them, and drop the frame. This is like someone yelling in a room, "Hey Marta, can I buy you an ice cream?" Only Marta will respond.

- **Source MAC (SMAC)**. Identifies the NIC of the sending computing system. This is how Host A knows where the message came from, and therefore to whom it should respond.
- **EtherType**. Defines the layer 3 protocol inside the frame, for example IPv4 or IPv6. Host A may be running an IPv4 protocol stack and an IPv6 protocol stack. The type field tells Host A which stack to pass the payload to for processing.
- **Payload**. Contains the data that to be transmitted. FCS (frame check sequence). Allows detection of corrupted data using the CRC or checksum method. As Host B creates this frame, it performs a mathematical calculation on most of the header and payload bits. It places the result of this calculation in the FCS field. When Host A receives the frame, it performs the same calculation. If it comes up with the same result that is stored in the FCS field, the frame is not corrupted. But if you have bad wiring, or strong electromagnetic interference (EMI) the data may become corrupted. In this case, Host A's calculation will not match the FCS, and so the frame is discarded – no need to accept bad data.
- **Frame checksum (FCS)**. Provides error correct to detect a frame that became corrupted during transmission from the source to the destination.

It is important to understand that the Ethernet payload not only carries application data – it also includes the header information from the upper layers. Remember, TCP adds a header, which IP considers part of the payload. Then IP adds a header, which Ethernet considers part of the payload. Let us look now at this IP header information.

Note

Figure 1-25 only shows the first and last bytes of each 6-byte MAC address, to ease the learning experience.

IPv4 header

Figure 1-26 shows the IPv4 header information added to a packet at layer 3, with a special emphasis on the most important fields in this header, as described below:

- **Time-to-Live (TTL)**: This field mitigates layer 3 route loops. If routers are misconfigured, IP packets could be bounced around between two or more routers forever. To prevent this, the originator places a number in this field from 0 to 255. For example, suppose that the TTL is 15. Each router that routes this packet decrements this field by 1. So, after this packet "hops" through the first router, TTL = 14. After the next router, TTL = 13, and so on. The router that decrements the TTL field to zero discards the packet.
- **Protocol**: This field indicates whether this IP packet carries TCP traffic or UDP traffic. Recall that application data is passed down to TCP or UDP, which adds a header and passes this on down to IP. From IP perspective, TCP, UDP, and other protocols are at higher layers in the model, and so can be collectively referred to as upper layer protocols (ULP).

- **Source IP Address**: This originator of this IP packet adds its IP address to the source address field. In the figure, host 10.1.1.1 sends a packet to host 10.2.2.2. Thus, the source IP address field = 10.1.1.1, and the destination IP address field = 10.2.2.2.
- **Destination IP Address**: The ultimate destination of the packet, as described above.

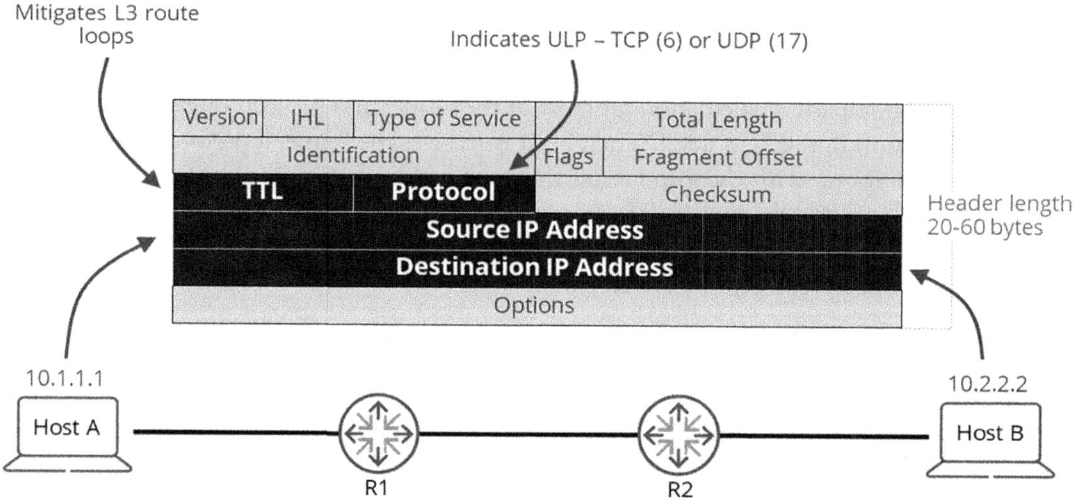

Figure 1-26: IPv4 header

The other fields are shown here in gray, because they are not relevant to your learning in this and several of the chapters that follow. You will learn about these other fields as they become relevant. For more detail, please see RFC 791: https://www.rfc-editor.org/rfc/rfc791#page-11.

However, you may want to understand the variable IP header length – variable due to the *Options* field. The IP header length is specified in the Internet Header Length (IHL) field. Four bits specify the number of 32-bit words, with a minimum value of 5. So, 5 x 32 = 160 bits = 20 bytes long. If any of the options are used, the size might be larger. The maximum IHL value = 15, and 15 x 32 = 480 bits = 60 bytes.

Remember, this is the header that layer 3 routers use to do their job. They receive an inbound packet, analyze the destination IP address field, and forward packets along the best path toward that destination. In Figure 1-26, router R1 receives this packet from host 10.1.1.1 and forwards it to R2. R2 receives it and forwards it to destination host 10.2.2.2.

TCP three-way handshake

Recall that TCP acts as a kind of *employee* for applications, which rely on TCP to get data to some destination host. Before TCP processes any application data, it must first contact the target device and establish a reliable, flow-controlled connection.

In Figure 1-27, Host B's HTTP application has data–sitting in a kind of *out box*, waiting to be transmitted to Host A.

- It is as if HTTP says, "Hey TCP, I need you to get this data to Host A".
- TCP says, "OK boss, but first I must establish a connection with Host A to ensure that your data is sent in a controlled, reliable manner".

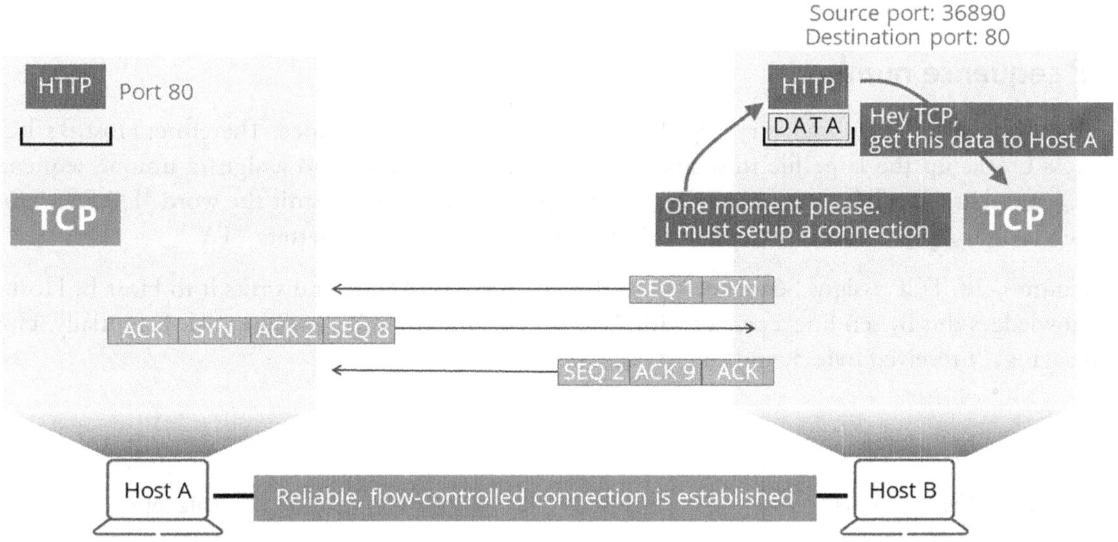

Figure 1-27: TCP three-way handshake

There are four primary fields in the TCP header that are used to establish this connection–the sequence (SEQ) number, the acknowledgement (ACK) number, the synchronize (SYN) flag, and the acknowledgement (ACK) flag.

- Host B starts by creating a TCP packet. This packet carries no data: it only signals Host A, "We should establish a connection".
- In this example, Host B sets SEQ = 1, and raises its SYN flag (sets it to a binary 1).

Host A receives this and knows, "This packet came from Host B with the SYN flag raised – Host B wants to connect". Host A must inform Host B that it received SEQ = 1, and so expects SEQ = 2 next.

- To do this, Host A sends a packet with ACK = 2, and both the SYN and ACK flags are raised. Understand that Host A is independent of Host B. Host B tagged its packet with SEQ = 1. Host

A chooses to tag its packet with SEQ = 8. Host A is saying, "I'm using my packet number 8 to acknowledge receipt of your packet number 1".

- Host B receives and responds to this packet, "You expected SEQ=2 next, so here it is...I am using my packet number 2 to acknowledge receipt of your packet number 8. I expect packet number 9 next. As you can see, both the SYN and ACK flags are raised in this packet.
- At this point, a reliable, flow-controlled connection is established. Each host has successfully received and acknowledged a sequence of packets from the other. Now TCP is ready to start transferring that data for the application.

HTTP uses a specific port to transmit the information; in this case the destination port number is 80; the source port in this case is randomly selected on Host B; in this case Host B selects the port 36,890 as the source port.

TCP sequence numbers

TCP may be asked to transfer very large files–too big to transfer all at once. Therefore, Host B's TCP process breaks up the large file into smaller pieces called segments and assigns a unique sequence (SEQ) number to each byte segment. In this example you want to transmit the word "DATA"; Host B then creates 2 pieces, one for the letters "DA" and the other for the letters "TA".

In Figure 1-28, TCP assigns Sequence = 3 to the first piece (DA) and transmits it to Host B. Host A acknowledges this by sending a packet with Acknowledgement (ACK) number = 4. Essentially, Host A is saying, "I received byte 3, and so I expect byte 4 next".

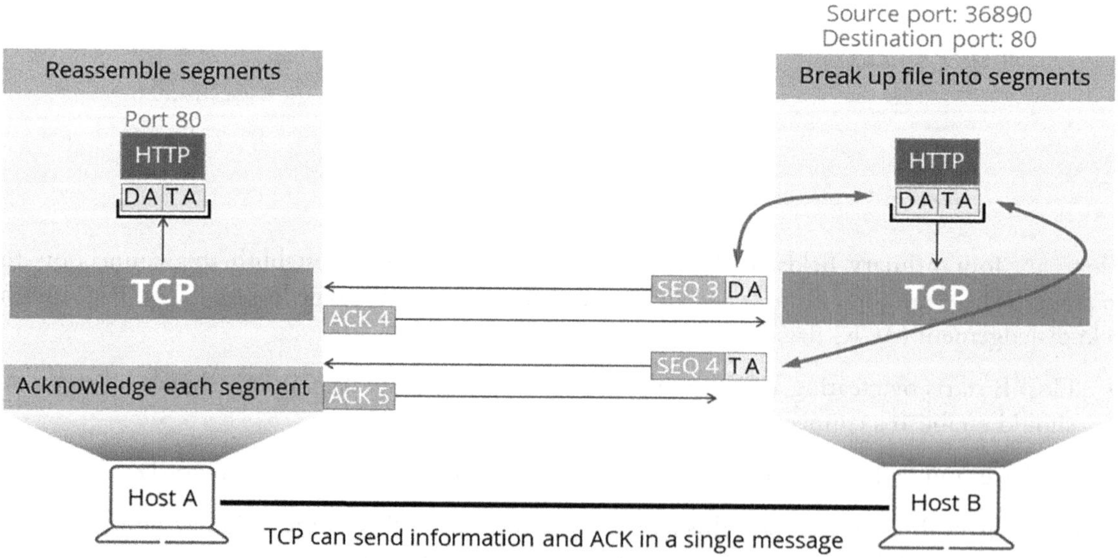

Figure 1-28: TCP sequence numbers

Host B then takes the next piece (TA), adds Sequence = 4, and sends it to Host A. Host A acknowledges this by sending ACK = 5. As Host A receives this data, it reassembles the pieces back into a complete file and passes it up to the application.

Note
TCP can send a block of information and acknowledge receipt of data blocks using a single ACK message.

TCP port numbers

Recall that TCP acts as a kind of "employee" for applications. TCP receives requests from HTTP, FTP, and many other applications. It is as if the application says, "Hey TCP – establish a connection with the target host, and make sure my data gets there reliably". TCP uses the destination port number to keep track of each application it serves (see Figure 1-29).

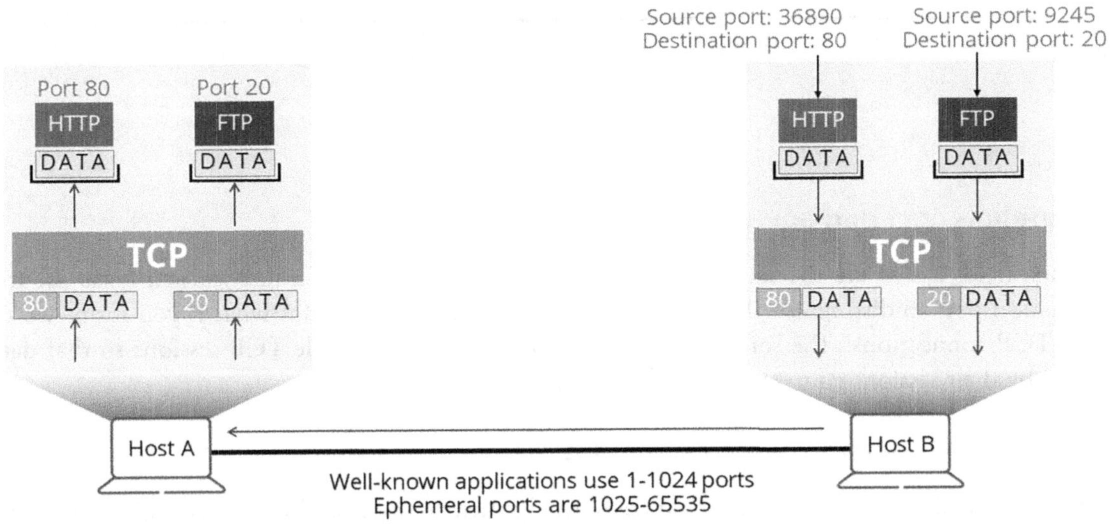

Figure 1-29: TCP port numbers

To accommodate this, each application is assigned a standard port number. FTP uses port 20 for data transfer, HTTP uses port 80, and so on. When Host B's TCP process accepts data from HTTP, it sets Destination Port = 80. Thus, when Host A receives this data, it knows to pass this data up to its HTTP application, as opposed to DNS or FTP.

In TCP you can use up to 65,535 ports; however, the well-known applications use the first 1,024 ports. From 1,025 to 65,535 ports are known as ephemeral port numbers. Applications use these ports to establish connections by randomly picking a port from that range and use that as source port for a TCP session.

TCP header

Figure 1-30 shows the entire TCP header information added to a layer 4 segment, with a special emphasis on the most important fields in this header, as described in the following sections.

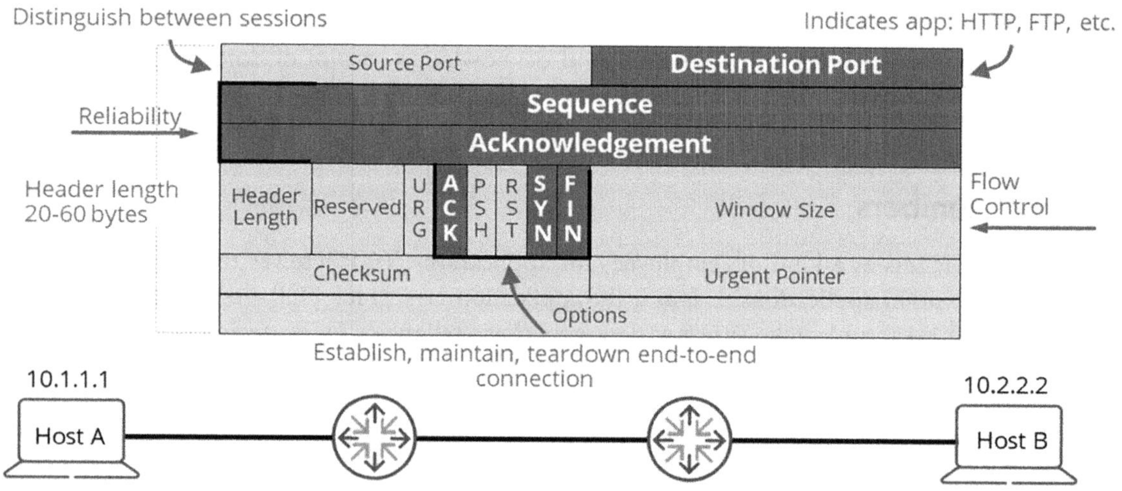

Figure 1-30: TCP header

Destination port and source port

The first fields added are the source and destination port numbers. You just learned hosts use the destination port to distinguish the various applications that it serves. But what if you form two or more TCP connections? The source port is used to distinguish multiple TCP sessions so that data carried by their sessions go to the right application.

Sequence and acknowledgement numbers

You also learned about the sequence and acknowledgement numbers. They are used along with the ACK and SYN flags to ensure reliability and flow control. Host B sends SEQ=5, Host A responds with ACK=6, and so on. But what if Host A responded with ACK = 5? Host B says, "Hmm. I guess host A did not receive segment 5, so I will resend it". In this way, TCP ensures reliability, resending any packets that are lost in transmission.

Flags

You learned how the SYN and ACK flags are used to formally synchronize TCP packet (sequence) numbers and acknowledge receipt of packets. There is also a Final (FIN) flag, which tells the receiver, "Hey Host A, this is the last piece of the file."

There is the urgent (URG) and push (PSH) flags to indicate that certain data is urgent and should be quickly pushed up to the receiving application. The URG flag is used in conjunction with the urgent pointer field, which indicates where the urgent data is located. There is also a reset (RST) flag, used to reset the connection.

Window Size

There's a checksum for reliability, and a window size field, used for flow control. If Host B is sending too much data at a time, Host A may not be able to handle the load. Host A may lower the window size, telling Host B, "Slow down, please do not send so much data at a time".

The remaining fields in the TCP header are not either rarely used or not relevant to this initial discussion. You will learn about these fields as you continue your learning, as appropriate. For more information, see RFC 793: https://www.ietf.org/rfc/rfc793.txt.

Again, you may want to understand the Header Length field, to understand the overall number of bytes a TCP header adds. Like in IP there is a minimum value of 5, and 5 x 32 = 160 bits = 20 bytes. If the Max Header Length value is 15, and 15 x 32 = 480 bits = 60 bytes.

UDP header

Figure 1-31 shows the UDP header information on a layer 4 segment. There are only source and destination ports, which you use exactly like TCP. There is a Length field and an optional checksum. For applications, TCP establishes a connection with the other side, retransmits lost packets, and makes sure not to overwhelm the receiver.

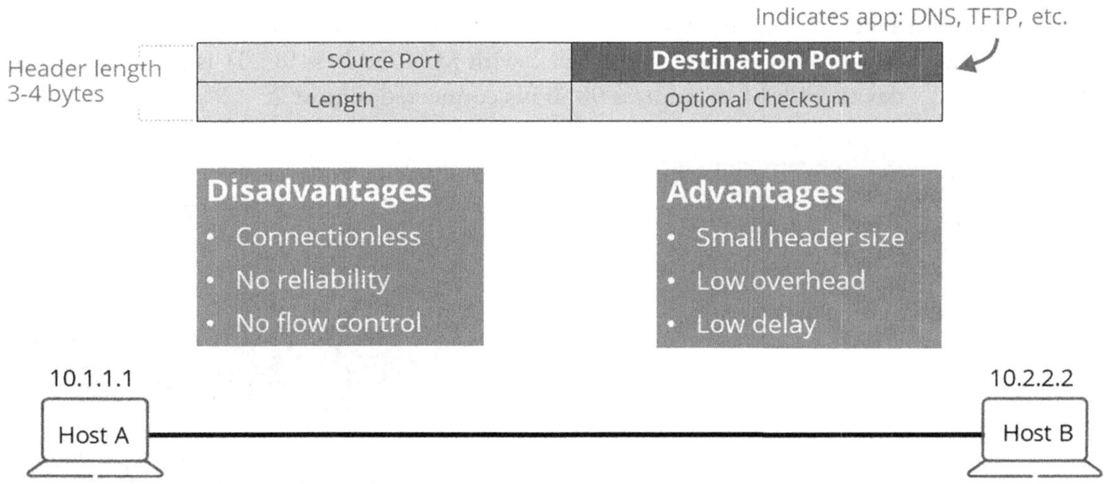

Figure 1-31: UDP header

In contrast, UDP offers no connection, no reliability, and no flow control. Does the destination receive the packets? Is the data I'm transmitting overwhelming the network or destination host? UDP does not care. Understand that this does not mean that reliability is sacrificed: the upper-layer application implements the reliability it needs.

UDP is less sophisticated but comes with a small header size meaning a lower overhead. The lack of flags and flow-control services means UDP transfers data with lower delay. It is not reliable but is quick. This is perfect for applications like voice over IP (VoIP) or video streaming. These applications do not need TCP's reliability. Programmers build the needed reliability into applications.

In UDP, you use up to 65,535 ports. Remember, well-known applications use the first 1,024 ports and ports 1,025 to 65,535 are ephemeral port numbers, same as TCP.

Networking devices

A functional society is composed of good people. And a functional network is composed of network devices. This section introduces you to networking devices.

Networking devices: Switches

Switches have multiple ports that are used to connect computing devices into one or more local area networks (LAN). This can include PCs, printers, video cameras, voice-over-IP (VoIP) phones, and more. You may encounter switches that have 8, 24, 48, or more ports. Understand that switches are *transparent* to endpoints. Connected devices are not aware of the existence of the switch: they perceive themselves as being directly connected to each other.

A switch is a layer 2 network device that forwards Ethernet frames, based on destination MAC addresses. Recall that MAC addresses are part of the data link layer, and so switches are considered layer 2 network devices. The figure shows the Aruba icon used for a switch, and below that a physical switch example. In Figure 1-32, you also see a simplified MAC address table that is maintained in switch memory. The switch has learned that a device with MAC address 90::01 is connected to port 1 of a switch and a device with MAC address 90::03 is connected to port 2.

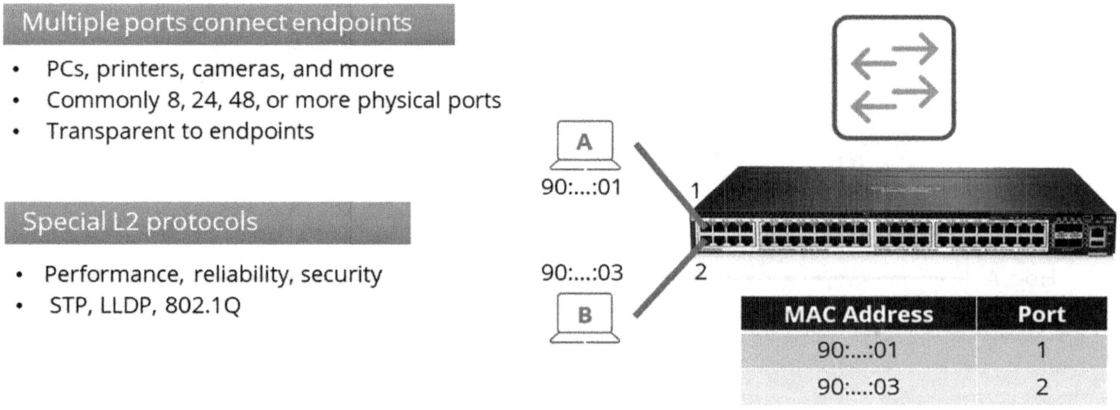

Figure 1-32: Networking devices: switches

Note

Switches forward Ethernet frames based on destination *MAC* addresses.

Switches also have special languages or protocols that they use to increase network performance, reliability, and security. Examples of these protocols include the following:

- Spanning Tree Protocol (STP)
- Link Layer Discovery Protocol (LLDP)
- 802.1Q

You will learn more about these protocols later in this book.

Networking devices: Routers

A router is a layer 3 network device that forwards packets based on the destination IP address. Since IP addresses are part of the network layer, routers are considered layer 3 devices.

Note

Routers forward packets based on destination *IP* addresses.

You know that switches connect computing devices into one or more networks. It is a router's job to connect those separate networks together, to create an inter-network. The Internet is the largest, most common example of an inter-network. Figure 1-33 shows four routers, used to interconnect five networks (links between the routers are also considered networks).

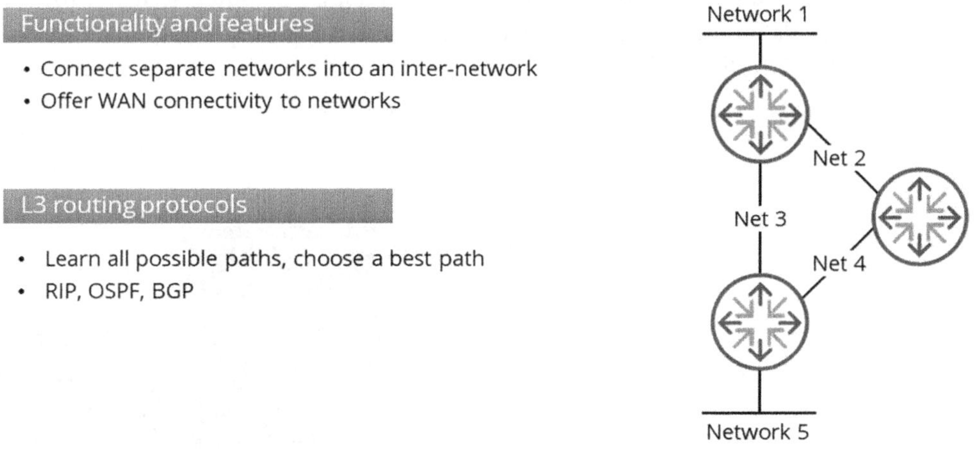

Figure 1-33: Networking devices: routers

While switches might have a few dozen or a couple hundred ports, routers have a relatively small port count—often between 2 and 6—which are used to connect to WAN networks.

When you take a long trip in your car, there are often several paths you could take to arrive at your destination. You consider all possible paths, and then choose the best path. You judge the best path based on some criteria that is important to you, such as whether the route is the fastest, or the shortest, or whether the route is particularly scenic. Then you drive along that path. Similarly, routers run protocols between them to learn all possible paths between all available networks. They then choose a best path for each destination, and forward packets along those best paths.

Examples of these routing protocols are:

- Routing Information Protocol (RIP)
- Open Shortest Path First (OSPF)
- Border Gateway Protocol (BGP)

You will learn more about these protocols later in this book.

Networking devices: Multi-layer switches

The OSI model sets clear distinctions between a layer 2 switch and a layer 3 router. However, networking professionals know that it is often good to have a single device that can perform both functions – a hybrid between a router and a switch.

A multi-layer switch, shown in Figure 1-34, has all the functionality of a layer 2 switch.

- They support typical layer 2 functionality, such as STP, LLDP, and VLANs. Multiple ports connect various endpoint devices together into one or more networks.
- A multilayer also has all or most of the functionality of a layer 3 router, so it can route between those separate LANs internally – an internetwork in a box! Like any router, it supports routing protocols like RIP, OSPF, and BGP.

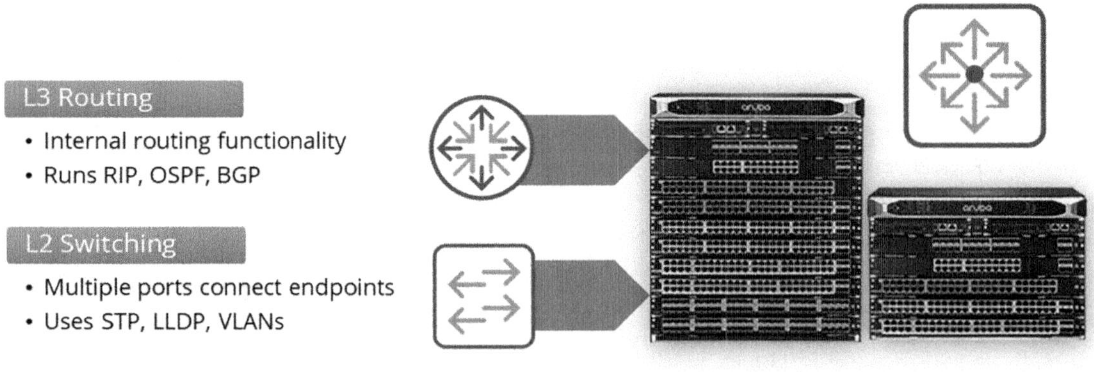

Figure 1-34: Networking devices: Multi-layer switches

Of course, multi-layer switches support layer 1 to actually transmit and receive data, and it can even perform some layer 4 functions – especially as relates to certain security features. These devices help to build a secure and flexible network.

Note

A multi-layer switch performs both layer 2 (L2) switching and layer 3 (L3) routing.

Networking devices: Access points

Access points (APs) enable wireless users to access wired resources. Without being tethered to Ethernet cables, users are free to roam about. APs are translational bridges: they accept Wi-Fi frames from endpoints, translate them into Ethernet frames, and forward them on to some wired resource. This could be an internal corporate server, storage, application, or an Internet resource. When that wired resource responds, the AP accepts that Ethernet frame, translates it to a Wi-Fi frame, and then transmits it over the air to the wireless host, as shown in Figure 1-35.

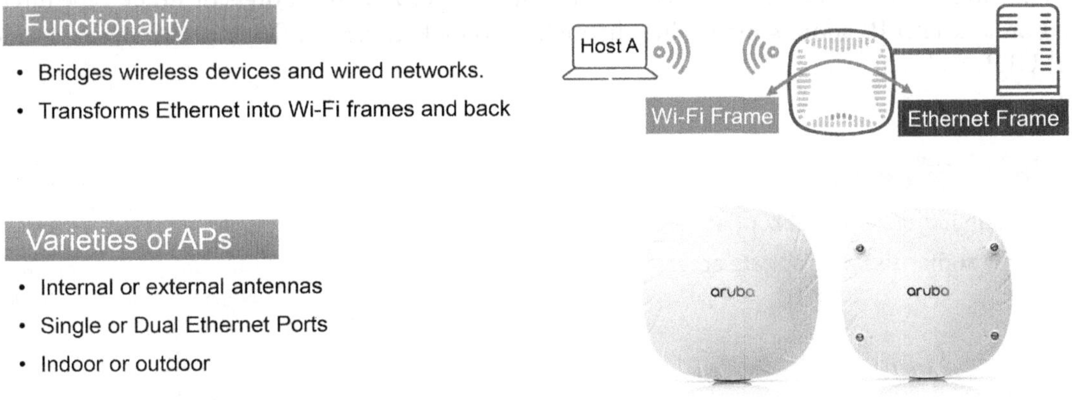

Figure 1-35: Networking devices: Access points

There are many varieties of APs to meet your technical and budgetary needs. They may be built with internal or external antennas, with one or dual Ethernet ports. Some may be for indoor use only, while others may be mounted outside.

Wi-Fi systems may be designed to use autonomous APs or gateway-based AP. Autonomous APs can perform all necessary functions to create a functional system. They process wireless and Ethernet frames and provide a certain amount of manageability and control. As the name implies, they work autonomously, without control from some external device. This type of deployment is relatively simple and easy, but it is not very scalable. It is suitable for smaller environments.

With gateway-based AP solutions, the APs send and receive wireless and Ethernet frames, much like autonomous APs. However, much of the processing and management functions are off-loaded to one or more centralized gateways.

With cloud-based APs, the APs are managed in a similar way to a gateway-based solution, but there is no gateway. Instead, all management and monitoring functions are offloaded to software that runs as a cloud service like Aruba Central.

These centralized configuration and control features might increase the initial complexity of the deployment, but once in place they provide far superior visibility into network health, enable more proactive network management, and can scale from medium to very large deployments.

Networking devices: Firewalls

A firewall is a security device. It denies unauthorized access and permits appropriate, authorized access.

Firewall functionality may operate at nearly any layer, from layer 2 to layer 7 (see Figure 1-36).

Firewalls might permit or deny access based on a simple layer 2 MAC address or Ether Type filter, or based on a layer 3 IP address/mask filter. It might also look at certain fields in the layer 4 headers: TCP, UDP, and etcetera.

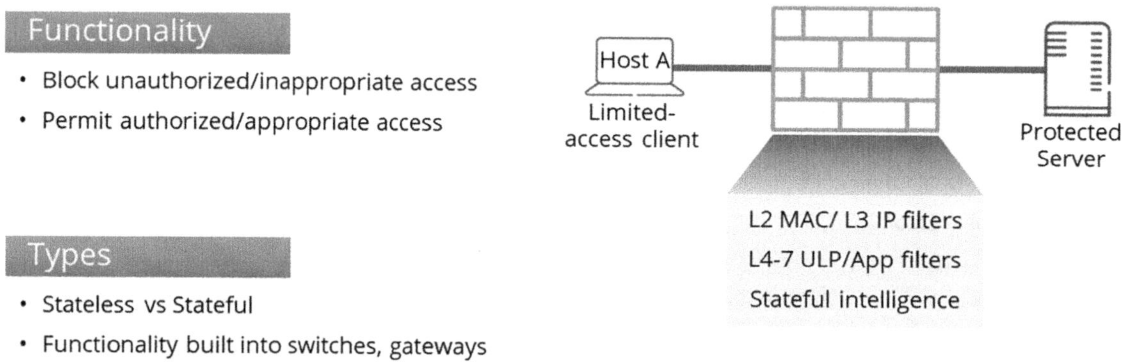

Figure 1-36: Networking devices: Firewalls

Some firewalls have the intelligence to be *application-aware*. They understand how certain applications are supposed to work, and block any nefarious activity that seeks to circumvent this normal operation. Or it might allow a particular user to access a corporate-authorized application on the web – perhaps a vendor catalog, but deny access to inappropriate applications and/or web sites.

Stateless firewalls lack any of this intelligence. They simply compare each layer 2, 3, and/or 4 header information to access control lists of permitted and denied traffic. Stateful firewalls are aware of normal TCP/IP communications. For example, they understand that all TCP connections begin with a three-way handshake, as you just learned. If this procedure is not followed, the firewall intelligently denies traffic.

Aruba builds firewall features into many of their products: switches, multi-layer switches, and especially into their gateway and AP products.

Lab 1: Testing remote lab connectivity

In this lab, you will test access and verify connectivity with the lab equipment.

Note

Lab access is not included with this book and must be purchased separately on Aruba Education's web portal. Purchasing lab access is optional, but highly recommended since you will be tested on your configuration skills and knowledge on the actual certification exam.

Objectives

After completing this lab, you will have all needed information to support the hands-on labs in this book.

Lab course topology

The official Aruba Training Lab provides the equipment you need to complete several lab activities. You should know the purpose and access procedures for this equipment. The layout of the equipment is shown in Figure 1-37 and includes the following:

- **PC-MGMT**: This client is used for remote lab management access, traffic analysis, and to access the switch's CLI via SSH.
- **PC-1:** This client is used for connectivity testing. It has a wired and wireless NIC.
- **PC-4**: This client is used for connectivity testing. It has a wired and wireless NIC.
- **6300-A switch**: This is one of your access switches; it will be named sw-edge1.
- **6300-B switch**: This is one of your Access switches, named sw-edge2.
- **8325-A switch**: This is the primary aggregation switch, named sw-agg1.
- **8325-B switch**: This is the secondary aggregation switch, named sw-agg2.
- **AP1**: This is the first AP, connected to your sw-edge1 on port 2.
- **AP2**: This is the second AP, connected to your sw-edge2 on port 2.

- **OOBM switch**: You have no access to this switch.
- **Core router**: Your aggregation switches connect with a routed connection to the Core router.
- **Windows Server**: This system will provide DHCP, DNS, and NTP services for the lab devices. You have NO access to this server.
- **ClearPass server**: It is used as an authentication, authorization, and accounting (AAA) RADIUS server for the network environment.

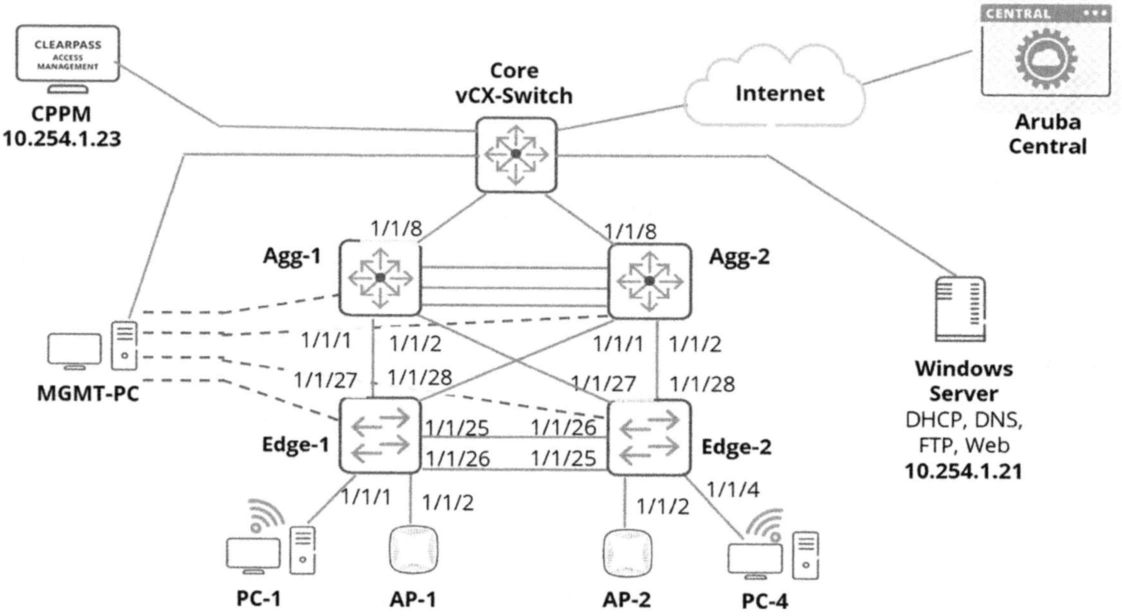

Figure 1-37: Lab topology

Task 1: Aruba Training Remote Lab access

Objectives

- Validate remote lab connectivity and ability to log in.
- Ensure that you have remote lab access during this training.

Steps

1. On your local computer, launch an HTML5 web browser, and access the Aruba Training Lab web portal at the URL: https://arubatraininglab.computerdata.com.
2. Enter your *username* and *password* (assigned to you from the registration email for the self-directed labs), and click the **Sign in** button.

Task 2: Testing connectivity

Objectives

- Test connectivity and authentication credentials for each of the devices.
- Working from the Aruba Training Lab diagram (shown previously in Figure 1-37), connect and log into the lab devices and your client PCs.

sw-edge1 (Edge-1)

This device will be the first edge switch in your lab environment, to be named **sw-edge1**.

1. To connect to the console of the Edge-1 switch, right-click on the icon in the lab diagram and select "Open Console." A new browser tab should open with a blank, black screen.
2. Press **[Enter]** a couple times, and you will see a user prompt. Log in by entering a username of **admin** and no password (leave it blank).
3. The switch will ask you to define a new password; hit **[Enter]** twice.

```
6300 login: admin
Password:

Please configure the 'admin' user account password.
Enter new password:
Confirm new password:
6300#
```

Note
This switch is factory default at the start of the labs. A factory default switch prompts the administrator to change the password after the first login.

4. Take note of the switch sw-edge1 *serial number* and *MAC address*. You may also save this in a text file on your local system. When using Aruba Central, this information will be used to identify the devices.

```
6300# show system
Hostname           : 6300
System Description : FL.10.09.1040
System Contact     :
System Location    :

Vendor             : Aruba
Product Name       : JL666A 6300F 24G CL4 PoE 4SFP56 Sw
```

```
Chassis Serial Nbr : SG00KN500Z
Base MAC Address   : 64e881-3f6540
ArubaOS-CX Version : FL.10.09.1040

Time Zone          : UTC

Up Time            : 4 days, 20 hours, 29 minutes
CPU Util (%)       : 10
Memory Usage (%)   : 23
```

sw-edge2 (Edge-2)

This device will be the second (indicated by B) edge switch in your lab environment, to be named **sw-edge2**.

1. To connect to the console of the 6300-B switch, right-click on the icon in the lab diagram and select "Open Console." A new browser tab should open with a blank, black screen.
2. Press **[Enter]** a couple times, and you will see a user prompt.

Note

This switch has been pre-configured in the remote lab environment; you must connect using the correct credentials—see the next step.

3. Login using **admin** and password **Aruba123!**

```
login: admin
Password:
```

Important

For ease of use we use a simple password in the lab (Aruba123!); please ***never*** use a simple password in real life!

4. Take note of the switch sw-edge2 *serial number* and *MAC address*. You may also save this in a text file on your local system.

```
6300# show system
```

8325-A and 8325-B

These devices will be the primary (A) and secondary (B) aggregation switches in your campus lab environment. They will be named *sw-agg1* and *sw-agg2*.

1. To connect to the console of the 8235-A switch, right-click on the icon in the lab diagram and select "Open Console." Press **[Enter]** a couple times; login with username **admin**, password **Aruba123!**.

2. This device has been pre-configured. Verify the prompt shows sw-agg1.

```
sw-agg1 login:admin
Password:
sw-agg1#
```

3. Repeat the previous 3 steps for the 8325-B, where the hostname should be **sw-agg2**.

```
sw-agg2 login:admin
Password:
sw-agg2#
```

Note

You don't need to record the serial and MAC address of the 8325 switches: they will be automatically provisioned with the correct configuration.

AP1 and AP2

There are two APs in the lab setup. These APs are factory default at the start of the training labs. In the next steps you will make an inventory with the MAC and serial number of both APs. This will make it easier in later labs to identify each AP either on the switch or in Aruba Central.

You will take note of the AP MAC address using the AP console connections. Right after an AP starts to boot, you can press **[Enter]** to access the *apboot* environment. In the apboot mode, you can execute the **mfginfo** command (manufacturing information) to see the AP MAC address and serial number.

In the next procedure you will reboot the AP. Make sure to switch quickly to the AP console upon rebooting: you will need to press **[Enter]** to access the apboot mode within a few seconds after the AP starts to boot.

1. To connect to the console of the AP1, right-click on the icon in the lab diagram and select "Open Console." Press **[Enter]** a couple times; you should see a login prompt.

2. Return to the lab dashboard; right-click **AP1** again and select **Reboot**.

3. Quickly switch back to the web page with the AP1 console access; press **[Enter]** when you see the option to access the apboot mode.

```
APBoot 2.4.0.8 (build 64221)
Built: 2018-03-28 at 20:30:14

Model: AP-303H
DRAM:  512 MiB
Flash: Detected W25Q32FV_SPI: total 4 MiB
NAND:  Detected MX35LFxGE4AB: total 128 MiB
Power: 802.3af POE
Net:   eth0
Radio: ipq4029#0, ipq4029#1
Reset: warm
FIPS:  passed

Hit <Enter> to stop autoboot:  0
apboot>
```

4. In the apboot context, execute the **mfginfo** command and press **[Enter]**. Take note of the MAC address and serial number of the AP. Here is an example output, where your MAC address and serial number will likely be different from that shown in this output.

```
apboot> mfginfo
Inventory:
Card 0: System
        Wired MAC              : 20:4c:03:c5:fc:34
        Wired MAC Count        : 4
        Date Code              : 052520
        Serial                 : CNKCK2R7R0
        Wireless MAC           : 24:62:ce:c5:b4:70
        Wireless MAC Count     : 2
        Country                : CCODE-US-b69c719895e67525a096729
                                 da53abcb37ee4b837
Card 1: CPU
        Assembly               : 2010258C
        Serial                 : Y105810DA
        Date Code              : 051320
        Major Rev              : 02
        Minor Rev/Variant      : 00
Card 2: Power
        Assembly               : 2010259C
        Serial                 : Y105803B8
        Date Code              : 051320
        Major Rev              : 02
        Minor Rev/Variant      : 00
apboot>
```

5. After you have noted the serial and MAC, enter the **reset** command to reboot the AP.

```
apboot> reset
resetting ...
```

Note

Even if you don't execute any command, the AP will reboot automatically after a few minutes.

6. Repeat the previous procedure for AP2 and take note of the *serial number* and *MAC address* of AP2.

PC-MGMT, PC-1, and PC-4

The PC-MGMT is used for device management access. The PC1 and PC4 devices will be used as lab test-hosts for wired and wireless access.

1. To access the desktop PC-MGMT, right-click on the icon in the lab diagram and select **Open Desktop**. A new browser tab will open with the remote desktop.
2. Repeat the previous 2 steps on PC-1 and PC-4.

Core-router (via PC-MGMT)

Unlike the access and aggregation switches, the core router in the remote lab is not a physical switch; it is running the AOS-CX simulator software developed by Aruba.

1. Switch back to the PC-MGMT desktop.
2. On the desktop, you can open PutTTY or MTPuTTY (multi-tab PuTTY). Either application is fine. Use what works best for you (separate Putty windows or multiple tabs in MTPutty).
3. You will find saved sessions to **rtr-core1**, like that shown in Figure 1-38. Double-click **rtr-core1** saved session to open the connection.

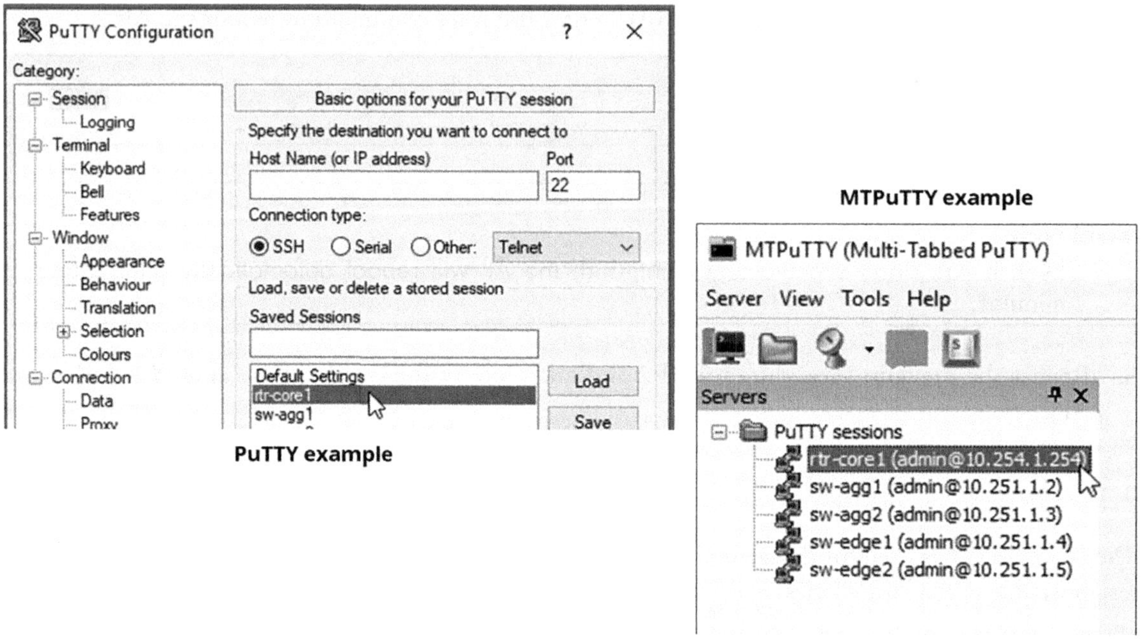

Figure 1-38: PuTTY and MTPuTTY access

4. Log in using the username **admin** and password **Aruba123!**

ClearPass (via PC-MGMT)

ClearPass is the AAA policy management solution used in the Aruba lab environment. The ClearPass admin web user interface (UI) can be accessed using the PC-MGMT desktop.

5. Switch back to the PC-MGMT desktop. On PC-MGMT, open a web browser, Google Chrome for example, and navigate to: **https://10.254.1.23**.
6. You will be presented a security certificate warning. Accept the warning. You will be presented with the login page.
7. Click the **Policy Manager** icon.
8. Log in using the username **admin** and password **Aruba123!**

Aruba Central access

Aruba Central is a cloud-based solution. It can be accessed with any system that has internet access. You can access Aruba Central using your local PC internet connection.

1. On your local PC, open a web browser and navigate to: **https://common.cloud.hpe.com**. Click **Sign in with SSO**.

2. In the Sign In With Single Sign-On box, enter the username for HPGLCP (with the @aruba-traininglabs.net suffix) and click **Next**. The username is given to you with your lab credentials when you purchase lab access separately. You will now be redirected to the remote lab login page based on the arubatraininglabs.net domain name.
3. Enter the username and the password for HPE GLCP (@arubatraininglabs.net) as provided with the self-directed lab options. Press **Log In**.
4. After the login has completed successfully, the GLCP will present the Aruba Central application in the Featured Applications list. Click **Launch** for the Aruba Central tile.
5. This will take you to the Aruba Central **Global** > **Overview** page (see Figure 1-39).

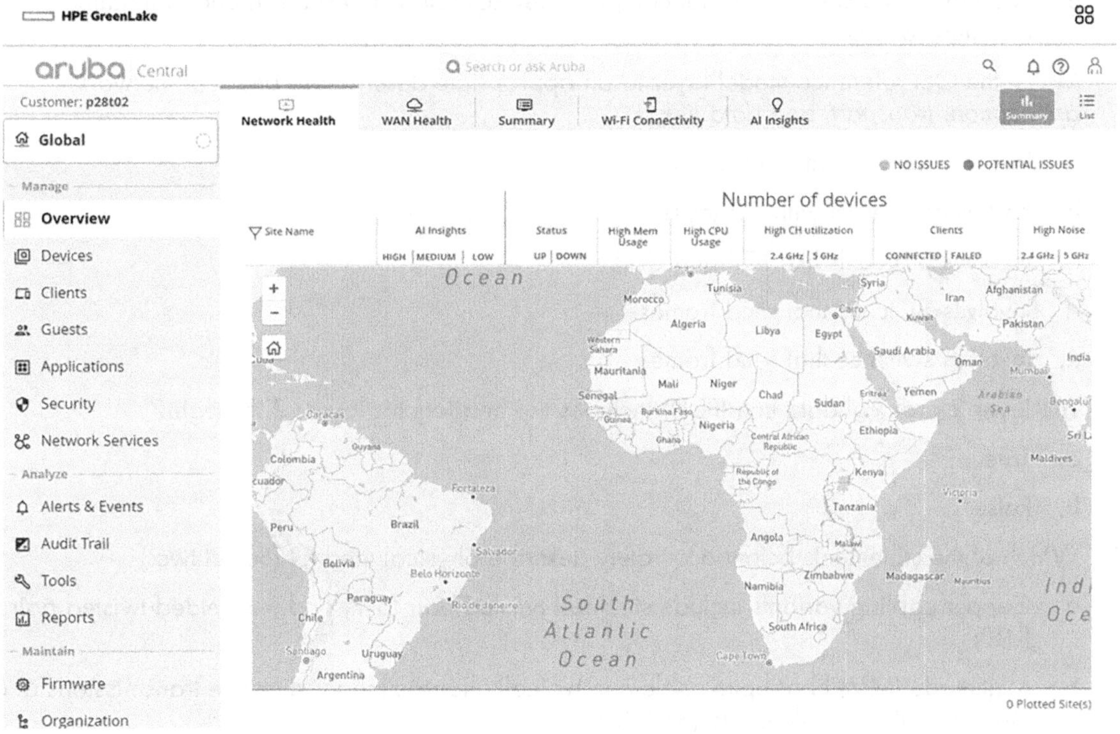

Figure 1-39: Aruba Central Overview window

Note

In these steps you have seen how to access Aruba Central. In the remainder of the lab guide these steps will not be repeated. You will simply be instructed to access Aruba Central. Please refer to this task if you would need assistance with the login steps.

Learning check

Discuss these questions.

1. Which of the statements below accurately describe basic network concepts? (Select two.)
   a. A computing network is a group of computing resources that permit digital data exchange between routers.
   b. A LAN is a group of computing devices in the same place, all interconnected via a router.
   c. A WAN is a group of computer resources that can communicate over large geographical distances.
   d. A protocol is a set of rules that computer devices follow to establish and maintain communications.
2. Match the OSI reference model layer to an appropriate description: *physical*, *network*, *application*, *transport*, and *data link*.
   a. Involves routers that send packets
   b. End users interact with this layer
   c. Involves protocols like TCP and UDP
   d. Involves switches that send frames
   e. Involves switches that send frames
3. The layer 2 protocol data unit (PDU) includes the headers for layers 3 through 7.
   a. True
   b. False
4. Which of the statements below accurately describe physical media? (Select two.)
   a. Copper cabling variants include shielded twisted pair (STP) and unshielded twisted pair (UTP).
   b. Multimode (MM) fiber optic cable can typically handle longer distance transmissions at a lower cost than single mode (SM) fiber.
   c. You typically use direct-attached copper (DAC) cables to directly connect two servers or other end systems to each other.
   d. Wi-Fi devices transmit data by modifying aspects of radio waves, such as amplitude, phase, or frequency.
   e. Multicast traffic is used to transmit data from one host to all other hosts on a LAN.

5. Match the TCP/IP stack layer (*network interface, application, transport,* and *internet*) to the appropriate description.
   a. Maps to three specific layers of the OSI model
   b. You might have an Ethernet or wireless NIC at this layer
   c. Deals with IP addresses
   d. Passes data to the application layer
6. Your colleague is new to networking and asks you about network devices. Which statements below would help your colleague understand this topic? (Select two.)
   a. You use switches to connect many end systems together, at layer 3 of the OSI model.
   b. Routers are often used in WAN environments, and forward packets based on IP address information.
   c. Multi-layer switches perform functions of a firewall and a router.
   d. Stateful firewalls have more intelligence than stateless firewalls, perhaps including application awareness features.

Summary

Congratulations! You have completed the first chapter. You should now be able to:

- Define networking, LAN, WAN, and their components.
- Explain the OSI model and encapsulation.
- Discuss different types of physical media.
- Compare unicast, multicast, and broadcast.
- Explain the TCP/IP stack.
- Discuss different types of networking devices.

2 AOS-CX Switching Fundamentals

OBJECTIVES

This chapter lays the foundation on which much of your future studies relies. After you take that new switch out of the box and connect it, then what?

After completing this chapter, you will be able to connect to and access switches and perform initial switch setup and validation. Then you configure VLANs and make certain port members of those VLANs, configure some ports as trunk ports, and configure other ports as layer 3 ports with IP addressing. Now that the network is functional, you can use LLDP and ICMP for network discovery, documentation, and diagnosis. Then you will learn to configure link aggregation to improve performance and resiliency.

After completing this chapter, you should be able to:

- ✓ Connect to and access the CLI of an AOS-CX switch.
- ✓ Perform an initial switch setup.
- ✓ Describe and configure VLANs, VLAN tagging (trunking), and IP addressing.
- ✓ Use LLDP and ICMP for network discovery and diagnosis.
- ✓ Configure link aggregation to improve performance and resiliency.

Initial switch setup

Let's begin with accessing and performing the initial setup of the AOS-CX switches.

Out-of-band management network

In-band management (left-side of Figure 2-1) is when you remotely connect to and manage switches via a normal Ethernet port. This seems convenient; in certain environments, like data centers, it could be a bad idea. A switch or router misconfiguration breaks your administrative access. There is also a security concern. Without tight security controls, hackers can access and control your devices.

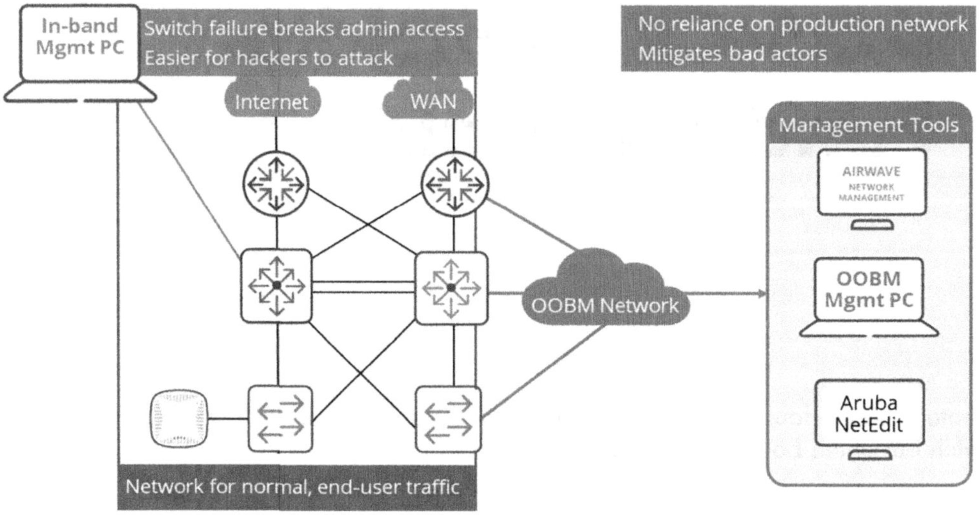

Figure 2-1: In-band and out-of-band management

Out-of-band (right-side of Figure 2-1) management (OOBM) uses alternate, dedicated connections to network devices, separate from the end user data network.

You can monitor and configure network devices without relying on the production network. Plus, you establish a trust boundary; with only one entry point for the management interface, it is far more difficult for bad actors to attack you. But how do you establish this connection?

Accessing the serial console port: OOBM

You typically perform initial switch configuration using out-of-band management (OOBM) on a console port for the initial switch configuration (see Figure 2-2). This special port exists on all switch models and types, to facilitate configuration, troubleshooting, and management.

Depending on the switch model, the console can be a USB-C port or an RJ-45 port; however, for both port types you must establish a serial connection between the management station (perhaps your laptop) and the switch. To do this, you need software on your laptop or management station that emulates a serial session. PuTTY, TeraTerm, and SecureCRT are common examples. Make sure to configure your terminal emulation software with the parameters shown in Figure 2-2.

If the switch has an RJ-45 port, you also need the serial cable that ships with the switch, and a USB-to-serial converter – modern laptops no longer include a serial port. If the switch has a USB-C port, you need a USB-C cable – connected directly from the switch to your laptop. All AOS-CX switches also ship with a Bluetooth USB dongle to enable easy initial setup with the CX mobile app. For more information, see: https://community.arubanetworks.com/community-home/digestviewer/viewthread?MID=35150. If the URL changes, do an Internet search on “How to: Using the Bluetooth USB Dongle for a Console Connection on a CX Switch”.

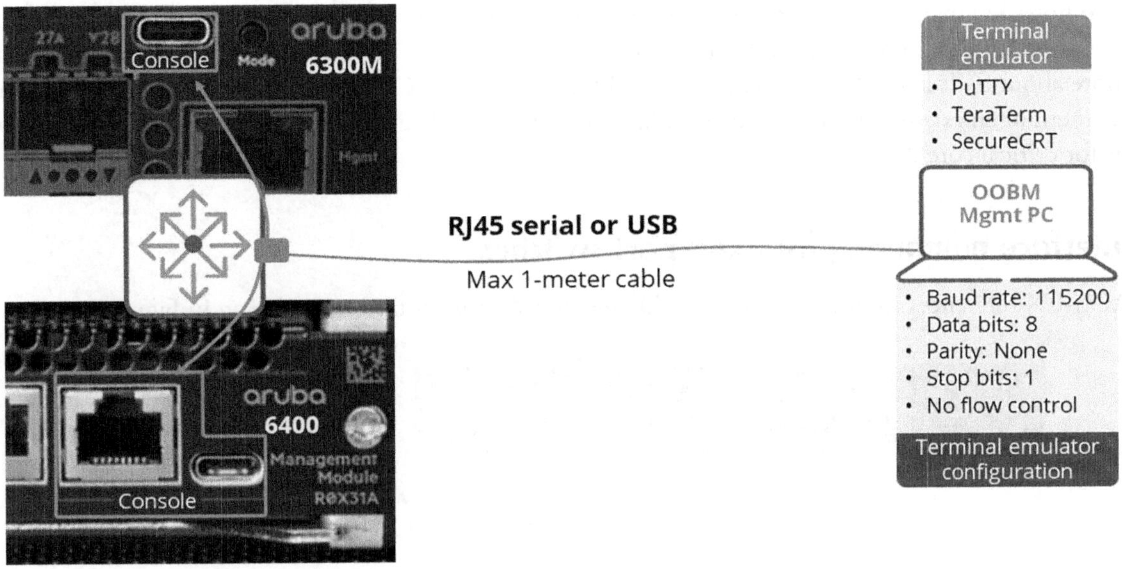

Figure 2-2: Accessing the console port

Management (mgmt) port

Another way to make an OOBM connection is via the special Ethernet management port. All AOS-CX switches include an out-of-band management (*mgmt*) port to completely isolate management traffic from all other traffic (see Figure 2-3). This isolation is achieved using VRF (virtual routing and forwarding), which is like a dedicated IP network stack separate from the default stack. So, management traffic does not use bandwidth in the data links, and data traffic cannot contact this interface. You must assign an IP address to this interface.

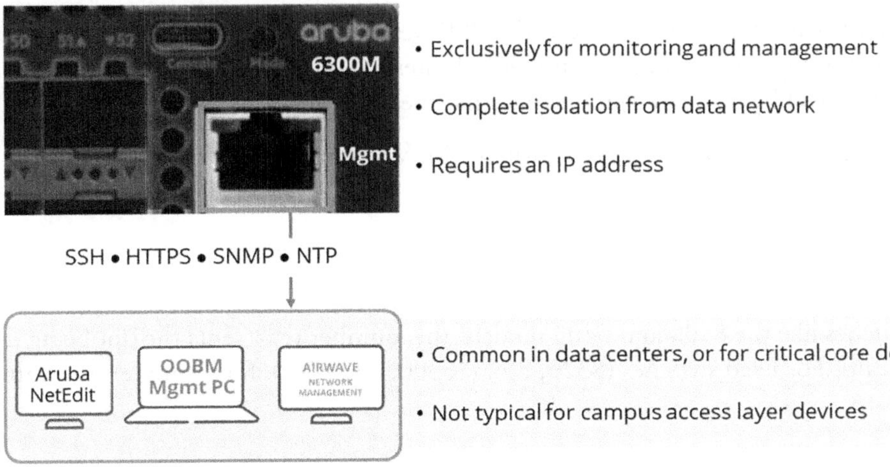

Figure 2-3: Ethernet management port

Also note that many access switches have a so-called *switch virtual interface* (SVI) enabled as a DHCP client, to support automatic connections to Aruba Central for Cloud management. You will learn more about SVIs later. Aruba tools like NetEdit, AirWave, and Central are great for centralized management of an existing network and compliment CLI access. The common use case is in a data center, or for critical core devices.

Interface numbering for fixed-port switches

AOS-CX references interfaces following the *member/slot/port* notation, as shown in Figure 2-4.

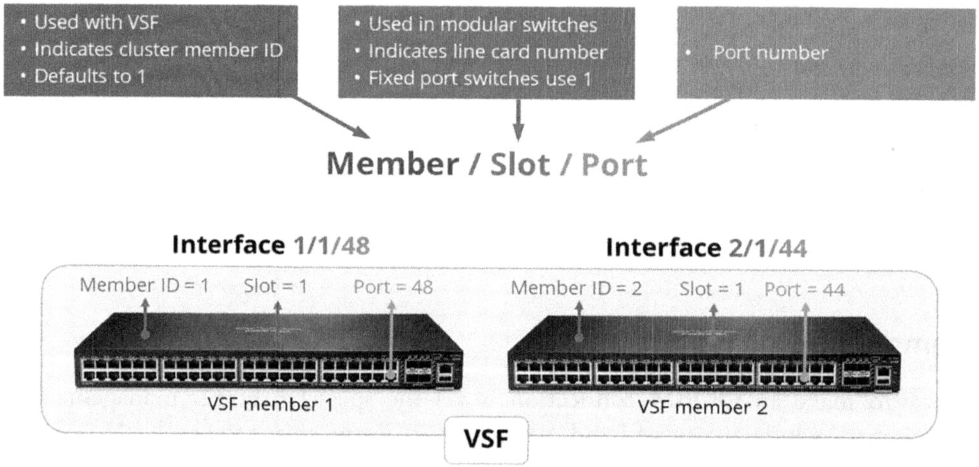

Figure 2-4: Interface numbering for fixed-port switches

Member

When the Aruba Virtual Switching Framework (VSF) feature is running, multiple switches can be seen as one virtual or logical switch; this number indicates the member ID in the cluster. By default, an AOS-CX switch will be member 1, indicating that these features are not running.

Figure 2-4 shows an example of a fixed-port switch with VSF enabled. If VSF is disabled all switches will have default member ID 1.

Slot

In modular switches like the 8400 and 6400 models, this number represents the slot being used by a particular line card. For fixed switches (8320, 8325, 6300, and 6200, for example) this number will be always 1.

Port

This number refers to the individual interface in the line card (modular switches) or in the chassis (fixed-port switches). Unlike some vendors, port numbers always start at *1* on AOS-CX switches.

> **Note**
> VSF is available for the 6200 and 6300 switch series.

Port numbering for chassis-based switches

Here's a couple of typical Aruba OS-CX switches–lots of Ethernet ports right (see Figure 2-5)? How do you identify them? You use the typical *member/slot/port* notation.

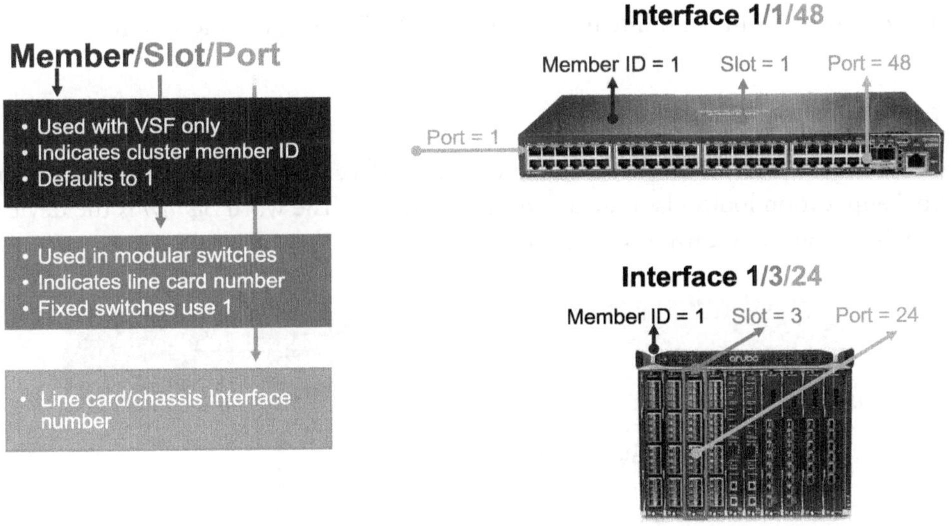

Figure 2-5: Port numbering for chassis-based switches

The member number is only used with a special feature—the Aruba Virtual Switching Framework (VSF), which makes multiple switches appear as one virtual or logical switch. The member number indicates a physical switch's unique member ID. The default member ID is 1, indicating that the feature is not running.

In modular switches like the 8400 and 6400, the *slot* number represents the slot a particular line card uses. For fixed switches like 8320, 8325, and 6300, this number is always 1. And the *port* number refers to the individual interface in the line card for modular switches, or in the chassis for fixed switches.

Look at the fixed-port switch in the top-right corner of Figure 2-5. It does not run VSF, so the member ID is always 1. It doesn't have slots because it is a fixed-port switch, so the slot number is always 1. Then the individual ports are numbered, from port 1 at the upper left, to port 48 at the lower right. So, you refer to this last port as *interface 1/1/48*.

The modular switch in bottom-right of Figure 2-5 does not run VSF, so its member ID is always 1. But modular switches have slots, into which you install special *line cards*. The third slot in from the left would be identified as slot three. The individual ports are numbered on each card; therefore, in Figure 2-5, the port highlighted for the switch at the bottom right is port 24. Thus, you refer to it as *interface 1/3/24*.

Now that you know how to refer to switch ports, so let's move on to the switch configuration.

AOS-CX prompts

This section covers the command line prompts for the AOS-CX operating system.

Default prompt

When you access the AOS-CX switch command line interface (CLI) for the first time, the prompt in the terminal application looks like that shown in Figure 2-6. The word *Switch* is the device's default hostname, which you soon learn how to change.

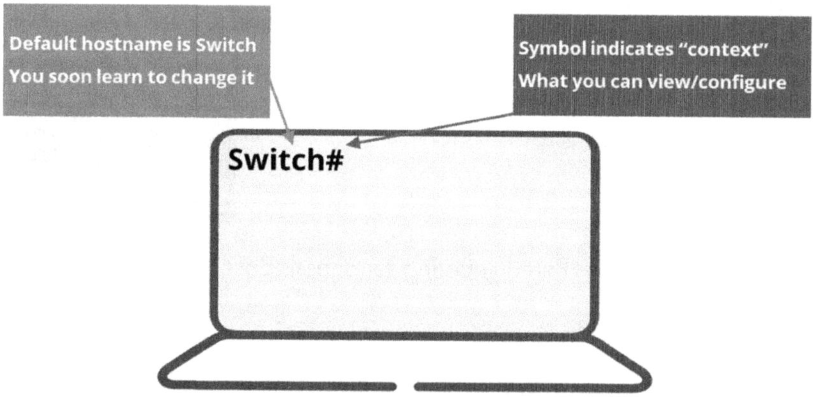

Figure 2-6: AOS-CX configuration prompts

The CLI is organized into various *contexts* or levels, as indicated by the symbol following the host-name. Each context determines which parts of the switch you manage, and which commands are available to users. Let's explore the contexts in more detail.

Prompt modes: Overview

The operating system defines the contexts shown in Figure 2-7. The *operator* context level is indicated by a greater-than (>) symbol. When you see this, know that you can run a sub-set of **show** or view commands, but cannot change configuration.

The *manager* context has the hashtag symbol, where you can run nearly all **show** commands to view status and configuration, but you cannot modify that configuration.

The *global configuration* context ends with the "(config)#" in the prompt. This is where you run commands that change the switch's configuration. These commands have a *global* effect on the device itself. But to configure specific components or features, you must access a sub-config mode.

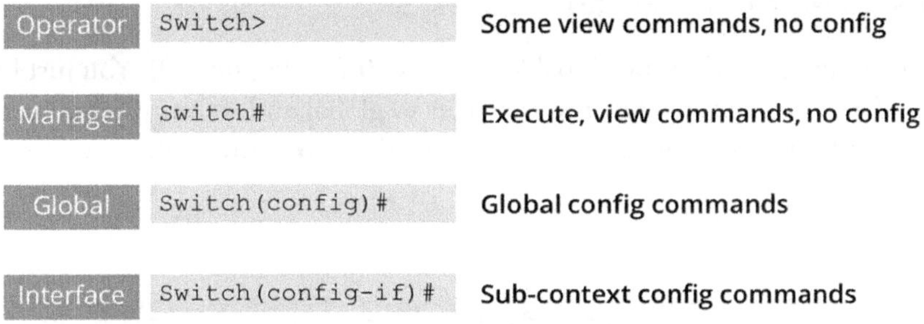

Figure 2-7: Prompt modes—overview

What you see in the prompt depends on what you are configurating. Figure 2-7 shows *config-if*– short for configure interface. You will see different prompts when you configure other aspects of the device. You'll see many sub-contexts throughout the material in this book. Mainly, you must understand how to navigate through these contexts. As you'll soon see, it's easy!

Prompt modes: Move to other modes

You begin as a lowly operator. To access manager mode, you enter the **enable** command. Your prompt changes from > to # (see Figure 2-8).

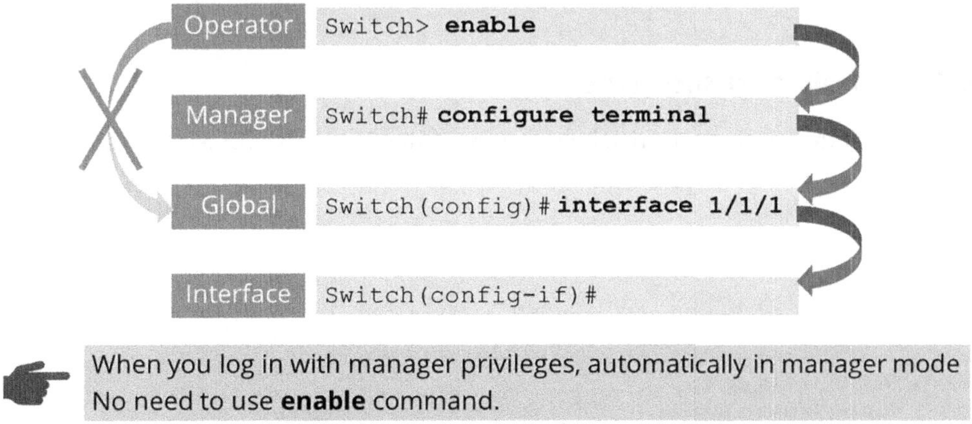

Figure 2-8: Prompt modes—move to other modes

To move from manager to the global config context, enter **configure terminal.** You see the "*config*" prompt displayed. To move from global to the interface context, use **interface** followed by the port identifier. Note that you cannot move directly from operator to global or interface using a single command. You must progress in the order shown in Figure 2-8.

Note

When you log in with manager privileges, the AOS-CX switches automatically places you in manager mode. Thus, there is no need to use the **enable** command in this situation.

Prompt modes: Return to previous modes

To move from the *interface* level to the global level, use **exit** (see Figure 2-9). You just learned that you cannot *skip a level* moving from operator to manager to global and so on. But you can move directly from any sub-config level to the manager context with the **end** command.

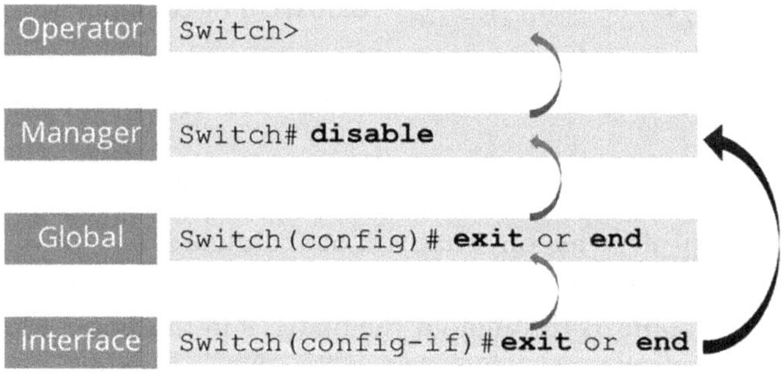

Figure 2-9: Prompt modes—return to previous modes

So, you can move from the global to manager level with either **exit** or **end**. You can also use **<Ctrl-z>** instead of **end**. To move from the manager to the operator level, use **disable**. To move from the interface to the manager level, use **end**. And from the operator level, if you want to log out of the switch, use the **exit** command.

AOS-CX: CLI help and shortcuts

This section covers the context-sensitive help feature of the AOS-CX CLI as well as CLI shortcuts.

Context sensitive help

The AOS-CX CLI provides help features. To see the available commands that you can run in the current context, enter the question mark (?) symbol (see Figure 2-10). Note that the question mark does not display on the screen when you enter it.

Or perhaps you know you want to use some variation of the **show access-list** command, but you are not sure what options you can specify. Just enter the command followed by a space and a question mark. So, if you wanted to see all access lists configured on the device, you would enter **show access-list all**, and so on. Pretty cool!

Show available commands in the current context

```
Switch# ?
boot                 Reboot all or part of the system
checkpoint           Checkpoint information
<<<Omitted Output>>>
Switch#
```

Show the available parameters for a command

```
Switch# show access-list ?
  all           All access-lists
  ip            Internet Protocol v4 (IPv4)
  ipv6          Internet Protocol v6 (IPv6)
  log-timer     Set ACL log timer length (frequency)
  mac           Ethernet MAC Protocol
Switch(config)# access-list
```

Figure 2-10: Context-sensitive help

Command abbreviation and completion

AOS-CX CLI supports command abbreviation–just enter enough letters to uniquely specify any command, and the CLI accepts it. For example, you can simply enter **conf** instead of typing **configure terminal** to access global config mode (see Figure 2-11).

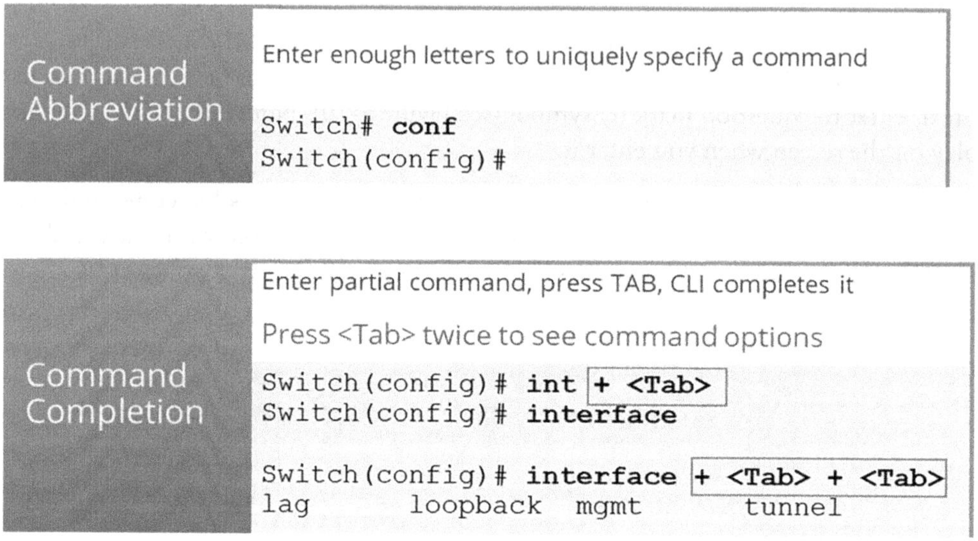

Figure 2-11: Command abbreviation and completion

The CLI also supports command completion. If you enter enough letters of a command to uniquely identify it, then press **<Tab>**, the CLI completes the command for you. If you have not entered enough letters to match a command, the CLI does not complete it.

In the example in Figure 2-11, someone has entered **int**, then pressed the **<Tab>** key. The CLI completed the command, so you see **interface** spelled out.

If you press **<Tab>** twice after a word, the CLI displays command options. For example, enter **interface**, then **<Tab><Tab>**. Notice that you see options (parameters) that you can enter after the **interface** command in Figure 2-11.

Common *show* commands

This section briefly covers some common **show** commands you can use to verify the operation of your AOS-CX switch.

show version command

The **show version** command displays version information about the network operating system software, service operating system software, and BIOS. In Figure 2-12, you see that this system is running AOS-CX version GL.10.04.0003. Near the bottom you see the service OS and BIOS versions. This is useful for your general knowledge, and to help to determine if upgrades are needed.

Version info for network OS, service OS, and BIOS

```
Switch# show version
-----------------------------------------------------------------
ArubaOS-CX
(c) Copyright 2017-2022 Hewlett Packard Enterprise Development LP
-----------------------------------------------------------------
Version      : FL.10.09.1031
Build Date   : 2022-06-14 04:21:23 UTC
Build ID     : ArubaOS-CX:FL.10.09.1031:dea0ef74d35b:202206140343
Build SHA    : dea0ef74d35b187ab79b41ffba4d68ec38cb5a63
Active Image : primary

Service OS Version : FL.01.09.0002
BIOS Version       : FL.01.0002
```

Figure 2-12: *show version* command

Context and authorization information for this command include:

- **Command context**: operator (>) or manager (#)
- **Authority**: operators or administrators

show images command

AOS-CX switches have two flash memory modules or *partitions* for storing switch software image files. AOS-CX refers to those modules as primary and secondary images. Having two flash memory locations helps to manage the switch more easily, and to help with planning firmware upgrades. For example, you can download a new firmware code in the flash memory that is not currently in use and then instruct the switch to boot from it during a maintenance window.

The **show images** command shows information about the software in the primary and secondary images (see Figure 2-13).

Version info for network OS, service OS, and BIOS

```
Switch# show images
---------------------------------------------------------------------------
ArubaOS-CX Primary Image
---------------------------------------------------------------------------
Version : FL.10.09.1031
Size    : 872 MB
Date    : 2022-06-14 04:21:23 UTC
SHA-256 : e91e7f1ade68ab54cd00918673501d6067cdbc428d3f4bf6f2beb31f5f8ef322
---------------------------------------------------------------------------
ArubaOS-CX Secondary Image
---------------------------------------------------------------------------
Version : FL.10.05.0021
Size    : 642 MB
Date    : 2020-10-29 10:36:02 PDT
SHA-256 : 4c795e8c9eec5952645ded19cf9a2018deb545c7ed0221f32a1a5bd0d64ee5f6
Default Image : primary
Boot Profile Timeout : 5 seconds
```

Figure 2-13: *show images* command

Context and authorization information for this command include:

- **Command context**: operator (>) or manager (#)
- **Authority**: operators or administrators

show system command

The **show system** command displays general status information about the system. During remote access, the device platform and version information may not be obvious. In the example shown in Figure 2-14, you are attached to device named *Switch*, which is an Aruba model 8325-48YC. It has 48x25 Gbps ports, and 8x100 Gbps ports. You see the serial number, base MAC address, and uptime, as well as CPU and memory utilization. This can give you some idea of the general health of the system. By baselining these parameters, you can begin to establish what *normal* is for your network.

Context and authorization information for this command include:

- **Command context**: operator (>) or manager (#)
- **Authority**: operators or administrators

General system status info

```
Switch# show system
Hostname               : Switch
System Description     : GL.10.04.0003
System Contact         :
System Location        :

Vendor                 : Aruba
Product Name           : JL635A Aruba 8325-48Y8C 48p 25G 8p 100G swch
Chassis Serial Nbr     : TW98KM003P
Base MAC Address       : 9020c2-bced00
AOS-CX Version         : GL.10.04.0003

Time Zone              : UTC
Up Time                : 1 week, 1 day, 11 hours, 20 minutes
CPU Util (%)           : 4
Memory Usage (%)       :19
```

Figure 2-14: *show system* command

Switch configuration files

This section introduces you to viewing and manipulating switch configuration files.

show running-config command

The **show running-config** command displays the current configuration running on the switch (see Figure 2-15). Note that no user information is displayed.

Current configuration, no user info

```
Switch# show running-config
Current configuration:
!
!Version AOS-CX GL.10.04.0003
!export-password:   default
Hostname Switch
!
Router ospf 1
      area 0.0.0.0
Vlan 1,1101,1103,1105,1107
Interface mgmt
      no shutdown
      ip static 10.250.1.3/24
      default-gateway 10.250.1.254
Interface 1/1/29
      no shutdown
      description To-8325-Core-Table01-Port47
      <<<Omitted output>>>
```

← Add **all** keyword to also see default values

Figure 2-15: *show running-config* command

Context and authorization information for this command include:

- **Command context**: manager (#)
- **Authority**: administrators

The **show running-config all** command includes configuration commands and values for all features, including the default configuration that is normally not visible in the configuration output.

Configuration file management

An AOS-CX switch uses RAM and flash memory to save configuration files. You should be aware of RAM and flash memory characteristics, as they relate to the storage of device configuration files (see Figure 2-16).

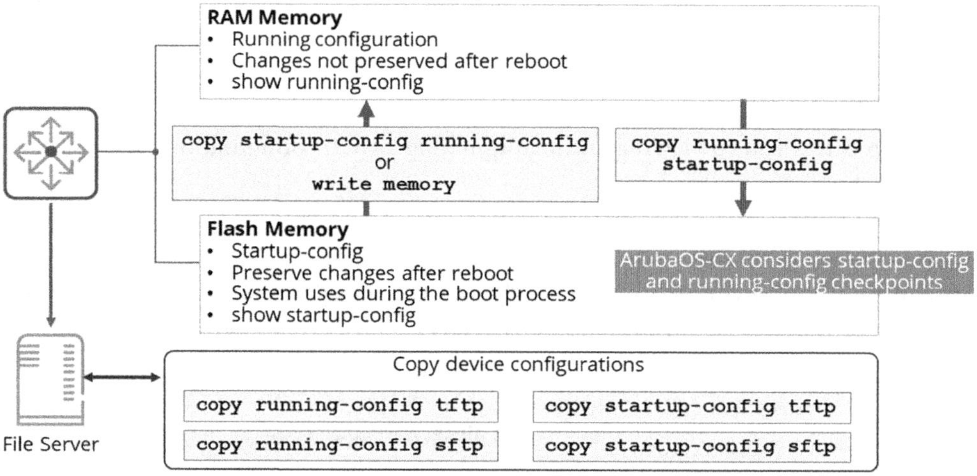

Figure 2-16: Configuration file management

RAM memory is used to save the configuration that is currently running on the device: the running-config. Remember that RAM is not permanent storage; the contents of RAM are lost when a device reboots or powers down. If you modify the running-config and the switch reboots, your changes are lost. Use the command **show running-config** to view this file.

Flash memory is used to permanently save a configuration file. The contents of flash memory survive during a reboot or power off. Use the **show startup-config** command to verify the contents on this file.

Of course, any time you make changes to the running-config, you should copy those changes to the startup-config file by using the **copy** command: to save the configuration that is running in the device, use the command **copy running-config startup-config.**

Note

Alternatively, you can use the command **write memory** command which does the exact same thing as the **copy running-config startup-config** command.

If you need to *restore* a configuration from the startup-config file, use the command **copy startup-config running-config.**

As a best practice, you should export the running and the startup configuration files to an external file server. This will help you to recover from catastrophic device failures or administrative mistakes. To accomplish this task, you can use the **copy** command, as shown here:

```
Switch# copy {startup-config | running-config}
{tftp://|sftp://USER@}{IP-address|host-
name}[:PORT] [;blocksize=VAIL]/file_name
```

For example, to copy the running-config to a TFTP server with and IP address of 10.253.1.21, use the command **copy running-config tftp://10.253.1.21/switch_config.cfg**

AOS-CX switches also allow you to copy files to a USB flash drive, which can be directly connected to the switch. To copy your running or startup-config to a USB device, use the following syntax:

```
Switch# copy {running-config | startup-config} usb:/file_name
```

AOS-CX considers startup and running config checkpoints, which you'll learn about in the next section.

Checkpoint overview

Using the **copy** commands that you just learned allows you to store configuration files and recover lost configurations in case of a power outage or an administrative mistake. This is a good thing, but these files do not include any additional data about the state of networking processes running on the switch. For a true recovery, a new approach is needed: *checkpoints*.

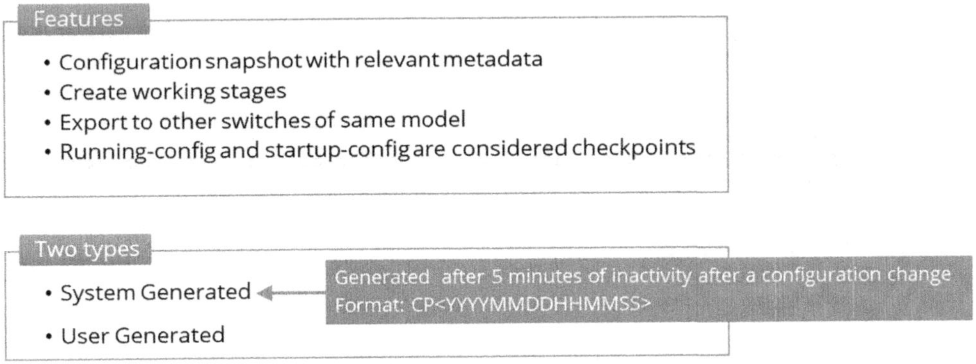

Figure 2-17: Checkpoint overview

A checkpoint (see Figure 2-17) is a snapshot of the switch running configuration and its relevant metadata during the time of creation of the configuration. You can use checkpoints to apply the switch configuration stored within a checkpoint whenever needed, such as revert to a previous working configuration. A checkpoint is flexible since it can be applied to other switches of the same model.

There are two types of checkpoints:

- System generated checkpoints are generated after 5 minutes of inactivity after a configuration change. By default, a checkpoint is taken every 300 seconds or 5 minutes. The naming format that this checkpoint type uses is CP<YYYYMMDDHHMMSS>. For example, *CPC20200111000038*
- User generated checkpoints are manually generated by the administrator and can be assigned whatever name the administrator deems relevant at the time of creation.

You'll learn how to configure these in the next section.

Checkpoint configuration

To create a checkpoint, use the **copy** command (see Figure 2-18):

```
Switch# copy {running-config | startup-config} checkpoint
<checkpoint-name>
```

To verify the checkpoint list, use the command **show checkpoint list all**.

Create manual checkpoint

```
Switch# copy {running-config | startup-config} checkpoint <checkpoint-name>
```

Verify checkpoint list

```
Switch# show checkpoint list all
```

Copy checkpoint

```
Switch# copy checkpoint <checkpoint-name> {running-config | startup-config
| <external-file-server>}
```

Restore to running-config

```
Switch# checkpoint rollback <checkpoint-name>
```

Figure 2-18: Checkpoint configuration

To revert the switch's configuration to a previously saved checkpoint use the following command:

```
Switch# copy checkpoint <checkpoint-name> {running-config |
startup-config}
```

Note

You can export a checkpoint file to an external server.

Alternatively, you can use the **checkpoint rollback** command. This command only restores the checkpoint configuration to the running-config file of the switch.

```
Switch# checkpoint rollback <checkpoint-name>
```

Note

The **checkpoint rollback** command performs the same task as **copy checkpoint running-config** command.

Checkpoint auto mode

There are situations when applying a configuration could lock out your access to the switch. This leaves no other option but to locally connect to the device using the console port and remove the offending configuration lines.

The checkpoint auto mode feature addresses this problem in an elegant way. The solution includes a timer that you set (see Figure 2-19). The system then creates a checkpoint with the current state. You then make your desired configuration changes, which the system applies, after a confirmation. If the configuration lines did not affect your access, you will be able to confirm the change. If your changes lock you out of the switch, confirmation does not occur. The switch thus automatically reverts to the latest snapshot after the timer expires.

1. Administrator enables checkpoint auto mode.
   `Switch# checkpoint auto <minutes>`
2. The system generates a checkpoint and starts a timer.
3. Administrator configures the device.
4. Administrator must confirm the changes to be permanent.
   `Switch# checkpoint auto confirm`

If no confirmation received, system rolls back to checkpoint.

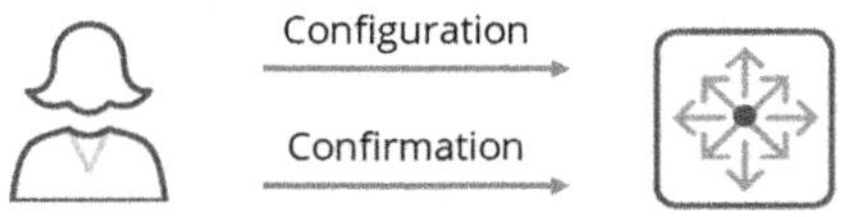

Figure 2-19: Checkpoint auto mode

To use the checkpoint auto mode, issue the following configuration.

```
Switch# checkpoint auto <minutes>
Switch# checkpoint auto confirm
```

Important

Do not execute the checkpoint auto command until you have verified that your configuration changes are functioning correctly.

Switch operation verification

This section introduces you to three commands that you can use to verify the operation of your AOS-CX switch.

top CPU

The **top cpu** command shows detailed CPU utilization information for the AOS-CX switch. You learned how to see high-level CPU utilization with the command **show system**. The **top cpu** command provides more detailed information (see Figure 2-20).

CPU utilization info

```
Switch# top cpu
top - 20:40:46 up 17 days, 23:45, 1 user, load average: 0.32, 0.26, 0.24
Tasks: 238 total, 1 running, 194 sleeping, 0 stopped, 0 zonbie
%Cpu(s): 2.5 us, 1.6 sy, 0.0 ni, 95.9 id, 0.0 wa, 0.0 hi, 0.0 si, 0.0 st
KiB Mem : 7868164 total, 3013360 free, 13999328 used, 3455476 buff/cache
KiB Swap: 2097148 total, 2095868 free,     1280 used. 6079420 avail Mem

  PID USER  PR  NI   VIRT   RES   SHR S %CPU %MEM     TIME+ COMMAND
26308 admin 20   0   3344  2064  1612 R 11.8 0.0    0:00.03 top -b -n 1 -s -c -o
  629 root  20   0  33928 17084  6124 S  5.9 0.2  777:57.14 /usr/sbin/ovsdb-server
 3047 root  20   0 791184 84416 32232 S  5.9 1.1  189:43.13 /usr/bin/prometheus
    1 root  20   0   7600  6312  2960 S  0.0 0.1    0:28.14 /sbin/init
```

Figure 2-20: **top cpu** command

Context and authorization information for this command include:

- **Command context**: operator (>) or manager (#)
- **Authority**: operators or administrators

Likewise, the **top memory** command shows memory utilization information for processes running on the switch.

show events command

The **show events** command displays the event logs generated by the switch modules since the switch last rebooted (see Figure 2-21). Log analysis is a powerful tool to investigate and troubleshoot system and protocol-related problems. Use the **-r** parameter to list the most recent log events first (ordered from newest to the oldest events).

Event logs since last reboot

```
Switch# show events -r
---------------------------------------------------
Event logs from current boot
---------------------------------------------------
2019-12-06T10:24:18.679695+00:00 Spine-8325-C hpe-sysmond[3212]:
Event|6303|LOG_INFO|AMM|1/1|Current system memory usage is 27%
2019-12-06T09:09:19.067316+00:00 Spine-8325-C lldpd[2165]:
Event|106|LOG_INFO|AMM|1/1|LLDP neighbor 90:20:c2:bc:7c:00 deleted  on 1/1/35
```

Use **-r** parameter to list most recent events first

Figure 2-21: **show events** command

Context and authorization information for this command include:

- **Command Context**: manager (#)
- **Authority**: administrators or auditors

Note

The **show event** command has other parameters that you can use like **-e**, **-s**, **-a**, **-n**, **-c**, and **-d**. You can learn more about how to use these parameters in the Command-Line Interface Guide AOS-CX document.

Note

The event log is a file on the switch SSD drive, where there is option to see events *before* the last reboot.

---

show interface brief command

The **show interface brief** command helps the administrator to see what interfaces are available and the current status of these interfaces. This command also briefly shows layer 2 and layer 3 configurations applied to the interface. This is a frequently used command for many network engineers. At a glance, you can tell the status of every interface on the entire device, with a specific focus on the *Enabled* and *Status* columns (see Figure 2-22).

In the *Enabled* column, *yes* means that the interface is not disabled in the switch configuration – no administrator has disabled the interface. *up* in the status column means that something is attached to that interface, and there is at least a good layer 1 and layer 2 connection between them.

Context and authorization information for this command include:

- **Command context**: operator (>) or manager (#)
- **Authority**: operators or administrators

Available interfaces and status

```
Switch# show interface brief
--------------------------------------------------------------------------------
Port    Native  Mode    Type        Enabled Status  Reason  Speed    Description
        VLAN                                                (Mb/s)
--------------------------------------------------------------------------------
1/1/1   1       routed SFP28DAC3   yes     up               25000    --
1/1/2   1       trunk  SFP28DAC3   yes     up               25000    --
1/1/3   1       trunk  SFP28DAC3   yes     up               25000    --
<<<Omitted Output>>>
```

Command output filtering

```
Switch# show interface brief | include up
1/1/1    11     access 1GbT        yes     up               1000     pc1
1/1/2    4      trunk  1GbT        yes     up               1000     ap1
1/1/3    1      access 1GbT        yes     up               1000     --
1/1/26   1      access SFP+DAC1    yes     up               10000    --
1/1/27   1      trunk  SFP+DAC1    yes     up               10000    --
1/1/28   1      trunk  SFP+DAC1    yes     up               10000    --
```

Figure 2-22: **show interface brief** command

The pipe (|) command is a CLI session command that filters the output of **show** commands according to the criteria specified by the parameter include, exclude, count, begin, or redirect.

- The pipe (|) command is supported for use with the **show** command only.
- You can use multiple pipe commands with a single show command.
- You can use the pipe command with the **page** command.
- Command completion by pressing the **<Tab>** key is not supported for pipe commands.

Here is an example where you see every line in the output of the running-config file that has the word "vlan" in it, with the exceptions of those output lines that contain the word "vlan2":

```
Switch# show running-config | include "vlan" | exclude "vlan2" | count
```

The **count** parameter causes the switch to display a count (the number of occurrences) instead of the actual lines that match the criteria in the pipes.

Basic switch configuration

This section introduces you to the basic configuration tasks you'll typically perform on an AOS-CX switch, like defining a hostname and configuring and verifying a switch's ports.

Configuring a hostname

You should name each network device. This ensures that you can easily identify them, especially in large networks. In a real environment this is one of the first configuration steps that you perform.

Many corporations have documented, standardized, naming conventions. For example, one standard might be *Building-Floor-Rack_Name*. For example, the switches in rack 2, on the 2nd floor of building 9, might be named *9-2-2_SWA*, *9-2-2_SWB*, *9-2-2_SWC*, and so on and so forth.

Figure 2-23 shows how to name devices using the **hostname** command.

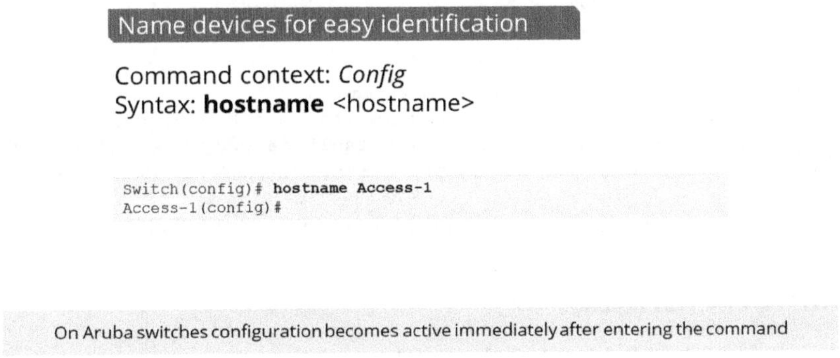

Figure 2-23: Configuring a hostname

Context and syntax information for this command include:

- **Command context**: config
- **Syntax**: **hostname** <hostname>

Here's an example:

```
Switch(config)# hostname Access-1
Access-1(config)#
```

Configuring the management port

The AOS-CX mgmt interface is enabled by default. It receives its IP parameters using DHCP. Use the **show interface mgmt** command to verify the interface's state (see Figure 2-24).

If you prefer, you can manually configure IP parameters on the management interface, as shown in Figure 2-24 (DHCP is the default setting for the interface addressing).

```
Switch(config)# interface mgmt
Switch(config-if-mgmt)# ip static <IP-address/mask>
Switch(config-if-mgmt)# default-gateway <default-gateway-IP>
Switch(config-if-mgmt)# nameserver <DNS-primary-server-IP>
```

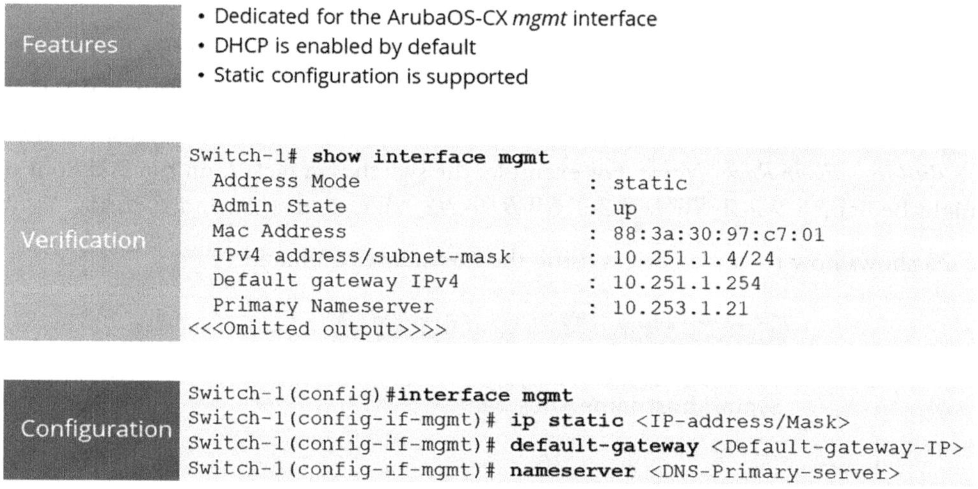

Figure 2-24: Configuring the management port

Managing Ethernet interfaces

In AOS-CX, all ports are disabled by default on all core and aggregation switches. On access layer switches, the port status is enabled by default. You can modify this state using the **no shutdown** and **shutdown** commands, as shown in Figure 2-25.

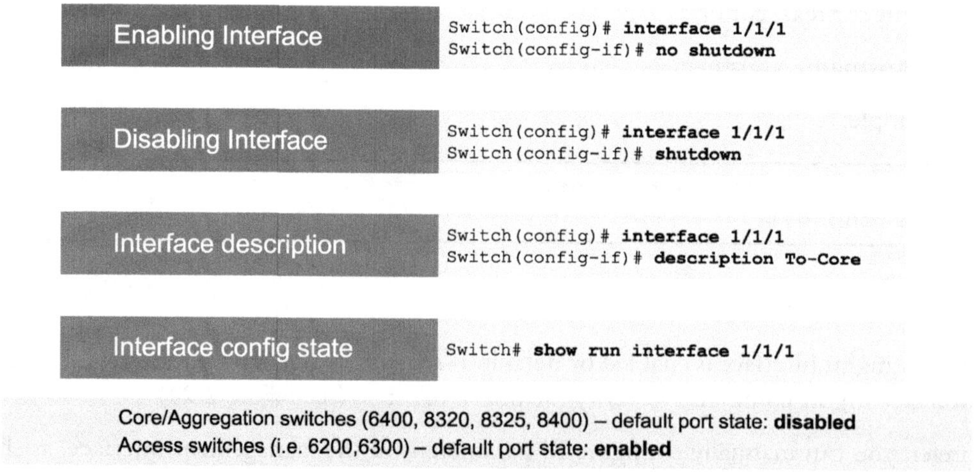

Figure 2-25: Managing Ethernet interfaces

Context and syntax information for this command include:

- **Command context**: config-if (interface level)
- **Syntax**: **shutdown** and **no shutdown**

Here is an example of enabling an interface:

```
Switch(config)# interface 1/1/1
Switch(config-if)# no shutdown
```

Here is an example of disabling an interface:

```
Switch(config)# interface 1/1/1
Switch(config-if)# shutdown
```

A description can be configured for each interface, which facilitates management.

Context and syntax information for the **description** command include:

- **Command context**: config-if (interface level)
- **Syntax**: **description** <description>

Here is a configuration example of defining a description for an interface:

```
Switch(config)# interface 1/1/1
Switch(config-if)# description To-Core-switch
```

You can view the current configuration state for an interface with the command **show run interface 1/1/1**, for example, as shown Figure 2-25.

You can also revert an interface to its default configuration from config-if mode with **default int 1/1/1**, for example.

Verifying interface statuses

The **show interface** command shows status and configuration for all interfaces on the switch. In Figure 2-26's top example, *Administratively down* means that a **shutdown** command was configured on that interface (or the interface is in its factory-default state in case of the AOS-CX core and aggregation switches). You can also see that an interface description was configured on that interface with the **description** command.

In the bottom example in Figure 2-26, notice that the interface is now up. Somebody has entered the **no shutdown** command. The interface description remains unmodified.

Context and syntax information for the **show interface** command include:

- **Command context**: manager (#)
- **Syntax**: **show interface** <interface-ID>

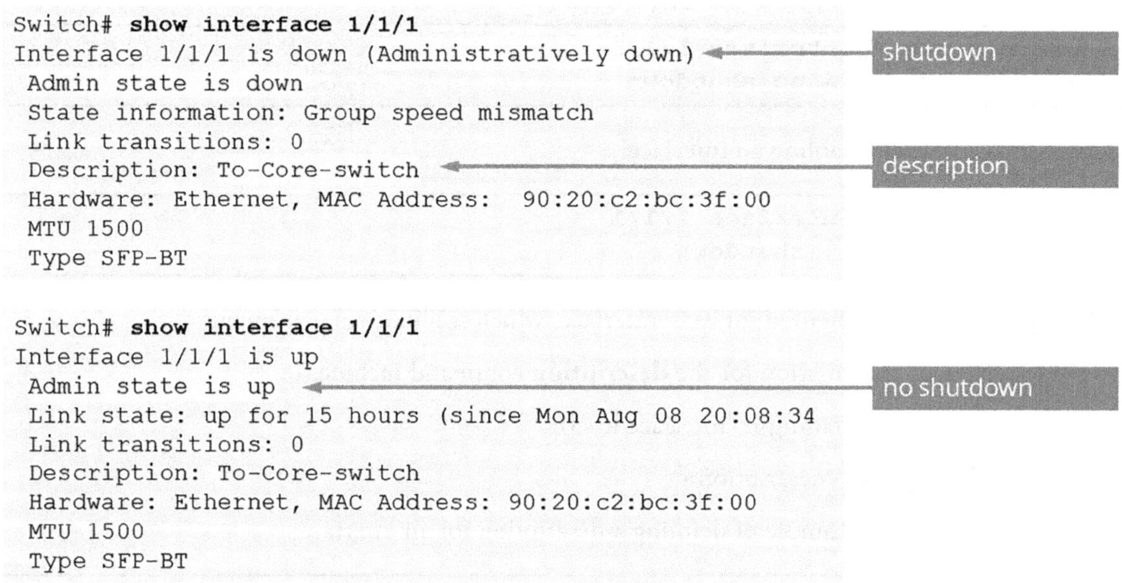

Figure 2-26: Verifying interface statuses

Introduction to VLANs

Virtual LANs (VLANs) are a fundamental networking component that help improve security, scalability, and many more features. This section introduces VLANs and their configuration and verification.

VLAN overview

A local area network (LAN) is a group of devices in the same broadcast domain. Figure 2-27 shows two LANs, labeled LAN-A and LAN-B. LAN-A switch has two connected devices. Host A connects to Port 1 and Host B connects to port 2. The other switch has similar connections.

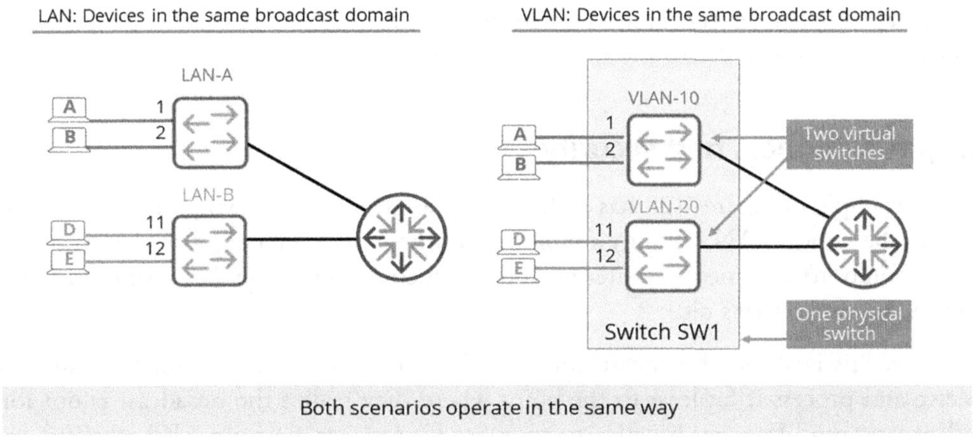

Figure 2-27: LANs and VLANs

Recall that if a switch receives a broadcast on any port, it floods it out all other ports, except the ingress port. Therefore, if Host A sends a broadcast, the LAN-A switch forwards it out all other ports, and so Host B receives the frame. Of course, Hosts D and E do not receive this frame – there is no connection between LAN-A and LAN-B. They are physically separate LANs, connected to physically separate switches.

What if you wanted to connect these three LANs into an internetwork? Simply add a routing device, like a traditional router, which can router unicast and multicast traffic between the LANs.

Now consider a VLAN, which is, just like a physical LAN, a group of devices in the same broadcast domain. Suppose that you have a single, physical Aruba switch named SW1. Hosts A–E connect to ports 1, 2, 11, and 12 on this switch (see the right side of Figure 2-27). By default, all these devices are in the same broadcast domain. If Host A sends a broadcast, all other hosts receive the frame. Now suppose that you learned about creating VLANs and created a VLAN named *VLAN10*. It is as if you have created a small virtual switch inside the physical switch. This virtual switch exists, but it is not connected to any of the physical switch ports.

And therefore, you define or *map* physical ports 1 and 2 as being members of the red VLAN 10. Similarly, you create VLAN20, and assign ports 11 and 12 as members of the blue VLAN 20. You have now created two separate broadcast domains on a single physical switch. When Host A sends a broadcast, the switch knows to only forward that frame out all ports that are in the same VLAN, with the exception of the ingress port. Only Host B receives the broadcast. If Host D broadcasts, only E receives it.

You have effectively recreated the scenario on the left of Figure 2-27, using a single physical switch. Just like the physical scenario on the left, there is *no* connectivity between the separate VLANs. No unicast, multicast, nor broadcast traffic shall pass between the VLANs. Of course, you could always connect a router to create an internetwork.

But if the scenarios operate in the same way, why bother with VLANs? Let's talk about the advantages of VLANs and learn how to create and implement them.

VLANs, performance, and security

The left-side example in Figure 2-28 has all hosts in a single VLAN. Having hundreds or thousands of hosts all in the same VLAN certainly simplifies your network design, and reduces your configuration burden. You don't even need a router for inter-host communication. You just need the router to get you to the internet. That's nice.

But remember, this is all one broadcast domain. When one host sends a broadcast all other hosts must receive and process it (at least to the point where they realize the broadcast is not for them). Most end systems send frequent broadcasts: to query for services, perform ARP requests, and much more. So now you have all devices receiving frequent broadcasts from all the other devices in the broadcast domain. This wastes bandwidth on the network, as well as CPU resources on each device.

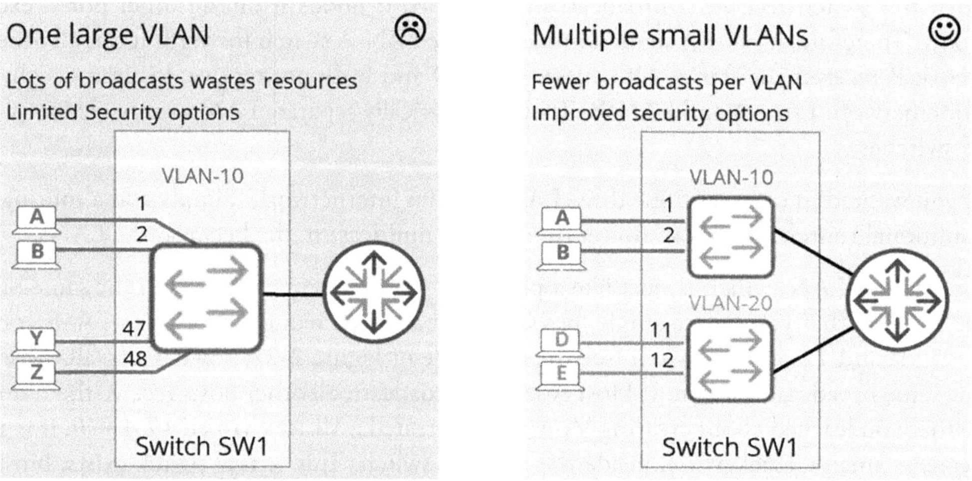

Figure 2-28: LANs and VLANs

Furthermore, simple layer 2 switches have relatively limited security options. By default, everyone in a VLAN can send packets to every other device in that VLAN. Of course, they can all see each other's broadcasts: bad actors might glean useful information from these broadcasts to instigate an attack.

Compare this to the significantly improved design on the right of Figure 2-28. Yes, you have a bit more complexity configuring various VLANs, IP addressing, and routing. However, that same large node network is now split into multiple smaller VLANs, say a couple hundred nodes per VLAN. (The actual number of nodes per VLAN depends on network design details and host application usage details that are beyond the scope of this book).

Now each host only sees broadcasts from other members of the VLAN, as opposed to all the hosts in the LAN. Far few CPU and bandwidth resources are wasted. You also get better isolation of problems. An issue on one VLAN will be less likely to affect other VLANs, depending on the nature of the issue, of course.

You know that by default folks in VLAN 10 cannot communicate with those in other VLANs. You must configure your routing services to route between those VLANs. And those router functions can include access control lists (ACLs) and other security features to secure that network. As you progress in your training, you will learn how Aruba makes this security much easier, with centralized policies pushed down to appropriate layer 3 devices, and other automated processes that can help maximize your security with minimal effort.

So that's *why you use* VLANs. Let's explore *how to configure* them.

Basic VLAN configuration

This section shows you how to create VLANs and associate them to edge ports.

Creating VLANs

AOS-CX refers to VLANs by their VLAN identifier (ID)—a number between 1 and 4,094. In AOS-CX, VLAN 1 is created by default and cannot be removed; and by default, all ports are members of this VLAN. This is a common default for many switches in the marketplace.

The **vlan** command creates a VLAN, which is enabled by default. Figure 2-29's top example shows the creation of VLAN 10. You can also create a series of VLANs with a single command, as shown in the example below it.

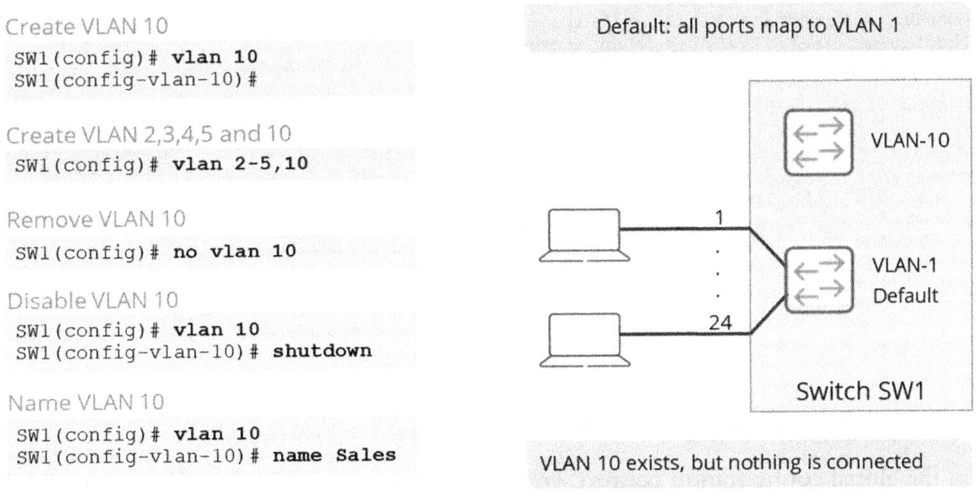

Figure 2-29: Creating VLANs

To remove the VLAN, use the command **no vlan** command, like no **vlan 10** to delete VLAN 10. Or, instead of deleting it, you could use the **shutdown** command to disable the VLAN.

It is also a good idea to name the VLAN, as shown in Figure 2-29, with the **name** command.

Remember that when you use the command **vlan 10** to create a VLAN, it is as if you have created a virtual switch inside of the actual, physical switch. Although it exists, no ports have yet been defined as members of this VLAN. No devices are yet attached to this virtual switch, as shown in the right-hand side of Figure 2-29.

The next thing you need to do is associate or *map* physical interfaces (edge ports) as members of the VLAN.

Access ports

Figure 2-30 shows SW1, with four connected devices, all in VLAN 1 by default.

- You define VLAN 10, with a name of Sales (top part of Figure 2-30).
- You define VLAN 20, with a name of Service (top part of Figure 2-30).

Define VLANs

```
SW1(config)# vlan 10
SW1(config-vlan-10)# name Sales
SW1(config-vlan-10)# exit
SW1(config)# vlan 20
SW1(config-vlan-20)# name Service
SW1(config-vlan-20)# exit
```

Map ports to VLANs

```
SW1(config)# interface 1/1/1-1/1/2
SW1(config-if-<1/1/1-1/1/2)# vlan access 10
SW1(config-if-<1/1/1-1/1/2)# exit
SW1(config)# interface 1/1/11-1/1/12
SW1(config-if-<1/1/4-1/1/5)# vlan access 20
SW1(config-if-<1/1/4-1/1/5)# exit
```

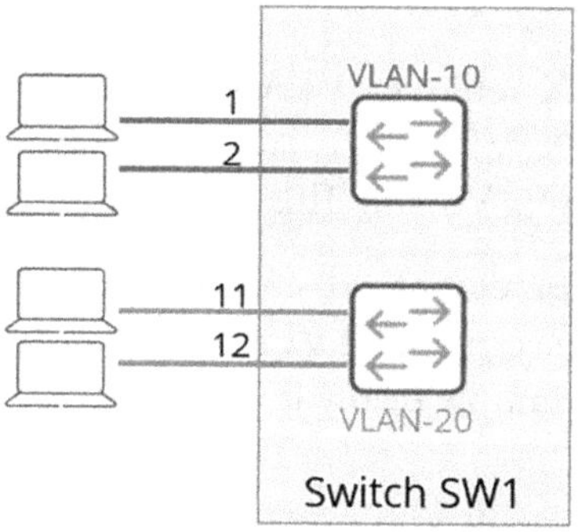


```
SW1# show vlan
---------------------------------------------------------------------
VLAN  Name            Status  Reason  Type      Interfaces
---------------------------------------------------------------------
1     DEFAULT_VLAN_1up        ok      default   <output omitted>
10    Sales             up    ok      static    1/1/1-1/1/2
20    Service           up    ok      static    1/1/11-1/1/12
```

The rest of the ports are still mapped to VLAN1

Figure 2-30: Assigning interfaces to VLANs

Now you need to assign or *map* ports to these VLANs.

- From the global configuration context, you choose to configure a range of interfaces at the same time – ports 1/1/1 through 1/1/2. Then you assign these ports to VLAN 10 (note that you can perform the assignment with just a single port, a range of ports, or a combination of ranges and individual ports) (middle part of Figure 2-30).

- You repeat these steps for ports 1/1/11 – 12 (middle part of Figure 2-30).
- Use **show vlan command** to verify VLAN creation and mapping (bottom part of Figure 2-30).

Note
You can perform the VLAN assignment with just a single port, a range of ports, or a combination of ranges and individual ports.

Understand that only one VLAN ID can be assigned to an interface (edge port). For example, interface 1/1/1 can't be a member of both VLAN 10 and VLAN 20 at the same time. This would be like you trying to be in the sales room for a meeting and in the engineering room for another meeting at the same time.

Note
AOS-CX switch interface modes can be layer 2 or layer 3. To convert an interface to layer 3 mode, use the command **routing** in the interface context. If you want to convert an interface to layer 2 mode, use the command **no routing**.

Extending VLANs across multiple switches

Figure 2-31 shows two switches, each with ports in VLANs 10 and 20. Now you want to connect the two physical switches together (SW1 and SW2), so you connect port 24 of SW1 to port 24 of SW2. What happens? Will all devices be able to communicate?

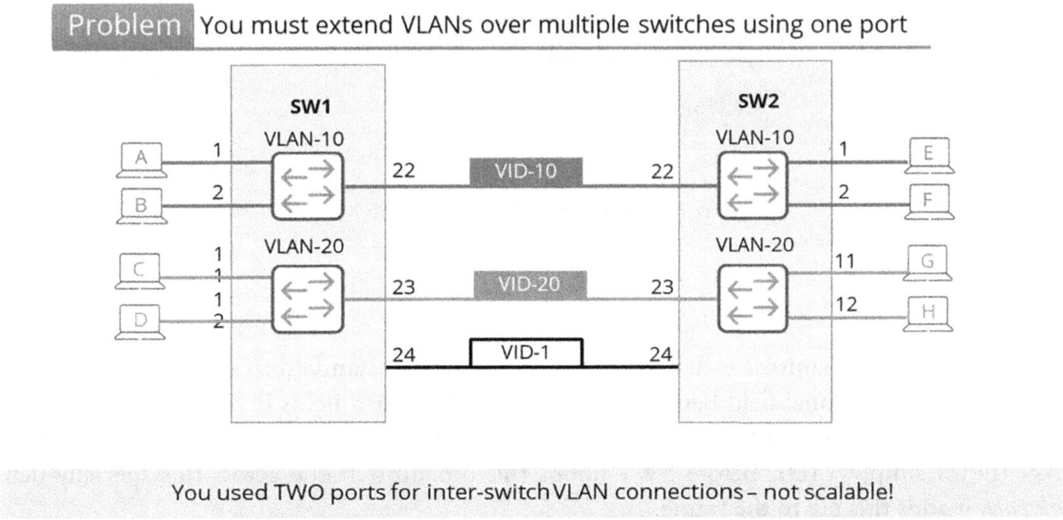

Figure 2-31: Extending VLANs across multiple switches

Recall that by default ports are members of VLAN 1, including port 24 on both switches. This means that port 24 is not connected to VLANs 10 and 20, and therefore cannot carry that traffic. Effectively, you still do not have connectivity between the switches for VLANs 10 and 20. How can you fix this?

One way is to connect two more pairs of port:

- Map port 22 on both switches to VLAN 10, and port 23 as to VLAN 20. Now all VLANs can communicate. A broadcast from Host A is flooded out all ports in VLAN 10 (except ingress port), and so all members of VLAN 10 receive the broadcast. This is good, but there is a problem.
- You used up 2 physical ports on each switch. What if you had 100 VLANs? We'd need 100 physical switch ports just for interconnects–that is not scalable and typically not feasible.

We need a way to use one single physical port to connect multiple VLANs.

IEEE 802.1Q tagging

Look at a standard Ethernet frame. Suppose that this was a broadcast frame from Host A, flooded out all ports in VLAN 10, including port 24. There is no field in a standard 802.3 Ethernet frame to inform receiving switches about the intended destination VLAN. If SW2 could talk, it might ask, "What do I do with this? Is this for VLAN 10? VLAN 20? VLAN1?"

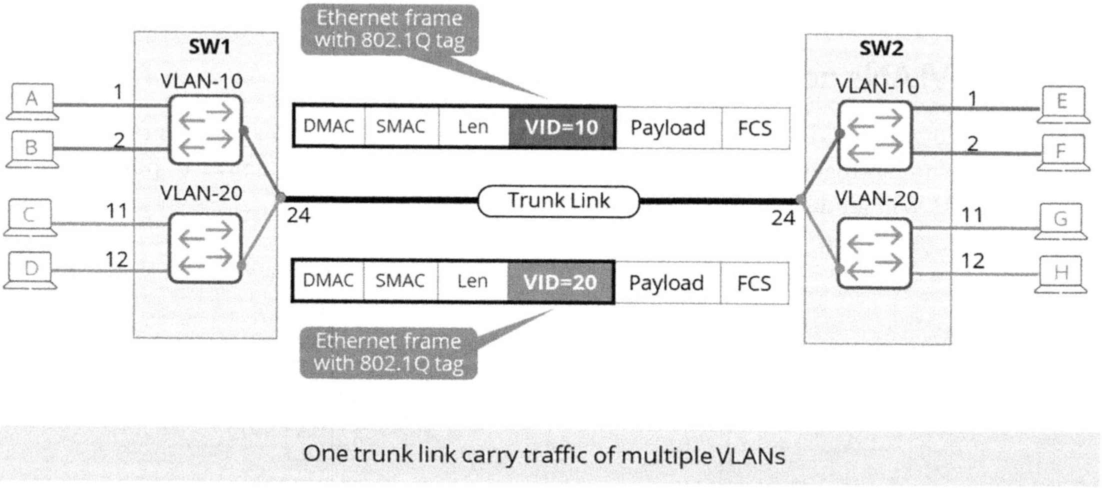

Figure 2-32: IEEE 802.1Q tagging

We need a tagging mechanism, as defined in the IEEE 802.1Q standard to deal with this issue. This standard adds an additional field between the length and payload fields in the Ethernet header: the 802.1Q tag field, as shown in Figure 2-32. The most important part of this field is the VLAN tag or VLAN ID (or simply VID). Before SW1 floods this broadcast frame across this specially-defined *trunk port*, it adds this tag to the frame.

Thus, SW2 receives the frame, looks at the tag and knows, "Ahhh, this frame is for VLAN 10". It strips off the special tag, and forwards a standard Ethernet frame out all ports in VLAN 10 to Hosts E and F. From Hosts A, E, and F's perspective, the tagging is transparent since they only deal with or see the normal 802.3 Ethernet frame.

Next Host G sends broadcast. SW2 floods a standard Ethernet frame out its local ports. Before flooding the frame out port 24, it's tagged: "This frame is for VLAN 20". SW1 receives this frame, sees that it is for VLAN 20, strips off the tag, and forwards a standard Ethernet frame out all ports in VLAN 20 (ports 11 and 12).

A trunk connection is like a virtual cable inside a physical cable, similar to a VLAN being virtual switch inside a physical switch. Thus, you can now extend the VLAN concept across multiple switches, using a single physical port.

Note

Many vendors use the term *trunk* when discussing a port that carries multiple VLANs; however, this is not true with all vendors and their switching products.

VLAN trunk ports and the native VLAN

The 802.1Q standard allows you to select a single VLAN on which no tag or VLAN ID (VID) is included in the layer 2 header (see Figure 2-33). This means that a standard, untagged Ethernet frame is sent out the port. This VLAN is known as the native VLAN or untagged VLAN. By default, in AOS-CX, VLAN 1 is the native VLAN. You can modify this by using this command: **vlan trunk native** <vlan-ID>.

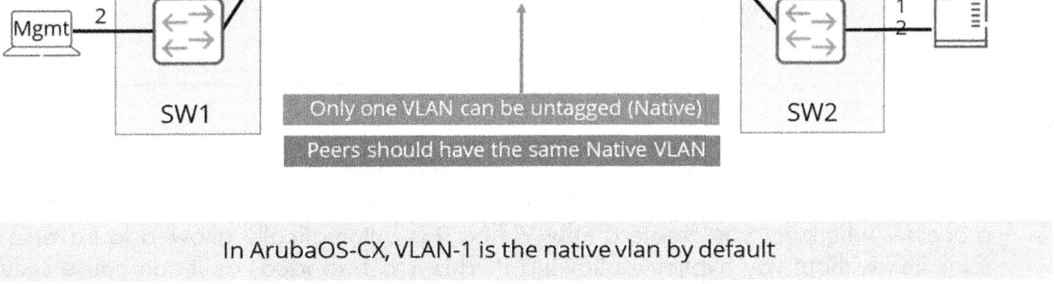

Figure 2-33: Native VLAN on trunk ports

Note

The VLAN-ID field is 12 bits long; this length allows a switch to carry traffic for up to 4,094 VLANs. To avoid inconsistencies Aruba recommends you use the same native VLAN in both sides of a trunk connection.

Configuring VLAN trunks: Allowed VLANs

In AOS-CX, use the command **vlan trunk allow** to allow a VLAN ID(s) to traverse a trunk interface. Multiple VLAN-IDs can be assigned to a trunk interface. In Figure 2-34, interface 1/1/25 is configured as a trunk link, and only allows VLANs 1, 10, and 20. Other VLANs will not traverse this link.

To verify the trunk interface, use the command **show vlan port** command.

Note

Figure 2-34 is focused on a trunk port configuration. It is assumed that the VLANs are already created, and access ports are mapped to them.

Configure trunk port

```
SW1(config)# interface 1/1/24
SW1(config-if)# vlan trunk allow 1,10,20
```

Validate trunk port

```
SW1# show vlan port 1/1/24

-----------------------------------------------------------------------
VLAN   Name             Mode                Mapping
-----------------------------------------------------------------------
1      DEFAULT_VLAN_1   native-untagged     port
10     Sales            trunk               port
20     Service          trunk               port
```

Figure 2-34: Allowing VLANs on a trunk

Note

The command **vlan allow all** allows all configured VLANs to traverse the trunk port. You may encounter folks who like to use this command, because it simplifies the configuration process – whenever you create a new VLAN, it is automatically allowed to traverse the trunk link without you explicitly allowing it. This is a bad idea, as it can cause security weaknesses and enable bad actors to potentially probe deeper into your network. Unless you are in a non-production lab environment, avoid using this command.

Changing the native VLAN

Recall that by default, all VLAN traffic is tagged to traverse a trunk link except the native VLAN traffic, where the native VLAN defaults to 1. However, you can change the native VLAN as desired. As in the example shown in Figure 2-35, use the **vlan trunk native** command to change the untagged VLAN: in this case it was changed to 10. This configuration must match on both switches, so be sure to apply this command to the appropriate port of the attached switch as well.

Verify your configuration with the command **show vlan port**.

Note

Some customers require all VLANs to be tagged, including the native VLAN, and AOS-CX can support this configuration. As typical, peer switch trunk configurations should match, or connections between devices might fail.

Modify the native VLAN

```
SW1(config)# interface 1/1/24
SW1(config-if)# vlan trunk native 10
```

Verifying Trunk port

```
SW1# show vlan port 1/1/24

-------------------------------------------------------------------
VLAN   Name               Mode                 Mapping
-------------------------------------------------------------------
1      DEFAULT_VLAN_1     trunk                port
10     Sales              native-untagged      port
20     Service            trunk                port
```

Peers must have the same Native VLAN

Figure 2-35: Changing the native VLAN

Note

You can use the **show lldp neighbor-info** command to see which switch is connected to port 1/1/24 of the switch you are currently configuring. This command is covered in a later section of this chapter.

Verifying the VLAN operation: MAC address table

Layer 2 switches use a MAC address table to make forwarding decisions. The switch builds this table automatically, based on the source MAC address of the frames that it receives from connected devices. How does the switch do this?

Suppose that you power on switch SW1, shown in Figure 2-36. Its MAC address table is initially empty. You can verify this with the **show mac-address** command.

Host A then transmits a frame to Host B. SW1 receives this frame, which contains a source MAC address of 90:...:00, on port 1; and port 1 is defined as a member of VLAN 10. Thus, the switch adds the entry to the MAC address table (the first entry as shown in Figure 2-36). The switch basically mapped the MAC address to the switch's VLAN and port as an entry in the MAC address table.

At this point, this is the only entry in the table. The switch does not yet know where to forward frames to other devices, like destination 00:...:37, so it floods this frame out all the remaining ports in VLAN 10. All hosts in VLAN 10 receive this frame and see the destination MAC address – most discard the frame since the destination MAC address doesn't match their own MAC address, with the exception of Host B: "This is my MAC address, and I will respond to host A's message".

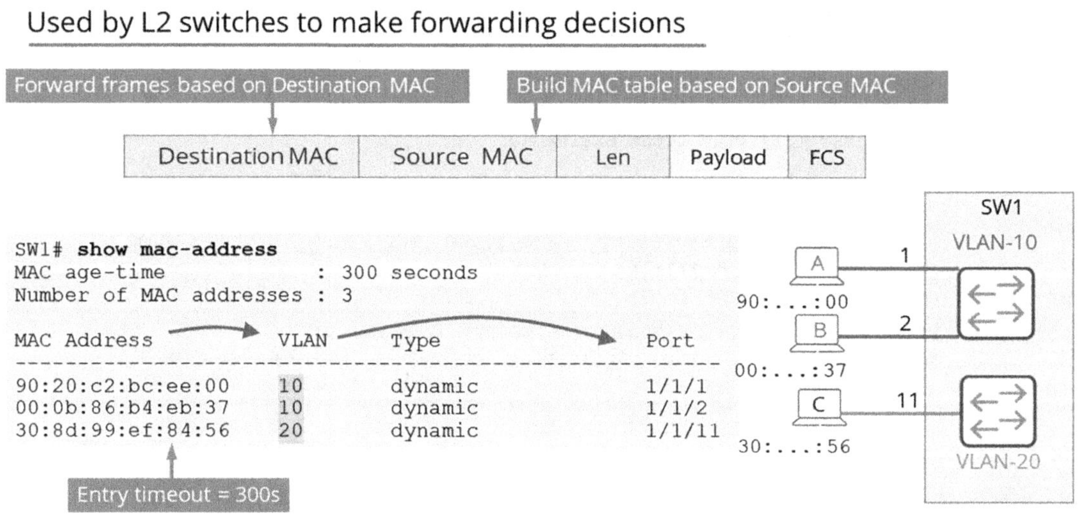

Figure 2-36: Building a MAC address table

And so Host B responds to Host A with a source MAC of 00:...:37, and a destination MAC of 90:...:00. The switch then adds a second entry to the MAC address table: 00:...:37 is in VLAN 10, connected to port 2. A similar process will happen with Host C on port 11.

Therefore, switches automatically build the MAC table based on the source MAC addresses that they see entering the switch ports, and forward frames based on the destination MAC addresses in these frames. By default, table entries are maintained for 300 seconds (5 minutes)—if an entry isn't referenced in this time frame, it is aged out of the table. You can view the entries in the MAC address table by using the **show mac-address** command.

Multi-layer interface types

You just learned about layer 2 switch ports and how to assign VLANs to them, often to connect to edge devices (access ports) or to interconnect switches (trunk ports). Because AOS-CX switches are multi-layer switches, you can create logical layer 3 interfaces for those VLAN broadcast domains. These interfaces are referred to as switch virtual interfaces (SVIs), which is basically a virtual layer 3 routed interface that exists only inside the switch and is specific to the respective VLAN.

Layer-3 switch ports

Suppose you create the logical interface SVI 10 by executing the **interface vlan 10** command in the switch's global context. Once you configure an IP address on this SVI interface, it connects to the internal routing construct for VLAN 10 (see Figure 2-37). The SVI interface then services routable traffic from VLAN 10 to other destination networks. You define SVI 20 similarly. With some additional configuration (discussed later in the book), your switch can route traffic between your VLANs.

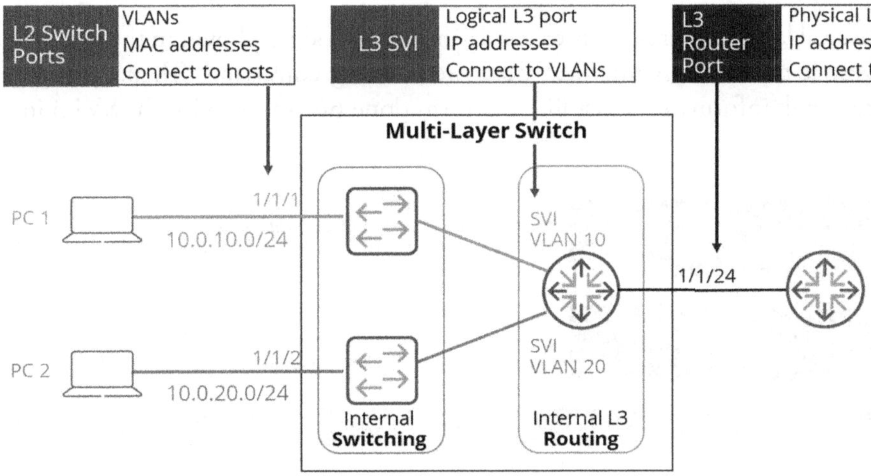

Figure 2-37: Multi-layer interface types

Suppose you need to connect your multi-layer switch to an external router, perhaps using port 1/1/24, as shown in Figure 2-37, and you do not want to use a trunk port between the switch and the router. By executing the **routing** command on the switch interface, this turns the physical port (1/1/24) into a layer 3 interface. This allows you to assign an IP address directly to the switch's physical interface; and the AOS-CX switch automatically adds the IP network address as a directly connected route in its routing table.

Note

Interfaces on the Aruba 6400 and higher series switches are configured as routed interfaces, by default. 6300 and smaller switches' interfaces are configured as layer 2 switch ports, by default. The Aruba 6200 supports SVIs, but its interfaces cannot be natively configured as routed ports.

Let's now look at configuration of SVI interfaces in more depth.

Two methods of IP address configuration

The top part of Figure 2-38 shows a typical SVI addressing configuration. You create the SVI interface by simply executing the **interface vlan** command from the switch's global config mode. This places you in the interface context where you can configure the interface properties, like the IP address of 10.1.3.1 with a /24 subnet mask (255.255.255.0). By default, creating an SVI interface automatically enables it.

To configure a physical layer 3 interface, enter config interface mode as shown in the middle of Figure 2-38. First, convert it to a layer 3 interface with the **routing** command. Then, configure the IP address and subnet mask information, just like what was done previously with the SVI 3 interface in Figure 2-38.

Assign IP address to a VLAN interface (SVI)

```
Switch(config)# interface vlan 3
Switch(config-if-vlan)# ip address 10.1.3.1/24
```

Assign IP address to a physical port

```
Switch(config)# interface 1/1/5
Switch(config-if)# routing
Switch(config-if)# ip address 10.1.5.1/24
```

Verification

```
Switch(config)# show ip interface brief
Interface          IP Address            Interface Status
                                          link/admin
1/1/5             10.1.5.1/24             up/up

vlan3             10.1.3.4/24             up/up
```

Figure 2-38: Configuring IP addressing on switch interfaces

Validate your configuration by executing the **show ip interface brief** command, which can be seen at the bottom of Figure 2-38.

Address Resolution Protocol (ARP)

Imagine that you are in a conference room and are leading a discussion with people that you have just met. You don't yet know everyone's name at this point. Then, someone enters the room and hands you a note that says, "Important message for Alvin Rogers". You do not know which person is Alvin, so you say, "Excuse me everyone. Who is Alvin Rogers?". Everyone hears your request, but only Alvin responds, "I'm Alvin". You can thus associate the name Alvin to an actual person seated at a particular chair in the room. This is very much how the Address Resolution Protocol (ARP) in IP functions, which maps layer 3 IP addresses to layer 2 MAC addresses.

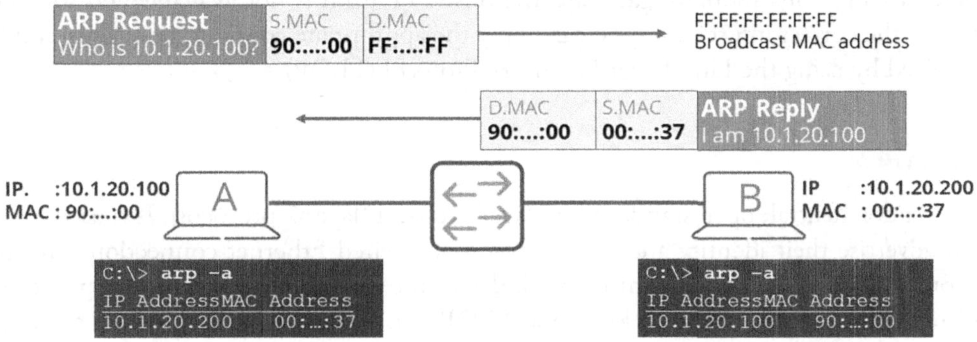

Figure 2-39: How ARP functions

In Figure 2-39, Host A knows that it must communicate with the host at IP address 10.1.20.200. To do this, it must build an Ethernet frame, which requires its own source MAC address (90::00), and Host B's MAC address for the destination, which it does not yet know.

To learn the destination's MAC address, Host A broadcasts an ARP request, where the destination MAC address is ff:ff:ff:ff:ff:ff, a layer 2 broadcast address. (In the IP header, the destination IP address is 255.255.255.255.) You previously learned that switches flood broadcasts out all ports in the same VLAN (except for the ingress port). Thus, most hosts in the VLAN receive this frame and ignore it, "I'm not 10.1.20.200", with the exception of Host B.

Host B responds with an ARP reply, which is a unicast message sent specifically to Host A's MAC address, "I am 10.1.20.100, and my MAC address is 00::37". Host A receives this reply and creates an entry in its ARP table, sometimes called an ARP cache. Host A maps the IP address of 10.1.20.200 to the MAC address of 00:0b:86:b4:eb:37. Thus, the next time Host A must communicate with Host B; it doesn't need to broadcast an ARP message, but instead uses the information it learned previously and placed in its ARP cache.

It is also important to point out that when Host B received the ARP request from Host A, it learned Host A's MAC and IP addresses. Thus, Host B created an entry in its own ARP table, mapping the IP address of 10.1.20.100 to the MAC address of 90:20:c2:bc:ee:00.

From a Windows PC's command prompt, you can execute the **arp –a** command to view the mappings of IP to MAC addresses that the PC has learned.

Discovering the network

Let's explore a few key tools to make it easy to discover, diagnose, and document nearly any network. This section covers the LLDP and ICMP protocols as well as the use of the ping and traceroute tools.

Link Layer Discovery Protocol (LLDP)

Imagine that you are remotely configuring an AOS-CX switch in a production environment. This switch is connected to multiple networking devices, and you must configure only the interface that is connected to an Aruba mobility gateway. The onsite IT team is not available and somehow you must identify the correct interface so you can apply the appropriate configuration. This problem can easily be solved by using the Link Layer Discovery Protocol (LLDP).

LLDP overview

LLDP is a vendor-neutral, open standard link layer protocol (layer 2 protocol). It is used by network devices to advertise their identity and capabilities over a wired Ethernet connection. This protocol enables you to discover and document network device interconnections. Media endpoint discovery (MED) is an enhancement of LLDP known as LLDP-MED. LLDP-MED provides auto-discovery of LAN policies such as the voice VLAN ID, and the layer 2 priority and differentiated services for quality of service (QoS), as well as many other types of information. LLDP and LLDP-MED also help to enable plug-and-play networking.

Devices send LLDP information on a regular interval and this information is encapsulated in Ethernet frames. Directly connected devices receive these frames and store the received LLDP information in a table in local memory. On AOS-CX switches, you can view this information with commands like **show lldp neighbor-info**, as shown in Figure 2-40.

Layer 2, IEEE standard protocol

- Network devices advertise identity/capabilities
- Helpful to discover and document

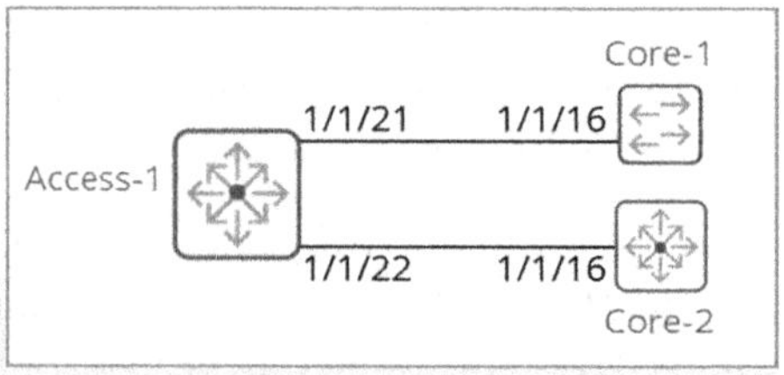


```
Access-1# show lldp neighbor-info

LLDP Neighbor Information
=========================
 <output omitted>

LOCAL-PORT  CHASSIS-ID         PORT-ID    PORT-DESC     TTL    SYS-NAME
---------------------------------------------------------------------------------
1/1/21      90:20:c2:bc:a8:00  1/1/16     1/1/16        120    Core-1
1/1/22      90:20:c2:bb:8b:00  1/1/16     1/1/16        120    Core-2
```

Figure 2-40: LLDP overview

In the diagram in Figure 2-40, Access-1's port 1/1/21 connects to port 1/1/16 of Core-1 and Access-1's port 1/1/22 connects to port 1/1/16 of Core-2. If someone has already taken the time to create a diagram like the one shown in Figure 2-40, you wouldn't need to use LLDP for the discovery process. However, network documentation is often incorrect, out-of-date, or was simply never created from the beginning. You can easily verify existing documentation and create new documentation using LLDP information.

Note

Many people do not believe that it is worth the time and/or money to create and maintain good and accurate network documentation: "The network is fine; why waste time making documents?" However, when the network fails, instead of having to discover the network (or rely on that one person who has the network memorized), everyone can look at the network documentation and be far more effective in troubleshooting the issue. Documentation is an insurance policy against slow, ineffective troubleshooting. This kind of documentation, for example, can be the difference between a 4-hour network outage and a 15-minute outage.

AOS-CX switches have LLDP enabled by default. The **show lldp neighbor-info** command displays information about neighboring devices for all interfaces or for a specific interface (see Figure 2-40).

To get more information about a specific entry in the LLDP table, you can append the interface at the end of the **show lldp neighbor-info** command. Here is an example:

```
Access-1# show lldp neighbor-info 1/1/21

Port                            : 1/1/21
Neighbor Entries                : 1
Neighbor Entries Deleted        : 0
Neighbor Entries Dropped        : 0
Neighbor Entries Aged-Out       : 0
Neighbor Chassis-Name           : Core-1
Neighbor Chassis-Description    : Aruba JL635A  GL.10.04.0001
Neighbor Chassis-ID             : 90:20:c2:bc:a8:00
Neighbor Management-Address     : 10.250.1.3
Chassis Capabilities Available : Bridge, Router
Chassis Capabilities Enabled    : Bridge, Router
Neighbor Port-ID                : 1/1/16
Neighbor Port-Desc              : 1/1/16
Neighbor Port VLAN ID           : 1107
TTL                             : 120
```

LLDP and LLDP-MED

LLDP is an open-standard data link layer protocol used by network devices to advertise their identity, capabilities, and neighbors on a local area network (see Figure 2-41). Based on IEEE 802 technology, you can use it to verify layer 2 connectivity between devices, discover an unknown network topology, and help build network documentation.

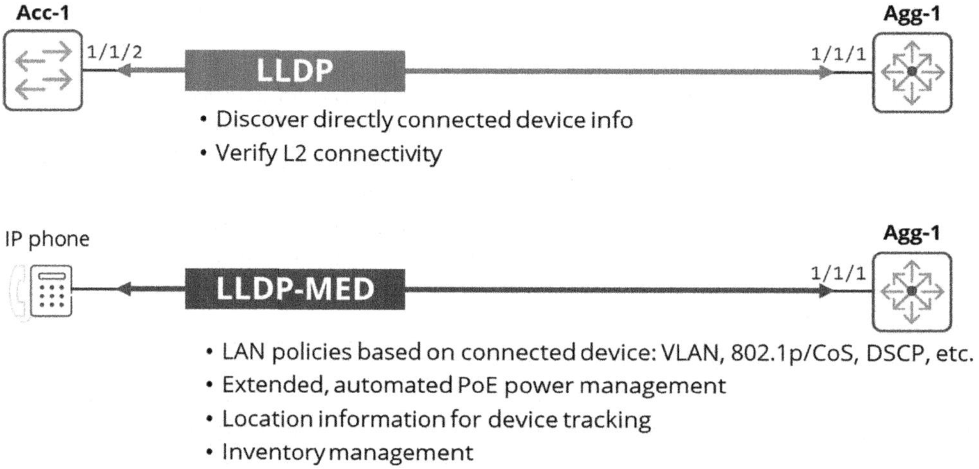

Figure 2-41: LLDP and LLDP-MED

The MED is an enhancement of LLDP, known as LLDP-MED. It helps to provision devices: determine correct power settings for connected VoIP phones and APs, auto-assign VLAN and QoS settings, and thus enable plug-and-play networking.

Features of LLDP and LLDP-MED include:

- You can define LAN policies based on the connected device: VLAN, layer 2 IEEE 802.1p class of service (CoS) values, layer 3 differential service code points (DSCP) settings, and more.
- You can discover the location of the device for VoIP and enhanced 911 services.
- Extended and automated power management provides the appropriate amount of power for power-over-Ethernet (PoE) edge devices.
- Inventory management allows you to rack your network devices, and determine device-specific information, like the vendor manufacturer, software and hardware versions, the serial or asset number, and layer 3 addressing information.
- Zero-touch provisioning (ZTP) provides support for quickly deploying new devices in a network (many of Aruba's products support ZTP).

LLDP operation

The IEEE 802.1AB LLDP protocol provides a standards-based method for network devices to discover each other and exchange information about their capabilities. Devices send LLDP data frames on all interfaces where outbound LLDP is enabled. Devices accept LLDP advertisements from layer 2 connected neighbors on ports where inbound LLDP is enabled. See Figure 2-42 for an illustration of this process.

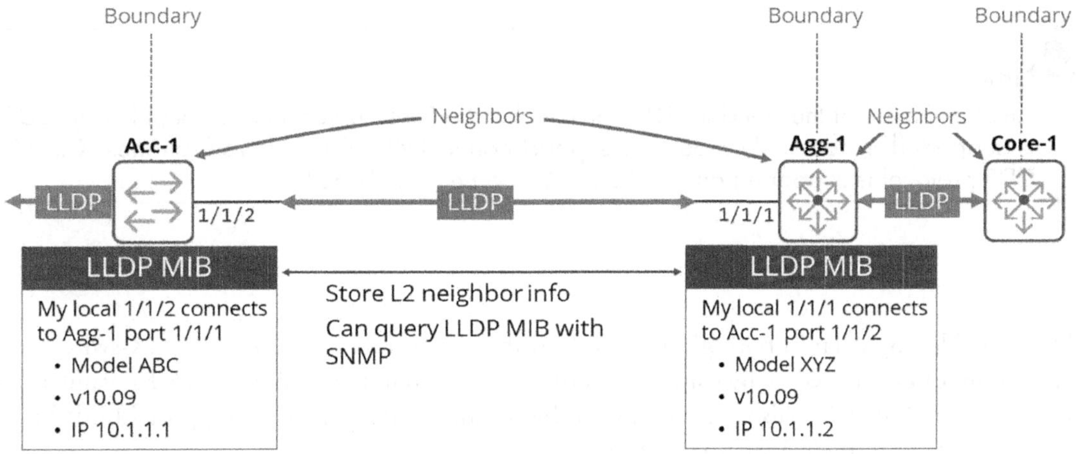

Figure 2-42: LLDP operation

Inbound packets from neighbor devices are stored in a special LLDP management information base (MIB). This information can then be queried by other devices through SNMP. Network management tools can use LLDP information to create accurate physical network topologies. LLDP helps to determine which devices are neighbors and through which interfaces they connect. LLDP operates at layer 2 and requires an LLDP agent to be active on each interface that sends and receives LLDP advertisements.

LLDP advertisements can contain a variable number of TLV (type, length, value) information elements. Each TLV describes a single attribute of a device such as: system capabilities, the management IP address, the device ID, and the port ID, to name a few.

Packet boundaries

LLDP frames travel only to directly-connected peers. These frames are never forwarded to any other devices, regardless of whether they are LLDP-enabled. An intervening hub or repeater forwards the LLDP packets it receives in the same manner as any other multicast packets it receives. Therefore, two LLDP switches joined by a hub or repeater handle LLDP traffic in the same way that they would if directly connected.

Any intervening IEE 802.1D device or layer 3 device that is either LLDP-unaware or has disabled the operation of LLDP will not forward (will drop) the LLDP frames.

LLDP-MED

LLDP-MED uses the standard LLDP process described earlier in this section, with some additional extensions that were not in the original LLDP standard, and also introduces new features unique to LLDP-MED operation.

Note

Cisco's version of the standard LLDP protocol is the Cisco Discovery Protocol (CDP). CDP is supported in read-only mode, to support connectivity of Cisco VoIP phones. Cisco's CDP protocol is supported on the AOS-CX switches, by default.

LLDP configuration

LLDP is enabled by default, both globally and on the switch interfaces for AOS-CX switches. You can configure LLDP, if needed, by using the commands shown in Figure 2-43, as well as validate what you configured. Aruba APs also support LLDP by default. Aruba gateways support LLDP, but it is disabled by default in their ArubaOS software.

A portion of the output is not shown in Figure 2-43. At the bottom of the **show lldp configuration** command, you would see the list of LLDP advertised TLVs. But you can also use the **show lldp tlv** command to also view this information. It is good that the switch advertises information about itself; however, with few exceptions, the switch does not advertise TLVs for LLDP-MED. You would have to manually enable the TLV values that are of interest to you.

```
lldp
interface 1/1/1
  lldp receive
  lldp transmit
```

Default on AOS-CX

```
Switch# show lldp configuration 1/1/1
LLDP Global Configuration
=========================
LLDP Enabled : Yes
LLDP Transmit Interval : 8
LLDP Hold Time Multiplier : 4
LLDP Transmit Delay Interval : 2
LLDP Reinit Time Interval : 2
LLDP Port Configuration
=======================
PORT TX-ENABLED RX-ENABLED
------------------------------------
1/1/1 Yes Yes
```

```
switch# show lldp tlv

TLVs Advertised
===============
Management Address
Port Description
Port VLAN-ID
System Capabilities
System Description
System Name
```

Figure 2-43: LLDP configuration and verification

Note

With LLDP-MED, you can use the port VLAN ID (native VLAN) to detect native VLAN configuration inconsistencies between connected networking devices.

LLDP discovery and verification

You can easily use LLDP for neighbor discovery as shown in Figure 2-44. You should examine the *Port VLAN ID* (native VLAN) to detect native VLAN inconsistencies between device.

```
Access-1# show lldp neighbor-info 1/1/21

Port                               : 1/1/21
Neighbor Entries                   : 1
Neighbor Entries Deleted           : 0
Neighbor Entries Dropped           : 0
Neighbor Entries Aged-Out          : 0
Neighbor System-Name               : Core-1
Neighbor System-Description        : Aruba JL635A  GL.10.04.0030
Neighbor Chassis-ID                : 90:20:c2:c0:ae:00
Neighbor Management-Address        : 90:20:c2:c0:ae:00
Chassis Capabilities Available     : Bridge, Router
Chassis Capabilities Enabled       : Bridge, Router
Neighbor Port-ID                   : 1/1/20
Neighbor Port-Desc                 : 1/1/20
Neighbor Port VLAN ID              : 1
TTL                                : 120
```

Figure 2-44: LLDP discovery and verification

ICMP

Suppose that a new printer is added into the network and a laptop wants to use this device. The first troubleshooting step is to verify connectivity between these two devices in order to ensure that the network is properly configured and functioning. Internet Control Message Protocol (ICMP) is one IP protocol that helps you address this issue.

ICMP is used by network devices to send error and operational information related to IP communications: it's basically a support protocol for IP. In Figure 2-45, Device B sends an ICMP message to Device A. The IP header of the IP packet has a "1" in the Protocol field. Each IP protocol has a number assigned to it, just like TCP (6) and UDP (17) are assigned protocol numbers. From this perspective, ICMP seems to *sit* or operate at layer 4, but its purpose is largely to function at layer 3. Think of ICMP as a kind of extension to the IP protocol at layer 3. Obviously, ICMP uses neither TCP nor UDP headers: however, IP and ICMP communicate directly. In other words, if IP has a problem, ICMP informs other devices about this problem.

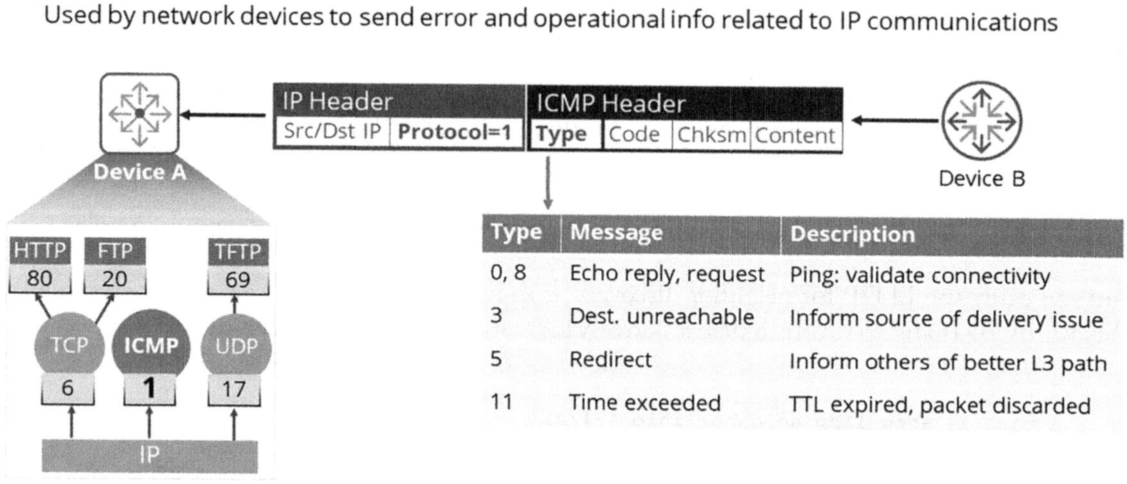

Figure 2-45: ICMP overview

What type of problem or information might this be? The ICMP *Type* field (with help from the *Code* field) indicates the type of message, as indicated by the table in Figure 2-45.

With most operating systems, including AOS-CX switches, the **ping** command can be used to test network connectivity. When executing ping, you specify a destination to test. The resulting ICMP packet consists of an ICMP Type 8 echo request, and the corresponding reply a Type 0 echo reply. A Type 3 response indicates that a router received an IP packet to be routed, but lacked a matching route in its route table: a *destination unreachable* message is displayed. Or perhaps the router knows that another router has a better path, and so indicates this information with a *redirect* message.

Every IP packet sent includes a time-to-live (TTL) value. Each router along the packet's path decrements the TTL by 1. The router that decrements the TTL field to 0 will discard the packet and inform the original source that its packet was discarded because its time-to-live value had expired. This ensures that if a routing loop exists, the packet doesn't continually rotate in the loop for infinity.

ping command

The **ping** command sends ICMP echo request messages to a destination device. The remote device responds to each received echo *request* with an echo *reply* message. When the source device receives the messages, it understands that the network was properly configured since the communication succeeded between the two devices.

On AOS-CX switches, the **ping** command must specify the IPv4 or hostname (see Figure 2-46).

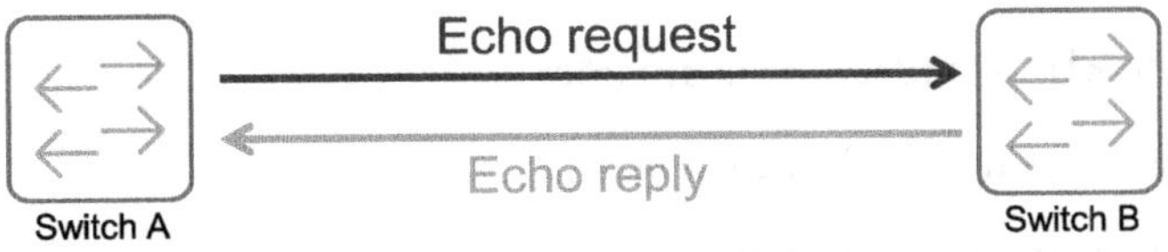


```
Switch(config)# ping 10.254.1.23
PING 10.254.1.23 (10.254.1.23) from 10.12.1.2 : 100(128) bytes of data.
108 bytes from 10.254.1.23: icmp_seq=1 ttl=61 time=4.72 ms
108 bytes from 10.254.1.23: icmp_seq=2 ttl=61 time=3.10 ms
108 bytes from 10.254.1.23: icmp_seq=3 ttl=61 time=2.98 ms
```

Figure 2-46: **ping** command

Here is more information concerning the syntax, context, and authorization information for the **ping** command:

- **Syntax**: **ping** <IPv4-address or hostname>
- **Command context**: operator (>) or manager (#)
- **Authority**: operators or administrators

The **ping** command is supported in most of the operating systems in the market today. The example ping session shown in Figure 2-46 is from the command prompt of an AOS-CX switch.

traceroute command

Traceroute is a variant of the ICMP protocol and it is a useful troubleshooting tool when you deploy a layer 3 feature, such as an IP routing protocol. This command tracks the path that a packet takes to reach its destination, listing the layer 3 devices that are in the path (the hops the packet takes to reach the destination).

In AOS-CX the traceroute command must specify an IPv4 address or hostname (see Figure 2-47).

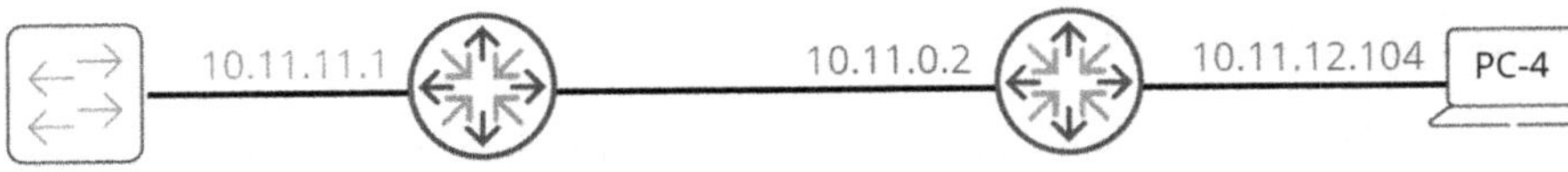


```
Switch# traceroute 10.11.12.104
traceroute to 10.11.12.104 (10.11.12.104), 1 hops min, 30 hops max, 3 sec.
timeout, 3 probes
  1   10.11.11.1     0.202ms  0.099ms  0.504ms
  2   10.11.0.2      0.188ms  0.101ms  0.099ms
  3   10.11.12.104   0.142ms  0.087ms  0.122ms
```

Figure 2-47: **traceroute** command

Here is more information concerning the syntax, context, and authorization information for the **traceroute** command:

- **Syntax**: **traceroute** <IPv4-address or hostname>
- **Command context**: operator (>) or manager (#)
- **Authority**: operators or administrators

In the example shown in Figure 2-47, the switch establishes connectivity to the destination 10.11.12.104, via two intermediate routers (10.11.11.1 and 10.11.0.2). You will learn more about routing and layer 3 in the upcoming chapters.

Note

Many firewalls block ping and traceroute commands. If you are attempting to use ping and traceroute commands to test connectivity through a firewall, the tests may fail. This does not necessarily mean that anything is wrong; it just means that the firewall is dropping your ping and trace packets as a security measure.

Note

An intermediate route control plane (software) in the path must generate an ICMP TTL expired packet when the TTL is decremented to 0. This is because the *data plane* drops the packet, while the *control plane* sends the notification.

Link aggregation

Link aggregation can be used to improve switch-to-switch performance and reliability. You might also leverage this feature for switch-to-server connections. This section will introduce you to what link aggregation is, its requirements, the different link aggregation types, how to configure and verify its configuration on AOS-CX switches, and how the load sharing feature works when multiple connections exist with an aggregated link between two networking devices.

Link aggregation overview

Ethernet link aggregation bundles multiple *physical* Ethernet links into one *logical link*, called a link aggregation group, or LAG for short. Figure 2-48 shows two switches connected by three physical Ethernet links, bundled into one virtual or logical link called LAG 1. Link aggregation is supported on many products, like servers, storage, routers, and firewalls, but is also commonly used to interconnect switches.

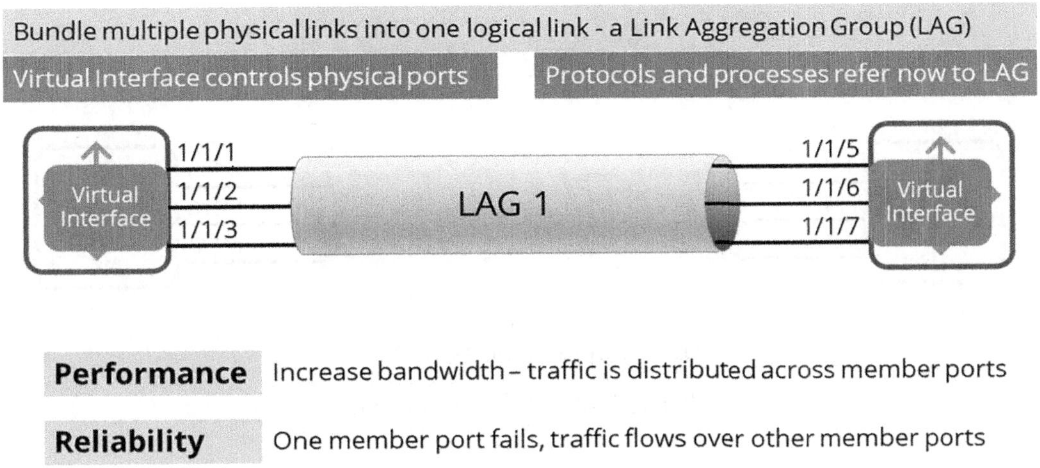

Figure 2-48: An overview of link aggregation

The benefits of link aggregation include:

- **Performance**: You get increased bandwidth because all traffic is distributed across the member ports.
- **Reliability:** Member ports dynamically back each other up. When a member port fails, its traffic automatically flows over other member ports.

Enabling link aggregation on a switch creates a virtual interface. You then assign physical ports to be members of that virtual interface. The switch's various protocols and processes will only perceive and refer to the virtual interface and not the actual physical interfaces of the LAG: they no longer perceive the individual physical interface members.

Because it is seen as a logical interface, understand that broadcast and multicast traffic is sent across a physical link in the bundle, but the remote peer sees this as coming from a logical interface, and thus ensures that the broadcast or multicast is not sent back to the source on one of the other physical links in the LAG in order to prevent the creation of layer 2 loops in the LAG itself.

LAG requirements

Configuring LAG interfaces with matching parameters is necessary in order to avoid errors and misconfigurations. Things that need to match between the two devices comprising a LAG include the duplexing (full or half), the link speed, and the media type (the type of cable connector). See Figure 2-49 as an example.

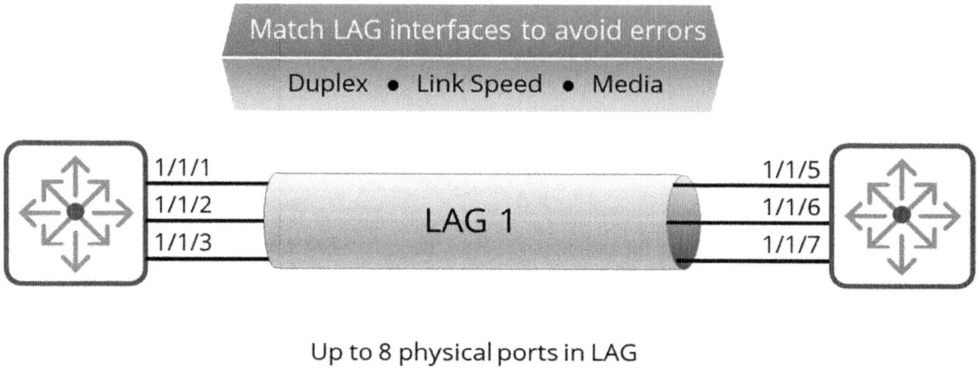

Figure 2-49: LAG requirements

AOS-CX displays a warning message if you attempt to map mismatched interfaces to a LAG: for example, you should not add a 10 Gb/s link to a LAG group of 1 Gb/s links.

AOS-CX allows each LAG to have up to 8 individual member ports. Use the command **show capacities** to verify your switch capacity.

LAG types

Now that you understand what a LAG is, you will learn about the two operating modes used for LAGs:

- *Static LAG*, which uses no protocol
- *Dynamic LAG*, which uses the Link Aggregation Control Protocol (LACP)

Static LAG

When using the static LAG mode, devices do not exchange any control information (left-side of Figure 2-50). In other words, there is no signaling between switch peers about the LAG. You simply configure the LAG independently on each peer. If your configuration is good, then the LAG will work. However, the switch peers have no knowledge of who they are connected to, or whether they are connected to the same peer, or connected correctly.

This mode is not recommended because a misconfiguration on one side cannot be detected by the remote peer. This can lead to unexpected behaviors, which can be challenging to detect and troubleshoot. Another scenario that could cause a problem is shown in the figure's right-hand example. Switch Access-1 is not aware that you have erroneously connected LAG member ports to two separate physical switches.

- Devices do not exchange control messages
- LAG configured independently on each peer

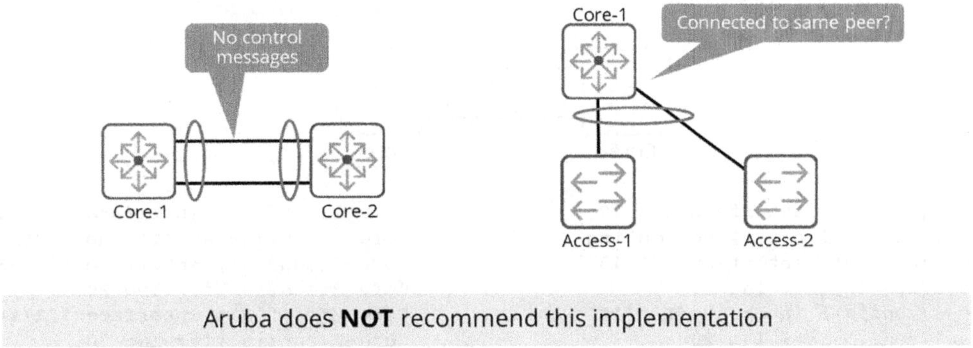

Figure 2-50: Static LAG

The following steps create a Layer 2 Dynamic Aggregation Group:

1. Create a LAG interface, represented by an identifier (any unused value between 1 and 256):

```
Switch(config)# interface lag <id>
```

2. Disable routing (Layer 3 capabilities) in the new LAG interface. This step limits the LAG to only process Layer 2 frame headers:

```
Switch(config-lag-if)# no routing
```

3. Map a port member to the LAG:

```
Switch(config)# interface <member/slot/port>
Switch(config-if)# lag <id>
```

Note

Member ports lose their port VLAN configuration when added to a LAG. A new LAG interface is an access port in VLAN 1, by default. Any protocol information from here onward would be done on the virtual LAG interface and not the physical interfaces that comprise the LAG.

See Figure 2-51 for the LAG command syntax and a configuration example.

```
Switch(config)# interface lag <id>               <- Create LAG interface
Switch(config-lag-if)# no routing                <- Disable routing on LAG
Switch(config)# interface <member/slot/port>
Switch(config-if)# lag <id>                      <- Map the interface to LAG
```

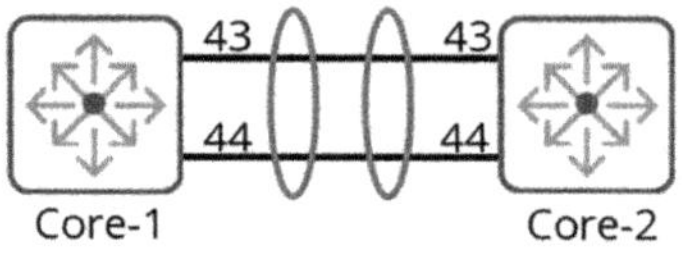


```
Core-1(config)# interface lag 20
Core-1(config-lag-if)# no routing
Core-1(config)# interface 1/1/43
Core-1(config-if)# lag 20
Core-1(config)# interface 1/1/44
Core-1(config-if)# lag 20
```

```
Core-2(config)# interface lag 20
Core-2(config-lag-if)# no routing
Core-2(config)# interface 1/1/43
Core-2(config-if)# lag 20
Core-2(config)# interface 1/1/44
Core-2(config-if)# lag 20
```

Figure 2-51: Static LAG configuration

Dynamic LAG with 802.3ad

With IEEE 802.3ad, peer devices that implement a dynamic LAG exchange control messages to establish and maintain the LAG link. This mechanism will also detect link failures and ensure that LAG port members terminate on the same device and meet the LAG requirements (see Figure 2-52).

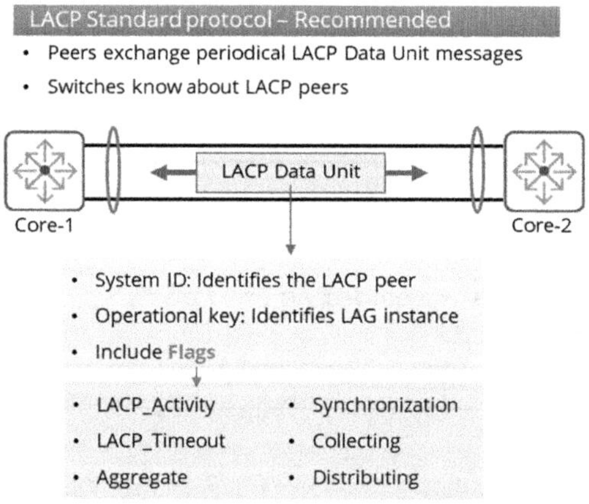

Figure 2-52: Dynamic LAG operation

The 802.3ad implementation is commonly known as the Link Aggregation Control Protocol (LACP). The two peers periodically exchange LACP messages called LACP data units (LACP-DUs). These messages include the following:

- **System ID**: uniquely identifies the switch.
- **Operational key**: uniquely identifies the LAG.

Note

Dynamic LAG or LACP is the recommended method to implement link aggregation, to avoid unexpected network problems.

There are some flags that are included in the LACP-DUs. The bottom content section of Figure 2-52 displays the meaning of these flags.

With 802.3ad standard you can bundle up to 16 ports; however, only 8 active ports can be actively used at any given time.

Dynamic LAG operational modes

LACP can be configured in one of two modes, which controls how peer negation proceeds (see Figure 2-53):

- *Passive* mode
- *Active* mode

Passive Device waits to receive a LACP Data Unit

Active Device transmits LACP Data Unit

Device A port mode	Device B port mode	LAG formed?
Passive	Passive	No
Passive	Active	Yes
Active	Passive	
Active	Active	

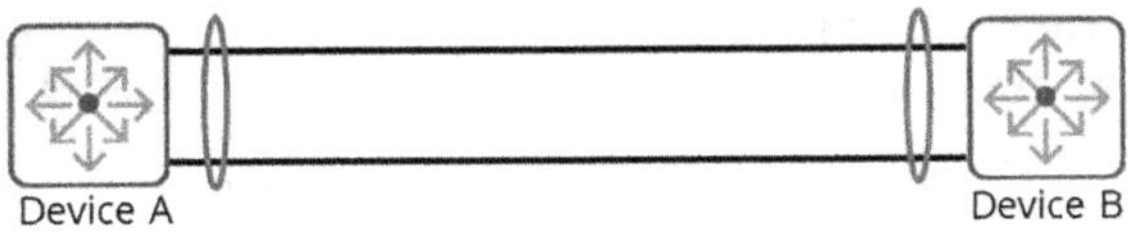

Figure 2-53: Dynamic LAG modes

Passive mode

With passive mode, the device passively waits to receive LACP-DU messages from the peer on order to dynamically create the LAG. This mode places the LAG in a listening state. It is as if the switch is thinking, "I'll just sit here and wait until I hear from my peer." If you configure the peer switch to also be in passive mode, then it is thinking the same thing. Both switches wait to hear from the other, and so a functional LAG connection is never formed. Therefore, at least one peer must be in active mode.

Active mode

With active mode, the device actively transmits LACP DU messages over its member ports: "Hey! Let's form a LAG". Whether the peer is passive or active, the peer will respond, negotiation will continue, and the LAG should successfully form if the requirements are met. Figure 2-53 has a table that illustrates what modes can be used in order to establish a LAG between two peers using LACP.

The rate LACP Data Unit messages are sent is configurable. There are 2 settings: fast and slow. Slow is the default and is the recommended setting.

Important

Dynamic link aggregation using LACP is the recommended configuration; as opposed to static link aggregation, which is not recommended.

Configuring a dynamic aggregation group

The configuration for dynamic LAG is the same as the one used for a static LAG, except for an extra line where you need to specify the dynamic LACP mode (see Figure 2-54). Fairly simple, yes?

```
Switch(config)# interface lag <id>                      <- Create LAG interface
Switch(config-lag-if)# no routing                       <- Disable routing on LAG
Switch(config-lag-if)# lacp mode {active | passive}     <- LACP mode
Switch(config)# interface <member/slot/port>
Switch(config-if)# lag <id>                             <- Map the interface to LAG
```

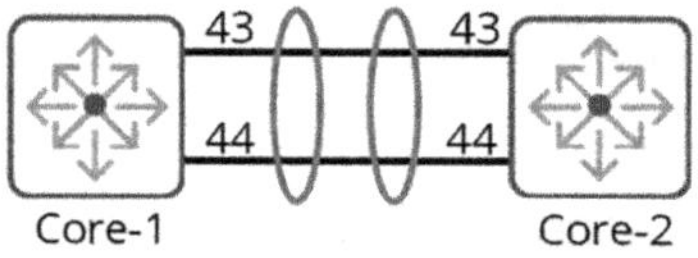


```
Core-1(config)# interface lag 20
Core-1(config-lag-if)# no routing
Core-1(config-lag-if)# lacp mode active
Core-1(config)# interface 1/1/43
Core-1(config-if)# lag 20
Core-1(config)# interface 1/1/44
Core-1(config-if)# lag 20
```

```
Core-2(config)# interface lag 20
Core-2(config-lag-if)# no routing
Core-2(config-lag-if)# lacp mode active
Core-2(config)# interface 1/1/43
Core-2(config-if)# lag 20
Core-2(config)# interface 1/1/44
Core-2(config-if)# lag 20
```

Figure 2-54: Dynamic LAG configuration

The top of Figure 2-54 shows the commands and syntax for setting up a dynamic LAG. The bottom part of the figure shows a configuration example.

Verifying the LAG operation

The following sections will illustrate how you can verify if your LAG is functioning correctly and how to identify issues that might cause a LAG to fail.

show interface lag command

Use the **show interface lag** command to verify your LAG configuration and status (see Figure 2-55). You can also use **show interface lag brief** command to see the LAG's total available bandwidth. The speed reflects the actual active member ports. This could be 20,000, 30,000, or 40,000 depending on active 10 GbE ports in the LAG for the example shown in Figure 2-56 (4x10 GbE links).

```
Switch# show interface lag

 Aggregate lag1 is up
 Admin state is up
 Description :
 MAC Address                    : 88:3a:30:97:59:c0
 Aggregated-interfaces          : 1/1/25 1/1/26 2/1/25 2/1/26
 Aggregation-key                : 1
 Aggregate mode                 : active
 Speed                          : 40000 Mb/s
 L3 Counters: Rx Disabled, Tx Disabled
 <<<<Omitted output>>>>
```

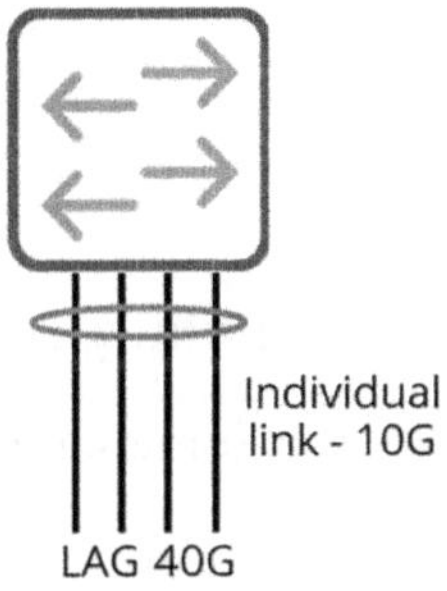


```
Switch# show interface lag brief
--------------------------------------------------------------------------
Port      Native  Mode    Type        Enabled Status  Reason          Speed
          VLAN                                                        (Mb/s)
--------------------------------------------------------------------------
lag1      1       trunk   --          yes     up      --              40000
```

Figure 2-55: **show interface lag** command

show lacp interfaces Command

You can obtain detailed information on the LAG by using the **show lacp interfaces** command. The output will provide valuable troubleshooting information such as the port forwarding state, the partner system mac-address, and the LACP flags. Flags indicate the state of link aggregation on each physical interfaces bundled into LAG. See Figure 2-56 for an example output of this command.

Note

The *Partner* output in the **show lacp interfaces** command is based on the received LACP information from the partner.

```
Switch(config)# show lacp interfaces

Actor details of all interfaces:
--------------------------------------------------------------------------------
Intf     Aggr     Port  Port  State    System-ID          System Aggr Forwarding
         Name     Id    Pri                               Pri    Key  State
--------------------------------------------------------------------------------
1/1/27   lag1     28    1     ASFNCD   88:3a:30:a5:d5:80 65534  1    up
1/1/28   lag1     29    1     ASFNCD   88:3a:30:a5:d5:80 65534  1    up

Partner details of all interfaces:
--------------------------------------------------------------------------------
Intf     Aggr     Port  Port  State    System-ID          System Aggr
         Name     Id    Pri                               Pri    Key
--------------------------------------------------------------------------------
1/1/27   lag1     28    1     ASFNCD   88:3a:30:a5:26:80 65534  1
1/1/28   lag1     29    1     ASFNCD   88:3a:30:a5:26:80 65534  1
```

Figure 2-56: **show lacp interfaces** command

Troubleshooting LAG problems

Let's look at some sample output of the **show lacp interfaces** command to troubleshoot problems with a LAG.

Figure 2-57 shows the first example. In this example, the second local port in LAG 256 (port 1/1/28) is not connected to the same logical LAG on the partner side as the local port 1/1/27. In other words, the *Partner* has two LAG interfaces, and this switch's physical ports in LAG 256 are connected to two different LAGs on the partner, which is a misconfiguration.

```
Switch(config)# show lacp interfaces

Actor details of all interfaces:
--------------------------------------------------------------------------------
Intf     Aggr     Port  Port  State    System-ID          System Aggr Forwarding
         Name     Id    Pri                               Pri    Key  State
--------------------------------------------------------------------------------
1/1/27   lag256   28    1     ALFNCD   8c:85:c1:65:0e:c0 65534  256  up
1/1/28   lag256   29    1     ALFO     8c:85:c1:65:0e:c0 65534  256  lacp-block

Partner details of all interfaces:
--------------------------------------------------------------------------------
Intf     Aggr     Port  Port  State    System-ID          System Aggr
         Name     Id    Pri                               Pri    Key
--------------------------------------------------------------------------------
1/1/27   lag256   1     1     ALFNCD   02:01:00:00:01:00 65534  1
1/1/28   lag256   1001  1     ALFN     02:01:00:00:01:00 65534  2
```

Figure 2-57: Port 2 connects to an invalid LAG ID on the partner side

See Figure 2-58 for the second example. In this example, local port 1/1/25 is connected to a different LACP system ID; in other words, the local LAG has one port connected to one partner and another port, in the same LAG, connected to a completely different partner. This is obviously a misconfiguration.

```
Switch(config)# show lacp interfaces

Actor details of all interfaces:
------------------------------------------------------------------------------
Intf    Aggr     Port  Port  State    System-ID          System Aggr Forwarding
        Name     Id    Pri                               Pri    Key  State
------------------------------------------------------------------------------
1/1/25  lag256   26    1     ALFO     8c:85:c1:65:0e:c0 65534  256  lacp-block
1/1/27  lag256   28    1     ALFNCD   8c:85:c1:65:0e:c0 65534  256  up
1/1/28  lag256   29    1     ALFO     8c:85:c1:65:0e:c0 65534  256  lacp-block


Partner details of all interfaces:
------------------------------------------------------------------------------
Intf    Aggr       Port  Port  State    System-ID          System Aggr
        Name       Id    Pri                               Pri    Key
------------------------------------------------------------------------------
1/1/25  lag256     27    1     ALFN     8c:85:c1:66:f0:40 65534  256
1/1/27  lag256     1     1     ALFNCD   02:01:00:00:01:00 65534  1
1/1/28  lag256     1001  1     ALFN     02:01:00:00:01:00 65534  2
```

Figure 2-58: Port 1/1/25 connects to a different LACP system ID

Figure 2-59 shows the third example. This example is similar to the last one. In this example, ports 1/1/27 and 1/1/28 are connected to one switch, where the ports are connected to *two* different LAGs on the same partner. On top of this, port 1/1/25 is connected to a completely different partner, but the key (LAG ID) matches that of what is configured on the local switch. Obviously, someone really messed up the wiring when trying to cable the switches together.

```
Switch(config)# show lacp interfaces

Actor details of all interfaces:
------------------------------------------------------------------------------
Intf    Aggr     Port  Port  State    System-ID          System Aggr Forwarding
        Name     Id    Pri                               Pri    Key  State
------------------------------------------------------------------------------
1/1/25  lag256   26    1     ALFO     8c:85:c1:65:0e:c0 65534  256  lacp-block
1/1/27  lag256   28    1     ALFNCD   8c:85:c1:65:0e:c0 65534  256  up
1/1/28  lag256   29    1     ALFO     8c:85:c1:65:0e:c0 65534  256  lacp-block


Partner details of all interfaces:
------------------------------------------------------------------------------
Intf    Aggr     Port  Port  State    System-ID          System Aggr
        Name     Id    Pri                               Pri    Key
------------------------------------------------------------------------------
1/1/25  lag256   27    1     ALFN     8c:85:c1:66:f0:40 65534  1
1/1/27  lag256   1     1     ALFNCD   02:01:00:00:01:00 65534  1
1/1/28  lag256   1001  1     ALFN     02:01:00:00:01:00 65534  2
```

Figure 2-59: Port connects to a different LACP system ID with the same key ID

Figure 2-60 shows the fourth and last example. In this example, the Partner port does not have LACP enabled.

```
Switch(config)# show lacp interfaces

Actor details of all interfaces:
---------------------------------------------------------------------------------
Intf     Aggr     Port  Port  State    System-ID           System Aggr Forwarding
         Name     Id    Pri                                Pri    Key  State
---------------------------------------------------------------------------------
1/1/25   lag256   26    1     ALFOE    8c:85:c1:65:0e:c0 65534  256  lacp-block
1/1/27   lag256   28    1     ALFO     8c:85:c1:65:0e:c0 65534  256  lacp-block
1/1/28   lag256   29    1     ALFNCD   8c:85:c1:65:0e:c0 65534  256  up

Partner details of all interfaces:
---------------------------------------------------------------------------------
Intf     Aggr        Port  Port  State    System-ID           System Aggr
         Name        Id    Pri                                Pri    Key
---------------------------------------------------------------------------------
1/1/25   lag256      0     0     PLFOEX   00:00:00:00:00:00 0      0
1/1/27   lag256      1     1     ALFN     02:01:00:00:01:00 65534  1
1/1/28   lag256      1001  1     ALFNCD   02:01:00:00:01:00 65534  2
```

Figure 2-60: Peer port does not have LACP enabled

Note

When the peer does not send LACP messages, AOS-CX switches will block the port, by default. This behavior may be different on other vendors' switches.

LAGs and load sharing

Peer LAG devices must balance the traffic load across multiple physical interfaces. How does the switch decide which member port should be used for a given packet? It uses a hashing algorithm.

Load sharing algorithm

A hash algorithm is a mathematical function that is applied to an input (x). Interestingly, if you see the output (y) of this function, you cannot derive or infer input (x). Therefore, it is referred to as a one-way function. Another characteristic of a hash is that if the same input (x) has been entered, the resulting output will be always the same. So, what does the switch use as inputs to this function when determine which link to use in the LAG?

The switch uses frame and/or packet header information as hash function inputs, and the result is the port member to be used for that particular frame. Depending on the switch algorithm used, inputs for the hash could be:

- Layer 4 TCP/UDP ports
- Layer 3 source and destination IP addresses
- Layer 2 source and destination MAC addresses

Note

Layer 3 is the default on AOS-CX switches, and is typically recommended to use in most situations.

Figure 2-61 shows how the layer 2 source and destination MAC addresses are taken as inputs for the algorithm. The output results in the switch using physical port 1 for all frames with this source and destination MAC address combination: always. Only a link failure will cause another port in the LAG to be used for these frames.

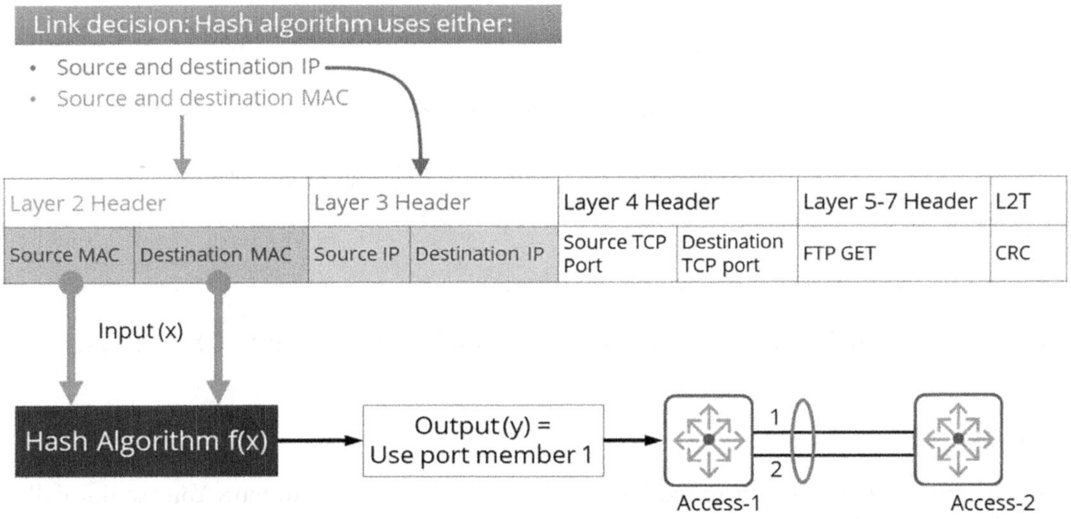

Figure 2-61: Load sharing algorithm

Configuring link aggregation load sharing

With AOS-CX switches, the default input information for the hash algorithm are the source and destination IP addresses. You can verify with the **show lacp aggregates** command (see Figure 2-62). Notice that the highlighted hash information is *l3-src-dst*. *l3* means layer 3, which uses the IP source and destination addresses.

In some cases, the hash algorithm's use of source and destination IP addresses may not properly distribute the load equally among port members. This could lead to traffic congestion and poor performance.

In that case, you can modify the hash input data. As shown in the figure, you have decided to use layer 4 source and destination port numbers as inputs to the hash function. This might help to avoid port member oversubscription when there are multiple connections between the same two devices, for example.

Verifying Hash algorithm

```
Switch# show lacp aggregates

Aggregate name    : lag1
Interfaces        : 2/1/26 1/1/25 2/1/25 1/1/26
Heartbeat rate    : Fast
Hash              : l3-src-dst
Aggregate mode    : Active
```

Changing algorithm

```
Switch(config)# interface lag <ID>
Switch(config-lag-if)# hash l4-src-dst
```

Default setting is `l3-src-dst`

Figure 2-62: Changing and verify the load sharing algorithm

Lab 2.01: Switching fundamentals

In this lab, you will apply your knowledge by placing configuration on AOS-CX switches.

Overview

In this lab activity, you will start to work on the new customer environment. Your senior colleague has already configured the aggregation switches in the new setup. You must learn how to add a factory default switch to this network.

The sw-edge1 is currently at factory default state. Your colleague wants you to understand how you must perform the initial switch and VLAN configuration on an Aruba AOS-CX access switch.

In the current stage of the labs, there is no need to consider redundancy. The new switch will be connected with a single link in this lab.

Objectives

After completing this lab, you will be able to:

- Gain initial access to a lab switch.
- Perform basic switch configuration and validation.

- Configure and verify an uplink connection to an aggregation switch.
- Configure a switch port as a VLAN access port.
- Configure a VLAN trunk on an uplink port.
- Get ample practice with VLAN port configurations.

Task 1: Initial access to the edge switch

Your colleague has asked you to add a new switch to the network. In this task, the initial configuration will be applied to a switch, sw-edge1, in its factory default state. The second edge switch (sw-edge2) and the aggregation switches have already been configured for you, so this task only applies to sw-edge1.

Objectives

- Configure an admin password.
- Configure the OOBM interface.
- Make an SSH connection to the switch's OOBM IP address.

Steps

1. Use the lab dashboard to open the console to **sw-edge1** and press **<Enter>**. Login with username **admin**, blank password (no password).

Important
Make sure to follow the requested passwords of the lab guide; some configurations and scripts depend on the correct password. For ease of use we use a simple password in the lab (Aruba123!): please, *never* use a simple password in a production network!

---

2. Review the default running configuration with the **show running-configuration** command.

```
6300# show running-config
Current configuration:
!
!Version ArubaOS-CX FL.10.09.1040
!export-password: default
user admin group administrators password ciphertext
AQBapUO8pKgfd1UOrOKxkHe65hSaTMZPhmtPrSdYPwREeA6UYgAAAH-
wShgpkD5jJYm9p1nV5y1aJBwiVIm18UsTiISeScMvi-
USwi8ON2rDxfNm+Gb+wWcL6nk+Qara/y1FnO/K/2Bgvqvq8QRkzzXNpn-
SAQaEWJ1M4BsHa-SAI8krMucrAIPa/89F
ntp server pool.ntp.org minpoll 4 maxpoll 4 iburst
ntp enable
!
!
...
```

3. Access configuration mode.

```
6300# configure
6300(config)#
```

4. Change the admin password to **Aruba123!**

```
6300(config)# user admin password plaintext Aruba123!
```

5. Change the hostname to **sw-edge1**.

```
6300(config)# hostname sw-edge1
sw-edge1(config)#
```

6. Access the management interface context (OOBM). The PC-MGMT system has an IP interface in this subnet. Enable the interface and configure a static IP address of 10.251.1.4/24.

```
sw-edge1(config)# interface mgmt
sw-edge1(config-if-mgmt)# no shutdown
sw-edge1(config-if-mgmt)# ip static 10.251.1.4/24
sw-edge1(config-if-mgmt)# end
sw-edge1#
```

Note

The OOBM port is not a commonly used feature in a campus deployment, but it is convenient in a training remote lab environment since you can break the 'normal' network and still have OOBM SSH access to the switches from the PC-MGMT desktop in the lab.

7. Review the interface mgmt configuration. (Example output: your output may be different.)

```
sw-edge1# show interface mgmt
  Address Mode                      : static
  Admin State                       : up
  Link State                        : up
  Mac Address                       : 64:e8:81:dd:81:81
  IPv4 address/subnet-mask          : 10.251.1.4/24
  Default gateway IPv4              :
  IPv6 address/prefix               :
  IPv6 link local address/prefix    : fe80::66e8:81ff:fedd:8181/64
  Default gateway IPv6              :
  Primary Nameserver                :
  Secondary Nameserver              :
  Tertiary Nameserver               :
```

8. Verify connectivity using a ping to PC-MGMT (10.251.1.90) in the mgmt VRF.

```
sw-edge1# ping 10.251.1.90 vrf mgmt
PING 10.251.1.90 (10.251.1.90) 100(128) bytes of data.
108 bytes from 10.251.1.90: icmp_seq=1 ttl=128 time=0.582 ms
108 bytes from 10.251.1.90: icmp_seq=2 ttl=128 time=0.639 ms
108 bytes from 10.251.1.90: icmp_seq=3 ttl=128 time=0.560 ms
108 bytes from 10.251.1.90: icmp_seq=4 ttl=128 time=0.465 ms
108 bytes from 10.251.1.90: icmp_seq=5 ttl=128 time=0.550 ms

--- 10.251.1.90 ping statistics ---
5 packets transmitted, 5 received, 0% packet loss, time 4094ms
rtt min/avg/max/mdev = 0.465/0.559/0.639/0.056 ms
```

Note

The mgmt interface is automatically bound to a dedicated VRF (routing table) so it is completely isolated from the front-facing network ports of the switch.

9. Make a local backup (checkpoint) of the current configuration.

```
sw-edge1# copy running-config checkpoint oobm
Copying configuration: [Success]
```

10. Review the list of checkpoints (local configurations); make sure there is a checkpoint named *oobm*.

```
sw-edge1# show checkpoint
NAME                  TYPE        WRITER   DATE(YYYY/MM/DD)
IMAGE VERSION
oobm                  latest      User     2022-10-15T17:52:30Z
FL.10.09.1040
```

11. In the next steps you will review the configured IP address on PC-MGMT. You then connect to sw-edge1, via SSH, to the management IP address you configured in the previous steps. Open a connection to the **PC-MGMT** in the lab topology.
12. On PC-MGMT, open a command prompt (**cmd**).
13. Run **ipconfig** to review the current interfaces and IP address assignments. You should see these 2 interfaces (see Figure 2-63):
    - OOBM
        - 10.**251**.1.90 (only for OOBM access)
        - no default gateway, isolated subnet to the management ports of the switches.
    - Lab NIC Core
        - 10.**254**.1.90
        - In-band access to the devices via the core router.

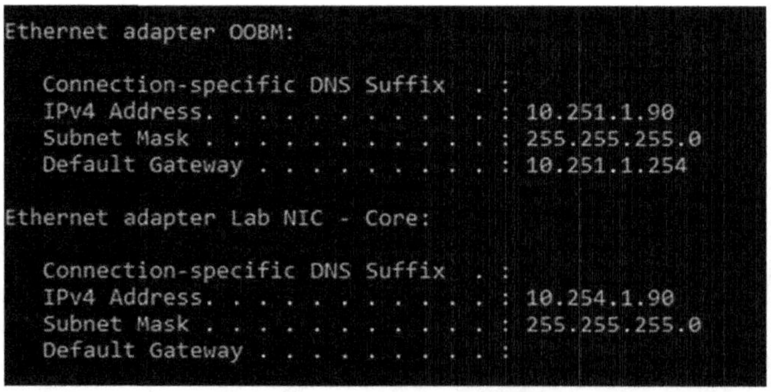

Figure 2-63: **ipconfig** output

14. Use PuTTY or MTPuTTY to open an SSH connection to the OOBM IP address of sw-edge1 (**10.251.1.4**) (see Figure 2-64).

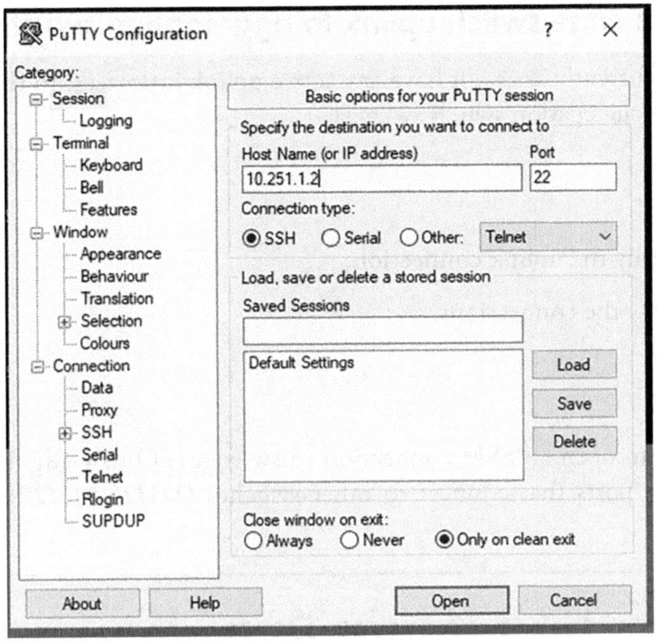

Figure 2-64: PuTTY connection

15. **Accept** the SSH key warning (see Figure 2-65).

Important

Be careful with the SSH key warning. You should only get this warning when the remote system is new or when it has re-generated its SSH host keys. Otherwise, some intermediate system may be trying to attack your systems.

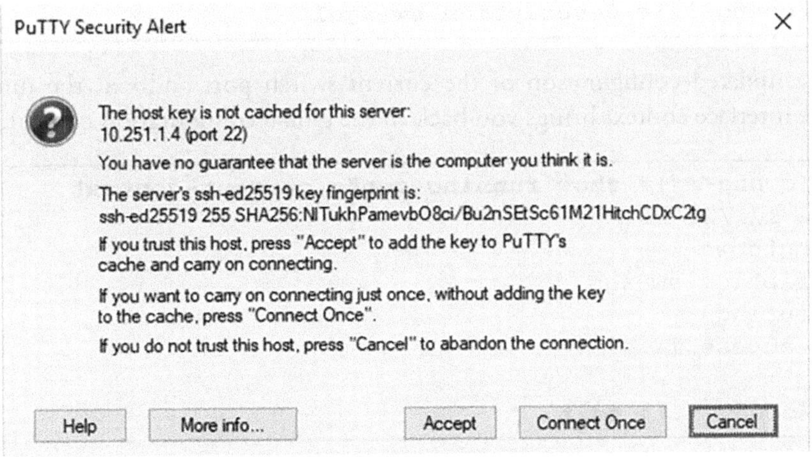

Figure 2-65: PuTTY security warning

16. Login with the **admin / Aruba123!** credentials.

Task 2: Configure edge switch uplink to aggregation switch

The sw-edge1 switch currently does not have any active uplink interfaces. In this task you will enable the uplink port to the aggregation switch sw-agg1.

Objectives

- Configure and verify the uplink connection.
- Use LLDP to verify the connections.

Steps

1. Use PC-MGMT to open an SSH connection to sw-edge1. On sw-edge1, access configuration mode and disable ports that connect to other switches (1/1/25-1/1/28) using the **shutdown** command.

```
sw-edge1# config
sw-edge1(config)# interface 1/1/25-1/1/28
sw-edge1(config-if-<1/1/25-1/1/28>)# shutdown
sw-edge1(config-if-<1/1/25-1/1/28>)# exit
```

2. On sw-edge1, enable port 1/1/27, which connects to sw-agg1.

```
sw-edge1(config)# interface 1/1/27
sw-edge1(config-if)# no shutdown
```

3. Configure port 1/1/27 with a description of **sw-agg1.**

```
sw-edge1(config-if)# description sw-agg1
```

4. Review the updated configuration of the current switch port and exit the interface context. Exiting the interface context brings you back to the global configuration context.

```
sw-edge1(config-if)# show running-config current-context
interface 1/1/27
    no shutdown
    description sw-agg1
    no routing
    vlan access 1
    exit
sw-edge1(config-if)# exit
```

5. Review the brief interface status. Port 1/1/27 should now have the "up" status and the updated description.

```
...
1/1/26        1       access   SFP+DAC1   no    down     Administratively
down   --           --
1/1/27        1       access   SFP+DAC1   yes   up
10000      sw-agg1
1/1/28        1       access   SFP+DAC1   no    down     Administratively
down   --           --
vlan1         --               --         yes   up
--         --
...
```

6. Verify that sw-agg1 is now listed as an LLDP neighbor.

```
sw-edge1(config)# show lldp neighbor-info

LLDP Neighbor Information
=========================

Total Neighbor Entries          : 6
Total Neighbor Entries Deleted  : 2
Total Neighbor Entries Dropped  : 0
Total Neighbor Entries Aged-Out : 2

LOCAL-PORT  CHASSIS-ID         PORT-ID                      PORT-DESC
TTL       SYS-NAME
-----------------------------------------------------------------------------
----------------------------------
1/1/1        68:b5:99:a3:a1:c0  11                          11
120       P28-PC1-Fanout
1/1/3        ec:b1:d7:1b:07:00  ec:b1:d7:1b:07:02           25
120
1/1/27       b8:d4:e7:da:80:00  1/1/1                       1/1/1
120       sw-agg1
mgmt         00:23:89:d6:7d:e7  GigabitEthernet2/0/21 T11-6300-A-
OOBM               120       P28-OOBM-Fanout
```

- **Question**: On what port is sw-edge1 connected on sw-agg1?
- **Answer**: Port 1/1/1

Note

There may be LLDP neighbors with a sys-name that contains 'Fanout'. These are switches that connect the lab VMs to the switch ports of your lab devices.

Note

Since there is no direct connection between the lab PC systems and the switch ports, shutting down the switch port will not be visible to the Lab VM PC, so you may need to manually refresh the IP configuration.

7. Verify the LLDP neighbor details for port 1/1/27.

```
sw-edge1(config)# show lldp neighbor-info 1/1/27

Port                              : 1/1/27
Neighbor Entries                  : 1
Neighbor Entries Deleted          : 0
Neighbor Entries Dropped          : 0
Neighbor Entries Aged-Out         : 0
Neighbor System-Name              : sw-agg1
Neighbor System-Description       : Aruba JL635A  GL.10.09.1040
Neighbor Chassis-ID               : b8:d4:e7:da:80:00
Neighbor Management-Address       : 10.1.1.1
Chassis Capabilities Available    : Bridge, Router
Chassis Capabilities Enabled      : Bridge, Router
Neighbor Port-ID                  : 1/1/1
Neighbor Port-Desc                : 1/1/1
Neighbor Port VLAN ID             : 1
TTL                               : 120

Neighbor EEE information          : DOT3
Neighbor TX Wake time             : 0 us
Neighbor RX Wake time             : 0 us
Neighbor Fallback time            : 0 us
Neighbor TX Echo time             : 0 us
Neighbor RX Echo time             : 0 us
```

- **Question**: What is the system description for the sw-agg1?
- **Answer**: Aruba switches report their product code as part of the system description. In this example JL635A (8325).

8. PC1 is connected to port 1/1/1; configure description 'pc1' on port 1/1/1.

```
sw-edge1(config)# interface 1/1/1
sw-edge1(config-if)# description pc1
sw-edge1(config-if)# exit
```

9. AP1 is connected to port 1/1/2. Configure a description of "ap1" on port 1/1/2.

```
sw-edge1(config)# interface 1/1/2
sw-edge1(config-if)# description ap1
sw-edge1(config-if)# exit
```

Task 3: Configure the PC port as a VLAN access port

In this task you will create a new VLAN on sw-edge1 and assign the PC connected to port 1/1/1 to this new VLAN.

Objectives

- Create a VLAN.
- Assign an access port to a VLAN.
- Verify the configuration.

Steps

1. Using PC-MGMT, open an SSH connection to sw-edge1 and access configuration mode.

> **Note**
> Use the command **config** to access the configuration mode; this will not be repeated for every task in the guide.

2. On sw-edge1, review the default list of VLANs.

```
sw-edge1(config)# show vlan
--------------------------------------------------------------------------------
----------------
VLAN  Name                            Status  Reason           Type
Interfaces
--------------------------------------------------------------------------------
-----------------
1     DEFAULT_VLAN_1                  up      ok               default
1/1/1-1/1/28
```

- **Question**: What VLANs exist by default on the switch?
- **Answer**: Only VLAN 1 exists by default on the switch. It is the default VLAN and it cannot be removed.

3. Examine the **show interface brief** output and review the default port status for port 1/1/1, which connects to PC1.

```
sw-edge1(config)# show interface brief
--------------------------------------------------------------------------
------------------------------------
Port       Native  Mode   Type              Enabled Status   Reason
Speed    Description
           VLAN
(Mb/s)
--------------------------------------------------------------------------
------------------------------------
1/1/1       1      access 1GbT             yes      up
1000     pc1
1/1/2       1      access 1GbT              yes      up
1000     ap1
```

- **Question**: What is the VLAN mode on a port by default?
- **Answer**: Access
- **Question**: What is the Native VLAN on a port by default?
- **Answer**: VLAN 1

4. Create VLAN 11, assign the VLAN name **v11-employee,** and revert to the global context.

```
sw-edge1(config)# vlan 11
sw-edge1(config-vlan-11)# name v11-employee
sw-edge1(config-vlan-11)# exit
```

5. List the VLANs to verify that the new VLAN was created.

```
sw-edge1(config)# show vlan
--------------------------------------------------------------------------
----------------
VLAN  Name                         Status  Reason             Type
Interfaces
--------------------------------------------------------------------------
----------------
1     DEFAULT_VLAN_1               up      ok                 default
1/1/1-1/1/28
11    v11-employee                 down    no_member_port     static
```

6. Review member ports of VLAN 11.

```
sw-edge1(config)# show vlan 11

------------------------------------------------------------------
----------------
VLAN  Name                            Status  Reason             Type
Interfaces
------------------------------------------------------------------
----------------
11    v11-employee                    down    no_member_port     static
```

7. Review the MAC-address table for VLAN 11.

```
sw-edge1(config)# show mac-address-table vlan 11
No MAC entries found.
```

8. Assign the port connected to PC1 (1/1/1) to VLAN 11 as an access port.

```
sw-edge1(config)# interface 1/1/1
sw-edge1(config-if)# vlan access 11
sw-edge1(config-if)# exit
```

9. Verify your configuration by reviewing the port membership of VLAN 11. Port 1/1/1 should now be member of VLAN 11.

```
sw-edge1(config)# show vlan 11

------------------------------------------------------------------
----------------
VLAN  Name                            Status  Reason             Type
Interfaces
------------------------------------------------------------------
----------------
11    v11-employee                    up      ok                 static
1/1/1
```

10. Verify that port 1/1/1 is no longer a member of VLAN 1.

```
sw-edge1(config)# show vlan 1

---------------------------------------------------------------------
----------------
VLAN  Name                        Status  Reason              Type
Interfaces
---------------------------------------------------------------------
----------------
1     DEFAULT_VLAN_1              up      ok                  default
1/1/2-1/1/28
```

11. Alternatively, you can also review the VLAN port configuration.

```
sw-edge1(config)# show vlan port 1/1/1

---------------------------------------------------------------------
------
VLAN  Name                            Mode            Mapping
---------------------------------------------------------------------
------
11    v11-employee                    access
```

Task 4: Configure an uplink port as VLAN trunk port

In this task the sw-edge1 uplink port to sw-agg1 will be configured to support multiple VLANs. This is known as a VLAN trunk. By default, VLAN 1 is the native (untagged) VLAN on the trunk. You will add VLAN 11 as a tagged VLAN to the trunk configuration.

Objectives

- Configure a VLAN trunk.
- Allow VLANs on a VLAN trunk.
- Verify the configuration.

Steps

1. On sw-edge1, verify that you are in configuration mode. Review the current configuration of uplink port 1/1/27.

```
sw-edge1(config)# interface 1/1/27
sw-edge1(config-if)# show running-config current-context
interface 1/1/27
    no shutdown
    description sw-agg1
    no routing
    vlan access 1
    exit
```

2. Configure the uplink as VLAN trunk; allow VLAN 11.

```
sw-edge1(config-if)# vlan trunk allowed 11
```

3. Review the updated port configuration.

```
sw-edge1(config-if)# show running-config current-context
interface 1/1/27
    no shutdown
    description sw-agg1
    no routing
    vlan trunk native 1
    vlan trunk allowed 11
    exit
```

Note

You may also check the interface configuration from any other context using the command: **show running-config interface** <member/slot/port>.

4. Review the list of allowed VLANs on the trunk port.

```
sw-edge1(config-if)# show vlan port 1/1/27
-------------------------------------------------------------------
------
VLAN  Name                                  Mode              Mapping
-------------------------------------------------------------------
------
11    v11-employee                          trunk             port
```

- **Question**: What VLANs are currently allowed on port 1/1/27?
- **Answer**: Only VLAN 11 is allowed on the port.

Note

When converting a port from access mode to trunk mode using the **vlan trunk allow** command, only the VLANs in the command are allowed on the port.

5. Allow VLAN 1 on the trunk.

```
sw-edge1(config-if)# vlan trunk allow 1
```

6. Review the list of allowed VLANs on the trunk port.

```
sw-edge1(config-if)# show vlan port 1/1/27
------------------------------------------------------------------
------
VLAN  Name                              Mode              Mapping
------------------------------------------------------------------
------
1     DEFAULT_VLAN_1                    native-untagged   port
11    v11-employee                      trunk             port
```

- **Question**: Did the allow VLAN 1 configuration overwrite the existing list of allowed VLANs?
- **Answer**: No

7. Simplify the allowed VLAN configuration by allowing all VLANS.

```
sw-edge1(config-if)# vlan trunk allowed all
```

- **Question**: Does the **allowed all** configuration imply that any VLAN will be passing on this VLAN trunk?
- **Answer**: No, it means that all VLANs that have been defined on this switch will be allowed. When a VLAN doesn't exist in the switch's global configuration, traffic marked with this VLAN ID will not be accepted on the VLAN trunk.
- **Question**: Will this apply to future VLANs?
- **Answer**: Yes, any VLAN that would be created on this switch would automatically be allowed on this VLAN trunk.

Note

Allowing all VLANs is easy for a lab environment, but it is typically not a best practice for a production environment: only the required VLANs should be allowed on the VLAN trunk ports in a production environment.

8. Review and compare the VLAN port configurations of ports 1/1/1 and 1/1/27. Notice the different modes (trunk and access) between the port VLANs.

```
sw-edge1(config-if)# show vlan port 1/1/1
--------------------------------------------------------------------------
------
VLAN  Name                                 Mode              Mapping
--------------------------------------------------------------------------
------
11    v11-employee                         access            port
sw-edge1(config-if)# show vlan port 1/1/27
--------------------------------------------------------------------------
------
VLAN  Name                                 Mode              Mapping
--------------------------------------------------------------------------
------
1     DEFAULT_VLAN_1                       native-untagged   port
11    v11-employee                         trunk             port
```

- **Question**: What does the 'native-untagged' mode represent for VLAN 1?
- **Answer**: Traffic in VLAN 1 will traverse this VLAN trunk as untagged ethernet frames, without a VLAN tag, just like an access port with VLAN 1.

9. On PC1, bounce the Lab NIC (disable and re-enable it) to renew its IP address.
10. On sw-edge1, review the MAC-address table for VLAN 11.

```
sw-edge1(config-if)# show mac-address-table vlan 11
MAC age-time              : 300 seconds
Number of MAC addresses : 2

MAC Address          VLAN     Type                       Port
--------------------------------------------------------------
00:50:56:b1:cd:26    11       dynamic                    1/1/1
b8:d4:e7:da:80:00    11       dynamic                    1/1/27
```

- **Question**: Do you see the MAC addresses associated with the uplink port? How many MAC addresses?
- **Answer**: Yes, the sw-agg1 MAC address.

Task 5: Practice VLAN port configuration

In this task you will practice the VLAN port configuration. You will configure these items:

- Prepare a VLAN trunk between sw-edge1 port 1/1/25 and sw-edge2 port 1/1/26 and enable the two ports.
- On sw-edge1, set the port 1/1/25 description to 'sw-edge2'.
- Ensure that only VLANs 1 and 11 are passing on this link.
- On sw-edge2, assign the port that connects to PC4 (1/1/4) to VLAN 11. Verify the port is enabled and up.

You will verify the configuration using these steps:

- Sw-edge1 port 1/1/25 is connected to port 1/1/26 on edge2 using LLDP.
- PC4 is getting an IP address in VLAN 11.
- Verify the MAC address tables on sw-edge2, sw-edge1, sw-agg1.
- Save the configuration after successful validation.

Objectives

- Practice the VLAN port configuration.

Steps

There are no steps in this task; use the commands you learned in the previous tasks to configure the objectives.

Configuration solution

1. sw-edge1:

```
int 1/1/25
 description sw-edge2
 vlan trunk allowed 1,11
 no shutdown
 exit
```

2. sw-edge2:

```
vlan 11
int 1/1/26
 description sw-edge1
 vlan trunk allowed 1,11
 no shutdown
 exit
```

```
int 1/1/4
 description pc4
 vlan access 11
 exit
```

3. Verify your configuration (both switches):

```
show lldp neighbor
show lldp neighbor 1/1/26
show mac-address-table vlan 11
```

4. Save the configurations on both edge switches.

```
write memory
```

5. You have completed this lab!

Lab 2.02: Link aggregation

In this lab you will configure a LAG and verify its connectivity.

Overview

In the previous lab you have added a factory default switch to the existing network. Your senior colleague has now asked you to learn about the bundling of links between devices.

This lab will introduce you to the aggregation of links, also known as a LAG. Link aggregation provides additional bandwidth capacity and redundancy in case of link failure. Your sw-edge1 and sw-edge2 have two links between each other. You will aggregate these two links into a LAG.

Objectives

After completing this lab, you will be able to:

- Configure link aggregation.
- Verify link aggregation.

Task 1: Configure a LAG between edge1 and edge2

In this task, you will configure sw-edge1 with a link aggregation configuration to aggregate ports 1/1/25 and 1/1/26 into a logical port called *LAG 50*.

First, a new logical or parent interface must be created, and then the physical member ports must be linked to this logical interface. The parent interface is referred to as a LAG interface.

On sw-edge1, you will configure two ports to the sw-edge2 switch as a LAG bundle.

Objectives

- Verify and configure link aggregation.

Steps

1. Using PC-MGMT, open an SSH connection to **sw-edge1** and enter configuration mode.
2. Define a new Link-Aggregation (LAG) interface with a LAG ID of 50.

```
sw-edge1(config)# interface lag 50
```

3. Review the interface list.

```
sw-edge1(config-lag-if)# show interface brief
--------------------------------------------------------------------
-----------------
Port       Native  Mode   Type            Enabled Status  Reason
Speed   Description
           VLAN
(Mb/s)
--------------------------------------------------------------------
-----------------
1/1/1      11      access 1GbT            yes     up
1000    pc1
1/1/2      1       access 1GbT            yes     up
1000    ap1
1/1/3      1       access 1GbT            yes     up
1000    --
...
lag50      1       access --              no      down    --
auto    --
```

- **Question**: Did the new interface show up in the list?
- **Answer**: Yes, the LAG interface is listed.
- **Question**: Is the LAG interface enabled, by default?
- **Answer**: No, by default it is disabled.

4. Enable the interface LAG 50.

```
sw-edge1(config-lag-if)# no shutdown
```

5. Review the aggregation mode.

```
sw-edge1(config-lag-if)# show lacp aggregates

Aggregate name    : lag50
Interfaces        :
Heartbeat rate    : N/A
Hash              : l3-src-dst
Aggregate mode    : Off
```

- **Question**: Based on what you have learned in this chapter, what are the types of LAG aggregation modes?
- **Answer**: A LAG can be static (no protocol) or dynamic (the LACP protocol is used to validate the link connections).

6. Configure LACP **active** as the aggregation mode.

```
sw-edge1(config-lag-if)# lacp mode active
```

7. Review the aggregation mode and verify that it now shows the LACP mode as active.

```
sw-edge1(config-lag-if)# show lacp aggregates

Aggregate name    : lag50
Interfaces        :
Heartbeat rate    : Slow
Hash              : l3-src-dst
Aggregate mode    : Active
```

8. At this point, the LAG does not contain any physical ports. In the next steps, you will configure port 1/1/25 as member port of the LAG 50.

9. Access the port 1/1/25 context and review the current configuration of the port. In the next steps you will link the port to the LAG 50 and check the impact on the current configuration of the port.

```
sw-edge1(config-lag-if)# interface 1/1/25
sw-edge1(config-if)# show running-config current-context
interface 1/1/25
    no shutdown
    no routing
    vlan access 1
    exit
```

10. Now link the port to LAG 50.

```
sw-edge1(config-if)# lag 50
```

11. Again, review the current port configuration.

```
sw-edge1(config-if)# show running-config current-context
interface 1/1/25
    no shutdown
    lag 50
    exit
```

- **Question**: What configuration changed after assigning the port to a LAG?
- **Answer**: A LAG member port does not have a VLAN port configuration anymore; this is configured at the LAG parent port context.

12. Review the LACP interface status.

```
sw-edge1(config-if)# show lacp interfaces
State abbreviations :
A - Active          P - Passive      F - Aggregable I - Individual
S - Short-timeout L - Long-timeout N - InSync      O - OutofSync
C - Collecting    D - Distributing
X - State m/c expired                E - Default neighbor state

Actor details of all interfaces:
-------------------------------------------------------------------
---------
Intf     Aggr    Port  Port    State    System-ID          System Aggr
Forwarding
         Name    Id    Pri                                 Pri    Key
State
-------------------------------------------------------------------
---------
1/1/25   lag50   26    1       ALFOE    64:e8:81:dd:81:80  65534  50
lacp-block

Partner details of all interfaces:
-------------------------------------------------------------------
---------
Intf     Aggr    Port  Port    State    System-ID          System Aggr
         Name    Id    Pri                                 Pri    Key
-------------------------------------------------------------------
---------
1/1/25   lag50   0     0       PLFOEX   00:00:00:00:00:00  0      0
```

- **Question**: The output shows *lacp-block*, why?
- **Answer**: On AOS-CX switches, LACP requires the peer device to send LACP PDUs. In this case, no LACP PDUs arrive on the port and therefore the port remains in LACP-block state. In this lab, the peer switch sw-edge2 has not been configured yet, so this state is expected at this point in the lab.

13. Now add port 1/1/26 as member port of the LAG 50 and enable the port.

```
sw-edge1(config-if)# interface 1/1/26
sw-edge1(config-if)# lag 50
sw-edge1(config-if)# no shutdown
sw-edge1(config-if)# exit
```

14. Review the LACP interface status.

```
sw-edge1(config)# show lacp interfaces

State abbreviations :
A - Active        P - Passive       F - Aggregable I - Individual
S - Short-timeout L - Long-timeout  N - InSync     O - OutofSync
C - Collecting    D - Distributing
X - State m/c expired                E - Default neighbor state

Actor details of all interfaces:
-----------------------------------------------------------------------------
-------
Intf      Aggr    Port  Port  State   System-ID         System Aggr
Forwarding
          Name    Id    Pri                             Pri    Key
State
-----------------------------------------------------------------------------
1/1/25    lag50   26    1     ALFOE   64:e8:81:dd:81:80 65534  50
lacp-block
1/1/26    lag50
down

Partner details of all interfaces:
-----------------------------------------------------------------------------
--------
Intf      Aggr    Port  Port  State   System-ID         System Aggr
          Name    Id    Pri                             Pri    Key
-----------------------------------------------------------------------------
--------
1/1/25    lag50   0     0     PLFOEX  00:00:00:00:00:00 0      0
1/1/26    lag50
```

- **Question**: What is the *Forwarding State*?
- **Answer**: lacp-blocked and down.
- **Question**: What is the result for clients connected to Edge2?
- **Answer**: PC4, which is connected to sw-edge2 on port 1/1/4, cannot reach the default gateway anymore.

15. You have now completed one side (sw-edge1) of the configuration; in the next task you will configure the sw-edge2 side.

Task 2: Practice and verify the LAG configuration on edge2

Objectives

In this task you will practice the LAG configuration steps on sw-edge2. Configure ports 1/1/25 and 1/1/26 in a LAG with ID 50 with LACP mode active. Make sure all ports are enabled and verify the status. Try to complete this task by yourself and only use the solution below if required.

Solution

1. On Sw-edge2:

```
interface lag 50
 no shutdown
 lacp mode active
 exit

interface 1/1/25,1/1/26
 lag 50
 no shutdown
 exit
```

Verification steps

1. On sw-edge2, verify 1/1/25, 1/1/26, and LAG 50 are enabled.

```
sw-edge2(config)# show interface brief
--------------------------------------------------------------------
---------------
Port       Native  Mode   Type        Enabled Status Reason
Speed   Description
           VLAN
(Mb/s)
--------------------------------------------------------------------
---------------
...
1/1/25     1       access SFP+DAC1    yes     up
10000   sw-edge1
1/1/26     1       access SFP+DAC1    yes     up
10000   sw-edge1
1/1/27     1       trunk  SFP+DAC1    no      down  Administratively
down   --       sw-agg1
1/1/28     1       access SFP+DAC1    no      down  Administratively
down   --       sw-agg2
vlan1      --              --         yes     up
--      --
lag50      1       access --          yes     up    --
20000   --
```

- **Question**: What is the reported speed for the LAG interface?
- **Answer**: The speed should be 20000Mbps: this is the combined speed of the 2 member ports, running at 10Gbps each.

2. Review the configuration of port 1/1/25.

```
sw-edge2(config)# show running-config interface 1/1/25
interface 1/1/25
    no shutdown
    description sw-edge1
    lag 50
    exit
```

3. Review the configuration of port 1/1/26.

```
sw-edge2(config)# show running-config interface 1/1/26
interface 1/1/26
    no shutdown
    description sw-edge1
    lag 50
    exit
```

4. Review the configuration of port LAG50.

```
sw-edge2(config)# show running-config interface lag 50
interface lag 50
    no shutdown
    no routing
    vlan access 1
    lacp mode active
    exit
```

5. Review the aggregate interfaces.

```
sw-edge2(config)# show lacp aggregates

Aggregate name    : lag50
Interfaces        : 1/1/25 1/1/26
Heartbeat rate    : Slow
Hash              : l3-src-dst
Aggregate mode    : Active
```

6. Review the LACP status of the interfaces; verify they are all UP.

```
sw-edge2(config)# show lacp interfaces

State abbreviations :
A - Active          P - Passive         F - Aggregable  I - Individual
S - Short-timeout L - Long-timeout  N - InSync      O - OutofSync
C - Collecting      D - Distributing
X - State m/c expired                   E - Default neighbor state

Actor details of all interfaces:
--------------------------------------------------------------------
-------
Intf      Aggr  Port  Port  State   System-ID          System Aggr
Forwarding
          Name  Id    Pri                               Pri    Key
State
```

```
------------------------------------------------------------------------
-------
1/1/25   lag50   26    1     ALFNCD   8c:85:c1:49:20:c0 65534   50   up
1/1/26   lag50   27    1     ALFNCD   8c:85:c1:49:20:c0 65534   50   up

Partner details of all interfaces:
------------------------------------------------------------------------
-------
Intf     Aggr    Port  Port  State    System-ID         System Aggr
         Name    Id    Pri                              Pri    Key
------------------------------------------------------------------------
-------
1/1/25   lag50   27    1     ALFNCD   64:e8:81:dd:81:80 65534  50
1/1/26   lag50   26    1     ALFNCD   64:e8:81:dd:81:80 65534  50
```

Task 3: Configuring the LAG as a VLAN trunk

In this task you will configure a LAG between sw-edge1 and sw-edge2, and configure it as a VLAN trunk. This ensures that traffic for VLAN 11, for example the PC4 on sw-edge2, can reach its default gateway via this new LAG.

Objectives

- Configure the LAG as a VLAN trunk.

Steps

1. On **sw-edge1**, review the list of interfaces.

```
sw-edge1(config)# show interface brief
------------------------------------------------------------------------
-----------------
Port       Native  Mode    Type        Enabled Status   Reason
Speed    Description
           VLAN
(Mb/s)
------------------------------------------------------------------------
-----------------
...
1/1/27     1       trunk   SFP+DAC1    yes     up
10000    sw-agg1
1/1/28     1       access  SFP+DAC1    no      down     Administratively
down   --        --
```

```
vlan1      --              --          yes     up
--     --
lag50      1       access --           yes     up     --
20000      --
```

- **Question**: What is the current VLAN port type of the LAG?
- **Answer**: The LAG is a new interface on the system. New interfaces are access ports and are placed in VLAN 1 by default.

2. Review the VLAN port status for LAG 50.

```
sw-edge1(config)# show vlan port lag50

-------------------------------------------------------------------------------
-------
VLAN   Name                              Mode              Mapping
-------------------------------------------------------------------------------
-------
1      DEFAULT_VLAN_1                    access            port
```

3. Configure LAG 50 as VLAN trunk and allow all VLANs.

```
sw-edge1(config)# interface lag 50
sw-edge1(config-lag-if)# vlan trunk allow all
sw-edge1(config-lag-if)# exit
```

4. Review the VLAN port status for LAG 50.

```
sw-edge1(config)# show vlan port lag50

-------------------------------------------------------------------------------
-------
VLAN   Name                              Mode              Mapping
-------------------------------------------------------------------------------
-------
1      DEFAULT_VLAN_1                    native-untagged   port1
1      v11-employee                      trunk             port
```

- **Question**: Is VLAN 11 now allowed on the LAG port?
- **Answer**: Yes.

5. On **sw-edge2**, apply the same configuration.

```
sw-edge2(config)# interface lag 50
sw-edge2(config-lag-if)# lan trunk allow all
sw-edge2(config-lag-if)# exit
```

6. Verify that PC4 now has access to the network again. Open a command prompt (**cmd**) and ping the default gateway IP address (10.1.11.1). This should be successful.

```
ping 10.1.11.1
```

7. Save the configuration on both edge switches.

```
write memory
```

Optional Task 4: Link-aggregation troubleshooting

The customer has had issues with link aggregations in the past due to incorrect cabling of the link aggregation member ports. You have been asked to explore how issues with cabling can be found on the switches. In this task you will review some of the common link aggregation issues.

Objectives

- Understand the LAG status.
- Troubleshoot a LAG.

Steps: LAG ports connected to two *different* LAGs on peer device

This scenario will show how to detect that the ports of a LAG are connected to 2 different LAGs on the peer side. This could occur due to a cabling error or a configuration error on the peer side.

In this example, a configuration error will be introduced on sw-edge2.

1. On sw-edge2, create a new LAG with ID 51 and set LACP to active mode.

```
sw-edge2(config)# interface lag 51
sw-edge2(config-lag-if)# no shutdown
sw-edge2(config-lag-if)# lacp mode active
```

2. Configure the LAG 51 as VLAN trunk and allow all VLANs.

```
sw-edge2(config-lag-if)# vlan trunk allow all
sw-edge2(config-lag-if)# exit
```

3. Configure port 1/1/26 as member of LAG 51. Note that you must un-assign the port from LAG 50 before you can assign it to the LAG 51.

```
sw-edge2(config)# interface 1/1/26
sw-edge2(config-if)# no lag 50
sw-edge2(config-if)# lag 51
sw-edge2(config-if)# exit
```

Troubleshooting the problem

4. Review the LACP status on both sides. Pay attention to the aggregation key ID.

```
sw-edge1(config)# show lacp interfaces

State abbreviations :
A - Active          P - Passive       F - Aggregable I - Individual
S - Short-timeout L - Long-timeout N - InSync      O - OutofSync
C - Collecting      D - Distributing
X - State m/c expired                 E - Default neighbor state

Actor details of all interfaces:
-------------------------------------------------------------------
-------
Intf     Aggr    Port  Port  State   System-ID          System Aggr
Forwarding
         Name    Id    Pri                              Pri    Key
State
-------------------------------------------------------------------
---------
1/1/25   lag50   26    1     ALFO    64:e8:81:dd:81:80 65534   50
lacp-block
1/1/26   lag50   27    1     ALFNCD  64:e8:81:dd:81:80 65534   50   up

Partner details of all interfaces:
-------------------------------------------------------------------
---------
Intf     Aggr    Port  Port  State   System-ID          System Aggr
         Name    Id    Pri                              Pri    Key
-------------------------------------------------------------------
---------
1/1/25   lag50   27    1     ALFN    8c:85:c1:49:20:c0 65534   51
1/1/26   lag50   26    1     ALFNCD  8c:85:c1:49:20:c0 65534   50
```

```
sw-edge2(config)# show lacp interfaces

State abbreviations :
A - Active         P - Passive      F - Aggregable I - Individual
S - Short-timeout L - Long-timeout N - InSync     O - OutofSync
C - Collecting    D - Distributing
X - State m/c expired              E - Default neighbor state

Actor details of all interfaces:
---------------------------------------------------------------------------
-------
Intf      Aggr   Port  Port  State   System-ID          System Aggr
Forwarding
          Name   Id    Pri                              Pri    Key
State
---------------------------------------------------------------------------
-------
1/1/25    lag50  26    1     ALFNCD  8c:85:c1:49:20:c0 65534  50 up
1/1/26    lag51  27    1     ALFN    8c:85:c1:49:20:c0 65534  51
lacp-block

Partner details of all interfaces:
---------------------------------------------------------------------------
-------
Intf      Aggr   Port  Port  State   System-ID          System Aggr
          Name   Id    Pri                              Pri    Key
---------------------------------------------------------------------------
-------
1/1/25    lag50  27    1     ALFNCD  64:e8:81:dd:81:80 65534  50
1/1/26    lag51  26    1     ALFO    64:e8:81:dd:81:80 65534  50
```

- **Question**: What do you observe?
- **Answer**: Different LAG Aggregation Key IDs, where the result is an LACP block.

5. On sw-edge2, revert the configuration.

```
sw-edge2(config)# interface 1/1/26
sw-edge2(config-if)# no lag 51
sw-edge2(config-if)# lag 50
sw-edge2(config-if)# exit
sw-edge2(config)# no interface lag 51
```

Optional Task 5: Verify link-aggregation failover

Your customer has heard that the failover of a link aggregation is very fast in case of a link failure. You want to be able to demonstrate this to the customer.

In this task you will prepare a failover test, verify link aggregation failover operation, and examine the impact on existing traffic.

Objectives

- Verify the LAG failover functionality.

Steps

1. On PC4, open a command prompt (cmd) and start a continuous ping to the default gateway.

```
ping 10.1.11.1 -t
```

2. On **sw-edge1**, shutdown the first port of the LAG.

```
sw-edge1(config)# interface 1/1/25
sw-edge1(config-if)# shutdown
```

3. Verify that the ping on PC4 continues.

> **Note**
> A single ping packet loss might be observed.

4. On sw-edge1, review the LACP interface state.

```
sw-edge1(config-if)# show lacp interfaces

State abbreviations :
A - Active        P - Passive      F - Aggregable I - Individual
S - Short-timeout L - Long-timeout N - InSync     O - OutofSync
C - Collecting    D - Distributing
X - State m/c expired              E - Default neighbor state

Actor details of all interfaces:
----------------------------------------------------------------------------
-------
```

```
Intf      Aggr   Port  Port  State   System-ID          System Aggr
Forwarding
          Name   Id    Pri                              Pri    Key
State
--------------------------------------------------------------------
-------
1/1/25    lag50
down
1/1/26    lag50  27    1     ALFNCD  64:e8:81:dd:81:80 65534  50
up
Partner details of all interfaces:
--------------------------------------------------------------------
-------
Intf      Aggr   Port  Port  State   System-ID          System Aggr
          Name   Id    Pri                              Pri    Key
--------------------------------------------------------------------
-------
1/1/25    lag50
1/1/26    lag50  26    1     ALFNCD  8c:85:c1:49:20:c0 65534  50
```

5. Review the port speed in the interface lag brief output.

```
sw-edge1(config-if)# show interface lag brief
--------------------------------------------------------------------
----------------
Port      Native  Mode   Type          Enabled Status  Reason
Speed   Description
          VLAN
(Mb/s)
--------------------------------------------------------------------
----------------
lag50     1       trunk  --            yes     up      --
10000   --
```

- **Question**: What happened with the reported interface LAG speed?
- **Answer**: Previously, with both links active, the reported link speed was 20 Gbps. With one link down, the reported link speed is 10 Gbps. The LAG speed is adjusted to the actual number of active interfaces and their respective speeds.

6. Enable port 1/1/25 again.

```
sw-edge1(config-if)# interface 1/1/25
sw-edge1(config-if)# no shutdown
```

7. Now simulate a failure of the second link by shutting down port 1/1/26.

```
sw-edge1(config-if)# interface 1/1/26
sw-edge1(config-if)# shutdown
```

8. On PC4, verify that the ping continues; this demonstrates the link aggregation failover test.
9. On sw-edge1, enable port 1/1/26 again.

```
sw-edge1(config-if)# interface 1/1/26
sw-edge1(config-if)# no shutdown
sw-edge1(config-if)# exit
```

10. Verify that both interfaces are in LACP UP state.

```
sw-edge1(config)# show lacp interfaces

State abbreviations :
A - Active          P - Passive          F - Aggregable I - Individual
S - Short-timeout L - Long-timeout     N - InSync     O - OutofSync
C - Collecting     D - Distributing
X - State m/c expired                  E - Default neighbor state

Actor details of all interfaces:
-------------------------------------------------------------------
-------
Intf      Aggr   Port  Port  State    System-ID          System Aggr
Forwarding
          Name   Id    Pri                               Pri    Key
State
-------------------------------------------------------------------
-------
1/1/25    lag50  26    1     ALFNCD   64:e8:81:dd:81:80 65534  50 up
1/1/26    lag50  27    1     ALFNCD   64:e8:81:dd:81:80 65534  50 up

Partner details of all interfaces:
-------------------------------------------------------------------
-------
Intf      Aggr   Port  Port  State    System-ID          System Aggr
          Name   Id    Pri                               Pri    Key
-------------------------------------------------------------------
-------
1/1/25    lag50  27    1     ALFNCD   8c:85:c1:49:20:c0 65534  50
1/1/26    lag50  26    1     ALFNCD   8c:85:c1:49:20:c0 65534  50
```

Optional Task 6: Static LAG configuration

The customer is aware that an LACP LAG is the recommended best practice for link aggregation. However, in some rare cases, they might work with a device or server that might not support LACP.

After consulting with your colleague, you learn that a static LAG can be configured for those devices. In this task you will practice the configuration of a static LAG. This is a LAG without a protocol, so LACP is not used.

To practice this configuration, the existing LACP LAG between the switches will be converted to a static LAG on sw-edge1.

Objectives

- Configure a static LAG.

Steps

1. On **sw-edge1**, remove the LACP mode active from the LAG 50.

```
sw-edge1(config)# interface lag 50
sw-edge1(config-lag-if)# no lacp mode active
sw-edge1(config-lag-if)# exit
```

2. Review the LACP interfaces.

```
sw-edge1(config)# show lacp interfaces

State abbreviations :
A - Active          P - Passive        F - Aggregable I - Individual
S - Short-timeout L - Long-timeout   N - InSync     O - OutofSync
C - Collecting     D - Distributing
X - State m/c expired                  E - Default neighbor state

Actor details of all interfaces:
----------------------------------------------------------------------------
-------
Intf     Aggr    Port  Port  State    System-ID          System Aggr
Forwarding
         Name    Id    Pri                               Pri    Key
State
----------------------------------------------------------------------------
-------
1/1/25   lag50
up
```

```
1/1/26   lag50
up

Partner details of all interfaces:
----------------------------------------------------------------------------
-------
Intf     Aggr    Port   Port   State    System-ID             System Aggr
         Name    Id     Pri                                   Pri    Key
----------------------------------------------------------------------------
-------
1/1/25   lag50
1/1/26   lag50
```

Note

Even though a static LAG has now been configured, the command used to verify the LAG is **show lacp interfaces**.

- **Question**: What is the peer ID reported for a static LAG?
- **Answer**: For a static LAG, no LACP peer ID will be reported.
- **Question**: What would happen if a static LAG was connected to an LACP LAG?
- **Answer**: This depends on the platforms that are used. For the AOS-CX switch platform, the LACP side will set all local interfaces to 'blocked'. This means no traffic would traverse the LAG. Other platforms may enable one interface of the LAG when no LACP partner is observed, or even enable all of the device's interfaces. You should become familiar with the behavior of the various platforms that you have in your network.
- **Question**: What would happen when a non-configured device is connected to an LACP LAG?
- **Answer**: The LACP side does not receive any LACP frames, so the result will be the same as the previous question (depends on the platform). During zero-touch-provisioning, a newly deployed access switch may not have the correct LAG configuration yet. This would result in an aggregation switch to block all traffic. To handle with this scenario, AOS-CX aggregation switches with VSX support a feature if it is not seeing LACP messages. This is known as *LACP fallback*, and this feature enables the ports when no LACP frames are received, as opposed to the 'blocking' default behavior and preventing a device from becoming provisioned (getting its configuration).

3. Revert the LAG mode to LACP active.

```
sw-edge1(config)# interface lag 50
sw-edge1(config-lag-if)# lacp mode active
sw-edge1(config-lag-if)# exit
```

4. Verify that the LACP status is *up*.

```
sw-edge1(config)# show lacp interfaces

State abbreviations :
A - Active        P - Passive       F - Aggregable I - Individual
S - Short-timeout L - Long-timeout  N - InSync     O - OutofSync
C - Collecting    D - Distributing
X - State m/c expired               E - Default neighbor state

Actor details of all interfaces:
--------------------------------------------------------------------
-------
Intf    Aggr    Port  Port State   System-ID         System Aggr
Forwarding
        Name    Id    Pri                            Pri    Key
State
--------------------------------------------------------------------
-------
1/1/25  lag50   26    1    ALFNCD  64:e8:81:dd:81:80 65534  50 up
1/1/26  lag50   27    1    ALFNCD  64:e8:81:dd:81:80 65534  50 up

Partner details of all interfaces:
--------------------------------------------------------------------
-------
Intf    Aggr    Port  Port State   System-ID         System Aggr
        Name    Id    Pri                            Pri    Key
--------------------------------------------------------------------
-------
1/1/25  lag50   27    1    ALFNCD  8c:85:c1:49:20:c0 65534  50
1/1/26  lag50   26    1    ALFNCD  8c:85:c1:49:20:c0 65534  50
```

5. You have completed this lab!

Learning check

Discuss these questions.

1. Which statements below accurately describe initial switch access and setup? (Select two.)
   a. You can make an out-of-band management (OOBM) connection to a switch using a serial connection.
   b. One advantage of making an out-of-band management (OOBM) connection to a switch using the management port is that it enhances redundancy for your data traffic.
   c. One advantage of making an out-of-band management (OOBM) connection to a switch using the management port is that it can mitigate bad actors trying to gain access to your switches.
   d. You connect to a switch via the console port, but you must configure that port with an IP address.
2. How would you properly identify the port indicated in Figure 2-66?

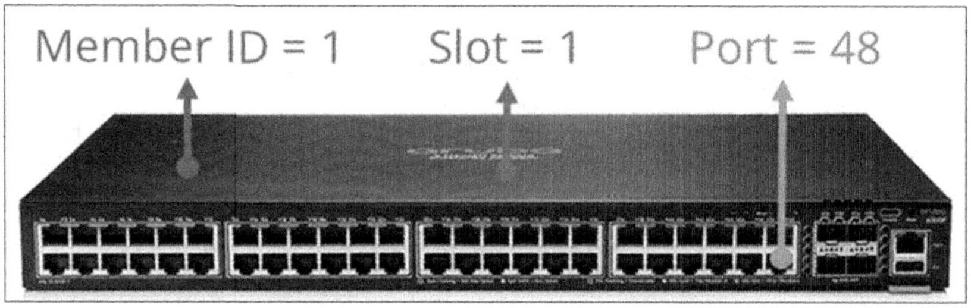

Figure 2-66: Question 2 diagram

   a. interface 1/48
   b. interface 1/1/48
   c. interface 48/1/1
   d. interface 1/48/1
3. When entering command line interface (CLI) commands on a switch, why might you use the <Tab> key?
   a. To move up one level in the CLI prompt mode hierarchy
   b. To access the context-sensitive help feature
   c. To exit out of global configuration mode
   d. To complete a CLI command that is partially entered

4. Your colleague asks you about managing configuration files on an AOS-CX switch. Which statements below would improve their understanding? (Select three.)
   a. The running-config is stored in RAM.
   b. The startup-config is stored in flash memory.
   c. When the switch reboots, unsaved changes to the startup-config are lost.
   d. AOS-CX considers both the startup-config and the running-config configurations to be "checkpoints".
5. Which of the options below are advantages of Virtual LANs (VLANs) as compared to physical LANs? (Select two.)
   a. VLANs segment the broadcast domain to improve performance.
   b. VLANs are more scalable.
   c. Routing between VLANs does not require a router.
   d. Multiple VLANs can be defined on a single switch.
6. You use trunk ports to extend the range of VLANs supported on a switch.
   a. True
   b. False
7. Which of the following are true about the Link Layer Discover Protocol (LLDP)? (Select two.)
   a. It helps you discover, document, and troubleshoot networks.
   b. LLDP has a Media Endpoint Discovery (MED) enhancement to help with "plug-and-play" networking.
   c. LLDP operates at Layers 2 and 3 of the OSI model.
   d. LLDP is enabled by default on both Aruba AOS-CX switches and Aruba Mobility gateways.
8. Choose the options below that accurately describe link aggregation. (Select three.)
   a. It enhances both performance and reliability.
   b. Member ports in a LAG should have matching duplex, speed, and media for their connected ports.
   c. A static LAG configuration is preferred over other methods, due to its simplicity.
   d. Peers that use the LACP can automatically form a LAG between them.

Summary

Congratulations! You have completed the second chapter. You should now be able to:

- Connect to and access the CLI of an AOS-CX switch.
- Perform an initial switch setup.
- Describe and configure VLANs, VLAN tagging (trunking), and IP addressing.
- Use LLDP and ICMP for network discovery and diagnosis.
- Configure link aggregation to improve performance and resiliency.

3 Basic IP Setup

OBJECTIVES

After completing this chapter, you will recognize fundamental concepts that are foundational for basic IP technologies, including IP addressing and routing. You will be able to:

- ✓ Discuss inter-VLAN routing.
- ✓ Explain DHCP relay.
- ✓ Discuss static IP routing.
- ✓ Configure a single-area OSPF network.

IP addressing

Proper IP addressing is vital for hosts, routers...and you. This section introduces you to IP addressing.

Introduction to IP addressing

The Internet Protocol (IP) provides an identification or address for each device in a network. Basic routing decisions are based on analyzing these layer 3 addresses. There are two versions of IP: IP version 4 (IPv4) and version 6 (IPv6). The main difference between them is the format of the IP address and the addressing space. Version 4 can allocate approximately 4.29 billion addresses, while version 6 can allocate 340 undecillion addresses. This book only focuses on IPv4 addressing and routing.

IPv4 addressing

An IPv4 address is comprised of 32-bits and is expressed in a dotted decimal notation. The dots divide the address in four 8-bit (1-byte) octets. This makes it easy for humans to work with IPv4 addresses. Eight bits enables values from 0 to 255.

Figure 3-1 shows three hosts in the same LAN, each with a unique IP address. For example, when Host C sends packets to Host A, the IP header contains a source IP address of 10.0.1.3 and a destination IP address of 10.0.1.1.

Layer 3 identifier for each network device

Feature	IPv4	IPv6
Size of address	32 bits – 4 octets	128 bits – 16 octets
Example address	192.168.10.250	fe80:0000:0000:0000:844d:a9ff:febb:62c3
Possible addresses	4.29 Billion	340 Undecillion (3.4 x 10^{38})

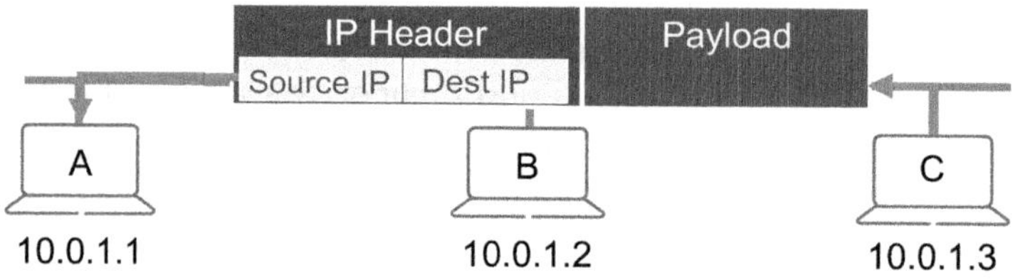

Figure 3-1: IP addressing

Each IP address has four octets, organized into two main or logical parts:

- Network portion
- Host portion

Let's explore IPv4 addressing in more depth.

Home address versus an IPv4 address

Your home address has a street name and a house number to help the mailperson deliver mail to your house. Likewise, IP addresses have a network portion and a host portion: this helps routers deliver data to a specific device. See Figure 3-2 for a comparison.

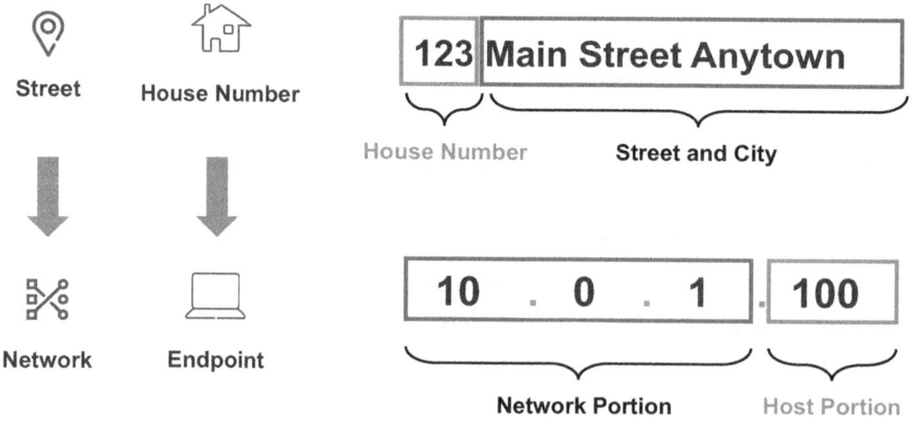

Figure 3-2: Home address versus an IPv4 address

Everyone on the same IP network has the same network number (like the street where you live). Perhaps everyone in VLAN 1 is associated with network 10.0.1, and one specific host on that network has a fourth octet of 100. But how do you know which portion of an IP address is for the network, and which portion is for the host?

Subnet masks

Every IP address has an associated subnet mask. This mask defines the line between the network portion and the host portion of the IP address. The subnet mask is a consecutive sequence of ones followed by a consecutive sequence of zeros. Where the ones end and the zeros begin is the line that divides the network and host portions of the address. Looking at the number in binary makes it very clear where this line is, but it is far more convenient for us humans to write this in decimal (see Figure 3-3). Regardless of how you write it, the mask is merely an instruction, telling devices how to perceive the associated IP address.

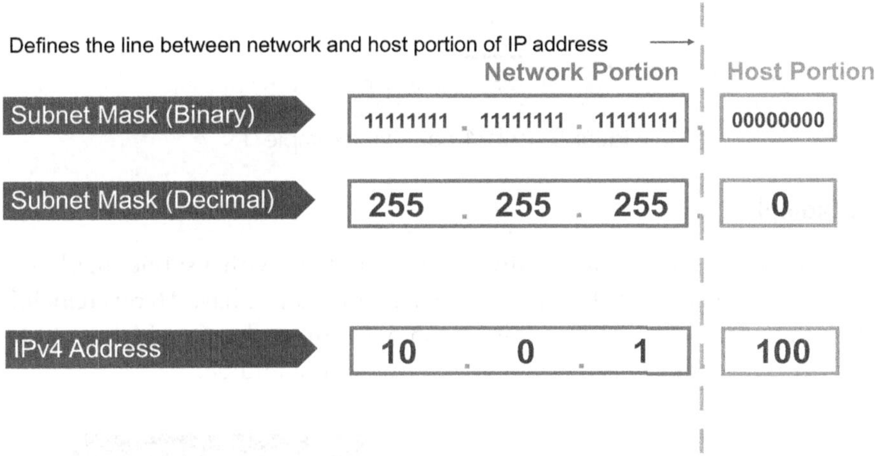

Figure 3-3: Subnet masks

Suppose an endpoint, switch, or router receives a packet with 10.0.1.100 as the destination IP address. It compares the address to the mask. Those portions of the IP address that line up with a binary 1 are part of the network number and those portions of the IP address that line up with the mask's binary 0s are part of the host portion. The subnet mask also defines the size of a network: the number of hosts it can contain. Let's look at a couple of examples.

IPv4 Network Example 1

The example shown in Figure 3-4 has IP address 10.0.1.100 with mask 255.255.255.0. Thus, the first three octets represent the network, and the last octet represents the host portion. So, how many hosts can you have on this subnet?

Well, 24 bits of this 32-bit IP address are used to specify the network, so only 8 bits remain to specify hosts on that network. Therefore, 2 to the power of 8 (2^8) equals number of available hosts. That's 256 unique numbers: from 0 to 255. Just to understand this a bit more, remember your binary to decimal conversion. If you have 11111111 in binary, then 128+64+32+16+8+4+2+1 = 255.

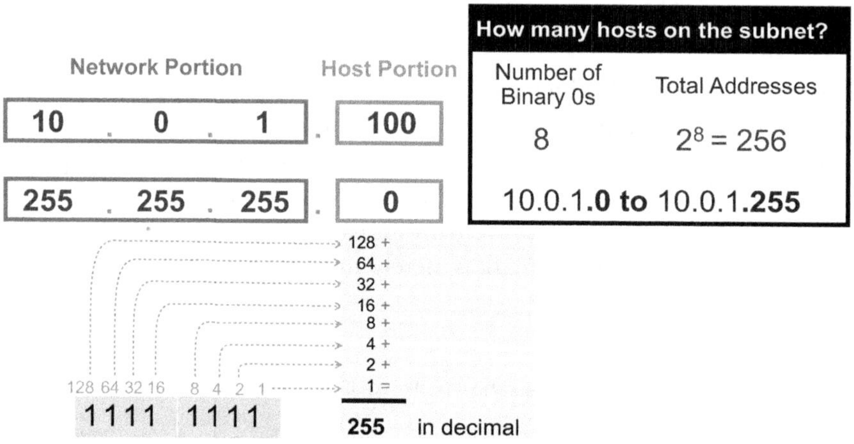

Figure 3-4: IPv4 network example 1

IPv4 Network Example 2

The example in Figure 3-5 shows a host address of 172.16.2.40, with a subnet mask of 255.255.0.0. In this example, only 16 bits are used to specify the network, so you have 16 bits remaining to specify hosts. And 2 to the 16th power (2^{16}) results in 65,536 unique hosts: addresses from 172.16.0.0 through 172.16.255.255, with 172.16.2. 40 somewhere in the middle.

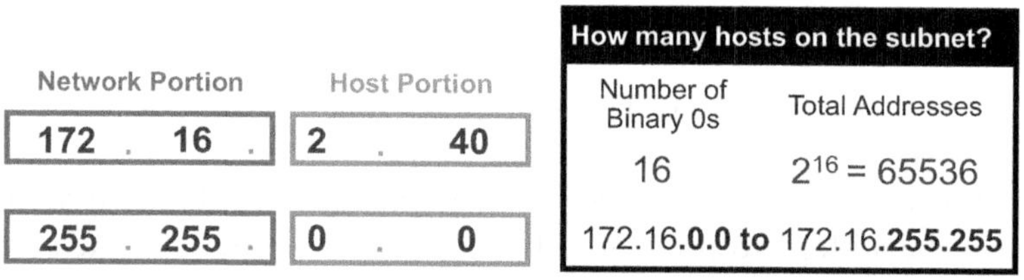

Figure 3-5: IPv4 network example 2

IPv4 address types

There are three types of IP addresses for a network, each with a specific function (see Figure 3-6):

- Network identifier (ID), which uniquely identifies the IP network
- Network broadcast, sometimes referred to as a directed broadcast when referring to a network's specific broadcast address
- Host addresses for the network

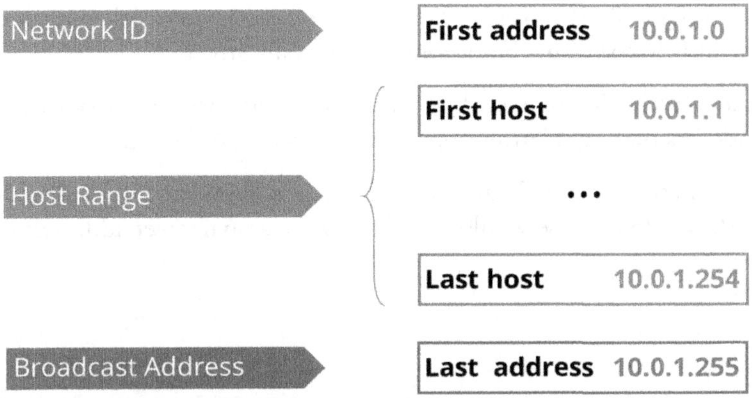

Figure 3-6: IPv4 address types

The network ID identifies the network itself, and all endpoints within it. This is always the first IP address in the host range, where all host bits are zeros.

The directed broadcast address is the last IP address in an address range, where all host bits are set to binary ones. This allows broadcast communication within that network, like saying, "Attention everyone on this subnet".

Therefore, since one number (first) is reserved as a kind of "name" for the network itself, and another (last) is reserved for the broadcast of that network, all numbers in between those two can actually be used to assign the hosts unique addresses. Remember, every endpoint in an IP network must have a unique host address.

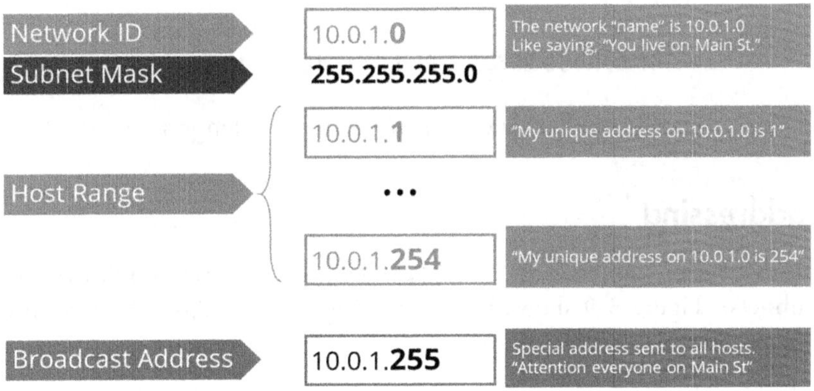

Figure 3-7: IPv4 address example

Let's look at an example, shown in Figure 3-7. Network 10.0.1.0 has a subnet mask of 255.255.255.0. The first three octets represent the network identifier and the last one represents the broadcast address of the network (directed broadcast).

The host octet's first value is 0. Therefore, 10.0.1.0 represents the network ID, which serves as a kind of "network name". It is like saying, "Your house is on Main Street".

The host octet's last value is 255, so 10.0.1.255 represents the directed broadcast address of the network. This is the address that says, "Attention, everyone on Main Street."

The addresses between the network ID and the broadcast address are assigned to devices on the network: 10.0.1.1 to 10.0.1.254. These would correlate to the actual street addresses for people that live on Main Street.

Prefix lengths

Let's look at some typical subnet masks, shown in Figure 3-8. Of the 32 bits in the first example, how many bits are binary 1s? How about the other two examples? That's right! 8, 16, and 24 bits respectively.

This gives us another, simpler way to specify the subnet mask. Instead of saying "The subnet mask is 255.255.255.0", you can simply say, "The mask is a /24". This is referred to as prefix notation.

The prefix length is written next to the IP address, as in the example shown at the bottom of Figure 3-8: IP address / prefix-length, or 10.0.1.100/24, for example.

Subnet Mask (Decimal)	Number of Ones	Prefix Length
255.0.0.0	8	/8
255.255.0.0	16	/16
255.255.255.0	24	/24

10.0.1.100 255.255.255.0 = 10.0.1.100/24

Figure 3-8: Subnet masks and prefix lengths

Using IPv4 addressing

You previously learned about single, large VLAN challenges, and how to split them up into multiple subnetworks (subnets). Figure 3-9 shows VLAN 1 using subnet 10.0.1.0/24, with over 250 hosts. Notice four things about this simple example:

1. The network ID and subnet mask is shown one time: 10.0.1.0/24.
2. The unique fourth octet for each host is shown next to the host. You know that every host on the same network has the same network number (10.0.1.0) and therefore you know that the entire address of the first host is 10.0.1.1, and the next host might be 10.0.1.2 and so on, perhaps up to 10.0.1.253.

3. A router is connected to this VLAN, and has been assigned the address of 10.0.1.254. Many seasoned network engineers assign either the first available address (.1) or the last available address (.254) to a connected router in the VLAN.

4. A layer 2 switch is not shown in this diagram. In discussions focused on layer 3 routing, layer 2 switches are depicted as "a wire". The switch typically exists, and each host connects to a unique port on the switch. But since layer 2 switches are transparent to layer 3 communications, it makes sense to omit it from the diagram, helping to simplify the figure and focus your mind on just the layer 3 concepts.

Similarly, you might show VLAN 2 using subnet 10.0.2.0/24, and so on and so forth.

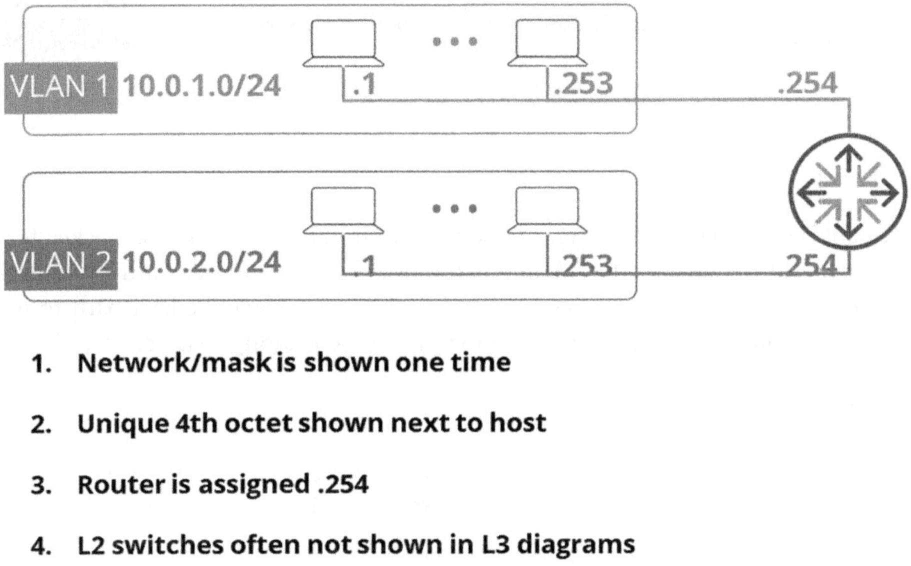

Figure 3-9: Subnet and VLANs

Let's now talk about how different VLANs, like VLAN 1 and 2 in Figure 3-9, can interact with each other.

IP routing

You have learned how devices communicate within the same network (i.e., broadcast domain), using layer 2 switching. Layer 2 switches forward frames to devices in the same LAN by processing layer 2 frame headers. Recall that switches build a MAC address table based on source MAC addresses and VLANs, and forward frames based on the destination MAC addresses.

Now you will learn how layer 3 routers transmit packets between different VLANs or broadcast domains, using the layer 3 IP destination addresses to make their forwarding decisions. An example is shown in Figure 3-10, where Host A in VLAN 10 wants to communicate with Host B in VLAN 20.

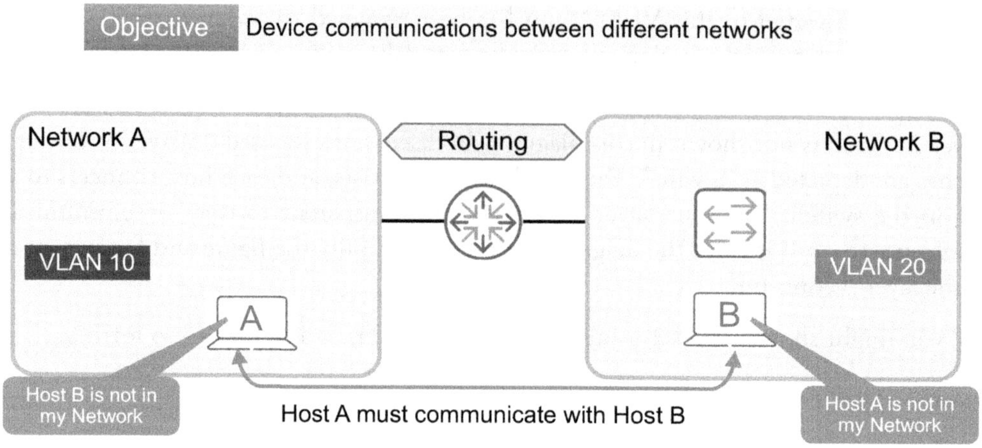

Figure 3-10: Routing between VLANs

Layer 3 analysis

The ability to analyze the layer 3 header and use the destination address for packet forwarding is known as *routing*. Layer 3 devices, like Aruba gateways, routers, firewalls, and multilayer switches, perform routing: they analyze layer 3 destination IP addresses, select the best path to get from the original source to ultimate destination, and the forward packets along that path (see Figure 3-11).

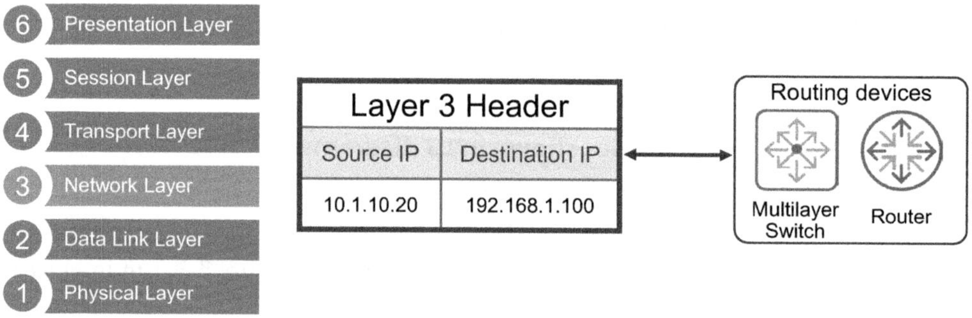

Figure 3-11: Routing examines layer 3 information to make routing decisions

IP routes

When a device communicates with others in a different network, it must know which local network device on its broadcast domain can route the traffic toward reach the remote destination network. This information is provided to computers by building an IP routing table.

For endpoints, this information must be manually added, in the form of so-called *static* or *connected* routes. Routers and multilayer switches can use manually-added static and/or connected routes or they can dynamically learn the best routes to each destination through the use of a dynamic routing protocol. You will soon learn about dynamic routing protocols in a later section of this chapter.

A static route might specify the following information, as shown in Figure 3-12 for Core-1 and Core-2:

- **Destination IP address**: where you want to go
- **Subnet mask**: How that destination IP address is split into a network portion and a host portion
- **Next-hop IP address**: The IP address of the router on your network that can route packets to next hop in order to reach the destination.

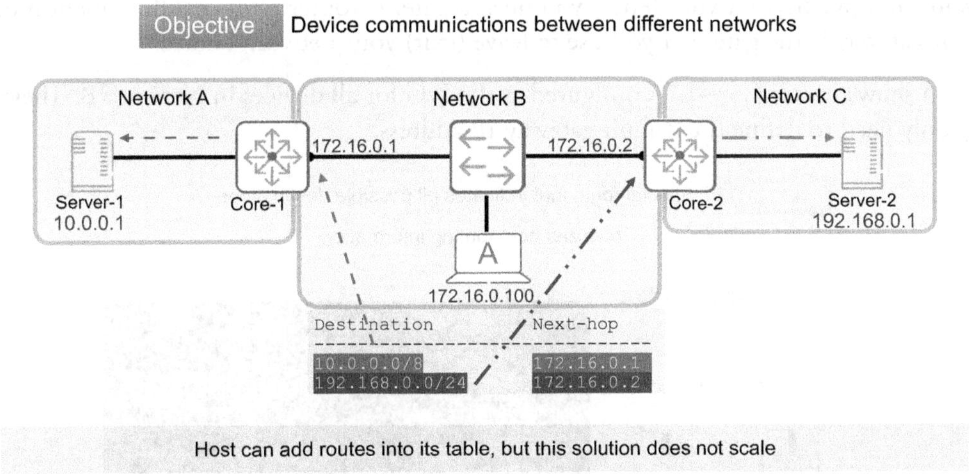

Figure 3-12: Device communications between different networks

Look at the scenario shown in Figure 3-12, which shows three networks connected by two routers. Core-1 connects Networks A and B, while Core-2 can route between networks B and C.

If Host A (in network B) needs to communicate with Server-1 in Network A, it must use the first route in its routing table: 10.0.0/8: "To get to destination 10.0.0.1, which has a mask of 255.0.0.0 (/8), you must use next-hop router at IP address 172.16.0.1. What if Host A wants to send packets to Server-2? The second route in its routing table, for 192.168.0.0/24, says to send packets to next-hop router at IP address 172.16.0.2, which is Core-2.

How did these routes get into Host A's routing table? For a connected route, the device examines the IP address and subnet mask configured for its Ethernet adapter and thus knows the local network to which it is connected. For remote networks, the destination network must be added manually. This is referred to as a *static* route.

Note

The problem with static routes is that they don't scale well: imagine a campus network that might have hundreds or thousands of destination networks. Manually adding these to every endpoint would be problematic and difficult to manage.

Default gateway

A default gateway (DG) is the device that routes to all network destinations for the endpoint devices in a broadcast domain or network. It is like telling a host, "To get to everywhere in the known universe, go to this next-hop address". The default gateway optimizes and simplifies endpoint routing decisions, since only a single, static route is required on the endpoint devices. This is commonly referred to as a *default* route and the routing device as a *default gateway*.

For example, assume that all the people in a neighborhood, including yourself, have fences around their homes. Inside the fence is your network. You can communicate with people inside your fence. If you want to move beyond your fence, you need a gate. A router performs the function of a gate: the default gateway is the gate that you use to leave (exit) your network choice.

Figure 3-13 shows router Core-1 is configured as the DG for all devices in Network B. These devices typically only need to define the default gateway IP address.

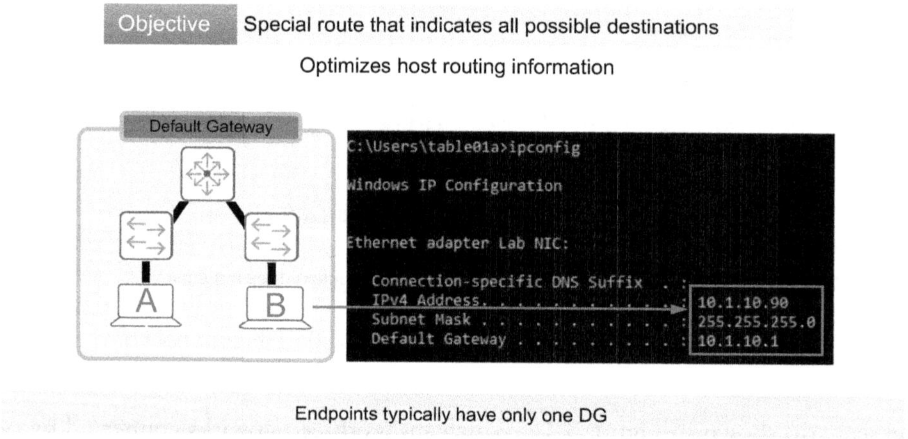

Figure 3-13: Default gateway on a Windows computer

Note

Different endpoints in the same subnet could have different default gateways, if there is more than one router on the network. However, this is very uncommon.

On a Microsoft Windows device, a static IP address and default gateway IP address can be defined at the same place. For Windows 10 devices you can follow these instructions:

1. Click the Windows **Start** button and type *Control Panel.*
2. Navigate to **Network and Internet → Network and Sharing Center → Change adapter settings**.

3. Right-click the Network Interface Card (NIC) that you want to configure and select **Properties**.
4. Double click *Internet Protocol Version 4 (TCP/IPv4)*.
5. Enter the proper information for the IP address, subnet mask, default gateway IP address. Optionally you can enter DNS server information.
6. Click **OK** twice.

To verify the previous set up, you perform the following:

1. Click the Windows **Start** button and type *command prompt* or *cmd*.
2. Execute the **ipconfig** command. Figure 3-13 shows an example of the output of this command.

Inter-VLAN routing

You must be able to connect your various VLANs together using a routing device, like a multilayer switch. Let's explore inter-VLAN routing with the AOS-CX switches.

Multilayer interface types

A switch port can only process layer 2 features and protocols, like creating and assigning VLANs, for example. Figure 3-14 shows ports 1 through 4 used as layer 2 interfaces. They connect to end systems and accept layer 2 frames as being a member of a layer 2 VLAN. The layer 2 frames from these ports and forward it based on the destination layer 2 MAC addresses in the frame headers. You learned in the last chapter that on some AOS-CX switches, the ports are layer 3 interfaces by default, and therefore you must configure those switch ports, like ports 1 to 5, with the command **no routing** on the respective interfaces.

But what if you wanted to route traffic between these VLANs. The AOS-CX switches are multilayer switches: they have both internal layer 2 switching functions and internal layer 3 routing capabilities. You need a way to connect each layer 2 VLAN to the internal routing functions. To accomplish this, you must either define the port as a routed port (assuming the switch has this capability, or to create switch virtual interfaces (SVIs) for the VLANs. An SVI is a virtual layer 3, routed interface that exists only inside the switch and is as a virtual (logical) construct.

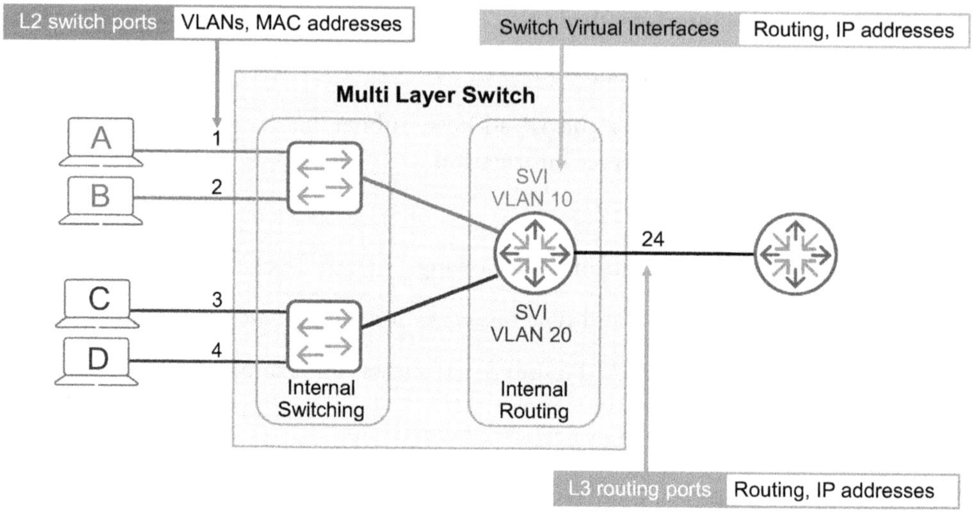

Figure 3-14: Multilayer interface types

Suppose that you create interface SVI 10. Because it is an SVI, by definition it connects to the internal routing construct. Because it is SVI "10", by definition, it connects to VLAN 10, and so services routable traffic from VLAN 10 to other destination VLAN networks. Similarly, you might create SVI 20. With some minimal routing configuration, which you will soon learn, your switch can now route traffic between your VLANs.

Now suppose that you need to connect your multilayer switch to an external router, perhaps using port 24, as shown in Figure 3-14. On some AOS-CX switches, ports are layer 3 interfaces by default. In this situation, port 24 connects to the internal routing functions by default. You merely need to configure it with typical layer 3 parameters, such as an IP address. You will soon learn about these concepts and the syntax for creating SVI interfaces.

The SVIs are virtual layer 3 interfaces, for internal routing, and port 24 in Figure 3-14 is a physical layer 3 interface, for external routing. *Both* are layer 3 interfaces, and so perform routing functions. They accept routable layer 3 packets and forward them based on their destination IP address.

At this point, you now know about three important interface types:

- Layer 2 switch ports
- Layer 3 SVIs
- Layer 3 physical routed ports

You are now ready to learn about another very important interface type when dealing with inter-VLAN routing: a trunk port.

Inter-VLAN routing solutions

You learned about VLANs in Chapter 2. A VLAN, as you recall, is a broadcast domain, with a unique IP network number. In other words, all devices in the same VLAN have the same network address. Examine Figure 3-15. All devices in the Sales VLAN are associated with network 10.1.10.x, where x is some unique host value. Therefore, IP addresses might be 10.1.10.100, 10.1.10.101, 10.1.10.102, and so on. Everyone in the HR VLAN might have host addresses like 192.168.20.100, 192.168.20.101, and so on.

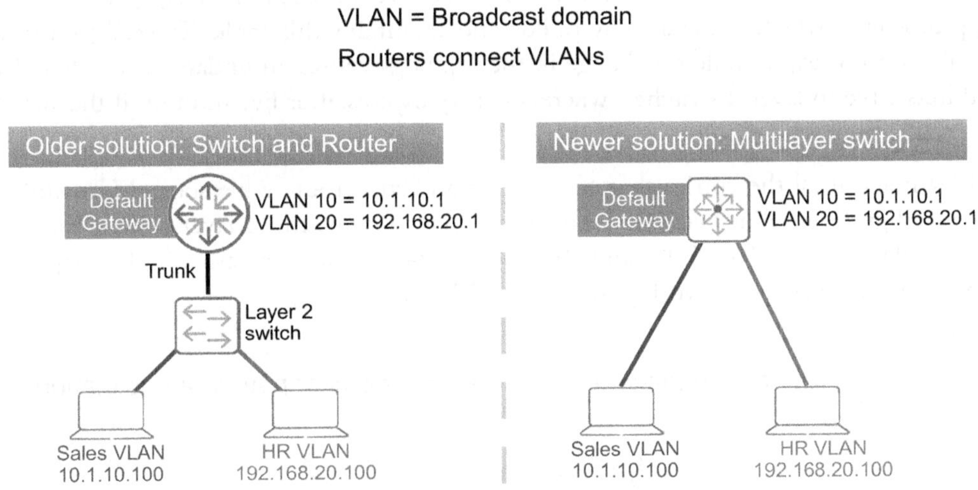

Figure 3-15: Inter-VLAN routing solutions

Recall that devices in different VLANs cannot communicate directly unless you connect them with a router. Inter-VLAN routing connects separate VLANs into a routed internetwork of communicating devices.

Many years ago (early 1990s), multilayer switches did not exist. In those days, layer 2 switches were used for host connectivity, and then routing was done with an external router solution, as shown in the right-hand side of Figure 3-15. This was called a *router-on-a-stick* configuration. You might still see this setup today, when a customer performs routing between network segments on an external firewall.

A potential problem with this deployment is that the switch-to-router link can become oversubscribed, although a LAG could alleviate this bandwidth problem to some extent. Performance can also be suboptimal, because sending frames to the router requires an additional switching operation.

Multilayer switches are more efficient devices to solve this issue. The switching and routing functions of the device are connected via a high-speed *internal* backplane. Initial routing decisions and other processes all happen *in the box*, which reduces latency and increases performance. Most AOS-CX switches are multilayer switches which have routing enabled, by default; the exception are the 6000/6100 series, which do not support routing—they are strictly layer 2 switches.

IP routing table

Routing devices (routers and multilayer switches) build and maintain a routing table that informs them of the best path to any given destination. One option to build the routing table is where you manually add entries to the route table in the form of static routes. Alternatively, you can configure a routing protocol(s), which automatically builds and maintains this table. Typically, entries in the routing table do not expire unless a change in the topology causes an update. This differs from the MAC address table in layer 2 switches, where an entry expires after five minutes if the switch stops receiving traffic from an endpoint.

You will learn about all the entries that can exist in an actual route table soon. Meanwhile, Figure 3-16 shows a slightly simplified view of a route table on the Core-1 multi-layer switch. There are three networks, 10.0.0.0/8, 172.16.0.0/16, and 192.168.0.0/24. Recall from earlier in the chapter that the subnet mask determines the network portion of an IP address.

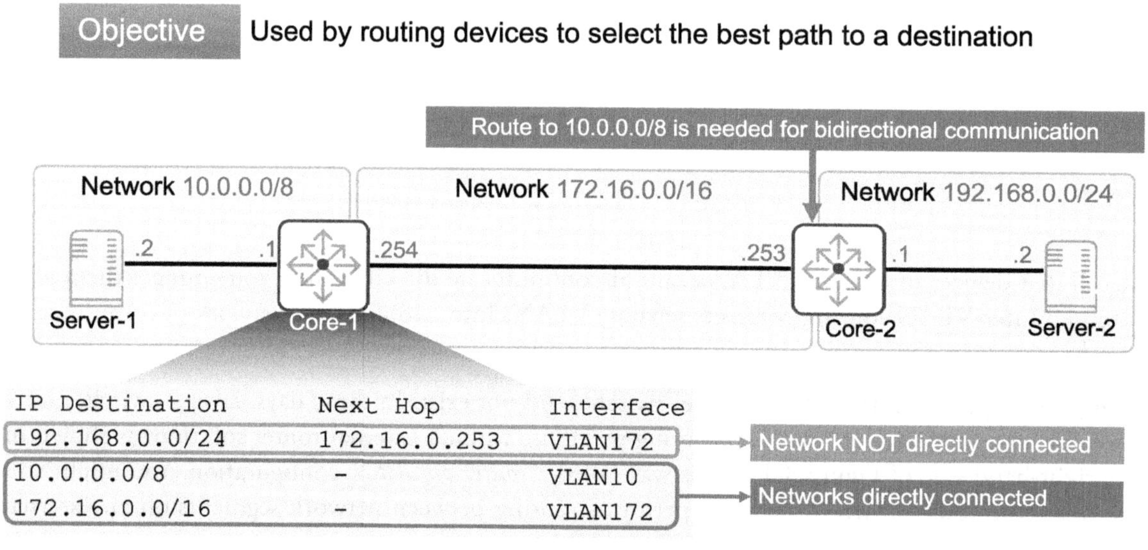

Figure 3-16: IP routing table

Let's analyze Core-1's routing table. The first entry equates to: "To get to any host on network 192.168.0.0/24, send packets to next-hop router 172.16.0.253 (Core-2). To get to that next-hop address, forward packets out interface VLAN 172".

The next entry says, "To get to any host on network 10.0.0.0/8, there is no next-hop...I'm directly connected to that network. Simply forward the packet out my local VLAN interface 10". Lastly, to connect to network 172.16.0.0/16, Core-1 is directly connected to this network and therefore there is no need to assign a next-hop IP address: simply forward the packet out the VLAN interface 172.

Consider at this moment that you configured only Core-1. Core-1 will properly route the packet and send it to Core-2. Core-2 will have no problem to send the packets to Sever-2, since this network is locally connected to Core-2. Therefore, communication from Server-1 to Server-2 is successful. However, the response from Server-2 is sent to Core-2. Core-2 receives the packet but since the destination (10.0.0.2) is not its routing table, the packet will be dropped. Therefore, bidirectional communication cannot be done by only configuring routing on Core1. To solve the problem, remember that Core-2 will need to be configured with a route to reach the non-direct connected network (10.0.0.0/8). Try to think about how this route would look like:

- Destination: 10.0.0.0/8
- Next-hop: 172.16.0.254
- Interface: VLAN that is connected to network 172.16.0.0 (in this example, VLAN 172)

This route would need to appear in Core-2's routing table (statically or dynamically) to allow for bi-directional communications between the two servers.

DHCP relay (IP helper)

You have previously learned how to configure static IP addressing parameters on an endpoint. However, configuring static IP address assignments is tedious and not scalable. Typically, most endpoints automatically get IP addressing information using the Dynamic Host Configuration Protocol (DHCP) (see the top part of Figure 3-17).

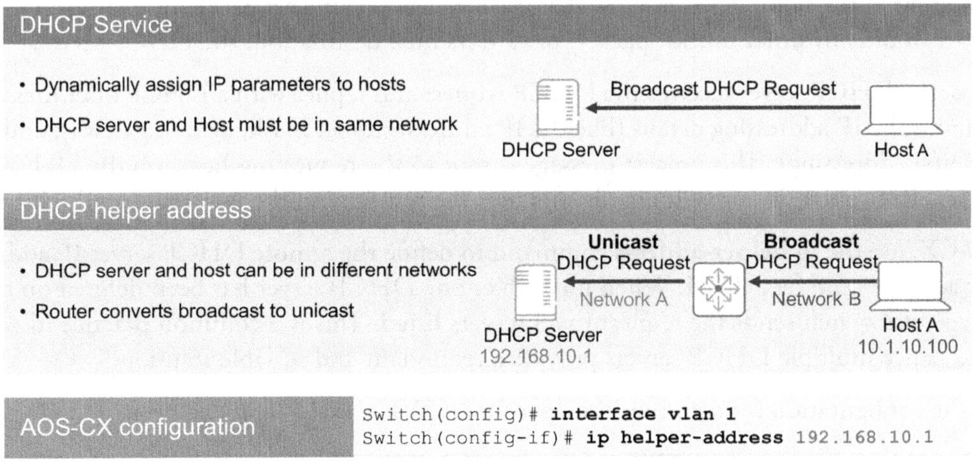

Figure 3-17: DHCP relay and DHCP helpers

To obtain IP addressing information dynamically, endpoints broadcast a DHCP request, "Hey everyone, I need an IP address, a subnet mask, a default gateway address, and a DNS server address". Because it is a broadcast, the host and server, by default, must be in the same subnet. Remember, a router defines the edge of a broadcast domain, and does not forward broadcast traffic. DHCP servers in different broadcast domains (different VLANs) do not receive the request, and so no address is assigned.

So, think about this for a minute. If an organization deploys thousands of broadcast domains (VLANs), then you would need thousands of DHCP servers–one per VLAN?! This obviously is not realistic. Instead, you need a centralized DHCP solution for all your VLANs.

From a routing perspective, the solution to the issue of endpoints being in separate VLANs from the DHCP servers is to configure a DHCP relay. Using DHCP relay, a DHCP helper address (or addresses if you have multiple DHCP servers) is configured on each router interface where DHCP request broadcasts would be received. That router interface typically serves as the default gateway for endpoints as well. The following process describes how the solution works, when a router is properly configured with a DHCP helper address (see the middle part of Figure 3-17):

1. The endpoint client broadcasts a normal DHCP query.
2. The router that is on the client's network (the client's default gateway) receives this broadcasted DHCP query.
3. After receiving the DHCP broadcast (which would be normally discarded), a new DHCP unicast packet is created and forwarded to the DHCP server. This new packet has source IP address set to the IP address of the SVI or routed interface on which the broadcast was originally received. This new DHCP unicast packet also sets the destination IP address to what is specified in the **ip helper-address** command (192.168.10.1, in this example). The destination address is the IP address of the DHCP server. Now that the message is a unicast packet, the router can forward it as it would any other unicast packet: toward its final destination, the DHCP server.
4. Thus, the DHCP server receives the DHCP request and replies with a DHCP offer message that contains the IP addressing details (like the IP address, subnet mask, default gateway, and a DNS server(s) addressing). This unicast message is sent to the requesting host, via the DHCP helper router. This router, in turns, forwards (proxies) the response to the requesting endpoint device.

In AOS-CX, use the **ip helper-address** command to define the remote DHCP server IP address. Up to eight addresses can be defined. When more than one DHCP server has been defined on the layer 3 interface, the switch sends the request to all servers listed. This is a common practice in company networks, since multiple DHCP servers provide a redundant and scalable solution.

Here is the configuration for defining IP helper for an SVI:

```
Switch(config)# interface vlan <VLAN-ID>
Switch(config-if)# ip helper-address <IP-address>
```

If you had a layer 3 routed interface, you would execute the **ip helper-address** command under the layer 3 interface context. The bottom of Figure 3-17 illustrates how to do this for the Network B SVI interface on the multilayer switch.

Route types and static IP routing

This section introduces you to the different types of routes and routing, but focuses specifically on one of those: static IP routing.

Route types in AOS-CX

You previously learned about the IP routing table, and that a routing device makes layer 3 forwarding decisions based on the best-path entries in this table for the destination IP addresses in packet headers. The entries in the routing table are populated manually or dynamically via a routing protocol. AOS-CX switches support multiple route types, shown in Figure 3-18.

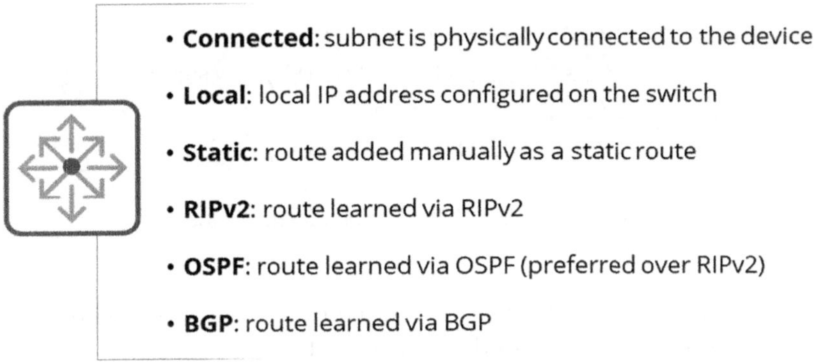

Figure 3-18: Route types in AOS-CX switches

Let's look at some examples of different route types.

Connected and local entries

Consider the SVI configuration in Figure 3-19. An SVI is configured for VLAN 1 with an IP address and it is assigned to port 1/1/1.

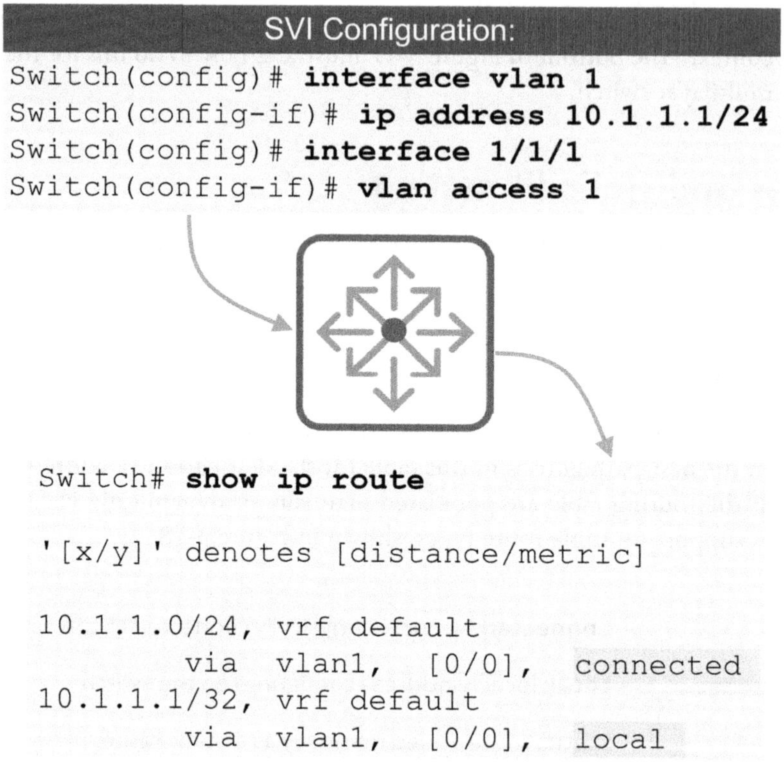

Figure 3-19: AOS-CX connected routes

This configuration generates some entries in the routing table, as seen with **show ip route** in Figure 3-19. In this output you see two entries in the routing table.

- **10.1.1.0/24** is a *connected* route type: the subnet physically connects to the switch and therefore there is no need for a next-hop device.
- **10.1.1.1/32** is a *local* route: the configured IP address participates in the switch's internal routing process. Recall that local route types indicate an SVI, as shown here, or a loopback interface, which you will see later in this chapter.

Note

You may notice that the local entry uses a /32 subnet mask. This is a special mask that indicates that subnetting cannot happen (no divisions are permitted) in the address space provided. This mask is useful to refer to individual host addresses.

Static route configuration and verification

You need a static route when the destination network is not directly or physically connected to the switch or router, and no routing protocol (like OSPF) advertises that route. With a static route, you manually define the destination network and the next hop in the path to the destination. This next-hop IP address must be directly connected to the device that you are configuring. Use the **ip route** command, as shown in Figure 3-20, to create a static route.

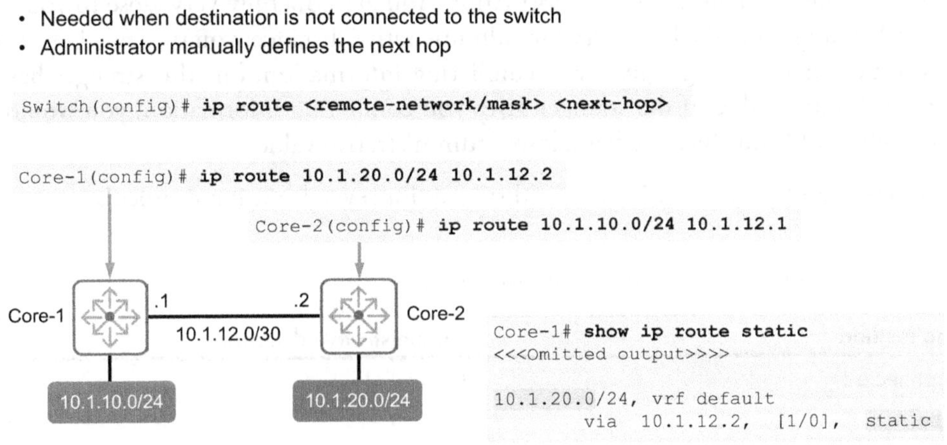

Figure 3-20: Static route configuration and verification

Suppose that endpoints in subnet 10.1.10.0 must communicate with devices in network 10.1.20.0. Core-1 switch has two physically connected networks: 10.1.10.0/24 and 10.1.12.0/30. However, the switch is not physically connected to the 10.1.20.0/24 subnet. Therefore, you add the route shown in Figure 3-20 for Core-1.

Core-2 switch also needs a static route to complete the path. It only has entries in its routing table for 10.1.20.0/24 and 10.1.12.0/30 networks. You must add a new route for Core-2 for network 10.1.10.0, as shown in Figure 3-20.

You can verify your configuration by executing the **show ip route static** command. An example is shown at the bottom right of Figure 3-20.

Note

Static routing is typically used for a default route, like how you would reach the Internet. A default route is represented with an IP address of 0.0.0.0 and a mask of 0.0.0.0 or "/0". It's like a *wild card*—if there is no specific entry in the routing table that matches the destination IP address, then use the default route to reach the destination.

Routing entries and administrative distance

Sometimes a router learns a route from two different sources. Maybe a router is running BGP and OSPF, and you have also configured some static routes. Perhaps all three methods have learned about the network 10.1.20.0/24. Which source for routing information should the router trust? The BGP route, the OSPF route, or the static route(s)? The protocol or method with the lowest *administrative distance* is used (a lower number is considered better or more reliable).

This is very much like humans learning information. You are hopefully very close to your mother... you trust her. You might say she has a very low administrative distance: you typically believe what she tells you. Imagine that a stranger gives you conflicting information, but this stranger has a higher administrative distance value. If the stranger gives you conflicting information, you would listen to what your mother told you since she has a lower administrative value.

Table 3-1 shows the AOS-CX administrative distances for several routing protocols.

Table 3-1. AOS-CX administrative distances for routes

Populating method	Administrative distance
Directly connected	0
Static route	1
External BGP	20
OSPF	110
Internal BGP	200
Unknown or unused	255

In Figure 3-21, Core-1 has learned about network 10.1.20.0 from OSPF and from a static route. Core-1 uses the path specified by the static route, due to the lower, more trusted administrative distance. You can see the administrative distance value in the routing entry: it is the first number located in the square brackets ("[ ]"). For example, the static entry has an administrative distance of 1 and OSPF a distance of 110.

- A trust rating for route entries
- Used if a route is learned by multiple methods
- Lower distance is preferred

```
Core-1# show ip route

Displaying ipv4 routes selected for
forwarding

'[x/y]' denotes [distance/metric]

10.1.20.0/24, vrf default
        via  10.1.12.2,  [1/0],  static
10.1.20.0/24, vrf default
        via  10.1.12.2,  [110/200],  ospf
```

Populating method	Admin Distance
Directly connected	0
Static Route	1
External BGP	20
OSPF	110
Internal BGP	200
Unknown or unused	255

Figure 3-21: Administrative distances

Note

The **show ip route** example output in Figure 3-21 is for illustration purposes. You would never see network 10.1.20.0 twice in the routing table with two different administrative distances: you would only see it once (i.e., the best route). Later, you will learn about equal cost multipath (ECMP), where you can have multiple entries in the routing table for a single destination; however, one requirement for ECMP is that the administrative distances must be the same for the multiple paths.

Note

Administrative distance is *not* defined in a standard. Different vendors might have different interpretations of what administrative distances should be defined for various routing methods. The ones used by AOS-CX switches in Table 3-1 are fairly common in the industry, but you should always verify with your vendor equipment what the defaults are. You can change the administrative distance for a specific route or a routing method, but that topic is beyond the scope of this book.

OSPF

This section introduces you to the most common dynamic routing protocol used internally by companies today: Open Shortest Path First (OSPF).

Introduction to dynamic routing

Configuring static routes doesn't scale very well and as mentioned earlier, static routing is typically used to configure default routes. Internally in your network, you might have hundreds or thousands of network destinations and you want to ensure that your routing devices learn about different ways of reaching the destination and populate the routing table with the best entry for each destination. In other words, you need a dynamic routing protocol.

Scalability issues

Working with static routes is simple and easy for small networks that only have a few subnets. However, when an organization has hundreds or even thousands of subnets, static routing is not efficient to configure or manage. Since there is no automatic route advertisement, you must manually configure all the routing. And on top of that, there's no dynamic failover mechanism if the primary path fails. With static routing, you would have to manually configure these policies.

The human factor is also a big consideration. You could easily cause serious problems by misconfiguring a static route, or adding it to the wrong router, which could result in layer 3 routing loops, traffic blackholing, and lost connectivity.

With static routes, recovering from routing device failures would be slow. You must manually configure new alternative paths. This is tedious and inconvenient, especially when dynamic routing would automatically do this for you.

Dynamic routing protocols are far more scalable, handling thousands of routes across multiple routing devices. Plus, they automatically failover to alternate paths with little to no downtime: in other words, they easily adapt to changes in the network topology. This is because routers running a dynamic routing protocol constantly exchange messages about changes in the network (a new network is added or an existing network is down), which minimizes your managerial workload.

See Figure 3-22 for a comparison of static and dynamic routing.

Static routes	Dynamic routes
• Suitable for simple networks	• Scale to any size
• No automated failover	• Offer automatic failover
• Human factor can cause outages	• Router exchanges improve availability
• Maintenance could be challenge	• Maintenance is simple

Figure 3-22: Static versus dynamic routing

Interior and exterior gateway protocols

An *autonomous system* (AS) is a collection of routing devices under a common administrative domain. Your Internet service provider (ISP) owns their own internal network: they have autonomy over this system and it is their AS. Each corporation has autonomy over their network: it is their AS.

To route packets inside an AS, each company chooses an interior gateway protocol (IGP). An IGP is simply a routing protocol or process that runs inside an AS. Examples of IGPs include the Routing Information Protocol (RIP) and Open Shortest Path First (OSPF), and even static routing. Each company chooses the IGP (or combination of IGPs) that works best for them.

In Figure 3-23, AS 1 has three routing devices. They use RIP to exchange network information to discover all paths inside the AS, and then find the best path for each network within the AS. AS 2 has a similar arrangement, but they have chosen to use OSPF instead of RIP.

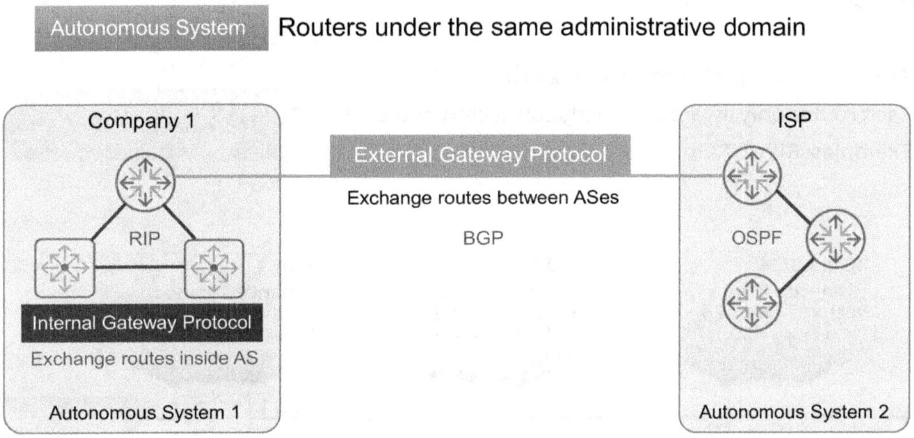

Figure 3-23: Autonomous systems and interior versus exterior gateway protocols

These companies might be partners and want to interact with each other. They connect to the Internet, via an ISP that owns completely different AS, like AS 3, for example. The two companies wish to route between AS 1 and AS 2 via the Internet. Robust, enterprise-class routing between these Autonomous Systems requires either static routing or an exterior gateway protocol (EGP). The only EGP currently in use today is the Border Gateway Protocol (BGP), which is beyond the scope of this book.

Note

Most companies will use a default route to connect to the Internet. BGP is commonly used by service providers connected to the Internet or the Internet backbone itself.

Distance vector routing protocols

The earliest methods devised to exchange IP information between routing devices are based on distance vector algorithms. Each router advertises its distance from each network, and the vector or direction packets must travel to get to that network. These algorithms were very simple in their construction and operation.

The distance is measured by number of *hops* or hop count. In Figure 3-23, how many routers must router R3 "hop over" to get to network 10.0.3.0/24? The answer is 0 hops – it is directly connected to that network.

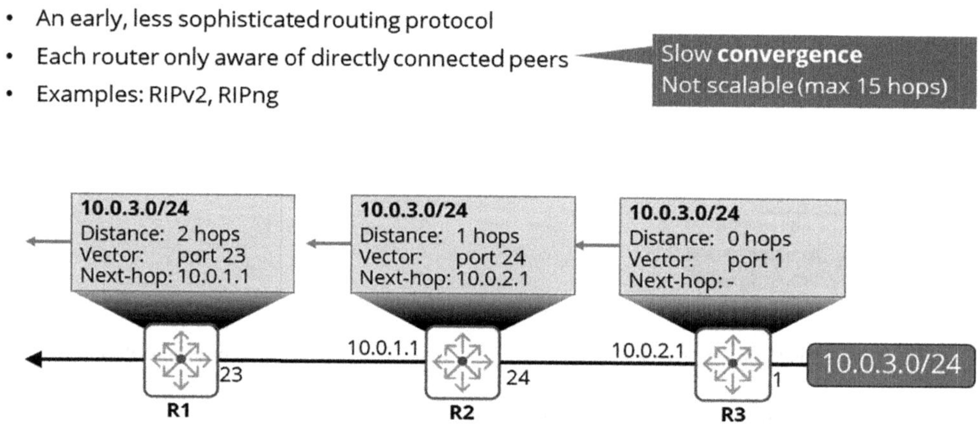

Figure 3-24: Distance vector routing protocols

What vector or direction must R3 forward packets to reach 10.0.3.0/24? The answer is: exit port 1. It is as if the router says, "to get to any host on destination network 10.0.3.0/24, which is zero hops away (I'm directly connected to the network), I must forward packets out my local port 1". Using something like RIP or RIPv2, R3 advertises its distance and vector to R2 (see Figure 3-24).

R2 receives this and knows, "If R3 is 0 hops away, and I must hop over R3 to get to that network, then I am one hop away from that route". To get to destination 10.0.3.0/24, which is one hop away, R2 must forward packets out local port 24 to the next-hop router 10.0.2.1. R2 then advertises its distance and vector to R1.

R1 receives this and knows, "If R2 is 1 hop away, then I must hop over R2 to get to the destination, and therefore I am two hops away". To get to 10.0.3.0/24, which is two hops away, R3 forwards packets out local port 23, to the next-hop router 10.0.1.1.

From this example, you can see that routers using a distance vector protocol are not aware of the entire network topology: they are only aware of what they are directly connected to, which includes their subnets/networks and their distance vector neighboring routers. This simplifies the configuration and route discovery process, but distance vector protocols, like RIP, historically have had two major limitations:

- **Slow convergence**: When a failure occurs, distance vector protocols can take minutes to converge (adapt to the change), depending on network size, complexity, and architecture.
- **Limited scalability**: Each router can be no more than 15 hops away from any other router. Understand that you could still have dozens, or even hundreds of routers, as long as it is a relatively flat architecture.

Common distance vector protocols include the Routing Information Protocol (RIP), RIP version 2 (RIPv2), and RIP Next Generation (RIPng). Note that RIP does not support a security feature to protect the advertisement messages that are exchanged between routing devices. And because of scalability issues with distance vector protocols in general, it is not recommended to use them, even in a small network.

Note

While RIP, RIPv2, and RIPng are supported on current AOS-CX switches, it is not recommended to use them. Therefore, even for small networks, you should use a routing solution that is more robust, like OSPF, which is *not* based on a distance vector implementation.

Link-state routing protocols

Link-state protocols resolve the limitations and disadvantages of distance vector protocols. Routers that run link-state protocols possess information about the network topology (not just what they are connected to but how their neighboring routing devices connect to devices beyond them). Thus, they can independently calculate their routing table using the sophisticated Dijkstra algorithm in order to select the best path for every network destination.

Actual implementations of link-state protocols include Open Shortest Path First version 2 (OSPFv2), which is used for IPv4, and OSPFv3, which is used for IPv6.These protocols provide faster convergence times compared with distance vector protocols. Another advantage is scalability: you can scale this to a much, much larger size (networks and routing devices) than what you could achieve with a distance vector solution (see Figure 3-25).

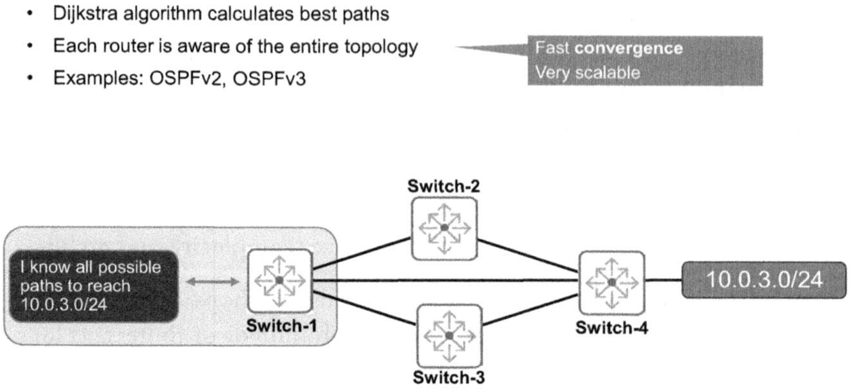

Figure 3-25: Link state protocols

Many years ago, one disadvantage of using protocols like OSPF is that they created more CPU, memory, and bandwidth utilization than distance vector protocols, like RIP. However, routing devices have become technologically faster and more capable over the years, and our approach to designing more efficient and scalable network architectures has greatly reduced this concern. In other words, you shouldn't even be concerned about the overhead involved with link state protocols because of the processing capability that exists in routing devices today, like the AOS-CX switches. The AOS-CX multilayer switches are designed to meet today's network requirements using protocols like OSPF. You view the OSPF capabilities on an AOS-CX switch with the **show capacities ospfv2** command:

```
Switch-6300# show capacities ospfv2

System Capacities: Filter OSPFv2
Capacities Name                                                   Value
-------------------------------------------------------------------------
Maximum number of OSPFv2 areas configurable in the s                128
Maximum number of OSPFv2 interfaces configurable in the system      128
Maximum number of OSPFv2 interfaces per area in the system          128
Maximum number of OSPFv2 neighbors allowed across all VRFs          128
Maximum number of OSPFv2 processes supported across each VRF          1
Maximum number of routes in OSPFv2 supported across all VRFs      64000
```

Note

This book focuses on OSPFv2 with a single area design. How to scale OSPF to very large sizes is covered in the Campus Access Professional-level course and book.

OSPF introduction

RFC 2328 defines OSPFv2, which describes how to route IPv4 packets. OSPF does not use TCP or UDP; instead OSPF advertisements are placed directly inside an IP packet. Therefore, it does not have a TCP or UDP port number, but has its own IP protocol number 89, as shown in Figure 3-26.

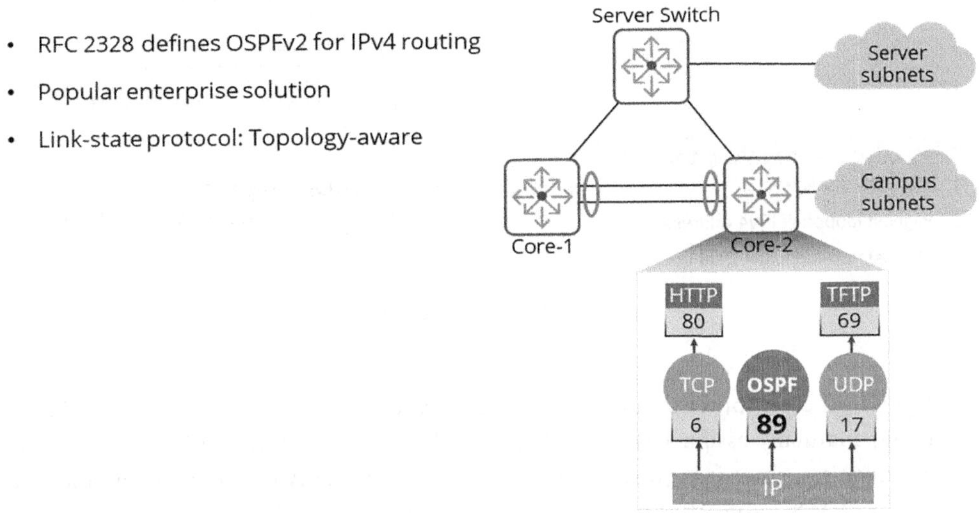

Figure 3-26: OSPF introduction

OSPF is a very popular enterprise-class IGP routing solution, due to its hierarchical scalability, quick convergence, and security mechanisms. For example, OSPF peers can authenticate packet exchanges.

As a link-state protocol, OSPF routers advertise information about their connected layer 3 interfaces and networks (links) and the cost associated with that interface. Let's see how it works, where we'll first discuss how each OSPF router is uniquely identified.

Router IDs

The OSPF configuration begins with the creation of the OSPF routing process. There can be multiple OSPF processes running on the AOS-CX switches, each having unique ID which is only locally significant. In Figure 3-27, the OSPF process ID is set to 1. In addition, all OSPF routers require a unique *router identifier* (RID). This is a 32-bit number which is typically associated with an actual IP address on the routing device. Routers include their ID in all OSPF packets that they generate. Humans might look at Figure 3-27 and see "Core-1, Core-2, and Server Switch". The routing switches, however, identify themselves based on their RIDs: 10.1.100.1, 10.1.100.2, and 10.0.100.0.

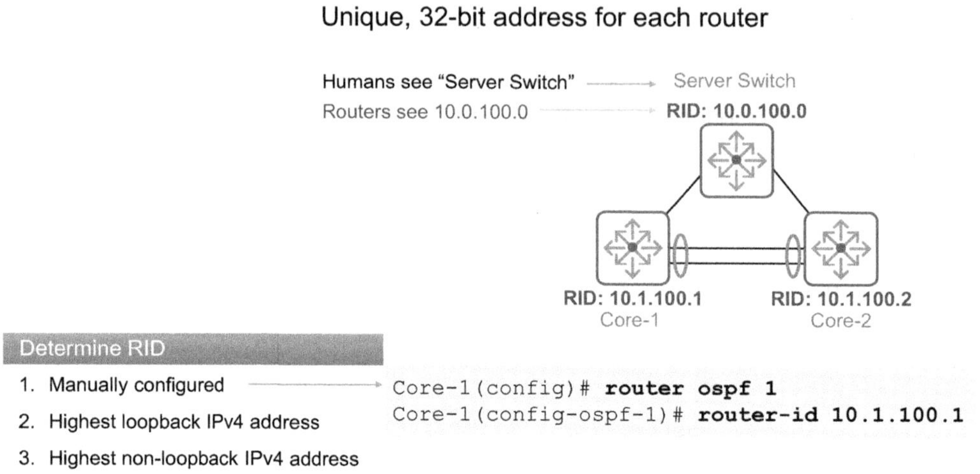

Figure 3-27: OSPF router IDs

You could depend on each router to automatically generate an RID itself, or you can take control of the situation and manually assign RIDs (see Figure 3-27) with the **router-id** command. Automatic assignment is tempting, since it requires little effort. However, many experienced engineers prefer to manually assign RIDs, due to certain documentation, troubleshooting, and management advantages.

AOS-CX switches use the following sequence to determine the router ID:

1. If you manually specify the RID, then that is what the router uses.
2. If you do not specify the RID, a loopback interface with the highest IP address becomes the RID.
3. If no loopback interfaces exist, the regular non-loopback, active interfaces with the highest IP address become the RID. Non-functional Interfaces that are in a down state are not considered.

Let's now explore using a loopback interface as a routing switch's RID.

Router IDs and loopback interfaces

A *loopback* interface is a logical interface that is always in an up state (unless you manually disable it). This interface enhances stability for processes and protocols that depend on the interface status to work, like OSPF. If you use a *physical interface* as the router ID, and that link fails, the router ID changes to an address on another alternate active interface, which is sub-optimal. Since loopbacks are always up, there is no danger of instability or wasted resources due to churning RIDs. Therefore, use loopbacks to "rid yourself" of this potential issue.

The IP address associated to a loopback interface can be routable, which means that external devices could initiate communication to it, assuming that you allow OSPF to advertise the network number of the loopback address (this doesn't happen by default). As you gain more education and experience, you will continue to learn about the advantages of loopback interfaces: scalability, troubleshooting, and network management.

In Figure 3-28, interface loopback 0 was created and assigned an IP address. Once this has been completed, you would then configure the OSPF routing process, specifying the loopback address as the RID. But you just learned that this would happen automatically anyway, so...why would you configure it?

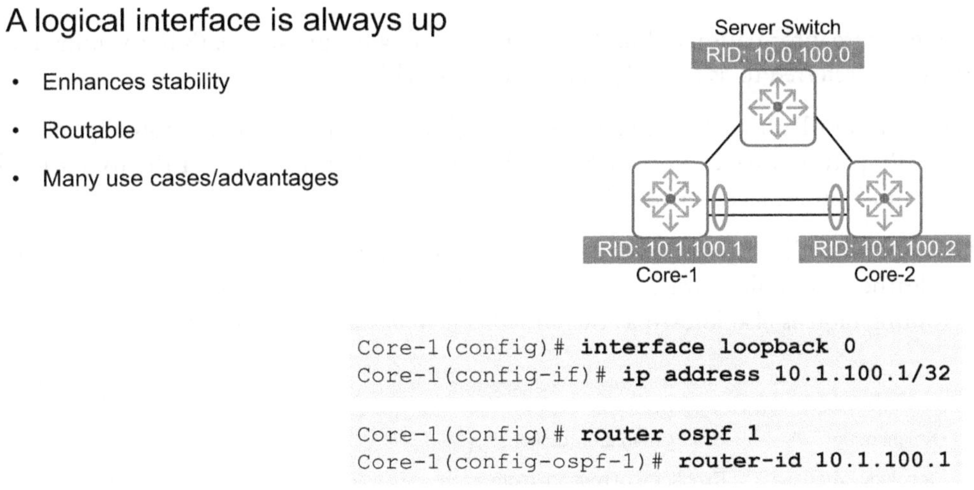

Figure 3-28: RIDs and loopback interfaces

What if someone on your team created a second loopback interface, with a higher IP address? At some point, when you would reboot the switch or restart the OSPF routing process, OSPF will then begin using the second loopback address instead of the loopback 0 address. This is not necessarily bad, but if you prefer to have a consistent, predictable behavior that matches your network documentation and facilitates quick troubleshooting, best practice is to not only configure the loopback interface, but also manually define that loopback IP address as the OSPF process's RID value as well.

OSPF general operation

This section will cover the general operation of OSPF. This includes the following topics:

- Building and validating an OSPF neighbor table.
- Building a list of every link (stored in a link state database, or LSDB for short).
- Share routing information using link state advertisements (LSAs).

- Use the LSDB to populate the routing table with the best paths.
- Learn how OSPF routers share messages.
- Understand the different states in which OSPF routers can be when establishing adjacencies with their OSPF routing peers.

Figure 3-29 provides an overview of the four key phases of OSPF's operations:

1. Form neighbor relationships with directly connected peers to build a neighbor table: a list of directly-connected OSPF peers.
2. Exchange route information to build an OSPF LSDB: a list of every path in the known topology, which is also referred to as a *topological* database or table.
3. Run the Shortest Path First (SPF) algorithm (Dijkstra's algorithm) on the topological database of *every* OSPF path to extract a list of *best* OSPF paths to destinations that the router has learned. This information is placed in an *OSPF* routing table.
4. Publish the best OSPF paths to the router's IP routing table. The IP routing table contains all routes learned from all routing protocols: OSPF, RIP, static, directly connected, and/or BGP. This routing table is also known as the *forwarding information base* (*FIB*).

Phase	Objective	Result
1. Neighboring	Build a neighbor relationship	Neighbor table
2. Topology database	Exchange of routing information	Link State Database
3. OSPF routing table	Run SPF to get the best path	OSPF routing table
4. IP routing table	OSPF publishes results to the IP routing table (FIB)	Routing table entries

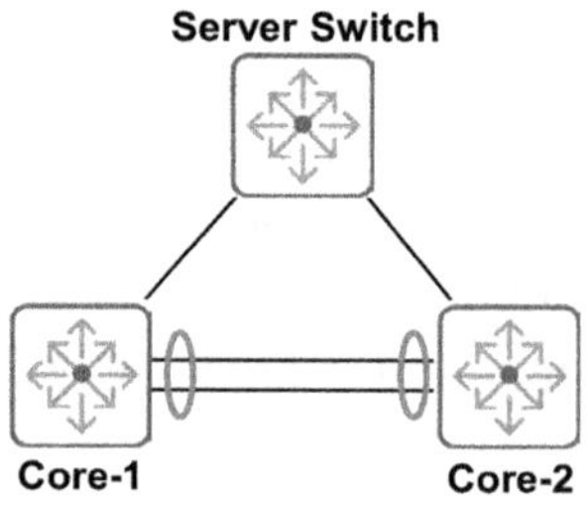

Figure 3-29: OSPF general operation overview

The following sections will discuss the four phases in more depth.

Phase 1: Building a neighbor table

First, OSPF neighbors introduce themselves to their directly connected neighbors on interfaces that OSPF enabled. During this process, they validate whether or not they are compatible and should share routing information. In this method the OSPF routers send OSPF *hello* packets out each active OSPF interface.

In Figure 3-30, each router is configured similarly (except for their IP addresses and perhaps interface numbers). First you create the OSPF process, and then you add the appropriate interfaces to that process: SVI 2 in the case of Core-1 and Core-2. You enter the interface context and use the **ip ospf** <process-ID> **area** <area-#> command to associate an interface to an OSPF process.

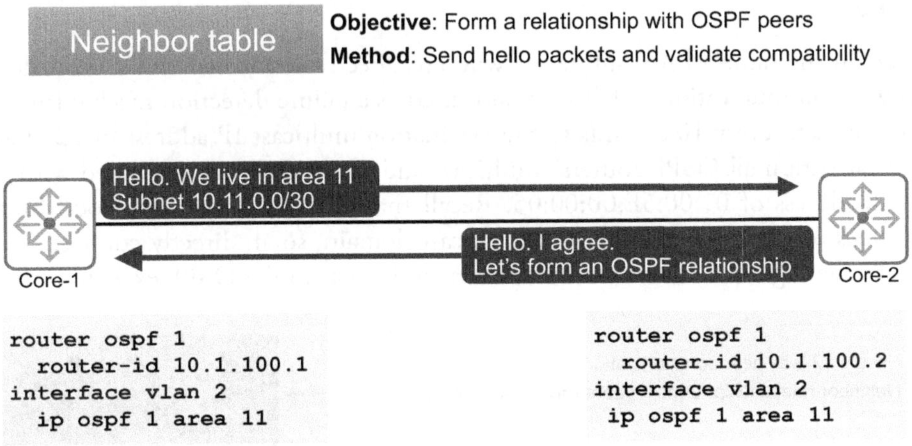

Figure 3-30: Building a neighbor table

Thus, Core-1 sends hellos out all OSPF-enabled interfaces, including the one shown here, in subnet 10.11.0.0/30 - "Hello. I am Core-1 (RID is 10.1.100.1) and I think that we live in Area 11, a normal area, and that we do not need to use secure authentication. We are connected on subnet 10.11.0.0/30".

Core-2 sends similar information in its hello packet. If the necessary OSPF criteria match, the OSPF routers agree to be *neighbors*; in OSPF terms, they form an OSPF neighbor relationship, which is sometimes referred to as an *adjacency*. If the required parameters do not match (typically due to a misconfiguration), the routers refuse to form a neighbor relationship, and will not share routes, resulting in your network being broken. Let us assume that all routers in Figure 3-30 successfully formed an adjacency with each directly-connected peer. This will be reflected in each router's neighbor table (we'll discuss the hello messages and adjacency states a bit later in the chapter).

Note

Only a small part of the hello packet contents is shown and discussed in this book: just enough information to convey the main idea of how the adjacency process works.

Note

The term *adjacency* and *neighbor* refer to the same relationship between two OSPF routers in a broadcast domain. As you will soon learn, just because two routers form an adjacency does not necessarily mean that the two routers will share link/routing information.

Hello messages

OSPF routers in the same VLAN/subnet or who are directly connected will send hello packets to ensure two-way communications. Hellos are also used as a failure detection mechanism. By default, these packets are sent every 10 seconds to the destination multicast IP address of 224.0.0.5. This is the reserved "attention all OSPF routers" multicast address, and this is associated with a multicast layer 2 MAC address of 01:00:5E:00:00:05. Recall that all layer 2 switches flood broadcast and multicast frames out all of its ports in the broadcast domain, so all directly connected routers will exchange hello messages (which are the first packets to be sent in the OSPF exchange process).

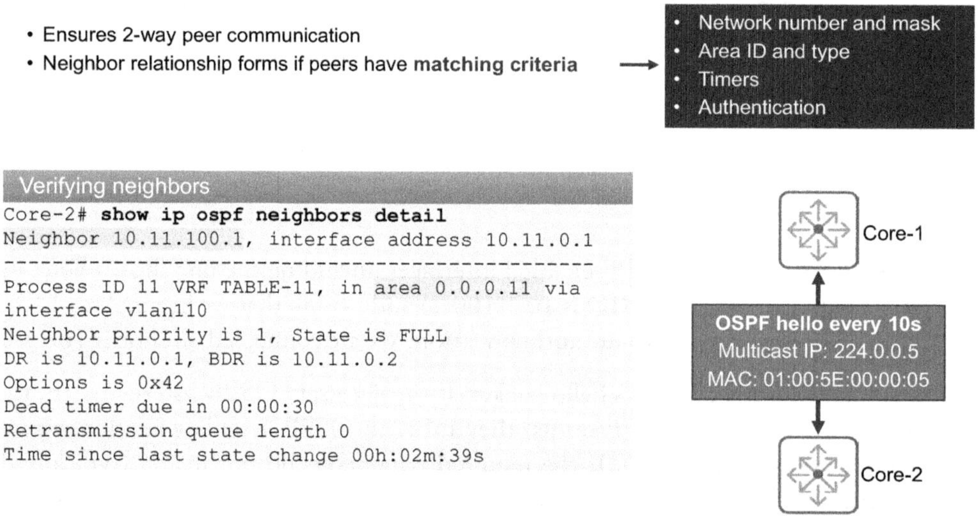

Figure 3-31: Becoming OSPF neighbors

Remember that the main purpose of hello packets is to build and maintain a neighbor table. Peers only form a neighbor relationship if they are compatibly configured (see Figure 3-31):

- They must agree that they are on the same subnet and have the same subnet mask.
- They must be in the same area and agree on the area type.
- Their timer values must match (a 10-second OSPF hello timer and a 40-second dead-interval timer, by default). You can verify the hello interval with the **show ip ospf neighbor detail** command.

- They must be configured to use the same authentication type.
- The MTU size should match (this is not a requirement, but if it doesn't match, you can experience issues when sharing a large amount of link/routing information).

The AOS-CX default hello interval is 10 seconds. You can customize the interval from 1 to 65553 seconds. A typical hello interval is 1 second.

```
Switch(config)# interface <interface-id>
Switch(config-if)# ip ospf hello-interval <seconds>
```

The dead interval is the time after which a neighbor is declared dead. Typically, this value is four times the hello interval, so 4 seconds for the dead interval if you use a 1-second hello timer. Note that the neighbor values must match for an adjacency to form. You can use the **show ip ospf neighbors** [**detail**] command to verify if the adjacencies (neighbor relationships) have successfully formed.

DRs and BDRs

An OSPF router will not form adjacencies to just any OSPF router. Instead, a client/server design is implemented in OSPF on each broadcast segment. For each multi-access broadcast segment, such as Ethernet, there is a designated router (DR) and a backup designated router (BDR) as well as other OSPF routers, called DROTHERs (designated router other). As an example, if you have 10 VLANs (subnets) in your switched area, you'll have 10 DRs and 10 BDRs. As you'll soon learn in a later section, if you have point-to-point links (only two OSPF routers), then there is no advantage for or against DR/BDRs. You'll learn how to configure network types shortly.

The OSPF router with the highest priority becomes the DR for the segment (see Figure 3-32). If there is a tie, the router with the highest router ID (not IP address on the segment) becomes the DR. By default, all routers have a priority of 1 (priorities can range from 0 to 255—it's an 8-bit value). If the DR fails, the BDR is promoted to DR and another router is elected as the BDR. If a router does not become a DR or BDR, it is referred to as a designated router other, or commonly DROTHER, for short. There is no preemption of the DR/BDR role. If a DR/BDR exists and a router with a higher router ID joins the segment, it will not preempt the existing DR/BDR. Election only occurs if there is no DR or BDR on a segment.

When an OSPF router comes up, it forms adjacencies with the DR and the BDR on each multi-access segment to which it is connected; if it is connected to three segments, it will form three sets of adjacencies. Any exchange of routing information is between these DR/BDR routers and the other OSPF neighbors on a segment (and vice versa). An OSPF router sends messages to a DR/BDR using the IP multicast address of 224.0.0.6. The DR and the BDR send messages to all OSPF routers using the 224.0.0.5 multicast IP address.

OSPF routers use link state advertisements (LSAs) to communicate with each other. One type of LSA is a hello, which is used to form neighbor relationships and as a keep-alive function. Hellos are generated every ten seconds. When sharing link information (directly connected routes), links are sent to the DR (224.0.0.6) and the DR disseminates this to everyone else (224.0.0.5) on the segment. In other words, two routers in a DROTHER relationship won't share routing information with each other; instead, routing information is shared with the DR/BDR and the DR/BDR disseminates the routing information to everyone else on the segment. On point-to-point links, since no DR/BDR is used, all OSPF packets are addressed to 224.0.0.5. You'll learn more about hello and LSA messages in the upcoming sections of this chapter.

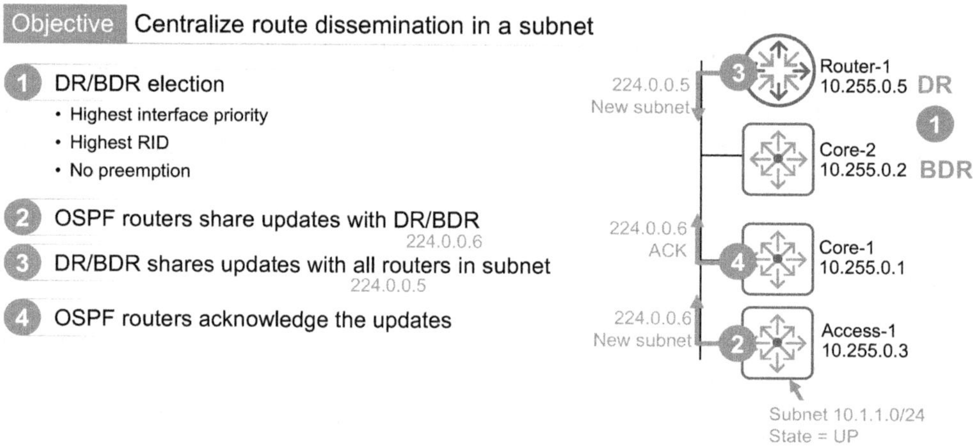

Figure 3-32: DRs and BDRs

Note

The DR/BDR feature was specifically designed based on the hardware (or lack thereof) capabilities of routers in the 1980s. Today, this is a moot point and you should not be concerned which of your routers play which role: DR, BDR, or DROTHER.

OSPF neighbor states

OSPF uses a finite state machine (FSM) to process the neighbor state transitions between routers, when certain conditions are satisfied. This process can be divided into two main phases:

- Establish neighbor adjacencies
- Synchronize OSPF databases

An OSPF router can be one of three types in a broadcast domain:

- Designated router (DR)
- Backup designated router (BDR)
- Not either of these two (DROTHER)

Is important to clarify that in a broadcast network, database synchronization only happens between the DR/BDR and DROTHER routers. In other words, DROTHER routers only establish neighbor adjacencies and stay in a 2-WAY state (which is considered a stable state), but never synchronize databases with other routers in a DROTHER state; however, they will move into a full adjacency with the DR and BDR, which allows them share information about links and networks.

As an example, consider that two core switches attempt to become OSPF neighbors over a broadcast network (see Figure 3-33):

1. Core-1 and Core-2 start in a DOWN neighbor state.
2. Core-1 transits to the INIT state when it sends its first hello message. This message includes its RID (10.1.100.1). This hello message also includes a seen field, which includes the routers that Core-1 has received hello messages previously. Since this is the first hello message, the field has a null value.
3. Core-2 receives Core-1's hello message and responds, indicating that the message has been seen and that the values are compatible. Core-2 then transits to an INIT state.
4. Core-1 receives the hello message and transits to a 2-WAY state. Core-1 again sends a hello message, but this time it includes any adjacently-compatible RIDs in the *seen* field.
5. Core-2 receives this message and then moves into a 2-WAY state.

Remember that all OSPF routers that meeting the matching criteria in a broadcast domain will move to a 2-way state with each other: the DR, the BDR, and all the DROTHERs.

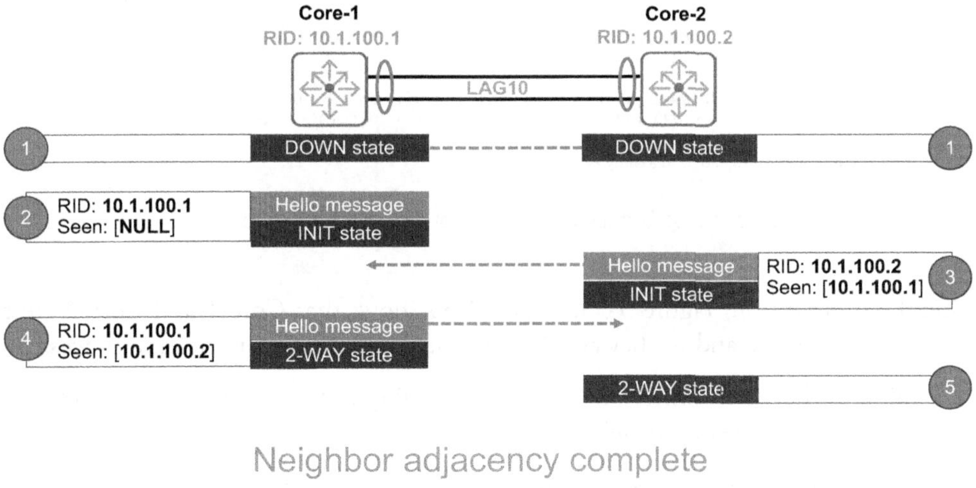

Figure 3-33: OSFP neighbor states (part 1)

Figure 3-34 shows an example of the neighbor process happening in a real-world environment, time stamped from the bottom, upwards (the newest message is first and the oldest is listed last based on using the **-r** option with the **show events** command).

- First, sw-agg1's neighbor, 10.1.0.3, moves from a *down* to an *init* state: a hello was sent to that neighbor.
- Next, sw-agg1 receives a hello from that neighbor, and so the state moves from *init* to *2-way*. Note that if the two neighbors are DROTHERs, they stay in a 2-way state and don't proceed any further.
- If this router or the neighboring router is either a DR or BDR, the two neighbors move to an *exstart* state: they begin to exchange LSDB information. This is covered shortly.
- These two neighbors will then move to the *exchange* state to share missing information with each other until each has a matching LSDB. This is covered shortly.
- Once this is complete, they are in a *full* state. You might say that these OSPF peers have formed a full adjacency. This is covered shortly.

```
sw-agg1# show events -r -d hpe-routing

T13:11:39.000761+00:00 sw-agg1 hpe-routing[7720]: Event2401LOG_INFOAMM1/1AdjChg:
Nbr  rtr ID 10.1.0.3 on IP addr 10.1.2.2( area ID 0.0.0.0): Exchange -> Full

T13:11:39.000105+00:00 sw-agg1 hpe-routing[7720]: Event2401LOG_INFOAMM1/1AdjChg:
Nbr  rtr ID 10.1.0.3 on IP addr 10.1.2.2( area ID 0.0.0.0): Exstart -> Exchange

T13:11:38.999138+00:00 sw-agg1 hpe-routing[7720]: Event2401LOG_INFOAMM1/1AdjChg:
Nbr  rtr ID 10.1.0.3 on IP addr 10.1.2.2( area ID 0.0.0.0): Two-way -> Exstart

T13:11:38.999093+00:00 sw-agg1 hpe-routing[7720]: Event2401LOG_INFOAMM1/1AdjChg:
Nbr  rtr ID 10.1.0.3 on IP addr 10.1.2.2( area ID 0.0.0.0): Init -> Two-way

T13:11:38.998994+00:00 sw-agg1 hpe-routing[7720]: Event2401LOG_INFOAMM1/1AdjChg:
Nbr  rtr ID 10.1.0.3 on IP addr 10.1.2.2( area ID 0.0.0.0): Down -> Init

<date stamp removed for clarity>
```

Figure 3-34: Sample event log of the adjacency process

Continuing the process from Figure 3-33, Figure 3-35 shows that Core-1 and Core-2 are the only two devices in the network, and so they continue to the database synchronization process (*exstart*).

6. Core-1 initiates the synchronization process by sending a database description (DBD) packet. The switch transits to the *exstart* neighbor state.
7. Core-2 does a similar process: it moves to the *exstart* state by sending a DBD packet.

The goal of the *exstart* state is to determine which router will become the *master* for the exchange process, based on the highest RID. Understand that the master role defines only which router will initiate the database exchange process. This role has nothing to do with DR/BDR: it is only for the

Link State Database (LSDB) exchange. In this example, Core-2 has a higher router-ID and becomes the MASTER.

8. Core-2 sends another Database Description packet as it transitions to the EXCHANGE state. Core-2 is now sharing a summary of the contents of its LSDB with Core-1.

9. Core-1 also sends a Database Description packet and moves to the EXCHANGE state, sharing its LSDB with Core-2.

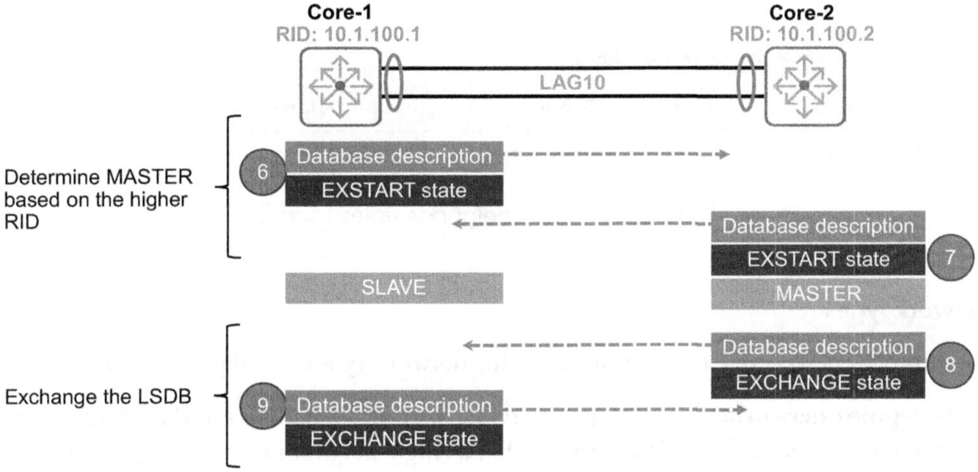

Figure 3-35: OSFP neighbor states (part 2)

After several exchanges are performed during the *exchange* state, each router will have a summary of the others' entire LSDB. They compare the information received and, in the next step, request the missing information (see Figure 3-36).

10. Core-1 and Core-2 request the missing information using link state update (LSU) packets. The other peer answers by sending the requested database information. Both switches transition to the *loading* neighbor state: this state is the actual routing information exchange.

11. Core-1 and Core-2 move to the *full* state when there is no more information to be exchanged. Now, both devices have the same LSDB information.

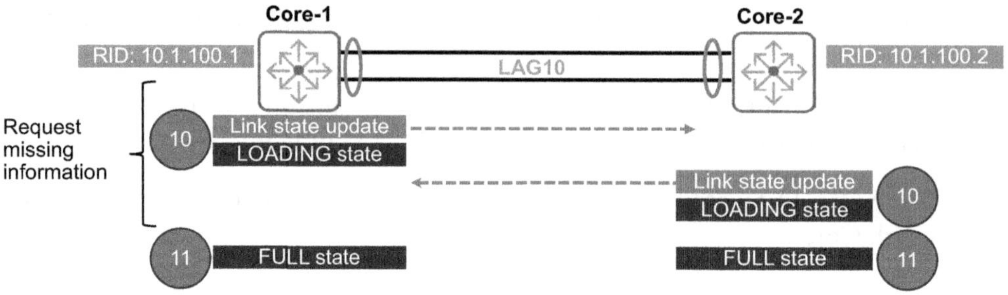


```
Core-1# show ip ospf neighbors

Neighbor ID      Priority  State              Nbr Address        Interface
--------------------------------------------------------------------------
10.1.100.2       1         FULL/DR            10.11.0.2          LAG10
```

Figure 3-36: OSFP neighbor states (part 3)

OSPF network types

From the protocol perspective two options exist for network types (see Figure 3-37):

- **Point-to-point network:** Only two peers are on the link. When an interface is configured to be part of a point-to-point link, OSPF knows that a single neighbor device is expected on the interface. PPP serial links (not used too much in networks now) or directly-connected Ethernet ports are examples of this network type.
- **Broadcast networks:** Two or more peers might be on the link. When an interface is configured to be part of a broadcast network, OSPF knows that more than one neighbor might be discovered on the interface.

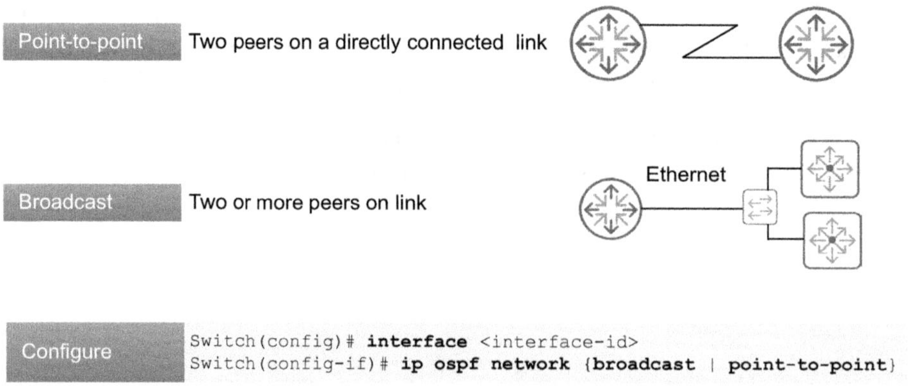

Figure 3-37: OSPF network types

AOS-CX interfaces default to the broadcast network type, but you can modify the network type using the **ip ospf network** command on the respective interface.

```
Switch(config)# interface <interface-id>
Switch(config-if)# ip ospf network {broadcast | point-to-point}
```

Unless you truly have multiple routers in the same broadcast domain, it is recommended to configure switch-to-switch links with the **point-to-point** network type. To verify the type of network that is in use you can use the **show ip ospf interface** command.

> **Note**
> Whenever possible, it is recommended to use point-to-point links to speed up the convergence of OSPF when it initially starts up.

> **Note**
> OSPF can also use the non-broadcast multiaccess (NBMA) network type. This network can support multiple devices (multi-access) but does not support the broadcast capability. Frame relay is an example of this network type. The use of this type of network is deprecated for modern networks, and so will not be covered in this book.

OSPF neighbor verification

The **show ip ospf neighbor** command allows you to verify the status of the neighbors (see Figure 3-38).

```
Core-1# show ip ospf neighbors

VRF : default                              Process : 1
===================================================

Total Number of Neighbors : 1

Neighbor ID      Priority  State         Nbr Address      Interface
-----------------------------------------------------------------------
10.1.100.2       255       FULL/DR       10.11.0.2        vlan2
```

Figure 3-38: Viewing the neighbor table

A state of *full* indicates that the router can share link information with the peer. In the example in Figure 3-38, the two routing switches are in a broadcast medium and the neighbor is the DR. If this was a point-to-point link, you would want to define the point-to-point link type for the network type to speed up the initial startup of OSPF (convergence), since no DR/BDR election process is necessary on a point-to-point link. If you see *2-way* as the state, this indicates that the neighbor is a DROTHER: all the required items match between the two peers, but no routing information is shared. In this instance, routing information is only shared with the DR/BDR peers.

Phase 2: Build a list of every link (LSDB)

A router shares known subnets with other routers if and only if a neighbor adjacency is fully formed. The objective is to build a topology database: a list of every single link (subnet), every single router, and exactly how those routers and subnets are interconnected. We humans look at a network diagram, in a pictorial format, to understand how router and links are interconnected. Routers, however, look at the topology database (in a numerical format) to understand this. The topology database is referred to as a link state database (LSDB) in OSPF.

The method involves sending special OSPF multicast messages called link state advertisements (LSAs). In Figure 3-39, the Server Switch sends LSAs out all interfaces participating in OSPF, by default: "Attention all OSPF routers: I am the Server Switch, and I am directly connected to 10.1.1.0/24, 10.1.2.0/24, 10.20.0.0/22". Every other OSPF router receives these LSAs and adds the information to its topological database. Soon, every router has received LSAs from every other router. Thus, each router has a full topology database. Again, this database is essentially the diagram you see in Figure 3-39, in a numerical format. There is only one reality from the routing switches' perspective, which is shown in Figure 3-39. Therefore, by default, every OSPF router has an identical topological database; and basically, the LSDB tells you *who* is connected to *what*.

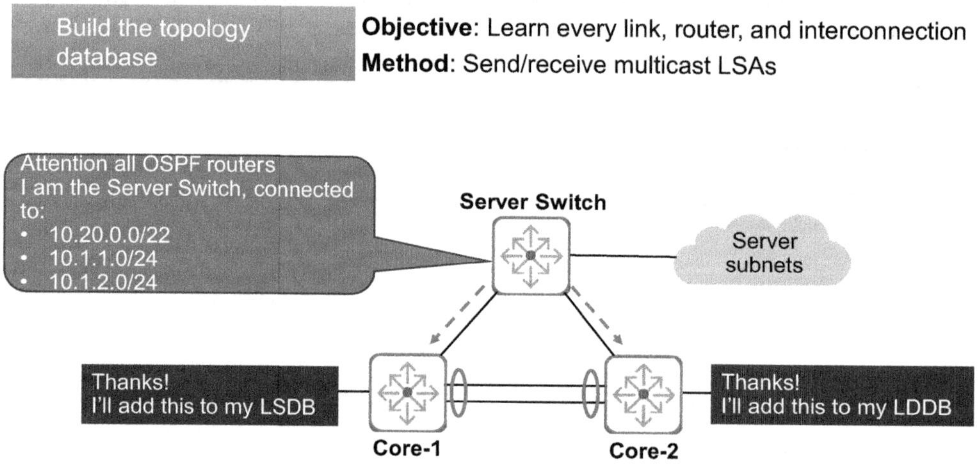

Figure 3-39: OSFP LSDB

Note

Figure 3-39 only shows the Server Switch advertising its directly-connected subnets, because that is all that it currently knows. However, once the Server Switch receives LSAs from the other routers (as they have just received from the Server Switch), it will advertise those routes as well. Routers advertise the entire contents of their topology database to all other routers (DRs and BDRs in a broadcast medium) in the same area. The topology database is also known as the link-state database (LSDB).

OSPF LSAs

OSPF routers generate different types of link state advertisements (LSAs), each with a different purpose and scope in the OSPF routing domain (see Figure 3-40). Each LSA includes information that describes the portion of the network that a router announces. LSAs are collected to make up the LSDB, which represents the topology of the network. This will be used by OSPF to calculate best OSPR routing path to every destination in the network. When OSPF routers establish adjacencies, they exchange their LSDB database entries so that every router in the network is aware of the entire topology.

In AOS-CX you can verify this information by using the **show ip ospf lsdb** command. This is a powerful troubleshooting command. If your network diagram shows that there are five routers, but you only see three listed, then you know that there is a problem.

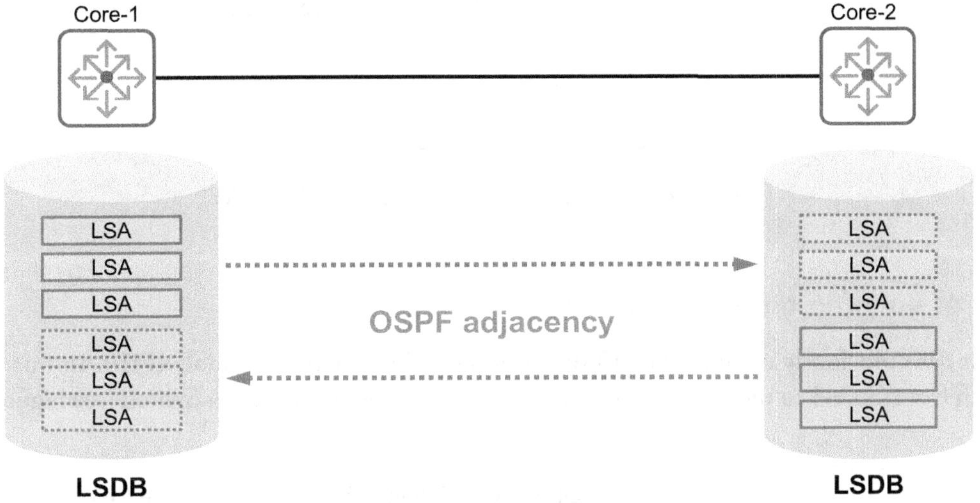

Figure 3-40: OSPF LSAs

Phase 3: Building a list of the best OSPF routes

Although there is only one topology, each router's position and connectivity within that topology is unique. Thus, each router's objective is to determine the best path to each link from where they sit within the topology. The method is to run the SPF (Dijkstra) algorithm on the topology database or LSDB.

For example, examine the network shown in Figure 3-41. Core-1's SPF algorithm analyzes the topology database and sees that Core-2 can reach destination 10.20.0.0/22 by routing packets out port LAG10. And Server Switch can reach the same destination via port 47. The algorithm considers the cost associated with each path (based on bandwidth) and chooses the lowest cost (fastest) path. OSPF's metric is *cost*. Cost is the inverse of the bandwidth of the interface. Therefore, faster links will have lower costs and the path with the *lowest* cost is preferred. This best path for a destination is added to the list of best paths, which is referred to as the OSPF routing table. Note that this is *not* the IP routing table, since the IP routing table, discussed in the next section, has the best possible paths from all route sources, not just OSPF. In the example shown in Figure 3-41, the best path for Core-1 to reach 10.20.0.0/22 is to forward traffic directly to the Server Switch.

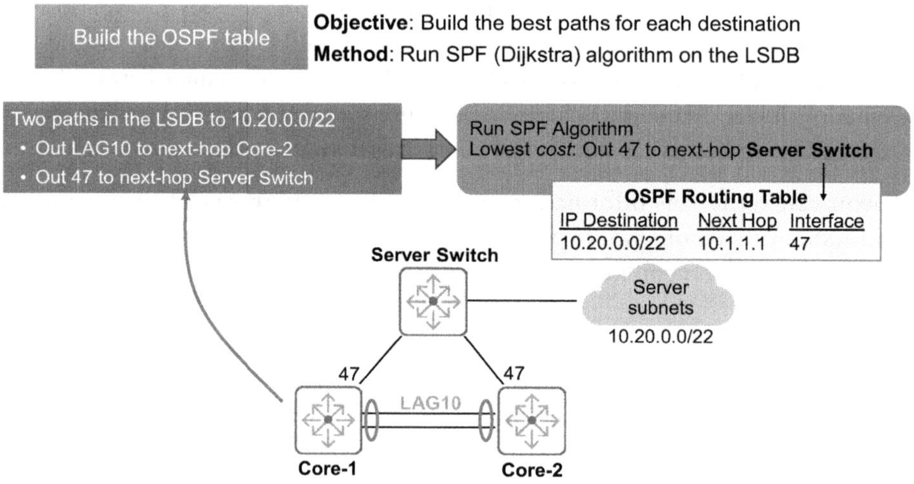

Figure 3-41: Building a list of the best OSPF routes

Phase 4: Building an IP routing table

The best path previously calculated by OSPF is published to the routing table (FIB). In this case the best path from Core-1 to the server subnets uses the path through the Server Switch (see Figure 3-42).

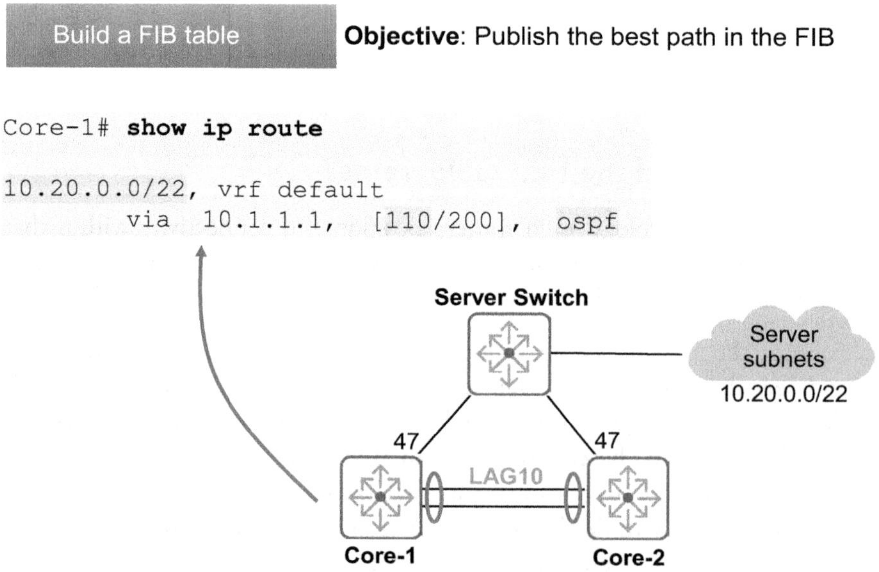

Figure 3-42: Building an IP routing table (FIB)

OSPF routing process

This section rounds out the rest of the OSPF configuration. Previously in the "OSPF introduction" section, you learned how routers obtain their RIDs and how to manipulate this. Without an RID, the OSPF process on an AOS-CX switch won't even start. Therefore, using the IP address on a switch's loopback is the preferred method of assigning an RID. The rest of the section will discuss an introduction to the rest of the OSPF configuration.

OSPF areas

An *area* is a group of OSPF routers that share the same LSDB. All routers must be part of an area. When you split a large network into separate areas, you reduce the size of the LSDB in each router, lower CPU utilization, and increase overall network stability and scalability. This is because each router in an area must only maintain the topology for *that* area.

For example, SW1 and SW2 in Figure 3-43 need not learn about the entire topology: they only need to know about area 10 topology. Any other destination is reached via the area border router(s), called ABRs. The ABRs in Figure 3-43 are connected to two areas—10 and 0—and thus have two LSDBs, one for each area.

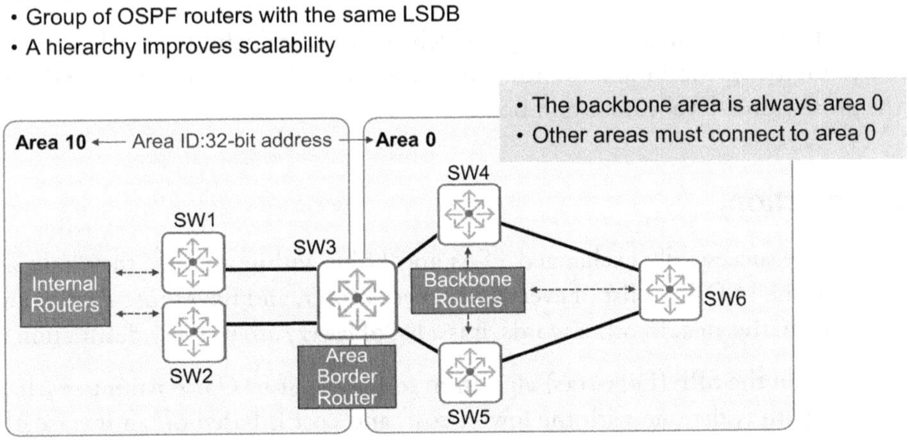

Figure 3-43: OSPF areas

Routers SW1 and SW2 are called *internal* routers: all their interfaces are in a single area. If an internal router must route outside of its area, it simply forwards packets to its ABR. An ABR is a router connected to two or more areas. SW3 is the ABR for area 10, with two interfaces in Area 10 and two interfaces in Area 0.

By definition, all areas must connect (preferably directly) to the special backbone area, called area 0. This is the most important area in the hierarchy, and so should have a redundant design. Communication between two non-backbone areas MUST communicate via the backbone area 0.

Routers with interfaces in area 0 are called *backbone* routers. Routers SW4, SW5, and SW6 have all interfaces in a single area, and are therefore called internal routers. Routers SW3, SW4, SW5, and SW6 are also backbone routers; however, SW3 is not an internal router since it has interfaces in two *different* areas.

The area ID is a 32-bit address that can be written in dot-decimal notation (area 0.0.0.0) or in decimal notation (area 0), as shown in Figure 3-43. AOS-CX supports both notations. The only area number reserved is 0 and it's used to define the backbone of the OSPF network.

First you need to create the area in the OSPF routing process:

```
sw-agg1(config)# router ospf 1
sw-agg1(config-ospf-1)# area <area-id>
```

Because this book focuses on a small campus design, only one OSPF routing process is used and it is identified by a process ID of 1 in the **router ospf** command.

Next, you assign interfaces (physical and/or logical) to an area by associating them to a specific area ID:

```
sw-agg1(config)# interface <interface-ID>
sw-agg1(config-loopback-if)# ip ospf 1 area <area-ID>
```

This guide is only focused on single-area designs which you'll typically find in small campus networks, where all router interfaces are in area 0. You can learn more about hierarchical and multiple-area designs in the professional-level course and book.

OSPF path selection

After all routers have successfully exchanged LSAs and LSUs within an area, they have an identical LSDB. Remember, the LSDB is a list of every link, every router, and how those routers and links are interconnected within the area. In other words, it is a list of *every path* to every destination in the area.

Next routers must run the SPF (Dijkstra's) algorithm to find the *best OSPF paths* to each destination network. The best path is the one with the lowest cost, and cost is based on an inverse of the bandwidth of the interfaces. Each interface has a cost associated with it, and as LSA routing information is received, the router increments the costs in the LSA messages by the interface cost on which the LSA arrived. The lower the cost, the more preferred the path to the destination. Therefore, if there are multiple paths to a single subnet, OSPF chooses the path with the lowest cumulative cost, which is the fastest path, by default.

Consider the topology shown in Figure 3-44. Core-1 has two paths to reach destination subnet 10.20.0.0/22. To determine each path's cost, simply add the cost values indicated for each link. Here's what you get for Core-1 to reach network 10.20.0.0/22:

- Using the Server Switch as the next hop, the cost is 100 = 100.
- Using Core-2 as the next hop, the cost is 100+50 = 150.

Obviously, using the Server-Switch results in the lowest *accumulated* path cost, and so that path is added to the *OSPF* routing table, as described earlier in the "Phase 3: Build a list of the best OSPF routes".

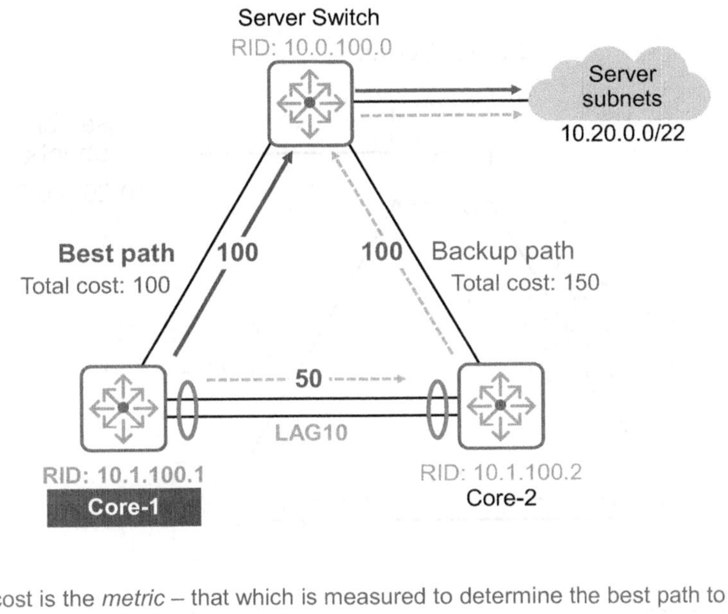

Figure 3-44: OSPF path selection

OSPF convergence

There are two components to OSPF routing convergence:

- Detecting topology changes
- Recalculating and selecting the next best OSPF route(s)

Topology change detection is supported in two ways by OSPF. The first, and the quickest, is a failure or change of status of the physical interface on a directly-connected link (point-to-point). The second is a timeout of the OSPF hello timer. An OSPF neighbor is deemed to have failed if the time to wait for a hello packet exceeds the dead timer, which defaults to four times the value of the hello timer. The default hello timer is 10 seconds, so the default dead timer is 40 seconds. The dead interval timer is necessary if the two OSPF routers are *not* directly connected together.

When a change is detected, an LSA is sent to all routers in the OSPF area to signal the topology change. In the example shown in Figure 3-45, the link between Core-1 and Sever Switch has failed. Server Switch and Core-1 detect this outage: assuming you've configured this as a point-to-point link, their links go from an up state to a down state. Thus, Server Switch and Core-1 originate the topology change LSA. Upon receiving the update, all the OSPF routers update their LSDB and rerun the SPF algorithm to calculate their new best paths for any affected network, such as 10.20.0.0/22. The Server Switch will not change anything since it is still directly connected to the best path. Core-1, on the other hand, will have to recalculate its best path for 10.20.0.0/22. In this case, it will use the path via Core-2.

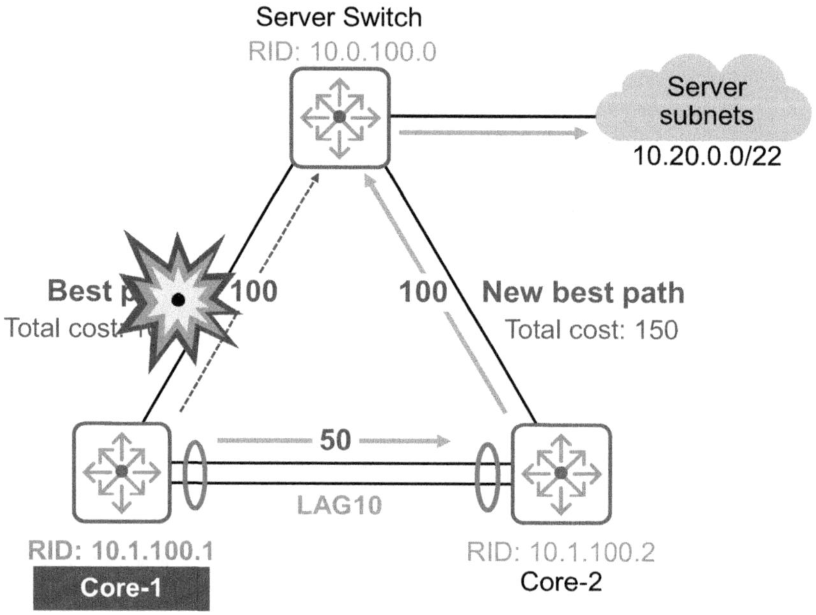

Figure 3-45: OSPF convergence

Ultimately, Core-2 receives LSAs about the change, but no recalculation is needed from its point of view since Core-2's best path is still it's connection to the Server Switch.

Each router performs route recalculation after a failure has been detected. This causes all routers to recalculate all their routes using the Dijkstra (SPF) algorithm when it is run against the routers' LSDBs. The result is a new OSPF routing table. The time it takes to do this is referred to as *convergence*.

Important

Because it is much quicker to detect a downed link than using a dead interval timer, it is highly recommended to use and configure OSPF point-to-point links whenever possible.

Equal Cost Multi-Path (ECMP)

Equal Cost Multi-Path (ECMP) is a mechanism that enables load sharing of two or more equal-cost paths to the same destination network. In Figure 3-46 there are two equal-cost paths Core-1 can use to reach network 10.1.0.0/24. Path 1 is via router with RID 10.0.100.2 and path 2 is via router with RID 10.0.100.3. When multiple paths exist to the same subnet that has the same cost AOS-CX router will perform load sharing using ECMP. ECMP is similar to link aggregation layer 3 based hashing: packets with different source/destination IP address are assigned different paths.

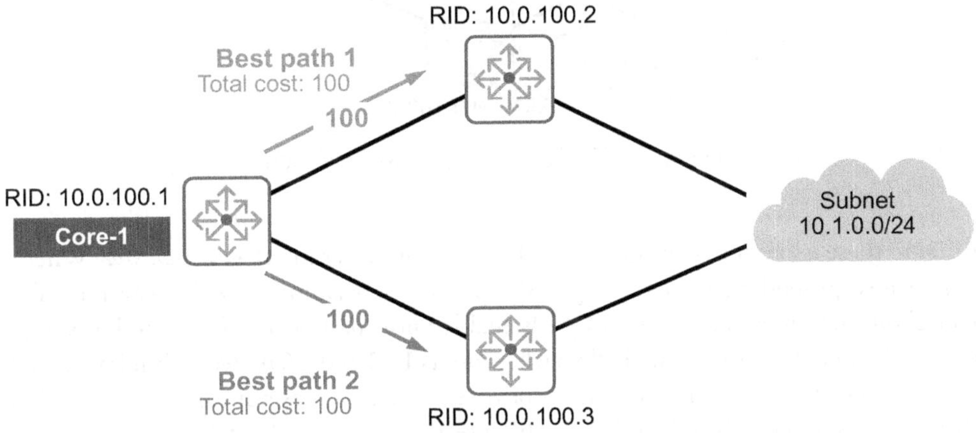

Figure 3-46: OSPF and ECMP

By default, AOS-CX switches load share across four equal-cost paths, but this can be changed; however, since most redundant designs typically only have two equal-cost paths, changing the ECMP maximum is not commonly done.

Passive interfaces

The OSPF configuration involves enabling the protocol on logical and physical interfaces. Thus, the router can generate LSAs to advertise subnets to other routers. This implies that the router sends periodic hello messages on all OSPF-enabled interfaces.

In some cases, this is not desired. The most common example is when only hosts exist on that subnet, commonly referred to as a stub link in OSPF. However, the OSPF router would still constantly send OSPF packets to the subnet when nobody (the user hosts) will process them. Hosts do not respond to the OSPF multicast 224.0.0.5 and 224.0.0.6 addresses. It is simply wasted processing cycles and bandwidth on the link. Also, if bad actors are on the link, they could learn information about your network, and potentially mount an attack.

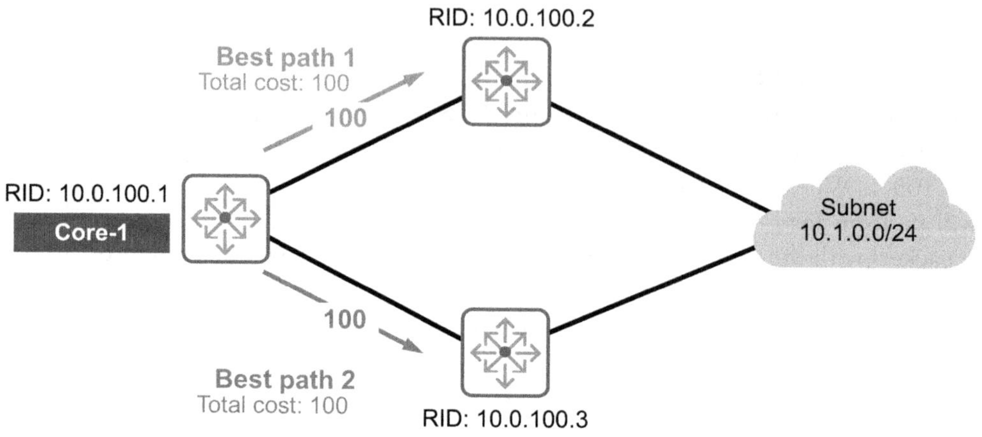

Figure 3-47: OSPF and passive interfaces

The solution to these issues is to configure these host-facing router interfaces as *passive*. When an interface is passive, it stops sending and receiving OSPF packets on that interface. However, it will continue to advertise about that interface out other OSPF-enabled non-passive interfaces. In the example shown in Figure 3-47, Switch-1 does not send hello packets on its 10.1.0.0/16 network, but it will tell Switch-2 about that network. Also, since the interface is passive, Switch-1 does not expect to see hello messages in the 10.1.0.0/16 subnet. If it did, and since the interface is passive, Switch-1 would basically ignore these hello (and LSA) messages. Switch-2 has a similar arrangement, as relates to the 10.2.0.0/16 link.

By default, OSPF-enabled interfaces on AOS-CX switches are non-passive (they fully participate in OSPF). In AOS-CX, you use the **ip ospf passive** command in the interface context to define an interface as a passive interface:

```
Switch(config)# interface <interface-ID>
Switch(config-if)# ip ospf passive
```

Lab 3.01: Basic IP configuration

In this lab, the senior network administrator has requested that you add a new VLAN to the network and prepare the IP routing configuration for the new VLAN on the aggregation switch.

Objectives

After completing this lab, you will be able to:

- Add a new VLAN IP interface, also known as an SVI, to the aggregation switch.
- Configure DHCP relay, also known as *IP helper*, on the aggregation switch.
- Verify routing between the new VLAN and an existing VLAN on the network.

Task 1: Prepare the Edge Switch Uplink to the Aggregation Switch

To ensure that each edge switch has a direct, single uplink to the aggregation switch, you will disable the LAG between the sw-edge1 and sw-edge2. Sw-edge1 has port 1/1/27 already connected to sw-agg1. This was done in the previous lab. Sw-edge2 needs to be configured.

Objectives

- Disable the LAG between the two edge switches.

Steps

1. On sw-edge1, disable the LAG 50 (that connects to sw-edge2).

```
sw-edge1(config)# interface lag 50
sw-edge1(config-lag-if)# shutdown
sw-edge1(config-lag-if)# exit
```

2. Verify that the interface LAG 50 is disabled.

```
sw-edge1(config)# show interface lag brief
----------------------------------------------------------------------------------
--------------------------------------
Port       Native  Mode    Type           Enabled Status  Reason
Speed    Description
           VLAN
(Mb/s)
----------------------------------------------------------------------------------
--------------------------------------
lag50      1       trunk   --             no      down    --
auto     --
```

3. Examine the LLDP neighbor list to verify that port 1/1/27 is connected to sw-agg1.

```
sw-edge1(config)# show lldp neighbor-info

LLDP Neighbor Information
=========================

Total Neighbor Entries          : 8
Total Neighbor Entries Deleted  : 0
Total Neighbor Entries Dropped  : 0
Total Neighbor Entries Aged-Out : 0
```

```
LOCAL-PORT  CHASSIS-ID          PORT-ID                     PORT-DESC
TTL       SYS-NAME
----------------------------------------------------------------------
------------------
...
1/1/25      8c:85:c1:49:20:c0  1/1/26                      sw-edge1
120       sw-edge2
1/1/26      8c:85:c1:49:20:c0  1/1/25                      sw-edge1
120       sw-edge2
1/1/27      b8:d4:e7:dAnswer:80:00  1/1/1                      1/1/1
120       sw-agg1
...
```

- **Question**: Do you still observe sw-edge2 in the LLDP neighbor list?
- **Answer**: Yes, even though the connection to sw-edge2 is down, the LLDP neighbor entries have their own aging timers (time-to-live); therefore, you may still notice the neighbor for some time. This is typically useful in troubleshooting, since you can still detect the connected device for a short period of time, even though the link is down.
- **Question**: What is the TTL shown for the sw-edge2?
- **Answer**: The TTL is 120 (seconds); this means that the neighbor will age out 120 seconds after the last received LLDP PDU.

4. On sw-edge2, enable port 1/1/27. This port connects to sw-agg1.

```
sw-edge2(config)# interface 1/1/27
sw-edge2(config-if)# no shutdown
sw-edge2(config-if)# exit
```

5. Verify that sw-agg1 is now listed as an LLDP neighbor.

```
sw-edge2(config)# show lldp neighbor-info

LLDP Neighbor Information
=========================

Total Neighbor Entries          : 5
Total Neighbor Entries Deleted  : 2
Total Neighbor Entries Dropped  : 0
Total Neighbor Entries Aged-Out : 2

LOCAL-PORT  CHASSIS-ID          PORT-ID                 PORT-DESC
TTL       SYS-NAME
---------------------------------------------------------------------
-----------------...
1/1/27      b8:d4:e7:dAnswer:80:00  1/1/2                   1/1/2
120       sw-agg1
...
```

Task 2: Configure a new Layer2 VLAN in the Edge Infrastructure

In this task you will prepare a new VLAN (VLAN 12) in the environment. Since the lab topology expects a loop-free setup for now, you will disable the LAG between sw-edge1 and sw-edge2. On sw-edge2, you will then enable the uplink to sw-agg1.

Objectives

- Practice the VLAN configuration.

Practice

- Define VLAN 12 on sw-edge1 and sw-edge2.
- Allow all VLANs on the uplink port 1/1/27 on both sw-edge1 and sw-edge2.
- On sw-edge2, configure port 1/1/4 as access port in VLAN 12.

Solution

1. Sw-edge1's configuration:

```
vlan 12
interface 1/1/27
 vlan trunk allow all
 no shutdown
 exit
```

2. Sw-edge2's configuration:

```
vlan 12

interface 1/1/27
 vlan trunk allow all
 no shutdown
 exit

interface 1/1/4
 vlan access 12
 exit
```

Task 3: Configure the Aggregation Switch with a new VLAN and SVI

In this task you will create the new VLAN on the aggregation switch. In a real deployment, the VLAN would be created on both aggregation switches. However, to keep the lab setup simple at this stage, you will only configure this on the first aggregation switch: sw-agg1.

In this task, you will configure a new IP interface for VLAN 12. A layer 3 interface configured on a VLAN is known as a switch virtual interface (SVI). This SVI will be the default gateway for clients in VLAN 12. You will also configure DHCP relay on the SVI, to ensure that clients in VLAN 12 can obtain a DHCP-based IP addressing information from the centralized DHCP server in the lab topology.

Objectives

- Configure an SVI on the aggregation switch.

Steps

Layer2 VLAN 12

1. On *sw-agg1*, create VLAN 12.

```
sw-agg1(config)# vlan 12
sw-agg1(config-vlan-12)# exit
```

2. Review the VLAN 12 port membership.

```
sw-agg1(config)# show vlan 12
-------------------------------------------------------------------
---------------
VLAN  Name                Status  Reason                Type
Interfaces
-------------------------------------------------------------------
---------------
12    VLAN12              up      ok                    static
1/1/1-1/1/2,lag256
```

- **Question**: Is port 1/1/1 member of VLAN12? Which ports are members of VLAN12?
- **Answer**: These ports are members of VLAN12: 1/1/1,1/1/2, LAG256.

3. Review the port configuration of port 1/1/1 and 1/1/2.

```
sw-agg1(config)# show running-config interface 1/1/1
interface 1/1/1
    no shutdown
    mtu 9198
    no routing
    vlan trunk native 1
    vlan trunk allowed all
    exit

sw-agg1(config)# show running-config interface 1/1/2
interface 1/1/2
```

```
    no shutdown
    mtu 9198
    no routing
    vlan trunk native 1
    vlan trunk allowed all
    exit
```

- **Question**: What do you observe?
- **Answer**: Ports 1/1/1 and 1/1/2 have been configured as VLAN trunk and allow all VLANs. This automatically includes the newly created VLAN12.

4. Now you will investigate what system connects to LAG256. Review the LACP interface status to determine the member ports.

```
sw-agg1(config)# show lacp interfaces

State abbreviations :
A - Active          P - Passive       F - Aggregable I - Individual
S - Short-timeout L - Long-timeout N - InSync      O - OutofSync
C - Collecting      D - Distributing
X - State m/c expired                 E - Default neighbor state

Actor details of all interfaces:
------------------------------------------------------------------------------
---------
Intf     Aggr     Port   Port   State    System-ID          System Aggr
Forwarding
         Name     Id     Pri                                Pri    Key
State
------------------------------------------------------------------------------
---------
1/1/46   lag256   47     1      ALFNCD   b8:d4:e7:dAnswer: 80:00 65534
256   up
1/1/47   lag256   48     1      ALFNCD   b8:d4:e7:dAnswer: 80:00 65534
256   up

Partner details of all interfaces:
------------------------------------------------------------------------------
---------
Intf     Aggr     Port   Port   State    System-ID          System Aggr
         Name     Id     Pri                                Pri    Key
------------------------------------------------------------------------------
---------
1/1/46   lag256   47     1      ALFNCD   b8:d4:e7:d9:e5:00 65534   256
1/1/47   lag256   48     1      ALFNCD   b8:d4:e7:d9:e5:00 65534   256
```

- **Question**: What are the member ports of the LAG 256?
- **Answer**: Ports 1/1/46 and 1/1/47 are member ports of the LAG 256.

3. Review the LLDP neighbor output to see what device connects to these ports.

```
sw-agg1(config)# show lldp neighbor-info

LLDP Neighbor Information
=========================

Total Neighbor Entries          : 6
Total Neighbor Entries Deleted  : 2
Total Neighbor Entries Dropped  : 0
Total Neighbor Entries Aged-Out : 2

LOCAL-PORT  CHASSIS-ID         PORT-ID                    PORT-DESC
TTL       SYS-NAME
--------------------------------------------------------------------
-----------------...
1/1/46      b8:d4:e7:d9:e5:00  1/1/46                      1/1/46
120       sw-agg2
1/1/47      b8:d4:e7:d9:e5:00  1/1/47                      1/1/47
120       sw-agg2
...
```

- **Question**: To what device is LAG 256 connected?
- **Answer**: LAG 256 consists of ports 1/1/46 and 1/1/47 and connects to the sw-agg2.

Layer3 SVI 12

You will now create an SVI for VLAN 12. An SVI is a layer 3 (IP) interface for a VLAN. When a switch doesn't have an SVI for a VLAN, it doesn't have an IP interface in that VLAN, so it cannot directly interact with devices, using IP, in that VLAN.

An access switch typically doesn't need an SVI in the endpoint VLANs; however, aggregation switches typically do need SVIs, since they will provide IP default gateway services for the VLANs.

1. On sw-agg1, review the current list of L3 interfaces. Make sure to check the list to the end.

```
sw-agg1# show ip interface brief
Interface          IP Address            Interface Status
                                           link/admin
...
vlan1             10.1.1.1/24               up/up
vlan2             10.1.2.2/24               up/up
vlan3             10.1.3.1/24               up/up
vlan11            10.1.11.1/24              up/up
```

- **Question**: Is there an SVI for VLAN 11 and for VLAN 12?
- **Answer**: In the existing network, only an SVI for VLAN 11 exists. Even though the layer 2 VLAN 12 was created on the switch, it doesn't have an IP interface in this VLAN yet.

2. Create a new SVI for VLAN 12.

```
sw-agg1(config)# interface vlan 12
```

3. Review the list of interfaces. This will now include SVI12, shown as *vlan12.*

```
sw-agg1# show ip interface brief
Interface            IP Address                 Interface Status
                                                  link/admin
...
vlan1               10.1.1.1/24                   up/up
vlan2               10.1.2.2/24                   up/up
vlan3               10.1.3.1/24                   up/up
vlan11              10.1.11.1/24                  up/up
vlan12              No Address                    up/up
```

4. Configure an IP address of 10.1.12.1/24 on this SVI.

```
sw-agg1(config-if-vlan)# ip address 10.1.12.1/24
```

5. Configure DHCP relay on this SVI. The IP address configured points to the centralized DHCP server.

```
sw-agg1(config-if-vlan)# ip helper-address 10.254.1.21
sw-agg1(config-if-vlan)# exit
```

6. Review the SVI 12 interface status. Take note of the interface MAC address of the interface. Note that your output may be different than what is displayed below.

```
sw-agg1(config)# show ip interface vlan12
Interface vlan12 is up
 Admin state is up
 Hardware: Ethernet, MAC Address: b8:d4:e7:dAnswer:80:00
 IP MTU 1500
 IP Directed Broadcast is Disabled
 IP Neighbor flood is Disabled
 IPv4 address 10.1.12.1/24
 active-gateway L3 source mac b8:d4:e7:dAnswer:80:00
```

- SVI 12 MAC address: ____________________

7. On sw-edge1, review the MAC-address table for VLAN 12.

```
sw-edge1(config)# show mac-address-table vlan 12
MAC age-time              : 300 seconds
Number of MAC addresses : 2
```

```
MAC Address              VLAN      Type                         Port
-------------------------------------------------------------------
00:50:56:b1:e3:6f        12        dynamic                      1/1/27
b8:d4:e7:dAnswer:80:00    12        dynamic                      1/1/27
```

- **Question**: Do you see the MAC address of the aggregation switch in the MAC table?
- **Answer**: Yes, the aggregation switch SVI is sending ARP broadcast packets. These contain the aggregation switch SVI MAC address as the source MAC, and this is how the edge switches learn the aggregation SVI MAC address.

SVI MAC on different VLANs

8. On sw-agg1, lookup the MAC address of the SVI11.

```
sw-agg1(config)# show ip interface vlan11

Interface vlan11 is up
 Admin state is up
 Hardware: Ethernet, MAC Address: b8:d4:e7:dAnswer:80:00
 IP MTU 1500
 IP Directed Broadcast is Disabled
 IP Neighbor flood is Disabled
 IPv4 address 10.1.11.1/24
 active-gateway L3 source mac b8:d4:e7:dAnswer:80:00
```

- **Question**: Is this MAC address different from the MAC address that is used on SVI12?
- **Answer**: No, the same MAC address is used on different SVIs.
- **Question**: Is it a problem to have the same MAC address on multiple places in the network?
- **Answer**: No, MAC addresses must only be unique within a VLAN. The fact that a MAC address is typically globally unique also ensures that it will be unique within a VLAN. It is typically no problem to re-use the same MAC address in different VLANs.

9. On sw-edge1, review the MAC address table for both VLAN 11 and VLAN 12. You should observe the aggregation switch MAC address in both VLANs, with the same MAC address (this is *okay* since the two VLANs are in different broadcast domains).

```
sw-edge1(config)# show mac-address-table vlan 11
MAC age-time              : 300 seconds
Number of MAC addresses : 2

MAC Address              VLAN      Type                         Port
-------------------------------------------------------------------
00:50:56:b1:cd:26        11        dynamic                      1/1/1
b8:d4:e7:dAnswer:80:00    11        dynamic                      1/1/27
```

```
sw-edge1(config)# show mac-address-table vlan 12
MAC age-time              : 300 seconds
Number of MAC addresses : 2

MAC Address          VLAN     Type                      Port
---------------------------------------------------------------
00:50:56:b1:e3:6f    12       dynamic                   1/1/27
b8:d4:e7:dAnswer:80:00    12       dynamic                   1/1/27
```

Verify VLAN 12 operation

With VLAN 12 configured on the edge and aggregation switch, and the SVI 12 configured with an IP address and IP helper active on the sw-agg1, your VLAN 12 should now be ready for use. Test the operation with your PC4; it should get an IP address in VLAN 12.

1. Open a connection to **PC4** to check the current IP address.
2. Right-click the **Network** icon in the Windows status bar.
3. Click **Open Network & Internet Settings** (see Figure 3-48).

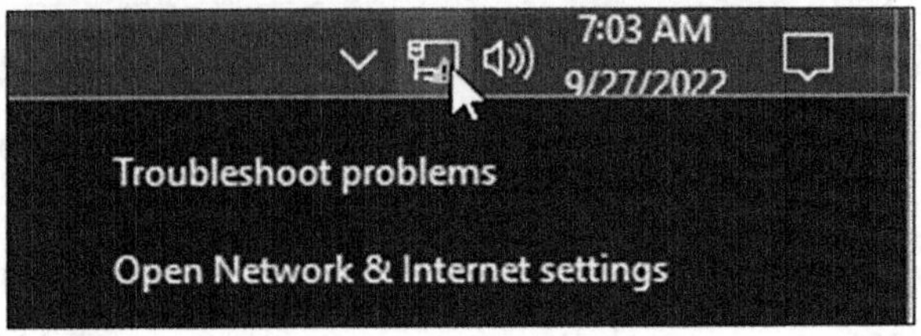

Figure 3-48: Windows task bar

4. Under Change your network settings, click **Change adapter options** (see Figure 3-49).

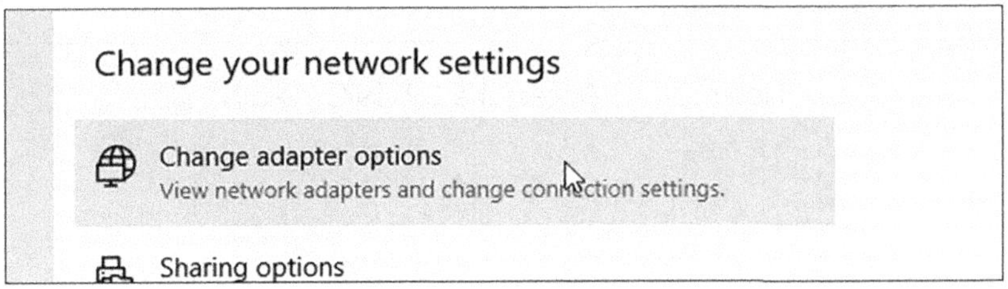

Figure 3-49: Selecting the change adapter options

5. Right-click the **Lab NIC** (this is the NIC connected to port 4 on sw-edge2). Click **Status** (see Figure 3-50).

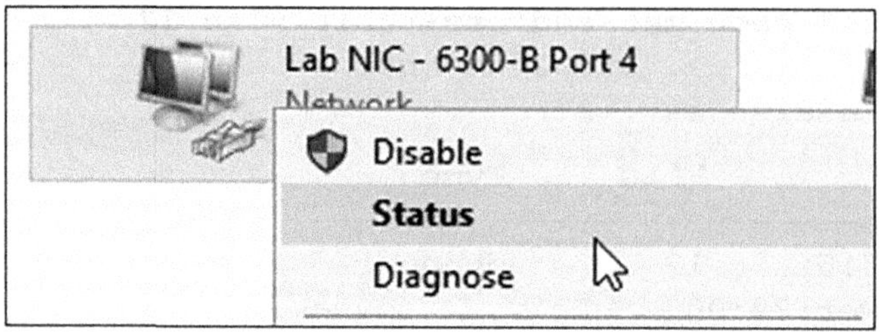

Figure 3-50: Access the Lab NIC status

6. Click **Details** (see Figure 3-51).

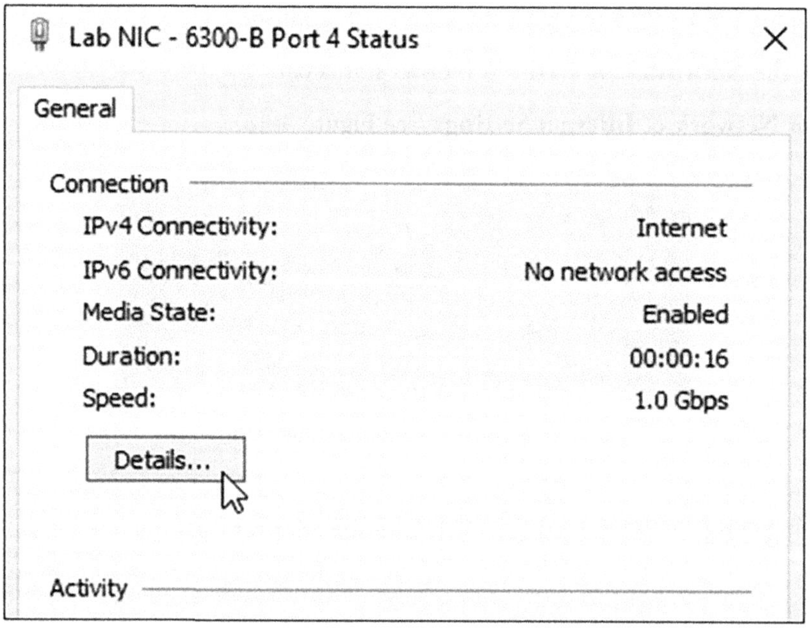

Figure 3-51: Access the Lab NIC details

7. Take note of the MAC address and the IP address of the NIC (see Figure 3-52).

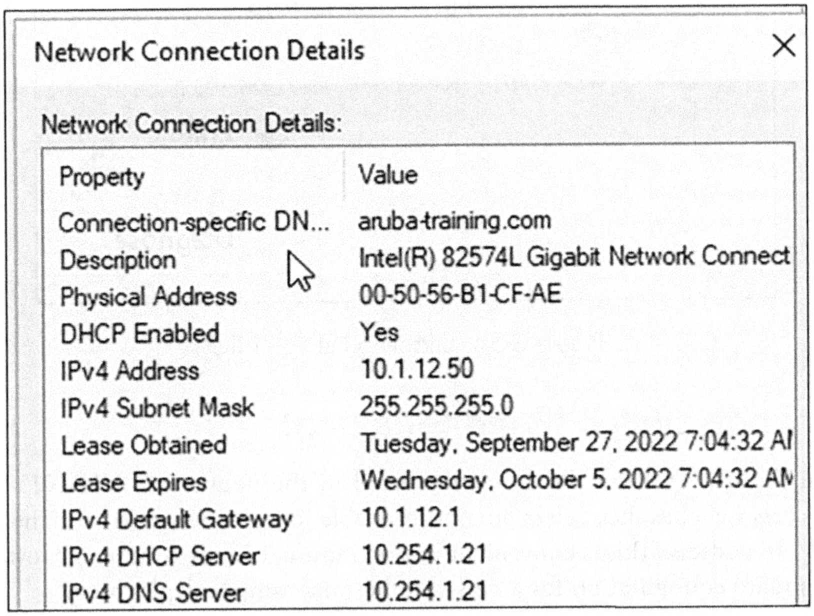

Figure 3-52: Lab NIC's MAC and IP addressing information

Note

You may need to **disable** and **enable** the *Lab NIC* interface to force a DHCP renewal. Make sure that you **don't disable the 'Do Not Touch'** interface (see Figures 3-53 and 3-54).

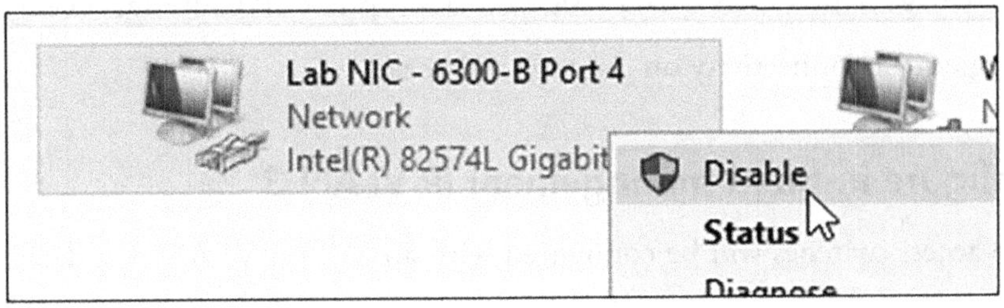

Figure 3-53: Disabling the LAB NIC

Figure 3-54: Enabling the Lab NIC

Lab 3.02: Edge switch IP management

At this stage in the lab, PC1 and PC4 are connected to the network in VLAN11 and VLAN12. However, the access switches themselves are not reachable in these networks yet. The access switches have an OOBM IP address. This is convenient in the training lab environment. However, this is not considered a standard configuration for access switches in a campus deployment.

In a typical campus deployment, the access switches will be configured through an in-band management VLAN. In this lab the access switches will be configured with an in-band management IP addresses and you will ensure that they can access the Internet. This will be required in later labs to provide cloud management functions using Aruba Central.

Objectives

After completing this lab, you will be able to:

- Configure management VLAN 3 and assign the appropriate IP addresses on edge switches.
- Verify management connectivity on the edge switches.

Task 1: Configure in-band management IP VLAN 3

In this task the access switches will be configured with an SVI on VLAN 3, a default gateway, and DNS server.

Objectives

- Configure an access switch management IP address.
- Configure DNS name lookup on the access switches.
- Configure time services using NTP.

Steps

Review the current management IP address

1. On sw-edge1, attempt to ping **8.8.8.8**.

```
sw-edge1(config)# ping 8.8.8.8
ping4: connect: Network is unreachable
```

2. Review the current IP interfaces.

```
sw-edge1(config)# show ip interface brief
Interface          IP Address               Interface Status
                                              link/admin
vlan1             No Address                  up/up
```

- **Question**: Did the ping work? Why?
- **Answer**: No, sw-edge1 currently does not have an in-band IP address.

3. Make sure you can distinguish in-band (**show ip interface brief**) from the OOBM management interface (**show interface mgmt**).

```
sw-edge1(config)# show interface mgmt
  Address Mode                   : static
  Admin State                    : up
  Link State                     : up
  Mac Address                    : 64:e8:81:dd:81:81
  IPv4 address/subnet-mask       : 10.251.1.4/24
  Default gateway IPv4           :
  IPv6 address/prefix            :
  IPv6 link local address/prefix: fe80::66e8:81ff:fedd:8181/64
  Default gateway IPv6           :
  Primary Nameserver             :
  Secondary Nameserver           :
  Tertiary Nameserver            :
```

Configure VLAN 3 in-band IP

1. On sw-edge1, create VLAN 3.

```
sw-edge1(config)# vlan 3
sw-edge1(config-vlan-3)# exit
```

2. Create SVI 3.

```
sw-edge1(config)# interface vlan 3
```

3. Assign an IP address of 10.1.3.4/24 and return to the global configuration context.

```
sw-edge1(config-if-vlan)# ip address 10.1.3.4/24
sw-edge1(config-if-vlan)# exit
```

4. Review the IP interface list.

```
sw-edge1(config)# show ip interface brief
Interface           IP Address              Interface Status
                                              link/admin
vlan1               No Address                up/up
vlan3               10.1.3.4/24               up/up
```

5. In VLAN 1, SVI 1 is enabled for DHCP by default. This is useful for ZTP deployments; however, if you are using a different VLAN ID for management, it is safe to disable the DHCP client and shutdown the interface SVI 1. Disable the DHCP client on SVI 1.

```
sw-edge1(config)# interface vlan 1
sw-edge1(config-if-vlan)# no ip dhcp
sw-edge1(config-if-vlan)# shutdown
sw-edge1(config-if-vlan)# exit
```

6. Attempt to ping sw-agg1 (IP **10.1.3.1**); this should now be successful.

```
sw-edge1(config)# ping 10.1.3.1

PING 10.1.3.1 (10.1.3.1) 100(128) bytes of data.
108 bytes from 10.1.3.1: icmp_seq=1 ttl=64 time=20.3 ms
108 bytes from 10.1.3.1: icmp_seq=2 ttl=64 time=0.173 ms
108 bytes from 10.1.3.1: icmp_seq=3 ttl=64 time=0.173 ms
108 bytes from 10.1.3.1: icmp_seq=4 ttl=64 time=0.224 ms
108 bytes from 10.1.3.1: icmp_seq=5 ttl=64 time=0.174 ms

--- 10.1.3.1 ping statistics ---
5 packets transmitted, 5 received, 0% packet loss, time 4079ms
rtt min/avg/max/mdev = 0.173/4.202/20.266/8.032 ms
```

7. Attempt to ping a host on the internet (IP **8.8.8.8**).

```
sw-edge1(config)# ping 8.8.8.8
ping4: connect: Network is unreachable
```

- **Question**: Did this work? Why?
- **Answer**: The ping did not work. The edge switch does not have a default route configured.

8. Review the IP routing table.

```
sw-edge1(config)# show ip route
Displaying ipv4 routes selected for forwarding

Origin Codes: C - connected, S - static, L - local
              R - RIP, B - BGP, O - OSPF
Type Codes:   E - External BGP, I - Internal BGP, V - VPN, EV - EVPN
              IA - OSPF internal area, E1 - OSPF external type 1
              E2 - OSPF external type 2

VRF: default

Prefix        Nexthop Interface  VRF(egress)  Origin/  Distance/
Age
                                                Type     Metric
---------------------------------------------------------------
10.1.3.0/24   -       vlan3      -            C        [0/0]      -
10.1.3.4/32   -       vlan3      -            L        [0/0]      -

Total Route Count : 2
```

9. Add a default route to pointing to a next-hop address of 10.1.3.1.

```
sw-edge1(config)# ip route 0.0.0.0/0 10.1.3.1
```

10. Verify the route is active in the IP routing table.

```
sw-edge1(config)# show ip route

Displaying ipv4 routes selected for forwarding

Origin Codes: C - connected, S - static, L - local
              R - RIP, B - BGP, O - OSPF
Type Codes:   E - External BGP, I - Internal BGP, V - VPN, EV - EVPN
              IA - OSPF internal area, E1 - OSPF external type 1
              E2 - OSPF external type 2
VRF: default

Prefix        Nexthop  Interface VRF(egress) Origin/   Distance/ Age
                                             Type      Metric
---------------------------------------------------------------
0.0.0.0/0     10.1.3.1 vlan3     -           S         [1/0]
00h:00m:14s
10.1.3.0/24 -          vlan3     -           C         [0/0]      -
10.1.3.4/32 -          vlan3     -           L         [0/0]      -

Total Route Count : 3
```

11. Attempt to ping a host on the internet (IP **8.8.8.8**). This should now be successful.

```
sw-edge1(config)# ping 8.8.8.8
PING 8.8.8.8 (8.8.8.8) 100(128) bytes of data.
76 bytes from 8.8.8.8: icmp_seq=1 ttl=114 (truncated)
76 bytes from 8.8.8.8: icmp_seq=2 ttl=114 (truncated)
76 bytes from 8.8.8.8: icmp_seq=3 ttl=114 (truncated)
76 bytes from 8.8.8.8: icmp_seq=4 ttl=114 (truncated)
76 bytes from 8.8.8.8: icmp_seq=5 ttl=114 (truncated)

--- 8.8.8.8 ping statistics ---
5 packets transmitted, 5 received, 0% packet loss, time 4005ms
rtt min/avg/max/mdev = 8.656/9.189/9.803/0.411 ms
```

Name resolution

You will now test and configure the DNS name lookup function in the network.

1. Attempt to ping the Aruba Path Quality Monitoring (PQM) host.

```
sw-edge1(config)# ping pqm.arubanetworks.com
ping4: pqm.arubanetworks.com: Temporary failure in name resolution
```

Note

The host **pqm.arubanetworks.com** is cloud-hosted by Aruba Networks for branch site connectivity and path quality checks.

- **Question**: Did the ping succeed? Why?
- **Answer**: No, the edge switch does not have any name resolution server (Domain Name Service or DNS) configured. While it can reach the internet, it currently cannot perform name resolution lookups.

2. Configure a DNS host 10.254.1.21.

```
sw-edge1(config)# ip dns server-address 10.254.1.21
```

3. Attempt to ping **pqm.arubanetworks.com** again, which should now be successful.

```
sw-edge1(config)# ping pqm.arubanetworks.com
PING pqm.arubanetworks.com (54.81.169.15) 100(128) bytes of data.
108 bytes from 54.81.169.15: icmp_seq=1 ttl=34 time=30.9 ms
108 bytes from 54.81.169.15: icmp_seq=2 ttl=34 time=30.9 ms
108 bytes from 54.81.169.15: icmp_seq=3 ttl=34 time=30.7 ms
108 bytes from 54.81.169.15: icmp_seq=4 ttl=34 time=30.7 ms
108 bytes from 54.81.169.15: icmp_seq=5 ttl=34 time=43.7 ms
```

```
--- pqm.arubanetworks.com ping statistics ---
5 packets transmitted, 5 received, 0% packet loss, time 4006ms
rtt min/avg/max/mdev = 30.659/33.379/43.708/5.165 ms
```

Note

Since it is cloud-hosted, *pqm.arubanetworks.com* runs on many different systems and therefore the IP address may be different for each lookup you perform.

Time services

The current release of the AOS-CX NTP client configuration includes a default NTP server definition. This can be overruled in the configuration in case a specific server is required (like one located within the campus network) or access to the default server is not allowed.

1. Review the default NTP host in the running-configuration.

```
sw-edge1(config)# show ntp servers
------------------------------------------------------
           NTP SERVER KEYID MINPOLL MAXPOLL OPTION VER
------------------------------------------------------
pool.ntp.org(default) --          4       4 iburst   4
------------------------------------------------------
```

2. Configure the server **10.254.1.21** as NTP server.

```
sw-edge1(config)# ntp server 10.254.1.21 iburst
sw-edge1(config)# ntp enable
```

3. Review NTP time updates. Your system time may have been using updates from the pool.ntp.org server previously.

```
sw-edge1(config)# show event -r | include ntp
2022-10-15T19:18:31.175905+00:00 sw-edge1 ntp-mgrd[4382]:
Event|1101|LOG_INFO|||NTP Association Add : 10.254.1.21 iburst
version 4
2022-10-15T19:18:03.342892+00:00 sw-edge1 ntpd[4669]: Event|System
date/time changed from 2022-10-15 19:20:48 to 2022-10-15 19:20:48
using 193.187.181.6
2022-10-15T12:17:48.336546+00:00 6300 ntp-mgrd[4382]:
Event|1107|LOG_INFO|||NTP enable state change : NTP is disabled ->
NTP is enabled
2022-10-15T12:17:43.173512+00:00 6300 ntp-mgrd[4382]:
Event|1101|LOG_INFO|||NTP Association Add : pool.ntp.org iburst
version 4 minpoll 4 maxpoll 4
```

4. Review the current time.

```
sw-edge1(config)# show clock
Sat Oct 15 19:20:14 UTC 2022
System is configured for timezone : UTC
```

5. Configure the correct time zone for the local system.

```
sw-edge1(config)# clock timezone us/michigan
```

6. Verify the current time now reflects the configured time zone.

```
sw-edge1(config)# show clock
Sat Oct 15 15:20:31 EDT 2022
System is configured for timezone : US/Michigan
```

7. The time on the switch should now be the same as the time on the PC-MGMT. This system has also been configured in this time zone.

Task 2: Practice configuring the sw-edge2 switch

Objectives

In this task you will practice applying the same settings on sw-edge2. You can use the solution if you are unsure about the steps or if you want to verify your steps.

On sw-edge2:

- Configure SVI 3 (VLAN 3) IP address 10.1.3.5/24 with default gateway to 10.1.3.1.
- Configure 10.254.1.21 as DNS server.
- Configure 10.254.1.21 as the NTP server and define the appropriate time zone.
- Verify the connectivity and the configuration.

Solution

1. Solution for sw-edge2:

```
vlan 3

interface vlan 3
 ip address 10.1.3.5/24
 exit
```

```
interface vlan 1
 no ip dhcp
 shutdown
 exit
ip route 0.0.0.0/0 10.1.3.1
ip dns server-address 10.254.1.21

ntp server 10.254.1.21 iburst
ntp enable
clock timezone us/michigan
```

2. Save the configuration on ***both*** edge switches!

```
write memory
```

Lab 3.03: Static routes

This lab will introduce you to static IP routes and static route summarization.

Objectives

After completing this lab, you will be able to:

- Configure static routes in the routing table.
- Understand the operation of static routes.
- Review the IP routing table.

Task 1: Configure a static route between aggregation and core

Your senior colleague has asked you to explore the IP routing table and make a static route configuration to reach the datacenter subnet 10.254.1.0/24. In this task the default route on the aggregation switch sw-agg1 will be removed and a static route for the Datacenter 1 subnet will be added. Datacenter 2 will be covered in the next task.

You will also explore the source IP address used when performing a ping on a routing switch.

The last step is to ensure there is bi-directional communication by configuring the core router with a static route for the return traffic.

Objectives

- Configure static routes.
- Verify static routes.

Steps

Remove the default route on sw-agg1

On sw-agg1, the default route must be removed to test the static route function in this lab. The default route was part of the initial configuration of sw-agg1.

1. On sw-agg1, remove the default route.

```
sw-agg1(config)# no ip route 0.0.0.0/0 10.254.101.254
```

2. Verify that the default route has been removed from the IP routing table.

```
sw-agg1(config)# show ip route | include 0.0.0.0
sw-agg1(config)#
sw-agg1(config)# show ip route 0.0.0.0

No ipv4 routes configured

sw-agg1(config)#
```

Review the routed uplink to core router

1. On sw-agg1, review the IP interface list. You should have port 1/1/8 configured with an IP address.

```
sw-agg1(config)# show ip interface brief
Interface            IP Address             Interface Status
                                              link/admin
...
1/1/8               10.254.101.2/24            up/up
...
```

2. Review the configuration of port 1/1/8 (uplink to core router). The core router has an IP address of 10.254.101.254.

```
sw-agg1(config)# show running-config interface 1/1/8
interface 1/1/8
    no shutdown
    mtu 9198
    description rtr-core1-1/1/1
    ip address 10.254.101.2/24
    exit
```

3. Review the IP routing table. There should be a directly connected route (*C*) for the 10.254.101.0/24 subnet.

```
sw-agg1(config)# show ip route

Displaying ipv4 routes selected for forwarding

Origin Codes: C - connected, S - static, L - local
              R - RIP, B - BGP, O - OSPF
Type Codes:   E - External BGP, I - Internal BGP, V - VPN, EV - EVPN
              IA - OSPF internal area, E1 - OSPF external type 1
              E2 - OSPF external type 2

VRF: default

Prefix           Nexthop           Interface       VRF(egress)   Origin/
Distance/     Age
                                                                 Type
Metric
------------------------------------------------------------------------
------------------
10.1.1.0/24       -                 vlan1           -             C
[0/0]          -
10.1.1.1/32       -                 vlan1           -             L
[0/0]          -
10.1.2.0/24       -                 vlan2           -             C
[0/0]          -
10.1.2.2/32       -                 vlan2           -             L
[0/0]          -
10.1.3.0/24       -                 vlan3           -             C
[0/0]          -
10.1.3.1/32       -                 vlan3           -             L
[0/0]          -
10.1.11.0/24      -                 vlan11          -             C
[0/0]          -
10.1.11.1/32      -                 vlan11          -             L
[0/0]          -
10.1.12.0/24      -                 vlan12          -             C
[0/0]          -
10.1.12.1/32      -                 vlan12          -             L
[0/0]          -
10.254.101.0/24   -                 1/1/8           -             C
[0/0]          -
10.254.101.2/32   -                 1/1/8           -             L
[0/0]          -
```

4. For troubleshooting, especially when dealing with a larger routing table, it can be useful to check if the IP routing table has a route available to reach a specific destination IP address. Check if there is a route available for IP 10.254.101.254. This should reveal the same directly connected route from the previous step.

```
sw-agg1(config)# show ip route 10.254.101.254
VRF: default
    Prefix              : 10.254.101.0/24
VRF(egress)            : -
    Nexthop             : -                                  Interface
: 1/1/8
    Origin               : connected                         Type
: -
    Distance             : 0                                 Metric
: 0
    Age                  : -                                 Tag
: 0
    Encap Type           : -                                 Encap
Details          : -
```

5. You may also use the pipe function to filter the command output. In this example, you will only receive the output lines that include the text **10.254.101**.

```
sw-agg1(config)# show ip route | include 10.254.101
10.254.101.0/24  -                 1/1/8          -              C
[0/0]          -
10.254.101.2/32  -                 1/1/8          -              L
[0/0]          -
```

6. Verify connectivity to the core router using the directly connected IP; this should be successful.

```
ping 10.254.101.254
```

7. Attempt to reach an IP in the server subnet (10.254.1.21).

```
sw-agg1(config)# ping 10.254.1.21
ping4: connect: Network is unreachable
```

- **Question**: Why did this ping fail?
- **Answer**: There is currently no IP route in the routing table for the requested destination IP.

8. Look up the requested destination IP in the local routing table to see if there are any matching routes.

```
sw-agg1(config)# show ip route 10.254.1.21
No ipv4 routes configured
```

- **Question**: What do you observe?
- **Answer**: There is currently no matching route for the requested address.

Configure static route to the datacenter subnet

1. On sw-agg1, add a static route for the 10.254.1.0/24 subnet with the core router as the next-hop IP address.

```
sw-agg1(config)# ip route 10.254.1.0/24 10.254.101.254
```

2. Look up the requested destination IP in the local routing table to see if there are any matching routes.

```
sw-agg1(config)# show ip route 10.254.1.21
VRF: default
   Prefix              : 10.254.1.0/24
VRF(egress)          : -
   Nexthop             : 10.254.101.254                  Interface
: 1/1/8
   Origin              : static                          Type
: -
   Distance            : 1                               Metric
: 0
   Age                 : 00h:01m:09s                     Tag
: 0
   Encap Type          : -                               Encap
Details        : -
```

3. Attempt to ping the host in the server subnet (10.254.1.21). This should now be successful.

```
ping 10.254.1.21
```

- **Question**: What is the source IP address used for this ping?
- **Answer**: By default, a system with multiple IP interfaces (such as the aggregation switch) will use the outbound interface IP address as the source IP.: in this example, it's the IP address configured on port 1/1/8.

4. Attempt to ping **10.254.1.21** using source IP address of SVI 1

```
sw-agg1(config)# ping 10.254.1.21 source vlan1
PING 10.254.1.21 (10.254.1.21) from 10.1.1.1 : 100(128) bytes of data.

--- 10.254.1.21 ping statistics ---
5 packets transmitted, 0 received, 100% packet loss, time 4081ms
```

- **Question**: Did this ping work?
- **Answer**: No.
- **Question**: In the first line of the **ping** command output, the system will report *PING … from …*. What is the from (source) IP address used for this ping?
- **Answer**: The source IP address is now 10.1.1.1. This is the SVI 1 IP address, as requested using the source command option.

5. Now try to identify what routing hop or hops (routers) is having trouble with this traffic flow. Use **traceroute** to set up an IP trace to destination IP **10.254.1.21** using source interface SVI1 (IP 10.1.1.1).

```
traceroute 10.254.1.21 source vlan1
```

- **Question**: Does any hop show a reply in the output? Why?
- **Answer**: When a different source IP is used, it is possible that there is no return route for the specified source subnet. In this example, it means that the ping packet did arrive at the destination host, but the return route may be missing on the return path. In this example, the fact that not even the first hop responds indicates that the issue may be on the next-hop (the core router).

6. Stop the traceroute using **<Ctrl>-c**.

```
sw-agg1(config)# traceroute 10.254.1.21 source vlan1
traceroute to 10.254.1.21 (10.254.1.21), 1 hops min, 30 hops max,
3 sec. timeout, 3 probes
^C
```

7. On the core router, review the IP routing table.

```
show ip route
```

8. First check if any routes are available for the 10.254.101.2 IP address. This is the IP address of sw-agg1 port 1/1/8.

```
rtr-core1# show ip route 10.254.101.2
VRF: default
    Prefix                : 10.254.101.0/24
```

```
VRF(egress)          : -
   Nexthop             : -                              Interface
: 1/1/1
   Origin              : connected                      Type
: -
   Distance            : 0                              Metric
: 0
   Age                 : -                              Tag
: 0
   Encap Type          : -                              Encap
Details          : -
```

- **Question**: Are there any matching routes for this IP? What is the type of route?
- **Answer**: Yes, this is a directly connected subnet on the core router.

9. Next, check if any routes are available for the **10.1.1.2** IP address; this is the sw-agg1 SVI 1 address that is used in the test ping and traceroute.

```
rtr-core1# show ip route 10.1.1.1

VRF: default

   Prefix             : 10.0.0.0/8
VRF(egress)           : -
   Nexthop              : -                               Inter-
face              : blackhole
   Origin               : static                          Type
: -
   Distance             : 1                               Metric
: 0
   Age                  : 00h:03m:23s                     Tag
: 0
   Encap Type           : -                               Encap
Details           : -
```

- **Question**: Are there any matching routes for this IP address? If so, why did the ping not succeed?
- **Answer**: There is a matching route on the core router, but this is a 10.0.0.0/8 summary route that points to the blackhole interface. At this moment, the core router simply drops (blackholes) traffic destined to 10.1.1.1, instead of sending it back to the sw-agg1.

Note

This route has been configured on the core router to ensure that traffic to any address in the 10.0.0.0/8 range that is not handled by a more specific subnet is dropped by the core router. This can help prevent routing loops when route summarization occurs at multiple places in the network. Route summarization is covered in the professional level course and book.

10. Add a specific static route for the **10.1.1.0/24** subnet with a next-hop IP of sw-agg1.

```
rtr-core1(config)# ip route 10.1.1.0/24 10.254.101.2
```

11. Review all the configured static routes in the routing table. Your new static route should be listed in the output.

```
rtr-core1(config)# show ip route static

Displaying ipv4 routes selected for forwarding

Origin Codes: C - connected, S - static, L - local
              R - RIP, B - BGP, O - OSPF
Type Codes:   E - External BGP, I - Internal BGP, V - VPN, EV - EVPN
              IA - OSPF internal area, E1 - OSPF external type 1
              E2 - OSPF external type 2

VRF: default

Prefix        Nexthop       Interface  VRF(egress) Origin/Distance/
Age

                                                    Type  Metric
-----------------------------------------------------------------
---------------------
0.0.0.0/0     10.254.1.253  1/1/9          -          S      [1/0]
00h:04m:07s
10.0.0.0/8    -             blackhole      -          S      [1/0]
00h:04m:07s
10.1.1.0/24   10.254.101.2  1/1/1          -          S      [1/0]
00h:00m:14s
10.1.3.0/24   10.254.101.2  1/1/1          -          S      [255/0]
00h:04m:07s
10.1.11.0/24  10.254.101.2  1/1/1          -          S      [255/0]
00h:04m:07s
10.1.12.0/24  10.254.101.2  1/1/1          -          S      [255/0]
00h:04m:07s

Total Route Count : 6
```

12. Verify in the routing table that the new route will be used to reach destination IP 10.1.1.2. Notice the difference with the previous command output three steps earlier.

```
rtr-core1(config)# show ip route 10.1.1.1

VRF: default
```

```
   Prefix              : 10.1.1.0/24
VRF(egress)          : -
   Nexthop              : 10.254.101.2                               Interface
: 1/1/1
   Origin               : static                                     Type
: -
   Distance             : 1                                          Metric
: 0
   Age                  : 00h:02m:05s                                Tag
: 0
   Encap Type           : -                                          Encap
Details        : -
```

13. On sw-agg1, verify that the ping to **10.254.1.21** works using SVI 1 as the source IP.

```
sw-agg1(config)# ping 10.254.1.21 source vlan1
PING 10.254.1.21 (10.254.1.21) from 10.1.1.1 : 100(128) bytes of data.
108 bytes from 10.254.1.21: icmp_seq=1 ttl=127 time=2.35 ms
108 bytes from 10.254.1.21: icmp_seq=2 ttl=127 time=2.22 ms
108 bytes from 10.254.1.21: icmp_seq=3 ttl=127 time=1.98 ms
108 bytes from 10.254.1.21: icmp_seq=4 ttl=127 time=1.98 ms
108 bytes from 10.254.1.21: icmp_seq=5 ttl=127 time=1.93 ms

--- 10.254.1.21 ping statistics ---
5 packets transmitted, 5 received, 0% packet loss, time 4004ms
rtt min/avg/max/mdev = 1.930/2.091/2.347/0.162 ms
```

- **Question**: Was the ping successful?
- **Answer**: Yes, routes are now available for both transmit and return traffic.

Task 2: Route aggregation with static routes

In this task you will explore route aggregation with static routes. The static route of the previous task only worked for SVI 1 communications. All other subnets connected to the aggregation layer would require their own static routes on the core router.

Your customer wants to know if, instead of configuring a static route for each IP subnet connected to the aggregation switches, a simplified static route configuration can be applied. You have heard that route aggregation can be used to aggregate, or summarize subnets into an aggregate static route.

In this task you will configure an aggregate route to handle all current and future subnets connected to sw-agg1. You will also ensure that both the current Datacenter 1 and the future Datacenter 2 can be reached using a single route. The data center subnet list can be found in Table 3-2.

Table 3-2.

Network	Subnet
Datacenter 1	10.254.1.0/24
Datacenter 2	10.254.2.0/24

Objectives

- Configure static route aggregation.

Steps

Prepare core router for this task

The core router has been pre-configured with some static routes in this lab environment to support the initial lab exercises with VLAN 11 and 12. To complete the current lab exercise, you must remove one of these static routes.

1. On the core router, remove the static route for the VLAN 12 subnet (10.1.12.0/24).

```
rtr-core1(config)# no ip route 10.1.12.0/24 10.254.101.2
```

2. Verify that SVI12 cannot access the datacenter subnet. On sw-agg1, attempt to ping **10.254.1.21** using source IP SVI 12: this should fail.

```
sw-agg1(config)# ping 10.254.1.21 source vlan12
PING 10.254.1.21 (10.254.1.21) from 10.1.12.1 : 100(128) bytes of data.

--- 10.254.1.21 ping statistics ---
5packets transmitted, 0 received, 100% packet loss, time 4088ms
```

Configure aggregate route

1. On the core router, attempt to find a route in the routing table for the sw-agg1 SVI 12's IP address (10.1.12.1).

```
rtr-core1(config)# show ip route 10.1.12.1

VRF: default

   Prefix              : 10.0.0.0/8                       VRF(egress)
: -
   Nexthop             : -                                Interface
: blackhole
   Origin              : static                           Type
: -
   Distance            : 1                                Metric
```

```
: 0
   Age                 : 00h:09m:42s                    Tag
: 0
   Encap Type          : -                              Encap Details
: -
```

- **Question**: Is there any matching route?
- **Answer**: Yes, the requested destination IP matches the 10.0.0.0/8 blackhole route, so traffic to 10.1.12.1 will be dropped by the core router. A destination of *blackhole* in a static route will cause the routing switch to *drop* all traffic that matches this route entry.

2. In the lab IP plan, all current and future campus IP subnets that connect to the aggregation switch belong to the 10.1.0.0/16 range.
3. On the core router, add an aggregate static route 10.1.0.0/16 for the campus subnets using sw-agg1 as the next-hop IP.

```
rtr-core1(config)# ip route 10.1.0.0/16 10.254.101.2
```

4. Again, check what route will be used to reach IP 10.1.12.1. The output should now show the 10.1.0.0/16 route.

```
rtr-core1(config)# show ip route 10.1.12.1

VRF: default

   Prefix              : 10.1.0.0/16                    VRF(egress)
: -
   Nexthop             : 10.254.101.2                   Interface
: 1/1/1
   Origin              : static                         Type
: -
   Distance            : 1                              Metric
: 0
   Age                 : 00h:00m:10s                    Tag
: 0
   Encap Type          : -                              Encap Details
: -
```

5. On sw-agg1, verify that the previous ping test that failed is now successful.

```
sw-agg1(config)# ping 10.254.1.21 source vlan12
PING 10.254.1.21 (10.254.1.21) from 10.1.12.1 : 100(128) bytes of data.
108 bytes from 10.254.1.21: icmp_seq=1 ttl=127 time=1.93 ms
108 bytes from 10.254.1.21: icmp_seq=2 ttl=127 time=2.18 ms
108 bytes from 10.254.1.21: icmp_seq=3 ttl=127 time=1.69 ms
```

```
108 bytes from 10.254.1.21: icmp_seq=4 ttl=127 time=1.25 ms
108 bytes from 10.254.1.21: icmp_seq=5 ttl=127 time=2.07 ms

--- 10.254.1.21 ping statistics ---
5 packets transmitted, 5 received, 0% packet loss, time 4004ms
rtt min/avg/max/mdev = 1.245/1.823/2.178/0.332 ms
```

Datacenter aggregation

Now you have connectivity with the first datacenter (10.254.1.0/24). The customer informs you that there will be a second datacenter in the future. This second datacenter will use IP subnet 10.254.2.0/24. The customer wants you to update sw-agg1, so a single static route is used to reach any IP address that starts with *10.254* in the first two octets in the IP address.

The customer has already activated a test IP address (10.254.2.1) on the core router to validate your configuration. This IP address is in the IP range of the second datacenter. The IP address of 10.254.2.1 can be used to test the connectivity for this second datacenter from the client PCs.

1. On sw-agg1, attempt to ping **10.254.2.1** using source IP SVI 12, which should fail at this point.

```
sw-agg1(config)# ping 10.254.2.1 source vlan12
ping4: sendmsg: Network is unreachable
PING 10.254.2.1 (10.254.2.1) from 10.1.12.1 : 100(128) bytes of data.
ping4: sendmsg: Network is unreachable
ping4: sendmsg: Network is unreachable
ping4: sendmsg: Network is unreachable
ping4: sendmsg: Network is unreachable

--- 10.254.2.1 ping statistics ---
5 packets transmitted, 0 received, 100% packet loss, time 4116ms
```

2. On sw-agg1, add an aggregate static route **10.254.0.0/16** for the datacenter subnets, using the core router as the next-hop IP.

```
sw-agg1(config)# ip route 10.254.0.0/16 10.254.101.254
```

3. On sw-agg1, verify that **10.254.2.1** is now reachable.

```
sw-agg1(config)# ping 10.254.2.1 source vlan12
PING 10.254.2.1 (10.254.2.1) from 10.1.12.1 : 100(128) bytes of data.
108 bytes from 10.254.2.1: icmp_seq=1 ttl=64 time=1.59 ms
108 bytes from 10.254.2.1: icmp_seq=2 ttl=64 time=1.55 ms
108 bytes from 10.254.2.1: icmp_seq=3 ttl=64 time=1.85 ms
108 bytes from 10.254.2.1: icmp_seq=4 ttl=64 time=1.76 ms
108 bytes from 10.254.2.1: icmp_seq=5 ttl=64 time=1.54 ms
```

```
--- 10.254.2.1 ping statistics ---
5 packets transmitted, 5 received, 0% packet loss, time 4005ms
rtt min/avg/max/mdev = 1.540/1.657/1.852/0.126 ms
```

Most specific route

It is possible that multiple routes could be used for a given destination, and so routers must decide which route to choose. The first decision is to pick the *most specific matching* route. This means that the routing device will compare the destination IP address in the packet header with the routes in the IP routing table: whichever route has the greatest number of bits that match the IP address will be the one that is used to route the packet. In these steps you will explore this behavior.

1. On the sw-agg1, review the route that would be used to reach 10.254.1.21.

```
sw-agg1(config)# show ip route 10.254.1.21
VRF: default
   Prefix          : 10.254.1.0/24          VRF(egress)
: -
   Nexthop         : 10.254.101.254         Interface
: 1/1/8
   Origin          : static                 Type
: -
   Distance        : 1                      Metric
: 0
   Age             : 00h:32m:22s            Tag
: 0
   Encap Type      : -                      Encap Details
: -
```

2. Now review the route that would be used to reach 10.254.2.1 (the second datacenter).

```
sw-agg1(config)# show ip route 10.254.2.1
VRF: default
   Prefix          : 10.254.0.0/16          VRF(egress)
: -
   Nexthop         : 10.254.101.254         Interface
: 1/1/8
   Origin          : static                 Type
: -
   Distance        : 1                      Metric
: 0
   Age             : 00h:01m:02s            Tag
: 0
   Encap Type      : -                      Encap Details
: -
```

- **Question**: Do you notice any difference? Why?
- **Answer**: For the 10.254.1.21 IP, there is a more matching route /24 (24 bits of the subnet mask match), for the 10.254.2.1 IP, the only available route is the /16 aggregate route.

3. Since the aggregate routes have now been added to the routing tables, the more specific routes can be removed.
4. On sw-agg1, *remove* the route for the **10.254.1.0/24** subnet.

```
sw-agg1(config)# no ip route 10.254.1.0/24 10.254.101.254
```

5. Again, review the route used to reach the 10.254.1.21 host; this should now show the /16 aggregate route instead of the more specific /24 route from the previous command output.

```
sw-agg1(config)# show ip route 10.254.1.21
VRF: default
   Prefix              : 10.254.0.0/16                 VRF(egress)
: -
   Nexthop             : 10.254.101.254                Interface
: 1/1/8
   Origin              : static                        Type
: -
   Distance            : 1                             Metric
: 0
   Age                 : 00h:01m:59s                   Tag
: 0
   Encap Type          : -                             Encap Details
: -
```

6. On the *core* router, remove the specific route for the 10.1.1.0/24 subnet.

```
rtr-core1(config)# no ip route 10.1.1.0/24 10.254.101.2
```

7. Verify that the core router is now using the aggregate route 10.1.0.0/16 for the 10.1.1.2 destination.

```
rtr-core1(config)# show ip route 10.1.1.1
VRF: default
   Prefix              : 10.1.0.0/16                   VRF(egress)
: -
   Nexthop             : 10.254.101.2                  Interface
: 1/1/1
   Origin              : static                        Type
: -
```

```
   Distance            : 1                       Metric
: 0
   Age                 : 00h:04m:19s             Tag
: 0
   Encap Type          : -                       Encap
Details: -
```

8. Save the configuration on *all* switches *and* the core router!

```
write memory
```

Lab 3.04: Basic OSPF configuration

Your customer is pleased to see that the static routes and route aggregation have simplified the routing table. Now they have explained that they may need to add several other sites with multiple links, and they are concerned that it may become hard to maintain the routing tables, using static routes, on all these devices.

After consulting with your senior colleague, you hear that OSPF is a routing protocol that can help with the distribution of routes and calculate the best paths in complex networks. In this lab you will introduce OSPF in the network and explore its features.

Objectives

After completing this lab, you will be able to:

- Start OSPF process on AOS-CX devices.
- Configure a single area OSPF.
- Enable interfaces for OSPF.
- Verify OSPF neighbor adjacencies.
- Verify OSPF exchange of routes.
- Advertise a default route using OSPF.

Task 1: Configure loopback interfaces on the aggregation switches

In this task you will configure an IP loopback interface on the aggregation switches.

Objectives

- Configure a loopback interface.

Steps

In the next steps you will configure a loopback interface on the sw-agg1 aggregation switch. A loopback interface is an internal, software interface. Since there is no physical interface and there is only one host (the device itself), you would typically configure a subnet mask of /32 for the IP address of a loopback interface.

A loopback interface is always up. Its state is independent of the state of any physical interface on the system. Loopback interfaces can be used as the source IP address for outbound connections. This ensures that, for example, all outbound SNMP, RADIUS, or packets to Aruba Central have the same, stable source IP address.

1. On *sw-agg1*, configure a new loopback interface with ID of **0**.

```
sw-agg1(config)# interface loopback 0
```

2. Configure an IP address of **10.1.0.2/32** on this new loopback interface.

```
sw-agg1(config-loopback-if)# ip address 10.1.0.2/32
sw-agg1(config-loopback-if)# exit
```

3. Reviewing the IP interface list shows the correct new loopback 0 and IP address of 10.1.0.2/32.

```
sw-agg1(config)# show ip interface brief
Interface          IP Address             Interface Status
                                            link/admin
...
loopback0         10.1.0.2/32               up/up
...
```

4. Review the IP routing table contains a /32 route for the loopback IP address.

```
sw-agg1(config)# show ip route

Displaying ipv4 routes selected for forwarding

Origin Codes: C - connected, S - static, L - local
              R - RIP, B - BGP, O - OSPF
Type Codes:   E - External BGP, I - Internal BGP, V - VPN, EV - EVPN
              IA - OSPF internal area, E1 - OSPF external type 1
              E2 - OSPF external type 2
```

```
VRF: default

Prefix                 Nexthop     Interface      VRF(egress)    Origin/
Distance/    Age
                                                                 Type

Metric
-------------------------------------------------------------------------
----------------
10.1.0.2/32            -           loopback0      -                   L
[0/0]           -
...
```

5. Verify you can locally ping the loopback IP address. This should be successful.

```
ping 10.1.0.2
```

6. On *sw-agg2*, configure loopback interface **0** with an IP address of **10.1.0.3/32**.

```
sw-agg2(config)# interface loopback 0
sw-agg2(config-loopback-if)# ip address 10.1.0.3/32
sw-agg2(config-loopback-if)# exit
```

7. Verify that the new loopback interface is listed in the IP interface list.

```
sw-agg1(config)# show ip interface brief
Interface              IP Address              Interface Status
                                                 link/admin

...
loopback0             10.1.0.3/32                up/up
...
```

Task 2: Configure an OSPF process on the aggregation switches

OSPF uses LSAs to advertise the topology and routes. These LSAs are added to the LSDB, but they require a unique identifier. The unique identifier for each OSPF router is configured using an OSPF router ID. You must also create an OSPF area to store the LSAs, and enable the layer 3 interfaces for OSPF in the new area.

Objectives

In this task you will:

- Enable OSPF on the aggregation switches.
- Configure OSPF to use the loopback IP address as the router ID.

- Enable OSPF on the SVI interface between the aggregation switches.
- Review the status of the OSPF routing process.

Steps

Configure OSPF process on sw-agg1

1. On *sw-agg1*, configure OSPF process **1**.

```
sw-agg1(config)# router ospf 1
```

2. Set the router ID to **10.1.0.2**.

```
sw-agg1(config-ospf-1)# router-id 10.1.0.2
```

3. Configure the area 0 (backbone area) and exit the OSPF context.

```
sw-agg1(config-ospf-1)# area 0
sw-agg1(config-ospf-1)# exit
```

4. Verify that OSPF is enabled.

```
sw-agg1(config)# show ip ospf
VRF : default                             Process   : 1
----------------------------------------------------

RouterID              : 10.1.0.2      OSPFv2                  : Enabled
BFD                   : Disabled      SPF Start Interval      : 200
ms
SPF Hold Interval     : 1000   m      SPF Max Wait Interval : 5000
ms
LSA Start Time        : 5000   ms     LSA Hold Time           : 0
 ms
LSA Max Wait Time     : 0      ms     LSA Arrival             : 1000
ms
External LSAs         : 0             Checksum Sum            : 0
ECMP                  : 4             Reference Bandwidth     : 100000
Mbps
Area Border           : false         AS Border               : false
GR Status             : Enabled       GR Interval             : 120
sec
GR State              : inactive      GR Exit Status          : none
GR Helper             : Enabled       GR Strict LSA Check     : Enabled
GR Ignore Lost I/F    : Disabled
```

```
Summary address:
Area        Total      Active
------------------------------
Normal       1          0
Stub         0          0
NSSA         0          0
Area   : 0.0.0.0
----------------
Area Type               : Normal     Status                : Inactive
Total Interfaces        : 0          Active Interfaces     : 0
Passive Interfaces      : 0          Loopback Interfaces   : 0
SPF Calculation Count   : 1
Area ranges         :
Number of LSAs          : 0          Checksum Sum          : 0
```

- **Question**: What is the router ID used by OSPF?
- **Answer**: The router ID should be 10.1.0.2

5. Examine the OSPF LSDB.

```
sw-agg1(config)# show ip ospf lsdb
No OSPF LSAs found on VRF default.
```

- **Question**: Why is the LSDB empty?
- **Answer**: Since there are no interfaces enabled for OSPF, no router LSAs exist or are learned.

6. Activate the loopback interface for OSPF process 1, area **0**.

```
sw-agg1(config)# interface loopback 0
sw-agg1(config-loopback-if)# ip ospf 1 area 0
sw-agg1(config-loopback-if)# exit
```

7. Examine the LSDB again.

```
sw-agg1(config)# show ip ospf lsdb
OSPF Router with ID (10.1.0.2) (Process ID 1 VRF default)
==========================================================

Router Link State Advertisements (Area 0.0.0.0)
------------------------------------------------

LSID            ADV Router      Age      Seq#        Checksum     Link
Count
--------------------------------------------------------------------
--------
10.1.0.2        10.1.0.2        13       0x80000001 0x00001c10    1
```

- **Question**: How can you recognize that the router LSA is generated by your own system?
- **Answer**: The LSA ID is based on the local OSPF router ID, in this example 10.1.0.2.
- **Question**: What is the link count for this router LSA?
- **Answer**: Only one link has been enabled for OSPF (the loopback). This is the only link that shows up as a router LSA. This also demonstrates that a router may have many IP interfaces or SVIs, but only the *OSPF-enabled* interfaces are included in the router LSA advertisements.

Adding additional interfaces

1. On sw-agg1, enable interfaces SVI **11** and SVI **12** for OSPF. This will ensure that these subnets are included in the sw-agg1 router LSAs. Routes that are included in the Router LSA are 'advertised' to the network. This means they will be reachable for other OSPF routers.

```
sw-agg1(config)# interface vlan11
sw-agg1(config-if-vlan)# ip ospf 1 area 0
sw-agg1(config-if-vlan)# exit
sw-agg1(config)# interface vlan12
sw-agg1(config-if-vlan)# ip ospf 1 area 0
sw-agg1(config-if-vlan)# exit
```

2. Review the LSDB and the link count.

```
sw-agg1(config)# show ip ospf lsdb
OSPF Router with ID (10.1.0.2) (Process ID 1 VRF default)
==========================================================

Router Link State Advertisements (Area 0.0.0.0)
-----------------------------------------------

LSID            ADV Router      Age      Seq#         Checksum    Link
Count
-----------------------------------------------------------------------------
10.1.0.2        10.1.0.2        16       0x80000001 0x0000ac69   3
```

- **Question**: How many links are included in the router LSAs?
- **Answer**: There are now three links in the router LSAs: the loopback subnet, the SVI 11, and the SVI 12 subnets.

Task 2: Configure and Tune OSPF Links to Core Router

In this task you will configure the connection between the sw-agg1 and core router. You will also observe the OSPF interface statistics and adjust the OSPF interface timers.

A link with only two routers is commonly configured as an OSPF point-to-point (P2P) link. P2P links speed up the OSPF neighbor adjacency setup and simplify the LSDB topology. You will also configure a P2P link.

Objectives

- Adjust OSPF timers.
- Configure an OSPF P2P link.

Steps

1. On sw-agg1, assign port **1/1/8** (connected to the core router) to OSPF process **1** area **0**.

```
sw-agg1(config)# interface 1/1/8
sw-agg1(config-if)# ip ospf 1 area 0
sw-agg1(config-if)# exit
```

2. Review the LSDB link count for the new router LSA.

```
sw-agg1(config)# show ip ospf lsdb
OSPF Router with ID (10.1.0.2) (Process ID 1 VRF default)
==========================================================

Router Link State Advertisements (Area 0.0.0.0)
------------------------------------------------

LSID            ADV Router      Age    Seq#       Checksum    Link
Count
-------------------------------------------------------------------
--------
10.1.0.2        10.1.0.2        12     0x80000001 0x0000d8b4  4
```

Note

Between OSPF routers on the same link (subnet), the OSPF hello and dead timers must match. A mismatch will prevent OSPF routers from establishing an adjacency. It is common practice to lower the hello and dead timer values from the defaults of 10 and 40 seconds, respectively, to 1 and 4 seconds. This change was already made on the core router side. If you want to review the existing OSPF configuration (including the timer configuration) on the core router, access the core router and run this command: **show running-configuration ospf**.

1. Review the OSPF interface statistics for port **1/1/8** on sw-agg1.

```
sw-agg1(config)# show ip ospf statistics interface 1/1/8
OSPF Process ID 1 VRF default, interface 1/1/8 statistics
(cleared 0h0m38s ago)
==================================================================
========

Tx Hello Packets       : 4          Rx Hello Packets       : 0
Tx Hello Bytes         : 256        Rx Hello Bytes         : 0
Tx DD Packets          : 0          Rx DD Packets          : 0
Tx DD Bytes            : 0          Rx DD Bytes            : 0
Tx LS Request Packets  : 0          Rx LS Request Packets  : 0
Tx LS Request Bytes    : 0          Rx LS Request Bytes    : 0
Tx LS Update Packets   : 0          Rx LS Update Packets   : 0
Tx LS Update Bytes     : 0          Rx LS Update Bytes     : 0
Tx LS Ack Packets      : 0          Rx LS Ack Packets      : 0
Tx LS Ack Bytes        : 0          Rx LS Ack Bytes        : 0

Total Number of State Changes : 39
Number of LSAs                : 0
LSA Checksum Sum              : 0
Total Transmit Failures       : 0
Total OSPF Packets Discarded  : 76

Reason                        Packets Dropped
---------------------------------------------
Invalid type                  0
Invalid length                0
Invalid checksum              0
Invalid version               0
Bad or unknown source         0
Area mismatch                 0
Self-originated               0
Duplicate router ID           0
Interface standby             0
Total Hello packets dropped   38
  Network Mask mismatch       0
  Hello interval mismatch     38
  Dead interval mismatch      0
  Options mismatch            0
  MTU mismatch                0
  Neighbor ignored            0
Authentication errors         0
  Type mismatch               0
```

```
  Authentication failures      0
Wrong protocol                 0
Resource failures              0
Bad LSA length                 0
Bad DD packets                 0
Others                         38

Total LSAs Ignored : 0
Bad Type           : 0
Bad Length         : 0
Invalid Data       : 0
Invalid Checksum   : 0
```

- **Question**: Are there any non-zero statistics for packets dropped?
- **Answer**: Yes, the 'Hello interval mismatch' shows several dropped packets.

2. Adjust the local timers of port 1/1/8 to match the core router side. Set the hello timer to **1** second and the dead timer to **4** seconds.

```
sw-agg1(config)# interface 1/1/8
sw-agg1(config-if)# ip ospf hello-interval 1
sw-agg1(config-if)# ip ospf dead-interval 4
```

3. Review the OSPF interface statistics again for port 1/1/8. The packets dropped should no longer be increasing and the *Rx Hello Packets* should be incrementing now.

```
sw-agg1(config-if)# show ip ospf 1 statistics interface 1/1/8
OSPF Process ID 1 VRF default, interface 1/1/8 statistics
(cleared 0h1m42s ago)
===========================================================================
========

Tx Hello Packets      : 25         Rx Hello Packets      : 17
Tx Hello Bytes        : 1664       Rx Hello Bytes        : 1152
Tx DD Packets         : 2          Rx DD Packets         : 2
Tx DD Bytes           : 124        Rx DD Bytes           : 144
Tx LS Request Packets : 1          Rx LS Request Packets : 1
Tx LS Request Bytes   : 68         Rx LS Request Bytes   : 56
Tx LS Update Packets  : 3          Rx LS Update Packets  : 2
Tx LS Update Bytes    : 320        Rx LS Update Bytes    : 288
Tx LS Ack Packets     : 2          Rx LS Ack Packets     : 3
Tx LS Ack Bytes       : 148        Rx LS Ack Bytes       : 192

Total Number of State Changes : 87
Number of LSAs                : 0
```

```
LSA Checksum Sum              : 0
Total Transmit Failures       : 0
Total OSPF Packets Discarded  : 170

Reason                        Packets Dropped
----------------------------------------------
Invalid type                  0
Invalid length                0
Invalid checksum              0
Invalid version               0
Bad or unknown source         0
Area mismatch                 0
Self-originated               0
Duplicate router ID           0
Interface standby             0
Total Hello packets dropped   85
  Network Mask mismatch       0
  Hello interval mismatch     85
  Dead interval mismatch      0
...
```

Tip

The change in statistics can be easier to read when you reset them to 0. If you want to reset the OSPF statistics, you can use the command: **clear ip ospf statistics**.

OSPF link type point-to-point

1. On sw-agg1, review if you have any active OSPF adjacencies.

```
sw-agg1(config-if)# show ip ospf neighbors
VRF : default                          Process : 1
===================================================

Total Number of Neighbors : 1

Neighbor ID      Priority  State          Nbr Address      Interface
---------------------------------------------------------------------
10.254.0.1       1         FULL/BDR       10.254.101.254   1/1/8
```

2. Examine the OSPF states in the event log for the **hpe-routing** daemon (background process). This daemon is responsible for routing protocols such as OSPF and BGP in AOS-CX.

```
sw-agg1(config-if)# show event -r -d hpe-routing -n 6
--------------------------------------------------
Event logs from current boot
--------------------------------------------------
2022-10-15T20:32:59.378180+00:00 sw-agg1 hpe-routing[6468]:
Event|2401|LOG_INFO|AMM|1/1|AdjChg: Nbr  rtr ID 10.254.0.1 on IP
addr 10.254.101.2( area ID 0.0.0.0): Loading -> Full
2022-10-15T20:32:59.375870+00:00 sw-agg1 hpe-routing[6468]:
Event|2401|LOG_INFO|AMM|1/1|AdjChg: Nbr  rtr ID 10.254.0.1 on IP
addr 10.254.101.2( area ID 0.0.0.0): Exchange -> Loading
2022-10-15T20:32:59.372921+00:00 sw-agg1 hpe-routing[6468]:
Event|2401|LOG_INFO|AMM|1/1|AdjChg: Nbr  rtr ID 10.254.0.1 on IP
addr 10.254.101.2( area ID 0.0.0.0): Exstart -> Exchange
2022-10-15T20:32:59.372375+00:00 sw-agg1 hpe-routing[6468]:
Event|2401|LOG_INFO|AMM|1/1|AdjChg: Nbr  rtr ID 10.254.0.1 on IP
addr 10.254.101.2( area ID 0.0.0.0): Two-way -> Exstart
2022-10-15T20:32:59.372276+00:00 sw-agg1 hpe-routing[6468]:
Event|2401|LOG_INFO|AMM|1/1|AdjChg: Nbr  rtr ID 10.254.0.1 on IP
addr 10.254.101.2( area ID 0.0.0.0): Init -> Two-way
2022-10-15T20:32:58.690624+00:00 sw-agg1 hpe-routing[6468]:
Event|2401|LOG_INFO|AMM|1/1|AdjChg: Nbr  rtr ID 10.254.0.1 on IP
addr 10.254.101.2( area ID 0.0.0.0): Down -> Init
```

3. Examine the OSPF LSDB.

```
sw-agg1(config-if)# show ip ospf lsdb
OSPF Router with ID (10.1.0.2) (Process ID 1 VRF default)
==========================================================


Router Link State Advertisements (Area 0.0.0.0)
------------------------------------------------


LSID            ADV Router       Age   Seq#        Checksum    Link
Count
-----------------------------------------------------------------
--------
10.1.0.2        10.1.0.2         138   0x80000002 0x000074a6  4
10.254.0.1      10.254.0.1       139   0x80000003 0x000058a6  5


Network Link State Advertisements (Area 0.0.0.0)
------------------------------------------------
```

```
LSID             ADV Router      Age       Seq#        Checksum
-----------------------------------------------------------------
10.254.101.2     10.1.0.2        143       0x80000001  0x00000cb9

AS External Link State Advertisements
--------------------------------------

LSID             ADV Router      Age       Seq#        Checksum
-----------------------------------------------------------------
0.0.0.0          10.254.0.1      907       0x80000002  0x00009234
```

- **Question**: Are there any new entries in the LSDB?
- **Answer**: Yes, there is one new router LSA, one new network LSA, and one AS external LSA. This indicates that routing information can now be calculated by OSPF.

4. Review the learned OSPF routes (from OSPF's point of view).

```
sw-agg1(config-if)# show ip ospf routes
Codes: i - Intra-area route, I - Inter-area route
       E1 - External type-1, E2 - External type-2

OSPF Process ID 1 VRF default, Routing Table
---------------------------------------------

Total Number of Routes : 3

10.1.11.0/24        (i) area: 0.0.0.0
     directly attached to interface vlan11, cost 100 distance 110
10.1.12.0/24        (i) area: 0.0.0.0
     directly attached to interface vlan12, cost 100 distance 110
10.254.101.0/24     (i) area: 0.0.0.0
     directly attached to interface 1/1/8, cost 10 distance 110
```

Note

Make sure to enter the correct command. There is also a command **show ip route ospf**. (**route ospf** versus **ospf route**). This command would show you the OSPF routes from the *IP* routing table.

- **Question**: Although there is an active OSPF adjacency, it seems that no OSPF routes are learned from the OSPF core router. What could be causing this issue?
- **Answer**: When only two OSPF routers exist on an Ethernet link, it is common practice to make the OSPF link a P2P connection. By default, OSPF links using Ethernet are a broadcast

(BCAST) network type. Your sw-agg1 believes it is connected to a BCAST network (the network LSA), while the core router has been pre-configured with a *P2P* link. It is important to configure both sides of the OSPF link with the same network type, otherwise OSPF cannot use this link when calculating the best path (the two routers will not form an adjacency).

5. On sw-agg1, configure the OSPF link on port **1/1/8** as OSPF **point-to-point**.

```
sw-agg1(config-if)# interface 1/1/8
sw-agg1(config-if)# ip ospf network point-to-point
sw-agg1(config-if)# exit
```

6. Verify that OSP routes can now be calculated.

```
sw-agg1(config)# show ip ospf routes
Codes: i - Intra-area route, I - Inter-area route
       E1 - External type-1, E2 - External type-2

OSPF Process ID 1 VRF default, Routing Table
---------------------------------------------

Total Number of Routes : 7

0.0.0.0/0             (E2)
     via 10.254.101.254 interface 1/1/8, cost 1 distance 110
10.1.11.0/24          (i) area: 0.0.0.0
     directly attached to interface vlan11, cost 100 distance 110
10.1.12.0/24          (i) area: 0.0.0.0
     directly attached to interface vlan12, cost 100 distance 110
10.254.0.1/32         (i) area: 0.0.0.0
     via 10.254.101.254 interface 1/1/8, cost 10 distance 110
10.254.1.0/24         (i) area: 0.0.0.0
     via 10.254.101.254 interface 1/1/8, cost 110 distance 110
10.254.101.0/24       (i) area: 0.0.0.0
     directly attached to interface 1/1/8, cost 10 distance 110
10.254.102.0/24       (i) area: 0.0.0.0
     via 10.254.101.254 interface 1/1/8, cost 110 distance 110
```

Review the OSPF routes

Based on the new LSAs, sw-agg1 has completed a topology calculation and has discovered new routes. The OSPF process will propose these routes to the IP routing table.

1. On sw-agg1, review the full IP routing table.

```
sw-agg1(config)# show ip route

Displaying ipv4 routes selected for forwarding

Origin Codes: C - connected, S - static, L - local
              R - RIP, B - BGP, O - OSPF
Type Codes:   E - External BGP, I - Internal BGP, V - VPN, EV - EVPN
              IA - OSPF internal area, E1 - OSPF external type 1
              E2 - OSPF external type 2

VRF: default

Prefix         Nexthop          Interface VRF(egress)  Origin/Dis-
tance/   Age
                                                       Type      Metric
--------------------------------------------------------------------
-----------
0.0.0.0/0      10.254.101.254   1/1/8         -        O/E2      [110/1]
00h:00m:42s
10.1.0.2/32    -                loopback0     -        L         [0/0]
-
10.1.1.0/24    -                vlan1         -        C         [0/0]
-
10.1.1.1/32    -                vlan1         -        L         [0/0]
-
10.1.2.0/24    -                vlan2         -        C         [0/0]
-
10.1.2.2/32    -                vlan2         -        L         [0/0]
-
10.1.3.0/24    -                vlan3         -        C         [0/0]
-
10.1.3.1/32    -                vlan3         -        L         [0/0]
-
10.1.11.0/24   -                vlan11        -        C         [0/0]
-
10.1.11.1/32   -                vlan11        -        L         [0/0]
-
10.1.12.0/24   -                vlan12        -        C         [0/0]
-
10.1.12.1/32   -                vlan12        -        L         [0/0]
-
```

```
10.254.0.0/16        10.254.101.254   1/1/8        -    S        [1/0]
00h:35m:07s
10.254.0.1/32        10.254.101.254   1/1/8        -    O      [110/10]
00h:00m:42s
10.254.1.0/24        10.254.101.254   1/1/8        -    O     [110/110]
00h:00m:42s
10.254.101.0/24      -                1/1/8        -    C        [0/0]
-
10.254.101.2/32      -                1/1/8        -    L        [0/0]
-
10.254.102.0/24      10.254.101.254   1/1/8        -    O     [110/110]
00h:00m:42s
Total Route Count : 18
```

- **Question**: How can you identify the OSPF routes in the IP routing table?
- **Answer**: The OSPF routes have an origin code *O*.
- **Question**: What is the administrative distance assigned to the OSPF routes?
- **Answer**: OSPF routes have a distance of 110 by default.
- **Question**: What is the administrative distance assigned to directly connected routes?
- **Answer**: Directly connected routes have a distance of 0 by default.
- **Question**: What is the administrative distance assigned to static routes?
- **Answer**: Static routes have a distance of 1 by default.

2. Now examine the list of OSPF routes in the IP routing table. Then check the output of the IP routes from the OSPF point of view.

```
sw-agg1(config)# show ip ospf routes
Codes: i - Intra-area route, I - Inter-area route
       E1 - External type-1, E2 - External type-2

OSPF Process ID 1 VRF default, Routing Table
---------------------------------------------

Total Number of Routes : 7

0.0.0.0/0            (E2)
     via 10.254.101.254 interface 1/1/8, cost 1 distance 110
10.1.11.0/24         (i) area: 0.0.0.0
     directly attached to interface vlan11, cost 100 distance 110
10.1.12.0/24         (i) area: 0.0.0.0
     directly attached to interface vlan12, cost 100 distance 110
10.254.0.1/32        (i) area: 0.0.0.0
     via 10.254.101.254 interface 1/1/8, cost 10 distance 110
10.254.1.0/24        (i) area: 0.0.0.0
```

```
     via 10.254.101.254 interface 1/1/8, cost 110 distance 110
10.254.101.0/24     (i) area: 0.0.0.0
     directly attached to interface 1/1/8, cost 10 distance 110
10.254.102.0/24     (i) area: 0.0.0.0
     via 10.254.101.254 interface 1/1/8, cost 110 distance 110
```

- **Question**: Is there any difference in the list of routes that were calculated by OSPF and the list of routes in the IP routing table?
- **Answer**: Some routes in the IP routing table, such as the static route for example, have a better administrative distance (1) than the OSPF routes (110). In this case, the static route will make it to the IP routing table instead of the proposed OSPF route.

Most specific matching route

During the previous lab, you have configured a static aggregate route for the 10.254.0.0/16 datacenter subnets.

1. On sw-agg1, now verify which route will be used when trying to reach 10.254.1.0.

```
sw-agg1(config)# show ip route 10.254.1.0
VRF: default
   Prefix                : 10.254.1.0/24                   VRF(egress)
: -
   Nexthop               : 10.254.101.254                  Interface
: 1/1/8
   Origin                : ospf                            Type
: ospf_intra_area
   Distance              : 110                             Metric
: 110
   Age                   : 00h:17m:29s                     Tag
: 0
   Encap Type            : -                               Encap
Details       : -
```

- **Question**: What route will be used for the 10.254.1.0 destination?
- **Answer**: The most matching route for the 10.254.1.0 subnet is now an OSPF route.
- **Question**: Do you still need the static aggregation route that was made in the previous lab activity?
- **Answer**: No, in this lab setup, OSPF advertises the exact routes from the datacenter. These OSPF routes are now used to reach the datacenter subnets.

2. On **sw-agg1**, review the configured static routes in the running configuration.

```
sw-agg1(config)# show running-config | include "ip route"
ip route 10.254.0.0/16 10.254.101.254
```

Note

Make sure to use double quotes to enclose filter strings that contain a *space*, like "ip route" in the previous example.

- **Question**: How many static routes are currently defined on sw-agg1?
- **Answer**: There is one static route for 10.254.0.0/16.
- **Question**: Is there a default static route in the configuration?
- **Answer**: No, it was removed during the "Static routes" lab activity.

3. Remove the aggregate route for the datacenter subnets.

```
sw-agg1(config)# no ip route 10.254.0.0/16 10.254.101.254
```

4. Verify you can still reach **10.254.1.21**.

```
ping 10.254.1.21
```

Default route

The core router uses OSPF to advertise the default route (0.0.0.0/0) to the OSPF neighbors.

1. On sw-agg1, check the IP routing table for the default route (0.0.0.0).

```
sw-agg1(config)# show ip route
Displaying ipv4 routes selected for forwarding
Origin Codes: C - connected, S - static, L - local
              R - RIP, B - BGP, O - OSPF
Type Codes:   E - External BGP, I - Internal BGP, V - VPN, EV - EVPN
              IA - OSPF internal area, E1 - OSPF external type 1
              E2 - OSPF external type 2
VRF: default
Prefix      Nexthop         Interface VRF(egress)  Origin/   Distance/
Age
                                                   Type      Metric
------------------------------------------------------------------
-----------------
0.0.0.0/0 10.254.101.254  1/1/8        -       O/E2      [110/1]
00h:19m:35s
...
```

```
sw-agg1(config)# show ip route 0.0.0.0
VRF: default
    Prefix              : 0.0.0.0/0
VRF(egress)           : -
    Nexthop             : 10.254.101.254                        Interface
: 1/1/8
    Origin              : ospf                                  Type
: ospf_type2_ext
    Distance            : 110                                   Metric
: 1
    Age                 : 00h:20m:35s                           Tag
: 0
    Encap Type          : -                                     Encap
De-tails        : -
```

- **Question**: Is there a default route present in the IP routing table?
- **Answer**: Yes, the default route is present, even when no local static default route is defined in the sw-agg1 configuration.
- **Question**: What is the advantage of a dynamic static route versus a static default route?
- **Answer**: A dynamic default route can still be advertised using alternate paths in case of a link failure. Doing this with manually configuring static routes can be more challenging.

2. Verify sw-agg1 can reach the Internet. Perform a ping to **8.8.8.8**; this should be successful.

```
sw-agg1(config)# ping 8.8.8.8
PING 8.8.8.8 (8.8.8.8) 100(128) bytes of data.
76 bytes from 8.8.8.8: icmp_seq=1 ttl=115 (truncated)
76 bytes from 8.8.8.8: icmp_seq=2 ttl=115 (truncated)
76 bytes from 8.8.8.8: icmp_seq=3 ttl=115 (truncated)
76 bytes from 8.8.8.8: icmp_seq=4 ttl=115 (truncated)
76 bytes from 8.8.8.8: icmp_seq=5 ttl=115 (truncated)

--- 8.8.8.8 ping statistics ---
5 packets transmitted, 5 received, 0% packet loss, time 4004ms
rtt min/avg/max/mdev = 8.593/8.849/9.168/0.188 ms
```

Task 3: Configure passive interface for campus access subnets

After the successful setup of OSPF routing in the network, your customer contacts you regarding a security concern. The IP interfaces (typically the SVIs) that connect clients to the network must be advertised using OSPF; this has been done by enabling the interface for OSPF. The side effect of the previous configuration is that OSPF will start sending hello messages and attempt to find OSPF neighbors on those interfaces as well. This means that any 'rogue' OSPF routers on the client subnets would be able to establish an adjacency with the customer OSPF router and could create problems.

Your senior colleague informs you that it is possible to disable OSPF hellos and the adjacency process on an AOS-CX interface, while keeping the interfaces enabled in the router LSA advertisement to other *networks*. This feature is known as an OSPF *passive* interface. In this task you will configure OSPF passive interfaces.

Objectives

- Configure an IP interface as passive.
- Configure the OSPF process to use passive interfaces by default.
- Enable the IP uplink interface for the OSPF process when passive interface default is active.

Steps

Passive single interface

1. On PC1, start a Wireshark capture of the lab-facing NIC, wait about 10-15 seconds, and stop the capture.
    - **Question**: Are there any OSPF packets seen by the client?
    - **Answer**: Yes, by default, all OSPF-enabled interfaces are active. Any client or rogue router in the client network could attempt to establish an OSPF adjacency with the sw-agg1.
    - **Question**: What is the interval of the hello packets?
    - **Answer**: By default, OSPF hello packets are sent every 10 seconds on a broadcast network type, such as Ethernet.

Note

An example network trace file of OSPF hello packets can also be found in the ACAF Student Files on the PC-MGMT desktop: lab03-04-ospf-hello-pc4.pcapng.

---

4. On sw-agg1, make the SVI 11 interface an OSPF passive interface.

```
sw-agg1(config)# interface vlan 11
sw-agg1(config-if-vlan)# ip ospf passive
sw-agg1(config-if-vlan)# exit
```

2. Review the list of OSPF interfaces and check the *Flags* column for active (A) and passive (P).

```
sw-agg1(config)# show ip ospf interface brief
OSPF Process ID 1 VRF default
===============================

Total Number of Interfaces: 4

Flags: P - Passive  A - Active

Interface      Area         IP Address/Mask   Cost     State
Status   Flags
-------------------------------------------------------------------
----------------------
1/1/8          0.0.0.0      10.254.101.2/24   10       Point-to-point
Up             A
loopback0      0.0.0.0      10.1.0.2/32       0        Loopback
Up             A
vlan11         0.0.0.0       10.1.11.1/24      100      DR-other
Up             P
vlan12         0.0.0.0      10.1.12.1/24      100      DR
Up             A
```

3. On PC1, start the Wireshark trace again, wait about 10-15 second, and then stop the trace.

- **Question**: Are there any OSPF packets in the trace?
- **Answer**: No, the sw-agg1 is no longer sending or processing OSPF packets on the interface SVI 11.

Passive interface default

When the aggregation switch has *many* client-facing IP subnets (SVIs), the configuration can be simplified by making all OSPF interfaces passive, by default. Then, the network administrator only needs to enable OSPF on the uplink ports during the initial OSPF configuration. Any SVIs that are added in the future will automatically be passive, though.

1. On sw-agg1, remove the SVI 11 OSPF passive configuration.

```
sw-agg1(config)# interface vlan 11
sw-agg1(config-if-vlan)# no ip ospf passive
sw-agg1(config-if-vlan)# exit
```

2. On sw-agg1, configure passive interface default for the OSPF process.

```
sw-agg1(config)# router ospf 1
sw-agg1(config-ospf-1)# passive-interface default
sw-agg1(config-ospf-1)# exit
```

3. On sw-agg1, verify with the IP OSPF interface details that all SVI interfaces are now passive.

```
sw-agg1(config)# show ip ospf interface brief
OSPF Process ID 1 VRF default
==============================

Total Number of Interfaces: 4

Flags: P - Passive  A - Active

Interface      Area            IP Address/Mask   Cost   State
Status       Flags
-------------------------------------------------------------------
----------------------
1/1/8          0.0.0.0         10.254.101.2/24   10     Point-to-point
Up           P
loopback0      0.0.0.0         10.1.0.2/32       0      Loopback
Up           P
vlan11         0.0.0.0         10.1.11.1/24      100    DR-other
Up           P
vlan12         0.0.0.0         10.1.12.1/24      100    DR-other
Up           P
```

Note

You may also use Wireshark on the PC again to verify that no more OSPF packets are advertised to SVI11.

4. However, by performing this configuration, *all* OSPF-enabled interfaces on sw-agg1 are passive, and thus adjacency with the core router has been lost. Examine the OSPF neighbor list.

```
sw-agg1(config)# show ip ospf neighbors
No OSPF neighbor found on VRF default.
```

5. Adjust the configuration by making port 1/1/8 OSPF non-passive.

```
sw-agg1(config)# interface 1/1/8
sw-agg1(config-if)# no ip ospf passive
sw-agg1(config-if)# exit
```

6. Verify with the IP OSPF interface details that interface 1/1/8 is now an active OSPF interface.

```
sw-agg1(config)# show ip ospf interface brief
OSPF Process ID 1 VRF default
==============================

Total Number of Interfaces: 4

Flags: P - Passive  A - Active

Interface      Area          IP Address/Mask    Cost    State
Status       Flags
-------------------------------------------------------------------
---------------------
1/1/8          0.0.0.0       10.254.101.2/24    10      Point-to-point
Up           A
loopback0      0.0.0.0       10.1.0.2/32        0       Loopback
Up           P
vlan11         0.0.0.0       10.1.11.1/24       100     DR-other
Up           P
vlan12         0.0.0.0       10.1.12.1/24       100     DR-other
Up           P
```

7. Verify that the OSPF neighbor list contains a FULL adjacency with the core router again.

```
sw-agg1(config)# show ip ospf neighbors
VRF : default                             Process : 1
===================================================

Total Number of Neighbors : 1

Neighbor ID       Priority  State         Nbr Address       Interface
-------------------------------------------------------------------
10.254.0.1        n/a       FULL          10.254.101.254    1/1/8
```

Task 4: Practice OSPF Configuration on sw-agg2

In this task you will add sw-agg2 to the OSPF topology. You will practice the OSPF configuration by applying it on agg2. Note that passive interfaces ensure no OSPF peering is possible on endpoint subnets. This also means that peering is not possible between agg1 and agg2 on any of the endpoint subnets. The uplink should be the only interface that is active for OSPF peering.

If the sw-agg1 or sw-agg2 uplink to the core router fails, they would not have an alternative path to reach the core router. To solve this, a direct IP (routed) link is required between the two aggregation switches. In the lab topology, SVI 2 (VLAN 2) has been configured on the LAG between the aggregation switches for this purpose.

It will be used as the routed link between the aggregation switches and, if an uplink fails, it can be used as an alternate path via the peer aggregation switch to reach the core router.

Complete these configuration steps on sw-agg2:

- Verify loopback 0 with IP 10.1.0.3/32.
- Define OSPF process 1 with RID of the IP address on of loopback 0.
- Activate uplink 1/1/8 with the correct OSPF interface timers and OSPF network type.
- Activate link SVI 2 to agg1 with the adjusted OSPF timers and OSPF network type.
- Ensure all future interfaces are passive by default.
- Verify uplink adjacency with core router and sw-agg1 is *full.*

Adjust the configuration on sw-agg1 as needed to support the above requirements.

Objectives

- Practice configuring OSPF on the two aggregation switches.

Solution

Sw-agg2

```
# sw-agg2 solution

router ospf 1
 router-id 10.1.0.3
 passive-interface default
 area 0.0.0.0

interface 1/1/8
 ip ospf 1 area 0.0.0.0
 no ip ospf passive
 ip ospf hello-interval 1
 ip ospf dead-interval 4
 ip ospf network point-to-point

interface loopback 0
 ip ospf 1 area 0.0.0.0
```

```
interface vlan2
 ip ospf 1 area 0.0.0.0
 ip ospf network point-to-point
 no ip ospf passive
 exit
```

Sw-agg1

```
# sw-agg1 solution

interface vlan2
 ip ospf 1 area 0.0.0.0
 ip ospf network point-to-point
 no ip ospf passive
 exit
```

Validation steps

```
show ip ospf neighbors
show ip ospf interface brief
show ip ospf route
```

1. Verify sw-agg1 has two OSPF peers (sw-agg2 and rtr-core1).
2. Verify sw-agg2 has two OSPF peers (sw-agg1 and rtr-core1).
3. Verify sw-agg2 has OSPF routes with next-hop to both sw-agg1 and rtr-core1.

Troubleshooting steps

In case you would have issues, check your OSPF configuration and the OSPF interface configuration and compare these settings with the peer device (sw-agg1 or rtr-core1).

```
show run ospf
show run interface vlan2
```

Important

Once you have completed and validated the OSPF configuration, save the configuration on *all* the switches with the **write memory** command!

Optional Task 5: Verify routing failover with OSPF

Now that you have completed the OSPF setup, the customer would like to see a demonstration of how OSPF handles a link failure and re-routes traffic in the network. In this task you will explore how OSPF responds to a link failure in the network.

Objectives

- Understand how OSPF handles link failures.

Steps

1. On PC1, open a command prompt (**cmd**) and start a continuous ping to **10.254.1.21**.
2. Start a second command prompt and perform a traceroute to **10.254.1.21**. Take note of the IP addresses in the trace output.
   - Trace route IP list: __________________

     ____

     ____

Tip

On the Windows client, use **tracert -d 10.254.1.21** (**-d** says to not resolve host names) to speed up the traceroute execution.

---

3. On sw-agg1, check the current number of SPF calculations in the **show ip ospf** command output.

```
sw-agg1(config)# show ip ospf
VRF : default                    Process  : 1
----------------------------------------------------

RouterID                : 10.1.0.2      OSPFv2                 : En-
abled
BFD                     : Disabled      SPF Start Interval     : 200
ms
SPF Hold Interval       : 1000  ms      SPF Max Wait Interval : 5000
ms
LSA Start Time          : 5000  ms      LSA Hold Time          : 0
ms
LSA Max Wait Time       : 0     ms      LSA Arrival            : 1000
ms
```

```
External LSAs              : 1          Checksum Sum          : 36917
ECMP                        : 4          Reference Bandwidth    :
100000 Mbps
Area Border                : false      AS Border             : false
GR Status                   : Enabled    GR Interval            : 120
sec
GR State                    : inactive   GR Exit Status         : none
GR Helper                   : Enabled    GR Strict LSA Check    : En-
abled
GR Ignore Lost I/F         : Disabled
Summary address:

Area        Total     Active
------------------------------
Normal       1         1
Stub         0         0
NSSA         0         0
Area   : 0.0.0.0
----------------
Area Type                  : Normal     Status                : Active
Total Interfaces           : 5          Active Interfaces     : 2
Passive Interfaces         : 3          Loopback Interfaces   : 1
SPF Calculation Count      : 20
Area ranges          :
Number of LSAs             : 3          Checksum Sum          : 126962
```

4. On sw-agg1, shutdown the uplink port to the core router (1/1/8).

```
sw-agg1(config)# interface 1/1/8
sw-agg1(config-if)# shutdown
```

5. On PC3, verify that the ping continued.
   - **Question**: Do you notice any difference in the ping output?
   - **Answer**: Yes, the TTL has changed because the path now requires one additional hop (via sw-agg2).
6. In the second command prompt, run the traceroute again. Compare this with the previous output.
   - **Question**: What do you observe?
   - **Answer**: The path takes sw-agg1, then is routed to sw-agg2 (via SVI 2), and finally is routed via the connection to the core router.

7. On sw-agg1, review the SPF count. This should have increased compared to the previous check.

```
sw-agg1(config-if)# show ip ospf
VRF : default                          Process  : 1
----------------------------------------------------

RouterID                : 10.1.0.2           OSPFv2          : Enabled
...
Area  : 0.0.0.0
----------------
Area Type               : Normal          Status                 : Active
Total Interfaces        : 5               Active Interfaces      : 2
Passive Interfaces      : 3               Loopback Interfaces : 1
SPF Calculation Count   : 22
Area ranges          :
Number of LSAs          : 3               Checksum Sum           : 45623
```

8. On sw-agg1, enable the port to the core router again (1/1/8).

```
sw-agg1(config-if)# interface 1/1/8
sw-agg1(config-if)# no shutdown
sw-agg1(config-if)# exit
```

9. Cleanup: On the PC close the command prompt windows.

Optional Task 6: Verify ECMP routing with OSPF

In the network setup, there is one link from the core router to each aggregation switch. For traffic that comes from the datacenter, two paths seem to be available to reach the client subnets that are connected to the aggregation switches. Your customer would like to understand if both links will be used and how this process works.

You learn from your colleague that both links can indeed be used thanks to a system that is known as ECMP. In this task, you will observe the operation of ECMP with regards to OSPF. ECMP is enabled by default. When multiple paths with the same cost exist for a destination network, all the paths will be placed in the routing table. The routing switch will use a load-balancing algorithm to distribute the load over these multiple paths.

Objectives

- Understand the ECMP operation.
- Configure ECMP.

Steps

1. Open an SSH connection to the core router.
2. In the IP routing table, lookup the entries for the 10.1.2.0 network.

```
rtr-core1(config)# show ip route

Displaying ipv4 routes selected for forwarding

Origin Codes: C - connected, S - static, L - local
              R - RIP, B - BGP, O - OSPF
Type Codes:   E - External BGP, I - Internal BGP, V - VPN, EV - EVPN
              IA - OSPF internal area, E1 - OSPF external type 1
              E2 - OSPF external type 2

VRF: default

Prefix              Nexthop            Interface  VRF(egress)
Origin/  Distance/   Age
Metric                                                    Type
---------------------------------------------------------------------------
0.0.0.0/0      10.254.1.253     1/1/9        -        S     [1/0]
01h:14m:34s
10.0.0.0/8     -                blackhole    -        S     [1/0]
01h:14m:34s
10.1.0.0/16    10.254.101.2     1/1/1        -        S     [1/0]
01h:04m:04s
10.1.0.2/32    10.254.101.2     1/1/1        -        O
 [110/100]    00h:01m:38s
10.1.0.3/32    10.254.102.3     1/1/2        -        O
  [110/100]    00h:03m:21s
10.1.2.0/24    10.254.101.2     1/1/1        -        O
 [110/200]    00h:01m:38s
               10.254.102.3     1/1/2        -
[110/200]
00h:01m:38s
10.1.3.0/24    10.254.101.2     1/1/1        -        S     [255/0]
01h:14m:34s
10.1.11.0/24   10.254.101.2     1/1/1        -        O
[110/200]    00h:01m:38s
...

rtr-core1(config)# show ip route 10.1.2.0

VRF: default

   Prefix          : 10.1.2.0/24            VRF(egress)        : -
   Nexthop         : 10.254.101.2           Interface          :
1/1/1
```

```
   Origin              : ospf                    Type              :
ospf_intra_area
   Distance            : 110                     Metric            :
200
   Age                 : 00h:04m:50s             Tag               : 0
   Encap Type          : -                       Encap Details     : -

   Prefix              : 10.1.2.0/24             VRF(egress)       : -
   Nexthop             : 10.254.102.3            Interface         :
1/1/2
   Origin              : ospf                    Type              :
ospf_intra_area
   Distance            : 110                     Metric            :
200
   Age                 : 00h:04m:50s             Tag               : 0
   Encap Type          : -                       Encap Details     : -
```

- **Question**: How many routes are available for the 10.1.2.0/24 network?
- **Answer**: The core router has two ECMP paths for the 10.1.2.0/24 network. The first path via sw-agg1, the second path via sw-agg2.

3. Review the OSPF default maximum paths for ECMP routing.

```
rtr-core1(config)# show ip ospf
VRF : default                         Process  : 1
------------------------------------------------

RouterID              : 10.254.0.1        OSPFv2                 : En-
abled
BFD                   : Disabled          SPF Start Interval     :200
ms
SPF Hold Interval     : 1000  ms          SPF Max Wait Interval :
5000  ms
LSA Start Time        : 5000  ms          LSA Hold Time        : 0
ms
LSA Max Wait Time     : 0     ms          LSA Arrival            :1000
ms
External LSAs         : 1                 Checksum Sum         : 36917
ECMP                     : 4                Reference Bandwidth    :
100000 Mbps
Area Border           : false             AS Border              :true
GR Status             : Enabled           GR Interval            : 120
sec
GR State              : inactive        GR Exit Status           :none
GR Helper             : Enabled         GR Strict LSA Check      : En-
abled
GR Ignore Lost I/F    : Disabled
...
```

4. Adjust the OSPF maximum paths to 1.

```
rtr-core1(config)# router ospf 1
rtr-core1(config-ospf-1)# maximum-paths 1
rtr-core1(config-ospf-1)# exit
```

5. In the IP routing table, examine the route entries for the 10.1.2.0 network.

```
show ip route
```

- **Question**: How many routes are available?
- **Answer**: Only one route is present in the IP routing table.

6. Review the 10.1.2.0 network in the OSPF routing table.

```
rtr-core1(config)# show ip route

Displaying ipv4 routes selected for forwarding

Origin Codes: C - connected, S - static, L - local
              R - RIP, B - BGP, O - OSPF
Type Codes:   E - External BGP, I - Internal BGP, V - VPN, EV - EVPN
              IA - OSPF internal area, E1 - OSPF external type 1
              E2 - OSPF external type 2

VRF: default

Prefix       Nexthop     Interface    VRF(egress)    Origin/ Distance/
Age
                                                     Type    Metric
----------------------------------------------------------------------
-------------------
0.0.0.0/0      10.254.1.253     1/1/9        -       S       [1/0]
01h:20m:01s
10.0.0.0/8     -                blackhole    -       S       [1/0]
01h:20m:01s
10.1.0.0/16    10.254.101.2     1/1/1        -       S       [1/0]
01h:09m:31s
10.1.0.2/32    10.254.101.2     1/1/1        -       O       [110/100]
00h:07m:05s
10.1.0.3/32    10.254.102.3     1/1/2        -       O       [110/100]
00h:08m:48s
10.1.2.0/24    10.254.102.3     1/1/2        -       O       [110/200]
00h:00m:21s
10.1.3.0/24    10.254.101.2     1/1/1        -       S       [255/0]
01h:20m:01s
```

- **Question**: How many routes are available?
- **Answer**: Only one route has been calculated by OSPF as the best route.

7. Restore the OSPF configuration by removing the maximum paths command.

```
rtr-core1(config)# router ospf 1
rtr-core1(config-ospf-1)# no maximum-paths
rtr-core1(config-ospf-1)# exit
```

8. Verify that both routes are available again in the routing table.

```
show ip route
```

Note

If you want to see the default values for any setting or command, you can use the command **show running-configuration all**. When you want to show the default OSPF commands on an interface, you can use **show running-config all | begin 50 1/1/8 | include ospf**. This will show the running-configuration beginning at the line containing **1/1/8** and showing the previous **50** lines. Out of those lines, only the lines that include the string **ospf** will be shown. See the next step for an example.

9. Here is an example output using filtering with the **show running-configuration all** command:

```
rtr-core1(config)# show running-config all | begin 50 1/1/8 | include
ospf
    ip ospf 1 area 0.0.0.0
    ip ospf dead-interval 4
    ip ospf hello-interval 1
    ip ospf retransmit-interval 5
    ip ospf transit-delay 1
    ip ospf priority 1
    ip ospf cost 100
    ip ospf network point-to-point
    ip ospf 1 area 0.0.0.0
    ip ospf dead-interval 4
    ip ospf hello-interval 1
    ip ospf retransmit-interval 5
    ip ospf transit-delay 1
    ip ospf priority 1
    ip ospf cost 100
    ip ospf network point-to-point
```

Learning check

Discuss these questions.

1. Given the IP address 172.16.3.42, and a subnet mask of 255.255.255.0, what is the host portion of the address? (Fill in the blank.) ______________________________

2. Match the TCP/IP stack layer to an appropriate description: *Switching Port*, *Switch Virtual Interface (SVI)*, and *Routing Port.*

   a. Involves IP addresses

   b. Used by multi-layer switches for internal IP routing between VLANs

   c. Assigned as a member of a VLAN

   d. A physical interface with an assigned IP address

   e. Involves MAC addresses

3. Refer to the diagram in Figure 3-55. Server-1's ping packet reaches Server-2. However, Server-2's reply packet does not reach Server-1. Which of the options below is a likely cause of this issue? (Select two.)

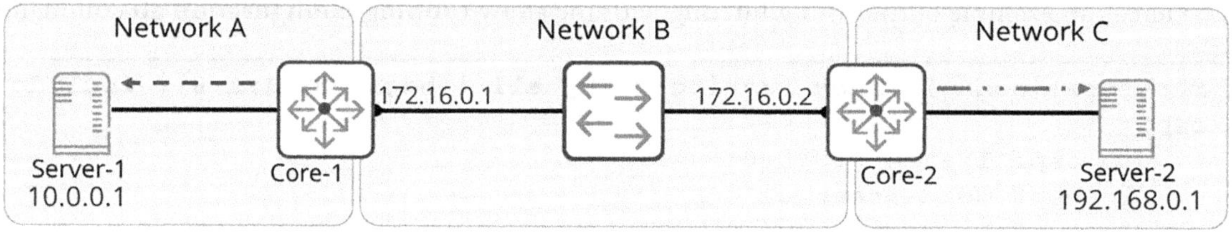

Figure 3-55: Question 3

   a. Server-1's default-gateway is not properly configured.

   b. Server-2's default-gateway is not properly configured.

   c. Core-1 has a route to reach Server-2 but lacks a return path to Server-1.

   d. Core-2 has a route to reach Server-2 but lacks a return path to Server-1.

4. Refer to the diagram in Figure 3-56. On router Core-1, your colleague has entered the command **ip route 10.1.20.0/24 10.1.12.2**. Which of the following options might you use to discuss this situation with your colleague?

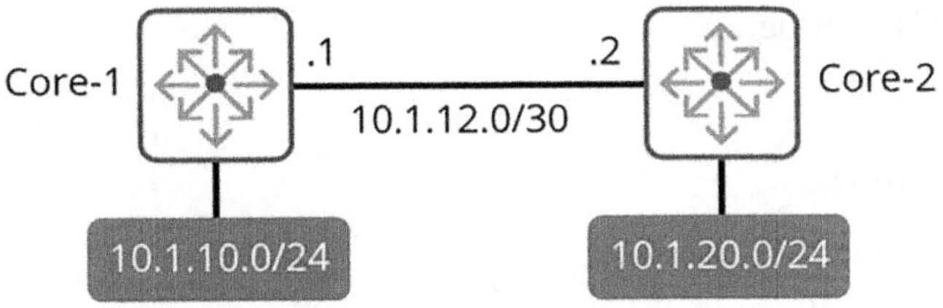

Figure 3-56: Question 4

   a. It appears that Core-1 will be able to route packets to subnet 10.1.20.0, which uses a mask of 255.255.224.0

   b. If devices on subnets 10.1.10.0/24 and 10.1.20.024 cannot communicate, it might be because you specified an incorrect next-hop route.

   c. If devices on subnets 10.1.10.0/24 and 10.1.20.024 cannot communicate, it might be because you need to configure a return path on Core-2.

   d. It appears that host 10.1.12.2 will be able to reach devices on subnet 10.1.10.0/24.

5. Which statements below are true about the Open Shortest Path First (OSPF) routing protocol? (Select two.)

   a. It is more scalable than RIP.

   b. For simplicity, it is best to let the router choose its own router ID (RID).

   c. You can leverage loopback interfaces to enhance OSPF stability.

   d. If you do not configure a router ID (RID), the OSPF router prefers to use the numerically highest IPv4 address assigned to a loopback interface.

6. Match an appropriate description to each of the OSPF entities below (*neighbor table, topology database, routing table*).

   a. A list of directly connected OSPF routers

   b. A list of the best paths to each destination

   c. A list of every possible path in the OSPF area

   d. Is the same for every OSPF router in the area

Summary

Congratulations! You have completed Chapter 3. You should now be able to:

- Discuss inter-VLAN routing.
- Explain DHCP relay.
- Discuss static IP routing.
- Configure a single-area OSPF network.

4 Network Redundancy

OBJECTIVES

Of course, you must maximize the resiliency of your vital corporate network systems. After completing this chapter, you will be able to describe the open standards for spanning tree and VRRP, as well as Aruba's VSX virtual switch feature.

After completing this chapter, you should be able to:

✓ Discuss spanning tree.

✓ Explain VRRP.

✓ Have a basic understanding of VSX.

Spanning tree

This chapter begins with a discussion of one of the first redundancy protocols, the spanning tree protocol.

Redundant networks

Network communications is often mission critical. Layer 2 switch or link failures cannot be tolerated. For example, the left-hand side of Figure 4-1 shows a single path to connect two access switches via Core-1. This design has single points of failure. If Core-1 fails, the network is down: Hosts A and B cannot communicate.

One common way to mitigate this is by adding a redundant Core switch, as in the Figure 4-1's right-hand side illustrates. In this scenario, if Core-1 fails, the network remains connected. Hosts A and B can still communicate over the redundant link using Core-2. While this redundant link mitigates a single point of failure issue, it creates a new challenge. The next three sections cover the main issues of layer 2 redundancy:

- Layer 2 loops
- Broadcast storms
- Multiple frame copies

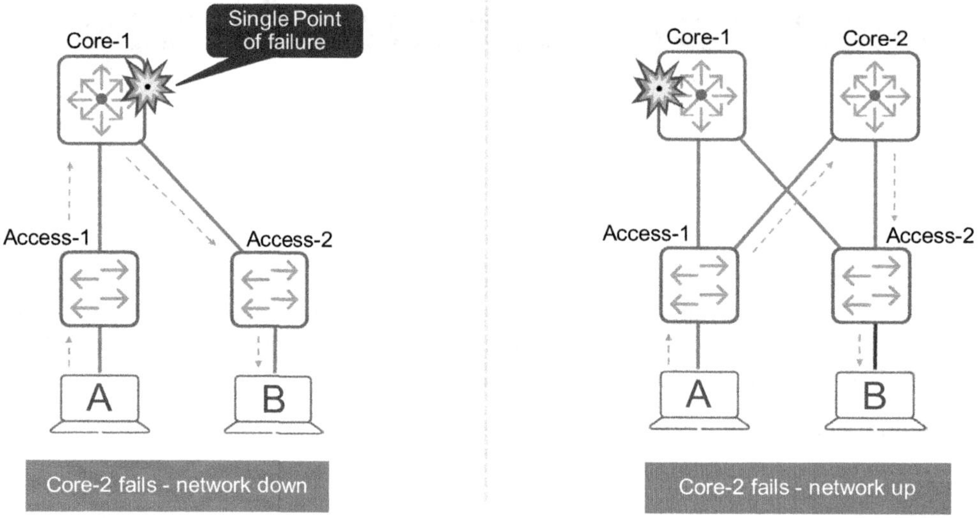

Figure 4-1: Redundant networks

Layer 2 loops

Connecting layer 2 switches with redundant links creates a loop at layer 2. A loop is also created if you connect an Ethernet cable from one port on a switch to another port on that same switch. Figure 4-2 shows two different ways that result in layer 2 loops. If not properly handled, these loops cause serious problems that can effectively disable a network. These loops can cause broadcast storms and multiple frame copies, which can, in turn, cause instability of the MAC address table of the switches and CPU utilization issues.

Figure 4-2: Layer 2 loops

Broadcast storms

You learned in Chapter 3 that switches flood broadcasts, as well as multicasts and unknown destinations, out all interfaces in the same VLAN, except for the ingress port. This flooding nature can cause problems in a redundant, looped network like the one shown in Figure 4-3. Let's look at this example more closely:

1. Suppose that Host A generates a broadcast frame.
2. Switch Access-1 receives this frame, and floods it out other ports in the same VLAN (copy 1).
3. Core-1 and Core-2 receive this and flood it out their connections to each other (copy 2).
4. Core-1, Core-2, and Access-2 receive the second copy and flood them out their other ports (copy 3).
5. Access-1 receives the third copy, and the cycle continues.
6. This frame(s) circles around the network for all of eternity.

And remember that Access-1 flooded this broadcast to Core-2. So, a second copy circles the network in the other direction, forever. And on top of that, each loop that the frame circles, it doubles in number.

As a natural part of network and endpoint operation, nearly all devices send broadcast frames, often many times per minute. Every broadcast from every device circulates around the network forever. Soon, all available bandwidth and CPU cycles are used up processing broadcasts. No resources are available to process normal data communication frames. Eventually, this will consume all your bandwidth and drastically affect the performance of the devices in the network: the network is effectively unusable!

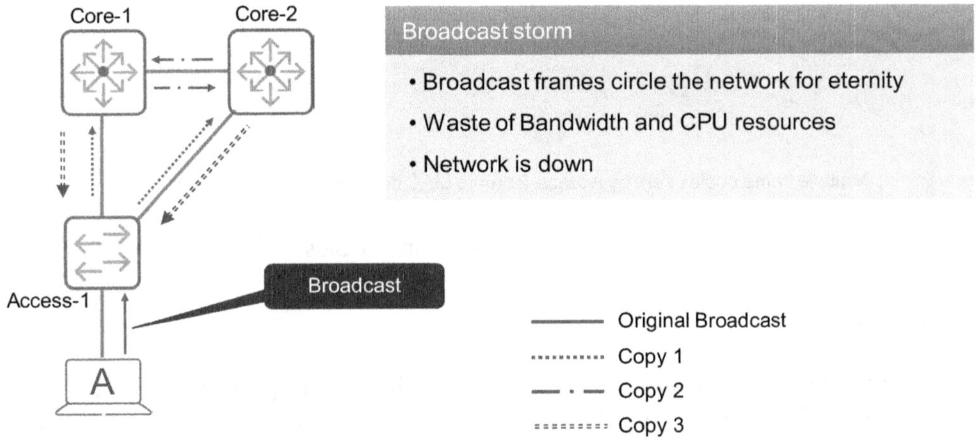

Figure 4-3: Broadcast storms

Multiple frame copies

You may recall that switches not only flood broadcasts out all ports in the VLAN (except the ingress port), but they do the same thing with unicasts to unknown destinations and multicasts. If a switch receives a unicast packet to some destination MAC address, and that address has not yet been learned in the MAC address table, then the switch floods the frame out all ports in the VLAN (except ingress port). These can also circulate and waste bandwidth and CPU cycles. These frames also cause other issues.

Suppose that Host B sends a frame to Host A (see Figure 4-4). Access-2 has not yet learned Host A's MAC address, and so it floods this frame out ports 21 and 22. Thus, Core-1 and Core-2 both switches receive a frame with a source MAC address of 90:...:00, inbound on port 2. *Both* cores send a copy to Access-1 switch, which receives it on ports 21 and 22. Access-1 now believes that it can reach Host B via ports 21 and 22. Access-1 is thus confused: Host B can only exist in one place, but it appears to exist in two places. Basically, Host B's MAC address will constantly switch in the MAC address table from ports 21 to 22 and back...and so on and so forth as Host B's frames arrive on Access-1 from Core-1 and Core-2. The multiple frame copies generated by Access-2 create MAC database instability on Access-1. As you can see, this is a serious problem. You need redundant links for reliability, but this redundancy can also bring your network to its knees.

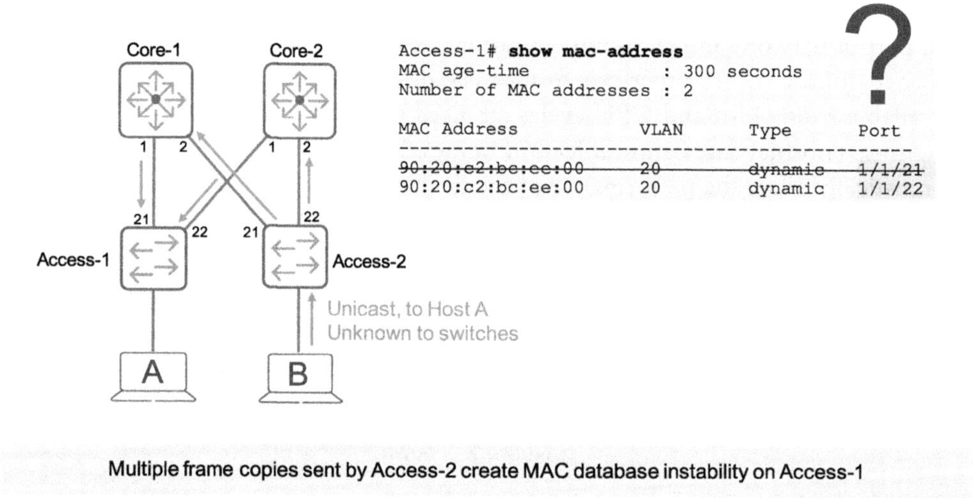

Figure 4-4: Multiple frame copies

Introduction to STP

It is time to learn about the solution to these challenges: the spanning tree protocol (STP). STP was invented in 1984 by the brilliant Miss Radia Perlman, working for DEC at the time. Legend has it that she created the algorithm for this protocol in a few hours. She was so pleased with her creation that she then wrote a poem about it, and then went home for the day. The implementation was known as DEC STP, and is only one of many solutions developed over the years to deal with layer 2 loop issues. The next section focuses on the open-standard implementation of STP.

Operational overview

The IEEE 802.1D standard version of the STP was developed to build and maintain redundant, yet loop-free networks. With STP, you eliminate single points of failure, while avoiding loops and MAC table instability. STP creates a loop-free topology by automatically disabling redundant links.

Figure 4-5 shows a highly redundant network for resiliency and fault tolerance. These redundant links could cause loops. But once STP is engaged, certain redundant links are automatically disabled: there are no loops, no broadcast storms, no MAC address table instability. These disabled links remain available to provide redundancy as needed. In STP, logically the redundant paths are disabled, but physically they are up. Originally, STP referred to the redundant ports (and paths) as being placed in a *blocked* state. This changed in the STP standard and these ports, today, are referred to be in a *discarded* state.

To do this, one switch in the spanning tree domain is elected as a *root bridge* or root switch. Many trees grow out of the ground, from its roots, as a single trunk, and then branch out from there. Likewise, the root bridge is the reference from which the spanning tree grows. All other switches are non-root bridges (switches), sometimes known as designated switches. A loop-free path "grows" from this root switch out to all non-designated (non-root) switches.

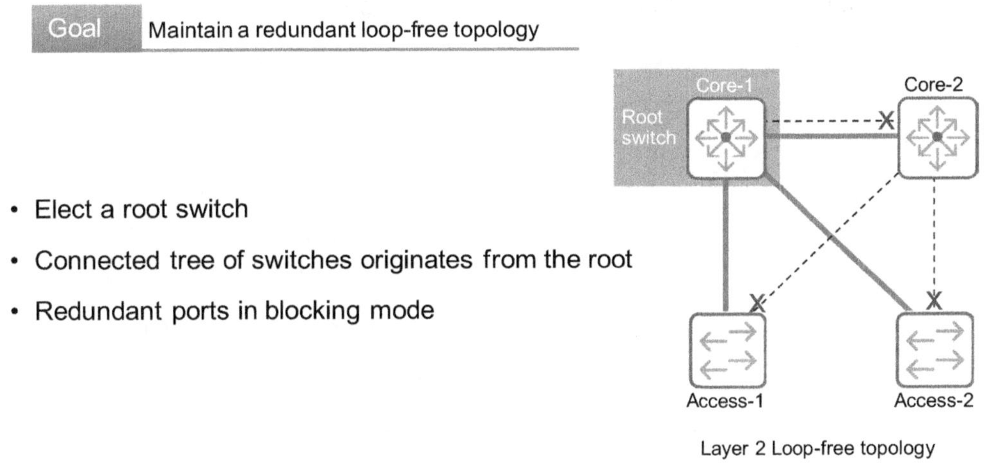

Figure 4-5: STP operational overview

Note

What is now called a *switch* is a very fast, multiple-port version of what used to be called a *bridge* (a few decades ago). When discussing STP, you will often see the term bridge being used. In this context, *bridge* and *switch* can be thought of as the same thing. When you see the term *bridge*, know that the discussion is actually about a switch in today's world. Both the terms root bridge and root switch may be used in this book, as they are considered interchangeable.

Convergence

The STP algorithm should run on all layer 2-connected switches, and STP will logically disable redundant links to avoid loops (see the left side of Figure 4-6). However, if a failure occurs, STP converges on a new topology of active links, which are used to forward frames (see the right side of Figure 4-6). A loop-free topology is thus maintained.

Note

The word *converge* can mean something like "to meet at some point". Convergence, in the context of networking, means that the devices all come to an agreement about a network topology. All devices converge on a new set of active paths to be used for frame forwarding in regards to layer 2 forwarding.

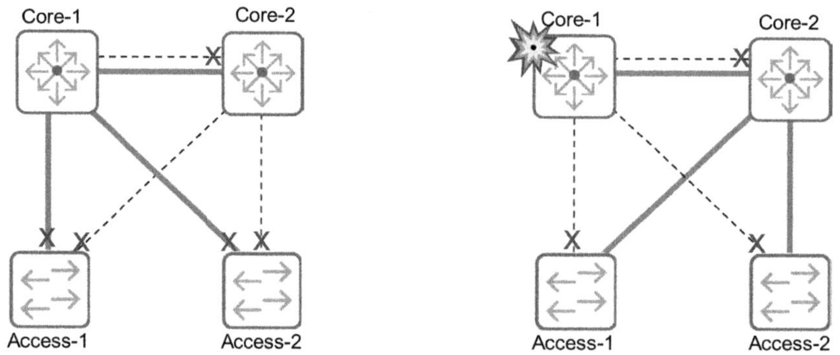

Figure 4-6: STP convergence

Important

STP uses a loop *avoidance* mechanism. In other words, STP attempts to avoid loops; however, typically because of misconfiguration, you could have loops in your network even though STP is enabled.

STP components

This section introduces the components of the STP standard, including the different STP versions, rapid STP (RSTP), bridge identifiers (BIDs), bridge protocol data units (BPDUs), port states, path costs, and port roles.

STP versions

There are three main versions of the IEEE STP protocol. See Figure 4-7 for a quick overview of these protocols.

Failure detection in the original IEEE STP (802.1d) standard was based on timers. With the IEEE 802.1d default configuration, the root switch (and only the root switch) originates a *hello* frame every 2 seconds. These hello frames are forwarded to the rest of the switches in the domain. If some switch downstream (farther away from the root switch) ceases to receive hello frames for 20 seconds (the default max age timer for STP), an outage is assumed, and all switches begin to converge on a new topology. The duration of this process is defined by the forward delay timer, which is 15 seconds by default. This protocol (802.1d) is considered obsolete and its use is no longer recommended in modern networks.

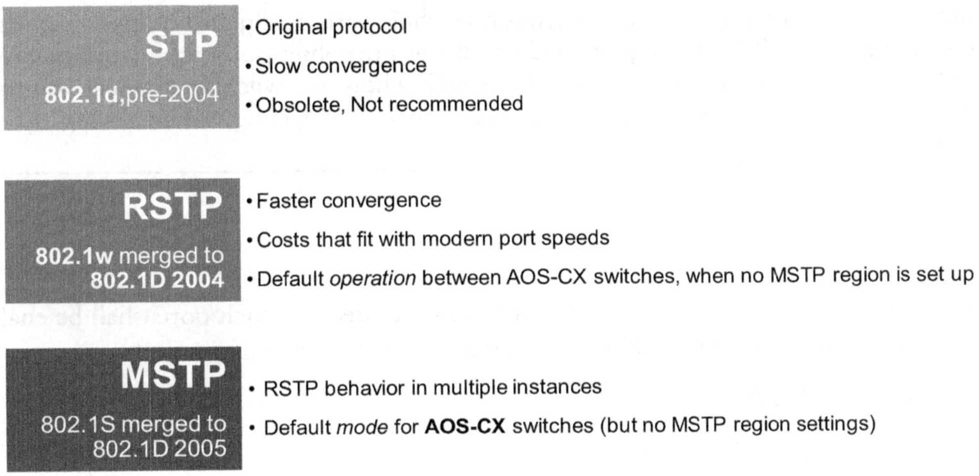

Figure 4-7: STP versions

The rapid spanning tree protocol (RSTP) was developed in 1998 to speed up convergence (IEEE 802.1w). Instead of only the root switch originating hello packets, *all* switches originate them. This means that RSTP now has a true keep-alive mechanism that can respond in seconds (or less). No need for old and slow max age and forward delay timers. You will soon learn the details of this new convergence process.

Note

RSTP is backwards compatible with the original 802.1d standard. However, to maintain this capability, some benefits of RSTP are lost.

IEEE 802.1s or the multiple spanning tree protocol (MSTP) was developed to improve the performance of STP for implementing multiple loop-free topologies or *instances* to load balance the traffic across *all* links. This protocol helps to create optimal paths and increase bandwidth. MSTP runs on AOS-CX switches by default. However, if no setting is configured, then this protocol behaves like RSTP. You will understand why this happens later in this chapter.

Note

802.1d, 802.1w, and 802.1s have all been integrated into the IEEE 802.1D standard. This book focuses on the additions that 802.1w and 802.1s have brought to STP.

Important

With AOS-CX 6200 switches and lower models, STP is globally *enabled* by default; however, on the 6400 and higher switches, STP is *disabled* by default. This makes sense since the 6200 and lower model switches have their ports configured as layer 2 ports, by default, and the 6400 and higher model switches have their ports configured as layer 3 ports, by default. Never make assumptions with whatever switching product you use in your network: always verify what is running on it, if anything, when setting up your network.

RSTP key components

Let's explore key elements that the RSTP algorithm uses to decide which ports shall be enabled for traffic forwarding, and which ones will not (see Figure 4-8). This includes:

- Bridge identifiers (BIDs)
- Bridge protocol data unit (BPDUs)
- RSTP port states
- RSTP port costs
- RSTP port roles

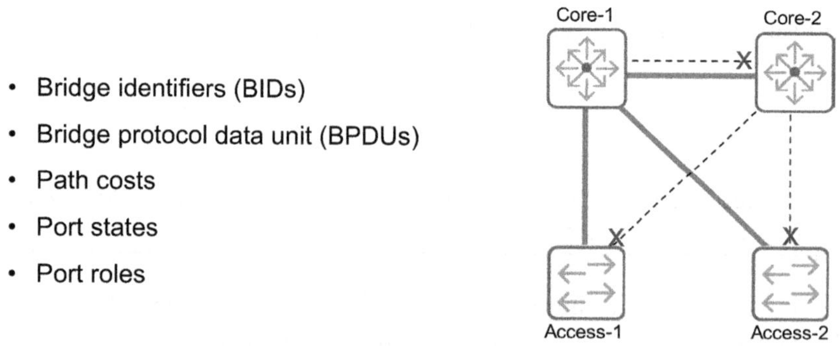

Figure 4-8: RSTP key components

Bridge Identifiers (BIDs)

STP assigns each switch a unique identifier called a bridge ID (BID). This identifier is composed of a 2-byte priority and 6-byte MAC address. The priority defaults to 32,768, so by default, all switches have the same priority. However, each switch has a unique MAC address, and so each BID will be unique. The switch with the lowest *ID* wins the election to be the root switch.

So, which switch would be the root in Figure 4-9? Remember, you have to look at both the combined priority *and* MAC address of each switch, which becomes its BID. In Figure 4-9, the Core-1 switch is elected as the root bridge. Obviously, an administrator changed the priority on Core-1, since on AOS-CX switches, the priority defaults to 32,768.

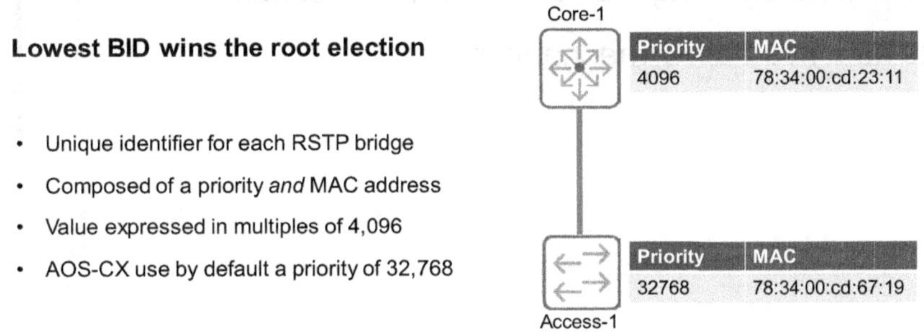

Figure 4-9: BIDs and electing a root bridge

Important
When defining the priority on AOS-CX switches, the number you define is *multiplied* by 4,096, resulting in its priority. For example, if you set a value of 2, the resulting priority would be: 2 x 4,096 = 8,192. Best practice is to configure one core switch with the lowest priority, like a multiplier of 0, and the other core switch with the second lowest priority, like a multiplier of 1. Then leave the access switches as their default (32,768). This allows you to understand what the resulting STP topology will be, help you tune it for your traffic patterns, and greatly assist you in troubleshooting issues.

Bridge protocol data units (BPDUs)

All switch participants in the STP algorithm exchange control messages called bridge protocol data units (BPDU), sometimes referred to as *hello* frames. In the original 802.1d standard, BPDUs were generated only by the root switch, and then had to *trickle down* to other switches. This led to the need for the long max age and forward delay timers, which *greatly* slowed convergence.

With RSTP, all switches originate BPDUs with their current information, every 2 seconds (the default hello-time period (see Figure 4-10). Thus, if a port stops receiving BPDUs for three consecutive hello timers, the switch quickly knows that it has lost connectivity to its neighbor. It ages out the protocol information and begins to converge.

Because each switch originates BPDUs, this becomes a true keep-alive mechanism. Failure detection will take no longer than 6 seconds; and if you're using point-to-point links, detecting link changes, like a link failure, is almost instantaneous.

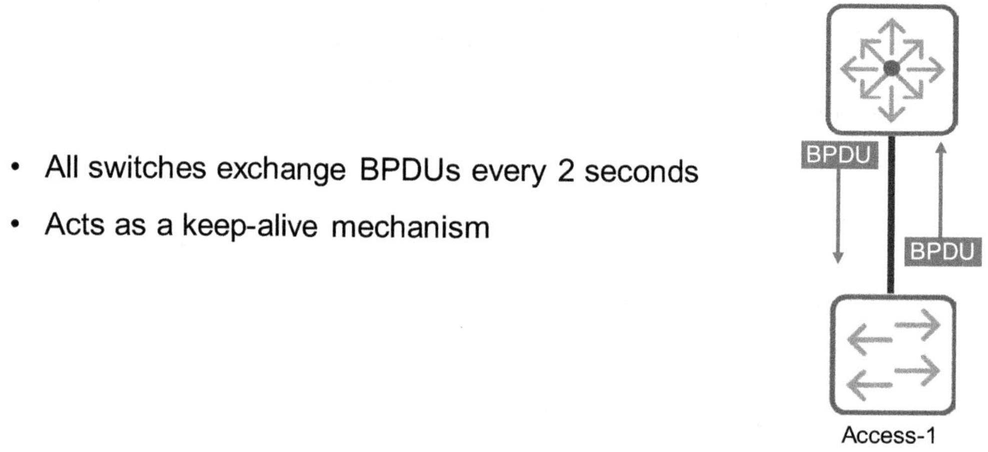

Figure 4-10: BPDUs

Port states

During the original tree establishment and any ensuing convergence, switches must decide which ports will forward data and which ports must be disabled (logically speaking) to prevent layer 2 loops. STP uses port states to transit from a blocking/disabled port state to a *forwarding* port state. The table in Figure 4-11 summarizes the port states and their specific task. You can see the states that were used on 802.1D as compared to RSTP 8021.1w port states.

STP (802.1D) port state	RSTP (802.1w) port state	Port included in active topology	Port learns MAC addresses
Disabled	Discarding	No	No
Blocking			
Listening			
Learning	Learning	Yes	Yes
Forwarding	Forwarding		

Figure 4-11: STP port states

To transition from blocking to forwarding, the original 802.1d standard takes at least 30 seconds: 15 seconds are spent in a listening state, and 15 in the learning state. This assumes that there is a change in the topology and not a failed root bridge. One of the reasons RSTP is more efficient than 802.1d is because the port transitions much more quickly from discarding to forwarding.

Note

The blocking and listening states are deprecated in RSTP.

Note

In RSTP the learning state is transitory and is only used during re-convergence upon a topology change. Discarding or forwarding states are stable. Also, a discarding port is not physically disabled. It can exchange STP, LLDP, and CDP BPDUs, but blocks almost all other traffic, by default.

Path costs

RSTP may have several possible paths to get from a non-root switch to some root switch. The non-root switch chooses the best path based on accumulated cost; and cost is based on an inverse of the link speed (faster ports have lower costs). Figure 4-12 shows the AOS-CX default port costs.

Link speed	Default port cost in AOS-CX
100 Mbps	200,000
1Gbps	20,000
10 Gbps	2,000
100 Gbps	200
1 Tbps	20

Direct Path	Indirect Path
0 + 20,000 = **20,000**	0 + 2,000 + 2,000 = **4,000**

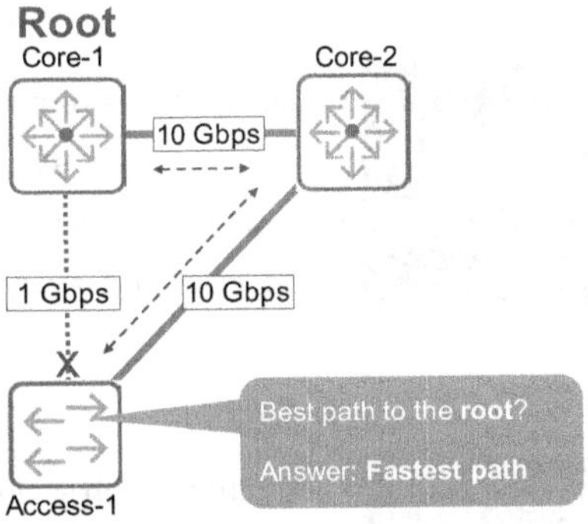

BPDUs advertise the path cost

Figure 4-12: STP path costs

Consider the example in Figure 4-12. Intuitively, you might think the Access-1's best path to the root switch's is the direct path; however, this is a 1 Gbps link, which has a cost of 20,000.

The indirect path from Access-1, via Core-2, to the root switch includes two 10 Gbps links, which only have a cost of 2,000 each. 2,000 + 2,000 = 4,000. Therefore, since this cost is less than 20,000, it becomes the preferred path. (Strictly speaking the root switch will add its cost to itself in this case the cost is equal to 0; in other words, for the root to reach itself, it has a cost of 0).

Thus, Access-1 prefers this path to reach the root (since it is the best and least-cost path), and any other redundant path(s) is/are disabled to avoid loops. As you can see in Figure 4-12, Access-1's left uplink is left in a discarding state while its right uplink is in a forwarding state.

Port roles

Switches that use 802.1w RSTP do not have a complete view of the topology. They build and maintain the loop-free topology by exchanging BPDUs, which indicate how close they are to the root switch. The BPDUs help switches to calculate the correct port role for each of their ports.

Designated and root ports

All ports on the root switch are *designated* ports (see Figure 4-13). The root switch is like the *boss* of the domain. It does not typically block its ports: only non-designated ports must worry about that. Looking at the non-root switches, what is the difference between the root port (RP) and designated port (DP)? They both forward frames. The RP is simply the port that the local switch uses to the root switch: the best path toward the root switch. The DP is designated to accept traffic from downstream root ports. Based on this topology, ports marked as DP or RP will forward traffic and other ports (non-designated) are placed in a discarding state.

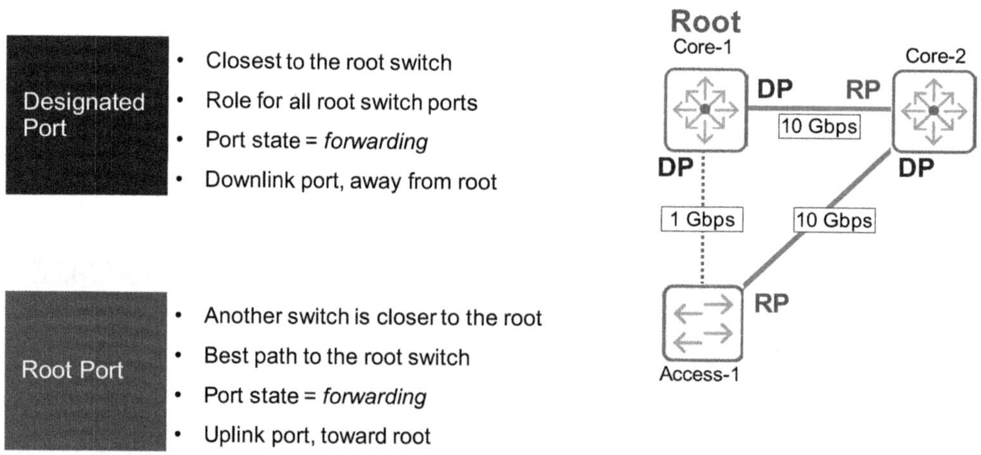

Figure 4-13: Port roles

Alternate and backup ports

To speed up convergence, RSTP introduces two new port states (see Figure 4-14):

- *Alternate* ports
- *Backup* ports

The following are true concerning alternate ports:

- The port is not the closest to the root on this link.
- This link does not offer the switch its best path to the root.
- The port is not connected to the same switch and has no loop to itself.
- This port becomes the root port if the active RP on itself fails or is no longer valid.

As an example, consider the following situation in Figure 4-14. Access-1 port 1/1/21 can become an alternate port since it fulfills the requisites to it.

The following are true concerning *backup* ports:

- The port is not the closest to the root in this link.
- This link does not offer the switch its best path to the root.
- The port is connected to the same switch. The switch has a looped connection to itself or it has a loop created by a layer 1 hub.
- The backup port becomes a designated port if the existing designated port fails.

A loop could be caused if a layer 1 hub is placed on the topology. As an example, take into consideration the network in Figure 4-14. In this situation ports 1/1/26 and 1/1/27 on Access-1 are in a loop situation because a hub or an unmanaged switch (no STP support) was introduced to the topology. If RTSP was running on Access-1, one of these ports is marked as the DP port and the other a backup port. Likewise, since neither of these ports have a path to the root, they will not become an alternate port.

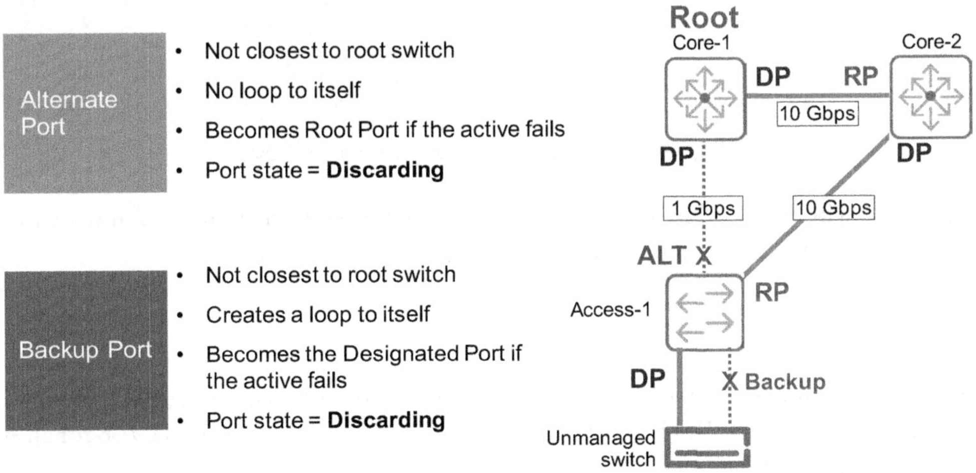

Figure 4-14: Alternate and backup ports

Note

Introducing layer 1 hubs or unmanaged switches into the topology should be avoided.

STP operation and building a loop-free topology

This section pulls together all that you learned and shows you how the entire process works. The RSTP algorithm converges on a loop-free topology by performing the following steps (see Figure 4-15):

- Elect the root switch.
- Select the root port (RP) on all non-root switches or non-designated bridges.
- Select the designated port (DP) on each switch-to-switch links.
- All remaining ports are placed in a discarded state.

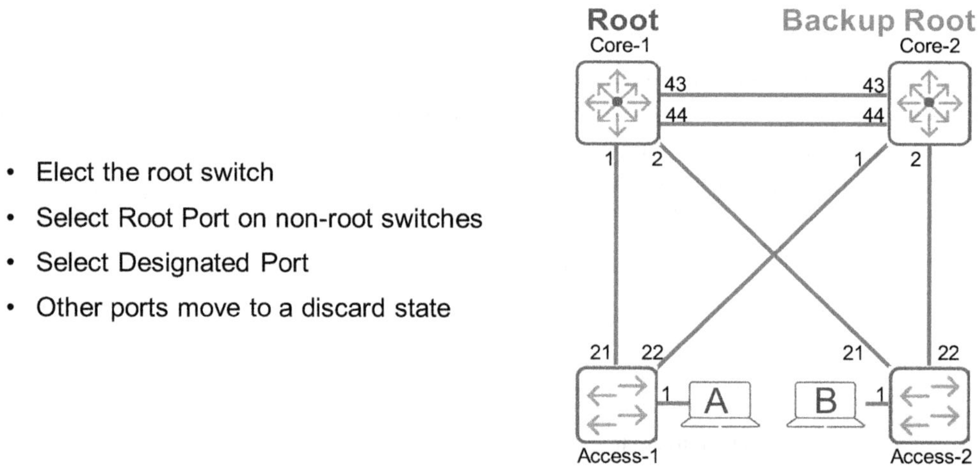

Figure 4-15: Building a loop-free topology

RP and DP ports will be placed in a forwarding state while other ports (alternate and backup) move to a discarding state to prevent loops. Let's look at this process in more depth.

Root switch election

For both the 802.1d and 802.1w standards, root switch election is based on the BID: the lowest value wins. Recall that the two-part BID is composed of a priority, which defaults to 32,768 for all switches, and a globally unique MAC address.

You can rely on this default, rather random behavior, for the election process, but it is not recommended. Some small, low-powered switch on the edge of your network might win the election. This makes for a less stable tree, with sub-optimal paths and poor resiliency. Consider the topology shown in Figure 4-16. If the priority is at their defaults, then Access-2 would become the root switch, since it would have the lowest BID: Since the priorities are all the same, it's the switch with the lowest MAC address.

- Based on BID
- The lowest BID wins
- Priority value helps define the root bridge
- AOS-CX default priority is 32,768

Device	Priority	MAC
Core-1	**4096**	**00:...:04**
Core-2	8192	00:...:03
Access-1	32768	00:...:02
Access-2	32768	00:...:01

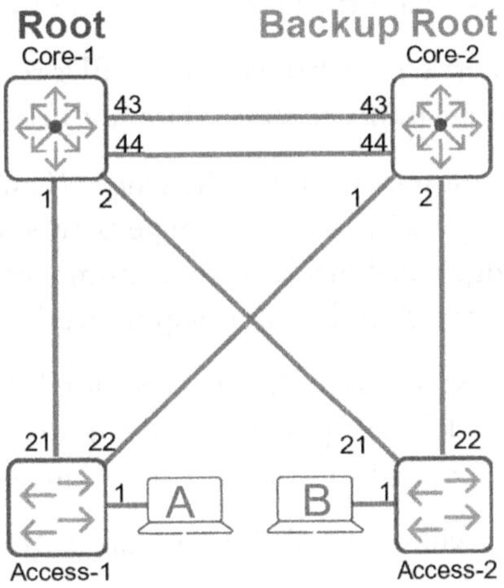

Figure 4-16: Root switch election

You want to ensure that one of the more reliable, more centrally-located core switches wins the election. You have a more robust, resilient, and optimally pathed tree structure. To do this, you simply lower the priority value of the preferred switches, to something lower than the 32,768 (the default). In this example, Core-1 becomes the root switch simply by setting its priority to 4,096 (multiplier of 1). However, if Core-1 fails, then Access-2 would become the root switch, which would not be optimal. Therefore, it is important that Core-2 becomes the second-best option. In this case Core-2 is defined with a priority of 8,192 (multiplier of 2). Recall that the priority value must be configured in increments of 4,096 via a configured multiplier value.

Tip

You might wonder why Core-1 wasn't assigned a multiplier of 0 and Core-1 a value of 1. Many network engineers prefer not to use 0 as a multiplier so that when troubleshooting STP issues, it is easy to force a particular switch to become the root by using a multiplier of 0. If two switches have a multiplier of 0, then the tie breaker is the one with the lowest MAC address, which might not be the switch you want to become root.

Non-root switches: Selecting root ports

Non-root switches or non-designated bridges must select the best path to the root switch, by selecting its root port: the port connected to this best path. Root port selection criteria are shown in

Figure 4-17. For example, Core-2 will receive BPDUs on ports 1, 2, 43, and 44. Core-2 must decide which port is the best one:

- Lowest path cost to the root bridge. Assuming that all links are the same speed then, indirect paths on ports 1 and 2 are discarded simply because the cost is higher than direct paths. Ports 43 and 44 have the same cost; the next criteria must be used.
- Lowest sender Bridge ID. BDPUs received on ports 43 and 44 are generated by the same device (Core-1); this criterion does not help to break the tie.
- Lowest port priority. The sender (Core-1) will include a port priority on the advertisement. The port priority comes in range between 0-240 and in AOS-CX the default value is set to 128; a bridge will consider a lower number the winner. In this case assuming that default values are set this criterion does not help to break the tie.
- Lowest port ID. The final decision is taken on the port ID, where the lowest the value the winner. Core-1 lowest port number is 43, so the port on Core-2 that is connected to this port becomes the root port (in this case it is Core-2's port 43).

Access-1 will receive BPDUs on ports 21 and 22. Access-1 must decide which port is the best one:

- Lowest root bridge ID. All devices agree that the root bridge is Core-1; this criterion does not help to select the best path.
- Lowest path cost to the root bridge. Assuming that all links are the same speed then, indirect paths on port 22 are discarded simply because the cost is higher than the direct path. Port 21 becomes the root port.

And Access-2 will receive BPDUs on ports 21 and 22. Access-2 must decide which port is the best one. A similar decision will take place as with Access-1: port 21 the direct path offers a best path so it will become the root port. RPs will be placed in a forwarding state.

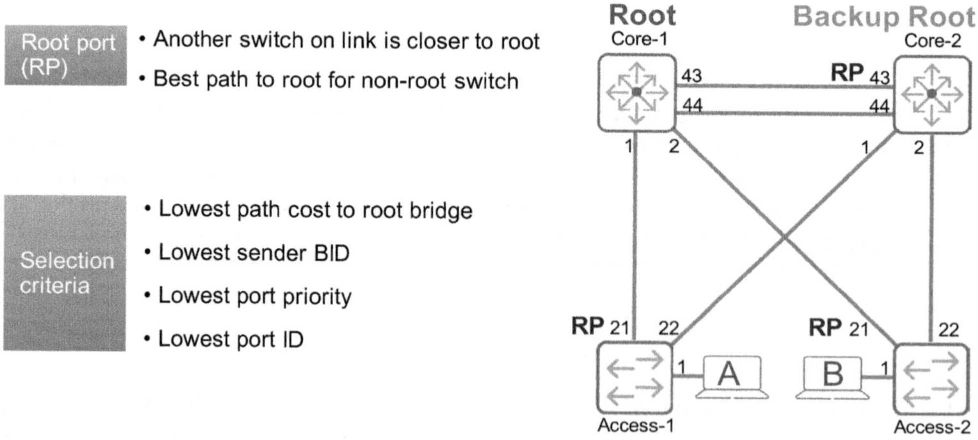

Figure 4-17: RP selection

Selecting designated and alternate ports

The criteria to select a designated port (DP) is the same as root port.

- Lowest root bridge ID.
- Lowest path cost to the root bridge.
- Lowest sender bridge ID.
- Lowest port priority.
- Lowest port ID.

The following statements are a shortcut to determine quickly the designated port on a link.

- All ports in the root switch are always designated ports.
- In a switch-to-switch link where previously a root port was elected on one side, the other side must always be a designated port.

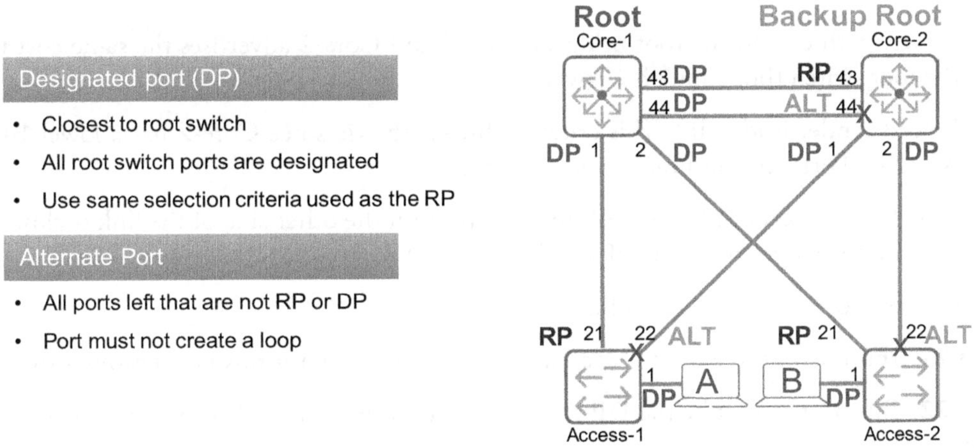

Figure 4-18: DP and alternate port selection

Examine Figure 4-18. Core-1 is the *root* switch and its ports will always be the closest to the root bridge; therefore ports 1, 2, 43, and 44 on Core-1 become *designated* ports.

On Core-2, let's focus on ports 1, 2, and 44.

- **Port 1**. Core-2 will evaluate if this port is closest to the root; this process is done by comparing the BPDU received from Access-1 on this port and the one that it sends out. This process follows the same rules as the root port selection, so let's start the analysis:
    - Lowest root bridge ID. All devices in the topology agree that Core-1 is the root; this criterion does not break the tie.
    - Lowest path cost to the root bridge. Access-1 and Core-2 advertises the same cost to reach the root switch (both are 1 hop away).
    - Lowest sender bridge ID. This criterion breaks the tie since Core-2 has a lower BID than Access-1. Therefore, port 1 in this link becomes the DP.
- **Port 2**. Core-2 will evaluate if this port is closest to the root; this process is done by comparing the BPDU received from Access-2 on this port and the one that it sends out.
    - Lowest root bridge ID. All devices in the topology agree that Core-1 is the root; this criterion does not break the tie.
    - Lowest path cost to the root bridge. Access-2 and Core-2 advertises the same cost to reach the root switch (both are 1 hop away).
    - Lowest sender bridge ID. This criterion breaks the tie since Core-2 has a lower BID than Access-2. Port 2 in this link becomes the DP.
- **Port 44**. This port cannot be the DP since the port on the other side of the link is closest to the root (the root itself provides the DP for the segment).

For Access-1 let's focus on ports 21 and 22:

- **Port 21**. This port is the only RSTP speaker in the link; therefor this port becomes the DP.
- **Port 22**. This port cannot be the DP since the port on the other side of the link is closest to the root.

For Access-2 let's focus on ports 21 and 22:

- **Port 21**. This port is the only RSTP speaker in the link; this port becomes the DP.
- **Port 22**. This port cannot be the DP since the port on the other side of the link is closest to the root.

In this example, the topology does not include any layer 1 hub which implies that there will not exist a backup port, but only alternate ports. Simply all ports that were not elected as DP or RP ports will become alternate ports and their state will be marked as discarding. In Figure 4-18 you can see that Access-1's port 22, Access-2's port 22, and Core-2's port 44 become *alternate* (ALT) ports.

STP and topology changes

In RSTP, a topology change occurs when non-edge ports move to a different state, or when BPDUs are no longer received. A switch detects this as topology change. The switch actively informs other switches of the topology change (TC) (see Figure 4-19). It sets the TC bit in BPDUs and transmits them. It flushes the MAC address table entries associated with all non-edge ports.

Other switches receive a BPDU with the TC bit set from a neighbor. The switch clears the MAC address table entries learned on all its ports, except the ingress port for the TC BPDU. These switches in turn send BPDUs with TC set (see Figure 4-20).

Note
In the original 802.1d standard, any switch could send TC BPDUs to notify other switches of a change, but the instruction to clear the MAC address table had to always come from the root switch. This two-step process took more time to complete.

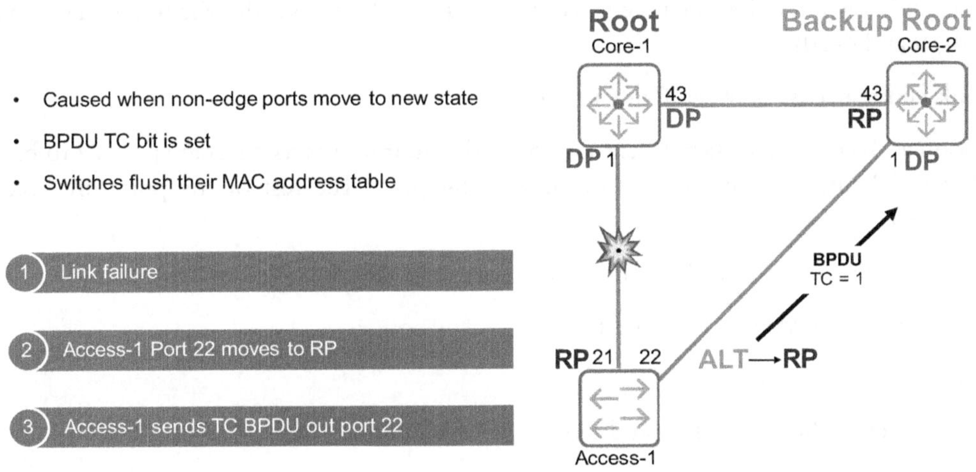

Figure 4-19: STP and topology changes (part 1)

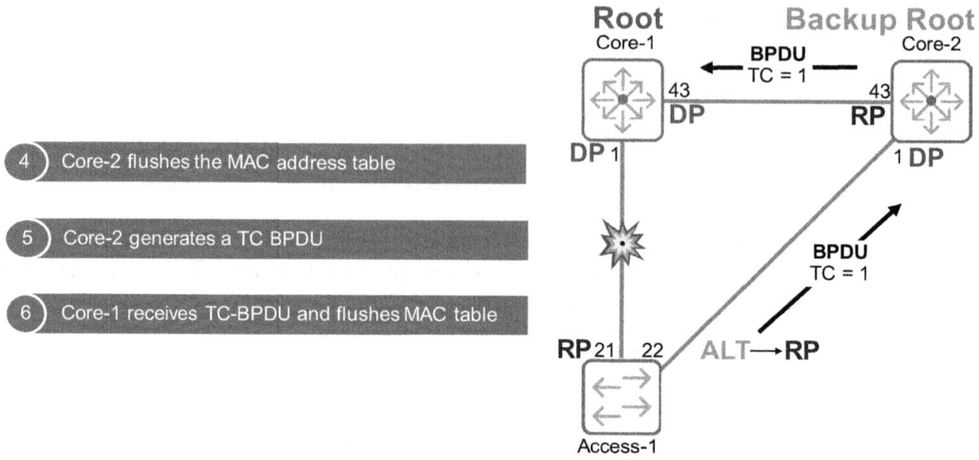

Figure 4-20: STP and topology changes (part 2)

To understand how a topology change is handled by Rapid STP consider the following scenario (see Figures 4-19 and 4-20):

1. A link failure occurs between Core-1 and Access-1.
2. As soon as Access-1 detects that its root port is down, it moves its alternate port 22 to be the new RP. Port 22 immediately moves to a forwarding state. The rest of the ports on Access-1 are unaffected.
3. Access-1 generates a TC BPDU and sends it out port 22.
4. Core-2 flushes its MAC address tables.
5. Core-2 sends a TC-BPDU.
6. Core-1 receives the TC-BPDU and flushes its MAC address table.

Note

Core-1 also detects the link failure and does the same process as Access-1, generating a TC and advertising it to its neighbor, Core-2.

Note

Edge port link state changes will not trigger a TCN.

Edge ports

Edge ports connect to an endpoint (like a PC, AP, printer, server, and etcetera), and so they do not expect to receive BPDUs. They need not participate in the STP algorithm because they cannot cause a loop: they are *leaf on the spanning tree*, so to speak. Since they cannot cause loops, they can quickly transition to a forwarding state, with no intermediate steps (see Figure 4-21).

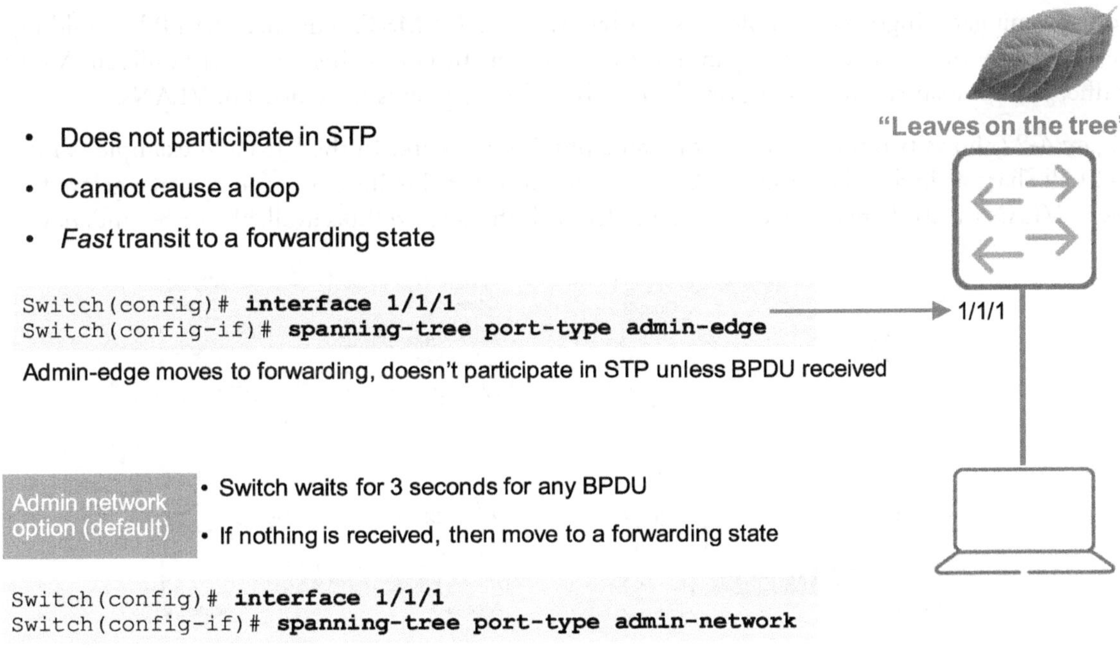

Figure 4-21: Edge ports

If BPDUs are received on an edge port with AOS-CX switches, then the port will act as a normal STP port and participate in the algorithm to prevent layer 2 loops.

There are two options for edge ports: **admin-edge** and **admin-network; admin-network** is the default. You can manually configure edge ports, as shown here.

```
Switch(config)# interface 1/1/1
Switch(config-if)# spanning-tree port-type admin-edge
```

admin-edge hardwires the port as an edge port. An alternative to the **admin-edge** option is the AOS-CX **admin-network** option. With this option, the port looks for BPDUs for the first 3 seconds after the link is up. If no BPDUs are received, the port becomes an edge port and immediately forwards frames. If BPDUs are detected, the port becomes as a non-edge port. In other words, it participates in normal STP operations.

Multiple STP (MSTP): 802.1s

This section introduces you to another part of today's 802.1D standard, multiple STP (MSTP), which was introduced originally in the 802.1s standard.

MSTP introduction

You can mitigate single STP topology issues by using MSTP. MSTP enhances RSTP by building a new topology or instance of the spanning tree protocol. Instances then can carry different VLAN traffic: you get load-sharing and optimal paths for different groups (instances) of VLANs.

Figure 4-22 shows two topologies on the same physical network. In the left-hand example, VLANs 1-20 all share a single STP instance. Core-1 is the root for this instance, and the tree grows from there. VLANs 1-20 share these forwarding links, with the unused links available for redundancy.

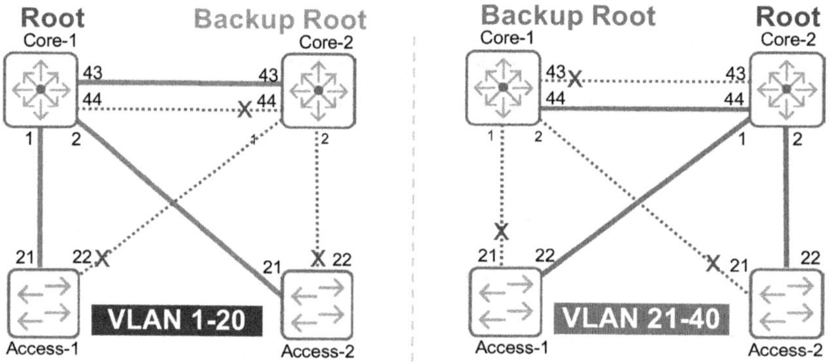

Figure 4-22: MSTP introduction

On the right, VLANs 21-40 share a different STP instance. Core-2 is the root, and the topology emanates from that root. The solid links are shared by VLANs 21-40, and the other links are available for backup. The dormant links for VLANS 1-20 are the active links for VLANs 21-40, and vice versa. So not only does this topology provide redundancy, it also allows you to actively use all the uplinks. Basically, the number of uplinks that you have determines, minimally, the number of instances that you'll need for your VLANs.

Common spanning tree (CST)

Common spanning tree (CST) defines one instance of spanning tree for the entire network, regardless of the number of VLANs. This is the original scenario that you just explored in the discussion of RSTP, which only has *one* instance. All BPDUs use untagged in the native VLAN (VLAN 1, by default), and so VLANs are not considered (see Figure 4-23).

There is no load balancing, so you get the sub-optimal pathing, poor link utilization, and potentially slower performance. You will now learn about these issues and how to resolve them with MSTP. So, what is the importance of CST? It is used for compatibility with other implementations of STP, like switches that only support RSTP or other MSTP switches that are misconfigured. When implementing MSTP, instance 0 is used to represent the CST topology and the native or the untagged VLAN on a trunk port *must* be a member of this instance.

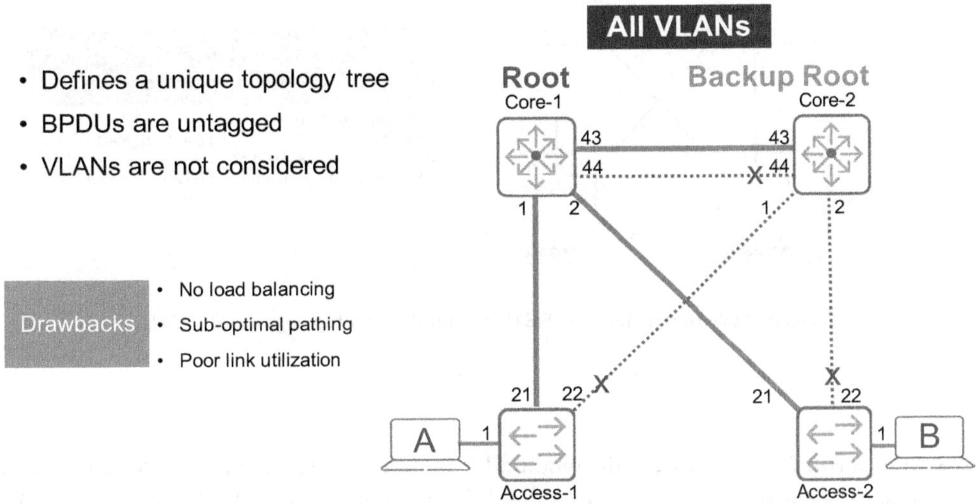

Figure 4-23: Common spanning tree

MSTP regions

MSTP introduces the concept of a *region*, which is a group of switches that have been defined with the same MSTP configuration. The items that must match between the switches in the region include the following (see Figure 4-24):

- Region name
- Region revision number
- VLAN-to-instance mappings

The above information is hashed and is sent in a switch's BPDUs. Neighboring MSTP switches create the hash and compare their hash (digest) with the incoming hash to determine if they match: i.e., they are in the same region, which is mapped or associated with a group of VLANs.

Note

A hash algorithm is a mathematical function that is applied to an input (x). A hash function is interesting because from the output (y) you cannot derive or infer what the input was. Thus, it is referred to as a one-way function. Another characteristic of a hash is that if the same input (x) has been entered, the result will be always the same. MSTP takes advantage of this characteristic.

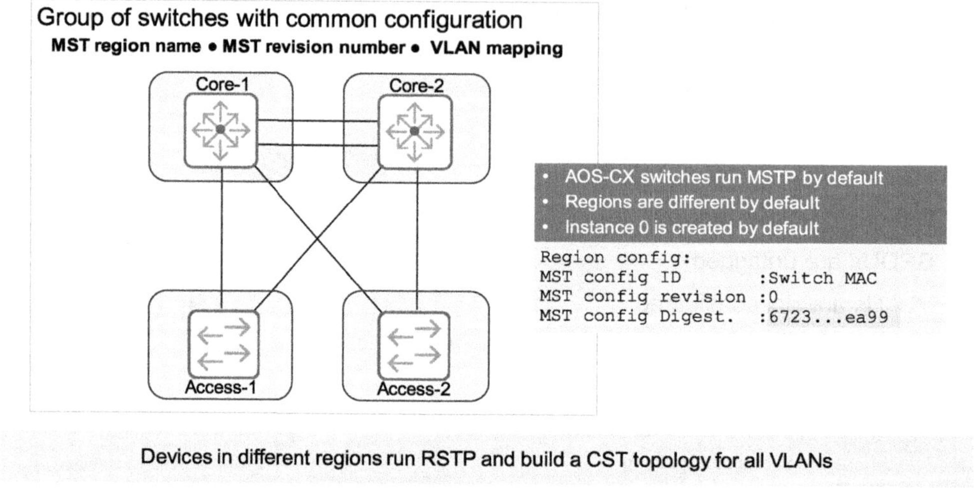

Figure 4-24: MSTP regions

AOS-CX switches run MSTP by default, when STP is enabled. At the factory default settings, each switch has the same revision number (0) and all VLANs are in instance 0. However, the region name (config ID) defaults to the switch's *system MAC address*. Thus, each AOS-CX switch operates in its own region since each switch has a different system MAC address. When this happens, the topology behaves like CST, creating a single topology or instance common for all VLANs. RSTP then exchanges messages between the individual devices, as you have learned in previous sections of this chapter.

Instance 0 or the internal spanning tree (IST) is used as both: a regular instance in MST and the creation of the CST in a multi-region deployment for backward compatibility with RSTP speakers; for this reason instance 0 is known as CIST (common internal spanning tree).

MSTP configuration

In Figure 4-25, notice that instances 1 and 2 are created. Along with the default instance 0, you have a total of 3 instances. The common configuration creates a single region where you can create multiple instances for optimal pathing and load balancing. Instance 1 is mapped VLANs 2-20 and uses Core-1 as the root switch. Instance 2 is mapped VLANs 21-40 and uses Core-2 as the root switch. Instance 0 has all other VLANs, including the untagged (native VLAN, which is 1 in this example).

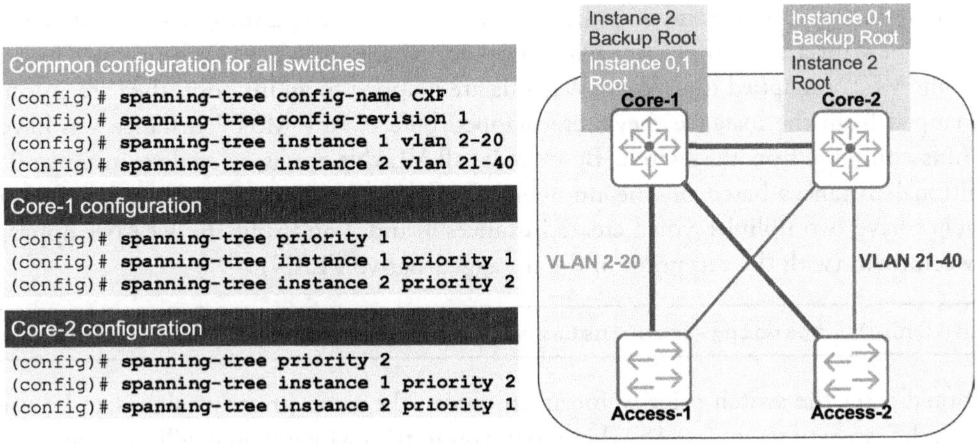

Figure 4-25: MSTP configuration

In AOS-CX, use the following steps to configure MSTP:

1. Configure the STP mode as MSTP. The command **spanning-tree mode** command sets the mode to either MSTP or Rapid per VLAN spanning tree (RPVST).

```
Switch(config)# spanning-tree mode mstp
```

Note

To change the spanning-tree mode, you must first disable STP with the command **no spanning-tree** command in the global context. RPVST is beyond the scope of this book.

2. Set up the MST region name. The **spanning-tree config-name** command sets the configuration name for the MST region in which the switch resides. All switches within an MST region must have identical configuration names, which is case *sensitive.*

```
Switch(config)# spanning-tree config-name <config-name>
```

3. Set up the configuration revision number. Use the revision number to differentiate between region configurations. This is valuable, for example, when changing configuration settings within a region, and you want to track the configuration version you use. It is also valuable when you create a new region from a subset of switches in a current region and you want to maintain the same region name. All switches within an MST region must have identical revision numbers.

```
Switch(config)# spanning-tree config-revision <revision-number>
```

4. Set up the MSTP instance and map the VLAN IDs to it. The **spanning-tree instance** command creates a new instance with VLAN mapped to an existing instance. Each instance must have at least one VLAN mapped to it. When VLANs are mapped to an instance, they are automatically unmapped from the instance they were mapped before. Any MSTP instance can have all the VLANs configured on the switch. By default, all VLANs are associated to instance 0. Create additional instances based on the number of unlinks you have. For example, if your access switches have two uplinks, you'd create instances 1 and 2 and split the VLANs, appropriately, between them (with the exception of the untagged/native VLAN).

```
Switch(config)# spanning-tree instance <instance-id> vlan <vlan-ids>
```

5. Optionally, set the switch priority for an instance. The same priority values can be used as in RSTP; the range to set up is 0-15. This allows you to turn which switch will become the root and backup root for the instance.

```
Switch(config)# spanning-tree instance <instance-id> priority <priority-
value>
```

6. And on the 6400 series of switches and higher, STP is disabled by default. To enable it, use the **spanning-tree** command.

```
Switch(config)# spanning-tree
```

Other protection features

This section covers two additional security features that most enterprise switching products support, but that are not part of the STP protocol standard itself. These two features operate within the AOS-CX switch and don't share information with other switches. How a vendor implements them is vendor-independent. These features are BPDU guard (sometimes referred to as BPDU protection) and loop protection, though other vendors or products might use a different name for them.

BDPU protection

BPDU protection (sometimes referred to as BPDU guard) is used to detect switches that are not supposed to be part of the network. Examine Figure 4-26. Notice that at the access layer, someone connected an unauthorized switch to Access-1. If this switch had STP enabled, it would participate in the STP topology, which would be undesirable (see Figure 4-26):

- It might create a security issue
- It might create an undesirable STP topology

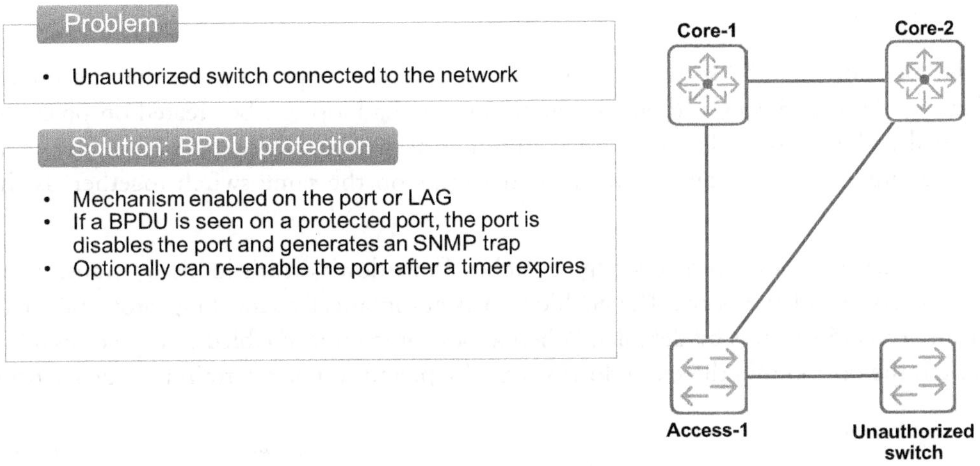

Figure 4-26: BPDU protection

To solve the issue the AOS-CX switches support a feature called *BPDU protection*, which is designed to protect the active STP topology by preventing spoofed or incidental BPDU frames from entering the STP domain. In a typical implementation, BPDU protection would be applied to edge ports connected to end user devices that do not run STP. If STP BPDU frames are received on a protected port, this feature disables that port and alerts the network manager using an SNMP trap.

BPDU protection is disabled by default and is enabled on the edge layer 2 or LAG interfaces:

```
switch(config)# interface <interface-ID>
switch(config-if)# spanning-tree bpdu-guard
switch(config-if)# exit
switch(config)# spanning-tree bpdu-guard timeout <seconds>
```

By default, the port is disabled when a BPDU is received on the port and an SNMP trap is generated. Optionally you can specify a timeout for the offense, where the port is re-enabled automatically after the timer expires; however, if another BPDU is seen on the port, it would be disabled again. Note that the timeout is defined globally for all ports.

Note

BPDU protection, sometimes referred to as BPDU guard, can only detect rogue switches that generate BPDU frames. If the rogue switch doesn't generate these frames, then a different security measure(s) should be used to detect its existence. Best practice, though, is to *disable* layer 2 ports that are not in use to prevent this kind of issue.

Loop protection

So far, the focus of the STP section has been on preventing loops on ports that are intentionally redundant–by design. However, there are situations when a loop can be created on ports that were not intended to be connected. This most commonly occurs on switch ports that face users: edge ports. For example, a user connects two Ethernet jacks on the same switch together, as shown in Figure 4-27.

The loop protection feature protects edge switches from downstream loops by transmitting loop protocol packets out of the ports. The enabled ports automatically send loop protection packets at specified intervals (5 seconds by default). When a loop-protection-enabled port receives a loop protocol packet, the port knows there is a downstream loop, and so takes certain remedial actions.

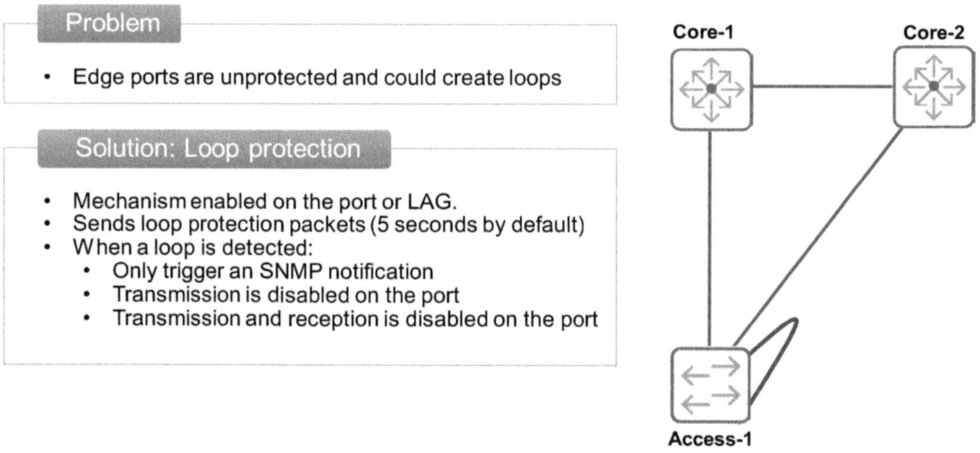

Figure 4-27: Loop protection

The actions supported on AOS-CX are:

- Do not disable. Only a notification is triggered and reported using SNMP.
- Transmission disabled. The port that transmitted the loop detection packet is disabled. This is the default action.
- Transmission and reception disabled.

In AOS-CX this feature can be enabled on a layer 2 interface or LAG. The AOS-CX configuration looks like this:

```
switch(config)# interface <interface-ID>
switch(config-if)# loop-protect
switch(config-if)# loop-protect action {do-not-disable | tx-disabled |
tx-rx-disabled}
```

VRRP

The second redundancy solution you'll cover in this chapter is the open standard Virtual Router Redundancy Protocol (VRRP).

Need for default gateway redundancy

In the previous chapter, you learned about the benefits of a default gateway (DG) for endpoints. Recall that an endpoint may only have one DG, and a single DG means a single point of failure. In the example in Figure 4-28, if Core-1 fails, PC-1 and all other hosts using it for the DG are now isolated. What can you do?

You could add another DG for redundancy. This seems like an easy solution...but maybe not. Each endpoint's DG is either configured manually or obtained via DHCP. When Core-1 fails, you must either manually reconfigure each host with a new DG, or reconfigure the DHCP scope, and then ask all end users to disconnect from the network and reconnect, or power cycle their PC, or teach them how to trigger a DHCP release and renew action. (In a Windows command prompt for example, use **ipconfig /release**, then **ipconfig /renew**). Whatever method you use, it is not very elegant or scalable, and is disruptive for end users. There must be a better way, right?

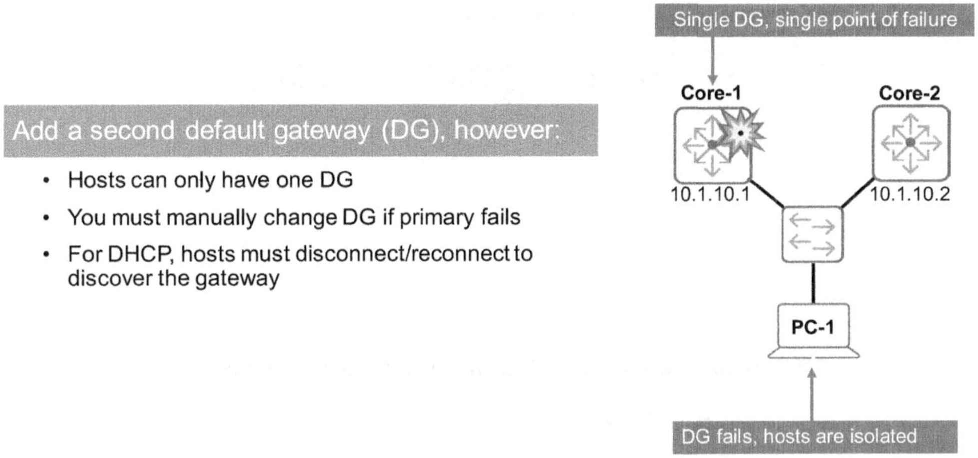

Figure 4-28: Need for default gateway redundancy

First-hop redundancy protocols

The real solution for this challenge can be resolved from the network side, instead from the endpoint side, by using some kind of first hop routing protocol (FHRP). This adds resiliency for endpoints by using a coordinated gateway solution: no change to endpoint IP configuration, no DHCP modifications, and no end user disruption. It is automatic!

Note

The term resiliency refers to the ability of the network to adapt to changes and failures.

An FHRP solution creates a single coordinated gateway from two or more physical routers (see Figure 4-29). The two physical routers present themselves to endpoints as a single device, with a single virtual IP (VIP) address. This VIP acts as the endpoint's DG. Normally, the primary routing device serves the DG role, forwarding traffic for endpoints. The secondary router monitors the primary router state. If the primary fails, the secondary device takes over: it takes on the primary role and VIP and forwards endpoint traffic for the VLAN. From the endpoint perspective the virtual IP address is always available. There is no DG address change, and there is no disruption for end users.

Note

There have been many FHRP solutions introduced to the industry over the years; however, the only standard one is VRRP, on which this book focuses.

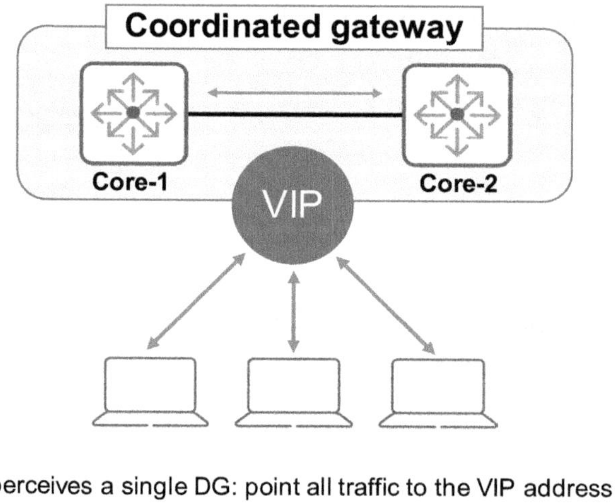

Figure 4-29: First hop redundancy protocols

VRRP components and operation

RFC 5798 defines the Virtual Router Redundancy Protocol (VRRP): a standards-based FHRP that enables two or more routing devices to provide gateway redundancy for endpoints in a VLAN/subnet (see Figure 4-30). VRRP uses an active-standby architecture. Only one gateway actively forwards traffic sent to the VIP address. This primary forwarding device is called the active, while the non-forwarding device is the standby.

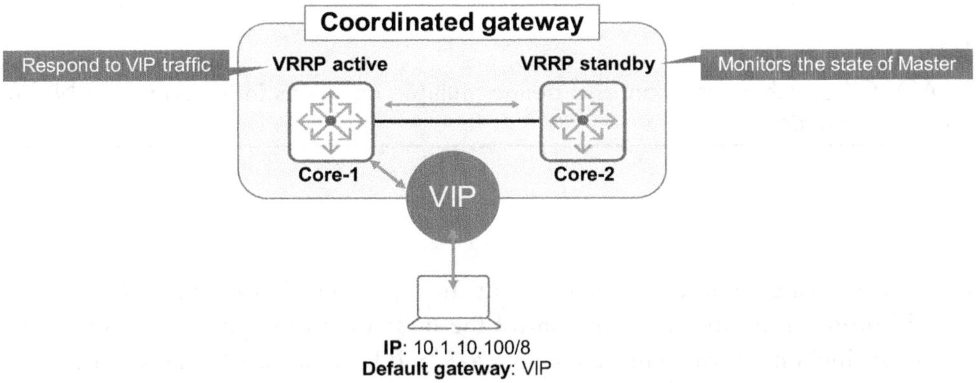

Figure 4-30: VRRP components and operation

VRRP instances

AOS-CX allows you to deploy multiple instances of VRRP, often to balance the load for VLANs, as shown in Figure 4-31. Each instance has a unique virtual router ID (VRID) number, as shown in the figure. VRRP instance 1 serves VLAN 10, while instance 2 serves VLAN 20.

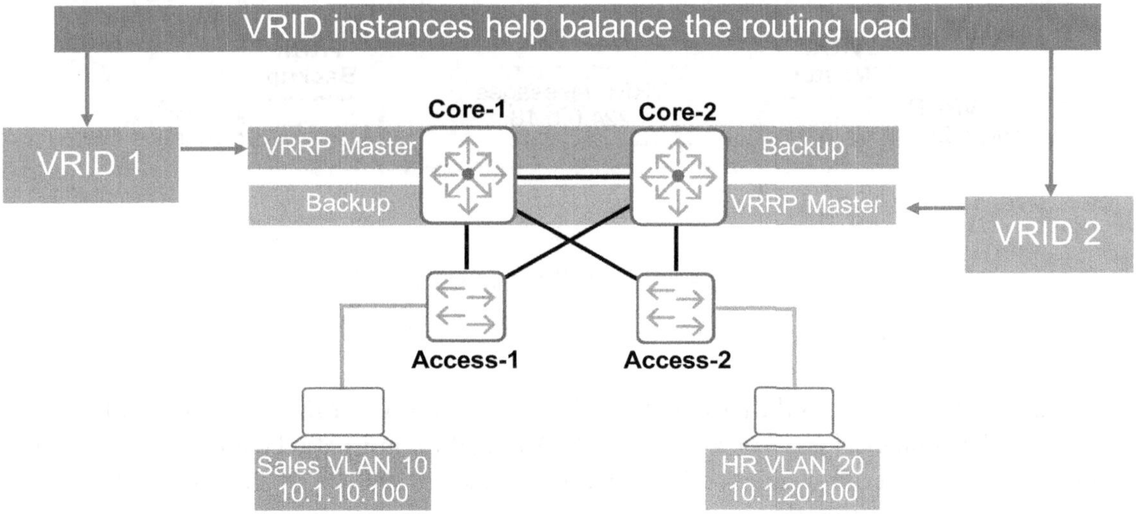

Figure 4-31: VRRP instances

Switch Core-1 is the master or active forwarder for VLAN 10, with Core-2 as the standby. Meanwhile, Core-2 is the master for VLAN 20, with Core-1 as the standby. This gives you a nice load balancing capability.

Note

On AOS-CX switches, you can also define multiple instances in the *same* VLAN, but this is uncommonly done.

master election

VRRP members exchange multicast advertisement messages to elect the master DG, using address 224.0.0.18, IP protocol number 112. To control the master election, you set a priority value from 1–255, where the highest priority wins (see Figure 4-32). If both devices have the same priority, then the gateway with the highest IP address in the VLAN wins the election. In AOS-CX the default priority value is 100.

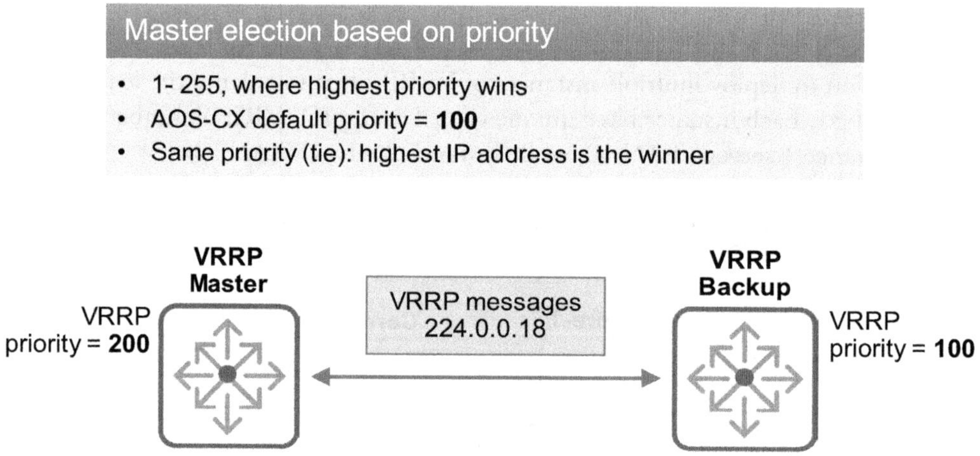

Figure 4-32: VRRP master election

Virtual IP (VIP) address

The virtual IP (VIP) address is the result of the DG coordination. You assign a unique, *real* IP address to each individual physical gateway, as normal. The VIP address must also be unique, but an *unused* IP address in the subnet. In Figure 4-33, the master is assigned 10.0.10.1, 10.0.10.2 to the backup, and 10.0.10.3 is used as the VIP.

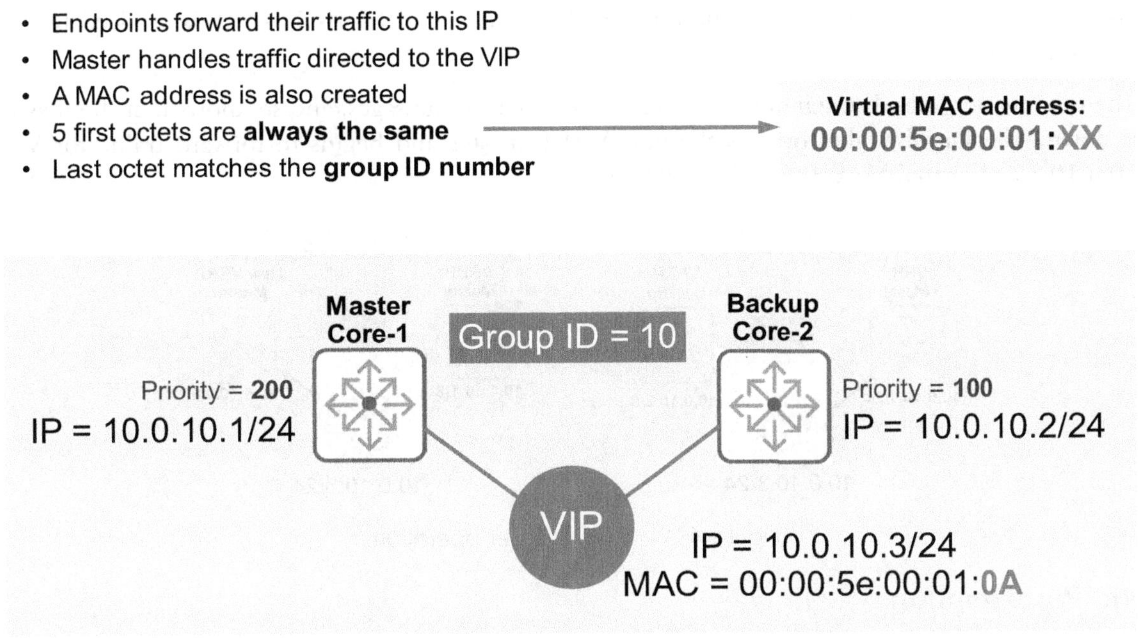

Figure 4-33: VIP address

Each VRRP instance has a unique VRRP address. 10.0.10.3 might be the VIP address for instance 10, used for VLAN 10. Meanwhile, 10.0.20.3 might be the VIP address for instance 20, used for VLAN 20. Note that the actual VLAN numbers and the VRRP instance or group numbers do not have to match, but it is a common practice by administrators to match them.

You must ensure that this VIP address is configured as the endpoint DG IP address. Thus, endpoints forward their traffic to the VIP. The VRRP master receives these packets and routes them. If the master fails, the standby router takes over. The edge devices do not learn about the routers' physical IP addresses, 10.0.10.1 and 10.0.10.2, in the example in Figure 4-33: they only need to know about 10.0.10.3, the VIP address.

A virtual MAC address (vMAC) is also automatically assigned to the VIP. As defined in the standard, this address is 00:00:5e:00:01:*xx*, where *xx* is the VRID group number, converted from decimal to hexadecimal. In Figure 4-31, the VRID is 10, and so the vMAC is 00:00:5e:00:01:0a (hex 0a is 10 in decimal). Thus, when endpoints ARP for the DG address of 10.0.10.3, they will learn the vMAC address, and add it to their ARP table. This simplifies operations, since in case there is a switchover from master to standby, the edge devices can still use the *same* IP and MAC address for the default gateway.

VRRP failover operations

Figure 4-34 shows a scenario where two Core switches run VRRP. Core-1 has a higher priority, and so is elected as the VRRP master. Core-2 is the VRRP standby, constantly monitoring Core-1 status via a keepalive mechanism. Endpoints forward their traffic to 10.0.10.3, which is serviced by the VRRP master.

Then the master fails. The standby stops receiving keepalive messages, and so knows that the master is down. The standby takes over as the new VRRP master and begins to forward traffic for VIP 10.0.10.3 and virtual MAC 00:00:5e:00:01:0a.

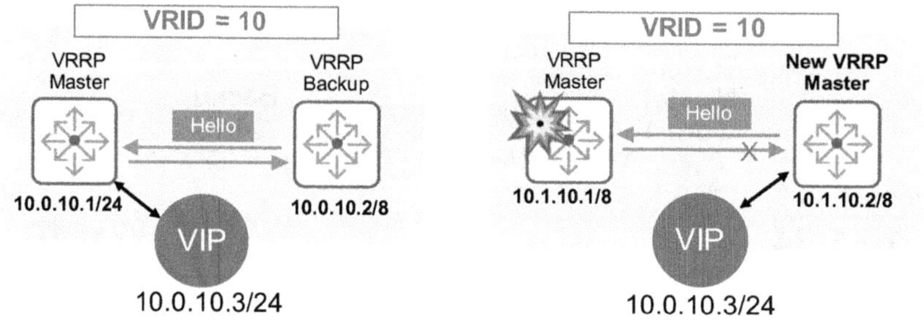

Figure 4-34: VRRP failover operations

VRRP preemption

If you continue to review Figure 4-34 and the failover operation, you can note that Core-1 failed, and so Core-2 took over as the new master. What happens when Core-1 comes back online? This depends on if you configure VRRP preemption.

If preemption is enabled (see the left side of Figure 4-35), then Core-1 will reassume its original master role and Core-2 reverts back to its standby role. This is the AOS-CX default setting. It is convenient because you know that under normal operating conditions, when all devices are up, the same router always acts as the master. This can be especially important if you are using multiple VRRP instances. If a master fails, the remaining router will carry the load for all endpoints. However, once the master comes back online, it resumes its original operational mode and processes traffic sent to the virtual addresses.

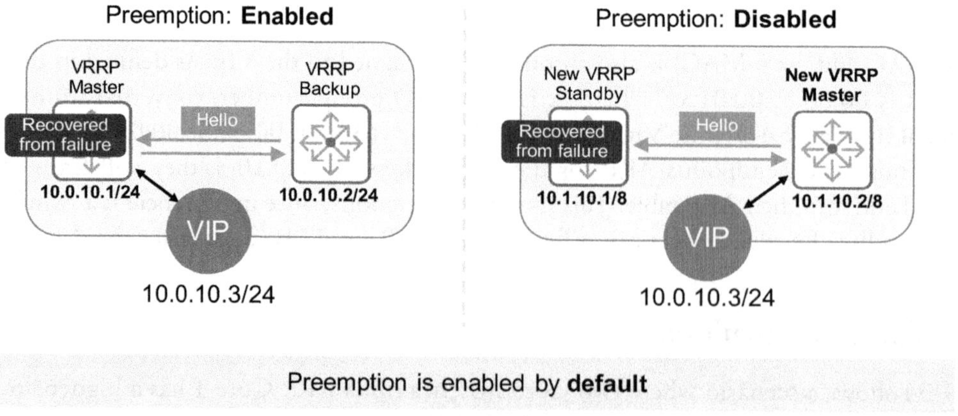

Figure 4-35: VRRP preemption

If preemption is disabled (see the right side of Figure 4-35), then Core-1 will not resume its original master role when it comes back online. Core-2 remains the master, and Core-1 takes on the standby role. You must manually disable preemption with the command **no preempt** if you want to use this mode. With preemption disabled, you lose the benefits described in the previous paragraph. Some administrators might choose to disable preemption if they are worried about the (very brief) time lag that might occur during the preemption process: while the routers switch back over to new operational states, potentially during the middle of a busy day. Due to the high-performance nature of AOS-CX switches, this is rarely a concern.

VRRP configuration

You can configure VRRP using commands presented in Figure 4-36. Globally you enable VRRP with the **router vrrp enable** command. Then you configure VRRP on each routed interface where you need DG redundancy. One requirement is that the VRRP group number (11, in the example in Figure 4-34) must match. If they don't match, then the resulting virtual MAC address will be different and will cause problems when a failover occurs, since edge devices only know about the original master vMAC address.

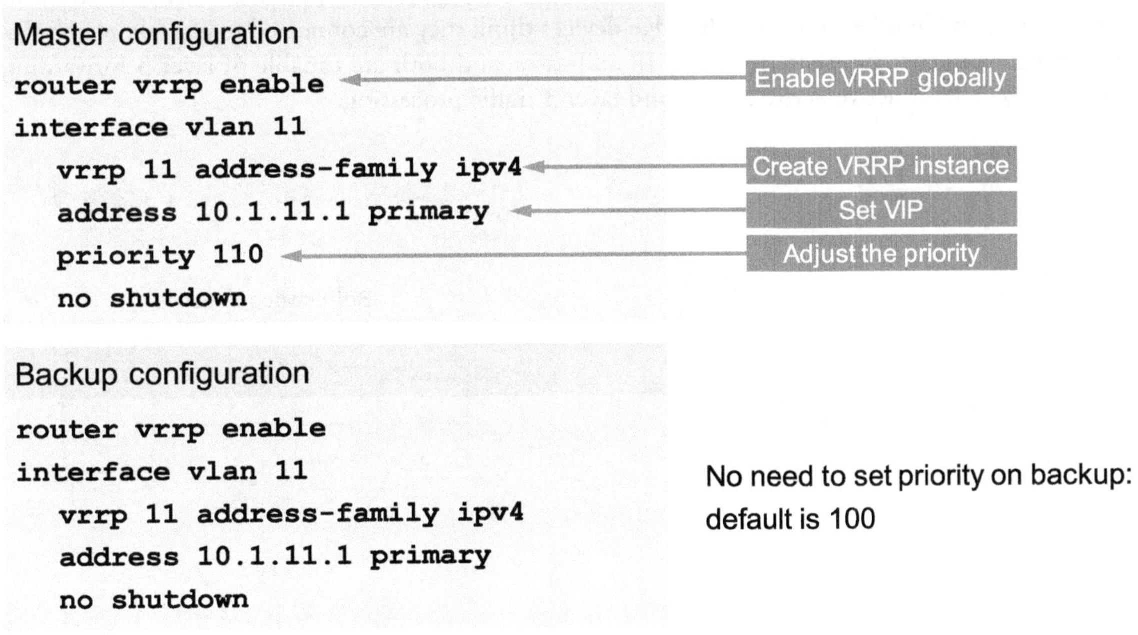

Figure 4-36: VRRP configuration

Introduction to VSX

The last part of this chapter explores one of Aruba's switching virtualization technologies. Aruba supports two virtual switching technologies: Virtual Switching eXtension (VSX) and Virtual Switching

Framework (VSF). This chapter provides a real brief introduction to VSX, where VSX is covered in depth in the Aruba professional-level course and book. Chapter 5 will cover VSF.

VRRP versus switch virtualization

Although VRRP and STP are supported protocols, Aruba has developed a *much* better solution. Look at Figure 4-37. The left side illustrates the implementation of VRRP and STP, which is active/standby solution. This means that only one of the switches is actively processing the traffic, and thus the solution is sub-optimal. Of course, you can enable multiple MSTP instances and configure multiple VRRP groups so one switch is a VIP master and the other is a VIP backup for a given VLAN. This, however, allows load-sharing only if the traffic originates from different VLANs—per VLAN load-sharing.

What if you want to use both switches and utilize all of the uplinks, even if the traffic comes from the clients connected to the *same* VLAN? The solution to the problem is virtualization that allows truly active/active traffic processing. You can combine two or more switches (depending on the virtual switching technology) to create *one* virtual switch. You can now use link aggregation instead of STP, which will use all the uplinks and allows load-sharing even if the traffic originates from the same VLAN. This is because when using LACP, the multiple switches that comprise the virtual switch send the same system identifier, and thus the edge devices think they are connected to a single switch. For IP routing, both switches share the same IP addresses, and both are capable of layer 3 forwarding: thus, you get a truly active/active layer 2 and layer 3 traffic processing.

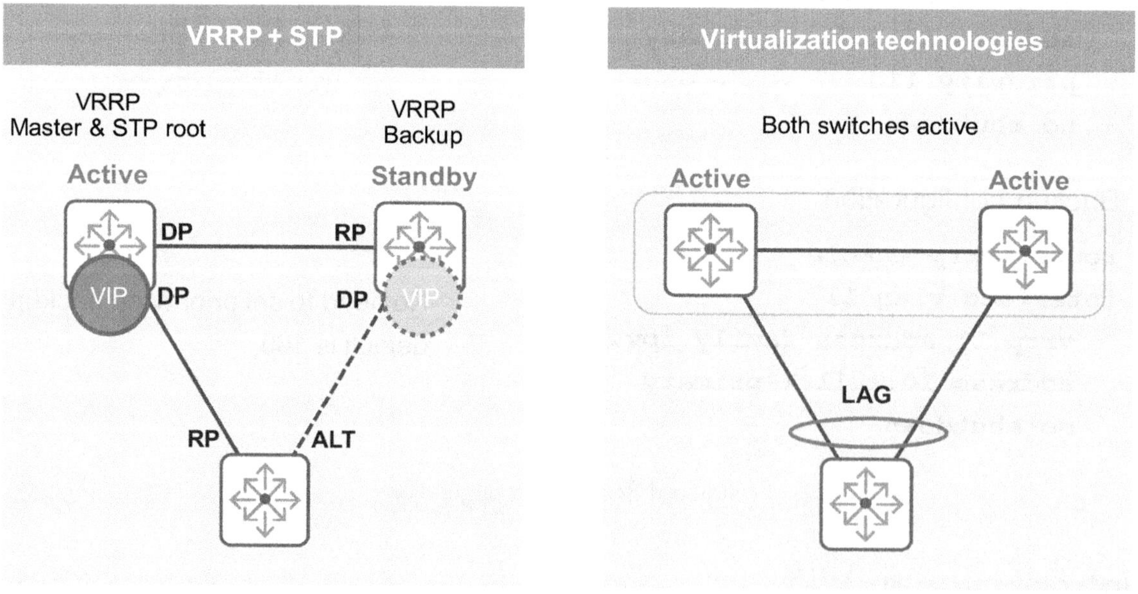

Figure 4-37: VRRP versus switch virtualization

AOS-CX switch virtualization: VSX versus VSF

Aruba's VSX and VSF are similar to each other, and even to earlier virtualization techniques like VRRP, where multiple devices are perceived as a single device. This feature improves resiliency while reducing configuration and management overhead. But there are distinct differences between VSX and VSF, as summarized in Figure 4-38.

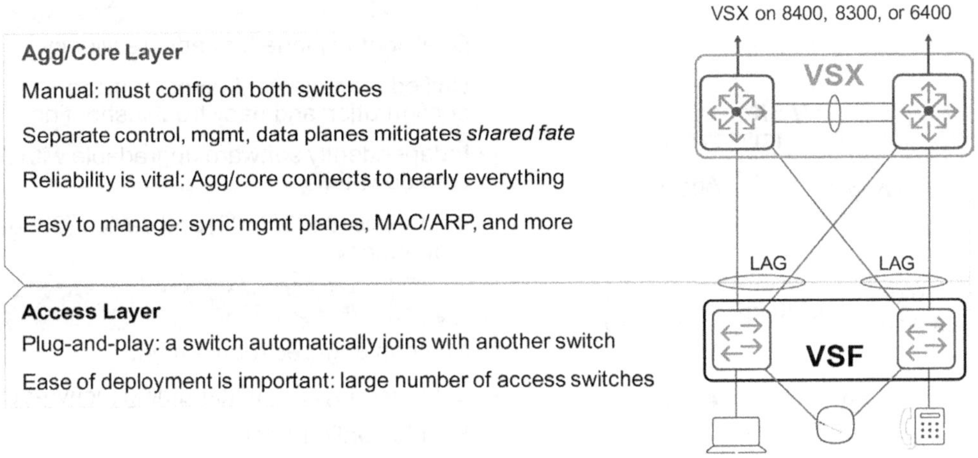

Figure 4-38: VSX versus VSF

You use VSX primarily at the campus network *aggregation* and *core* layers. You use VSF primarily at the campus *access* layer. The differences in features and function are intentional, since priorities differ between access and core/aggregation.

Configuring VSF (sometimes referred to as *stacking*) is a plug-and-play affair: a switch, at default settings, listens to discover another VSF switch on some designated ports. This ease of deployment is especially important at the access layer: you may have dozens or hundreds of access switches. VSF accommodates up to ten switches, depending on the switch model, while a VSX configuration has exactly two switches: primary and secondary.

You must manually configure VSX on both switches. VSX maintains separate control plane and management planes, and much of the data plane. This mitigates the *shared fate* nature of common planes, for improved uptimes: you get high availability *and* redundancy. This is especially important when aggregating all those access layer switches. And it still remains easy to manage, because you can synchronize the management planes, and optionally, other features like MAC and ARP addresses, and much more.

VSX benefits: VSX-LAG

Figure 4-39 shows an example of an Aruba-recommended topology in which you deploy VSX at the aggregation layer.

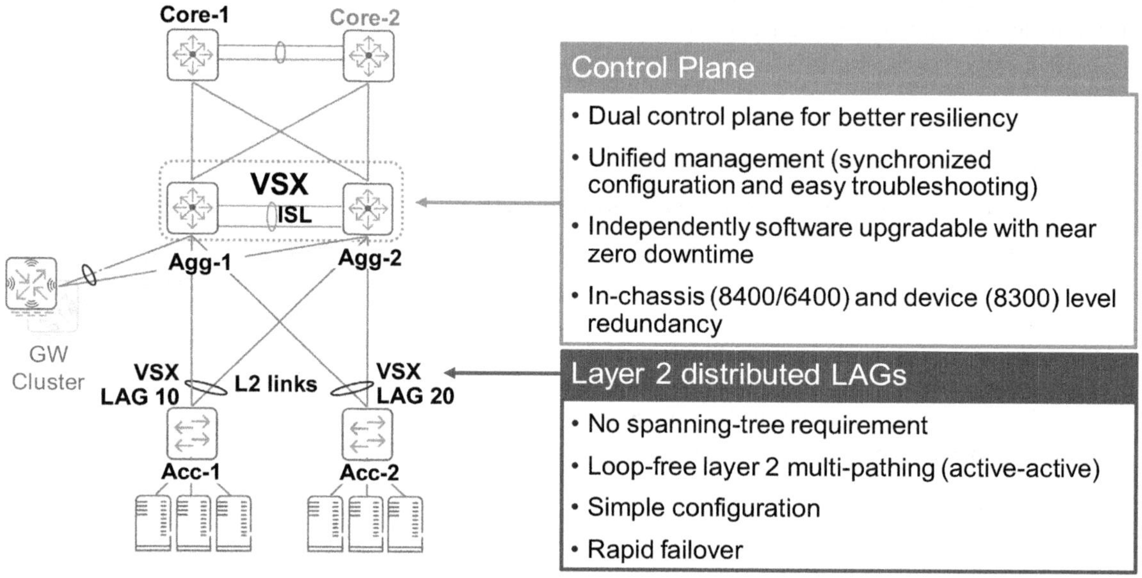

Figure 4-39: VSX benefits—VSX-LAG

While you can deploy VSX at the core if the topology requires, you may prefer to keep redundant core switches entirely independent. A dual control plane provides the best resiliency and software upgrades with near zero downtime. At the same time, VSX permits unified management on the features that administrators choose to synchronize.

The VSX pair uses link aggregations (LAGs) with links on both VSX switches to connect to access layer switches. This type of LAG is called a multi-chassis (MCLAG), distributed, or VSX LAG. These VSX LAGs operate at layer 2 because the access layer switches typically do not route, and the aggregation layer VSX pair acts as the default gateway for the access VLANs. The layer 2 LAGs eliminate the need for MSTP on the redundant links. They provide a loop-free path that permits the use of all the links in the aggregation (active-active design). Configuring a LAG-based topology is simple; and if a link fails, failover occurs very quickly.

Lab 04.01 STP

Your customer informs you that they have experienced network loops in the past. They would like to understand what options are available to deal with network loops. After contacting your senior colleague, you learn that STP and loop protect can be used for network loop detection.

STP can also be used to select the best path in a redundant layer 2 topology. However, in most networks today, this function will be performed by a VSX cluster at the core and/or aggregation and VSX-LAG connections to the edge switches. In this type of setup, the role of the STP is reduced to a loop detection protocol.

Since STP is considered a core network administrator competency, this lab will still provide you a brief introduction to STP layer 2 best path selection and path failover. In this lab you will explore the operation and basic configuration of STP and loop protection.

Objectives

After completing this lab, you will be able to:

- Review the STP default settings.
- Understand STP path cost and failover.
- Configure STP access security with BPDU protection (BPDU Guard).
- Configure and verify loop protection.

Task 1: Explore STP costs and path failover

In this task you will explore the default settings of STP on AOS-CX switches. You will explore the root path cost and topology change notifications (TCNs).

To get a redundant layer 2 path in the network, the link between sw-edge1 and sw-agg2 will be enabled. This will result in a loop in the network that will be resolved by STP. Note that this is only an example of a looped topology. In a later lab activity, you will configure a LAG between the access switch and the VSX aggregation pair of switches to achieve an active-active forwarding topology.

Objectives

- Explore default STP settings.
- Review the root path cost.
- Review the root port.

Steps

Verify STP on the root bridge

For this lab, STP should be enabled on all switches. This is the default state on the 6xxx series switches. In the next steps you will review the STP status on your lab switches.

1. On sw-agg1, review the spanning-tree status. STP should be enabled. Take note of the bridge ID priority and MAC-address.

```
sw-agg1(config)# show spanning-tree
Spanning tree status       : Enabled Protocol: MSTP
MST0
  Root ID     Priority    : 0
              MAC-Address: b8:d4:e7:da:80:00
              This bridge is the root
              Hello time(in seconds):2  Max Age(in seconds):20
              Forward Delay(in seconds):15

  Bridge ID   Priority   : 0
              MAC-Address: b8:d4:e7:da:80:00
              Hello time(in seconds):2  Max Age(in seconds):20
              Forward Delay(in seconds):15

Port     Role             State       Cost   Priority    Type   BPDU-Tx   BPDU-Rx
TCN-Tx   TCN-Rx
------------ --------------- ----------- --------------- ----------
----------------- --------
1/1/1    Designated       Forwarding  2000   128         P2P    75994     14
10       14
1/1/2    Designated       Forwarding  2000   128         P2P    64085     9
6        6
lag256   Designated       Forwarding  800    64          P2P    76128     2
22       2

Number of topology changes    : 19
Last topology change occurred : 128119 seconds ago
```

- **Question**: What is the priority configured on sw-agg1?
- **Answer**: sw-agg1 is the target root bridge. In the lab configuration it was configured with priority 0, the lowest priority value. The lowest value is the best value for STP.

Verify STP status and root port on the access switches

1. On sw-edge1, identify the switch base system MAC address.

```
sw-edge1# show system
Hostname                : sw-edge1
System Description : FL.10.09.1040
System Contact          :
System Location         :

Vendor                  : Aruba
Product Name            : JL666A 6300F 24G CL4 PoE 4SFP56 Sw
Chassis Serial Nbr : SG09KN500B
Base MAC Address        : 64e881-dd8180
ArubaOS-CX Version : FL.10.09.1040

Time Zone               : US/Michigan

Up Time                 : 1 day, 18 hours, 34 minutes
CPU Util (%)            : 1
Memory Usage (%)        : 22
```

2. Review the spanning-tree status. STP should be enabled. Take note of the bridge ID priority and MAC-address.

```
sw-edge1# show spanning-tree
Spanning tree status        : Enabled Protocol: MSTP
MST0
  Root ID      Priority    : 0
               MAC-Address: b8:d4:e7:da:80:00
               Hello time(in seconds):2  Max Age(in seconds):20
               Forward Delay(in seconds):15

  Bridge ID    Priority   : 32768
               MAC-Address: 64:e8:81:dd:81:80
               Hello time(in seconds):2  Max Age(in seconds):20
               Forward Delay(in seconds):15
...
```

- **Question**: Is there a relation between the bridge ID MAC and the system MAC?
- **Answer**: Yes, they are the same.
- **Question**: What is the (default) bridge priority on sw-edge1?
- **Answer**: The default priority for spanning tree is 32768.
- **Question**: Based on the bridge IDs you have seen on sw-agg1 and the edge switches, which switch will be elected by STP as the root bridge?
- **Answer**: Sw-agg1 has the lowest bridge ID and will be elected as the root bridge.

Verify the root path cost

In the next steps you will explore the path cost used by STP to measure the cost to reach the root bridge. This is described as the root path cost.

1. Use the **show spanning-tree mst** command to take note of the root path cost.

```
sw-edge1# show spanning-tree mst
#### MST0
Vlans mapped:  1-4094
Bridge         Address:64:e8:81:dd:81:80    priority:32768
Regional Root
Operational    Hello time(in seconds): 2  Forward delay(in seconds):15
Max-age(in seconds):20  txHoldCount(in pps): 6
Configured     Hello time(in seconds): 2  Forward delay(in seconds):15
Max-age(in seconds):20  Max-Hops:20
Root           Address:b8:d4:e7:da:80:00  Priority:0
               Port:1/1/27                Path cost:2000
Regional Root  Address:64:e8:81:dd:81:80  Priority:32768
               Internal cost:0            Rem Hops:20
...
```

- **Question**: What is the current root path cost?
- **Answer**: The root path cost is 2000.
- **Question**: What type of link will result in a cost of 2000?
- **Answer**: A single 10 Gbps link has a default cost value of 2000.
- **Question**: How does the STP cost 2000 link to 10 Gbps?
- **Answer**: As part of the STP standard, 2000000 has been set as the reference cost for a 10 Mbps connection. Other speeds are based on this reference cost in Table 4-1:

Table 4-1. STP port costs.

Speed	STP link cost
10 Mbps	2,000,000
100 Mbps	200,000
1 Gbps	20,000
10 Gbps	2,000
100 Gbps	200

2. Review the port list of the **show spanning-tree mst** command output.

- **Question**: For port 1/1/27, what is the Role, State, and Cost?
- **Answer**: Port 1/1/27 is the root port, it is forwarding, and the cost is 2000.

3. Review port 1/1/27 in the brief interface list.

```
sw-edge1# show int 1/1/27 brief
----------------------------------------------------------------------------
---------------------------------
Port       Native  Mode    Type        Enabled Status  Reason
Speed    Description
           VLAN
(Mb/s)
----------------------------------------------------------------------------
---------------------------------
1/1/27     1       trunk   SFP+DAC1    yes     up
10000    sw-agg1
```

- **Question**: What is the speed for the port 1/1/27?
- **Answer**: The speed is 10000 (Mb/s). This represents 10 Gbps and results in the STP cost of 2000.

4. On sw-edge2, review the spanning-tree status. STP should be enabled. Take note of the bridge ID priority and MAC-address.

```
sw-edge2# show spanning-tree
Spanning tree status       : Enabled Protocol: MSTP

MST0
  Root ID     Priority   : 0
              MAC-Address: b8:d4:e7:da:80:00
              Hello time(in seconds):2  Max Age(in seconds):20
              Forward Delay(in seconds):15

  Bridge ID   Priority  : 32768
              MAC-Address: 8c:85:c1:49:20:c0
              Hello time(in seconds):2  Max Age(in seconds):20
              Forward Delay(in seconds):15
...
```

Verify the root path cost

1. Use the **show spanning-tree mst** command to take note of the root path cost. This should also be 2000, based on a single 10 Gbps link to the sw-agg1.

```
sw-edge2# show spanning-tree mst
#### MST0
Vlans mapped:  1-4094
Bridge         Address:8c:85:c1:49:20:c0    priority:32768
Regional Root
Operational    Hello time(in seconds): 2  Forward delay(in seconds):15
Max-age(in seconds):20  txHoldCount(in pps): 6
```

```
Configured      Hello time(in seconds): 2  Forward delay(in seconds):15
Max-age(in seconds):20  Max-Hops:20
Root            Address:b8:d4:e7:da:80:00  Priority:0
                Port:1/1/27                Path cost:2000
Regional Root   Address:8c:85:c1:49:20:c0  Priority:32768
                Internal cost:0            Rem Hops:20
...
```

Investigate the interface cost for LAG ports

Each interface link speed has its own assigned cost. In the next steps you will explore what the impact of the LAG speed is with regards to the STP cost.

1. On sw-agg2, review the spanning-tree status. STP should be enabled. Take note of the bridge ID priority and MAC-address.

```
sw-agg2# show spanning-tree
Spanning tree status      : Enabled Protocol: MSTP

MST0
  Root ID     Priority    : 0
              MAC-Address: b8:d4:e7:da:80:00
              Hello time(in seconds):2  Max Age(in seconds):20
              Forward Delay(in seconds):15

  Bridge ID   Priority   : 4096
              MAC-Address: b8:d4:e7:d9:e5:00
              Hello time(in seconds):2  Max Age(in seconds):20
              Forward Delay(in seconds):15
```

2. On sw-agg2, use the **show spanning-tree mst** command to see the root path cost.

```
sw-agg2# show spanning-tree mst
#### MST0
Vlans mapped:   1-4040
Bridge          Address:b8:d4:e7:d9:e5:00    priority:4096
Regional Root
Operational     Hello time(in seconds): 2  Forward delay(in seconds):15
Max-age(in seconds):20  txHoldCount(in pps): 6
Configured      Hello time(in seconds): 2  Forward delay(in seconds):15
Max-age(in seconds):20  Max-Hops:20
Root            Address:b8:d4:e7:da:80:00  Priority:0
                Port:lag256                Path cost:800
Regional Root   Address:b8:d4:e7:d9:e5:00  Priority:4096
                Internal cost:0            Rem Hops:20
```

- **Question**: What is the current root port?
- **Answer**: The current root port is lag256.
- **Question**: What is the current root path cost?
- **Answer**: The current root path cost is 800.
- **Question**: What is the cost for port lag256?
- **Answer**: Port lag256 has a cost of 800.

3. Review the interface brief output, filter 'up' in the output. Pay attention to the speed.

```
sw-agg2# show interface brief | include up
1/1/8       --        routed  SFP+DAC3          yes      up
10000    rtr-core1-1/1/2
1/1/46      1         trunk   SFP28DAC0.65      yes      up
25000    --
1/1/47      1         trunk   SFP28DAC0.65      yes      up
25000    --
loopback0 --          routed  --                yes      up
--       --
vlan2       --                --                yes      up
--       --
lag256      1         trunk   --                yes      up         --
50000    --
```

- **Question**: What is the speed of ports 1/1/46 and 1/1/47?
- **Answer**: Ports 1/1/46 and 1/1/47 are connected at 25 Gbps.
- **Question**: What would be the standard STP cost for a 25 Gbps link?
- **Answer**: 25 Gbps is '4 times slower' than a 100 Gbps link, so 200 x 4 = 800.
- **Question**: What was the root path cost on sw-agg2?
- **Answer**: The root path cost is 800. This is the cost of a 25 Gbps link.
- **Question**: What is the speed of the LAG 256?
- **Answer**: The interface speed is 50 Gbps, based on 2 member ports of 25 Gbps.
- **Question**: What could you expect the LAG STP cost to be, based on a speed of 50 Gbps?
- **Answer**: 50Gbps is '2 times slower' than a 100Gbps, so 200 x 2 = 400. However, the LAG is using an STP cost of 800.
- **Question**: Then why is the LAG cost 800?

- **Answer**: This is a vendor choice. For AOS-CX, the choice was made to keep the link speed of a LAG to the individual member port speed. This ensures that a link down event for a LAG member will not impact STP, since the STP port speed did not change.
- **Conclusion**: On AOS-CX, the STP cost of a LAG is based on the member port speed, not the total reported speed of the LAG to ensure STP stability during LAG member port changes.

Add a redundant link to the network

In the next steps you will activate a new link in the network. This will result in a loop in the network topology. You will enable the link between sw-edge1 and sw-agg2.

1. On **sw-agg2**, enable the interface 1/1/1. It connects to sw-edge1.

```
sw-agg2(config)# interface 1/1/1
sw-agg2(config-if)# no shutdown
sw-agg2(config-if)# exit
```

2. On **sw-edge1**, enable the interface 1/1/28. It connects to sw-agg2.

```
sw-edge1(config)# interface 1/1/28
sw-edge1(config-if)# description sw-agg2
sw-edge1(config-if)# no shutdown
sw-edge1(config-if)# exit
```

3. Verify that interface 1/1/28 is up and connects to port 1/1/1 on sw-agg2.

```
sw-edge1(config)# show interface 1/1/28 brief
--------------------------------------------------------------------------
-------------------------------------
Port       Native  Mode   Type            Enabled Status  Reason
Speed   Description
           VLAN
(Mb/s)
--------------------------------------------------------------------------
-------------------------------------
1/1/28     1       access SFP+DAC1        yes     up
10000   sw-agg2

sw-edge1(config)# show lldp neighbor-info

LLDP Neighbor Information
=========================

Total Neighbor Entries          : 7
Total Neighbor Entries Deleted  : 4
Total Neighbor Entries Dropped  : 0
Total Neighbor Entries Aged-Out : 4
```

```
LOCAL-PORT  CHASSIS-ID          PORT-ID                      PORT-DESC    TTL
SYS-NAME
---------------------------------------------------------------------------------
-----------------
...
1/1/27      b8:d4:e7:da:80:00  1/1/1                        1/1/1        120
sw-agg1
1/1/28      b8:d4:e7:d9:e5:00  1/1/1                        1/1/1        120
sw-agg2
...
```

Review STP for the redundant link

1. On **sw-edge1**, review the spanning-tree MSTP status. Pay attention to the *State* column. This is the forwarding mode.

```
sw-edge1(config)# show spanning-tree mst
...
Port            Role            State       Cost Priority Type       BPDU-Tx Rx
TCN-Tx Rx
-------------- -------------- ---------- ---------- ----------
---------------- ----------
...
1/1/24          Disabled        Down        20000   128    P2P           0    0
0          0
1/1/27          Root            Forwarding 2000     128    P2P  Bound  16   76904
14         12
1/1/28          Alternate       Blocking    2000     128    P2P  Bound  2    63
0          2
...
```

- **Question**: Is there any interface in a blocking state?
- **Answer**: Yes, port 1/1/28 is shown with a *Blocking* state.

Task 2: Test STP path failover

In this task you will test the STP failover on the access switch sw-edge1 in case of an uplink failure. The uplink to sw-agg1 will be disabled. As a result, STP should enable the path to sw-agg2.

Objectives

- Verify STP path failover.

Steps

1. On **sw-edge1**, check the STP MST output and take note of the current root path cost. (your output may differ).

```
sw-edge1(config)# show spanning-tree mst
#### MST0
Vlans mapped:  1-4094
Bridge         Address:64:e8:81:dd:81:80    priority:32768
Regional Root
Operational    Hello time(in seconds): 2  Forward delay(in seconds):15
Max-age(in seconds):20  txHoldCount(in pps): 6
Configured     Hello time(in seconds): 2  Forward delay(in seconds):15
Max-age(in seconds):20  Max-Hops:20
Root           Address:b8:d4:e7:da:80:00  Priority:0
               Port:1/1/27                Path cost:2000
Regional Root  Address:64:e8:81:dd:81:80  Priority:32768
               Internal cost:0            Rem Hops:20
...
Topology change flag            : False
Number of topology changes      : 45
Last topology change occurred : 50 seconds ago
```

- **Question**: What is the current root path cost?
- **Answer**: The current root path cost is 2000. By default, this represents a single 10 Gbps connection to reach the root bridge.

2. Take note of the current number of *topology changes*.

- **Question**: What is the current root port?
- **Answer**: The current root port is port 1/1/27, which connects to sw-agg1.

3. On **PC1**, open a command prompt and start a continuous ping to server IP 10.254.1.21.

```
C:> ping 10.254.1.21 -t
```

4. On **sw-edge1**, shut down the uplink to sw-agg1; this is port 1/1/27.

```
sw-edge1(config)# interface 1/1/27
sw-edge1(config-if)# shutdown
sw-edge1(config-if)# exit
```

5. On **sw-edge1**, review the spanning-tree MSTP output (your output may be different).

```
sw-edge1(config)# show spanning-tree mst
#### MST0
Vlans mapped:  1-4094
Bridge         Address:64:e8:81:dd:81:80    priority:32768
Regional Root
Operational    Hello time(in seconds): 2  Forward delay(in seconds):15
Max-age(in seconds):20  txHoldCount(in pps): 6
Configured     Hello time(in seconds): 2  Forward delay(in seconds):15
Max-age(in seconds):20  Max-Hops:20
Root           Address:b8:d4:e7:da:80:00  Priority:0
               Port:1/1/28                Path cost:2800
Regional Root  Address:64:e8:81:dd:81:80  Priority:32768
               Internal cost:0            Rem Hops:20
...
Topology change flag           : False
Number of topology changes     : 47
Last topology change occurred  : 2 seconds ago
```

- **Question**: What port is the root port now?
- **Answer**: Port 1/1/28
- **Question**: What is the root path cost now?
- **Answer**: Previously it was 2000 (single 10 Gbps connection); now it shows 2800 (10 Gbps connection to reach sw-agg2, then the 25 Gbps connection to reach sw-agg1).
- **Question**: What is the number of topology changes? Did this change from the previous output?
- **Answer**: The number of topology changes has now increased, due to the uplink port change.

6. On PC1, check your ping status.
   - **Question**: Is the ping successful?
   - **Answer**: No, although it seems from the previous output that STP successfully calculated a new best path to the root bridge, the user traffic has stopped.

7. Let's investigate. What could be the issue?

8. On **sw-edge1**, review the MAC address table for VLAN 11. You should expect to see a client MAC address and a MAC address for the default gateway (sw-agg1) on the uplink port (your output may be different).

```
sw-edge1(config)# show mac-address-table vlan 11
MAC age-time               : 300 seconds
Number of MAC addresses : 1

MAC Address          VLAN     Type                     Port
--------------------------------------------------------------
00:50:56:b1:cd:26    11       dynamic                  1/1/1
```

- **Question**: Do you see a MAC address for the client and the default gateway?
- **Answer**: No, only a MAC address on port 1/1/1 for the client is shown. This indicates there is something wrong with the uplink port VLAN.

9. Review the VLAN 11 port membership.

```
sw-edge1(config)# show vlan 11
-----------------------------------------------------------------------------
-----------------------------------
VLAN  Name                                    Status  Reason
Type        Interfaces
-----------------------------------------------------------------------------
-----------------------------------
11    v11-employee                            up      ok
static      1/1/1,1/1/27,lag50
```

- **Question**: What are the port members of VLAN 11?
- **Answer**: Ports 1/1/1, 1/1/27, and LAG 50
- **Question**: What port is missing from this list?
- **Answer**: Port 1/1/28 is now the uplink port, but it is not member of VLAN 11.

10. Review the port 1/1/28 configuration.

```
sw-edge1(config)# show running-config interface 1/1/28
interface 1/1/28
    no shutdown
    description sw-agg2
    no routing
    vlan access 1
    exit
```

- **Question**: What do you observe?
- **Answer**: The port is an access port in VLAN 1. This is still the default configuration.

11. Adjust the configuration. Make port 1/1/28 a VLAN trunk that allows all VLANs.

```
sw-edge1(config)# interface 1/1/28
sw-edge1(config-if)# vlan trunk allowed all
sw-edge1(config-if)# exit
```

1. Verify your configuration. Check the VLAN 11 port membership.

```
sw-edge1(config)# show vlan 11
--------------------------------------------------------------------------
------------------------------------------
VLAN  Name                                Status  Reason
Type        Interfaces
--------------------------------------------------------------------------
------------------------------------------
11    v11-employee                        up      ok
static      1/1/1,1/1/27-1/1/28,lag50
```

12. Check the MAC-address table for VLAN 11 again.

```
sw-edge1(config)# show mac-address-table vlan 11
MAC age-time            : 300 seconds
Number of MAC addresses : 2

MAC Address          VLAN     Type                    Port
--------------------------------------------------------------
00:50:56:b1:cd:26    11       dynamic                 1/1/1
b8:d4:e7:da:80:00    11       dynamic                 1/1/28
```

- **Question**: Were any MAC addresses learned on the uplink port?
- **Answer**: Yes, a MAC address has been learned on the port 1/1/28 now.

a. On PC1, verify that the ping has resumed.

13. On **sw-edge1**, repeat the uplink failure to confirm the STP failover. Enable the original uplink again.

```
sw-edge1(config)# interface 1/1/27
sw-edge1(config-if)# no shutdown
```

14. Wait about 10 seconds, then shut down the link again.

```
sw-edge1(config-if)# shutdown
```

15. On **PC1**, verify that the ping continues. This demonstrates the STP failover.

Note
It is expected to see the loss of a ping packet during a topology change.

16. On **sw-edge1**, restore the uplink connection by enabling port 1/1/27.

```
sw-edge1(config-if)# no shutdown
```

17. On **sw-edge1**, review the STP MST output.

```
sw-edge1(config-if)# show spanning-tree mst
#### MST0
Vlans mapped:  1-4094
Bridge         Address:64:e8:81:dd:81:80    priority:32768
Regional Root
Operational       Hello  time(in  seconds):  2   Forward  delay(in  seconds):15
Max-age(in seconds):20  txHoldCount(in pps): 6
Configured        Hello  time(in  seconds):  2   Forward  delay(in  seconds):15
Max-age(in seconds):20  Max-Hops:20
Root           Address:b8:d4:e7:da:80:00  Priority:0
               Port:1/1/27                Path cost:2000
Regional Root  Address:64:e8:81:dd:81:80  Priority:32768
               Internal cost:0            Rem Hops:20
...
```

- **Question**: What is the current root port?
- **Answer**: The root port should have reverted to port 1/1/27.

2. On **PC1**, verify the ping has continued when the original uplink was restored.

Cleanup

3. On **PC1**, close the command prompt.

Task 3: Configure STP access security

In this task you will configure access ports that connect to clients with STP BPDU protection. BPDU protection will shut down a port when it receives an STP BPDU. This could occur when two wall outlets are connected to each other (network loop) or when a rogue switch is introduced into the network. In such cases, BPDU protection will shut down the ports to contain the problem. A BPDU protection timeout can be configured to automatically restore the link after a set amount of time.

In the lab, the LAG between sw-edge1 and sw-edge2 will be used to test the feature. Since sw-edge2 is advertising BPDUs, sw-edge1 will shut down the link.

Objectives

- Configure STP access security with BPDU guard.
- Configure automatic recovery for BPDU guard disabled ports.

Steps

1. On sw-edge1, configure the BPDU Guard recovery timer. Set it to 60 seconds.

```
sw-edge1(config)# span bpdu-guard timeout 60
```

2. Configure a port range with ports 1/1/1 to 1/1/24 and enable BPDU Guard.

```
sw-edge1(config)# interface 1/1/1-1/1/24
sw-edge1(config-if-<1/1/1-1/1/24>)# spanning-tree bpdu-guard
sw-edge1(config-if-<1/1/1-1/1/24>)# exit
```

Note

In the training lab, there are no STP switches connected to the access ports, so the previous step is only an example configuration.

To test the BPDU protection feature, you will enable it on the LAG connected to sw-edge2; this is LAG50. Since sw-edge2 generates STP BPDUs, sw-edge1 will consider it a violation of the BPDU protection and shut down the LAG port.

3. On sw-edge1, enable BPDU Guard on the LAG 50 (link that connects to sw-edge2).

```
sw-edge1(config)# interface lag 50
sw-edge1(config-lag-if)# spanning-tree bpdu-guard
```

4. Enable the LAG 50. Now STP BPDU frames that are generated by sw-edge2 will be received on a protected port.

```
sw-edge1(config-lag-if)# no shutdown
```

5. Review the STP output.

```
sw-edge1(config-lag-if)# show spanning-tree
...
lag50         Disabled        Bpdu-Error 20000            64          P2P
913          8            8           7
...
```

- **Question**: What is the status for port lag50?
- **Answer**: Port lag50 is reported with *Bpdu-Error*.

6. To see more details, review the STP inconsistent-ports output.

```
sw-edge1(config-lag-if)# show spanning-tree inconsistent-ports
Instance ID  Blocked Port   Reason
------------ -------------- ------------
0            lag50          BPDU Guard
```

- **Question**: Do you see any inconsistent ports?
- **Answer**: The LAG 50 is reported as inconsistent port with reason BPDU Guard.

7. Wait 60 seconds and check the event log in reverse order.

```
sw-edge1(config-lag-if)# show events -r -d hpe-mstpd
---------------------------------------------------
Event logs from current boot
---------------------------------------------------
2022-10-17T03:14:55.024057-04:00 sw-edge1 hpe-mstpd[3243]:
Event|2007|LOG_WARN|CDTR|1|Port lag50 disabled - BPDU received on protected
port
2022-10-17T03:14:55.022192-04:00 sw-edge1 hpe-mstpd[3243]:
Event|2012|LOG_INFO|CDTR|1|CIST - Topology Change generated on port lag50
going in to forwarding
2022-10-17T03:13:49.594762-04:00 sw-edge1 hpe-mstpd[3243]:
Event|2007|LOG_WARN|CDTR|1|Port lag50 disabled - BPDU received on protected
port
2022-10-17T03:13:47.506076-04:00 sw-edge1 hpe-mstpd[3243]:
Event|2012|LOG_INFO|CDTR|1|CIST - Topology Change generated on port lag50
going in to forwarding
...
```

- **Question**: Did the port LAG 50 come back up?
- **Answer**: Yes, after the BPDU Guard timeout expired (60 seconds), sw-edge1 enabled the port LAG 50 again, but since it received another BPDU, the port was also disabled again.

Cleanup

1. On sw-edge1, remove the STP BPDU protection command.

```
sw-edge1(config)# interface lag 50
sw-edge1(config-lag-if)# no spanning-tree bpdu-guard
```

2. Shut down the port LAG 50.

```
sw-edge1(config-lag-if)# shutdown
sw-edge1(config-lag-if)# exit
```

Task 4: Configure loop protection

After the successful STP setup, your customer informs you that they have had some issues with loops in the past that were not handled by STP. Some of their contractors had connected their own switches to some switch ports, and it turned out that these switches were blocking the STP BPDU frames, while at the same time, they were not sending out STP BPDU frames themselves. Since STP did not receive its own BPDUs back on the looped connection, the loop went unnoticed. They have asked if you have a solution for this scenario.

After consulting with your senior colleague, you learn that AOS-CX switches have a feature that is known as loop protect. It is an Aruba-specific technology that can also detect loops. The Aruba best practice is to have loop protect enabled on the access ports. In this task you will enable loop protect and verify its operation.

Objectives

- Configure loop protection.
- Verify the loop protection operation.

Steps

Prepare the setup

Introducing a loop in the network can have significant impact on the network. STP and loop protect are designed to handle loops in the network. However, if you make a configuration mistake while configuring loop protect or STP, the loop will remain effective in the remote lab environment. To limit the impact of the loop, you will apply a rate-limiter on the looped port. This will ensure that broadcasts, multicast, and unknown unicasts will only pass at 100 packets per second (pps).

1. On sw-edge1, configure port lag50 with a rate-limiter.

```
sw-edge1(config)# interface lag 50
sw-edge1(config-lag-if)# rate-limit unknown-unicast 100 pps
sw-edge1(config-lag-if)# rate-limit broadcast 100 pps
sw-edge1(config-lag-if)# rate-limit multicast 100 pps
```

2. To demonstrate loop protection, you will disable STP on this port: **lag50**. If you would leave STP enabled, STP may be faster with the loop detection, and you may not have a chance to see loop protect in action. A BPDU filter can be used to disable STP on a port. No STP BPDUs will be transmitted, and any received BPDUs will be dropped and ignored.

```
sw-edge1(config-lag-if)# span bpdu-filter
This filter command allows the port to go into a continuous
forwarding mode and spanning tree will not interfere, even if
the port would cause a loop to form in the network topology.
If you suddenly experience high traffic load, shutdown the
port and remove the applicable filter configuration under
interface context using the following CLI command(s):
no spanning-tree bpdu-filter
no spanning-tree rpvst-filter
sw-edge1(config-lag-if)# exit
```

Configure loop protect

Now that STP has been disabled, you can enable loop protection on the port and enable the port. This will introduce the loop in the network.

1. On the LAG 50, enable loop protection.

```
sw-edge1(config)# interface lag 50
sw-edge1(config-lag-if)# loop-protect
```

2. Enable the port LAG 50 to activate the loop.

```
sw-edge1(config-lag-if)# no shutdown
```

Verify the loop protect operation

1. Review the event log file. Pay attention to the *hpe-lpd* daemon events.

```
sw-edge1(config-lag-if)# show events -r
---------------------------------------------------
Event logs from current boot
---------------------------------------------------
2022-10-17T03:18:32.409479-04:00 sw-edge1 lacpd[3150]:
Event|1321|LOG_INFO|CDTR|1|LAG 50 State change for interface 1/1/25: Actor
state: ALFO, Partner state ALFNCD
2022-10-17T03:18:32.398850-04:00 sw-edge1 lacpd[3150]:
Event|1321|LOG_INFO|CDTR|1|LAG 50 State change for interface 1/1/26: Actor
state: ALFO, Partner state ALFNCD
2022-10-17T03:18:32.377134-04:00 sw-edge1 intfd[718]:
Event|404|LOG_INFO|UKWN|1|Link status for interface 1/1/26 is down - Network
loop detected
2022-10-17T03:18:32.376924-04:00 sw-edge1 intfd[718]:
Event|404|LOG_INFO|UKWN|1|Link status for interface 1/1/25 is down - Network
loop detected
2022-10-17T03:18:32.348222-04:00 sw-edge1 hpe-lpd[3131]:
Event|2808|LOG_INFO|CDTR|1|Ports TX lag50 and RX 1/1/27 are involved during
TX port disabling
2022-10-17T03:18:32.348056-04:00 sw-edge1 hpe-lpd[3131]:
Event|2801|LOG_WARN|CDTR|1|Port lag50 is disabled by Loop-protection after
loop detection on VLAN 1
```

- **Question**: What are the TX and RX ports involved in the loop protect port disabling?
- **Answer**: Ports lag50 and 1/1/27 (the uplink port to sw-agg1) are involved.
- **Question**: Did you enable port 1/1/27 for loop protect?
- **Answer**: No. Only the transmitting port (lag50) was enabled. When the looped packet comes back on any interface (in this example, via sw-edge2 > sw-agg1, back to port 1/1/27), the switch can disable the transmitting (TX) interface. In this example, this is the lag50.

2. Check the operational status for the ports.

```
sw-edge1(config)# show interface brief
--------------------------------------------------------------------------
------------------------------------
Port      Native  Mode   Type           Enabled Status  Reason
Speed   Description
          VLAN
(Mb/s)
--------------------------------------------------------------------------
------------------------------------
...
```

```
1/1/25    1        access SFP+DAC1        yes     down    Network loop
detected  --       --
1/1/26    1        access SFP+DAC1        yes     down    Network loop
detected  --       --
...
```

3. Check the operational status for the individual port 1/1/25 (this is the member port of the lag 50).

```
sw-edge1(config)# show interface 1/1/25

Interface 1/1/25 is down
 Admin state is up
 State information: Network loop detected
 Link state: down for 3 minutes (since Thu Sep 22 05:25:00 EDT 2022)
 Link transitions: 192
 Description:
```

4. Check the loop protect status.

```
sw-edge1# show loop-protect loop-detected

Status and Counters - loop protection Information

Transmit Interval                : 5 (sec)
Port Re-enable Timer             : Disabled
Loop Detected Trap               : Disabled

Interface 1/1/25
  Loop-protect enabled           : Yes
  Action on loop detection       : TX disable
  Loop detected count            : 1
  Loop detected                  : Yes
    Detected on VLAN             : 1
    Detected at                  : 2022-09-22T05:25:00
  Interface status               : down
```

5. Review the interface 1/1/25.

```
sw-edge1(config-lag-if)# show interface 1/1/25

Interface 1/1/25 is down
 Admin state is up
 State information: Network loop detected
...
```

- **Question**: What is the state information?
- **Answer**: The state information reports network loop detected.

6. Check the current configuration of the looped port.

```
sw-edge1(config-lag-if)# show running-config int lag50
interface lag 50
    no shutdown
    no routing
    vlan trunk native 1
    vlan trunk allowed all
    lacp mode active
    rate-limit unknown-unicast 100 pps
    rate-limit broadcast 100 pps
    rate-limit multicast 100 pps
    rate-limit icmp ip-all 100 pps
    loop-protect
    spanning-tree bpdu-filter
    exit
sw-edge1(config-lag-if)#
```

- **Question**: Is the port disabled in the configuration with a **shutdown** command?
- **Answer**: No, the port is operationally disabled, not through the **shutdown** configuration command.

Cleanup

First you will recover the looped status on the ports; the last step will be disabling the LAG 50.

1. On sw-edge1, remove the STP BPDU filter on port lag 50.

```
sw-edge1(config)# interface lag 50
sw-edge1(config-lag-if)# no spanning-tree bpdu-filter
```

4. Remove loop protect on the port lag 50.

```
sw-edge1(config-lag-if)# no loop-protect
sw-edge1(config-lag-if)# exit
```

2. Bounce the physical ports to clear the loop status; STP is now active again and it will set the port in Discarding state.

```
sw-edge1(config)# interface 1/1/25-1/1/26
sw-edge1(config-if-<1/1/25-1/1/26>)# shutdown
sw-edge1(config-if-<1/1/25-1/1/26>)# no shutdown
sw-edge1(config-if-<1/1/25-1/1/26>)# exit
```

3. Verify with **show interface brief** that the ports 1/1/25, 1/1/26, and the LAG 50 are **up** (This is physical up; STP is now blocking traffic on the ports.)

```
sw-edge1(config)# show interface brief
------------------------------------------------------------------------------
--------------------------------------
Port       Native  Mode   Type     Enabled Status   Reason
Speed    Description
           VLAN
(Mb/s)
------------------------------------------------------------------------------
-----------------...
1/1/27     1       trunk  SFP+DAC1 yes     up
10000    sw-agg1
1/1/28     1       trunk  SFP+DAC1 yes     up
10000    sw-agg2
vlan1      --             --       no      down     Administratively down
--       --
vlan3      --             --       yes     up
--       --
lag50      1       trunk  --       yes     up       --
20000    --
...
```

4. As a last step, shut down the LAG 50.

```
sw-edge1(config)# interface lag 50
sw-edge1(config-lag-if)# shutdown
sw-edge1(config-lag-if)# exit
```

5. Save the configurations on **all** switches.

```
Sw-edge1(config)# write memory
```

Lab 04.02 VRRP

In your customer deployment, the final setup will use Aruba VSX and active gateway as the first hop redundancy protocol. This will offer an active-active default gateway function for the endpoints.

The active gateway feature is specific for the AOS-CX switches with VSX. In some environments, such as an SD-Branch gateway deployment, it may not be available and an alternative default gateway solution can be used, such as VRRP. Your senior colleague wants you to become familiar with the VRRP technology and has asked you to configure it in one of the VLANs.

The VRRP configuration will be replaced with the VSX active gateway configuration in a later lab activity.

Objectives

After completing this lab, you will be able to:

- Configure VRRP between two Aruba switches.
- Review the status of VRRP.
- Verify the VRRP failover.

Task 1: Reconfigure sw-agg1 with VRRP on SVI11

In this task you will change the sw-agg1 SVI11 configuration. You will set the local IP address to 10.1.11.2/24 and you will activate VRRP with IP Address 10.1.11.1. This is not a redundant configuration yet, sw-agg2 will be added as the redundant VRRP peer in the next task.

Objectives

- Configure VRRP.
- Verify VRRP Operation.

Steps

1. On **PC1**, open a command prompt and review the ARP cache entry for 10.1.11.1.

```
C:> arp -a 10.1.11.1

Interface: 10.1.11.50 --- 0xe
  Internet Address       Physical Address      Type
  10.1.11.1              b8-d4-e7-da-80-00     dynamic
```

   - Take note of the current MAC address of 10.1.11.1: ______________________________

2. Using PC-MGMT, open an SSH connection to sw-agg1.
3. On **sw-agg1**, check the current SVI 11 MAC address.

```
sw-agg1(config)# show interface vlan11

Interface vlan11 is up
 Admin state is up
 Description:
 Hardware: Ethernet, MAC Address: b8:d4:e7:da:80:00
 IPv4 address 10.1.11.1/24
    active-gateway L3 source mac b8:d4:e7:da:80:00
 L3 Counters: Rx Disabled, Tx Disabled
```

- **Question**: Does the SVI 11 MAC address match the ARP cache entry of PC1?
- **Answer**: Yes

4. Check the global VRRP status.

```
sw-agg1(config)# show vrrp

VRRP is not enabled
```

5. Enable VRRP globally and verify the status again.

```
sw-agg1(config)# router vrrp enable
sw-agg1(config)# show vrrp

VRRP is enabled
```

6. Change the SVI11 IP address to 10.1.11.2/24.

```
sw-agg1(config)# interface vlan 11
sw-agg1(config-if-vlan)# ip address 10.1.11.2/24
```

7. Create a VRRP VRID instance on SVI11; this will put you to the VRID context.

```
sw-agg1(config-if-vlan)# vrrp 11 address-family ipv4
```

8. Set the VRRP primary address to 10.1.11.1.

```
sw-agg1(config-if-vrrp)# address 10.1.11.1 primary
```

9. Set the VRRP VRID priority to 110.

```
sw-agg1(config-if-vrrp)# priority 110
```

10. Enable the VRRP VRID 11 and return to the SVI configuration context.

```
sw-agg1(config-if-vrrp)# no shutdown
sw-agg1(config-if-vrrp)# exit
```

11. Review the current SVI configuration. Verify you have the updated SVI IP address and the VRRP configuration.

```
sw-agg1(config-if-vlan)# show running-config current-context
interface vlan 11
    ip address 10.1.11.2/24
    ip helper-address 10.254.1.21
    ip ospf 1 area 0.0.0.0
    vrrp 11 address-family ipv4
        address 10.1.11.1 primary
        priority 110
        no shutdown
        exit
```

12. Review the VRRP status.

```
sw-agg1(config-if-vlan)# show vrrp

VRRP is enabled

Interface vlan11 - Group 11 - Address-Family IPv4
  State is ACTIVE
  State duration 31.386 secs
  Virtual IP address is 10.1.11.1
  Virtual MAC address is 00:00:5e:00:01:0b
  Advertisement interval is 1000 msec
  Version is 2
  Preemption is enabled
   min delay is 0 sec
  Priority is 110
  Active Router is 10.1.11.2 (local)
  Active Advertisement interval is 1000 msec
  Active Down interval is 3570 msec
```

- **Question**: Do you see interface vlan11 in the VRRP list?
- **Answer**: Yes
- **Question**: What is the current state?
- **Answer**: Active
- **Question**: What is the virtual MAC Address for this VRID?
- **Answer**: 00:00:5e:00:01:0b

13. Connect to **PC1** and check the ARP cache entry for 10.1.11.1 again.

```
C:> arp -a 10.1.11.1

Interface: 10.1.11.50 --- 0xe
  Internet Address      Physical Address      Type
  10.1.11.1             00-00-5e-00-01-0b     dynamic
```

- **Question**: What do you observe?
- **Answer**: The MAC address for the entry has changed to the VRRP Virtual MAC address 00:00:5e:00:01:0b.

Task 2: Configure sw-agg2 with VRRP on SVI11

In this task you will configure sw-agg2 with VRRP on SVI11. Sw-agg2 will act as the standby VRRP host, while sw-agg1 will be the active VRRP host.

On sw-agg2, you will first add SVI11 and assign it the .3 IP address in the subnet. Next you will activate VRRP with the default priority and with VIP 1. Finally, you will review the VRRP status on the active (sw-agg1) and standby (sw-agg2) systems.

Objectives

- Configure the standby VRRP host.
- Verify the standby VRRP status.

Steps

1. Use PC-MGMT to open an SSH connection to **sw-agg2.**
2. **sw-agg2** does not have an SVI for VLAN 11 yet. Review the layer 3 IP interfaces to verify that vlan11 is **not** listed.

```
sw-agg2(config)# show ip interface brief
```

3. Create SVI 11 and assign static IP address 10.1.11.3/24.

```
sw-agg2(config)# interface vlan 11
sw-agg2(config-if-vlan)# ip address 10.1.11.3/24
```

4. Enable the SVI for OSPF.

```
sw-agg2(config-if-vlan)# ip ospf 1 area 0
```

5. Now configure sw-agg2 SVI11 with the following VRRP properties:
   - VRID: **11**
   - Primary IP: **10.1.11.1**

```
sw-agg2(config-if-vlan)# vrrp 11 address-family ipv4
sw-agg2(config-if-vrrp)# address 10.1.11.1 primary
```

6. Review the VRRP status.

```
sw-agg2(config-if-vrrp)# show vrrp

VRRP is not enabled

Interface vlan11 - Group 11 - Address-Family IPv4
  State is INIT (Group Disabled)
  State duration
  Virtual IP address is 10.1.11.1
  Virtual MAC address is 00:00:5e:00:01:0b
  Advertisement interval is 1000 msec
  Version is 2
  Preemption is enabled
   min delay is 0 sec
  Priority is 100
  Active Router is 0.0.0.0
  Active Advertisement interval is 1000 msec
  Active Down interval is 3609 msec
```

   - **Question**: What is the state of VRRP on Interface vlan11?
   - **Answer**: The state is INIT (Group Disabled).
   - **Question**: What should you change to enable the VRID group?
   - **Answer**: Enable the VRID.

7. Enable the VRID.

```
sw-agg2(config-if-vrrp)# no shutdown
```

8. Review the VRRP status again.

```
sw-agg2(config-if-vrrp)# show vrrp

VRRP is not enabled

Interface vlan11 - Group 11 - Address-Family IPv4
  State is INIT
  State duration
  Virtual IP address is 10.1.11.1
  Virtual MAC address is 00:00:5e:00:01:0b
  Advertisement interval is 1000 msec
  Version is 2
  Preemption is enabled
   min delay is 0 sec
  Priority is 100
  Active Router is 0.0.0.0
  Active Advertisement interval is 1000 msec
  Active Down interval is 3609 msec
```

- **Question**: What is the VRRP state?
- **Answer**: Interface VLAN 11 state is INIT; the global state is not enabled.
- **Question**: What should you change to enable VRRP?
- **Answer**: Enable VRRP at the global level.

9. Return to the global configuration level.

```
sw-agg2(config-if-vrrp)# exit
sw-agg2(config-if-vlan)# exit
```

10. Enable VRRP at the switch global configuration level.

```
sw-agg2(config)# router vrrp enable
```

11. Verify the VRRP state again.

```
sw-agg2(config)# show vrrp

VRRP is enabled

Interface vlan11 - Group 11 - Address-Family IPv4
  State is STANDBY
  State duration 02.158 secs
  Virtual IP address is 10.1.11.1
  Virtual MAC address is 00:00:5e:00:01:0b
  Advertisement interval is 1000 msec
  Version is 2
  Preemption is enabled
   min delay is 0 sec
  Priority is 100
  Active Router is 10.1.11.2
  Active Advertisement interval is 1000 msec
  Active Down interval is 3609 msec
```

- **Question**: What is the VRRP state for interface vlan11?
- **Answer**: STANDBY
- **Question**: What is currently the active router?
- **Answer**: The active router is 10.1.11.2 (sw-agg1).
- **Question**: What is the default priority of VRRP?
- **Answer**: The default priority is 100. Since sw-agg1 has been configured with VRRP priority 110, it will be the active VRRP host.

Task 3: Test the VRRP Failover

In this task you will test the VRRP VIP failover. On the PC1 you will initiate some test traffic with a continuous ping. While this ping is active, you will disable the sw-agg1 SVI11 interface. VRRP on sw-agg2 will detect that the VRRP keepalive packets are missing and it will resume the active state on the configured VIP.

Objectives

- Verify the VRRP failover from active to standby.
- Verify the VRRP preemption feature.

Steps

1. Open a connection to PC1.
2. Open a command prompt and start a continuous ping to 10.254.1.21.

```
C:> ping 10.254.1.21 -t
```

3. Open a *second* command prompt and run a traceroute to 10.254.1.21.

```
C:> tracert -d 10.254.1.21

Tracing route to 10.254.1.21 over a maximum of 30 hops

  1    <1 ms    <1 ms    <1 ms  10.1.11.2
  2     2 ms     1 ms     1 ms  10.254.101.254
  3     2 ms     1 ms     1 ms  10.254.1.21

Trace complete.
```

- **Question**: What is the IP address of the first router in the path?
- **Answer**: Currently, sw-agg1 with IP 10.1.11.2 is the first router that responds in the traceroute path.

4. Use PC-MGMT to open an SSH connection to **sw-agg1.**
5. Enter the SVI11 context and shut down the SVI interface.

```
sw-agg1(config)# interface vlan 11
sw-agg1(config-if-vlan)# shutdown
```

6. On PC-1, verify that the ping continued after a few seconds.

```
Reply from 10.254.1.21: bytes=32 time=2ms TTL=126
Reply from 10.254.1.21: bytes=32 time=1ms TTL=126
Request timed out.
Reply from 10.254.1.21: bytes=32 time=2ms TTL=126
Reply from 10.254.1.21: bytes=32 time=2ms TTL=126
```

- **Question**: What is the failover time for VRRP?
- **Answer**: VRRP takes about 3-4 seconds to detect and take over the Virtual IP address by default.

7. On PC-1, run the traceroute to 10.254.1.21 again.

```
C:> tracert -d 10.254.1.21

Tracing route to 10.254.1.21 over a maximum of 30 hops

  1    <1 ms    <1 ms    <1 ms  10.1.11.3
  2     2 ms     1 ms     1 ms  10.254.102.254
  3     1 ms     1 ms     1 ms  10.254.1.21

Trace complete.
```

- **Question**: What is the first router in the path?
- **Answer**: The first router is now sw-agg2 with IP address 10.1.11.3.

8. On sw-agg1, review the VRRP state.

```
sw-agg1(config-if-vlan)# show vrrp

VRRP is enabled

Interface vlan11 - Group 11 - Address-Family IPv4
  State is INIT (Interface Down)
  State duration 01 mins 49.683 secs
...
```

9. On sw-agg2, review the VRRP state.

```
sw-agg2(config)# show vrrp

VRRP is enabled

Interface vlan11 - Group 11 - Address-Family IPv4
  State is ACTIVE
  State duration 02 mins 22.561 secs
...
```

- **Question**: What is the sw-agg2 state?
- **Answer**: sw-agg2 is now ACTIVE.

10. This demonstrates the VRRP failover function. Now you will explore what happens when the original sw-agg1 SVI 11 comes back again. On sw-agg1, review the VRRP state.

```
sw-agg1(config-if-vlan)# show vrrp

VRRP is enabled

Interface vlan11 - Group 11 - Address-Family IPv4
  State is INIT (Interface Down)
  State duration 01 mins 49.683 secs
  Virtual IP address is 10.1.11.1
  Virtual MAC address is 00:00:5e:00:01:0b
  Advertisement interval is 1000 msec
  Version is 2
  Preemption is enabled
   min delay is 0 sec
  Priority is 110
  Active Router is 0.0.0.0
  Active Advertisement interval is 1000 msec
  Active Down interval is 3570 msec
```

- **Question**: What is the preemption state for interface vlan11?
- **Answer**: Preemption is enabled by default.
- **Question**: What is the expected behavior when preemption is enabled?
- **Answer**: With preemption enabled, when a VRRP with a better priority returns to service, it will claim control of the Virtual IP address.

11. Verify the preemption behavior. On sw-agg1, enable SVI11 again.

```
sw-agg1(config)# interface vlan 11
sw-agg1(config-if-vlan)# no shutdown
```

12. Wait about 3 seconds, then review the VRRP state.

```
sw-agg1(config-if-vlan)# show vrrp

VRRP is enabled

Interface vlan11 - Group 11 - Address-Family IPv4
  State is ACTIVE
  State duration 10.691 secs
```

13. On sw-agg2, review the VRRP state.

```
sw-agg2(config)# show vrrp

VRRP is enabled

Interface vlan11 - Group 11 - Address-Family IPv4
  State is STANDBY
  State duration 35.574 secs
```

- **Question**: What is the state on sw-agg2?
- **Answer**: Sw-agg2 has now reverted to STANDBY, since sw-agg1 performed a preempt of the ACTIVE role.

14. Save the configuration on both aggregation switches.

```
sw-agg2(config)# write memory
```

Task 4: Optional – MAC address tracing of the aggregation switches

Trace the MAC address of traffic sent to the router and traffic from the router to the endpoint when a failover occurs.

Objectives

- Understand VRRP MAC address use in the network.

Steps

Prepare the setup

1. Use PC-MGMT to open an SSH connection to **sw-agg2**.
2. By default, OSPF ECMP would have the core router distribute traffic destined to 10.1.11.0/24 to both sw-agg1 and sw-agg2. In this lab we need all the traffic to return via sw-agg1. You will achieve this by increasing the cost of SVI11 on sw-agg2.
3. Enter the SVI 11 context and set OSPF cost 200. The default cost is 100.

```
sw-agg2(config)# interface vlan 11
sw-agg2(config-if-vlan)# ip ospf cost 200
```

Start test traffic

4. On PC1, start a continuous ping to 10.254.1.21.

```
C:> ping 10.254.1.21 -t
```

5. On PC-1, start Wireshark trace on the LAB NIC.
6. Set a display filter to **icmp**. Don't forget to press <**Enter**> to activate the filter (see Figure 4-40).

Figure 4-40: Wireshark—filtering ICMP traffic

7. Open an SSH connection to sw-agg1.
8. Enter the SVI 11 and shut down the SVI.

```
sw-agg1(config)# interface vlan 11
sw-agg1(config-if-vlan)# shutdown
sw-agg1(config-if-vlan)# exit
```

9. On PC-1, wait until the ping continues, then stop the Wireshark trace.

Analyze the return traffic

1. Explore **ICMP request** from the beginning of the trace; check the destination MAC. This should be the VRRP MAC (see Figure 4-41).

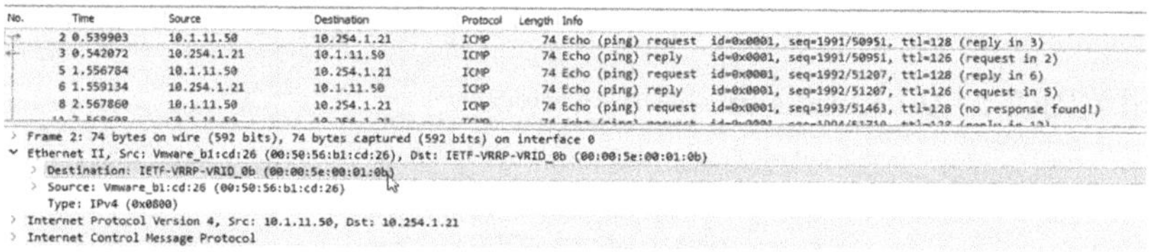

Figure 4-41: Wireshark—look for the VRRP vMAC address

2. Explore **ICMP reply** from the beginning of the trace; check the source MAC. This is the sw-agg1 SVI 11 MAC (see Figure 4-42).

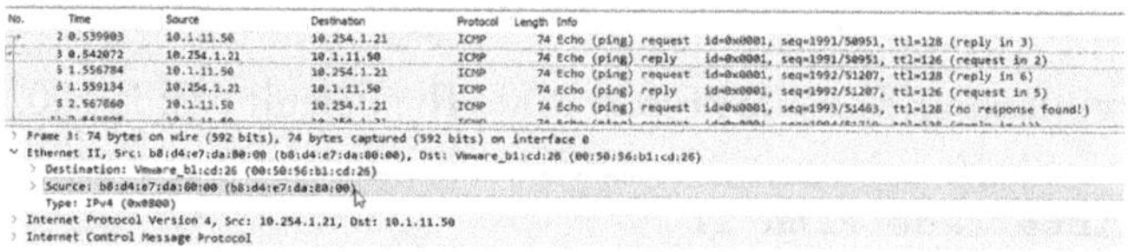

Figure 4-42: Wireshark—expand the ICMP reply

3. Now move to the end of the trace.

4. Explore the **ICMP request** from the end of the trace; check the destination MAC. This should be the VRRP MAC. This shows that from the endpoint point of view, the default MAC address does not change (see Figure 4-43).

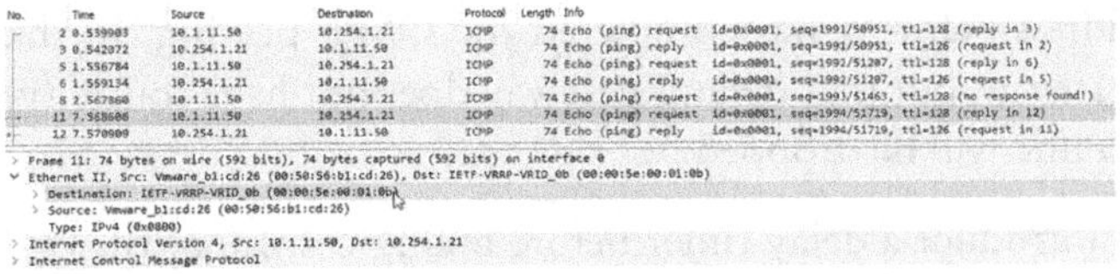

Figure 4-43: Wireshark—expand the ICMP request

5. Explore **ICMP reply** from the end of the trace - check the source MAC. This is the sw-agg2 SVI11 MAC (see Figure 4-44).

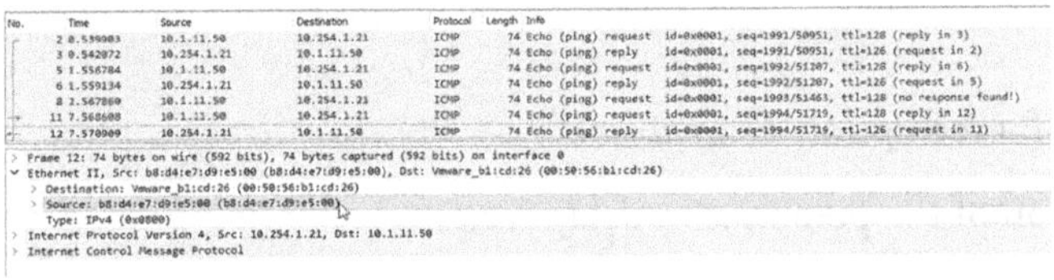

Figure 4-44: Wireshark—expand the ICMP reply and check the source MAC

Note

Example traces from the screenshots can be found in the ACAF Student Files on the desktop of PC-MGMT: lab04.02-vrrp-svi-mac-trace.pcapng.

Cleanup

1. On sw-agg2, restore the SVI11 OSPF cost to the default by removing the configured cost.

```
sw-agg2(config)# interface vlan 11
sw-agg2(config-if-vlan)# no ip ospf cost
sw-agg2(config-if-vlan)# exit
```

2. On sw-agg1, enable the SVI11 again.

```
sw-agg1(config)# interface vlan 11
sw-agg1(config-if-vlan)# no shutdown
sw-agg1(config-if-vlan)# exit
```

3. On PC1, close the command prompt and Wireshark.

Task 5: Optional – VRRP preempt delay

In some scenarios, performing immediate VRRP pre-emption may have a negative side-effect. When the router needs to establish upstream routing adjacency, for example with OSPF, the router may take control of the VIP address before it has completed the OSPF peering. In that case, the clients will have a (new) default gateway, but that default gateway does not have routes (yet) to forward the traffic. The result is that traffic will be dropped.

To solve this, you can introduce a delay timer before taking control of the endpoint default gateway function. This gives upstream routing a chance to learn routes before claiming control of the default gateway VIP.

In this lab activity, it applies to upstream OSPF adjacency, but in case of SD-WAN gateways, it may apply to establishing upstream IPsec connections and learning SD-WAN routes.

Objectives

- Configure a VRRP preemption delay timer.
- Verify the operation of the preemption delay timer.

Steps

Verify the problem

1. On **PC1**, open a command prompt and start a continuous ping to 10.254.1.21.

```
C:> ping 10.254.1.21 -t
```

2. Use PC-MGMT to open an SSH connection to **sw-agg1**.
3. Save the configuration and reboot the sw-agg1; confirm the reboot.

```
sw-agg1# write memory
sw-agg1# boot system
```

4. On **PC1**, verify the failover takes about 3 seconds (1 lost ping).

```
Reply from 10.254.1.21: bytes=32 time=1ms TTL=126
Reply from 10.254.1.21: bytes=32 time=3ms TTL=126
Request timed out.
Reply from 10.254.1.21: bytes=32 time=4ms TTL=126
Reply from 10.254.1.21: bytes=32 time=1ms TTL=126
```

5. Wait for the sw-agg1 to complete the reboot (about 2 minutes); keep checking the PC1 ping output.

```
Reply from 10.254.1.21: bytes=32 time=3ms TTL=126
Reply from 10.254.1.21: bytes=32 time=2ms TTL=126
Reply from 10.254.1.21: bytes=32 time=2ms TTL=126
Request timed out.
Reply from 10.1.11.2: Destination net unreachable.
Reply from 10.1.11.2: Destination net unreachable.
Reply from 10.1.11.2: Destination net unreachable.
Reply from 10.1.11.2: Destination net unreachable.
Request timed out.
Reply from 10.1.11.2: Destination net unreachable.
Reply from 10.1.11.2: Destination net unreachable.
Reply from 10.1.11.2: Destination net unreachable.
Reply from 10.1.11.2: Destination net unreachable.
Request timed out.
Reply from 10.1.11.2: Destination net unreachable.
Reply from 10.1.11.2: Destination net unreachable.
Request timed out.
Reply from 10.254.1.21: bytes=32 time=1ms TTL=126
Reply from 10.254.1.21: bytes=32 time=2ms TTL=126
Reply from 10.254.1.21: bytes=32 time=2ms TTL=126
```

- **Question**: What do you observe when sw-agg1 comes back online?
- **Answer**: Several ping packets are lost when the sw-agg1 returned to service.
- **Question**: What is the reported ping error and what IP address reports this error?
- **Answer**: The echo reply comes from 10.1.11.2, with the message "Destination net unreachable".
- **Question**: What does this error indicate?
- **Answer**: This indicates that 10.1.11.2 (sw-agg1) is receiving the ICMP request, since it is the active VRRP host. However, it does not have a route to the requested destination (10.254.1.21), since the OSPF adjacency was not established yet. Some seconds later, when the OSPF adjacency is established and the routing tables have been populated, the ping works again.
- **Question**: How could this timing problem be solved?
- **Answer**: The VRRP system should be delayed a little bit to give the upstream routing protocol (OSPF) the opportunity to learn the upstream routes. A typical delay is 60 seconds.

Configure Preempt Delay Timer on sw-agg1

1. On **sw-agg1**, configure a preempt delay timer of 60 seconds.

```
sw-agg1(config)# interface vlan 11
sw-agg1(config-if-vlan)# vrrp 11 address-family ipv4
sw-agg1(config-if-vrrp)# preempt delay minimum 60
sw-agg1(config-if-vrrp)# exit
sw-agg1(config-if-vlan)# exit
```

Verify the configured solution

1. On **PC1**, verify the ping to 10.254.1.21 is still running.
2. On **sw-agg1**, save the configuration and reboot the switch.

```
sw-agg1# write memory
sw-agg1# boot system
```

3. On **PC1**, verify the ping continues when sw-agg1 completes the reboot.

Cleanup steps

1. On PC1, close the command prompt.
2. There is no need to revert the preempt delay timer on sw-agg1.

Lab 04.03 Introduction to VSX

In this lab you will be introduced to an Aruba redundant aggregation switch configuration, known as VSX. Your senior colleague has completed the aggregation switch configuration for the customer setup and wants you to review the VSX configuration that will be active on the aggregation switches. The working VSX configuration will be loaded to the sw-agg1 and sw-agg2 systems, and you will be able to explore the status and the basic configuration of VSX.

As the associate engineer, you will then connect the sw-edge1 in a redundant way to the VSX aggregation using a VSX LAG. A VSX LAG is also known as multi-chassis link aggregation (MCLAG).

In the last task you will explore Aruba VSX active gateway. This is an active-active default gateway solution and replaces the active-standby VRRP protocol in a VSX cluster.

Objectives

After completing this lab, you will be able to:

- Review a working VSX system status.
- Review an MC-LAG port configuration.
- Verify the VSX MC-LAG operation from the access switch point of view.
- Understand the LACP fallback feature.
- Verify the active-active gateway configuration.

Task 1: Prepare the topology

In this task you will load the correct start configuration for this lab on

- sw-agg1
- sw-agg2

Your senior colleague has prepared the VSX configuration for the customer aggregation switches. You have been asked to apply the configuration to the sw-agg1 and sw-agg2 switches.

Objectives

- Apply a prepared configuration to the aggregation switches.

Steps

Important
Make sure you apply the *correct* configuration to *each* aggregation switch in the next steps. Please ***double-check*** that you are applying sw-agg1 configuration to the actual sw-agg1 host, and the sw-agg2 configuration to the sw-agg2 host.

sw-agg1

1. Use PC-MGMT to open an SSH connection to sw-agg1.
2. On PC-MGMT, open the folder **ACAF Student Files** on the desktop; copy the contents of the file named **lab04.03-vsx-to-load-on-sw-agg1.txt.**
3. Paste the commands in the SSH connection of **sw-agg1**.

sw-agg2

1. Use PC-MGMT to open an SSH connection to sw-agg2.
2. On PC-MGMT, open the folder **ACAF Student Files** on the desktop; copy the contents of the file named **lab04.03-vsx-to-load-on-sw-agg2**.
3. Paste the commands in the SSH connection of **sw-agg2**.

With this configuration applied, the aggregation switches have been configured with VSX. In the next task you will explore the status of VSX on the aggregation switches.

Task 2: Review global VSX configuration and status

In this task you will review the VSX configuration that was prepared by your senior network administrator colleague. This will allow you to understand the basic concept of VSX redundancy at the aggregation layer. You will review the global VSX configuration and status, the roles assigned to the switches in the VSX cluster, and the VSX system MAC.

Objectives

- Understand the VSX components.
- Describe the VSX ISL.
- Describe the VSX device roles.
- Understand the VSX system MAC.

Steps

1. Open an SSH connection to sw-agg1.
2. First review the connections to sw-agg2.

```
sw-agg1(config)# show lldp neighbor-info

LLDP Neighbor Information
=========================

Total Neighbor Entries          : 6
Total Neighbor Entries Deleted  : 0
Total Neighbor Entries Dropped  : 0
Total Neighbor Entries Aged-Out : 0
```

```
LOCAL-PORT   CHASSIS-ID           PORT-ID                    PORT-DESC
TTL       SYS-NAME
--------------------------------------------------------------------------
----------------------------------
1/1/1        64:e8:81:dd:81:80    1/1/27                     sw-agg1
120       sw-edge1
1/1/2        8c:85:c1:49:20:c0    1/1/27                     sw-agg1
120       sw-edge2
...
1/1/46       b8:d4:e7:d9:e5:00    1/1/46                     1/1/46
120       sw-agg2
1/1/47       b8:d4:e7:d9:e5:00    1/1/47                     1/1/47
120       sw-agg2
...
```

- **Question**: What ports are connected to sw-agg2?
- **Answer**: Ports 1/1/46 and 1/1/47

3. Review the LAG port members.

```
sw-agg1(config)# show lacp interfaces

State abbreviations :
A - Active          P - Passive       F - Aggregable I - Individual
S - Short-timeout L - Long-timeout N - InSync      O - OutofSync
C - Collecting      D - Distributing
X - State m/c expired                 E - Default neighbor state

Actor details of all interfaces:
--------------------------------------------------------------------------
---------
Intf         Aggr        Port    Port   State    System-ID          System Aggr
Forwarding
             Name        Id      Pri                                Pri    Key
State
--------------------------------------------------------------------------
---------
1/1/1        lag1(mc)    1       1      IE       02:01:00:00:01:00 65534  1      up
1/1/2        lag2(mc)    2       1      IE       02:01:00:00:01:00 65534  2      up
1/1/46       lag256      47      1      ALFNCD   b8:d4:e7:da:80:00 65534  256    up
1/1/47       lag256      48      1      ALFNCD   b8:d4:e7:da:80:00 65534  256    up

Partner details of all interfaces:
--------------------------------------------------------------------------
---------
```

```
Intf         Aggr         Port  Port  State    System-ID            System  Aggr
             Name         Id    Pri                                 Pri     Key
--------------------------------------------------------------------------------
---------
1/1/1        lag1(mc)     0     0     IE       00:00:00:00:00:00 0          0
1/1/2        lag2(mc)     0     0     IE       00:00:00:00:00:00 0          0
1/1/46       lag256       47    1     ALFNCD   b8:d4:e7:d9:e5:00 65534      256
1/1/47       lag256       48    1     ALFNCD   b8:d4:e7:d9:e5:00 65534      256
```

- **Question**: What LAG do ports 1/1/46 and 1/1/47 belong to?
- **Answer**: These ports belong to LAG256.

4. Review the VSX status output.

```
sw-agg1(config)# show vsx status
VSX Operational State
---------------------
  ISL channel              : In-Sync
  ISL mgmt channel         : operational
  Config Sync Status       : In-Sync
  NAE                      : peer_reachable
  HTTPS Server             : peer_reachable

Attribute           Local               Peer
------------        --------            --------
ISL link            lag256              lag256
ISL version         2                   2
System MAC          02:01:00:00:01:00   02:01:00:00:01:00
Platform            8325                8325
Software Version    GL.10.09.1040       GL.10.09.1040
Device Role         primary             secondary
```

- **Question**: What is the VSX role of sw-agg1?
- **Answer**: sw-agg1 has been configured as primary.
- **Question**: What is the Inter Switch Link (ISL) used for VSX communication between sw-agg1 and sw-agg2?
- **Answer**: LAG 256. Typically, a redundant link, such as a link aggregation, is used as the ISL.
- **Question**: What is the system MAC?
- **Answer**: On the lab VSX system, the system MAC has been set to 02:01:00:00:01:00.
- **Question**: What is the difference between the VSX system MAC and the base switch MAC?
- **Answer**: The base switch MAC is used as the *local* layer 2 identifier of the switch, for example in LLDP communication. The VSX system MAC is used in VSX *shared* layer 2 protocols, such as LACP on the VSX-LAG and the STP BID on a VSX system.

Review the VSX LAG configuration

In the previous labs, you have configured a LAG between sw-edge1 and sw-edge2. A traditional LAG requires all the member ports to be connected between the same two devices. With VSX, the adjacent devices can be connected with a LAG, using one link(s) to the primary VSX node and another link(s) to the secondary VSX node (multiple links are also possible).

Examples for adjacent device would be:

- An access switch with a LAG
- A server with NIC teaming
- A firewall with port bonding
- An Aruba gateway with a port-channel

The senior administrator has prepared the VSX system with VSX LAG 1 for sw-edge1 and VSX LAG 2 connecting to sw-edge2.

1. Review the VSX settings in the running configuration; look for the LAG interfaces.

```
sw-agg1(config)# show running-config vsx
vsx
    system-mac 02:01:00:00:01:00
    inter-switch-link lag 256
    role primary
    vsx-sync dhcp-relay mclag-interfaces ospf vsx-global
interface lag 256
    no shutdown
    no routing
    vlan trunk native 1
    vlan trunk allowed all
    lacp mode active
interface 1/1/47
    no shutdown
    lag 256
interface 1/1/46
    no shutdown
    lag 256
interface lag 1 multi-chassis
    no shutdown
    no routing
    vlan trunk native 1
    vlan trunk allowed all
    lacp mode active
...
```

- **Question**: Do you see anything special for interface lag 1?
- **Answer**: The interface lag 1 has a new option: *multi-chassis*. This indicates that the state will be shared with the VSX peer.

2. Review the LACP interfaces.

```
sw-agg1(config)# show lacp interfaces

State abbreviations :
A - Active          P - Passive      F - Aggregable I - Individual
S - Short-timeout L - Long-timeout N - InSync      O - OutofSync
C - Collecting     D - Distributing
X - State m/c expired                 E - Default neighbor state

Actor details of all interfaces:
------------------------------------------------------------------------------
-----------
Intf        Aggr        Port  Port  State    System-ID          System Aggr
Forwarding
            Name        Id    Pri                               Pri    Key
State
------------------------------------------------------------------------------
-----------
1/1/1       lag1(mc)    1     1     IE       02:01:00:00:01:00 65534  1     up
1/1/2       lag2(mc)    2     1     IE       02:01:00:00:01:00 65534  2     up
1/1/46      lag256      47    1     ALFNCD   b8:d4:e7:da:80:00 65534  256   up
1/1/47      lag256      48    1     ALFNCD   b8:d4:e7:da:80:00 65534  256   up

Partner details of all interfaces:
------------------------------------------------------------------------------
-----------
Intf        Aggr        Port  Port  State    System-ID          System Aggr
            Name        Id    Pri                               Pri    Key
------------------------------------------------------------------------------
-----------
1/1/1       lag1(mc)    0     0     IE       00:00:00:00:00:00 0      0
1/1/2       lag2(mc)    0     0     IE       00:00:00:00:00:00 0      0
1/1/46      lag256      47    1     ALFNCD   b8:d4:e7:d9:e5:00 65534  256
1/1/47      lag256      48    1     ALFNCD   b8:d4:e7:d9:e5:00 65534  256
```

- **Question**: What does *mc* represent in the lag1(mc) output?
- **Answer**: This indicates that LAG1 is a MCLAG (multi-chassis), as opposed to the LAG256, that is a 'local' LAG.

- **Question**: What do the sections 'Actor' and 'Partner' represent?
- **Answer**:
  - **Actor:** Represents the local LACP communication information: what this side of the LAG will advertise to the adjacent device.
  - **Partner:** Represents the LACP information that has been received from the partner adjacent device. In case no LACP information has been received from the partner, the System-ID will be shown as 00:00:00:00:00:00. In this lab environment, the sw-edge1 or sw-edge2 will be the LACP partner.
- **Question**: In the *Actor* section, why is the System-ID for the LAG 1 different from the LAG 256?
- **Answer**: LAG 1 is an MCLAG, so both sw-agg1 and sw-agg2 will use this VSX system MAC in their LACP communication with the sw-edge1 access switch. LAG 256 is a local LAG between sw-agg1 and sw-agg2, so each aggregation switch will use its own Layer 2 identifier (MAC address).

3. Since both sw-agg1 and sw-agg2 are part of the MC-LAG configuration and status, they need to be aware of each other's LACP states. This will be synchronized by VSX over the ISL. To see the LACP state for an MC-LAG, use the command **show lacp interfaces multi-chassis**.

```
sw-agg1(config)# show lacp interfaces multi-chassis

State abbreviations :
A - Active         P - Passive      F - Aggregable I - Individual
S - Short-timeout L - Long-timeout N - InSync     O - OutofSync
C - Collecting     D - Distributing
X - State m/c expired              E - Default neighbor state

Actor details of all interfaces:
----------------------------------------------------------------------------
----------
Intf         Aggregate  Port     Port       State   System-ID          System
Aggr
             name       id       Priority                              Priority
Key
----------------------------------------------------------------------------
----------
1/1/1        lag1(mc)   1        1          IE      02:01:00:00:01:00  65534
1
1/1/2        lag2(mc)   2        1          IE      02:01:00:00:01:00  65534
2
```

```
Partner details of all interfaces:
-----------------------------------------------------------------------------
----------
Intf       Aggregate  Partner Port     State   System-ID          System
Aggr
           name       Port-id Priority                            Priority
Key
-----------------------------------------------------------------------------
----------
1/1/1      lag1(mc)   0       0        IE      00:00:00:00:00:00  0
0
1/1/2      lag2(mc)   0       0        IE      00:00:00:00:00:00  0
0

Remote Actor details of all interfaces:
-----------------------------------------------------------------------------
----------
Intf       Aggregate  Port    Port     State   System-ID          System
Aggr
           name       id      Priority                            Priority
Key
-----------------------------------------------------------------------------
----------
1/1/1      lag1(mc)   1001    1        IE      02:01:00:00:01:00  65534
1

Remote Partner details of all interfaces:
-----------------------------------------------------------------------------
----------
Intf       Aggregate  Partner Port     State   System-ID          System
Aggr
           name       Port-id Priority                            Priority
Key
-----------------------------------------------------------------------------
----------
1/1/1      lag1(mc)   0       0        IE      00:00:00:00:00:00  0
0
```

- **Question**: What is different in this output compared to the basic **show lacp interfaces** command?
- **Answer**: This command also includes *Remote Actor* and *Remote Partner* output. These sections show the status on the VSX peer. Since you have executed this command on the sw-agg1, the Remote Actor shows the 'actor' state of sw-agg2.
- **Question**: Is the Remote Actor using a different system-ID compared to the Actor system-ID?
- **Answer**: No, for MC-LAGs, both VSX switches will identify themselves using the VSX System MAC, so the adjacent device believes it is connected to the same device.

LACP Fallback

A successful LACP connection, such as the LAG 256, has LACP State *ALFNCD* listed.

1. Check the output of **show lacp interfaces**.

```
sw-agg1(config)# show lacp interfaces

State abbreviations :
A - Active          P - Passive       F - Aggregable I - Individual
S - Short-timeout L - Long-timeout N - InSync      O - OutofSync
C - Collecting      D - Distributing
X - State m/c expired                 E - Default neighbor state

Actor details of all interfaces:
----------------------------------------------------------------------------
----------
Intf         Aggr         Port  Port  State   System-ID            System  Aggr
Forwarding
             Name         Id    Pri                                Pri     Key
State
----------------------------------------------------------------------------
----------
1/1/1        lag1(mc)     1     1     IE      02:01:00:00:01:00  65534   1     up
1/1/2        lag2(mc)     2     1     IE      02:01:00:00:01:00  65534   2     up
...
```

- **Question**: What is the LACP state for lag1(mc)?
- **Answer**: The state is *IE*.
- **Question**: Use the *Flags* legend from the output; what do the IE Flags indicate?
- **Answer**: I = Individual, E = Defaulted Neighbor. This state is shown when the peer does not have LACP enabled and the LACP Fallback feature is enabled. In the next steps you will explore what this LACP Fallback means.
- **Question**: What is the forwarding state?
- **Answer**: The Forwarding State is *up*.

In a previous lab, you configured the LAG between sw-edge1 and sw-edge2. When sw-edge1 was configured with LACP, but sw-edge2 was not configured yet, the forwarding state was *lacp-blocked*. In this activity, the peer is not configured yet, but the forwarding state is *up*. Let's investigate what is different now.

2. Check the lag1 interface configuration.

```
sw-agg1(config)# show running-config interface lag1
interface lag 1 multi-chassis
    no shutdown
    no routing
    vlan trunk native 1
    vlan trunk allowed all
    lacp mode active
    lacp fallback
    exit
```

- **Question**: What new command do you observe in this LAG configuration context?
- **Answer**: The command **lacp fallback** is a new command.
- **Question**: What is the purpose of this command?
- **Answer**: By default, an LACP LAG will block all traffic, unless it can successfully negotiate with an LACP partner. During new switch deployments with ZTP, when a factory default switch is connected, it does not have any LACP configuration yet. This means a factory default switch would never be able to connect to an existing LACP LAG. With the **lacp fallback** command, you can set the member ports of the LAG to forwarding when no LACP control frames are received. Instead of the status *lacp-blocked*, the status will be *up*.

3. Verify the LACP fallback is working by testing connectivity on sw-edge1. On sw-edge1, verify you can still reach 10.254.1.21.

```
C:> ping 10.254.1.21
```

Optional steps: Test default LACP behavior

Be sure to continue after this optional section!

1. On sw-agg1, disable the LACP fallback feature on LAG1.

```
sw-agg1(config)# interface lag 1
sw-agg1(config-lag-if)# no lacp fallback
sw-agg1(config-lag-if)# exit
```

2. On sw-agg1, verify the LACP state now changed to **lacp-blocked**.

```
sw-agg1(config)# show lacp interfaces

State abbreviations :
A - Active         P - Passive      F - Aggregable I - Individual
S - Short-timeout L - Long-timeout N - InSync      O - OutofSync
C - Collecting     D - Distributing
X - State m/c expired               E - Default neighbor state

Actor details of all interfaces:
------------------------------------------------------------------------------
----------
Intf        Aggr        Port  Port  State   System-ID          System Aggr
Forwarding
            Name        Id    Pri                              Pri    Key
State
------------------------------------------------------------------------------
----------
1/1/1       lag1(mc)    1     1     ALFOE   02:01:00:00:01:00 65534  1
lacp-block
1/1/2       lag2(mc)    2     1     IE      02:01:00:00:01:00 65534  2
up
```

3. On sw-edge1, verify that you can no longer reach the host 10.254.1.21.

```
C:> ping 10.254.1.21
```

A factory default switch attempts to send a DHCP request on SVI1. VLAN 1 is untagged on all ports by default. When the upstream aggregation switch has the ports in lacp-blocked state, the factory default switch would not be able to get a DHCP-based IP address.

4. On -sw-agg1, enable LACP fallback again on LAG1.

```
sw-agg1(config)# interface lag 1
sw-agg1(config-lag-if)# lacp fallback
sw-agg1(config-lag-if)# exit
```

5. On sw-agg1, verify the LACP state is *up* again.

```
sw-agg1(config)# show lacp interfaces

State abbreviations :
A - Active         P - Passive      F - Aggregable I - Individual
S - Short-timeout L - Long-timeout N - InSync      O - OutofSync
C - Collecting     D - Distributing
X - State m/c expired               E - Default neighbor state
```

```
Actor details of all interfaces:
------------------------------------------------------------------------------
----------
Intf       Aggr       Port  Port  State   System-ID           System  Aggr
Forwarding
           Name       Id    Pri                               Pri     Key
State
------------------------------------------------------------------------------
----------
1/1/1      lag1(mc)   1     1     IE      02:01:00:00:01:00   65534   1      up
1/1/2      lag2(mc)   2     1     IE      02:01:00:00:01:00   65534   2      up
```

End of optional section, the next steps are required again!

Task 3: Configure access switch with LAG to VSX

Your colleague has asked you to connect the access switch sw-edge1 with a redundant connection (two links) to the VSX aggregation cluster. In this task you will properly connect sw-edge1 to the VSX aggregation using uplink ports 1/1/27 and 1/1/28 with a LAG.

Objectives

- Configure an access switch with a LAG to connect to a VSX system.
- Verify the LAG status on a VSX peer device.

Steps

1. Use PC-MGMT to open an SSH connection to **sw-edge1**.
2. Review the current STP uplink port states for ports 1/1/27 and 1/1/28.

```
sw-edge1(config)# show spanning-tree mst
#### MST0
Vlans mapped:  1-4094
Bridge         Address:64:e8:81:dd:81:80    priority:32768
Regional Root
Operational    Hello time(in seconds): 2  Forward delay(in seconds):15
Max-age(in seconds):20  txHoldCount(in pps): 6
Configured     Hello time(in seconds): 2  Forward delay(in seconds):15
Max-age(in seconds):20  Max-Hops:20
Root           Address:02:01:00:00:01:00  Priority:0
               Port:1/1/27                Path cost:2000
```

```
Regional Root  Address:64:e8:81:dd:81:80  Priority:32768
               Internal cost:0            Rem Hops:20

Port           Role           State       Cost        Priority   Type
BPDU-Tx     BPDU-Rx     TCN-Tx      TCN-Rx
-----------------------------------------------------------------------------
---------------------------- ...
1/1/27         Root           Forwarding 2000        128        P2P  Bound
112         80235       51          34
1/1/28         Alternate      Blocking   2000        128        P2P  Bound
141         3720        30          33
...
```

- **Question**: What are the uplink port states?
- **Answer**: Ports 1/1/27 is root port and 1/1/28 is *blocking*.

3. Configure a new LACP LAG. Typically, the highest possible LAG ID is used for the uplink to the aggregation switch; in this example it will be LAG ID 256. Enable the LAG.

```
sw-edge1(config)# interface lag 256
sw-edge1(config-lag-if)# no shutdown
sw-edge1(config-lag-if)# lacp mode active
```

4. Configure the LAG as VLAN trunk; allow all VLANs.

```
sw-edge1(config-lag-if)# vlan trunk native 1
sw-edge1(config-lag-if)# vlan trunk allowed all
sw-edge1(config-lag-if)# exit
```

5. Assign ports 1/1/27 and 1/1/28 to the LAG 256 and enable them.

```
sw-edge1(config)# interface 1/1/27,1/1/28
sw-edge1(config-if-<1/1/27,1/1/28>)# lag 256
sw-edge1(config-if-<1/1/27,1/1/28>)# no shutdown
sw-edge1(config-if-<1/1/27,1/1/28>)# exit
```

6. Review the LLDP neighbors.

```
sw-edge1(config)# show lldp neighbor-info

LLDP Neighbor Information
=========================

Total Neighbor Entries          : 5
Total Neighbor Entries Deleted  : 7
```

```
Total Neighbor Entries Dropped   : 0
Total Neighbor Entries Aged-Out : 7

LOCAL-PORT  CHASSIS-ID          PORT-ID                        PORT-DESC
TTL       SYS-NAME
------------------------------------------------------------------------------
-----------------
...
1/1/27      b8:d4:e7:da:80:00  1/1/1                          1/1/1
120       sw-agg1
1/1/28      b8:d4:e7:d9:e5:00  1/1/1                          1/1/1
120       sw-agg2
...
```

- **Question**: What is the Chassis-ID on the ports 1/1/27 and 1/1/28? Are they the same or different?
- **Answer**: LLDP uses the base system MAC of sw-agg1 and sw-agg2. Each aggregation switch can be uniquely identified based on the LLDP information.

7. Review the LACP interface status; check the *Partner* section of the output.

```
sw-edge1(config)# show lacp interfaces

State abbreviations :
A - Active          P - Passive      F - Aggregable I - Individual
S - Short-timeout L - Long-timeout N - InSync      O - OutofSync
C - Collecting     D - Distributing
X - State m/c expired                 E - Default neighbor state

Actor details of all interfaces:
------------------------------------------------------------------------------
----------
Intf         Aggr        Port  Port  State   System-ID           System  Aggr
Forwarding
             Name        Id    Pri                               Pri     Key
State
------------------------------------------------------------------------------
----------
1/1/25       lag50
down
1/1/26       lag50
down
1/1/27       lag256      28    1     ALFNCD  64:e8:81:dd:81:80  65534   256  up
1/1/28       lag256      29    1     ALFNCD  64:e8:81:dd:81:80  65534   256  up
```

```
Partner details of all interfaces:
-----------------------------------------------------------------------------
----------
Intf        Aggr        Port  Port  State    System-ID          System Aggr
            Name        Id    Pri                               Pri    Key
--------------------------------------------------------------------------
----------
1/1/25      lag50
1/1/26      lag50
1/1/27      lag256      1     1     ALFNCD   02:01:00:00:01:00  65534  1
1/1/28      lag256      1001  1     ALFNCD   02:01:00:00:01:00  65534  1
```

- **Question**: What is the LACP System-ID on the ports 1/1/27 and 1/1/28? Are they the same or different?
- **Answer**: They are the same. Since the aggregation switches have been configured with a multi-chassis LAG, sw-agg1 and sw-agg2 will use the VSX system MAC in the LACP communication. sw-edge1 believes it is connected to the same LACP partner on both ports.
- **Question**: What is the Partner Port ID on the ports 1/1/27 and 1/1/28?
- **Answer**: Port 1/1/27 is connected to port 1/1/1 on sw-agg1; Port 1/1/28 is connected to port 1/1/1 on sw-agg2. Since a unique port ID is required for the different LACP port members, the ports of the secondary switch (sw-agg2) start with 1000 + the port number. Therefore port 1/1/28 is reported to be connected to port 1001 (base 1000 + port 1).

Task 4: Review VSX active gateway configuration and status

In this task you will review VSX active gateway configuration. This is the VSX active-active alternative to VRRP.

Objectives

- Review the active gateway configuration.
- Verify the active gateway operation.

Steps

1. Open an SSH connection to sw-agg1.
2. Review the configuration of SVI 11.

```
sw-agg1(config)# show running-config interface vlan 11
interface vlan11
    vsx-sync active-gateways
    ip address 10.1.11.2/24
    active-gateway ip mac 12:01:00:00:01:00
    active-gateway ip 10.1.11.1
    ip ospf 1 area 0.0.0.0
    ip helper-address 10.254.1.21
    exit
```

- **Question**: What is the interface IP address?
- **Answer**: 10.1.11.2/24
- **Question**: What is the configured active gateway IP?
- **Answer**: 10.1.11.1
- **Question**: What is the configured active gateway MAC?
- **Answer**: 12:01:00:00:01:00

3. VSX allows the administrator to easily run commands on the *peer* VSX node. While connected to sw-agg1, repeat the previous command, but add **vsx-peer** to the end. This will execute the requested command on the VSX peer device, in this case on sw-agg2.

```
sw-agg1(config)# show running-config interface vlan 11 vsx-peer
interface vlan11
    vsx-sync active-gateways
    ip address 10.1.11.3/24
    active-gateway ip mac 12:01:00:00:01:00
    active-gateway ip 10.1.11.1
    ip ospf 1 area 0.0.0.0
    ip helper-address 10.254.1.21
    exit
```

- **Question**: What is the interface IP address?
- **Answer**: 10.1.11.3/24. On sw-agg2, this is indeed the configured IP address.
- **Question**: Are there any differences in the active gateway configuration between sw-agg1 and sw-agg2?
- **Answer**: No, they both have the same virtual MAC and IP configured. Active gateway is not an active-standby protocol, such as VRRP. With active gateway, both aggregation switches actively forward traffic that is sent to the virtual MAC.

4. For active gateway, there is no standard MAC range available like what you saw for VRRP. Therefore, you manually configure a private MAC address as the active gateway MAC.

5. On PC1, check the ARP cache.

```
C:> arp -a 10.1.11.1
Interface: 10.1.11.50 --- 0xe
  Internet Address        Physical Address       Type
  10.1.11.1               12-01-00-00-01-00      dynamic
```

- **Question**: What is the MAC address of the default gateway IP 10.1.11.1?
- **Answer**: 12:01:00:00:01:00. This is the configured active gateway MAC of the aggregation switches.

5. Save the configuration on ***all*** switches.

```
switch# write memory
```

Optional Task 5: Verify the VSX failover

The customer is pleased with the active-active VSX cluster configuration, but they would like to see how a switch failure is handled by the VSX system.

In this task you will reboot each of the aggregation switches and verify the network connectivity using PC1. This will allow you to demonstrate:

- The MC-LAG failover for layer 2 traffic.
- The active gateway failover for the layer 3 traffic.
- The OSPF routing failover for the uplink routed traffic.

Objectives

- Verify the VSX system failover.

Steps

1. On sw-edge1, verify the LACP state of the LAG 256; both member ports should be up.

```
sw-edge1(config)# show lacp interfaces

State abbreviations :
A - Active          P - Passive      F - Aggregable I - Individual
S - Short-timeout L - Long-timeout N - InSync      O - OutofSync
C - Collecting     D - Distributing
X - State m/c expired               E - Default neighbor state
```

```
Actor details of all interfaces:
----------------------------------------------------------------------------
----------
Intf        Aggr        Port  Port  State   System-ID          System Aggr
Forwarding
            Name        Id    Pri                              Pri    Key
State
----------------------------------------------------------------------------
----------
1/1/25      lag50
down
1/1/26      lag50
down
1/1/27      lag256      28    1     ALFNCD  64:e8:81:dd:81:80 65534  256  up
1/1/28      lag256      29    1     ALFNCD  64:e8:81:dd:81:80 65534  256  up

Partner details of all interfaces:
----------------------------------------------------------------------------
----------
Intf        Aggr        Port  Port  State   System-ID          System Aggr
            Name        Id    Pri                              Pri    Key
----------------------------------------------------------------------------
----------
1/1/25      lag50
1/1/26      lag50
1/1/27      lag256      1     1     ALFNCD  02:01:00:00:01:00 65534  1
1/1/28      lag256      1001  1     ALFNCD  02:01:00:00:01:00 65534  1
```

2. On sw-edge1, enable terminal-monitor. Make sure you are connected using PC-MGMT and an SSH session, ***not*** the console connection!

```
sw-edge1(config)# terminal-monitor
Terminal-monitor is enabled successfully
```

3. On PC1, start a continuous ping to 10.254.1.21.

```
C:> ping 10.254.1.21 -t
```

4. On sw-agg1, save the configuration.

```
sw-agg1# write mem
```

5. Reboot the switch; confirm the reboot with **y**.

```
sw-agg1# boot system
```

6. On PC1, verify the ping continues.

7. On sw-edge1, review the events displayed on your terminal. You should see STP root failover occurs to sw-agg2 (priority 4096) and the LAG failover (due to port 1/1/27 link status down).

```
2022-10-17T05:36:10.586174-0400 hpe-mstpd[3243] <INFO>
Event|2008|LOG_INFO|CDTR|1|CIST starved for a BPDU Rx on port lag256 from
0:020100-000100
2022-10-17T05:36:10.586566-0400 hpe-mstpd[3243] <INFO>
Event|2006|LOG_INFO|CDTR|1|CST - Root changed from 0: 02:01:00:00:01:00 to
32768: 64:e8:81:dd:81:80
2022-10-17T05:36:15.978772-0400 intfd[718] <INFO>
Event|404|LOG_INFO|UKWN|1|Link status for interface 1/1/27 is down
2022-10-17T05:36:15.998591-0400 lacpd[3150] <INFO>
Event|1321|LOG_INFO|CDTR|1|LAG 256 State change for interface 1/1/27: Actor
state: ALFO, Partner state ALFNCD
2022-10-17T05:36:16.205704-0400 hpe-mstpd[3243] <INFO>
Event|2006|LOG_INFO|CDTR|1|CST - Root changed from 32768: 64:e8:81:dd:81:80
to 4096: 02:01:00:00:01:00
```

8. On sw-edge1, review the events during the sw-agg1 boot process.

```
2022-10-17T05:37:36.539034-0400 hpe-mstpd[3243] <INFO>
Event|2006|LOG_INFO|CDTR|1|CST - Root changed from 4096: 02:01:00:00:01:00
to 0: 02:01:00:00:01:00
2022-10-17T05:38:07.161997-0400 lldpd[3153] <INFO>
Event|106|LOG_INFO|CDTR|1|LLDP neighbor b8:d4:e7:da:80:00 deleted on 1/1/27

2022-10-17T05:38:49.809988-0400 intfd[718] <INFO>
Event|403|LOG_INFO|UKWN|1|Link status for interface 1/1/27 is up
2022-10-17T05:38:49.827259-0400 lacpd[3150] <WARN>
Event|1310|LOG_WARN|CDTR|1|Partner is out of sync for interface 1/1/27 LAG
sport: 2. Actor state: ALFOX, partner state PSFO
2022-10-17T05:38:49.828560-0400 lacpd[3150] <INFO>
Event|1321|LOG_INFO|CDTR|1|LAG 256 State change for interface 1/1/27: Actor
state: ALIO, Partner state PLIO
2022-10-17T05:38:49.833016-0400 lacpd[3150] <INFO>
Event|1321|LOG_INFO|CDTR|1|LAG 256 State change for interface 1/1/27: Actor
state: ALFO, Partner state PLIO
2022-10-17T05:38:49.836836-0400 lacpd[3150] <INFO>
Event|1321|LOG_INFO|CDTR|1|LAG 256 State change for interface 1/1/27: Actor
state: ALFO, Partner state PLFO
2022-10-17T05:38:49.839875-0400 lacpd[3150] <INFO>
Event|1321|LOG_INFO|CDTR|1|LAG 256 State change for interface 1/1/27: Actor
state: ALFOE, Partner state PLFO
2022-10-17T05:38:49.843065-0400 lacpd[3150] <INFO>
Event|1321|LOG_INFO|CDTR|1|LAG 256 State change for interface 1/1/27: Actor
state: ALFOE, Partner state PSFO
```

```
2022-10-17T05:38:49.846235-0400 lacpd[3150] <INFO>
Event|1321|LOG_INFO|CDTR|1|LAG 256 State change for interface 1/1/27: Actor
state: ALFOX, Partner state PSFO
2022-10-17T05:38:49.853395-0400 lacpd[3150] <INFO>
Event|1309|LOG_INFO|CDTR|1|Partner is detected for interface 1/1/27 LAG 256
: 65534,02:01:00:00:01:00. Actor state: ALFOX, partner state ALFOX
2022-10-17T05:38:49.853564-0400 lacpd[3150] <INFO>
Event|1321|LOG_INFO|CDTR|1|LAG 256 State change for interface 1/1/27: Actor
state: ALFOX, Partner state ALFOX
2022-10-17T05:38:49.856348-0400 lacpd[3150] <INFO>
Event|1321|LOG_INFO|CDTR|1|LAG 256 State change for interface 1/1/27: Actor
state: ALFO, Partner state ALFOX
2022-10-17T05:38:49.860439-0400 lacpd[3150] <INFO>
Event|1321|LOG_INFO|CDTR|1|LAG 256 State change for interface 1/1/27: Actor
state: ALFO, Partner state ALFO
2022-10-17T05:38:51.454255-0400 lacpd[3150] <INFO>
Event|1321|LOG_INFO|CDTR|1|LAG 256 State change for interface 1/1/27: Actor
state: ALFN, Partner state ALFO
2022-10-17T05:38:51.533436-0400 lacpd[3150] <INFO>
Event|1321|LOG_INFO|CDTR|1|LAG 256 State change for interface 1/1/27: Actor
state: ALFN, Partner state ALFN
2022-10-17T05:38:51.541340-0400 lacpd[3150] <INFO>
Event|1321|LOG_INFO|CDTR|1|LAG 256 State change for interface 1/1/27: Actor
state: ALFNC, Partner state ALFN
2022-10-17T05:38:51.548912-0400 lacpd[3150] <INFO>
Event|1321|LOG_INFO|CDTR|1|LAG 256 State change for interface 1/1/27: Actor
state: ALFNC, Partner state ALFNC
2022-10-17T05:38:51.555263-0400 lacpd[3150] <INFO>
Event|1321|LOG_INFO|CDTR|1|LAG 256 State change for interface 1/1/27: Actor
state: ALFNCD, Partner state ALFNC
2022-10-17T05:38:51.565517-0400 lacpd[3150] <INFO>
Event|1321|LOG_INFO|CDTR|1|LAG 256 State change for interface 1/1/27: Actor
state: ALFNCD, Partner state ALFNCD
2022-10-17T05:38:51.868979-0400 lldpd[3153] <INFO>
Event|104|LOG_INFO|CDTR|1|LLDP neighbor b8:d4:e7:da:80:00 added on 1/1/27
```

9. On sw-edge1, disable the terminal monitor logging.

```
sw-edge1# no terminal-monitor
```

Note

The remote lab uses the CX simulator for the Core router. Therefore, there is no direct, physical connection between the sw-agg1 and the core router. This means that the link down of the sw-agg1 (due to the reboot) will not be observed immediately by the core router and it needs to rely on OSPF timers to expire. Therefore, the failover time in this lab will be higher than a production environment with directly connected devices.

Learning check

Discuss these questions.

1. Which of the statements below accurately describe spanning tree protocols? (Select three.)
   a. They mitigate layer 2 loops, broadcast storms, and multiple frame copies.
   b. Switches elect a root bridge, from which the spanning tree emanates.
   c. All switch ports forward traffic, to improve resiliency.
   d. When a link failure occurs, switches converge on a new topology.

2. Match the descriptions to the correct STP version (*802.1D STP*, *802.1w RSTP*, and *802.1s MSTP*).
   a. You map certain VLANs to a particular STP instance.
   b. Default operation between AOS-CX switches.
   c. Default mode for AOS-CX switches.
   d. Improved convergence over older versions.
   e. Slow convergence.
   f. Not recommended.

3. See Figure 4-45. Assuming default settings and normal operation, which interface will switch Access-1 use as its root port, and why does it choose that interface?

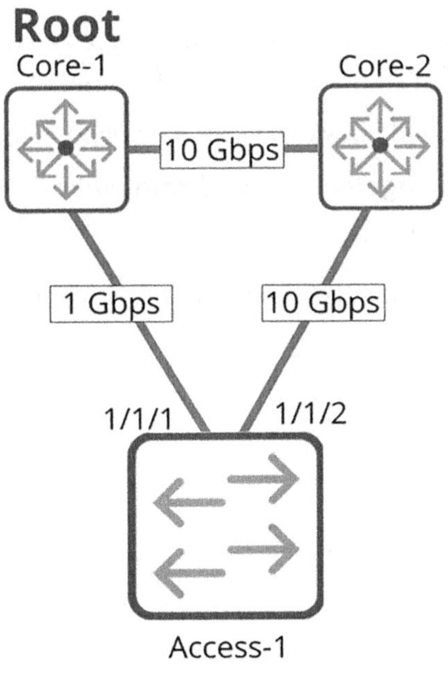

Figure 4-45: Question 3

a. Port 1/1/1, because it is directly connected to the root
b. Port 1/1/1, because it has a lower port number
c. Port 1/1/2, because it has a higher port number
d. Port 1/1/2, because it has a lower cost

4. Your colleague is confused about why a certain switch in your network was elected as the root switch. Which of the statements below would help your colleague understand this aspect of spanning-tree protocols? (Select two.)
   a. Root switch election is based on the lowest bridge ID (BID) and the highest priority.
   b. Root switch election is based on the lowest bridge ID (BID), which you can override by configuring a priority value on your preferred root switch.
   c. The bridge ID (BID) is derived from an IP address on the device.
   d. It is a best practice to manually configure a priority value on both your preferred root switch and backup root switch.

5. Which statements below are true about the Virtual Router Redundancy Protocol (VRRP)? (Select two.)
   a. It provides better first-hop redundancy as compared to configuring hosts with two default gateways (DGs).
   b. With VRRP, two routing devices actively forward traffic.
   c. You can define VRRP instances to improve load-sharing for VRRP deployments.

6. Which statements below provide accurate comparisons of Aruba's Virtual Switching eXtension (VSX) versus Aruba's Virtual Switching Framework (VSF)? (Select three.)
   a. Both solutions allow multiple devices to be perceived as a single device.
   b. You typically run VSX on aggregation and core switches.
   c. VSF can be easier to configure.
   d. Like VSF, VSX maintains separate control and management planes.

Summary

Congratulations! You have completed Chapter 4. You should now be able to:

- Discuss spanning tree.
- Explain VRRP.
- Have a basic understanding of VSX.

5 VSF

OBJECTIVES

One great way to optimize network resiliency and ease network management tasks is by using the Virtual Switching Framework (VSF) feature on access layer AOS-CX switches.

After completing this chapter, you will be able to describe VSF technology, operation, and use cases. You will be able to configure VSF, and describe aspects like auto-VSF and VSF Multi-Active Detection (MAD).

After completing this chapter, you should be able to:

✓ Describe VSF.

✓ Configure VSF.

✓ Describe Auto-VSF.

✓ Explain VSF MAD.

VSF introduction

Let's begin with an introduction to VSF.

Introduction to stacking technologies

Stacking technologies allow you to manage a group of switches as a single device, commonly referred to as a virtual switch (see Figure 5-1). Control and management plane functions are centralized in one group member, but each member in the stack runs its own independent data plane. The tight relationship between the control and data planes is maintained; it just happens on an inter-switch basis.

Stacking benefits include:

- **Ease of management:** You no longer need to connect to, configure, and manage each individual switch. You simply configure the primary switch. That configuration is then automatically distributed to other virtual switch members. This simplifies network setup, operation, and maintenance.
- **Network simplification:** Since multiple devices share a common control plane, routing protocols and STP are no longer needed inside the stacking group. Connected devices perceive the group as a single device.

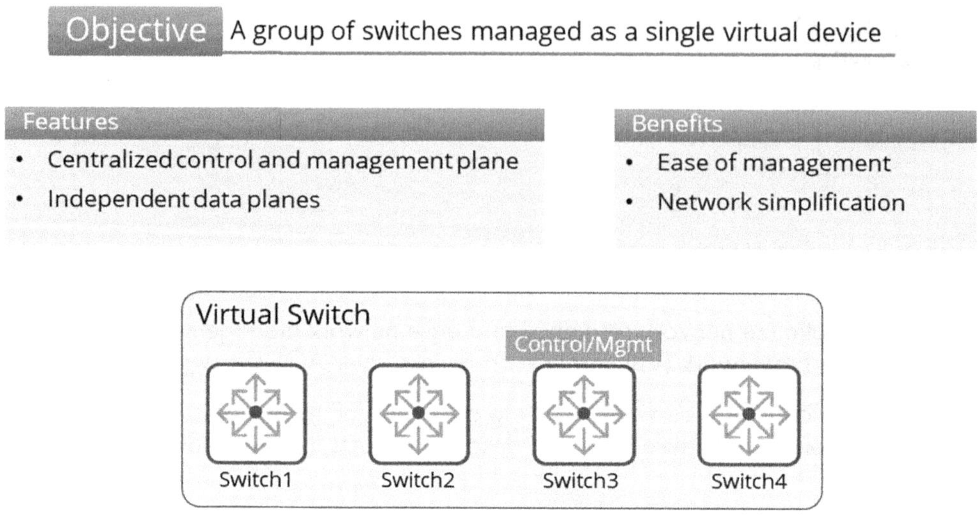

Figure 5-1: Features and benefits of VSF

Distributed data plane and distributed link aggregation

Stacking technologies maintain the data plane distributed across all the members; this implies that each physical device individually creates and populates forwarding tables like ARP and MAC. These tables are then shared across all members using the control plane.

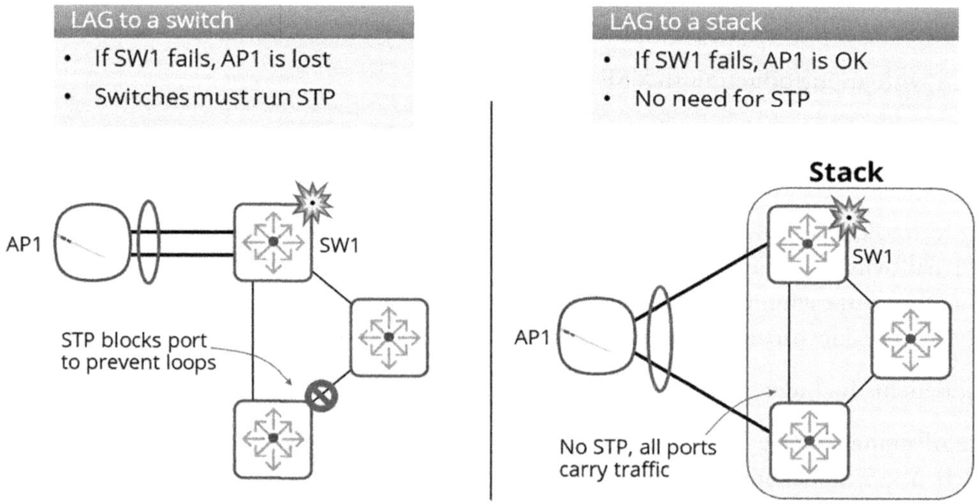

Figure 5-2: Distributed data plane and distributed link aggregation

Mobility gateways (formerly referred to as controllers), firewalls, APs, and other devices can benefit from stacking with link aggregation (LAG) enabled switches, since they perceive a LAG connection to a single device. However physical links terminate in different stack members (referred to as *distributed* link aggregation). If SW1 fails (see the example shown in Figure 5-2), the traffic can still use other LAG links.

LAG and stacking features enable the network to fully use all available links at the same time. Notice that STP is not needed because the stack operates from a single control plane. Aruba highly recommends this implementation since it provides resiliency and is easier to manage one virtual switch than multiple physical switches.

Aruba VSF stacking solution and platforms

VSF defines a *virtual switch,* which is composed of individual physical switches interconnected using Ethernet links. All member switches share a *single control plane.* Devices connected to the VSF perceive a single device. As you recently learned, this virtual switch behavior has significant benefits related to simplified management and improved connectivity.

In AOS-CX, VSF is supported on the 6200, 6300M, and 6300F access switches (see Figure 5-3). You can configure a maximum of 10 members in the stack. This feature is *enabled* by default.

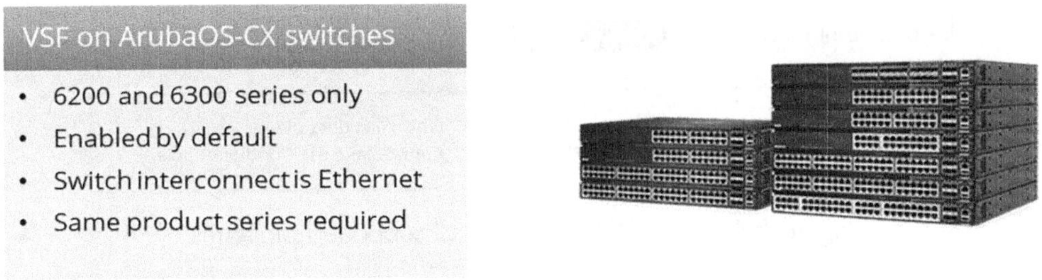

Figure 5-3: Aruba VSF solution and supported platforms

The VSF feature is also available in the older Aruba AOS switching family, including the 5400 and 2930 series, but the feature is disabled by default on those switches. Also, you cannot form a VSF stack with switches from different model families. This means that VSF can only be formed using any combination of 6200 switches or 6300 ArubaOS CX switching series; i.e., you cannot have both 6200 and 6300 switches in the same stack.

VSF member roles and links

Aruba VSF creates a single control plane that runs on a single VSF member. This is the *primary* member and always uses member ID 1. This device assumes the *conductor* role in the stack.

Note

A factory-default Aruba 6200 or 6300 switch boots up as VSF enabled switch with member ID 1. This implies that the switch behaves as the primary member.

A *secondary* member provides for high availability in case primary failure. You can choose any member in the stack to take the secondary role, but it must be explicitly defined. Just configure any member ID with the exception of 1, which is reserved for the primary switch.

The other devices in the VSF are *members*, which only perform data plane functions and cannot assume the primary/conductor role (see Figure 5-4).

The VSF switches are interconnected using uplink ports.

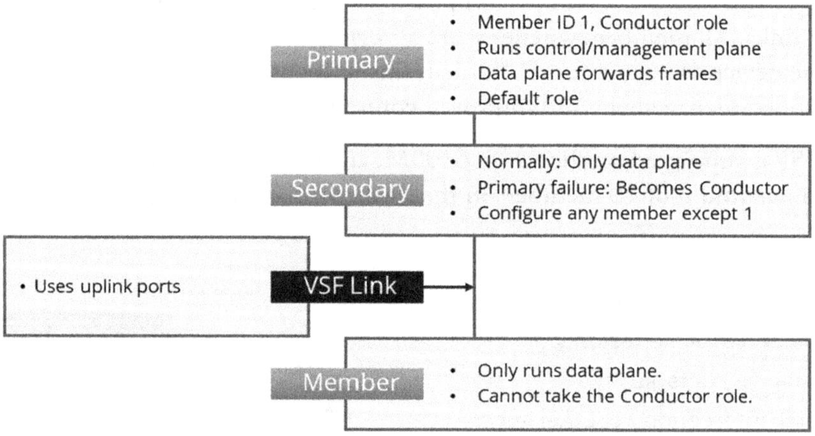

Figure 5-4: Aruba VSF solution and supported platforms

VSF Open Virtual Switch Database

You previously learned about the *Current State Database* (CSDB) when discussing checkpoints in Chapter 2. As the most important element of the ArubaOS CX architecture, it is contacted by all control and management protocols. In a VSF stack, this database runs on the primary/conductor switch (see Figure 5-5).

When VSF runs on a group of switches the Open Virtual Switch Database (OVSDB) is also created. This new database is created on the conductor switch and contains state and configuration data for the VSF stack itself. The conductor switch *synchronizes* the OVSDB content with the secondary switch to ensure that it can quickly take over the active role without interruption.

The OVSDB database includes six tables:

- VSF member table
- VSF link table
- System table: Includes the number of members, MAD status, fragment status, and topology type.
- Subsystem table: Boot time for each member
- Interface table
- Topology table

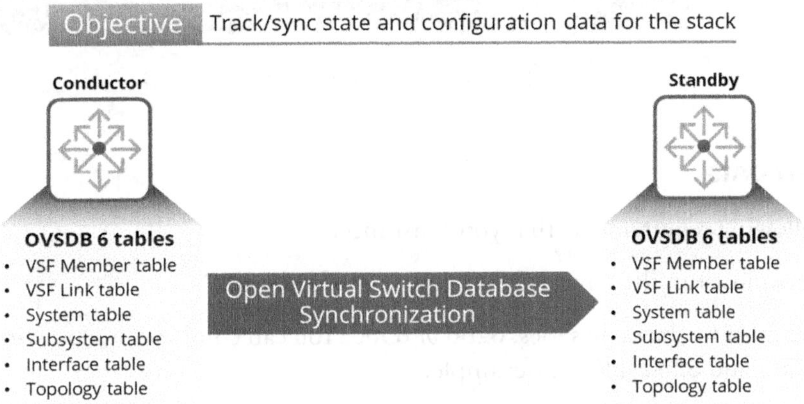

Figure 5-5: VSF Open Virtual Switch Database

VSF topologies

Of course, VSF members must be physically connected to form the switching stack, utilizing one of two topologies:

- **Daisy chain.** As the name implies, this topology interconnects VSF members with a single chain of Ethernet connections. From Figure 5-6 (top) you can easily see that a switch or link failure causes the stack to be split. This means that part of the stack is unable to provide endpoint connectivity.
- **Ring topology.** This topology is recommended since it offers a backup path in case of a switch or link failure (see the bottom of Figure 5-6).

Note

A full mesh topology is not supported with VSF.

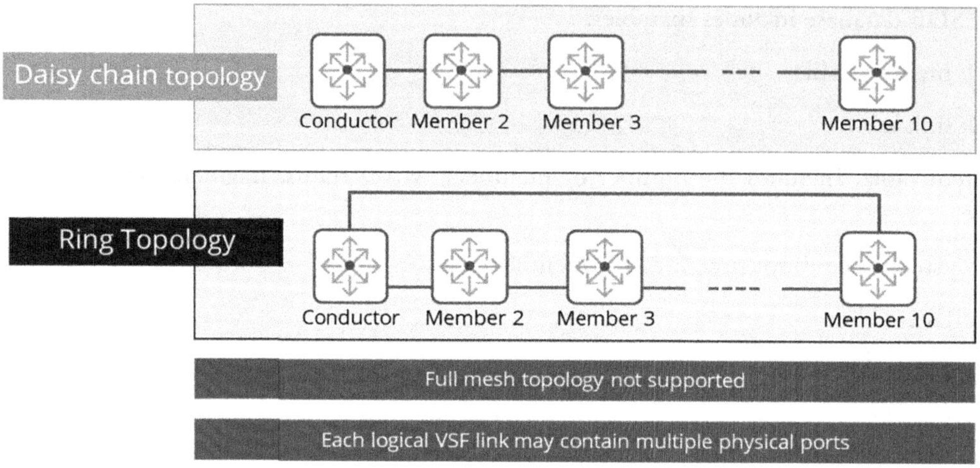

Figure 5-6: VSF topologies

VSF requirements

VSF has the following requirements that you must meet:

- All members must run the same AOS-CX version of software.
- All switches must be the same series: 6200 *or* 6300. You can combine models within a series, like the 6200-25G and 6200-48G, for example.
- The VSF link uses regular Ethernet ports.
- It is recommended to use the uplink ports which support 10, 25, or 50 Gbps speeds.
- A maximum of eight (8) members are supported in a 6200 stack and ten (10) members in a 6300 stack (see Figure 5-7).

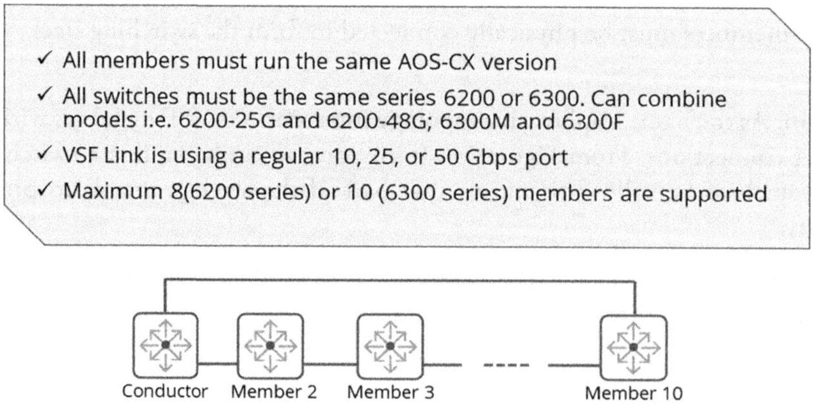

Figure 5-7: VSF requirements

Note

Firmware is automatically synchronized within the VSF stack by the conductor. Joining switches will automatically upgrade or downgrade to the switch's firmware to the firmware running on the conductor switch.

VSF configuration

Now that you understand some of the basics concerning VSF let's talk about how to set it up.

VSF member IDs and port numbers

When the VSF stack forms, all physical devices use a single control plane. This means that all switch interfaces are be available for configuration using the standard ArubaOS-CX *member/slot/port* notation, as shown in Figure 5-8. Use the command **show interface brief** to see all available ports for *all* switches within a stack.

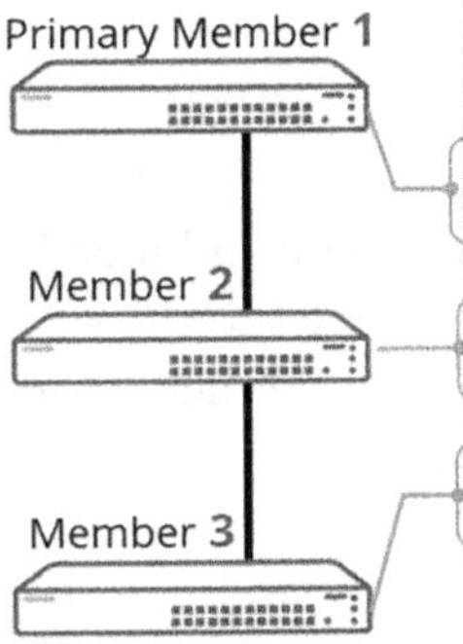


```
Access-6300# show interface brief
-----------------------------------------------------------------------
Port      Native  Mode    Type           Enabled Status      Speed
          VLAN                                                (Mb/s)
-----------------------------------------------------------------------
1/1/1     1       access  SFP+DAC1       yes     up          10000
1/1/2     1       access  SFP+DAC1       yes     up          10000
...
2/1/1     1       access  SFP+DAC1       yes     up          10000
2/1/2     1       access  SFP+DAC1       yes     up          10000
...
3/1/1     1       access  SFP+DAC1       yes     up          10000
3/1/2     1       access  SFP+DAC1       yes     up          10000
```

Figure 5-8: VSF member IDs and port numbers

Note

Since all 6200 and 6300 switch series models are fixed-port switches, the *slot* number is always 1.

VSF configuration

Consider the scenario where three 6300 switches are assigned to the same VSF stack, shown in Figure 5-9. You see the port configuration for all three switches. The **link** command specifies which physical port or ports are used to connect the switches. If you use a ring topology, then you would have two link interfaces **1** and **2**. After Access-2 reboots, you add the **vsf secondary-member 2** command, which will cause it to renumber its member ID from 1 to 2. The switch will have to reboot a second time. The same would be required of Access-3, where it's member number would be changed to 3. Note that since there is already a primary and secondary member, the third switch will be considered a normal member of the stack.

Once you have set up VSF, use the **show vsf** command to determine member status and role. As you can see setting up VSF is a very simple process; you'll be doing this in the lab later in the chapter.

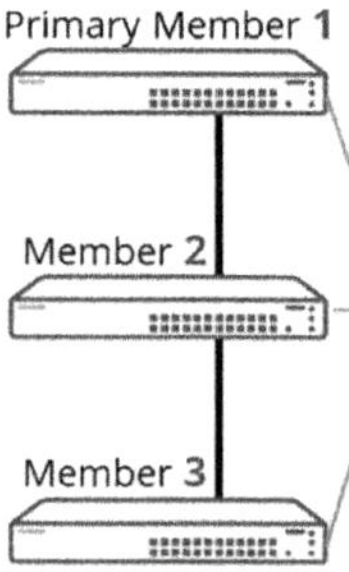


```
Access-6300# show interface brief
------------------------------------------------------------------------
Port      Native  Mode    Type            Enabled Status         Speed
          VLAN                                                   (Mb/s)
------------------------------------------------------------------------
1/1/1     1       access  SFP+DAC1        yes     up             10000
1/1/2     1       access  SFP+DAC1        yes     up             10000
...
2/1/1     1       access  SFP+DAC1        yes     up             10000
2/1/2     1       access  SFP+DAC1        yes     up             10000
...
3/1/1     1       access  SFP+DAC1        yes     up             10000
3/1/2     1       access  SFP+DAC1        yes     up             10000
```

Figure 5-9: VSF configuration example

VSF pre-provisioning

AOS-CX supports member pre-provisioning. This enables you to prepare the VSF link and member configuration for a specific 6200 or 6300 switch model before the switch is connected to the stack. This configuration is performed from the primary/conductor switch. Then, when the new switch boots up, it can automatically join the stack and obtain its proper configuration.

- Prepare VSF Link and member configuration
- Based on switch model
- VSF-connected switch boots with proper config

Example: VSF Provisioning for a 6300F 24-port switch

```
Switch-6300(config-vsf)# vsf member 4
Switch-6300(config-vsf)# type jl668a
Switch-6300(config-vsf)# link 1 1/1/25
Switch-6300(config-vsf)# link 2 1/1/26
```

P/N	Model
JL658A	6300M 24SFP+/4SFP56
JL659A	6300M 48SR PoE CLS 6/4SFP56
JL660A	6300M 24SR PoE CLS 6/4SFP56
JL661A	6300M 48G PoE CLS 4/4SFP56
JL662A	6300M 24G PoE CLS 4/4SFP56
JL663A	6300M 48G/4SFP56
JL664A	6300M 24G/4SFP56
JL665A	6300F 48G PoE CLS 4/4SFP56
JL666A	6300F 24G PoE CLS 4/4SFP56
JL667A	6300F 48G/4SFP56
JL668A	6300F 24G/4SFP56

Check Aruba Networks website for complete list of products and P/N

Figure 5-10: VSF pre-provisioning

Figure 5-10 shows a VSF pre-provisioning example for a 6300F 24-port switch done from a conductor switch. The table in Figure 5-10 indicates some available options for the switch model number. The **type** command, followed by a space and a "?", will show the available model options. The list will change when new products or models are released. A firmware update may be required to support these new models, however.

Auto-VSF

As you can see, setting up VSF manually is a fairly simple process. Let's see if we can make VSF deployments even easier with the auto-VSF feature.

Auto-VSF overview

The VSF auto-stacking feature provides a mechanism to automatically form a stack when the stack members are physically connected in a desired topology (see Figure 5-11). This reduces the number of user intervention touch points to form a VSF stack, simplifying the VSF deployment process.

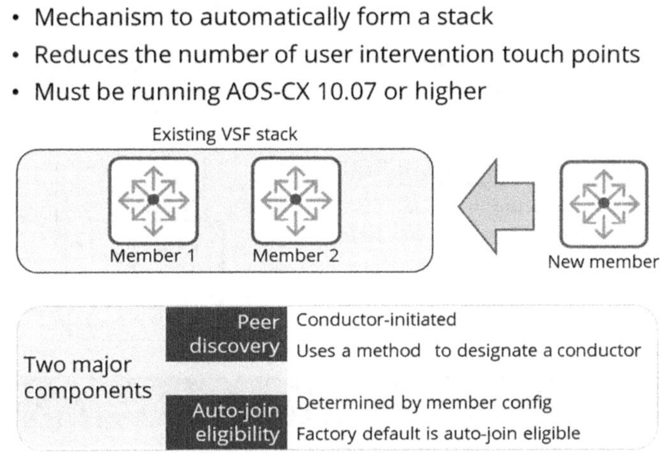

Figure 5-11: Auto-VSF feature

A manual stack formation procedure generally requires you to explicitly log in to each of the switches, configure the links, renumber the member ID, and then make the physical connection to form a VSF stack of the desired size and topology. This is error prone since there are multiple touch-points involved in the whole work-flow for each member.

The auto-stacking feature eases this problem by reducing the number of touch-points involved to just the physical connections of the links. A new factory default switch can be added into an existing stack by physically connecting it to a given member of the stack. The new switch is automatically assigned a member ID, then it reboots. After it reboots, the newly added member joins the stack.

There are two major components to the auto-stacking solution:

- **Peer discovery**: Initiated from the conductor using one of the methods of designating a conductor switch.
- **Auto-join eligibility**: Determined by the configuration state of each stack member. A switch with a factory default configuration is always auto-join eligible.

Peer discovery

Auto-stacking peer discovery is a unidirectional process. It starts on the conductor of the existing VSF stack with the VSF link containing the higher denomination VSF port. The conductor sends a VSF peer discovery protocol packet on that port (see Figure 5-12). The peer receives the packet, determines if it is valid, and sends a response with information including its auto-join eligibility, MAC address, and part (model) type. If the peer is auto-join eligible, the VSF member and link configurations are automatically added to the running-configuration of the active conductor.

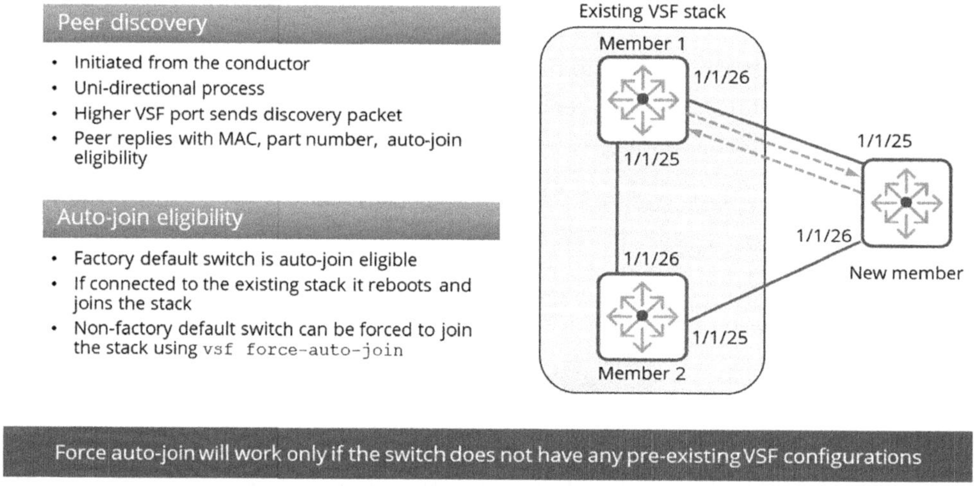

Figure 5-12: Peer discovery and auto-join eligibility

Auto-join eligibility

A switch in its factory default configuration state is considered to be auto-join eligible. If the auto-join eligible switch is connected to existing stack, it will automatically reboot and join the stack. Once it moves out of the factory default configuration state, it is not considered as auto-join eligible and cannot automatically join an existing stack. However, an administrator can still manually configure the links, renumber the device to make it part of a stack; or just reset a member back to its factory default state.

Force auto-join support

Only a switch with a *factory default* configuration is considered to be auto-join eligible. In order to support factory express deployments where the administrator wants to add a switch which is not in its non-factory default configuration, the force auto-join configuration support is provided. Use the **vsf force-auto-join** command to force the switch to join the stack automatically. Once the user sets **force auto-join** in the switch configuration, the switch will be considered as auto-join eligible and will join the stack even though the switch does not have the factory default configuration.

> **Note**
> Force auto-join will work only if the switch does not have any pre-existing VSF configurations. You would need either remove the VSF configuration or reset the switch back to its factory default state if you wanted it to join a stack.

Forming a VSF stack using the auto-stacking command

Remember that you need to meet these requirements for auto-stacking to work:

- All switches must be in the factory default configuration.
- All stack members must be connected in a desired topology on the reserved VSF link ports.

Figure 5-13 illustrates the steps to use auto-stacking. It's fairly simple since the only configuration command needed is done on the conductor: **vsf start-auto-stacking**.

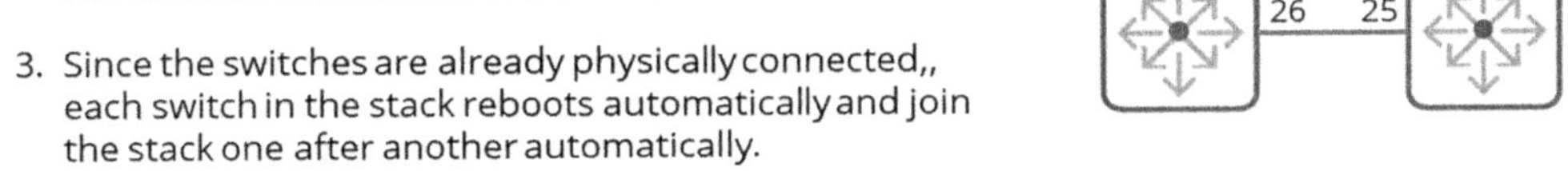

Figure 5-13: Peer discovery and auto-join eligibility

Reserved interfaces for auto-stacking

Based on the product type of a switch, the following two interfaces are reserved for the auto-stacking process (see Figure 5-14):

- 24-port switch models: 25 and 26
- 48-port switch models: 49 and 50

Users can physically connect the switch to an existing stack on one of these reserved auto-stacking interfaces.

For auto-stacking, it is required that the existing stack conductor's VSF link port have a higher value. For example, if you connect member 2 to member 1 (the existing VSF stack conductor), you must connect interface 26 of member 1 to interface 25 of member 2. Otherwise, the member 2 will not join in auto-stacking.

Platform	SKU part number	Ports reserved
6300	JL658A JL660A JL662A JL664A JL666A JL668A	25 and 26
6300	JL659A JL661A JL663A JL665A JL667A JL762A	49 and 50
6200	JL724A JL725A	25 and 26
6200	JL726A JL727A JL728A	49 and 50

Figure 5-14: Reserved interfaces for auto-stacking

VSF MAD

Don't get angry about potential stack system failures! You've got VSF MAD (Multiple Active Detection). This section briefly introduces what MAD is and what it does.

VSF link failures

VSF link failures could cause a fragmented stack. Figure 5-15 shows a fragmented VSF stack, due to link failures. Originally, SW1 was the primary and SW2 was the secondary in a stack. After fragmentation, there is no direct connection between the two switches. SW1 continues to think that it is the primary. SW2 no longer hears from SW1, and so now thinks that it is the primary switch in the stack (this assumes that SW2 is configured as a secondary switch). This is known as a *split-brain* condition: both fragments continue to function although there is no communication between them.

A split-brain can cause unstable and unexpected network behavior. A packet received in one fragment and destined for the other fragment is discarded. Even worse, both segments use the same IP and MAC addresses and the same routing information: external devices could start populating duplicated data, which can cause very strange network behaviors.

Note

A split-brain situation could also occur if a VSF member fails in a daisy chain topology. Therefore, Aruba always recommends deploying a stack using a *ring* topology.

The best way to solve a split-brain situation is to disable the ports on one of the segments: the segment that contained the original secondary switch.

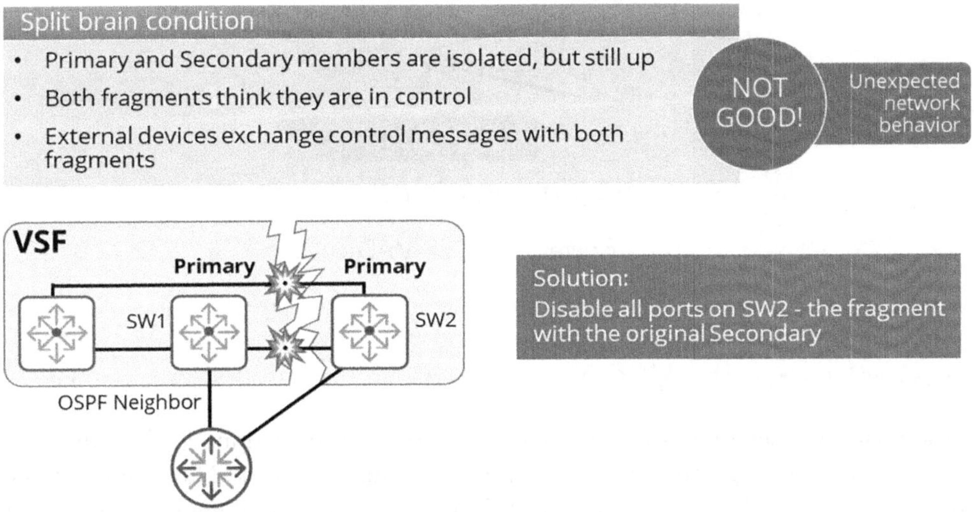

Figure 5-15: VSF link failures and fragmented stacks

Multi-Active Detection (MAD)

You can use Multi-Active Detection (MAD) to avoid split-brain situations (see Figure 5-16). Thus, if a VSF link failure occurs, the segmented switches that do not include the primary switch verifies the primary member status. If the original primary is up, then all members in the fragment that does *not* include the primary member will disable all their ports, with the exception of the VSL link ports.

VSF uses the following mechanism to detect and verify the status of the primary member: management interface split detection**.** This method requires you to connect out-of-band management (OOBM) interfaces to primary and secondary stack members. These interfaces must be in the same layer 2 broadcast domain (VLAN); thus, if you have more than two switches in a stack, you'll need to connect them to an intermediate switch that is not participating in the stack. This network is used to identify active stack fragments. Each member broadcasts Split Detection Protocol (SDP) packets to identify stack fragments that are currently operational.

MAD is disabled by default. To enable MAD on the OOBM port, use the **vsf split-detect mgmt** command.

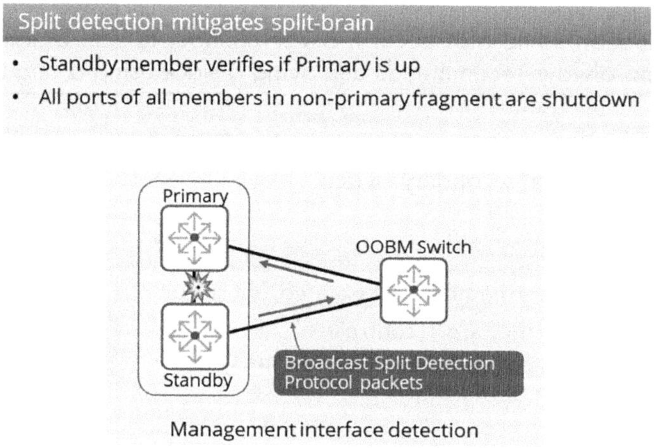

Figure 5-16: VSF MAD

Lab 5: Stacking with VSF

In the previous lab, your senior network colleague prepared the aggregation switches and you explored how VSX is used on the aggregation layer to build a redundant network. Now your colleague wants you to configure the access layer switches into a VSF stack and connect them with a LAG to the VSX aggregation layer.

Using a VSF stack in the access layer has several advantages:

- **Reduced number of management points:** When a customer requires four switches on a floor, they must remember to connect to the correct switch management IP when a configuration change is required. This may require an additional lookup to find the switch that is connected to a wall outlet. When the four switches on a floor are connected into a VSF stack, the number of management points is reduced by a factor four.

- **Reduced number of uplink ports:** A VSF stack of four switches operates as one switch. They can use two physical uplinks to the aggregation layer, if that is sufficient bandwidth. However, if four *individual* switches are used, each switch needs its own set of two uplinks or STP must be used to offer redundancy.
- **Easy device replacement:** In the unfortunate event of a switch failure, using a single switch approach means a new switch must be deployed and the configuration restore procedure must be followed.

On the other hand, if a VSF stack is used, the new switch only requires adding to the existing VSF stack. The configuration of the ports will automatically be assigned to the new member.

Overview

During this lab you will initially erase the configuration of sw-edge2 since it will join the sw-edge1 VSF stack. Later in the lab, you will also test the VSF automatic stacking. At that point, you will again, erase the switch configuration. This is okay since you will be connecting the switches to Aruba Central in the lab in the next chapter. You will then use Aruba Central to push the correct configurations to the switches again.

Objectives

- Configure VSF.
- Understand auto-VSF.
- Verify VSF operation.
- Configure split-brain protection with VSF.

Task 1: Prepare the topology

In this task you will apply the correct start-up configuration for this lab on sw-agg1 and sw-agg2. In the upcoming tasks, you will configure the two edge switches to form a single VSF stack. Once you do this, the switches can form a LAG with four member ports to the VSX cluster.

In this task you will also prepare the VSX cluster, so all ports to the edge switches belong to the same LAG (LAG 1).

You will make the sw-edge2 factory default in the next tasks. However, in the remote lab, a ZTP DHCP server provides the initial configuration file. This would break the auto-join VSF activity. Therefore, you will disable ZTP for the sw-edge2.

Objectives

- Reconfigure the VSX system to use a LAG with 4 member ports.
- Disable the ZTP for sw-edg2.

Steps

Configure aggregation VSX

The configuration in the next steps will assign ports 1/1/1 and 1/1/2 on *both* sw-agg1 and sw-agg2 to LAG1.

1. Use PC-MGMT to open an SSH connection to **sw-agg1**.
2. Configure port **1/1/2** to become member of LAG1 and enable the port.

```
sw-agg1(config)# interface 1/1/2
sw-agg1(config-if)# no lag 2
sw-agg1(config-if)# lag 1
sw-agg1(config-if)# no shutdown
sw-agg1(config-if)# exit
```

3. Use PC-MGMT to open an SSH connection to sw-agg2.
4. Configure port 1/1/2 to become member of LAG1 and enable the port.

```
sw-agg2(config)# interface 1/1/2
sw-agg2(config-if)# no lag 2
sw-agg2(config-if)# lag 1
sw-agg2(config-if)# no shutdown
sw-agg2(config-if)# exit
```

Disable ZTP for sw-edge2

On PC-MGMT you will rename the sw-edge2 ZTP file in the TFTP folder. This will prevent the sw-edge2 from loading the start configuration.

1. On **PC-MGMT** open the TFTP folder on the desktop.
2. Rename cx-ztp-sw-edge2.cfg file to ***no-cx-ztp-sw-edge2.cfg***. When sw-edge2 attempts to download the file, it will not find it anymore. The ZTP process will not be performed, and the switch will automatically join the VSF stack in this lab.

Task 2: Create a VSF stack using auto-join of a new member

In this task sw-edge1 and sw-edge2 will become part of the same VSF stack. sw-edge1 will be used as VSF member 1, while sw-edge2 will be configured to become VSF member 2.

Every AOS-CX access switch is by default part of a VSF stack (of 1 switch) and assigned member ID 1. The original member 1 (sw-edge1 in this setup) will keep its configuration in this task. The result will be that sw-edge1 simply stays in the network with the same configuration, but it will receive more ports than it previously had. This shows how an existing customer can extend any existing (VSF-capable) AOS-CX switch to a stack without impacting the existing configuration.

You will use the auto-join function to connect sw-edge2 to the VSF stack. This means a new member can be added to an existing stack without having to access the new member device. Factory default switches are enabled for the auto-join function by default. You only need to configure the VSF links on the existing VSF system. The existing VSF system will start sending VSF hello messages on these VSF links. The factory default switch is automatically listening for these special VSF hello messages on the predefined VSF ports and it will automatically reboot and join the VSF stack.

Objectives

- Understand the concept of VSF member IDs.
- Understand the concept of VSF links.
- Understand the auto-join VSF feature.
- Verify the auto-join VSF.
- Verify the VSF operation.

Steps

In the next steps you will review the default VSF status and configure sw-edge1 with a VSF link to support new member devices.

1. Open an SSH connection to **sw-edge1**.
2. Review the current VSF status.

```
sw-edge1# show vsf
Force Autojoin                : Disabled
Autojoin Eligibility Status: Not Eligible
MAC Address                   : 64:e8:81:dd:81:80
Secondary                     :
Topology                      : Standalone
Status                        : No Split
Split Detection Method        : None

Mbr Mac Address           type           Status
ID
--- ------------------- -------------- ---------------
1   64:e8:81:dd:81:80   JL666A         Conductor
```

- **Question**: What is the current member ID (Mbr ID)?
- **Answer**: 1.
- **Question**: What is the current device status?
- **Answer**: Conductor.
- **Question**: What does the secondary option mean in the output?
- **Answer**: In a VSF stack, one switch (the primary/conductor) controls the control plane and management plane processes, such as LACP, STP, or the SSH server. When the conductor goes offline, due to a power failure for example, the secondary switch should resume the conductor state. This other switch is configured using the secondary option. Without a configured secondary, the VSF stack would be offline when the conductor is not active, and therefore, you should always define a secondary switch!
- **Question**: Has any member been configured as the secondary?
- **Answer**: No, by default no secondary is configured.
- **Question**: What is the default *Split Detection Method*?
- **Answer**: None. If there is a split stack condition, both fragments would remain active.

3. Review the default VSF running-configuration. The example output, your output may be different.

```
sw-edge1# show run vsf
vsf member 1
    type jl666a
```

- **Question**: How many VSF members do you see in the configuration?
- **Answer**: Only one.
- **Question**: What does the JLxxx number in the line with "type jl.." represent?
- **Answer**: Each Aruba AOS-CX switch model has a unique product code. It indicates whether the switch has 24 or 48 ports, PoE ports, smart rate ports, etcetera. By setting the product code, AOS-CX knows how many ports to provision and which features, such as PoE, are available on the VSF member.

Note

Visit https://www.arubanetworks.com for a complete list of products. Visit https://www.arubanetworks.com/assets/ds/DS_6300Series.pdf for a list of the 6300 switch series.

4. On the **sw-edge1**, configure the *future* member **2** to become the **secondary** (standby) control plane in the VSF stack. Confirm the message with **y**.

Note

This step has no impact on your current setup, since there is no member 2 yet. This command will only take effect when member 2 joins the VSF stack!

```
sw-edge1(config)# vsf secondary-member 2
This will save the configuration and reboot the specified switch.
Do you want to continue (y/n)? y
```

5. Configure the split stack detection to use the OOBM mgmt interface.

```
sw-edge1(config)# vsf split-detect mgmt
```

Note

If there is no OOBM network available, you can connect a cable directly from the commander to the standby member OOBM port to support split stack detection.

6. Verify the configuration changes. A secondary should be present and the split method should be set to *mgmt* now.

```
sw-edge1(config)# show vsf
Force Autojoin              : Disabled
Autojoin Eligibility Status: Not Eligible
MAC Address                 : 64:e8:81:dd:81:80
Secondary                   : 2
Topology                    : Standalone
Status                      : Active Fragment
Split Detection Method      : mgmt

Mbr Mac Address           type           Status
ID
--- ------------------- -------------- ---------------
1   64:e8:81:dd:81:80   JL666A         Conductor
```

7. Enter the VSF member 1 context.

```
sw-edge1(config)# vsf member 1
```

8. Configure VSF link 1 with port 1/1/26.

```
sw-edge1(vsf-member-1)# link 1 1/1/26
```

9. Configure VSF link 2 with port 1/1/25.

```
sw-edge1(vsf-member-1)# link 2 1/1/25
```

10. Exit the VSF context.

```
sw-edge1(vsf-member-1)# exit
```

Note

You will use auto-VSF to automatically establish the VSF system. Auto-VSF requires dedicated VSF link-to-port mappings. Make sure you have mapped the ports correctly:

- VSF link 1 to port 1/1/26
- VSF link 2 to port 1/1/25

11. Verify the configured VSF link to port assignment. Previously ports 1/1/25 and 1/1/26 were configured as member ports of LAG 50.

```
sw-edge1(config)# show vsf link
VSF Member 1
     Link        Peer    Peer
Link State       Member  Link   Interfaces
---- ---------- ------- ------ ---------------------------
1    down        0       0      1/1/26
2    down        0       0      1/1/25
```

12. Review the port 1/1/25 configuration.

```
sw-edge1(config)# show running-config interface 1/1/25
interface 1/1/25
    no shutdown
    exit
```

- **Question**: What happened with the port configuration?
- **Answer**: When a port is assigned to a VSF link, all the existing settings are removed.

13. Review the interface brief output.

```
sw-edge1(config)# show interface brief
-------------------------------------------------------------
--------------------------
Port       Native  Mode   Type           Enabled Status  Reason
Speed   Description
VLAN
```

```
(Mb/s)
-------------------------------------------------------------
----------------------------------
1/1/1      11      access 1GbT        yes      up
1000     pc1
...
1/1/24     1       access 1GbT        yes      down Waiting for link
--       --
1/1/25     --      VSF    SFP+DAC1    yes      up
10000    sw-edge2
1/1/26     --      VSF    SFP+DAC1    yes      up
10000    --
1/1/27     1       trunk  SFP+DAC1    yes      up
10000    sw-agg1
1/1/28     1       trunk  SFP+DAC1    yes      up
10000    sw-agg2
...
```

- **Question**: What are the port modes you have seen so far?
- **Answer**: Access, trunk, and routed.
- **Question**: What is the new port mode for the VSF linked ports, such as 1/1/25 and 1/1/26?
- **Answer**: They are assigned a new port mode: *VSF*. This indicates the port is used for a stacking link.

14. Review the current VSF links.

```
sw-edge1(config)# show vsf link
VSF Member 1
     Link        Peer    Peer
Link State       Member  Link   Interfaces
---- ---------- ------- ------ ---------------------------
1    down        0       0      1/1/26
2    down        0       0      1/1/25
```

15. Review the current VSF topology with a single switch.

```
sw-edge1(config)# show vsf topology
Conductor
+-------+
|   1   |
+-------+
```

Perform a factory reset of sw-edge2

sw-edge2 will be factory reset. In the default state, it will listen for VSF frames and it can automatically join the VSF stack. The switch will now reboot and remove all locally stored configurations and log files.

1. Open a console connection to sw-edge2 and login using **admin / Aruba123!.**
2. Zeroize the sw-edge2 and confirm the operation.

```
sw-edge2# erase all zeroize
This will securely erase all customer data and reset the switch
to factory defaults. This will initiate a reboot and render the
switch unavailable until the zeroization is complete.
This should take several minutes to one hour to complete.
Continue (y/n)? y
The system is going down for zeroization.
Oct 17 06:50:47 hpe-mgmtmd[3776113]: RebootLibPh1: Reboot reason:
Reboot requested by user
```

Review the VSF join on sw-edge1

While sw-edge2 performs the zeroize operation, you can review the status on sw-edge1.

1. On sw-edge1, review the VSF status.

```
sw-edge2# show vsf
```

2. Start a repeat sequence. This will automatically repeat the previous command with the configured delay. This way, you don't need to repeat the command yourself while the sw-edge2 is rebooting and auto-joining the VSF system.

```
sw-edge1(config)# repeat delay 10

**************************************
Iteration : 1 Command : show vsf
**************************************
Force Autojoin                : Disabled
Autojoin Eligibility Status: Not Eligible
MAC Address                   : 64:e8:81:dd:81:80
Secondary                     : 2
Topology                      : Standalone
Status                        : Active Fragment
Split Detection Method        : mgmt
```

```
Mbr Mac Address          type           Status
ID
--- ------------------- -------------- ---------------
1   64:e8:81:dd:81:80   JL666A         Conductor
...
...
***************************************

Iteration : 26 Command : show vsf
***************************************
Force Autojoin                : Disabled
Autojoin Eligibility Status: Not Eligible
MAC Address                   : 64:e8:81:dd:81:80
Secondary                     : 2
Topology                      : Standalone
Status                        : Active Fragment
Split Detection Method        : mgmt


Mbr Mac Address          type           Status
ID
--- ------------------- -------------- ---------------
1   64:e8:81:dd:81:80   JL666A         Conductor
2                       JL666A         Not Present
...

...
***************************************
Iteration : 38 Command : show vsf
***************************************

Force Autojoin                : Disabled
Autojoin Eligibility Status: Not Eligible
MAC Address                   : 64:e8:81:dd:81:80
Secondary                     : 2
Topology                      : Ring
Status                        : Active Fragment
Split Detection Method        : mgmt


Mbr Mac Address          type           Status
ID
--- ------------------- -------------- ---------------
1   64:e8:81:dd:81:80   JL666A         Conductor
2   8c:85:c1:49:20:c0   JL666A         Standby Booting
```

```
*************************************
Iteration : 39 Command : show vsf
*************************************

Force Autojoin                 : Disabled
Autojoin Eligibility Status: Not Eligible
MAC Address                    : 64:e8:81:dd:81:80
Secondary                      : 2
Topology                       : Ring
Status                         : No Split
Split Detection Method         : mgmt


Mbr Mac Address         type           Status
ID
--- ------------------- -------------- ---------------
1   64:e8:81:dd:81:80   JL666A         Conductor
2   8c:85:c1:49:20:c0   JL666A         Standby
```

3. After a few minutes, the sw-edge2 should have completed the join. When you see two entries in the **show vsf** repeated output, you can stop the repeat using the <Ctrl>-c key sequence.

```
sw-edge1(config)# <Ctrl>-c
```

4. On sw-edge1, review the VSF running configuration.

```
sw-edge1(config)# show run vsf

vsf split-detect mgmt
vsf secondary-member 2
vsf member 1
    type jl666a
    link 1 1/1/26
    link 2 1/1/25
vsf member 2
    type jl666a
    link 1 2/1/25
    link 2 2/1/26
```

- **Question**: Do you see the member 2 in the configuration?
- **Answer**: Yes. Member 2 was automatically added with the correct type to the configuration.
- **Question**: Did you have to configure VSF links for member 2?
- **Answer**: No. These member 2 VSF links were automatically added to the configuration. Only the VSF links on the existing stack (the conductor in this case) had to be configured.

5. Check the VSF topology with member 2 active.

```
sw-edge1(config)# show vsf topology
 Conductor     Standby
 +-------+     +-------+
 |   1   |1==1|   2   |
 +-------+     +-------+
     2             2
     +=============+
```

6. Check the VSF links that have been configured.

```
sw-edge1(config)# show vsf link

 VSF Member 1

     Link       Peer    Peer
Link State      Member  Link   Interfaces
---- ---------- ------- ------ ---------------------------
1    up         2       1      1/1/26
2    up         2       2      1/1/25


 VSF Member 2

     Link       Peer    Peer
Link State      Member  Link   Interfaces
---- ---------- ------- ------ ---------------------------
1    up         1       1      2/1/25
2    up         1       2      2/1/26
```

7. Use the **show vsf detail** command to see serial and MAC details of the member devices.

```
sw-edge1(config)# show vsf detail
VSF Stack
      MAC Address                  : 64:e8:81:dd:81:80
      Secondary                    : 2
      Topology                     : ring
      Status                       : No Split
      Split Detection Method       : mgmt
      Software Version             : FL.10.09.1040
      Force Autojoin               : Disabled
      Autojoin Eligibility Status  : Not Eligible
      Autojoin Ineligibility Reason: Configuration changes detected
      Name                         : Aruba-VSF-6300
```

```
        Contact                             :
        Location                            :

Member ID                                   : 1
        MAC Address                         : 64:e8:81:dd:81:80
        Type                                : JL666A
        Model                               : 6300F 24-port 1GbE Class 4
PoE and 4-port SFP56 Switch
        Status                              : Conductor
        ROM Version                         : FL.01.09.0002
        Serial Number                       : SG09KN500B
        Uptime                              : 1 day, 22 hours, 43 minutes
        CPU Utilization                     : 3%
        Memory Utilization                  : 22%
        VSF Link 1                          : Up, connected to peer
member 2, link 1
        VSF Link 2                          : Up, connected to peer
member 2, link 2

Member ID                                   : 2
        MAC Address                         : 8c:85:c1:49:20:c0
        Type                                : JL666A
        Model                               : 6300F 24-port 1GbE Class 4
PoE and 4-port SFP56 Switch
        Status                              : Standby
        ROM Version                         : FL.01.09.0002
        Serial Number                       : SG11KN501V
        Uptime                              : 2 minutes
        CPU Utilization                     : 5%
        Memory Utilization                  : 11%
        VSF Link 1                          : Up, connected to peer
member 1, link 1
        VSF Link 2                          : Up, connected to peer
member 1, link 2
```

Console access to VSF member sw-edge2

In the next steps you will explore the console connection of the VSF stack members.

1. Use the lab dashboard to open a console connection to sw-edge2. Notice how the hostname prompt shows sw-edge1, even though you are connected to the console of the sw-edge2.

```
sw-edge1 login:
```

2. Login with **admin/Aruba123!**

```
sw-edge1 login: admin
Password:

standby#
```

- **Question**: Do you see the hostname in the prompt? What does the prompt show on the standby switch?
- **Answer**: No, the hostname is not shown, only the *standby* prompt.

3. Use **?** to see a list of commands.

```
standby# ?
  clear                Reset functions
  diagnostics          Change diagnostic commands availability
  exit                 Exit current mode and change to previous mode
  list                 Print command list
  member               VSF member selection
  no                   Negate a command or set its defaults
  page                 Enable page break
  show                 Show running system information
  start-shell          Start Bash shell
  terminal-monitor     Enables Terminal-monitor
  top                  Top command
  vsf-factory-reset    Erase all customer data and reset the switch
to factory defaults.
```

- **Question**: Do you have full CLI access on the standby device?
- **Answer**: No, only minimal diagnostics and **show** commands are available.

You may leave the console connection to sw-edge2 open for now. This concludes the member auto join VSF.

Task 3: Connect VSF access stack to VSX aggregation with an MCLAG

In this task you will connect the VSF stack with a LAG of four uplinks to the VSX aggregation switches. In the previous lab, you prepared the VSX system with a single MCLAG (LAG1 on the VSX) of four member ports.

Typically, an access VSF stack would only have two uplink ports: one uplink port on the conductor and one uplink port on the secondary. In this lab environment, each access switch connects with two ports to the aggregation layer. This is why four uplink ports are bundled into the link-aggregation:

- sw-edge1 and sw-edge2 have their port 1/1/27 connected to sw-agg1, on ports 1/1/1 and 1/1/2 respectively.
- sw-edge1 and sw-edge2 have their port 1/1/28 connected to sw-agg2, on ports 1/1/1 and 1/1/2 respectively.

Objectives

- Configure a VSF stack with LAG to VSX aggregation.
- Use a LAG with member ports of different VSF members.

Steps

1. Open an SSH connection to **sw-edge1**.
2. Review the interface brief list.

```
sw-edge1(config)# show interface brief
---------------------------------------------------------------------------
-------------------------------
Port       Native  Mode   Type           Enabled Status  Reason
Speed   Description
           VLAN
(Mb/s)
---------------------------------------------------------------------------
--------------------------------
1/1/1      11      access 1GbT           yes     up
1000    pc1
...
1/1/24     1       access 1GbT           yes     down    Waiting
for link        --      --
1/1/25     --      VSF    SFP+DAC1       yes     up
10000   sw-edge2
1/1/26     --      VSF    SFP+DAC1       yes     up
10000   --
1/1/27     1       trunk  SFP+DAC1       yes     up
10000   sw-agg1
1/1/28     1       trunk  SFP+DAC1       yes     up
10000   sw-agg2
2/1/1      1       access 1GbT           yes     down    Waiting
for link        --      --
...
2/1/24     1       access 1GbT           yes     down    Waiting
for link        --      --
```

```
2/1/25    --       VSF    SFP+DAC1        yes      up
10000   --
2/1/26    --       VSF    SFP+DAC1        yes      up
10000   --
2/1/27    1        access SFP+DAC1        yes      up
10000   --
2/1/28    1        access SFP+DAC1        yes      up
10000   --
vlan1     --              --              no       down
Administratively down   --       --
vlan3    --          --          yes    up                        --     --
lag50     1        trunk  --              no       down     --
auto    --
lag256    1        trunk  --              yes      up       --
20000   --
```

- **Question**: In the output, do you notice ports that start with 2/1/?
- **Answer**: Yes, these are the ports of the member 2 that are now under the management of this VSF system.

3. Check the LLDP neighbors.

```
sw-edge1(config)# show lldp neighbor-info

LLDP Neighbor Information
=========================

Total Neighbor Entries           : 10
Total Neighbor Entries Deleted   : 2
Total Neighbor Entries Dropped   : 0
Total Neighbor Entries Aged-Out  : 2

LOCAL-PORT  CHASSIS-ID          PORT-ID       PORT-DESC
TTL       SYS-NAME
------------------------------------------------------------------
-------------------
...
1/1/27      b8:d4:e7:da:80:00  1/1/1         1/1/1
120       sw-agg1
1/1/28      b8:d4:e7:d9:e5:00  1/1/1         1/1/1
120       sw-agg2
...
2/1/27      b8:d4:e7:da:80:00  1/1/2         1/1/2
120       sw-agg1
2/1/28      b8:d4:e7:d9:e5:00  1/1/2         1/1/2
120       sw-agg2
...
```

- **Question**: On what ports do you see sw-agg1 and sw-agg2 as LLDP neighbors?
- **Answer**: On ports 1/1/27 and 1/1/28, as well as 2/1/27 and 2/1/28.

4. Review the current LACP interface status.

```
sw-edge1(config)# show lacp interfaces


State abbreviations :
A - Active          P - Passive      F - Aggregable I - Individual
S - Short-timeout L - Long-timeout N - InSync      O - OutofSync
C - Collecting     D - Distributing
X - State m/c expired              E - Default neighbor state

Actor details of all interfaces:
-------------------------------------------------------------------
-------------
Intf     Aggr        Port  Port  State   System-ID         System
Aggr Forwarding
         Name        Id    Pri                             Pri
Key  State
-------------------------------------------------------------------
1/1/27   lag256      28    1     ALFNCD  64:e8:81:dd:81:80 65534
256  up
1/1/28   lag256      29    1     ALFNCD  64:e8:81:dd:81:80 65534
256  up


Partner details of all interfaces:
-------------------------------------------------------------------
---------------
Intf     Aggr        Port  Port  State   System-ID         System
Aggr
         Name        Id    Pri                            Pri
Key
-------------------------------------------------------------------
1/1/27   lag256      1     1     ALFNCD  02:01:00:00:01:00 65534
1
1/1/28   lag256      1001  1     ALFNCD  02:01:00:00:01:00 65534
1
```

- **Question**: What ports are missing from the LAG 256 configuration?
- **Answer**: The new ports 2/1/27 and 2/1/28 of member 2 are missing from the LAG 256.

5. Now add ports 2/1/27 and 2/1/28 to the LAG 256 and enable them.

```
sw-edge1(config)# interface 2/1/27,2/1/28
sw-edge1(config-if-<2/1/27,2/1/28>)# lag 256
sw-edge1(config-if-<2/1/27,2/1/28>)# exit
```

6. Verify the LACP interface status. The *Forwarding State* for all four ports should be *up* now.

```
sw-edge1(config)# show lacp interfaces
State abbreviations :
A - Active        P - Passive      F - Aggregable I - Individual
S - Short-timeout L - Long-timeout N - InSync     O - OutofSync
C - Collecting    D - Distributing
X - State m/c expired              E - Default neighbor state

Actor details of all interfaces:
------------------------------------------------------------------------------
--------------------
Intf       Aggr      Port  Port  State   System-ID           System Aggr
Forwarding
           Name      Id    Pri                                Pri   Key
State
------------------------------------------------------------------------------
--------------------
1/1/27     lag256    28   1    ALFNCD  64:e8:81:dd:81:80       65534  256
up
1/1/28     lag256    29   1    ALFNCD  64:e8:81:dd:81:80       65534  256
up
2/1/27     lag256    92   1    ALFNCD  64:e8:81:dd:81:80       65534  256
up
2/1/28     lag256    93   1    ALFNCD  64:e8:81:dd:81:80       65534  256
up


Partner details of all interfaces:
------------------------------------------------------------------------------
Intf     Aggr     Port  Port  State   System-ID         System Aggr
         Name     Id    Pri                             Pri    Key
------------------------------------------------------------------------------
1/1/27   lag256   1     1     ALFNCD  02:01:00:00:01:00 65534  1
1/1/28   lag256   1001  1     ALFNCD  02:01:00:00:01:00 65534  1
2/1/27   lag256   2     1     ALFNCD  02:01:00:00:01:00 65534  1
2/1/28   lag256   1002  1     ALFNCD  02:01:00:00:01:00 65534  1
```

Task 4: Split-brain detection test

Your customer is pleased with the VSF configuration and connectivity, but they have a concern about the VSF stacking cables. They would like to know what happens when one or more VSF links fail, and how that affects the network.

You plan a meeting with your senior colleague and the customer, where your colleague explains the impact:

- A VSF stack will typically be connected in a ring topology. The ring topology provides protection against a link failure on any of the VSF links. This ensures there is no single point of failure.
- However, in the unlikely event that a second VSF link fails, it is possible that there is a split, or fragmented stack, with a conductor in both fragments.
- The member 1 will be the conductor in one fragment, while the configured secondary member will be the conductor in the other fragment.
- This may be a challenge for the rest of the network, since there are two different instances of STP/ LACP operating, but they are using the same system MAC address.

The solution is to make sure that the fragment with the secondary switch as the conductor will disable its ports; this will effectively remove it from the network. The split detection will provide a dedicated connection between the member 1 and the secondary to support this detection. If there is a fragment without the member 1 or the secondary member, the fragment will not have a conductor. As such there is no split-brain, but it also means the switches in this fragment will not be operational, due to the lack of control plane.

In this task you will explore the configuration and operation of a split stack situation.

Objectives

- Configure the split stack detection.
- Verify the split stack operation.
- Verify the operation after reconnecting the fragments.

Steps

1. Open an SSH connection to **sw-edge1** and review the VSF status.

```
sw-edge1(config)# show vsf

Force Autojoin               : Disabled
Autojoin Eligibility Status: Not Eligible
MAC Address                  : 64:e8:81:dd:81:80
```

```
Secondary                    : 2
Topology                     : Ring
Status                       : No Split
Split Detection Method       : mgmt


Mbr Mac Address          type            Status
ID
--- ------------------- -------------- ---------------
1   64:e8:81:dd:81:80   JL666A         Conductor
2   8c:85:c1:49:20:c0   JL666A         Standby
```

2. Enable the terminal monitor to see live event logging.

```
sw-edge1(config)# terminal-monitor filter vsf
Terminal-monitor is enabled successfully
```

Note

The **terminal-monitor** command is not available on the console connection. If you receive an error when using **terminal-monitor**, make sure to use PC-MGMT and open an SSH connection to sw-edge1.

3. Disable the VSF links by shutting down the physical member ports.

```
sw-edge1(config)# interface 1/1/25,1/1/26
sw-edge1(config-if-vsf-<1/1/25,1/1/26>)# shutdown
sw-edge1(config-if-vsf-<1/1/25,1/1/26>)# exit
```

4. Review the events that are displayed on your session.

```
2022-10-17T07:06:25.278928-0400 vsfd[713] <INFO> Event|9924|LOG_
INFO|CDTR|1|VSF link 2 is down
2022-10-17T07:06:25.286822-0400 vsfd[713] <INFO> Event|9908|LOG_
INFO|CDTR|1|Topology is Standalone
2022-10-17T07:06:25.286981-0400 vsfd[713] <WARN> Event|9913|LOG_
WARN|CDTR|1|Lost member 2 with Loss of communication
2022-10-17T07:06:25.299888-0400 vsfd[713] <INFO> Event|9924|LOG_
INFO|CDTR|1|VSF link 1 is down
2022-10-17T07:06:26.459501-0400 vsfd[713] <INFO> Event|9927|LOG_
INFO|CDTR|1|Fragment with conductor 1 is Active
```

5. Check vsf status again.

```
sw-edge1(config)# show vsf

Force Autojoin                : Disabled
Autojoin Eligibility Status: Not Eligible
MAC Address                   : 64:e8:81:dd:81:80
Secondary                     : 2
Topology                      : Standalone
Status                        : Active Fragment
Split Detection Method        : mgmt


Mbr Mac Address          type           Status
ID
--- ------------------- -------------- ---------------
1   64:e8:81:dd:81:80   JL666A         Conductor
2                       JL666A         In Other Fragment
```

- **Question**: What is the status for member 2?
- **Answer**: In Other Fragment.

6. Open or switch to the console connection on sw-edge2; login with **admin/Aruba123!**

```
sw-edge1 login: admin
Password:
Last login: 2022-10-17 11:00:17 from the console
User "admin" has logged in 1 time in the past 30 days
sw-edge1#
```

- **Question**: Is the hostname prompt still *standby*?
- **Answer**: No, since the secondary lost the conductor, it became the active conductor and assumed the conductor hostname, in this example sw-edge1.

7. On the console of sw-edge2, review the interface list.

```
sw-edge1# show interface brief
--------------------------------------------------------------------
------------------------------------
Port      Native  Mode   Type          Enabled Status  Reason
Speed   Description
          VLAN
(Mb/s)
--------------------------------------------------------------------
------------------------------------
2/1/1   1        access 1GbT          yes     down    Disabled
by VSF
--      --
```

```
2/1/2    1      access 1GbT          yes      down     Disabled by VSF
--       --
2/1/3    1      access 1GbT          yes      down     Disabled by VSF
--       --
2/1/4    1      access 1GbT          yes      down     Disabled by VSF
--       --
2/1/5    1      access 1GbT          yes      down     Disabled by VSF
         --       --
2/1/6    1      access 1GbT          yes      down     Disabled by VSF
--       --
2/1/7    1      access 1GbT          yes      down     Disabled by VSF
--       --
2/1/8    1      access 1GbT          yes      down     Disabled by VSF
--       --
2/1/9    1      access 1GbT          yes      down     Disabled by VSF
--       --
2/1/10   1      access 1GbT          yes      down     Disabled by VSF
--       --
2/1/11   1      access 1GbT          yes      down     Disabled by VSF
--       --
2/1/12   1      access 1GbT          yes      down     Disabled by VSF
--       --
2/1/13   1      access 1GbT          yes      down     Disabled by VSF
--       --
2/1/14   1      access 1GbT          yes      down     Disabled by VSF
--       --
2/1/15   1      access 1GbT          yes      down     Disabled by VSF
--       --
2/1/16   1      access 1GbT          yes      down     Disabled by VSF
--       --
2/1/17   1      access 1GbT          yes      down     Disabled by VSF
--       --
2/1/18   1      access 1GbT          yes      down     Disabled by VSF
--       --
2/1/19   1      access 1GbT          yes      down     Disabled by VSF
--       --
2/1/20   1      access 1GbT          yes      down     Disabled by VSF
--       --
2/1/21   1      access 1GbT          yes      down     Disabled by VSF
--       --
2/1/22   1      access 1GbT          yes      down     Disabled by VSF
--       --
2/1/23   1      access 1GbT          yes      down     Disabled by VSF
--       --
2/1/24   1      access 1GbT          yes      down     Disabled by VSF
--       --
2/1/25   --     VSF    SFP+DAC1      yes      down     Waiting for link
--       --
```

```
2/1/26   --     VSF    SFP+DAC1      yes     down    Waiting for link
--       --
2/1/27   1      trunk  SFP+DAC1      yes     down    Disabled by VSF
--       --
2/1/28   1      trunk  SFP+DAC1      yes     down    Disabled by VSF
--       --
vlan1    --            --            no      down    Administratively
down     --            --
vlan3     --           --             yes     down
--       --
lag50      1   trunk   --             no      down     --
uto      --
lag256     1   trunk   --             yes     down     --
auto     --
```

- **Question**: What is the status and reason text for the interfaces?
- **Answer**: Disabled by VSF. This is the split-brain detection that has disabled the ports on this fragment.
- **Question**: What is the status for the VSF links?
- **Answer**: Waiting for link.
- **Question**: Why are the VSF links not disabled by VSF?
- **Answer**: When the VSF links would be restored, this fragment will automatically detect that the links come up again and it will automatically rejoin the VSF stack by rebooting the current fragment. If the VSF links would have been disabled, this fragment would not be able to detect that the VSF links would have been restored.

8. On the console of **sw-edge2**, review VSF status.

```
sw-edge1# show vsf

Force Autojoin               : Disabled
Autojoin Eligibility Status: Not Eligible
MAC Address                  : 64:e8:81:dd:81:80
Secondary                    : 2
Topology                     : Standalone
Status                       : Inactive Fragment
Split Detection Method       : mgmt


Mbr Mac Address          type            Status
ID
--- ------------------- -------------- ---------------
1                        JL666A          In Other Fragment
2   8c:85:c1:49:20:c0   JL666A          Conductor
```

- **Question**: What is the status of the current system?
- **Answer**: The secondary has detected using the OOBM mgmt interface that the conductor is still active; it has marked itself the *Inactive Fragment*.

9. You may close the console connection to sw-edge2.

Restore the connection.

1. On sw-edge1, enable the ports 1/1/25 and 1/1/26 again.

```
sw-edge1(config)# interface 1/1/25,1/1/26
sw-edge1(config-if-vsf-<1/1/25,1/1/26>)# no shutdown
sw-edge1(config-if-vsf-<1/1/25,1/1/26>)# exit
```

2. Review the events on the terminal. Sw-edge2 will now reboot to join the VSF stack again.

```
2022-10-17T07:14:04.971464-0400 vsfd[713] <INFO>
Event|9923|LOG_INFO|CDTR|1|VSF link 2 is up
2022-10-17T07:14:04.979681-0400 vsfd[713] <INFO>
Event|9923|LOG_INFO|CDTR|1|VSF link 1 is up
2022-10-17T07:14:07.978152-0400 vsfd[713] <INFO>
Event|9916|LOG_INFO|CDTR|1|Member 2 elected as standby reason-Des-
ignated standby
2022-10-17T07:14:08.117673-0400 vsfd[713] <INFO>
Event|9924|LOG_INFO|CDTR|1|VSF link 1 is down
2022-10-17T07:14:28.653101-0400 vsfd[713] <INFO>
Event|9924|LOG_INFO|CDTR|1|VSF link 2 is down
```

3. On sw-edge1, you may disable the terminal monitor.

```
sw-edge1(config)# no terminal-monitor
```

4. Wait a few minutes and confirm that member has joined the VSF stack as the standby system.

```
sw-edge1(config)# show vsf
```

Task 5: Conductor failover test

Your customer is very interested in the VSF stacking technology, and they would like to know what happens when the VSF conductor fails or reboots. In this task you will test this failure. This prepares you to respond to the customer questions.

The member 1 is by default the active conductor in the VSF stack. If member 1 is down due to a reboot or power loss, the secondary assumes the conductor role. It is important that the secondary member ID is set in the configuration, which must be done manually.

In the following tasks you will explore auto-stacking. With auto-stacking, member 2 is automatically configured as the secondary switch in the VSF stack, so there is no need to configure it manually.

Objectives

- Verify the impact of the conductor failover.

Prepare test connections

First you will initiate some ping tests from various systems to the VSF system.

There will be an out-of-band ping test and an in-band ping test that you will perform.

Out-of-band

1. On sw-edge1, use **show interface mgmt**, and take note of the MAC address. The example output, your output may be different.

```
sw-edge1(config)# show interface mgmt
  Address Mode                    : static
  Admin State                     : up
  Link State                      : up
  Mac Address                     : 64:e8:81:dd:81:81
  IPv4 address/subnet-mask        : 10.251.1.4/24
  Default gateway IPv4            :
  IPv6 address/prefix             :
  IPv6 link local address/prefix  : fe80::66e8:81ff:fedd:8181/64
  Default gateway IPv6            :
  Primary Nameserver              :
  Secondary Nameserver            :
  Tertiary Nameserver             :
```

- What is the OOBM MAC: ____________________

2. On PC-MGMT, perform a ping to the OOBM mgmt interface (**10.251.1.4**) of the sw-edge1.

```
C:> ping 10.251.1.4
```

3. On PC-MGMT, check the ARP table.

```
C:> arp -a 10.251.1.4
Interface: 10.251.1.90 --- 0xf
   Internet Address      Physical Address      Type
   10.251.1.4            64-e8-81-dd-81-81     dynamic
```

- **Question**: Does the ARP entry for 10.251.1.4 match the MAC address you have recorded in step 1?
- **Answer**: Yes, this is the same MAC.

4. On PC-MGMT, start a continuous ping to the OOBM IP 10.251.1.4.

```
C:> ping 10.251.1.4 -t
```

In-band

For the in-band test, you will use sw-agg1 and ping to the SVI 3 IP address (the access switch in-band management IP).

5. On sw-agg1 ping to **10.1.3.4.**

```
sw-agg1(config)# ping 10.1.3.4
```

6. Check the ARP table entry for MAC.

```
sw-agg1(config)# show arp | include 10.1.3.4
10.1.3.4          64:e8:81:dd:81:80  vlan3          lag1
reachable
```

- Question: Take note of the MAC address of the IP 10.1.3.4: ____________________.

7. On sw-agg1, now initiate a series of 100 pings. You need to disable paging to ensure the output does not stop after 1 page.

```
sw-agg1(config)# no page
sw-agg1(config)# ping 10.1.3.4 repetitions 100
PING 10.1.3.4 (10.1.3.4) 100(128) bytes of data.
108 bytes from 10.1.3.4: icmp_seq=1 ttl=64 time=0.167 ms
108 bytes from 10.1.3.4: icmp_seq=2 ttl=64 time=0.166 ms
...
```

Reboot the conductor

You are now ready to test the conductor failover and observe the impact on the MAC address of the stack.

1. On the VSF system (SSH connection to sw-edge1), reboot member 1.

```
sw-edge1(config)# vsf member 1 reboot
The conductor switch will reboot and the standby will become the
conductor.
Do you want to continue (y/n)? y
sw-edge1(config)#
Oct 17 07:19:39 vsfd[713]: RebootLibPh1: Reboot reason: Reboot of
conductor due to Reboot requested by user
```

2. On PC-MGMT, verify the ping continues to the OOBM IP using the same mgmt MAC.

```
Reply from 10.251.1.4: bytes=32 time<1ms TTL=64
Reply from 10.251.1.4: bytes=32 time<1ms TTL=64
Reply from 10.251.1.4: bytes=32 time<1ms TTL=64
Request timed out.
```

```
Request timed out.
Request timed out.
Reply from 10.251.1.4: bytes=32 time<1ms TTL=64
Reply from 10.251.1.4: bytes=32 time=10ms TTL=64
Reply from 10.251.1.4: bytes=32 time<1ms TTL=64
```

3. On PC-MGMT, stop the ping and check the ARP table again.

```
C:> arp -a 10.251.1.4

Interface: 10.251.1.90 --- 0xf
  Internet Address      Physical Address      Type
  10.251.1.4            64-e8-81-dd-81-81     dynamic
```

4. On sw-agg1, verify the ping continues on the in-band IP using the same SVI MAC. This demonstrates the conductor role failover to the secondary and the stability of the VSF MAC address during the failover.

Restore member 1 as the conductor

Once member 1 comes back online, it will not automatically claim the conductor role. This is fine in a production environment. In the lab environment, the next lab requires member 1 to be the conductor. Therefore, you will reboot the secondary. This will make member 1 the conductor again.

1. Re-open the SSH connection to sw-edge1 (this is the OOBM IP 10.251.1.4, but it is currently handled by the sw-edge2 as this is the current conductor).
2. Review the VSF status.

```
sw-edge1# show vsf

Force Autojoin                : Disabled
Autojoin Eligibility Status: Not Eligible
MAC Address                   : 64:e8:81:dd:81:80
Secondary                     : 2
Topology                      : Ring
Status                        : No Split
Split Detection Method        : mgmt


Mbr Mac Address        type           Status
ID
--- ------------------ -------------- ---------------
1   64:e8:81:dd:81:80  JL666A         Standby
2   8c:85:c1:49:20:c0  JL666A         Conductor
```

3. Reboot member 2 to make member 1 conductor. You may also use the **vsf switchover** or **vsf member 2 reboot** commands.

```
sw-edge1# vsf switchover
This will cause an immediate switchover to the standby
and the conductor will reboot.
Do you want to continue (y/n)? y
sw-edge1#
 Oct 17 07:23:34 vsfd[732]: RebootLibPh1: Reboot reason: Reboot of
conductor due to VSF switchover requested
```

Important

Wait for the member 2 to join the VSF stack again before moving to the next task.

Task 6: Automatic conductor VSF stack formation

Your customer is pleased with the auto-join function to add a factory default switch to an existing stack. However, the customer is thinking about deploying a lot of switches, and they are wondering if the initial configuration of the complete VSF stack can also be done by their technician when a new location needs to be deployed. This would include the initial conductor configuration, but without accessing the configuration of the factory default member 1. After consulting with your senior colleague, you learn that the auto-stacking feature can also be used to configure to the conductor.

In this task you will see an easy method to form a VSF stack, including the initial assignment of the conductor in the stack. This procedure must be applied to factory default switches. This means the first steps in this task will be to make the access switches factory default again. Auto-stacking requires the use of dedicated ports on the switches. The following is an extract from the VSF configuration guide:

https://www.arubanetworks.com/techdocs/AOS-CX/10.09/HTML/vsf/Content/fir-int.htm.

Reserved interfaces for auto-stacking

Based on the product type of a switch, the following two interfaces are reserved for the auto-stacking process:

- 24-port switch models: 25 and 26
- 48-port switch models: 49 and 50

Users can physically connect the switch to an existing stack on one of these reserved auto-stacking interfaces. Figure 5-14 shows the list of reserved auto-stacking interfaces based on the product type and platform. An example connection diagram for four switches can be seen in Figure 5-17.

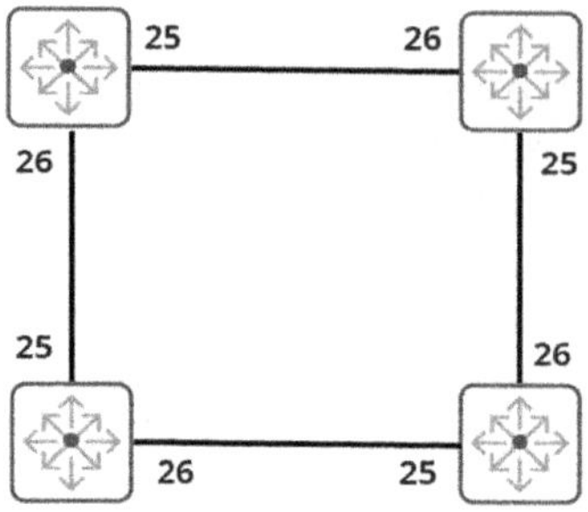

Figure 5-17: Example cabling diagram

Note

Use the higher port number on the member 1 (either 26 or 50) to connect it to the lower port number on the member 2 (either 25 or 49). The rest of the connections should automatically follow this logic of connecting the higher port to the next member lower port.

Objectives

- Understand VSF auto-stacking.
- Enable VSF auto-stacking on the conductor.
- Verify VSF auto-stacking.

Steps

Factory reset the devices

1. Open an SSH connection to **sw-edge1**.
2. Verify that member 2 is active as *Standby* in the VSF stack. It may still be booting after the last task; make sure it has completed the reboot before continuing.

```
sw-edge1# show vsf
Force Autojoin                 : Disabled
Autojoin Eligibility Status: Not Eligible
MAC Address                    : 64:e8:81:dd:81:80
Secondary                      : 2
Topology                       : Ring
Status                         : No Split
Split Detection Method         : mgmt


Mbr Mac Address          type          Status
ID
--- ------------------- ------------- ---------------
1   64:e8:81:dd:81:80   JL666A        Conductor
2   8c:85:c1:49:20:c0   JL666A        Standby
```

Important
Make sure to wait for both switches to be online (rebooted) after the previous task!

3. Zeroize the configuration. It will take *several minutes* to complete the factory reset.

```
sw-edge1# erase all zeroize
This will securely erase all customer data and reset the switch
to factory defaults. This will initiate a reboot and render the
switch unavailable until the zeroization is complete.
This should take several minutes to one hour to complete.
Continue (y/n)? y
The system is going down for zeroization.
sw-edge1#
 Oct 17 07:41:30 hpe-mgmtmd[34933]: RebootLibPh1: Reboot reason:
Reboot requested by user
```

Prepare the conductor switch

Note
This step can also be performed by pushing the LED button on the physical switch. In that case no console cable is required to initiate the creation of a VSF stack. Here is an excerpt from the AOS-CX 10.09 *Virtual Switching Framework (VSF) Guide*:
1. Physically connect the switches in a desired topology on the reserved VSF link ports.
2. Press the LED mode button on the conductor until the mode changes to "Stk". The stack members reboot one after another and join the stack.
   During stacking operation, the port LEDs are displayed in three different states:
   - Flashing green—Indicates that the member is the conductor.
   - Flashing orange—Indicates that the member is rebooting to join the stack or offline due to error condition.
   - Solid green—Indicates that the member joined the stack and is operational.

In this remote lab, you don't have access to the LED button, so you will use the CLI to initiate the auto-stacking.

1. Open a connection to the console of sw-edge1. Since the switch is now factory default, connect with **admin** / no password (blank password).

2. When prompted for a new password, leave it blank (just press <**Enter**>).

```
6300 login: admin
Password:


Please configure the 'admin' user account password.
Enter new password:
Confirm new password:
6300#
```

3. Review the default VSF configuration.

```
6300# show vsf

Force Autojoin                : Disabled
Autojoin Eligibility Status: Eligible
MAC Address                   : 64:e8:81:dd:81:80
Secondary                     :
Topology                      : Standalone
Status                        : No Split
Split Detection Method        : None


Mbr Mac Address          type           Status
ID
--- ------------------- -------------- ---------------
1   64:e8:81:dd:81:80   JL666A         Conductor
```

- **Question**: Is there any standby member configured?
- **Answer**: No, by default there is no standby member set.

4. Review the VSF running configuration.

```
6300# show running-config vsf
vsf member 1
    type jl666a
```

- **Question**: Is there a member 2 present?
- **Answer**: No. By default, only member 1 is present.
- **Question**: Are there any VSF links configured?
- **Answer**: No. By default, no VSF links are configured.

5. Enter the configuration mode and initiate the auto-stacking.

```
6300# config
6300(config)# vsf start-auto-stacking
This will configure links and secondary on conductor

Do you want to continue (y/n)? y
```

6. Review the updated VSF configuration.

```
6300(config)# show vsf

Force Autojoin                : Disabled
Autojoin Eligibility Status: Not Eligible
MAC Address                   : 64:e8:81:dd:81:80
Secondary                     : 2
Topology                      : Standalone
Status                        : Active Fragment
Split Detection Method        : None


Mbr Mac Address          type           Status
ID
--- ------------------- -------------- ---------------
1   64:e8:81:dd:81:80   JL666A         Conductor
2                       JL666A         Not Present
```

- **Question**: Is there a secondary member configured now?
- **Answer**: Yes, member 2 has been set as the standby.

7. Review the updated VSF running-configuration.

```
6300(config)# show running-config vsf
vsf secondary-member 2
vsf member 1
    type jl666a
    link 1 1/1/26
    link 2 1/1/25
vsf member 2
    type jl666a
    link 1 2/1/25
    link 2 2/1/26
```

8. Wait for member 2 to complete the reboot in the VSF stack with the role of secondary.

9. Review the VSF status.

```
6300(config)# show vsf

Force Autojoin              : Disabled
Autojoin Eligibility Status: Not Eligible
MAC Address                 : 64:e8:81:dd:81:80
Secondary                   : 2
Topology                    : Ring
Status                      : No Split
Split Detection Method      : None


Mbr Mac Address          type           Status
ID
--- ------------------- -------------- ---------------
1   64:e8:81:dd:81:80   JL666A         Conductor
2   8c:85:c1:49:20:c0   JL666A         Standby
```

- **Question**: What is the configured Split Detection Method?
- **Answer**: None. Split detection is not automatically configured by the auto-stacking feature.

This concludes the VSF auto-stacking task.

Task 7: Cleanup the configuration

In this last task you will:

- Revert the aggregation switch configuration to support two access switches with an MC-LAG for each.
- Factory reset the access switches. After the reset they will operate as two individual switches again.

Revert the aggregation switch configuration

1. On sw-agg1, assign port 1/1/2 back to LAG 2.

```
sw-agg1(config)# interface 1/1/2
sw-agg1(config-if)# no lag 1
sw-agg1(config-if)# lag 2
sw-agg1(config-if)# no shutdown
sw-agg1(config-if)# exit
```

2. On sw-agg2, assign port 1/1/2 back to LAG 2.

```
sw-agg2(config)# interface 1/1/2
sw-agg2(config-if)# no lag 1
sw-agg2(config-if)# lag 2
sw-agg2(config-if)# no shutdown
sw-agg2(config-if)# exit
```

Factory reset the edge switches

1. On sw-edge1, zeroize the configuration.

```
6300# erase all zeroize
This will securely erase all customer data and reset the switch
to factory defaults. This will initiate a reboot and render the
switch unavailable until the zeroization is complete.
This should take several minutes to one hour to complete.
Continue (y/n)? y
The system is going down for zeroization.
```

Note

The edge switches are left at factory default state at the end of this lab activity. Since you are *not* using the VSF **start-auto-stacking** command, the two switches will remain in their factory default, single switch state. This is the expected state for the next lab activity.

Learning check

Discuss these questions.

1. Which of the following statements accurately describe Aruba's Virtual Switch Framework (VSF)? (Select four.)
   a. VSF defines a virtual switch, composed of individual switches interconnected via a proprietary backplane link.
   b. You can configure a maximum of 10 members in a VSF stack.
   c. VSF primary and secondary roles can perform the conductor role, but a VSF member cannot.
   d. VSF conductor and standby maintain synchronized state and configuration data by leveraging an Open Virtual Switch Database (OVSDB).
   e. VSF supports daisy chain, ring, and full-mesh topologies.
   f. You can leverage VSF pre-provisioning to facilitate adding new members to a VSF stack.

2. Which VSF feature provides a mechanism to automatically form a stack?
   a. VSF pre-provisioning
   b. VSF priority configuration
   c. VSF LAG
   d. Auto-VSF
3. VSF Multi-Active Detection (MAD) mitigates a split-brain condition by shutting down all ports of all members in a non-primary fragment.
   a. True
   b. False

Summary

Congratulations! You have completed Chapter 5. You should now be able to:

- Describe VSF.
- Configure VSF.
- Describe Auto-VSF.
- Explain VSF MAD.

6 Introduction to Aruba Solutions

OBJECTIVES

After completing this chapter, you will recognize fundamental concepts that are foundational for mastering network technology. You will be able to:

✓ Explain Aruba Edge Services Platform (ESP).

✓ Describe various Aruba switching products.

✓ Describe various Aruba wireless LAN (WLAN) products.

✓ Understand the Aruba Central management platform.

✓ Describe the Aruba ClearPass product.

Aruba Edge Services Platform (ESP)

Let's start the chapter by introducing the Aruba Edge Services Platform (ESP).

Aruba ESP Architecture

ESP is based on four core attributes (see Figure 6-1):

- **Unified Infrastructure**: manages different types of devices in different locations using a single interface. It eliminates silos and streamlines operations.
- **Zero trust security**: means that every single device that plugs into the network gets authenticated and authorized. You are not inherently trusted and zero trust security includes role-based security to minimize the need for manual configuration.
- **Artificial intelligence operations**: gives the network a *sixth sense*, resolving issues *before* they impact the business or at least resolve them faster.
- **Aruba Central**: powers ESP, which is a *software as a service* (SaaS) management solution, ready to monitor and manage any size company. Aruba Central's flexibility gives you greater choice and control over how you consume and pay for technology to accomplish their goals.

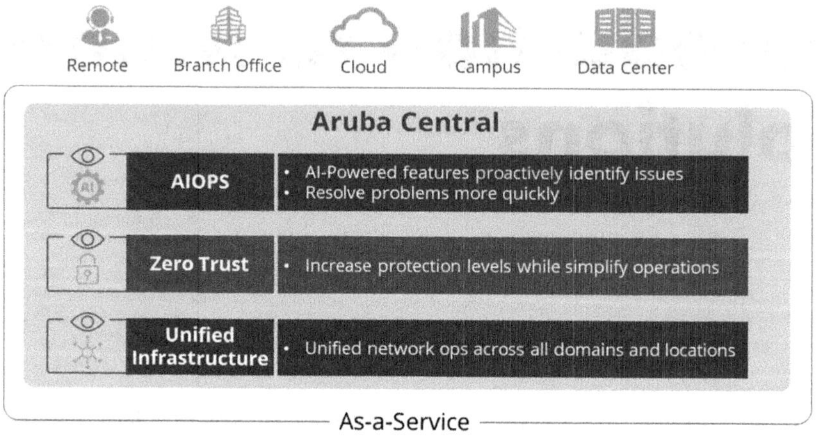

Figure 6-1: Aruba Edge Services Platform (ESP)

Reference architecture: Small campus

Figure 6-2 shows an example network of many of the different components that you will typically see in a small campus network.

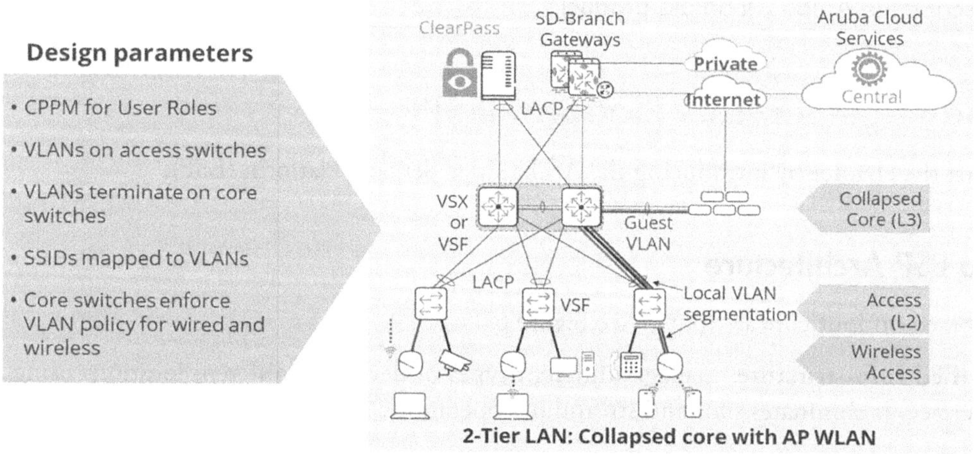

Figure 6-2: Reference architecture: Small campus

Aruba switch portfolio

Let's move on and introduce the Aruba CX switching portfolio.

Aruba CX switching overview

The Aruba CX switching advantages include the following (see Figure 6-3).

- A single operating system across your entire environment spanning from edge to data center.
- Aruba Central is a cloud management tool. It is a unified infrastructure, single pane-of-glass to manage and monitor your Aruba wired and wireless networking devices. Central supports a multi-editor mode for advanced configurations (based on their stand-alone NetEdit product), which allows you to orchestrate common configuration changes across one or multiple CX switches.
- Aruba Network Analytics Engine (NAE) is a tool to help you with monitoring and troubleshooting. NAE provides network operators with distributed, real-time insights into network health for a faster, more proactive way of monitoring, detecting, and troubleshooting IT- and user-impacting issues.
- Aruba Virtual Switch Framework (VSF) and Virtual Switching Extension (VSX): These technologies (covered in Chapters 4 and 5) bundle two or more physical switches into a single logical switch. This ensures high availability and support for live upgrades to eliminate maintenance windows and reduce downtime.

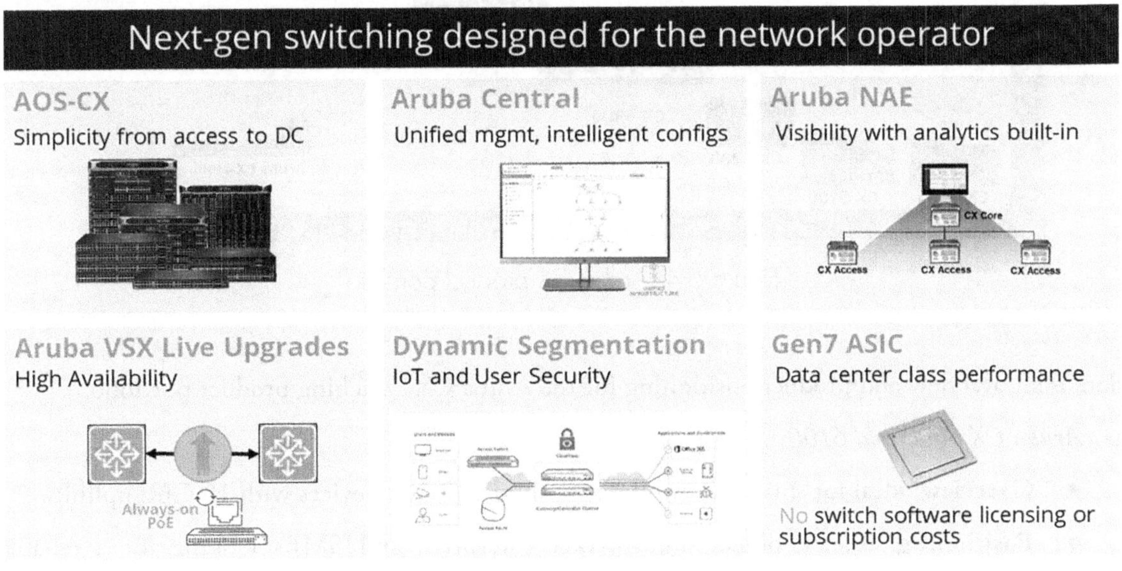

Figure 6-3: Aruba switch portfolio overview

- Dynamic segmentation is a way to unify network access and network security regardless of whether the user is using wired or Wi-Fi connection.
- The top-shelf, Gen7 ASIC is a microprocessor specialized in processing network traffic. It offers programmable, data center class performance with no bottlenecks and requires no additional switch software licensing or subscriptions.

Aruba CX product portfolio

The Aruba CX switching portfolio includes modular access, aggregation, and core switches (see Figure 6-4). At the access and aggregation layers you get power over Ethernet (PoE), multi-gigabit speeds, secure segmentation, and stacking capabilities. At the campus core and data center, you get high availability, bandwidth, and speed: with 40 and 100 GbE scalability.

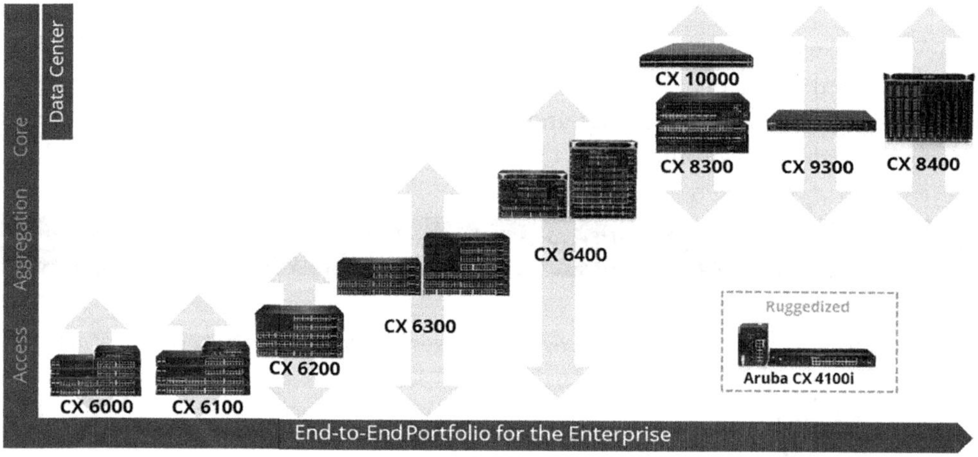

Figure 6-4: Aruba CX product portfolio

Here is an overview and product positioning for the Aruba CX switching product portfolio.

- *Aruba CX 6000 and 6100*:
  - **Overview**: ideal for smaller networks of around 100-500 devices with 10 GbE uplinks.
  - **Positioning**: branch offices, mid-market enterprise, and SMBs looking for a reliable, cost-sensitive entry-level access
- *Aruba CX 6200*:
  - **Overview**: These fixed-port access switches support 8-member stacking with built-in 1/10 GbE uplinks and PoE.
  - **Positioning**: Ideal for branch offices, mid-market enterprise, and SMBs looking for a reliable, simple to deploy, cost-sensitive access solution.

- *Aruba CX 6300*:
  - **Overview**: A top-of-rack (ToR) model with back-to-front airflow designed for cost-effective 1 GbE ToR and OOBM data center deployments.
  - **Positioning**: Ideal for use in business-critical networks with redundant power and fans, with IT initiatives driving to Wi-Fi 6, plans for Internet of Things (IoT), and looking for automation to help with rollouts. Supports bandwidth intensive applications, plus investment protection with future-ready smart rate multi-gigabit capabilities and high-power PoE.
- *Aruba CX 6400*:
  - **Overview**: Versatile 5- and 10-slot high availability (HA) chassis with up to 28 Tbps capacity, 1/10/25/40/50/100 GbE line rate interfaces, and high-density smart rate multi-gig and high-power 60w PoE port capability.
  - **Positioning**: Ideal for use from mid-large enterprise campus access to core and some data center use cases that require performance, scale, and high availability.
- *Aruba CX 8300*:
  - **Overview**: Compact 1U (1 unit) switches with 6 Tbps of capacity with high-speed 1/10/25 GbE, 1/10/40 GbE, and 40/100 GbE connectivity with redundant hot swappable power and fans and options for front-to-back or back-to-front cooling for flexible hot-cold aisle deployments.
  - **Positioning**: Ideal for enterprise aggregation, core, ToR and, end-of-rack (EoR) deployments. The Aruba Fabric Composer (AFC) product speeds up infrastructure provisioning and reduces risk by utilizing pre-engineered and tested Aruba and HPE integrations.
- *Aruba CX 8400*:
  - **Overview**: A carrier-class, 8-slot chassis switch which provides up to 19.2 Tbps of capacity with high-density 10/40/100 GbE connectivity with fully redundant fabric, management, power, and fans.
  - **Positioning**: Ideal for large enterprise campus core and data center networks where HA is a must and the largest scale (table sizes and buffers) is required.
- *Aruba CX 9300*:
  - **Overview**: The Aruba CX 9300 switch series is a next-generation 25.6 Tbps, 1U fixed configuration switch supporting 32 ports of 100/200/400 GbE. The switch is an ideal solution for flexible, cost-effective, high-density 400 GbE connectivity server, storage, and intra-fabric connectivity.
  - **Positioning**: Designed for spine, core, aggregation, ToR, leaf, or EoR in a data center.

- *Aruba CX 10000*:
  - **Overview**: Provides advanced services to the data center such as distributed stateful firewall, encryption services, and stateful network address translation (NAT), zero-trust networking (ZTN).
  - **Positioning**: On-premises enterprise data centers and private clouds.
- *Aruba CX 4100i*: Is a layer 2 switch with ruggedized design. It is a special use-case for challenging environments. Two fanless models can survive extended temperatures, minus 40 to 70 degrees Celsius.

Aruba Central for AOS-CX management

Aruba Central features an even simpler, more powerful user interface (UI), to ease routine changes to interfaces, ports, VLANs, and more on the AOS-CX switches (see Figure 6-5). You can easily toggle to MultiEdit mode for advanced configuration settings, with assistances like command completion, syntax highlighting, and built-in validation of network policies. You have improved flexibility over how to implement configuration changes, where you can flip back and forth of using the UI or MultiEdit mode depending on your needs and preferences.

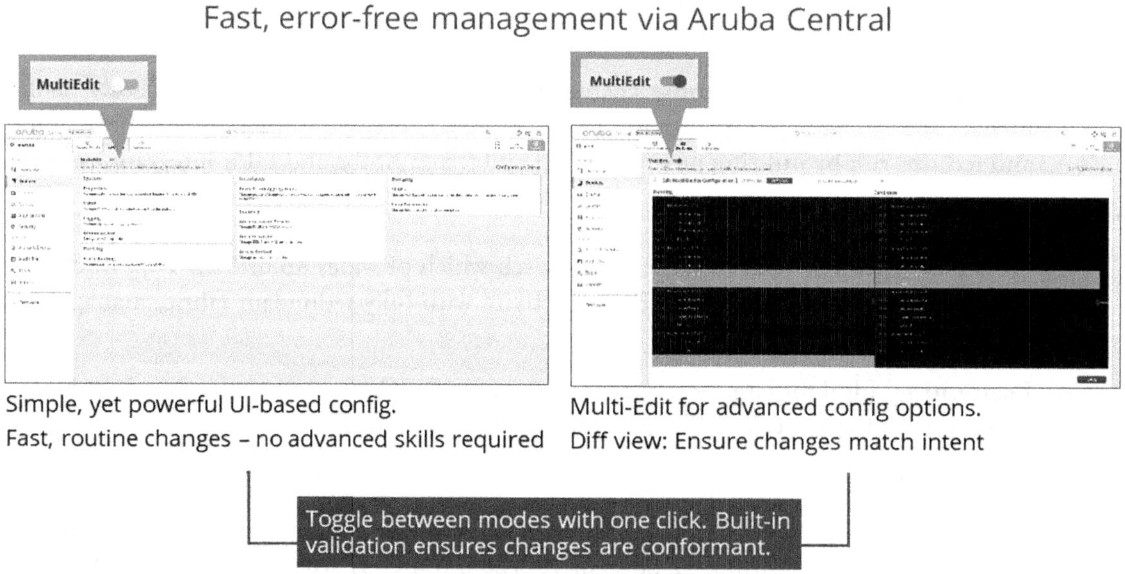

Figure 6-5: Aruba Central for AOS-CX management

Network automation with AOS-CX

The NAE engine is integrated with AOS-CX system configuration and time series databases (see Figure 6-6). This enables you to examine historical trends and predict future problems due to scale, security, and performance bottlenecks. With that information, you can create software modules that automatically detect issues, take appropriate actions, and reduce the time spent on manual tasks and troubleshooting.

NAE uses Python-based agents, developed by Aruba and the Aruba community, to automate network monitoring and troubleshooting operations. Agents monitor the state, settings, and statistics for system components in the current state database (CSDB), for example, the number of bytes transmitted on an interface or the switch's operating temperature. The agent then writes ongoing results to the time series database (TSDB) for ongoing analysis and tracking.

You can interact with the NAE through the switch's web UI's analytics dashboard. You can also automate the NAE through the REST API using tools such as Postman, Ansible, or another of your choice. You can create or obtain Python scripts for new agents and load the agents on the switch using the web UI or the REST API.

You can also deploy a pre-defined set of NAE scripts to CX switches using the Central configuration UI. Custom NAE scripts can be pushed via template configurations as well.

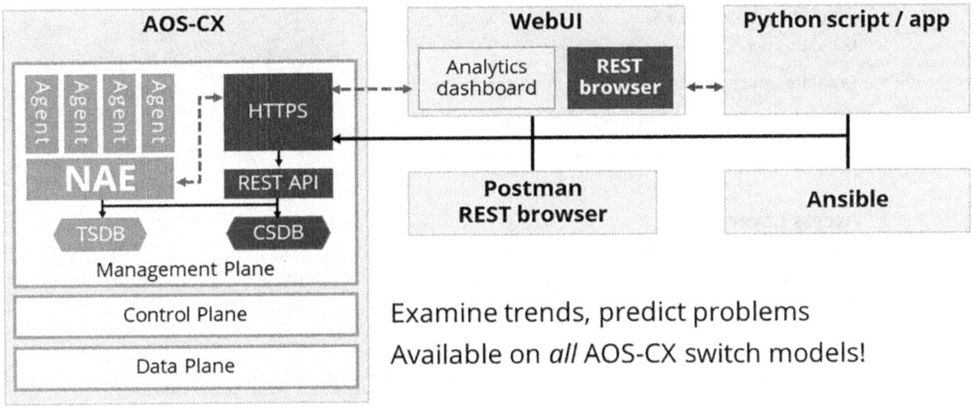

Figure 6-6: Network automation with AOS-CX

AOS-CX switch virtualization: VSX and VSF

VSX and VSF are similar to each other, and even to earlier virtualization techniques like the Virtual Router Redundancy Protocol (VRRP). With these virtualization techniques *multiple devices are perceived as a single device*. Switch virtualization allows you to eliminate the need for the Spanning Tree Protocol (STP) and supports active/active traffic forwarding so all uplinks are being utilized at the same time. This improves resiliency while reduces configuration and management overhead. But there are distinct differences between VSX and VSF, as summarized in Figure 6-7. You use VSX at the campus network aggregation and core layers. You use VSF primarily at the access layer. The differences in features and functions are intentional since priorities differ between the access and core layers.

Configuring VSF stacking is a plug-and-play affair: a switch, at default settings, automatically will join with another switch. The 6200 and 6300 CX series switches support VSF. This ease of deployment is especially important at the access layer: you may have dozens, or hundreds of access switches and this auto-join process simplifies the deployment of new switches.

VSX is supported on the 6400, 8300, 8400, and 9300 series switches. Unlike VSF, with VSX, you must manually configure it on both switches. VSX maintains separate control plane and management planes, and much of the data plane. This mitigates the *shared fate* nature of common planes, and thus provides for improved uptimes: you get high availability and redundancy. This is especially important when aggregating all those access layer and/or aggregation layer switches. VSX is easy to manage, since you can synchronize the management planes, and optionally, as well as other features, like the MAC and ARP addresses, table information, and configuration information. With few exceptions, it uses an opt-in process: you select what you want synchronized between the two switches, giving you complete control over the solution.

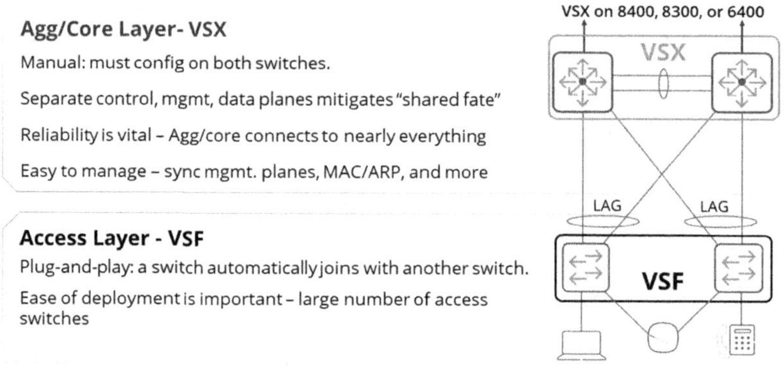

Figure 6-7: AOS-CX switch virtualization—VSX and VSF comparison

Dynamic segmentation

All modern desktops, laptops, tablets, and smart phones support robust security mechanisms like 802.1X/EAP (Extensible Authentication Protocol). But many devices lack these security mechanisms. This includes specialty devices, and some devices that rely on PoE, such as security cameras, payment card readers, Internet of Things (IoT), and medical devices. This lack of endpoint security can pose a serious risk to your infrastructure (see Figure 6-8).

A switch with dynamic segmentation improves your security stance while easing administration. It authenticates these devices using Aruba ClearPass, and tunnels the traffic to an Aruba gateway, which centralizes formerly disparate access control lists (ACLs) and firewall rules into a single, intuitive policy system. You can continue to use these user devices without compromising network security.

With dynamic segmentation, the switch authenticates the user by interacting with an authentication, authorization, and accounting (AAA) server, like ClearPass. ClearPass profiles each device, deems appropriate endpoints as *acceptable*, and sends accept or reject messages to the switch. The accept messages include tunnel information, the assigned VLAN, and secondary role. The switch uses this information to establish the tunnel to the gateway (mobility controller, or MC for short). The user's traffic is tunneled to the gateway where the gateway applies the appropriate policies to it, like firewall policies. In Figure 6-9, a user was profiled with the Finance role, and so their port is configured for VLAN 15, and their traffic is securely tunneled to the gateway. The gateway applies centralized policy enforcement to limit finance traffic as appropriate. Printers and cameras are similarly treated, as shown in Figure 6-9.

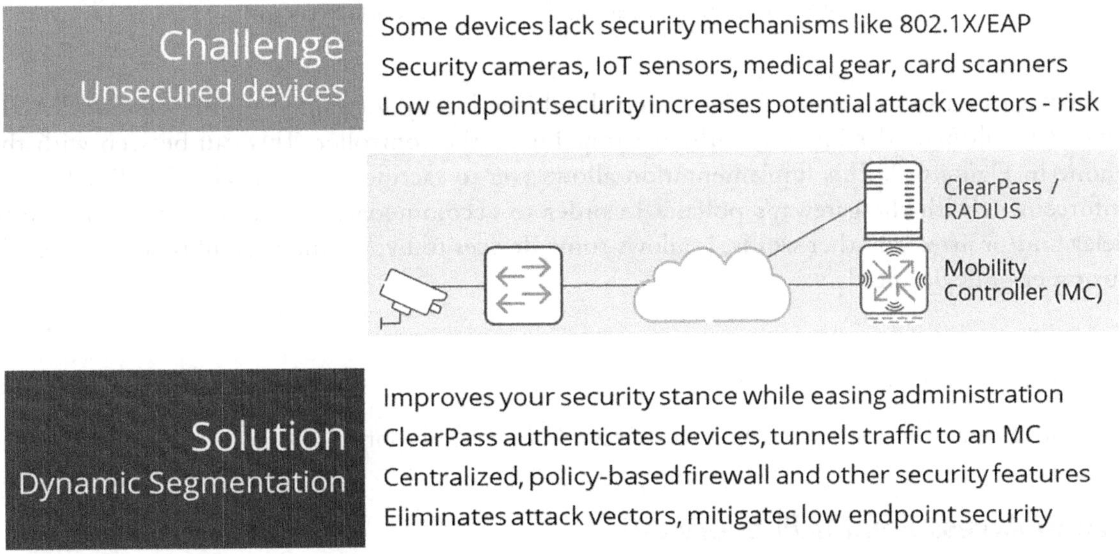

Figure 6-8: Dynamic segmentation

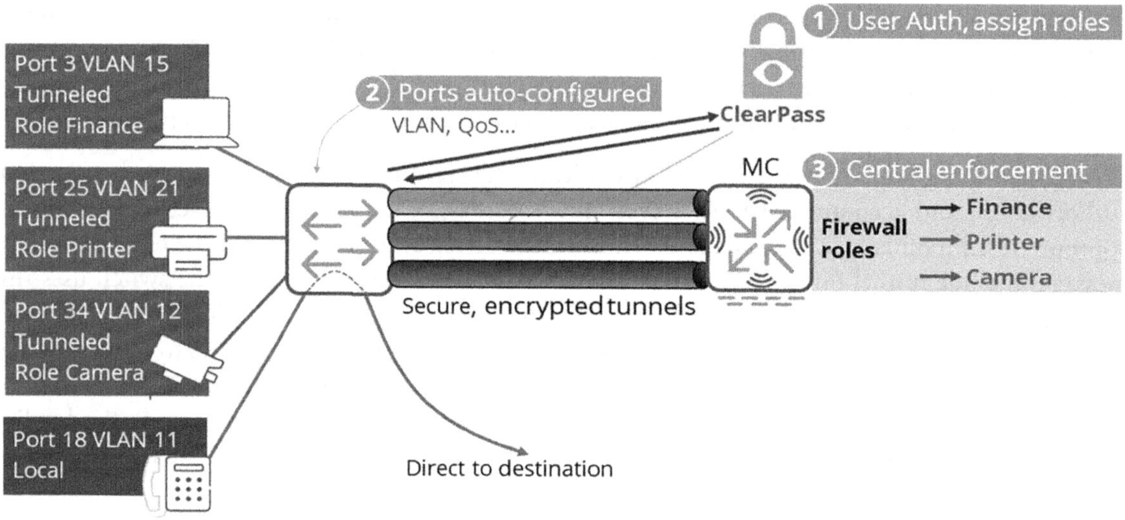

Figure 6-9: Dynamic segmentation operational overview

ClearPass can also indicate if device traffic should be locally forwarded by the switch, directly toward its ultimate destination, without tunneling to the controller. This can be seen with the phone in Figure 6-9. This implementation allows you to sacrifice the centralized policy firewall enforcement with the gateway's policies in order to accommodate devices that are sensitive to delay and/or jitter; in other words, it allows some devices to bypass the firewall based on specific customer or device needs.

Aruba WLAN portfolio

Let's move on and introduce the Aruba wireless LAN (WLAN) portfolio.

Aruba Access Point (AP) overview

Aruba WLAN products provide edge connectivity that's intelligent, safe, and fast, with AI-powered mobility and enterprise-class design. Aruba APs deliver superior services and segmentation without the complexity that other vendors add. And now, with Wi-Fi 6, Aruba delivers critical security and reliability to mobile users, IoT, and latency-sensitive applications, even in Wi-Fi crowded areas. Enterprises with 5G strategies also benefit from Aruba Wi-Fi 6, serving as an onramp to 5G (in place of small cell/DAS). The different Aruba AP models (see Figure 6-10) represent different levels of performance and capacity as well as different operational environment: indoor vs outdoor. The rest of this section will briefly introduce the Aruba AP models.

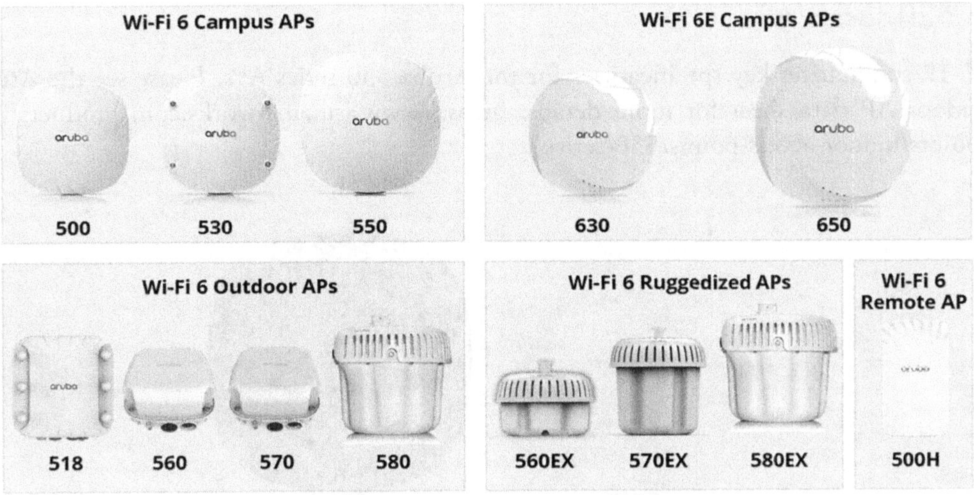

Figure 6-10: Aruba AP model overview

Aruba 500 series APs

Figure 6-11 summarizes the key specifications for the Aruba 500 series APs. Please see the Aruba 500 series indoor AP data sheet for more details: https://www.arubanetworks.com/products/wireless/access-points/indoor-access-points/500-series/.

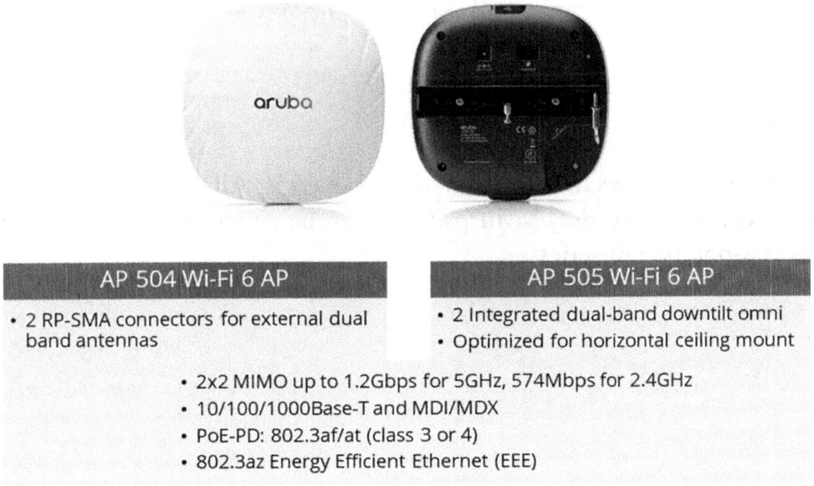

Figure 6-11: Aruba 500 series APs

Aruba 530 series APs

Figure 6-12 summarizes key specifications for the Aruba 530 series APs. Please see the Aruba 530 series indoor AP data sheet for more details: https://www.arubanetworks.com/products/wireless/access-points/indoor-access-points/530-series/.

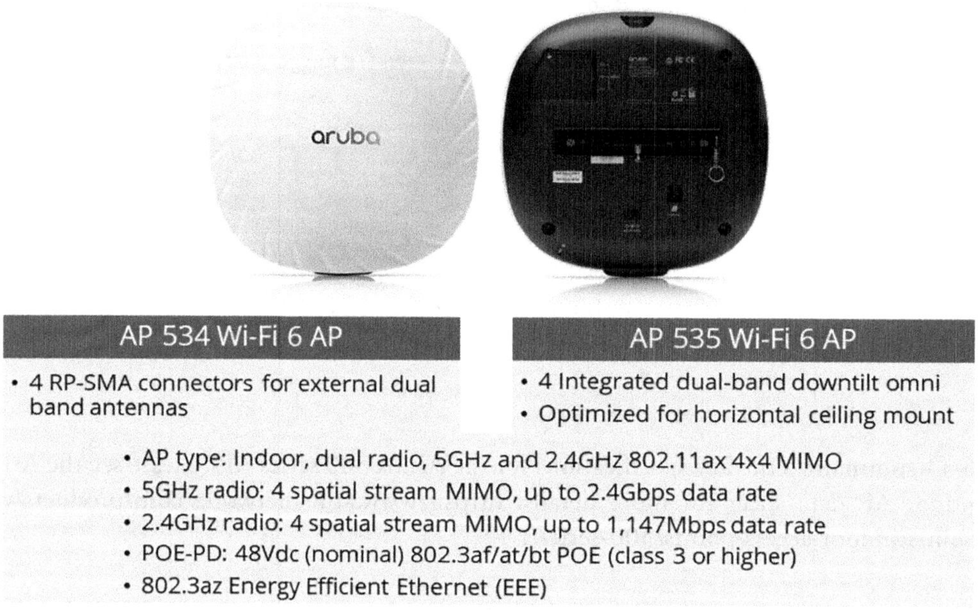

Figure 6-12: Aruba 530 series APs

Aruba 550 AP

Figure 6-13 summarizes key specifications for the Aruba 550 AP. Please see the Aruba 550 series indoor AP data sheet for more details: https://www.arubanetworks.com/products/wireless/access-points/indoor-access-points/550-series/.

Figure 6-13: Aruba 550 AP

Aruba 600 Series APs

Figure 6-14 summarizes key specifications for the Aruba 600 series APs.

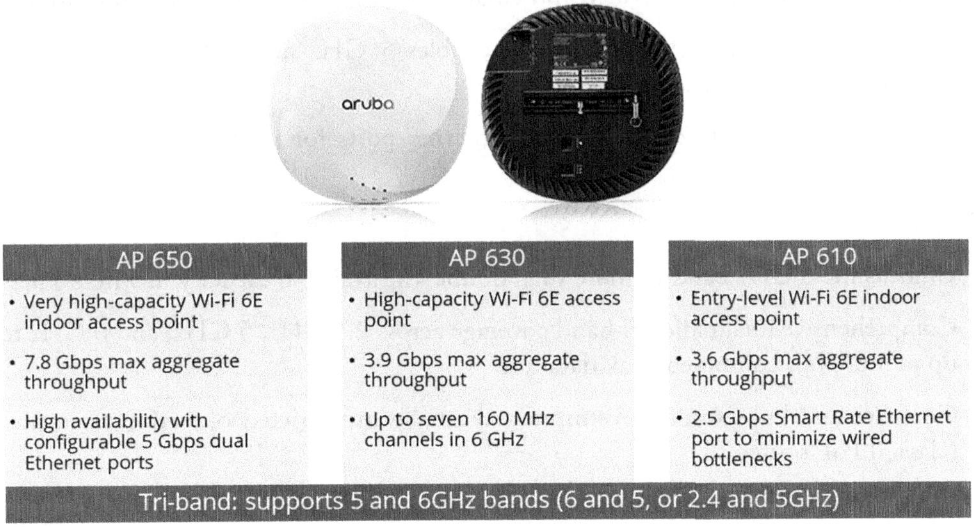

Figure 6-14: Aruba 600 series APs.

Here is a brief summary:

- *AP 650*:
    - Comprehensive tri-band coverage across 2.4 GHz, 5 GHz, and 6 GHz to deliver 7.8 Gbps maximum aggregate data rate
    - 4x4 MIMO radios to deliver peak performance and increased capacity using MU-MIMO and OFDMA (uplink and downlink for both)
    - Up to seven 160 MHz channels in 6 GHz support low-latency, bandwidth-hungry applications like high-definition video and augmented reality/virtual reality applications
    - Unique ultra tri-band filtering enables 5 GHz and 6 GHz to operate without restrictions or interference
    - High availability with configurable 5 Gbps dual Ethernet ports for hitless failover of Ethernet and power
- *AP 630*:
    - Unlocks the 6 GHz band to more than double the available capacity
    - Comprehensive tri-band coverage across 2.4 GHz, 5 GHz, and 6 GHz to deliver 3.9 Gbps maximum aggregate data rate
    - Up to seven 160 MHz channels in 6 GHz support for low-latency, bandwidth-hungry applications like high-definition video and augmented reality/virtual reality applications
    - Unique ultra tri-band (UTB) filtering enables 5 GHz and 6 GHz to operate without restrictions or interference
    - High availability with 2.5 Gbps dual Ethernet ports for hitless failover of Ethernet and power
- *AP 610*:
    - Unlocks the 6 GHz band to more than double the available capacity in wireless networks
    - Comprehensive dual radio/tri-band coverage across 2.4 GHz, 5 GHz, and 6 GHz to deliver up to 3.6 Gbps combined peak data rate
    - Compact and low power consumption: virtually unrestricted operation from an 802.3af (class 3) PoE source
    - Up to seven 160 MHz channels in 6 GHz to support low-latency, bandwidth-hungry applications like high-definition video, and augmented reality/virtual reality applications
    - 2.5 Gbps smart rate Ethernet port to minimize wired bottlenecks
    - Built-in GPS receivers and fine time measurement (802.11mc) to enable APs to automatically locate themselves and serve as reference points for accurate indoor location measurements

Aruba 300 series APs

Figure 6-15 summarizes key specifications for the Aruba 300 series APs.

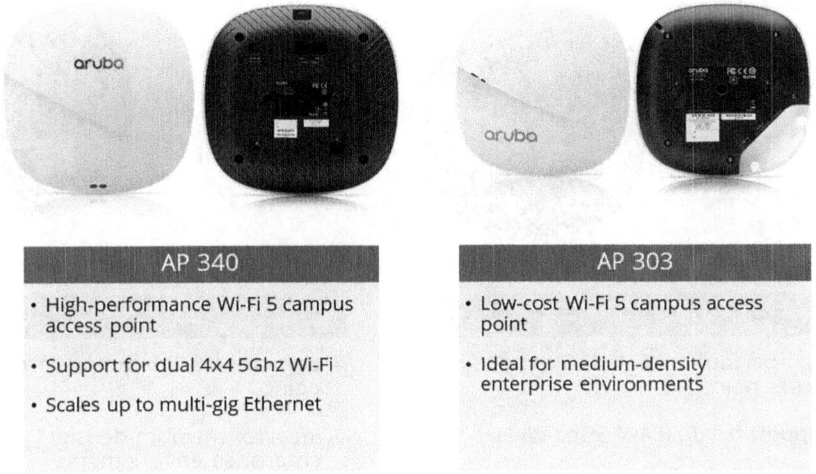

Figure 6-15: Aruba 300 series APs

Here is a brief summary:

- *AP 340*:
  - Designed with an integrated, 802.3bz compliant, smart rate multi-gigabit Ethernet port to eliminate wired bottlenecks
  - Maximum data rates of 2,166 Mbps in the 5 GHz band and 800 Mbps in the 2.4 GHz band (for an aggregate peak data rate of 3 Gbps in dual-band and 4.3 Gbps in dual-5GHz)
  - Includes an integrated Bluetooth low energy (BLE) radio for advanced location and indoor wayfinding
  - Participates in Aruba's dynamic segmentation solution
- *AP 303*:
  - Up to 1.2 Gbps aggregate peak data rate
  - Daisy-chain another AP using the secondary PoE-out port
  - Integrated Zigbee and Bluetooth 5 radio for IoT connectivity
  - Resolve sticky client issues with MU-MIMO-aware Aruba ClientMatch
  - Participates in Aruba's dynamic segmentation solution

Aruba remote/hospitality APs

Figure 6-16 summarizes key specifications for the Aruba remote/hospitality series APs.

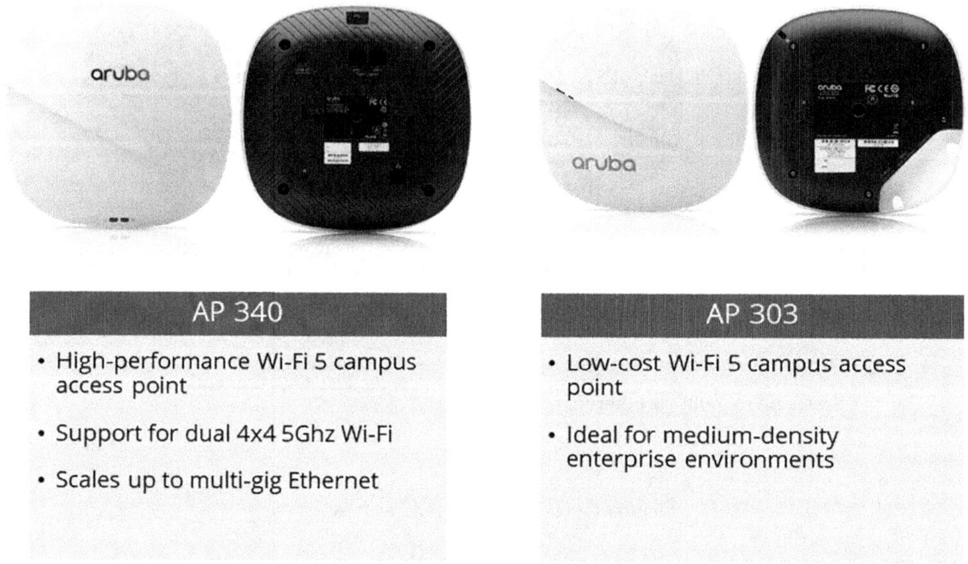

Figure 6-16: Aruba remote/hospitality series APs

Here is a brief summary:

- *AP 303H*:
    - Cost-effective Wi-Fi 5 (Wave 2) AP. A flexible, secure solution ideal for hospitality, branch offices, medical clinics, and remote workstations.
    - Unified dual-radio 802.11ac Wave 2 2x2:2 hospitality and branch AP with internal antennas, three local gigabit Ethernet ports, PoE out and USB host interface
    - Supports wall-box and desk mount deployments
- *AP 505H*:
    - High-end dual radio Wi-Fi 6 hospitality AP with 1+4 Ethernet ports, PSE, USB
    - High-performance, cost-effective, medium-density, 2X2 radios
    - 505H High-end dual radio, four Ethernet ports, 2X PoE-outdoor, 500 Mbps
    - 503H Mid-range radio, two Ethernet ports, PoE indoor, 100 Mbps cryptographic VPN performance

Aruba gateways

Now that you have some familiarity with Aruba's AP product portfolio, let's now look at Aruba's gateway portfolio. A quick overview of the portfolio is shown in Figure 6-17.

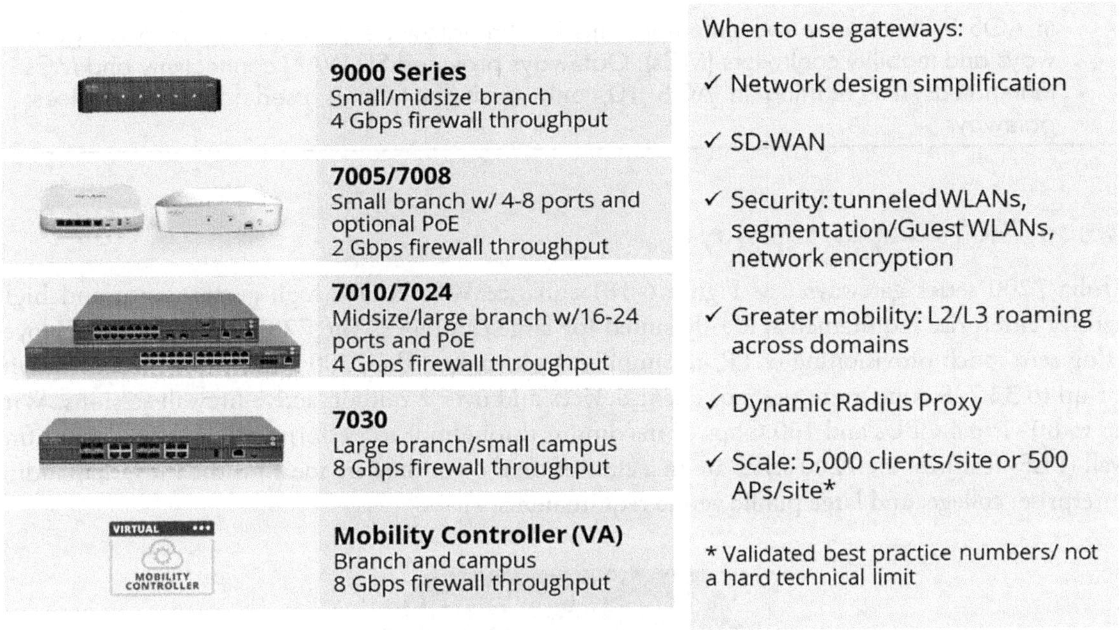

Figure 6-17: Aruba gateway product portfolio

The main reason you need gateways in a large-scale deployment is to simplify network design and provide scalability. The gateway is responsible for traffic termination, so all wireless traffic is being tunneled from the APs to the gateway, which then passes through the firewall function on the gateway and from there it is forwarded on its way to the traffic's final destination. This eliminates excessive span of VLANs. Aruba offers gateways to support the needs of larger enterprises who require additional scalability, 24x7 operations, and greater security. Gateways are optional: in other words, you can run the APs in standalone mode; but if you need scalability and simplicity, gateways are the solution of choice.

You should use a gateway deployment when any of the following is required:

- Simplified network design
- SD-WAN connectivity
- Security: tunneled WLANs, segmentation/guest WLANs, network encryption
- More scalable and robust mobility: layer 2 and/or layer 3 roaming across domains

- Dynamic RADIUS proxy
- Scalability: More than 5,000 clients per site or more than 500 APs per site

Note

In AOS 8, there were two different terms for the hardware shown in Figure 6-17: gateways and mobility controllers (MCs). Gateways provided SD-WAN connectivity and MCs terminated APs. Starting in AOS 10, only a single term is used for these devices: *gateways*.

Aruba 7200 campus gateway

Aruba 7200 series gateways (see Figure 6-18) enhance WLANs for high-performance and high-density enterprise requirements, ideally suited for large campuses. The 7200 series can be deployed using zero touch provisioning (ZTP) to simplify deployment. The 7200 series provides connectivity for up to 32,728 concurrent users or client devices and over 2 million active firewall sessions. With up to 80 virtual CPUs and 100 Gbps of maximum throughput to perform policy enforcement firewall (PEF) features, the 7200 series are in a class of their own: they are ideal for the most demanding enterprise, college, and large public venue requirements.

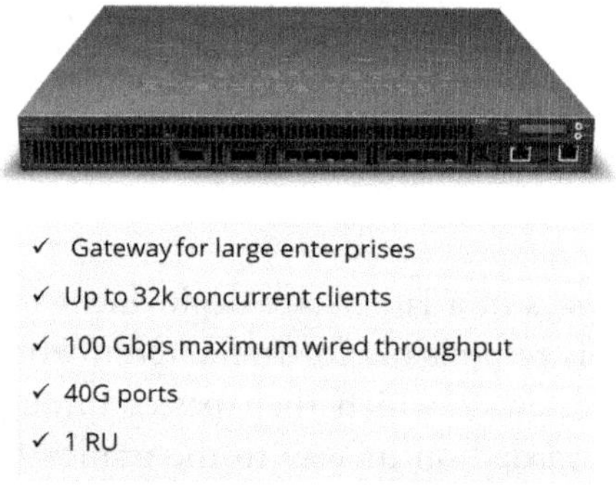

Figure 6-18: Aruba 7200 campus gateway

Aruba 9200 series campus gateway

The Aruba 9200 series campus gateway (see Figure 6-19) delivers enhanced Wi-Fi scalability and security to meet demands at the network edge. Engineered for mission-critical environments, the 9200 series delivers high-performance, seamless roaming, 24x7 reliability, and always-on connectivity: even during upgrades. Unlike other edge appliances, the 9200 series provides consistent policy enforcement across wired and wireless and dynamic segmentation for any user or device, even IoT.

The base unit provides 20 Gbps of throughput, 64K device, 500K IP addresses, 32K GRE tunnels, and 32K IPsec tunnels. Applying a hardware capacity license will increase the performance and scalability of the 9200. With the Silver license, the maximum throughput is 30 Gbps and 128K devices, 1M IP addresses, 64K GRE tunnels, and 64K IPsec tunnels. And finally, with the Gold license, you can achieve 40 Gbps throughput and 256K devices, 2M IP addresses, 128K GRE tunnels, and 128K IPsec tunnels.

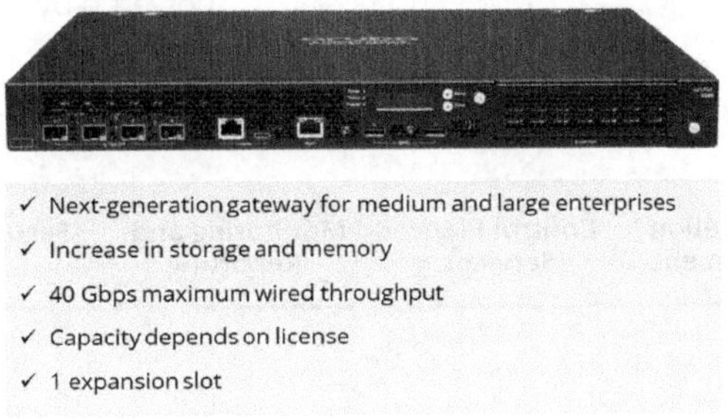

Figure 6-19: Aruba 9200 campus gateway

Running Aruba APs and gateways with AOS 10

Aruba offers APs and gateways running AOS 10.x. You can now deploy and manage your WLANs from a single and unified cloud-based network management system: Aruba Central. This combination of AOS 10.x-based devices and Aruba Central is powerful! It simplifies network administration with automated workflows, end-to-end visibility, AI-powered insights, and analytics to enhance and optimize the wireless experience for users, as shown in Figure 6-20.

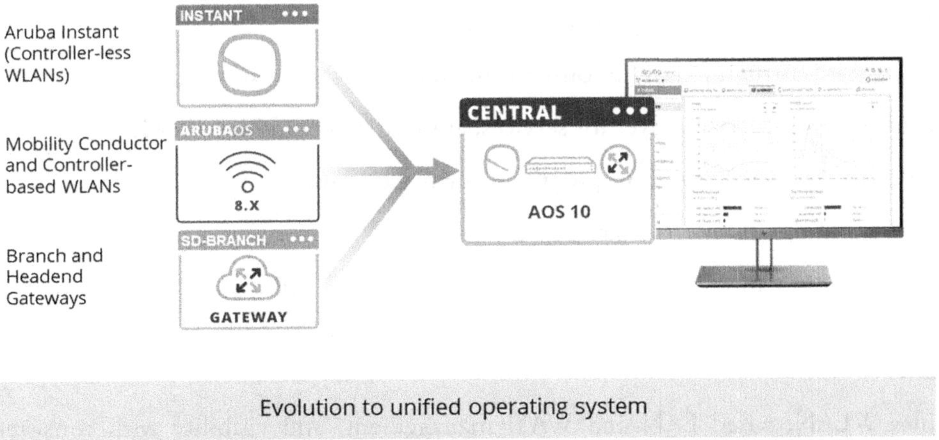

Figure 6-20: Running Aruba APs and gateways with AOS 10

Let's explore AOS 10 a bit more in the next sections.

ArubaOS 10: Overview

AOS 10.x provides a single architecture for customer WLAN access deployments (see Figure 6-21). It integrates access networks, cloud service offerings like Aruba ESP, which provides a unified architecture for branch and campus, wired and wireless, with a no sub-WLAN architectures such as AOS 6/8 and Instant 4/6/8: *only* AOS 10 is needed.

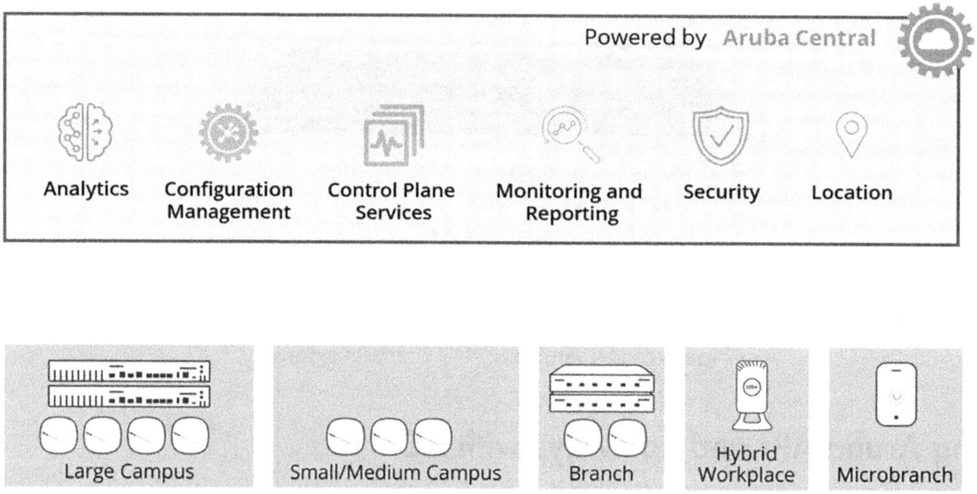

Figure 6-21: AOS 10 overview

ArubaOS 10: Architecture

AOS 10.x is comprised of three *layers* (see Figure 6-22).

- A required infrastructure layer of on-premise WLAN APs and switches
- A required central services layer for services, applications, and NMS/CMRT
- An optional gateway layer of on-premise security gateways

The advantages of this architecture are that it:

- Scales beyond the mobility conductor 8.x domain of 10K APs or instant clusters of 2K devices
- Enables cloud services including AI-powered insights, workflow automation, and robust security
- Unifies WLAN, wired LAN and WAN management, and visibility with consistent UI and workflows

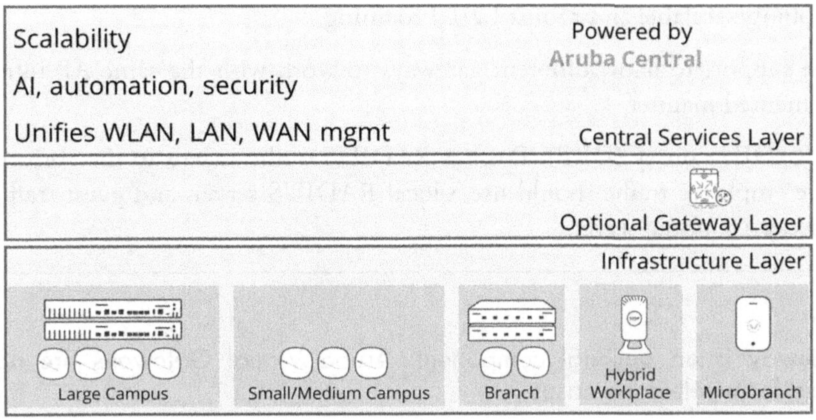

Figure 6-22: AOS 10 architecture

Let's zoom in on the optional gateway layer (see Figure 6-23).

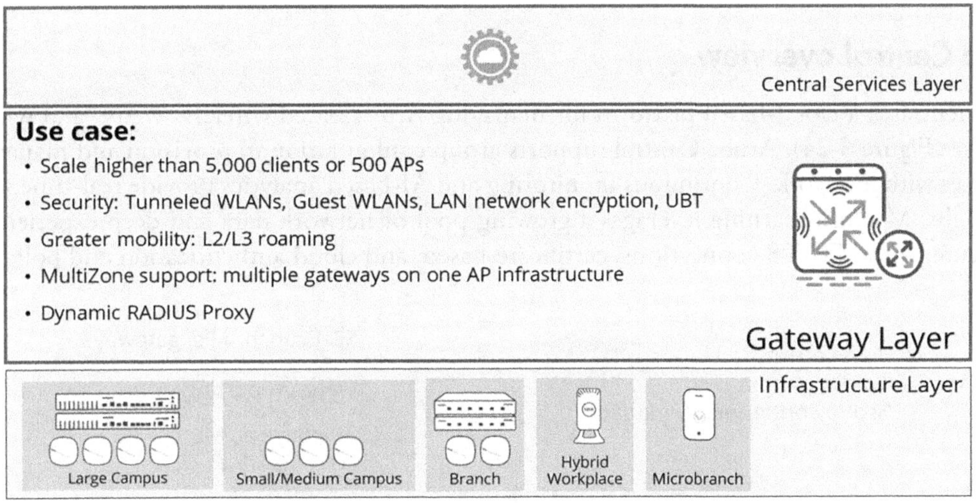

Figure 6-23: AOS 10 architecture—gateway layer

AOS 10 provides additional scalability and security for enterprise customers. Customers should consider a gateway if they have one or more of the following requirements:

- Scale higher than 5,000 clients or 500 APs (Note that this is due to management complexity and not a hard technical limit)
- Security: Tunneled WLANs, segmentation with guest WLANs, network encryption over their LAN, or user-based tunneling (UBT)

- Greater mobility: scalable and robust L2/L3 roaming
- Multi-zone support to allow different gateways to work with the same AP infrastructure in a secure, segmented manner
- Dynamic RADIUS proxy (DRP): Ensures RADIUS traffic is sent to the correct server. Useful if corporate employee traffic should use a local RADIUS server, and guest traffic should use a centralized captive portal server.

Note

The gateway is an optional component. Aruba Virtual Gateways are not currently supported for wireless use cases.

Introduction to Aruba Central

Let's take a look at Aruba Central in this section.

Aruba Central overview

Aruba Central is a cloud-based platform for managing Aruba-based wireless, WAN, and wired networks (see Figure 6-24). Aruba Central supports group configuration to provision and manage similar devices with less work. Continuous monitoring and AI-based analytics provide real-time visibility and insight. Machine learning leverages a growing pool of network data and deep experience. It is very secure, with HTTPS connections, certificate-based, and cloud authentication and policies.

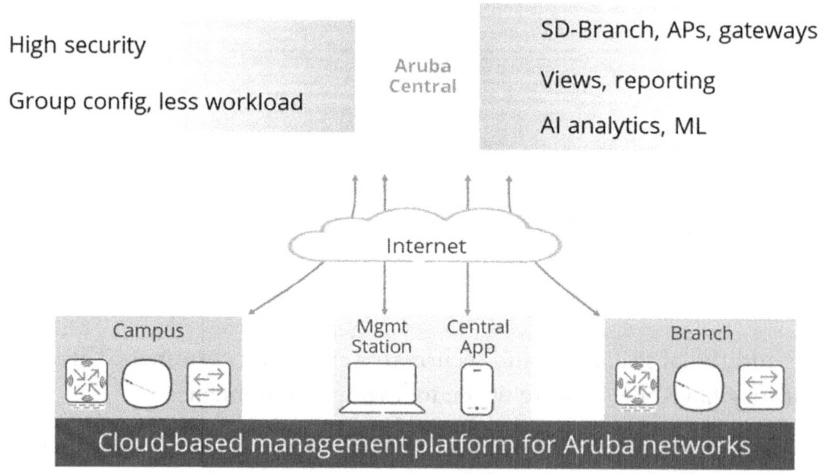

Figure 6-24: Aruba Central overview

You get comprehensive network views, device status and health information, and application usage. You can monitor, identify, and address issues with a data-driven dashboard, alerts, reports, and troubleshooting workflows.

Note

Central also runs part of the control plane for some network devices and solutions, using the Aruba SD-branch solution and Aruba APs and gateways running AOS 10.

Cloud-managed

Central is simple, with no hardware or software requirements on the customer's end (see Figure 6-25): It's in the cloud, with all the resources needed for performance and visibility. Central is reliable, with guaranteed uptime, and local management in case your connectivity to Aruba Central experiences issues or failures.

Plus, you get intelligent features to improve performance and visibility, with analytics to improve the business and user experience. And flexibility? It's easy to scale up as needed, to accommodate growing systems and ongoing innovations, with simple subscription options and anytime, anywhere access.

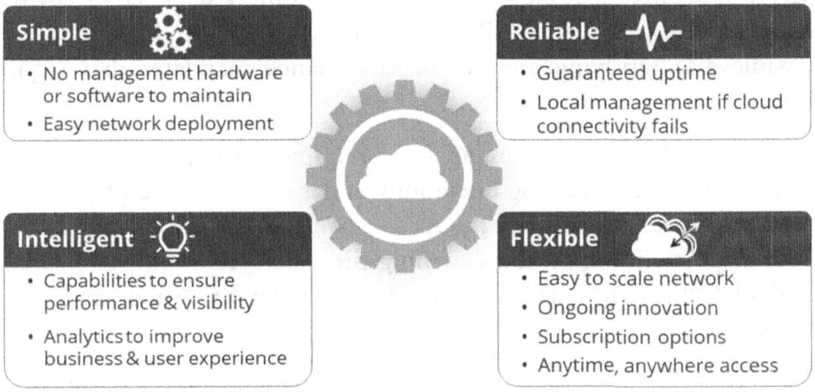

Figure 6-25: Central—cloud-managed

Flexible and cost-effective cloud networking

Central supports management, monitoring, and configuration for the Aruba IAPs, Aruba AOS and AOS-CX switches, SD-Branch gateways, and AOS 10 APs and gateways (see Figure 6-26). And since it's a cloud-based management system, it only requires a simple 1, 3, 5, 7, or 10-year subscription license per device, which includes tech support, with additional optional services available.

Figure 6-26: Flexible and cost-effective cloud networking

Aruba Central on Premise (COP)

Aruba wants to provide organizations with the choice and flexibility to deploy and manage their networks in ways that best align with their technical and security requirements. And that's why Aruba offers Central On Premises (COP) to complement their Aruba Central cloud offering.

Aruba COP delivers most of the same services and capabilities of Aruba Central (see Figure 6-27). It has the same user interface, same code, and same user experience and workflows: it's just powered by purpose-built server appliances that are deployed in a customer's data center environment. In this way, networking teams still gain a cloud-like management experience for better operational agility and efficiency, while also satisfying security or compliance mandates that require on-premises infrastructure.

Like Aruba AirWave, Aruba COP is designed to simplify the deployment and maintenance of wireless and wired networks by unifying operations under a single pane of glass, with support for Aruba APs: both gateway-based and instant APs, as well as AOS 8 mobility controllers and/or conductors and both portfolios of Aruba switching products (AOS and AOS-CX). But an added advantage of Aruba COP is that it delivers much higher scalability: supporting up to 25K devices in a seven-node server cluster.

COP has limits in terms of number of managed devices, depending on the number of servers hosting COP. With a three-node deployment, you can manage up to 8,000 devices, while a five-node deployment supports up to 16,000 devices, and, finally, a seven-node deployment supports up to 25,000 devices. In addition to this, there are limits for AI Insights that can only be used in a five-node and seven-node deployment. A five-node deployment supports up to 160,000 clients and a seven-node deployment supports up to 300,000 clients.

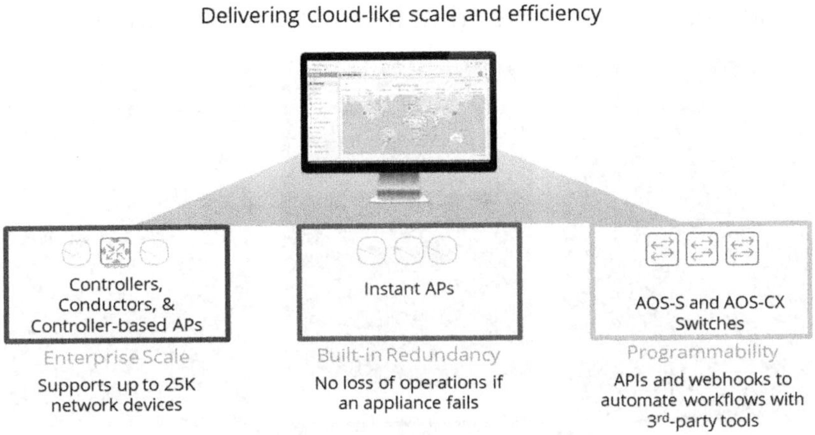

Figure 6-27: Aruba Central on Premise

Introduction to Aruba ClearPass

Aruba ClearPass helps to ease security operations and improve your security stance. This section introduces you to the ClearPass product.

Aruba ClearPass overview

Aruba ClearPass is an authentication server that can easily handle all network access control (NAC) functions. It is a decision-making point when clients connect to the network and go through an authentication and authorization process. One of the great things about ClearPass is that Aruba has included many features within the base software set (see Figure 6-28). Aruba has had TACACS and RADIUS servers for over 10 years so they're tried and tested. The policy engine, basic guest, and device registration tools work without the need for dedicated appliances or extra licenses.

The visibility dashboard, policy creation templates profiling, and troubleshooting services have been there since Aruba's been shipping the product. The exchange interoperability components (API and syslog messaging) are included. Customers are free to use built-in integration templates or create their own workflows. And, if customers want, they can try any of the three expandable applications because Aruba has included 25 enterprise licenses with the base platform. ClearPass is available as an appliance or a virtual machine (VM) installation.

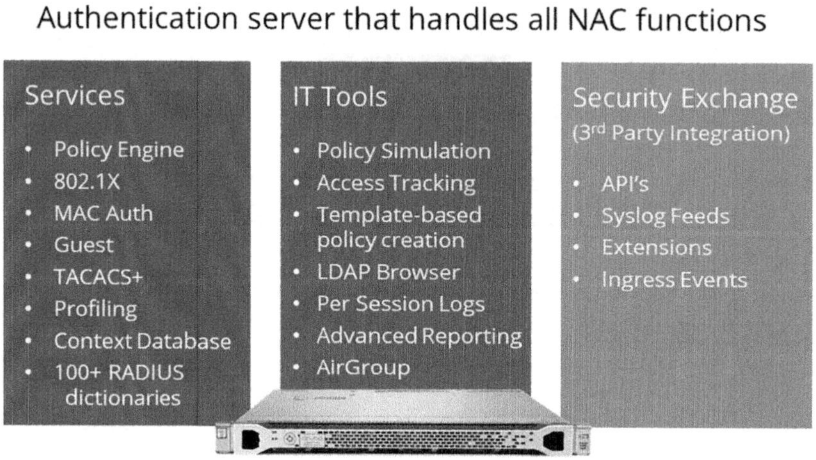

Figure 6-28: Aruba ClearPass overview

ClearPass applications

ClearPass Onboard is the easiest way to support a bring-your-own-device (BYOD) environment (Figure 6-29). Based on its own internal certificate authority (CA), Onboard provides an intuitive and easy to use portal for your users to securely provision their devices to connect to your company's network. To enforce the concept that every device should be granted access on its own unique merits, ClearPass Onboard assigns a unique identity to each device that goes through the Onboard process. For auditing capabilities, this identity includes information about the client device as well as the user that put the device on the network.

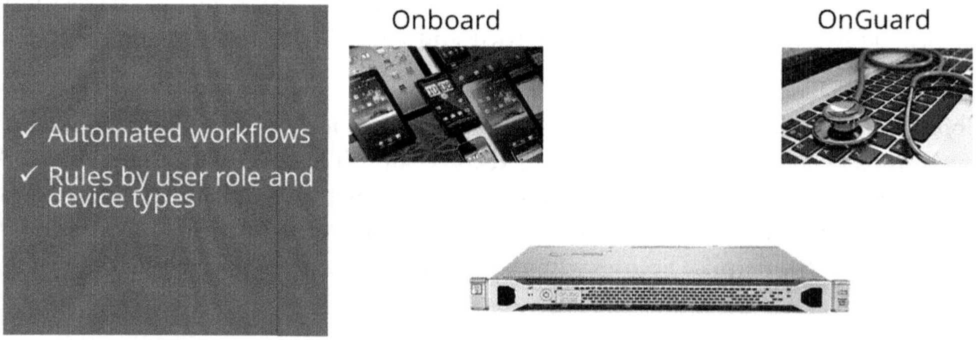

Figure 6-29: Aruba ClearPass applications

Posture checks and client system health validation are necessary for most networks. Using an agent, either deployed through a web browser to the client or as a permanent software applet, ClearPass

OnGuard can provide security status validation on your wired, wireless, and VPN networks. OnGuard makes it easy to enforce the organization's device compliance policies.

Learning check

Discuss these questions.

1. Which of the options below are the main aspects of Aruba's Edge Services Platform (ESP) architecture? (Select all that apply.)
   a. Artificial Intelligence Operations (AIOps)
   b. Zero Trust Security (ZTP)
   c. Aruba Central
   d. Unified infrastructure
2. Which AOS-CX switch models offer network automation features like the Network Analytics Engine (NAE) to examine trends and predict problems?
   a. Only the chassis-based AOS-CX switches
   b. All CX 6300 series switches and above
   c. All AOS-CX switches
   d. All AOS-CX switches except the ruggedized CX4100i
3. Which Aruba feature solves the challenge of tedious switch port configurations and securing devices that lack support for 802.1X/EAP authentication?
   a. IPsec tunneling
   b. Dynamic segmentation
   c. Analytics dashboard
   d. Third-party integration
4. Which Aruba Access Points (APs) support three RF bands: 2.4 GHz, 5 GHz, and 6 GHz?
   a. All Aruba APs
   b. All 500 and 600 series APs
   c. All APs but the older 300 series and hospitality APs
   d. Only the 600 series APs
   e. No Aruba APs support all three bands.

5. Which statements below accurately describe Aruba gateways? (Select two.)
    a. They can simplify network design by acting as a tunnel termination point for Generic Routing Encapsulation (GRE) tunnels.
    b. They are key components of both SD-WAN solutions and many Wi-Fi solutions.
    c. They are required components of all AOS-CX validated design deployments.
    d. Running Aruba APs and gateways with AOS 10 enables you to manage WLANs from Aruba Central.
6. Aruba Central enables you to manage APs, switches, and gateways at all locations from a single portal, supported only as a cloud-based solution.
    a. True
    b. False
7. Aruba ClearPass is an authentication server that handles all network access control (NAC) functions, with extensive third-party integration features.
    a. True
    b. False

Summary

Congratulations! You have completed Chapter 6. You should now be able to:

- Explain Aruba Edge Services Platform (ESP).
- Describe various Aruba wireless LAN (WLAN) products.
- Understand the Aruba Central management platform.
- Describe the Aruba ClearPass product.

7 Central for Device Management

OBJECTIVES

Aruba Central provides centralized and cloud-based management platform for Aruba's ESP. After completing this chapter, you will be able to perform Aruba Central device onboarding and create and use Central groups to ease administration. You will be able to use user interface (UI) and template configuration mode and describe Central licensing.

After completing this chapter, you should be able to:

✓ Perform device onboarding.

✓ Create Central groups.

✓ Use the UI configuration mode.

✓ Use template configuration mode.

✓ Describe Central licensing.

Device onboarding

Let's begin the chapter by discussing device onboarding.

Device onboard workflow

The onboard device process starts at the device inventory (see Figure 7-1). First, you must add any device that you want Aruba Central to manage to the device inventory list. Once you've added the devices, you must assign each device a subscription (license). This enables Aruba Central to manage the device. And third, you configure your devices.

Figure 7-1: Device onboarding workflow

Adding devices to the Aruba Central inventory

There are four ways to have devices communicate with Central (see Figure 7-2). Zero-touch is the best way as it simplifies installations. All devices receive a cloud activation key that you can use to direct devices to Central. Alternatively, you can also manually add a device to Central.

You can download the Aruba app to your smartphone to use the zero-touch option. This app allows you to simply take a picture of the device's serial number (SN) and MAC address. Then Central automatically updates Activate with the SN and MAC. If you have access to an Activate account, you can then add in Activate rules to assist in the zero-touch provisioning (ZTP) process. If not, you can call Aruba support and they can assist you in setting up the appropriate Activate rule or rules for your devices.

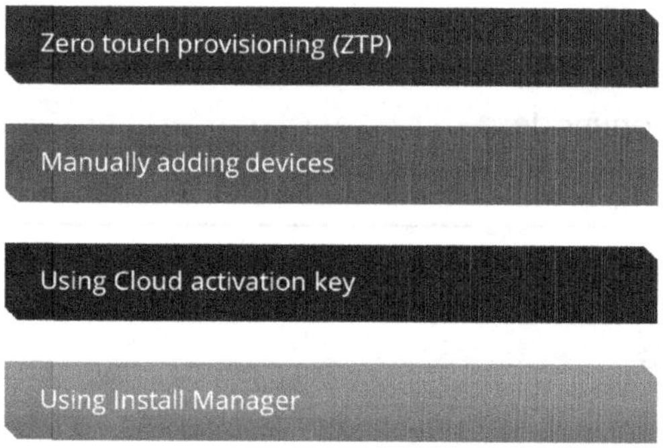

Figure 7-2: Adding devices to the Central inventory

Aruba Activate

Before starting the ZTP process, you need to understand what Aruba Activate is: it is a key component for provisioning Aruba devices. Aruba Activate is a cloud-based service designed to enable customers to efficiently manage their Aruba device inventory and facilitate the provisioning of their Aruba devices. Use of Aruba Activate is free to customers.

Aruba Activate provides two main services (see Figure 7-3):

- Zero-touch provisioning system for all Aruba networking devices.
- Firmware repository for Aruba networking devices.

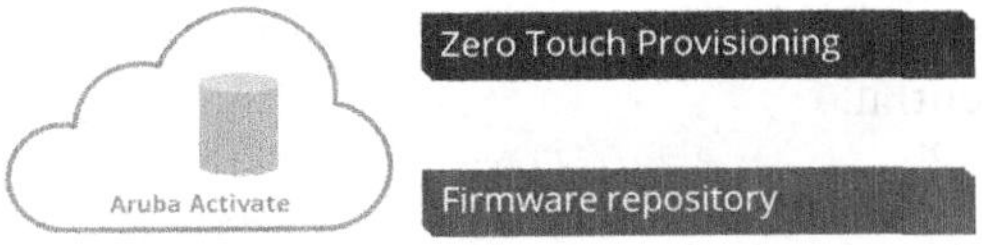

Figure 7-3: Aruba Activate

New customer process

Figure 7-4 depicts an example of how a new customer receives the Aruba Central key and joins the Central network using ZTP and Aruba Activate.

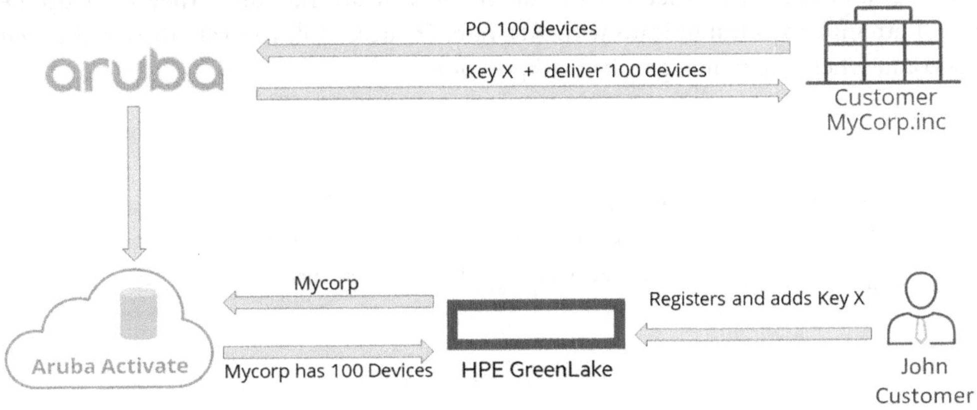

Figure 7-4: New customer process

Here are the steps for the process:

1. First, the new customer, in this case MyCorp Inc., sends Aruba a purchase order (PO) for 100 devices (Aruba APs, for example). Aruba delivers the 100 devices (APs) as well as the Aruba Central key to MyCorp Inc.
2. Next, Aruba updates Activate with the PO information including the customer's name, the subscription key ("X" in Figure 7-4), and the 100 APs they purchased.
3. Aruba uses Activate to dynamically provision APs and gateways to point them to Central. This is done by Aruba creating a rule in Activate that directs these devices to contact Central when they boot up at their factory default settings.
4. Now, John Customer, a MyCorp. Inc employee, registers for Aruba Central using his company email. Then John adds the Central subscription key that Aruba sent to him with the purchase confirmation.
5. The Central account communicates with Activate and passes the customer's name to Activate.
6. Activate then informs the Central account about all 100 APs that belong to this account. The APs will show up in MyCorp Inc's Central inventory under *My Devices*.

As you can see from this process, Aruba has automated most of the process of getting the newly purchased devices into Aruba Central.

ZTP: Device functionality

You might choose to assign an AP to an AOS 10 group with a firmware policy (see Figure 7-5). Therefore, after communication with Aruba Central, the AP automatically downloads the AOS 10 firmware and reboots. Once the upgrade completes, the AP again communicates with Aruba Central, which pushes the appropriate configuration to the AP.

Aruba switches and gateways in factory default mode will do the same. They boot up, get an IP address, and attempt to communicate with Activate. If successful, Activate directs the switch to a Central account where the switch gets its configuration.

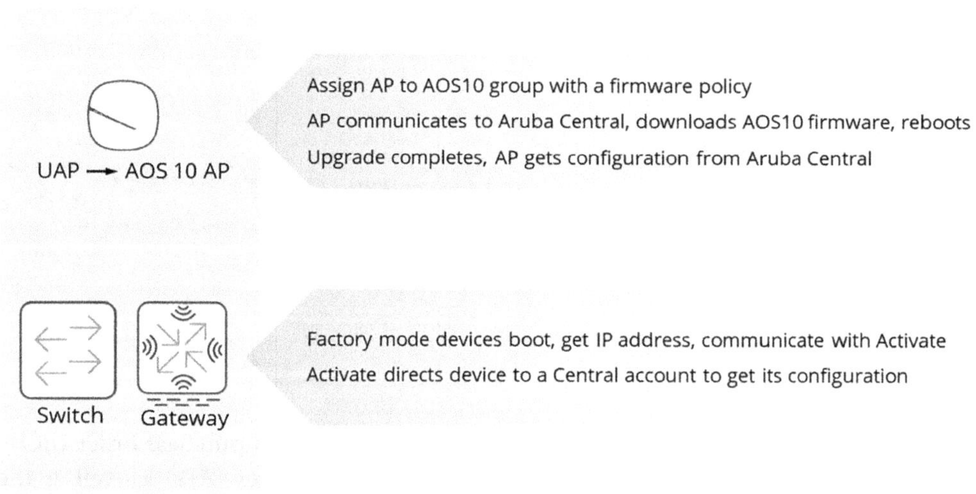

Figure 7-5: ZTP: Device functionality

HPE GreenLake management task overview

HPE GreenLake is a cloud management and monitoring platform from HPE. It hosts and provides not only Aruba Central services but many other HPE services as well. HPE GreenLake's device management feature lets you easily onboard, view, and manage all the devices in your account.

One of the first things you do is onboard your various network devices (see Figure 7-6). You can add them manually by using a .csv file or by manually entering each devise's serial number and MAC address. You can also onboard devices using an Aruba cloud activation key. This method is not supported by AOS-CX switches or AOS 10 APs.

After onboarding devices to HPE GreenLake, you must assign them to the correct HPE GreenLake application like Aruba Central. You will learn to perform all these activities later in the lab. For now, let's look at some items related to using that cloud activation key.

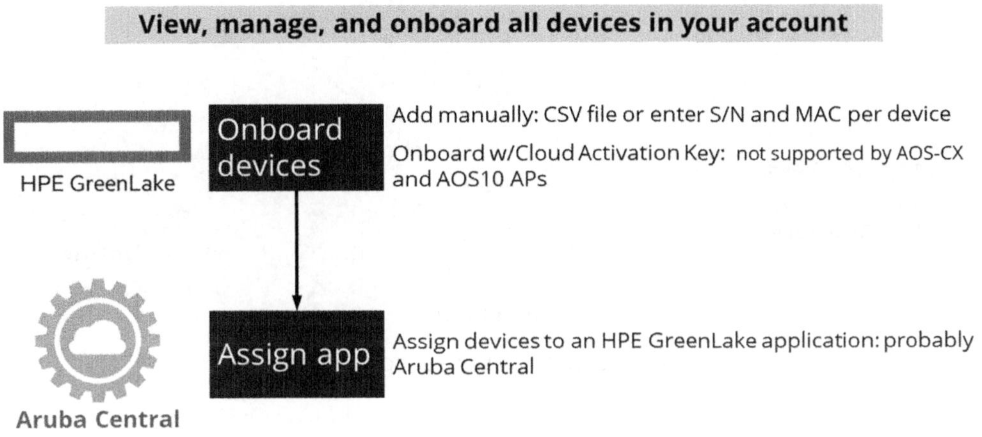

Figure 7-6: HPE GreenLake management task overview

Aruba cloud-based management flow

Getting Aruba devices to appear in Central involves the following steps (see Figure 7-7):

1. First, the customer registers the devices in HPE GreenLake as described earlier.
2. HPE GreenLake communicates information about the devices to Aruba Activate as those devices are assigned to the Aruba Central platform.
3. Now that Aruba Activates is aware of the devices' serial number and MAC address, it will accept ZTP requests coming from these devices.
4. Aruba Activate upgrades the devices firmware, if necessary, and directs the devices to the appropriate Central cluster.
5. The devices establish an HTTPS session to the Aruba Central cluster and now appear in the Central web interface.
6. The customer can now start configuring and monitoring the devices via Central.

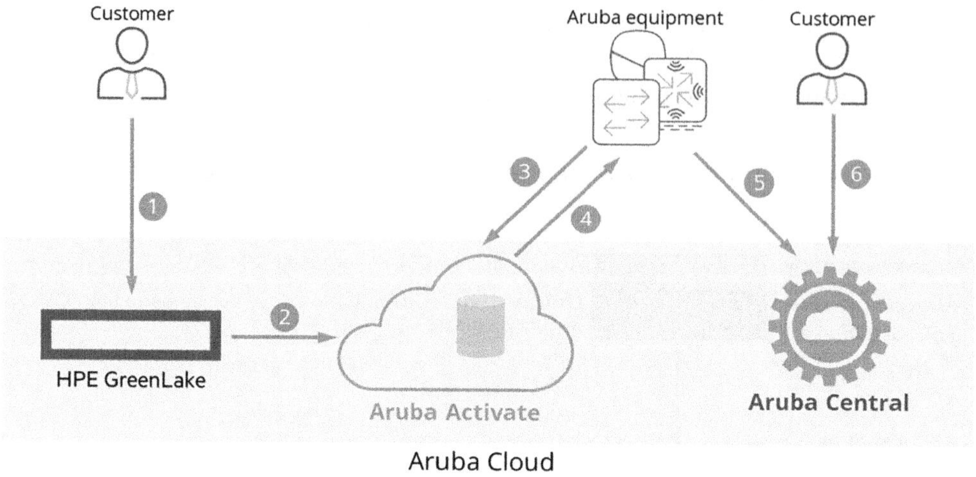

Figure 7-7: Aruba cloud-based management flow

Aruba Central licensing

This section explores Aruba Central's licensing.

Aruba Central 2.5.5 licensing

Aruba Central supports foundation and advanced license keys. These keys are for APs, switches, SD-Branch gateways, and gateways running AOS 10. Foundation keys enable you to onboard, monitor, and manage these Aruba devices. Advanced keys add functionality and features.

Every device Aruba Central manages requires one license (subscription). Licenses are based on 1, 3, 5, 7, and 10-year subscriptions. The licensing model provides the license types shown in Figure 7-8.

For gateways, there is also a Virtual Gateway (VGW) license, available for AWS, Azure, and ESXi platforms and is licensed based on the bandwidth required. The license types available for VGW are VGW-500M, VGW-2G, and VGW-4G. Security gateway licenses add security features related to IPS and IDS, security dashboard, and anti-malware. Base licenses have the same features as the respective licenses but are limited to 75 clients.

Note

There are different licenses for different device families: Switch foundation licenses for 6200 switch series and foundation licenses for 6400 switch series; and Advanced gateway licenses for the 7000 gateway family, and Advanced licenses for the 7200 gateway family.

To check the latest information on Aruba Central License keys, navigate to the Aruba Central online help (**Help and support icon > Documentation center**) and select **Aruba Central Licenses** under *Key Features.*

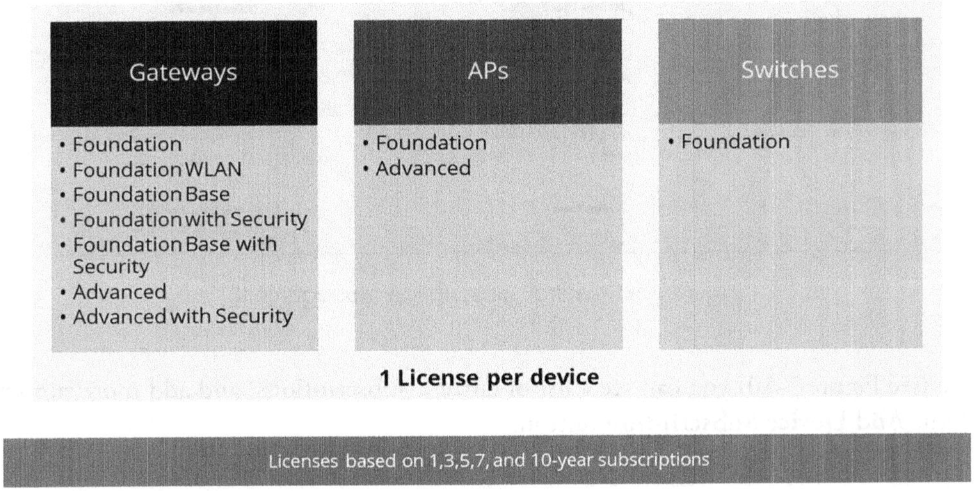

Figure 7-8: Aruba Central 2.5.5 licensing

Subscription management

You view and mange HPE GreenLake subscription keys at the Subscription Management page (**HPE GreenLake > Activities & Tasks > Subscriptions**). The Subscriptions page shows your current subscriptions and allows you to add more as needed. To access it, click the **Subscription** tile shown in Figure 7-9.

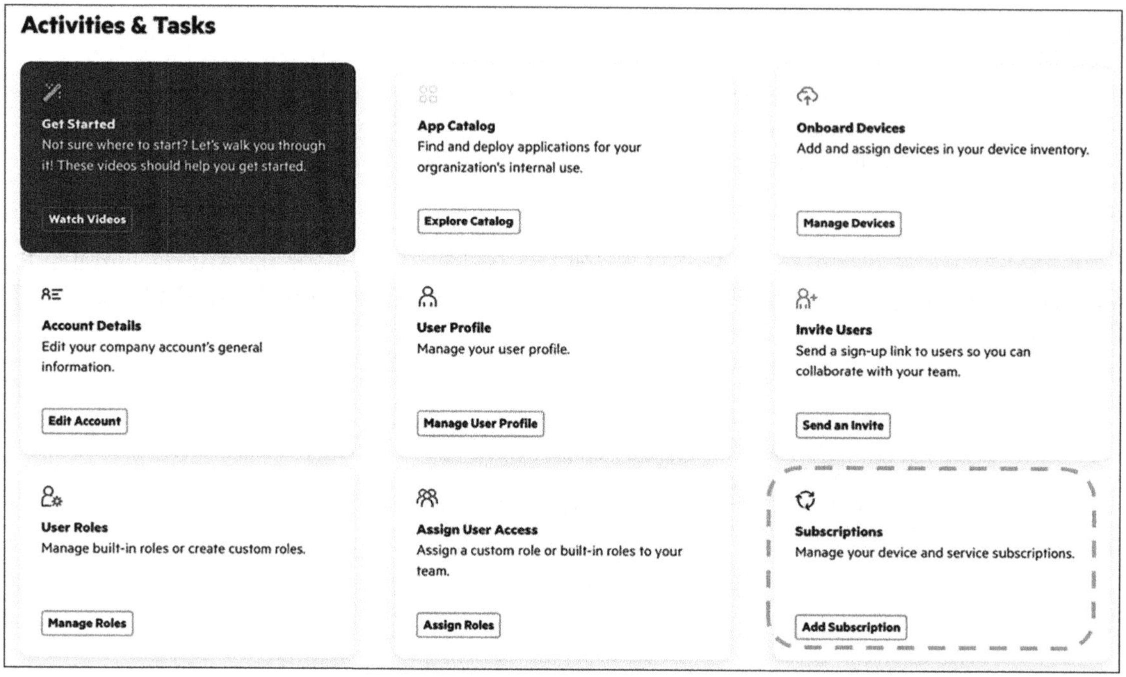

Figure 7-9: Central subscription management

From here (see Figure 7-10) you can see a list of current subscriptions, and add more subscriptions: just click the **Add Device Subscription** button.

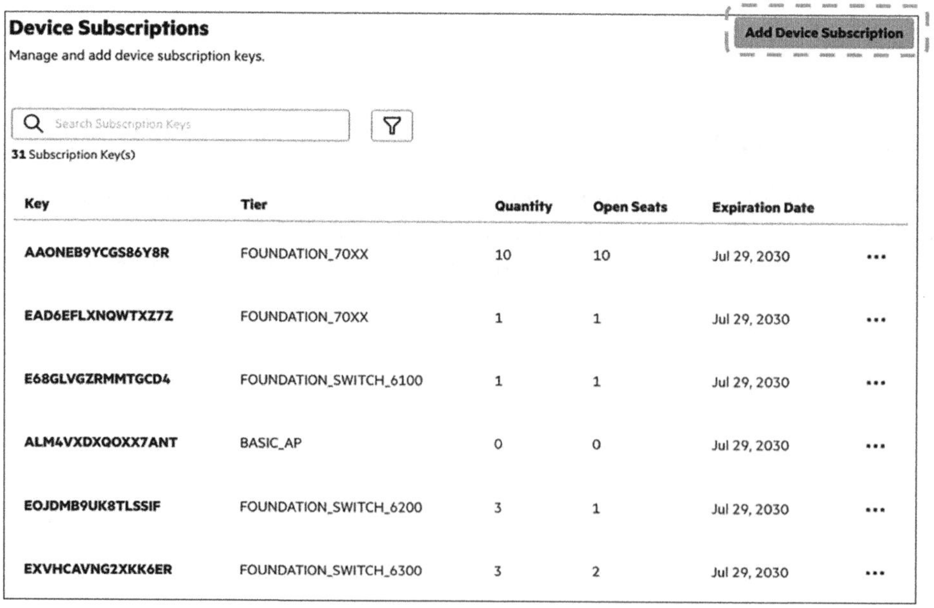

Figure 7-10: Viewing and adding device subscriptions

Subscription assignment

After adding subscription keys, you must assign them to Aruba devices. From the HPE GreenLake management page, select **Manage Devices**, which will take you to the page shown in Figure 7-11. Then click the **Onboard Devices** tile.

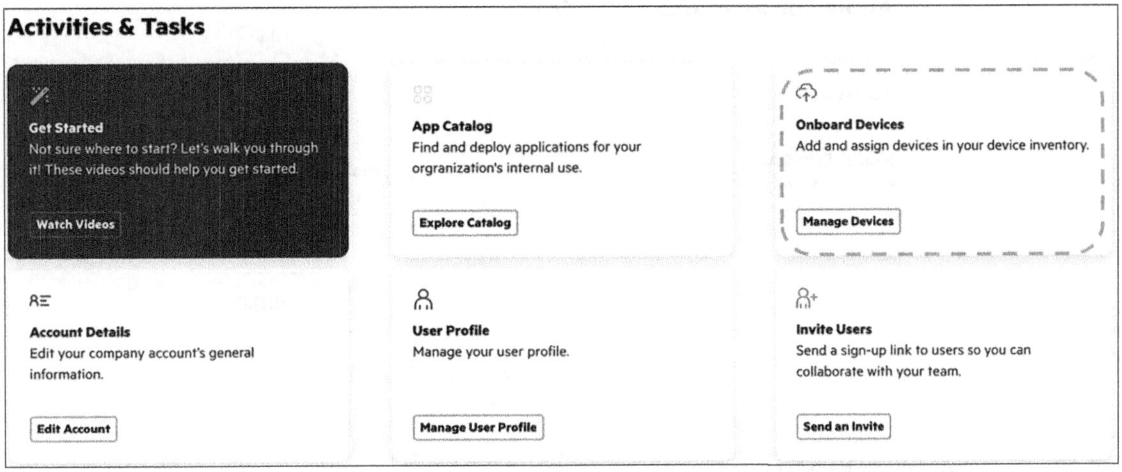

Figure 7-11: Subscription assignment

License assignment

You are then taken to the window shown in Figure 7-12. Complete the license assignment here (**HPE GreenLake > Activities & Tasks > Manage Devices**). Choose devices to which the subscription shall be assigned and click **Apply Subscription**. To update license assignments, select the device, click **Detach Subscription**, and assign a new subscription to the device. To avoid the manual process, you can select **Auto-Subscribe**. For now, let's assume you click **Apply Subscription**.

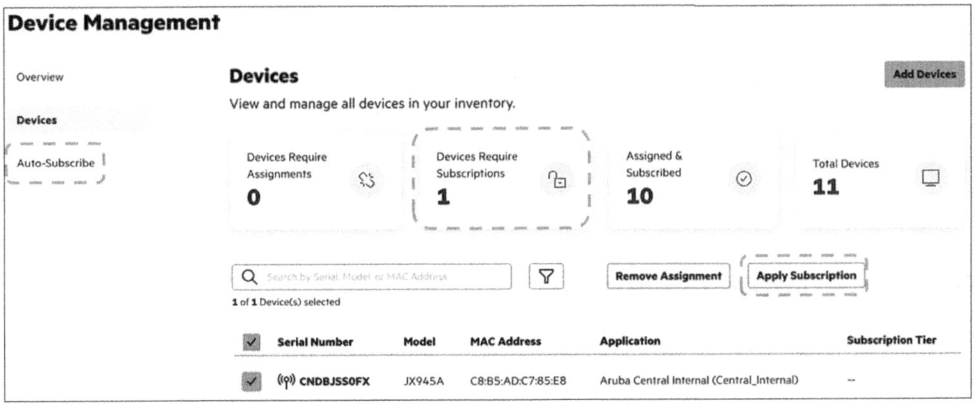

Figure 7-12: License assignment

Manual license assignment

You are taken to the Apply Subscriptions to Devices page, shown in Figure 7-13. For each selected device type, click **Apply Subscriptions**. Then, in Figure 7-14, select the type of license to assign.

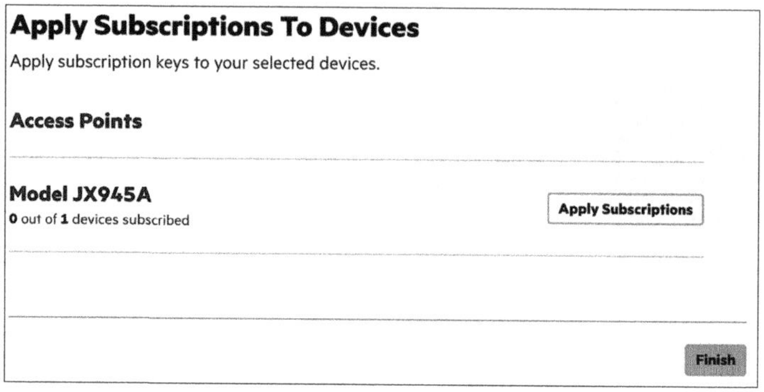

Figure 7-13: Manual license assignment

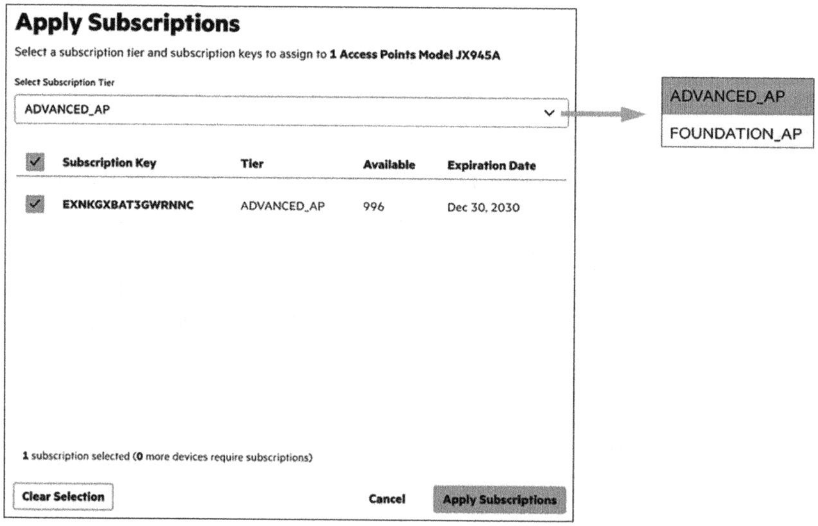

Figure 7-14: Manual license assignment (part 2)

Auto-Subscribe license assignment

With Auto-Subscribe you can assign licenses to all new devices automatically when they connect to Central (see Figure 7-15).

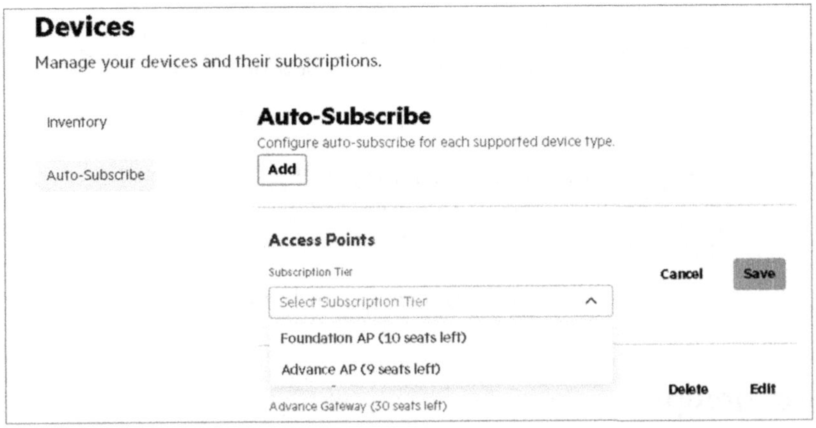

Figure 7-15: Auto-Subscribe assignment

Central groups, sites, and labels

Aruba Central's groups, sites, and labels can ease workflows and save you time (see Figure 7-16). *Groups* in Aruba Central are a primary configuration element: a container for device management, monitoring, and maintenance. Groups enable you to combine devices with identical or similar configuration requirements and manage these devices efficiently by using either a UI-based configuration workflow or CLI-based configuration template or editor.

Device groups enable you to:

- Provision multiple devices in a single group. For example, a group can consist of multiple APs. These APs can share common configuration settings. You can also apply a common security policy for the devices deployed in a specific geographical location.
- Provision different types of devices in a group. For example, a group can consist of APs, gateways, and switches.
- Create a base configuration and add devices as necessary. When you assign a new device to a group, it inherits the configuration that is currently applied to the group.
- Create a clone of an existing group. If you want to build a new group based on an existing group, you can create a clone of the group and customize it as per your network requirements.

A *site* refers to a physical location where a set of devices are installed: a campus, branch, or venue. In Aruba Central, you use sites as a primary navigation element. So, if your devices are deployed in a campus, you could create a site called CampusA, for example. You can also tag the devices in CampusA using labels. If the campus consists of multiple buildings, you could label the devices deployed in the campus as *Building1* or *Lobby*. If the devices in a specific location or an area within a specific location must have similar configuration, you can place them in the same group. You'll learn more about sites shortly.

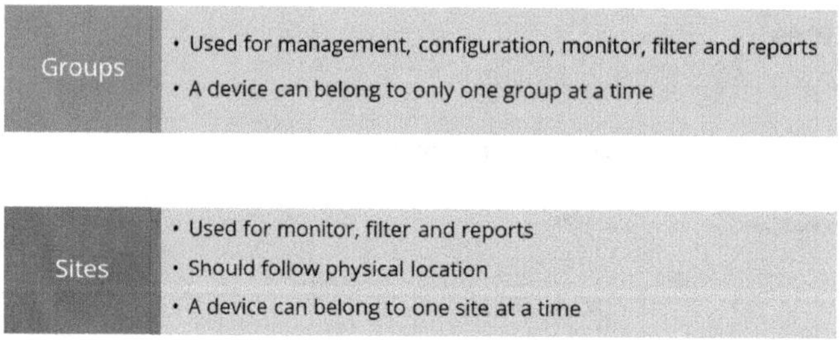

Figure 7-16: Groups and sites

Aruba Central groups

A group's main purpose is to simplify your device configuration. All devices in the same group will receive the same configuration. However, you can also use groups to monitor devices. When you select a group to monitor, you'll only see the devices in that group.

APs advertise SSIDs based on their group membership and configuration. APs will inherit the group configuration based on the group with which they are associated. A device can only belong to one group at a time. If you want all APs to advertise the same SSIDs, then you only need one group.

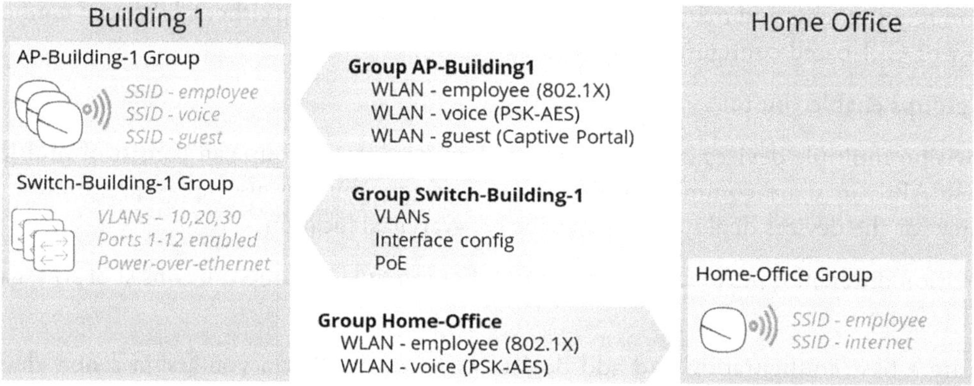

Figure 7-17: Groups

Figure 7-17 shows an AP group named *AP-Building-1*. In this example, these APs need to support three SSIDs: employee, voice, and guest. Therefore, all APs are placed in the same group. Certain employee homes have a Microbranch setup, so the scenario includes a Home-Office group. This group supports an SSID named employee. This SSID configuration securely tunnels corporate traffic back to the office. The internet SSID configuration bridges traffic directly to the user's local LAN, and on to the internet. This example also includes a group named Switch-Building-1 for the Aruba switches' configuration in the building.

Group types

Group types include:

- UI group
- Template group
- Default group
- Unprovisioned group

UI Group

You configure UI group device members using a web GUI interface within Central, which allows you to customize and manage configuration parameters using UI workflows. For example, you can configure APs in a UI group using the menu options under **Organization** > **Groups**.

Template Group

You manage template group device configurations with CLI-based configuration templates. You can combine devices with similar configuration requirements in a single template group and apply a common configuration template to these devices.

Default Group

The default group is a system-defined container that consists of devices running factory default configurations. When a new device with a factory default configuration connects to Aruba Central, it is automatically added to the *default* group. You can use the default group as a placeholder for new devices, with factory default configurations, allowing the network administrator to find and provision those devices easily.

Unprovisioned Group

If a device has a customized configuration and connects to Aruba Central, Central marks this device as *unprovisioned.* If you want to preserve the device configuration, create a new group and assign this device to the newly created group. If you want to overwrite the configuration, move the unprovisioned device to an existing group. You will learn to manage groups, sites, and labels during the lab activities.

Group personas

A device's persona represents the device's role in a network deployment. The group persona defines the type of devices and features that are to be available and managed by the group (see Figure 7-18). Creating personas for devices helps you to customize configuration workflows, automate parts of configurations, and see the default configuration and relevant settings for the device. It also helps you to customize the monitoring screens and troubleshoot workflows appropriate for the device.

- Represents the device's role in a network deployment
- Defines the type of devices and features to be available and managed by the group

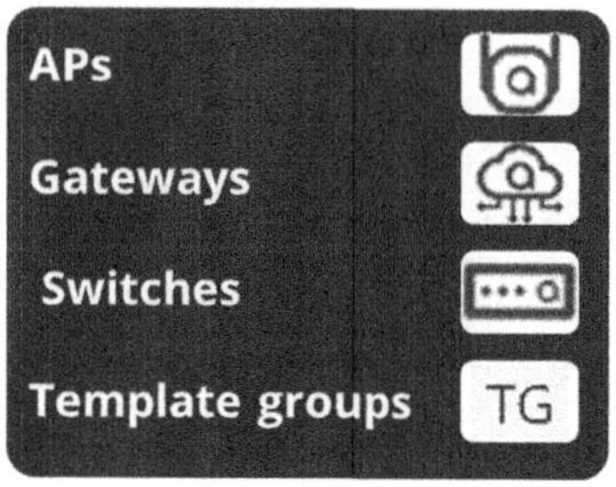

✓ Customize configuration workflows
✓ Automate parts of configurations
✓ see default config and relevant settings
✓ Customize monitoring screens
✓ Troubleshoot workflows

Figure 7-18: Group personas

Creating a Group Persona

You can create a persona when you create a group. Persona and architecture are set at the group level (see Figure 7-19). All devices within a group inherit the same persona from the group settings. While creating a group, you can mark the current group's architecture and persona settings as preferred settings for adding subsequent groups. For subsequent groups, you can either automatically apply the preferred settings or manually select settings for the new group.

Based on the device persona you select for a group, the device configuration page displays only particular device tabs for that group. For example, if a group has only access points persona assigned to it, then the device configuration page for that group displays only the access points tab.

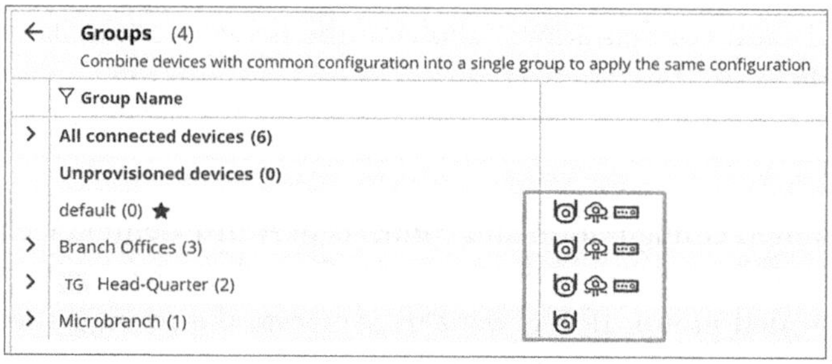

Figure 7-19: Creating a group persona

Group personas for devices

Group personas for devices is summarized in Figure 7-20.

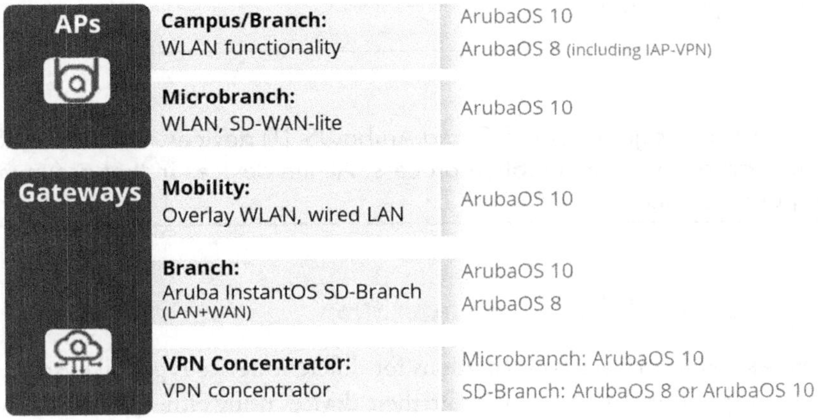

Figure 7-20: Group personas for devices

Access points can have the following persona:

- **Campus/Branch**: in this persona, the AP provides WLAN functionality. This persona applies to both ArubaOS 10 and legacy ArubaOS 8 (including IAP-VPN) architectures.
- **Microbranch**: in this persona, the AP provides SD-WAN-lite functionality in addition to the WLAN AP functions. This persona applies only to the ArubaOS 10 architecture.

Gateways can have the following persona:

- **Mobility**: in this persona, gateways provide overlay WLAN and/or wired (UBT) LAN functionalities. This persona applies only to the ArubaOS 10 architecture.
- **Branch**: in this persona, gateways provide Aruba InstantOS SD-Branch (LAN + WAN) functionality. This persona applies to both ArubaOS 10 and ArubaOS 8 architectures, with ArubaOS 10 being a superset of ArubaOS 8.
- **VPN Concentrator**: in this persona, gateways provide VPN concentrator functionality for ArubaOS 10 microbranch or ArubaOS 8 or ArubaOS 10 SD-Branch deployments.

The two architectures for creating groups are highlighted here:

- **ArubaOS 8**: Instant AP-based deployment, including Aruba InstantOS 6.x or Aruba InstantOS 8.x (IAP, IAP-VPN), or Aruba InstantOS 8.x SD-Branch deployments.
- **ArubaOS 10**: ArubaOS 10 deployment, including AP-only underlay, AP+gateway overlay, microbranch, or AOS 10.x SD-Branch deployments.

You can apply a device persona to both ArubaOS 10 and ArubaOS 8 deployments. All of the ArubaOS 8 personas apply to ArubaOS 10 deployments as well. The ArubaOS 10 deployment supports a couple of additional personas that do not apply to ArubaOS 8 deployments. The persona workflow differs based on the deployment type.

Note

A group *cannot* manage ArubaOS 8 *and* ArubaOS 10 devices. The ArubaOS version is selected as part of the group creation process. All the devices in that group must comply with the group persona.

Groups: Template and UI-based configuration

Aruba Central allows you to manage configurations for different device types, like Aruba APs, gateways, and switches, in your inventory. You can configure these devices using either UI workflows or configuration templates (see Figure 7-21). You define a preferred configuration method when creating the group.

Aruba Central allows you to create a single group with different configuration methods defined for each device type. For example, you can create a group with the name *Group1* and within this group, you can enable a template-based configuration method for switches and a UI-based configuration method for APs and gateways. Aruba Central identifies both groups under a single name (*Group1*). If a device type in the group is marked for the template-based configuration method, the group name is prefixed with *TG* (like *TG Group1*, for example). You can use *Group1* as the group ID for workflows such as user management, monitoring, reports, and audit trail.

After you assign devices to a group and access the configuration containers, Aruba Central automatically displays relevant configuration options based on the configuration method you defined for the device group. The Groups page allows you to create, edit, or delete a group, view the list of groups provisioned in Aruba Central, and assign devices to groups.

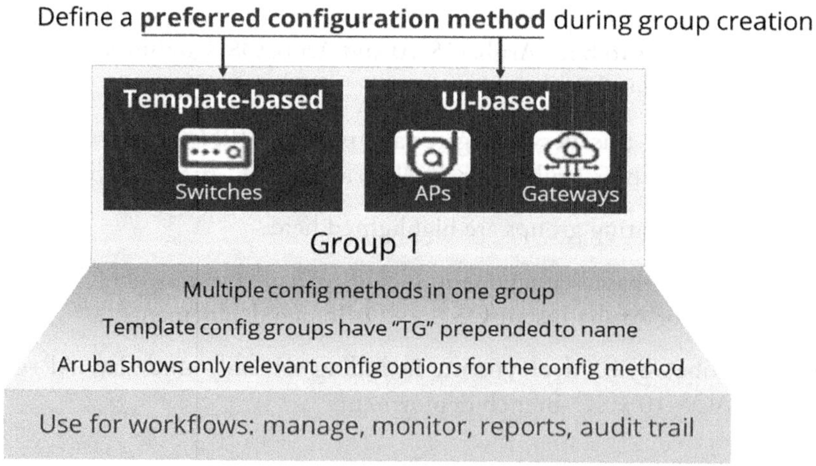

Figure 7-21: Template and UI-based configurations

Managing groups and personas: Task overview

This section covers creating a group, defining a group persona, and adding devices to a group (see Figure 7-22).

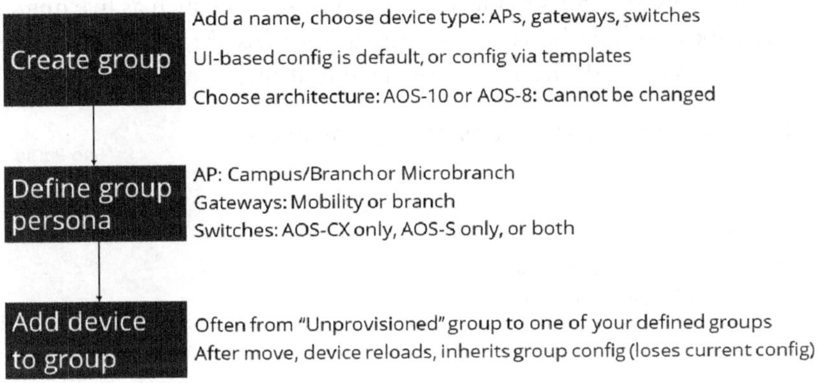

Figure 7-22: Managing groups and personas

Creating the group

Group creation starts with assigning a group name and the device types that include: access points, gateways, and switches. You then choose UI-based configuration (the default) or template-based configuration (and then choose which device types in the group shall be managed using templates).

You can mix and match the UI and template methods in a single group: for example, a UI configuration for access points and gateways, but use a CLI template configuration for the switches. This adds flexibility: you to choose the type of management interface to use per device type within a group. Keep in mind that you cannot change the type of the group between UI and template groups after creating the group.

Defining the group

You then define the group by choosing an architecture: ArubaOS 10 or ArubaOS 8. You choose the group persona by choosing the architecture for each device category. These selections affect the configuration options at the group and device levels. Also, you *cannot* change the group persona after you create the group. For example, if you selected ArubaOS 8 at the time of the creation, the only way to manage ArubaOS 10 devices is by creating a new group with the AOS 10 architecture.

Adding devices to the group

After you define groups, you move devices into those groups as members—perhaps devices from the *unprovisioned* group to one of your defined groups. Be careful: after a device is moved to a new group, it reloads and inherits the group configuration. This means that the device will erase the current configuration and download its new group configuration!

Sites

A *site* refers to a physical location where a set of devices are installed; for example, campus, branch, or venue (see Figure 7-23). Sites are more than just a tool to filter views, place devices on a map, and generate reports. You can use sites for several Aruba Central services such as live upgrades, AirMatch, install manager, network health dashboard, User Experience Insight (UXI), and etcetera. As a best practice, you should assign every device to an Aruba Central site.

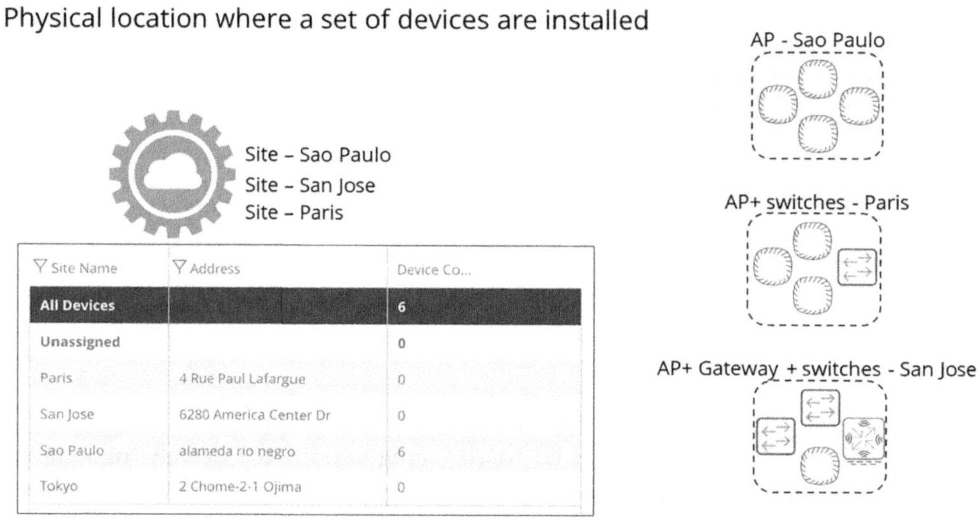

Figure 7-23: Sites

Note

Devices can only belong to one site.

When creating a new site, you need the site name, street address, city, country, and region. The more information you add, the more accurate the site is displayed on the map. Most information is mandatory, but the zip or postal code is not.

Note

VisualRF, which will be covered later in this book, currently does not support hierarchical arrangement for a location. If a campus has two or more buildings, it is recommended to create an Aruba Central site for *each* building.

Using an Open Location Code

An Open Location Code (OLC) is a geocode system for identifying an area anywhere on the earth. Location codes created by the OLC system are referred to as *plus codes*. Aruba Central supports plus codes when creating new sites. Simply use plus code as a street address (see Figure 7-24). The site will then be automatically positioned on the map in Central. A good example of why OLC is useful would be a campus that has multiple buildings. The street name and number for each building be the same since it is the same campus. Therefore, entering an address when creating the site would not be precise. Plus codes allow you to specify the exact location when setting up the site. You will learn how to add, modify, and manage sites during the lab activities.

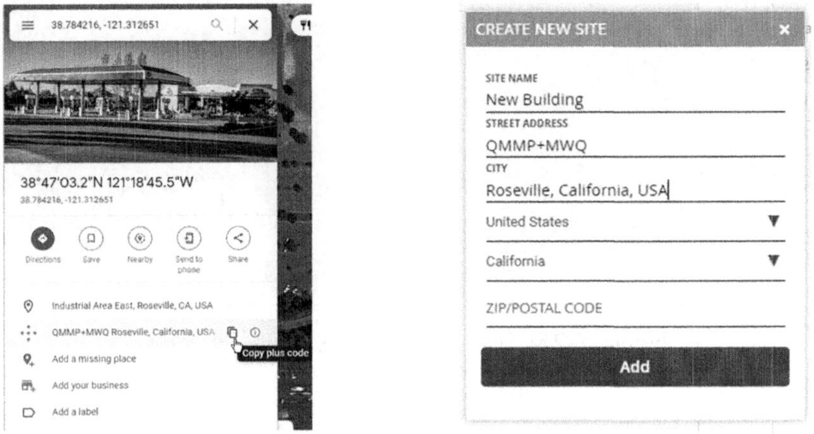

Figure 7-24: Open location codes

Keep in mind that many Aruba Central services use the site information for planning and managing the devices and their activities. Therefore, be sure that all devices are assigned to the right site to allow those services to run correctly!

Sites and the network health map

Sites with proper addresses will appear on the network health map (see Figure 7-25). The map provides a pictorial view of the various network sites. The sites are color-coded as red or green. Red indicates potential issues, and green indicates no issues. You can click the site on the map to view details of any issues at the selected site. You can also zoom into a site and get to a street level view.

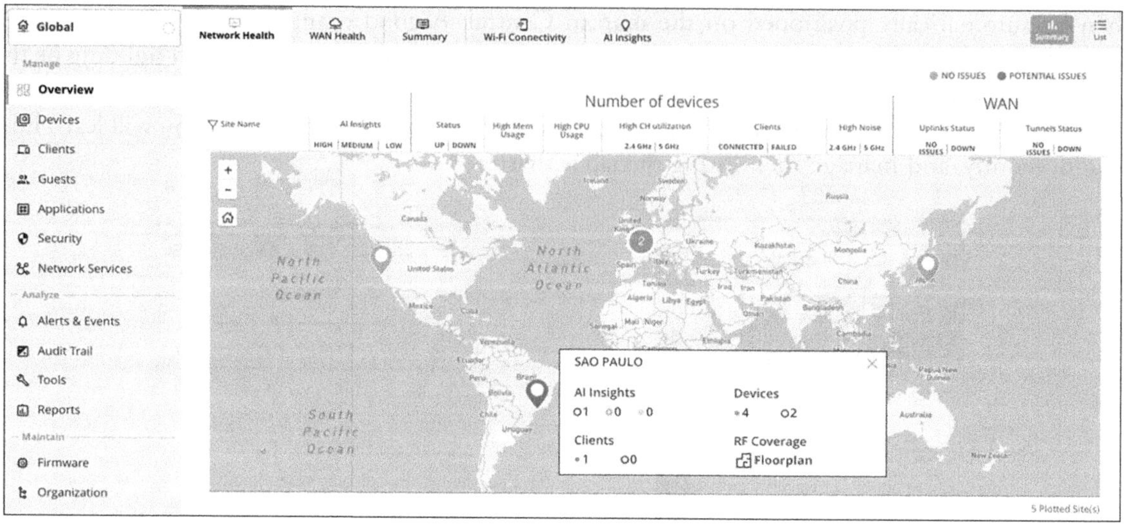

Figure 7-25: Sites and the network health map

Labels

Labels are tags that are used to filter devices for monitoring, managing, and reporting purposes. A device can have multiple labels, which can help you determine device ownership, departments, and functions. Suppose you have several buildings in a city, all with their own devices. Therefore, you could create a label of BLD-1-FLR-1, and all the devices in building 1, floor 1 would be associated to that label (for example, all APs and switches on that floor). Select the label to see only the devices associated with it and unselect the label to remove the filter.

You might create other labels as shown in Figure 7-26: for other buildings, floors, cities, and perhaps a conference room label for even more granularity. The noted AP in Figure 7-26 can be associated with four labels: conference rooms, city ABC, building 1, and BLD-1-FLR-1.

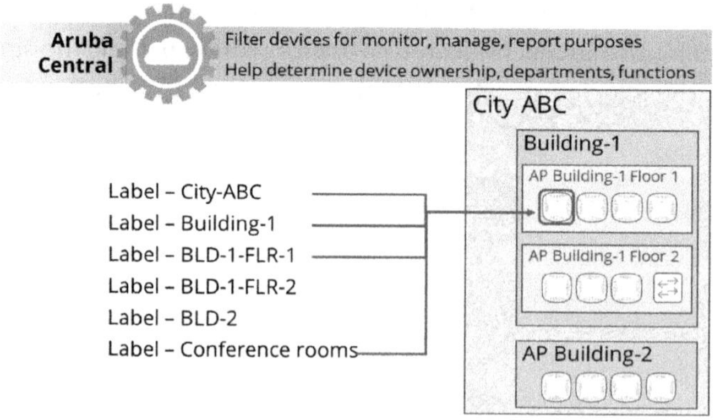

Figure 7-26: Labels

Lab 7.01: Managing aggregation switches with Aruba Central

Your senior network colleague is very happy with your progress in networking skills. Your next assignment is to learn to use the Aruba Central cloud-based management platform to manage and monitor the switches. This lab will first introduce you to the basic access and navigation in Aruba Central. This will include the configuration of sites and labels. Next you will migrate the aggregation switches to Aruba Central using a template configuration that has been prepared for you by your senior colleague.

Objectives

- Configure Aruba Central sites and labels.
- Understand the requirements for the cloud connection.
- Assign devices to a template-based group.
- Verify the deployment of a template configuration.

Task 1: Prepare the topology

In this task you will configure the aggregation switches with an IP helper address on SVI1. This ensures that the factory default edge switches can receive a DHCP address and attempt to connect to Aruba Central for their cloud management connection.

Objectives

- Configure IP helper on the aggregation switches.

Steps

1. Use PC-MGMT to open an SSH connection to sw-agg1.
2. Configure SVI1 with an IP Helper address of 10.254.1.21.

```
sw-agg1(config)# interface vlan 1
sw-agg1(config-if-vlan)# ip helper-address 10.254.1.21
sw-agg1(config-if-vlan)# exit
```

Task 2: Essential Aruba Central configuration

In this task you will perform some essential steps when configuring an Aruba Central environment for the first time.

Objectives

- Access Aruba Central using HPE GLCP.
- Understand the core navigation flow in Aruba Central.
- Configure sites and labels in Aruba Central.

Steps

Access HPE GreenLake Cloud Platform (GLCP)

1. On your local system, use a browser to open a connection to the HPE GreenLake Cloud Platform: **https://common.cloud.hpe.com**.
2. Click **Sign In with SSO**. Enter the email address provided by your instructor and click **Next**. You are redirected to the Aruba Training Labs SAML host.
3. Enter the email address and password provided by your instructor and click **Login**. You are now logged in to HPE GreenLake Cloud Platform.

Launch Aruba Central Application

1. In the **Aruba Central** tile, click **Launch**.

Aruba Central UI Navigation Instructions

In the next steps, you will be introduced to the core navigation structure in Aruba Central: using the context, navigation, and top areas. This will be used in the remainder of the labs to guide you to the correct Aruba Central screen.

Context filter

The context filter is used to narrow the scope of the Aruba Central UI.

1. By default, the Global context is selected (see Figure 7-27).

Figure 7-27: Global context

2. Click the context area to see a pop-up with context options. Currently, only the default group is listed in the context filter (see Figure 7-28).

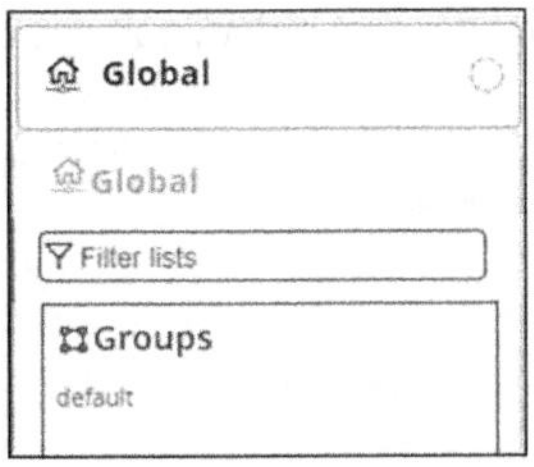

Figure 7-28: Context area

3. As you progress with the lab activities, additional groups, sites, and labels will show up in this context selection list. Figure 7-29 shows an example.

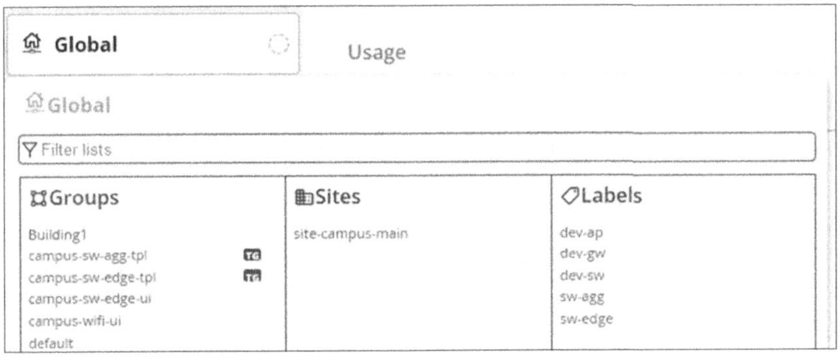

Figure 7-29: Context filter example with groups, sites, and labels

4. Click the **Context** button under the Groups heading. Then click **default** (see Figure 7-30). The context filter will now show the *group* icon (four squares connected as a larger square) and the name of the group: *default* (See Figure 7-31).

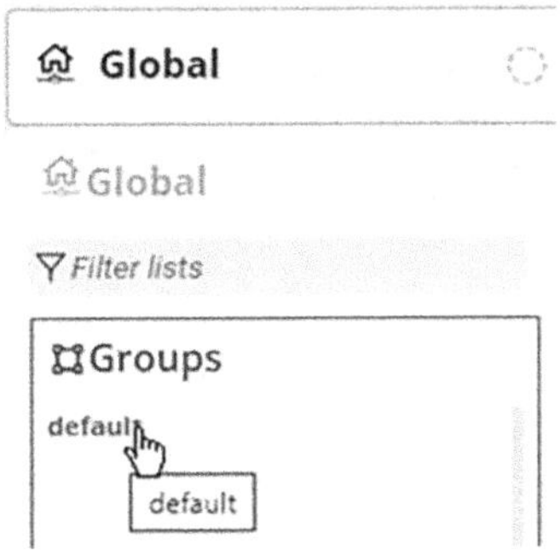

Figure 7-30: Context menu

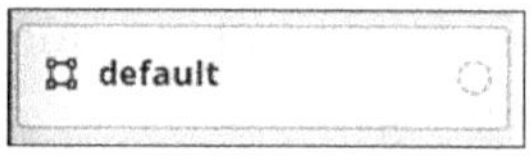

Figure 7-31: Default group

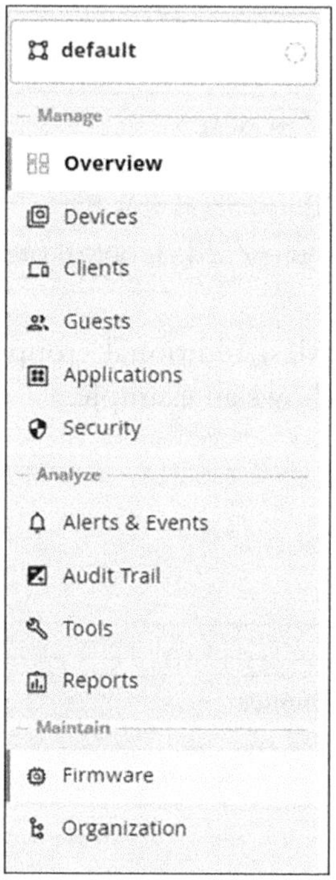

Figure 7-32: Navigation menu

Navigation menu

On the left side of the UI, you can see the navigation menu. This menu provides access to various status and configuration screens for a given context.

1. Figure 7-32 shows an example of the navigation menu with Manage, Analyze, and Maintain sections. You may see an instruction such as the following in the lab:
   - Navigate to: Context: **Groups / default** > Navigation: **Devices.**
2. You should first verify that you are at the correct context filter level (or change the context filter), then use the navigation menu on the left side and click **Devices** (see Figure 7-33).

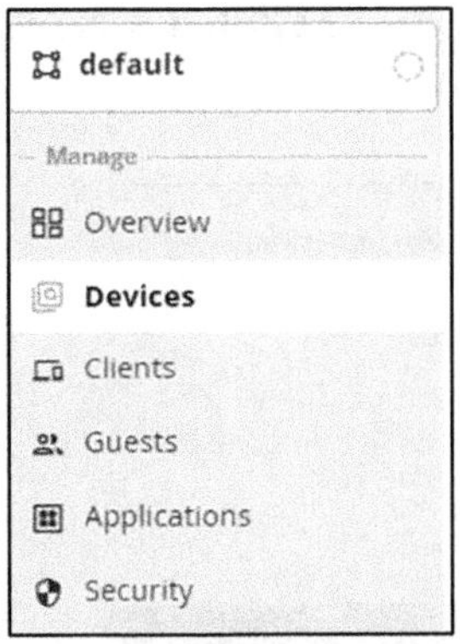

Figure 7-33: Select Devices from the context menu's default group

Top Menu

Each of the navigation menu's Aruba Central screens have one or more tabs at the top of the screen.

1. In the example shown in Figure 7-34 (Context: **Group / default** > Navigation: **Devices**), the top options are *Access Points* and *Switches*. The example instruction to reach this page would be: Context: **Groups / default** > Navigation: **Devices** > Top: **Access Points**.

Figure 7-34: Tabs at the top of the screen

Top Right

Many of the Aruba Central pages will have similar options at the top right of the screen.

1. In example shown in Figure 7-35 (Context: **Group / default** > Navigation: **Devices** > Top: **Access Points**), the top right options are Summary, List, and Config. To access this example page, the instruction would be: Context: **Group / default** > Navigation: **Devices** > Top: **Access Points** > Top right: **List**

Figure 7-35: Options at the top right

2. This concludes the introduction to the Aruba Central navigation.

Sites and Labels

Aruba Central uses sites to map devices to their physical locations. In Aruba Central, all devices in the same location should be mapped to the same site. A device can only belong to one site.

3. In Aruba Central, navigate to: Context: **Global** > Navigation: **Organization** > Top: **Network Structure** > **Sites** (see Figure 7-36).

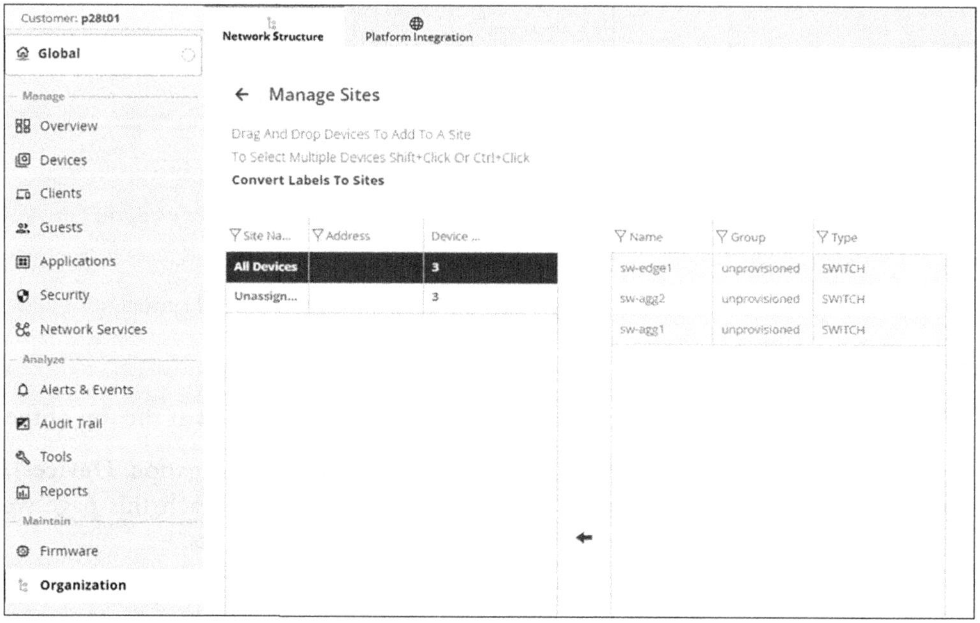

Figure 7-36: Manage sites screen

4. At the bottom of the page, click the **plus sign** icon to create a new site.

Note

This lab uses a fictitious example site's addressing information. Feel free to enter your own site address information, but please use the site name **site-campus-main**. See Table 7-1 for an example.

Table 7-1. Example site information

Field	Value
Name	site-campus-main
Street Address	378 Reed Grass Dr.
City	Oviedo
Country	USA
State	FL
ZIP code	32765

5. Click **Add** to save the site.

Device Labels

Labels can be used to assign logical markers to devices. Unlike a site, it is possible to assign multiple labels to a single device.

1. In Aruba Central, navigate to: Context: **Global** > Navigation: **Organization** > Top: **Network Structure** > **Labels.**
2. At the bottom of the page, click the plus sign to create a **New Label**. Set the label name to **sw-agg**.
3. Add another label and set the label name to **sw-edge**.

Verify Sites and Labels

4. In Aruba Central, click the **Context** filter. The new site and labels will be listed in the Sites and Labels columns (see Figure 7-37). This makes it convenient to jump to a site or location.

Task 3: Provide aggregation switches with static cloud connection

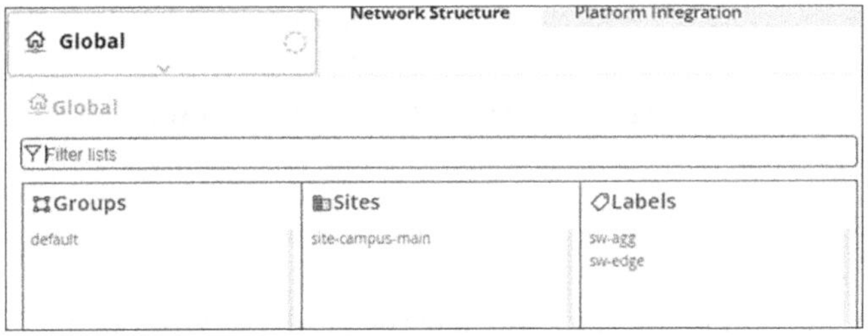

Figure 7-37: Example sites and labels configuration

In this task you will connect the existing aggregation switches to Aruba Central. The goal of this task is to show the connection requirements and how the connection is established for a switch that has an existing configuration. Aruba Central uses groups to organize device configurations. At this point in the lab, no group has been created. This also means that the aggregation devices will not receive a configuration from Aruba Central yet. This will be done in the upcoming tasks. For right now, only the connection to Aruba Central will be verified.

Objectives

- Connect an AOS-CX switch to Aruba Central.

Steps

1. Open a connection to Aruba Central. Navigate to: Context: **Global** > Navigation: **Devices** > Top: **Switches** (see Figure 7-38).
   - **Question**: Are there any switches listed?
   - **Answer**: Yes, your edge and aggregation switches are connected.
   - **Question**: What is the color of the small dot in front of the sw-agg1 device name?
   - **Answer**: This is a green dot; it represents an active connection.

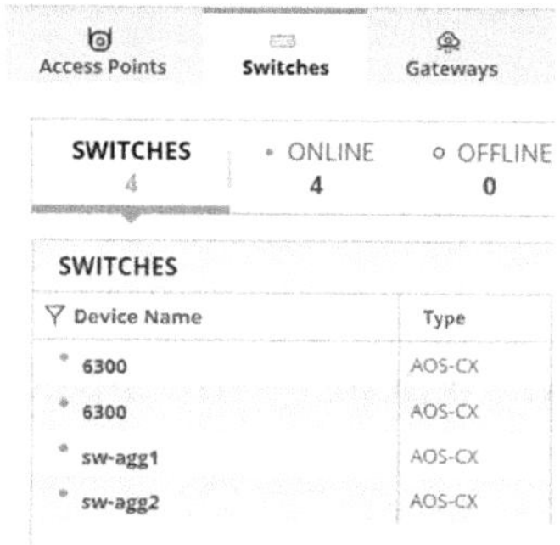

Figure 7-38: Example switches screen

Sw-agg1

1. Open an SSH Connection to sw-agg1. Review the Aruba Central connection status.

```
sw-agg1# show aruba-central
```

   - **Question**: What is the admin state?
   - **Answer**: Aruba Central support is enabled. This is the default state.
   - **Question**: What is the connection status?
   - **Answer**: The reported connection status is connected.

2. Aruba Central requires internet access and DNS functionality. You will now review these requirements. Attempt to reach an IP address on the internet to verify IP reachability.

```
sw-agg1# ping 8.8.8.8
```

   - **Question**: Was the ping successful?
   - **Answer**: Yes, in previous lab you have configured OSPF adjacency with the core router. The core router provides a default route to the internet. Behind the core router, there is router that provides NAT services for your lab.

3. Attempt to reach an internet system based on a DNS FQDN. You can use the Aruba Path Quality Monitoring (PQM) host for this test.

```
sw-agg1# ping pqm.arubanetworks.com
```

- **Question**: What do you observe?
- **Answer**: The name cannot be resolved. This indicates something is missing in the DNS configuration.

4. Review the DNS configuration.

```
sw-agg1# show ip dns
```

- **Question**: Are there any DNS servers configured?
- **Answer**: No, a DNS server is missing from the configuration.
- **Question**: How is it possible that the connection with Aruba Central is connected when no DNS servers have been configured?
- **Answer**: The Aruba Central client process on AOS-CX includes by default the DNS server 8.8.8.8. Only the Aruba Central client component on the switch uses this default DNS server. Other utilities, such as ping, use the global DNS configuration.

5. To override the default DNS, configure the host 10.254.1.21 as the DNS server address.

```
sw-agg1# ip dns server-address 10.254.1.21
```

6. Attempt to validate the change with a ping to pqm.arubanetworks.com. The ping should now be successful.

```
sw-agg1# ping pqm.arubanetworks.com
```

7. Check the status of the Aruba Central connection again; it should still show as connected.

Note

If something was wrong with the connection, you can speed up the connection retry process by disabling and re-enabling the Aruba Central feature on the switch.

```
aruba-central
  disable
  enable
  exit
```

- **Question**: What is the Aruba Central location?
- **Answer**: This is the URL of the Aruba Central region your account is hosted on.

```
sw-agg1(config)# show aruba-central
Central admin state                          : enabled

Central location                             : device-
uswest4-d2.central.arubanetworks.com
VRF for connection                           : default
Shared Token                                 : N/A
Central connection status                    : connected

Central source                               : activate
Central source connection status             : connected
Central source last connected on             : Tue Sep 27 16:02:10 UTC 2022
System time synchronized from Activate       : True

Activate Server URL                          : devices-v2.arubanetworks.com
CLI location                                 : N/A
CLI VRF                                      : N/A

Source IP                                    : 10.254.101.2
Source IP Overridden                         : False

Central support mode                         : disabled
```

Assign the Site

Now that the aggregation switches are connected to Central, you can assign them to a site. A device can only belong to a single site.

1. In Aruba Central, navigate to: Context: **Global** > Navigation: **Organization**> Top: **Network Structure** > **Sites.**
2. In the left pane, select **Unassigned**. The two aggregation switches should be listed in the pane on the right side.
3. Select both aggregation switches (you can use the control or command key to select multiple entries), then drag them to the site named **site-campus-main**.
4. Confirm the action with **Yes**.

Assign a Label

Next you can assign a label. Up to five labels can be assigned to a device.

1. In Aruba Central, navigate to: Context: **Global** > Navigation: **Organization**> Top: **Network Structure** > **Labels**
2. In the left pane, select **All Devices**. The two aggregation switches should be listed in the pane on the right side.

3. Select both aggregation switches, then drag them to the label named **sw-agg**.
4. Confirm the action with **Yes**.

Verify Site and Label

1. In Aruba Central, click the **Context** menu.
2. While this menu is open, your cursor is in the **Filter lists** field. You can start typing here to filter the context menu. Type **main** (see Figure 7-39).
   - **Question**: What do you notice?
   - **Answer**: Only the groups, sites, and labels that contain the string *main* will be displayed now. This is convenient when a customer has many groups, sites, or labels.

Figure 7-39: Example of filtering

Task 4: Aggregation switches template configuration group

Your senior colleague wants to ensure that the aggregation switches receive their full configuration from Aruba Central. This should be based on a configuration that has been prepared by your senior colleague.

In this task you will prepare a template group in Aruba Central for the aggregation switches and upload the prepared configuration. Next you will assign the aggregation switches to this group; this will make Aruba Central push the configuration to the switches. When Aruba Central is managing the configuration using a template, network administrators should not make any changes to the switch configuration anymore.

In this task you will see how the configuration will become read-only, and how you can still make exceptions to this to provide troubleshooting and support.

Objectives

- Create a switch template group.
- Assign a template to a group.
- Move a device to the group.
- Understand the configuration lockout feature.
- Understand how the support mode feature can be used.

Steps

Create Template Group

First you will create the switch template group.

1. In Aruba Central, navigate to: Context: **Global** > Navigation: **Organization**> Top: **Network Structure** > **Groups.**
2. At the right-top, click the **plus** sign to add a new group.
3. For the name, enter **campus-sw-agg-tpl**.
4. For the value *Group will contain*, only select **switches**.
5. Configure using **templates**: move the slider to the right (**enabled**). The checkbox for switches is automatically selected.
6. Click **Next**.
7. Types of switches used: Select **AOS-CX only**.

Note

You don't need to select *Make these the preferred settings*. This would make the current selection the default selection for future group additions.

8. Click **Add**.
9. Verify that the group **campus-sw-agg-tpl** is now listed (see Figure 7-40).

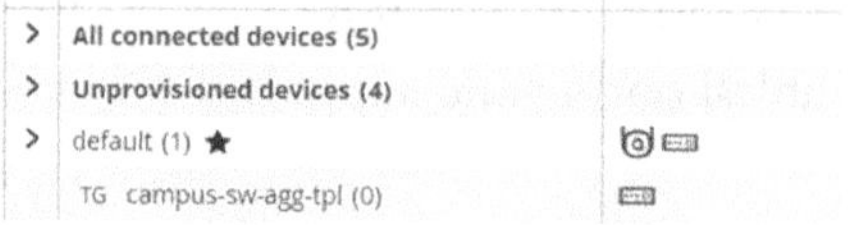

Figure 7-40: Verify the group listing

- **Question**: Why is there a TG label in front of the group name?
- **Answer**: This is how Central shows that the group configuration is based on a template group (TG).
- **Question**: What icon type do you see in the second column?
- **Answer**: Central shows what type of devices are enabled for the group. Since you only enabled switches, only a switch icon will be displayed for this group.

Assign a Template to a Group

You will now apply a configuration to this group. This configuration was prepared by your senior colleague and has been saved in the ACAF Student files folder on the PC-MGMT Desktop.

1. Navigate to Context: **Groups / campus-sw-agg-tpl** > Navigation: **Devices**> Top: **Switches** > **Config** (gear icon). See Figure 7-41, where the top navigation shows *Templates*, *Variables*, and *Configuration Audit*.

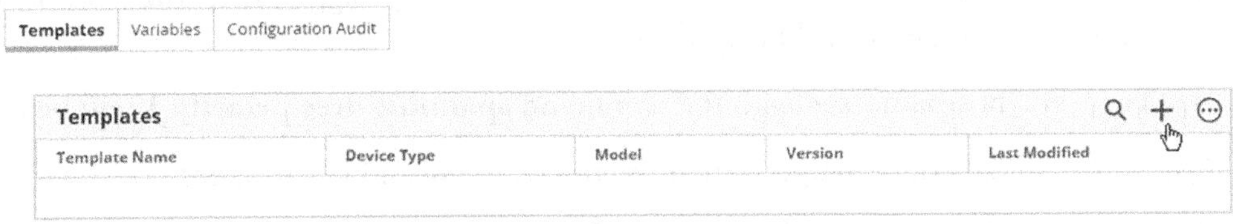

Figure 7-41: Top navigation view

2. Under **Templates**, at the right-top, use the + icon to add a template. The Add Template window appears.
3. For template name, enter **sw-agg**.
4. For device type, verify **Aruba CX** is selected (default based on the group settings).
5. You can leave the other fields at default and click **Next**.
6. From the PC-MGMT, navigate to **Desktop > ACAF student files** and open **lab07.01-sw-agg-tpl.txt**.
7. Copy all the contents of the template text file.
8. Paste the contents of the file in the Template screen.
9. Click **Save** to complete this step. The template will now be uploaded to Aruba Central.
10. Verify that the new template is now listed in the Templates list.
11. To view the contents of an existing template, move the mouse cursor over the **sw-agg** template entry. At the end of the line a pen icon and a delete icon will show up.

12. Click the **pen** icon to edit the template. The template contents will now be displayed.
13. Navigate to line 25 in the template text.
    - **Question**: What is the text you see on the lines 25-30?
    - **Answer**:

```
%if _sys_hostname=sw-agg1%
spanning-tree priority 0
    %endif%
%if _sys_hostname=sw-agg2%
spanning-tree priority 1
    %endif%
```

- **Question**: What does this mean?
- **Answer:** There is a condition:
    - If a variable named '_sys_hostname' has the value 'sw-agg1', then the command **spanning-tree priority 0** will be applied.
    - For '_sys_hostname' sw-agg2 the command **spanning-tree priority 1** will be applied.
- **Question**: What would happen when the hostname would be sw-agg3?
- **Answer**: Since the hostname does not match any condition, none of these two example commands would be applied. This template relies on the correct hostname of the switches. Next, you will double-check the hostname of your aggregation switches before moving them to this group.

Verify the Aggregation Switch Hostnames

1. In Aruba Central, click **Cancel** to close the template details.
2. In Aruba Central, navigate to Context: **Global** > Navigation: **Devices**> Top: **Switches** > Right-top: **List.**
3. Verify in the Device Name column that the listed name is **sw-agg1** and **sw-agg2**.
4. If the hostname is not correct, the next steps can be used to correct it.
    a. Only perform these steps if the system would have an incorrect hostname
    b. In the remote lab, the aggregation switches have initially been configured using a Zero Touch Provisioning file; this includes the correct hostnames sw-agg1 and sw-agg2.
    c. The next steps are only required if you would have changed the hostname yourself to some other hostname.
    d. Using PC-MGMT, open an SSH connection to sw-agg1 or sw-agg2.
    e. Verify that the hostname is exactly **sw-agg1**.

```
show hostname
```

Important

If the hostname is not exactly sw-agg1, change it with the configuration:

```
config
  hostname sw-agg1
  end
```

Move a Device to the Group

In the next steps you will start by moving *only* sw-agg1 to the target group, then you will verify the applied configuration. *After* these validation steps have been completed, you will move sw-agg2.

1. In Aruba Central, navigate to: Context: **Global** > Navigation: **Organization**> Top: **Network Structure**> **Groups.**
2. Expand *All connected devices* by clicking the > icon.
3. Under *All connected devices*, select **sw-agg1**.

Note

Please pay attention. This screen is a multi-select enabled by default; clicking on multiple devices will select all of them. To un-select, click again on the device.

4. On the right-hand side, a popup will be displayed with the Move Devices action button (see Figure 7-42). Click the **Move Devices** button.

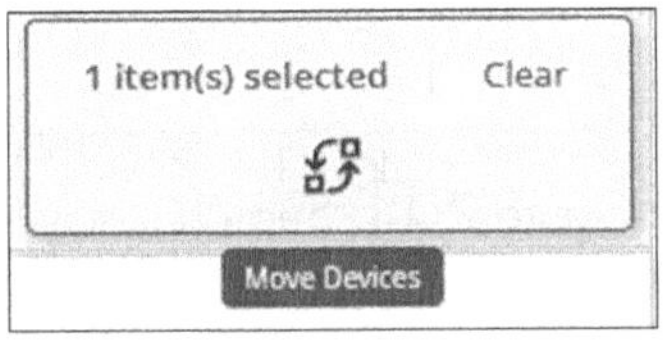

Figure 7-42: Select the Move Devices button

5. Click the **Destination Group** field.
6. You can either select the group **campus-sw-agg-tpl** or start typing a substring of the group name, such as **agg**, then select the group from the filtered list. This can be convenient when a lot of groups exist in Central.
7. Click the **Move** button to continue. With this action, Aruba Central will be in control of the sw-agg1 configuration. The existing configuration of sw-agg1 will be completely replaced with the template that you have previously uploaded to Central.

Verify the Configuration Sync Status

In these steps you will verify the status of the configuration sync. This can be done at the device or group level.

In the next steps you will check the configuration sync at the device level. This level also allows you to check the 'Attempted' configuration and the 'Running' configuration of the device.

The *Attempted* configuration is the configuration that Central attempts to push to the device, while the running configuration is the configuration that is currently running on the device. There may be a difference in the syntax (for example, port ranges) and command order between these two versions.

1. In Aruba Central, navigate to the AI search bar: Top : "Search or ask Aruba" (see Figure 7-43).

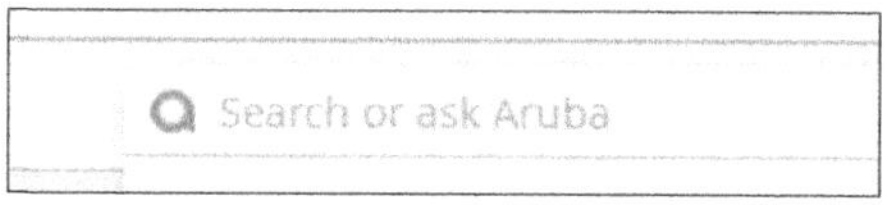

Figure 7-43: Search field

2. Enter **sw-agg1** in the search box and press **<Enter>**. After a few seconds, the results will be displayed in a popup; this will include sw-agg1 (see Figure 7-44).

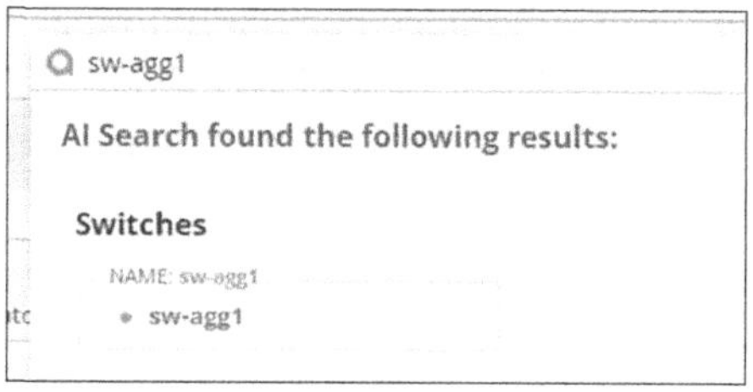

Figure 7-44: Search results

3. Click the name **sw-agg1** in the popup window. This will take you to the device level homepage. Notice that the context filter has been set to **sw-agg1** now (see Figure 7-45).

Note
Several shortcuts are available in the popup window. Just click the sw-agg1 text next to the green dot.

Figure 7-45: Context filter results

4. In the Navigation bar, click **Device**, under Top: **Switch**, select **Configuration Audit**.
5. Check the status in the Configuration Status box (see Figure 7-45).
   - **Question**: What is the status in the box?
   - **Answer**: The status should be *Config In Sync*. See Figure 7-46.

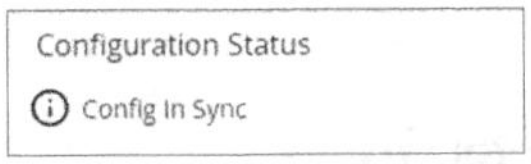

Figure 7-46: Configuration status results

Check Attempted and Running Configuration

1. Under this box, click the **View Details** link (see Figure 7-47). A new window will appear with configuration options.

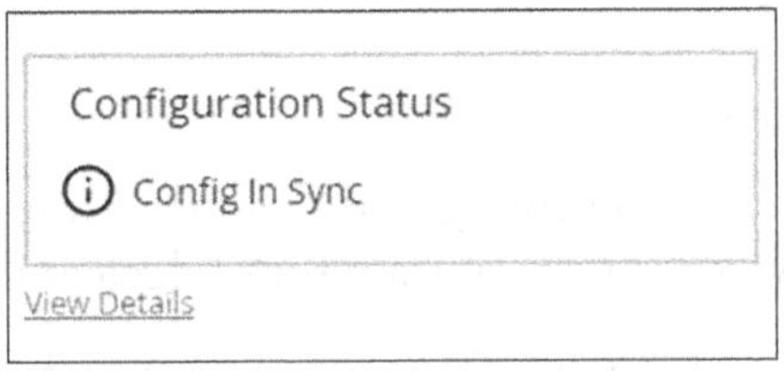

Figure 7-47: View Details link

2. To see the attempted configuration, click the **Attempted Configuration** link.
   - **Question**: Scroll through the attempted configuration. What is the spanning tree priority command you see?
   - **Answer**: The attempted configuration shows the command 'spanning-tree priority 0'.
   - **Question**: Do you see the %if .. conditional logic that was present in the template that you have uploaded?
   - **Answer**: No, this is the final (the attempted) configuration that Aruba Central has generated based on the template that you have uploaded, combined with variables (such as the host-name) and the conditional logic. This is a good place to detect any command issues with the template for a given device.
3. To see the actual device running configuration, click the **Device Running Configuration** link.
4. Click **Close** to close the configuration window.
5. Use PC-MGMT and open an SSH connection to **sw-agg1**.

6. Check the **show interface brief** output.

```
sw-agg1# show interface brief
```

- **Question**: Do you see an interface description for interface 1/1/1 and 1/1/2?
- **Answer**: Yes, the template has pushed the updated configuration; this includes interface descriptions.

Move sw-agg2 to the campus-sw-agg-tpl Group

After the successful migration of sw-agg1, you will now move sw-agg2 to the *campus-sw-agg* group.

You have learned in the previous section how to move a device to a new group. Try to move the device sw-agg2 to the group campus-sw-agg-tpl yourself. Refer to the previous section if you are unsure about the steps.

In the next section you will verify the configuration sync.

Verify sw-agg2 Configuration Sync

Previously you have used the device level configuration audit. This time you will use the group-level configuration audit. This is convenient to monitor a list of devices in a single operation.

1. In Aruba Central, navigate to: Context: **Groups / campus-sw-agg-tpl** > Navigation: **Devices>** Top: **Switches** > Right-top: **Config.**
2. Click **Configuration Audit**.
3. In the middle of the page, a **Configuration Status** box is shown.
   - **Question**: What is the count for *Not In Sync* devices?
   - **Answer**: The count will be 0 when the configuration was successfully applied. It may take a few moments (less than a minute) to push the configuration. If the count would be 1, refresh the page until it is 0.
4. Use PC-MGMT to open an SSH connection to sw-agg2.
5. Check the show interface brief command output.

```
sw-agg1# show interface brief
```

- **Question**: Are the interface descriptions updated for ports 1/1/1 and 1/1/2?
- **Answer:** Yes, the interface descriptions have been updated.

Lab 7.02: Managing edge switches with Aruba Central

Now that you have migrated the aggregation switches, you will explore options to deploy access switches using zero touch provisioning (ZTP). When the access switches have an active connection, you will explore how Central UI and MultiEdit can be used to configure the access switches. The last step is to apply a prepared template configuration to the access switches. This ensures that they have a 'known-good' configuration; it makes them ready for the rest of the lab activities.

Objectives

- Understand the ZTP process.
- Assign devices to a UI-based configuration group.
- Use MultiEdit to configure switches.
- Apply a template to edge switches and verify the template deployment.

Task 1: Review edge switch cloud connection with ZTP

In the previous lab, you connected the aggregation switches to Aruba Central using a static IP address and a template configuration. In this task you will use the edge switches. You will review how the factory default access switches can be onboarded with ZTP. This is convenient when performing a deployment of new devices.

Objectives

- Understand the ZTP process.
- Understand the infrastructure requirements for ZTP.

Steps

ZTP Untagged Native VLAN

To support ZTP, the access switches must get a DHCP IP address. By default, AOS-CX switches have a DHCP client enabled on their SVI 1 (VLAN 1). This is untagged on all ports.

When connecting the factory default switch to the network aggregation switches, the aggregation switches must be provisioned with an untagged native VLAN on their ports connecting to the access switches. This does not need to be VLAN 1 on the aggregation side.

When you loaded the template to the aggregation switches, the native VLAN on the LAG to the access switches was set to VLAN 1. This VLAN is used as the ZTP device onboarding VLAN in the lab.

1. Use **PC-MGMT** to open an SSH connection to **sw-agg1**.
2. Review the current configuration of LAG1. This LAG connects to sw-edge1.

```
sw-edge1# show running-config interface lag1
interface lag 1 multi-chassis
     no shutdown
     description sw-edge1
     no routing
     vlan trunk native 1
     vlan trunk allowed 1,3-4,11-15
     lacp mode active
     lacp fallback
     exit
```

   - **Question**: What is the configured trunk native VLAN?
   - **Answer**: The trunk native VLAN is set to VLAN 1.

3. Review the SVI 1 configuration.

```
sw-agg1# show running-config interface vlan1
interface vlan1
    ip address 10.1.1.2/24
    active-gateway ip mac 12:01:00:00:01:00
    active-gateway ip 10.1.1.1
    ip ospf 1 area 0.0.0.0
    ip helper-address 10.254.1.21
    exit
```

   - **Question**: Is there an IP helper address configured?
   - **Answer**: Yes, the IP helper points to 10.254.1.21. Any device connecting on SVI1 could receive a DHCP address.

4. Now check if there are any IP devices active on VLAN 1. The DHCP server scope IP range is 10.1.1.50-10.1.1.150. Check the ARP table for VLAN1 entries.

```
sw-agg1# show arp | include vlan1

10.1.1.51     64:e8:81:dd:81:80 vlan1    lag1
reachable
10.1.1.52     20:4c:03:c6:09:50 vlan1    lag1
reachable
10.1.1.54     20:4c:03:b1:d1:9a vlan1    lag256
reachable
10.1.1.55     20:4c:03:b7:70:1a vlan1    lag256
reachable
```

- **Question**: Are there any IP DHCP devices active?
- **Answer**: Yes. There are four devices active.
- **Question**: Why four devices?
- **Answer**: The APs are connected to the factory default access switches. Since all ports on the access switches belong to VLAN 1 (in the factory default state), the APs are also getting an IP address from SVI 1.

5. Optional: You can check the DHCP client status and the DHCP options, such as default gateway and DNS Servers. Open the console of sw-edge1 and check the DHCP client of the switch.

```
6300# show ip dhcp

INTERFACE-NAME ADDRESS      DEFAULT_GATEWAY  DOMAIN_NAME          VRF
DNS-SERVERS
---------------------------------------------------------------------------
--------------------
vlan1      10.1.1.51/24  10.1.1.1           aruba-training.com
default   10.254.1.21
```

ZTP LACP Fallback

Typically access switches connect via an LACP LAG to the aggregation switches. When a factory default switch connects to an LACP LAG, the aggregation switch does not receive any LACP communication from the access switch and the ports are set to a *lacp-blocked* state.

To support ZTP with factory default switches, the aggregation switch should enable LACP fallback on the LAG. This is enabled on the ports when no LACP communication is received.

1. On **sw-agg1**, review the running configuration of LAG1.

```
sw-agg1# show running-config interface lag1
interface lag 1 multi-chassis
    no shutdown
    description sw-edge1
    no routing
    vlan trunk native 1
    vlan trunk allowed 1,3-4,11-15
    lacp mode active
    lacp fallback
    exit
```

- **Question**: What LACP commands do you see?
- **Answer**: The LACP commands are:

```
lacp mode active
lacp fallback
```

2. Review the LACP interface state.

```
sw-agg1# show lacp interfaces
State abbreviations :
A - Active          P - Passive      F - Aggregable I - Individual
S - Short-timeout L - Long-timeout N - InSync     O - OutofSync
C - Collecting    D - Distributing
X - State m/c expired                E - Default neighbor state
Actor details of all interfaces:
------------------------------------------------------------------
-----------
Intf         Aggr     Port   Port   State System-ID          System Aggr
Forwarding
             Name     Id     Pri                             Pri    Key
             State
------------------------------------------------------------------
------------
1/1/1       lag1(mc)  1     1      IE    02:01:00:00:01:00 65534  1 up
. . .


Partner details of all interfaces:
------------------------------------------------------------------
-----------
Intf        Aggr      Port Port State  System-ID           System
Aggr
            Name      Id   Pri                              Pri      Key
------------------------------------------------------------------
----------
1/1/1       lag1(mc) 0     0    IE     00:00:00:00:00:00  0        0
. . .
```

- **Question**: What is the lag1(mc) LACP state?
- **Answer**: Since the access switch is not sending any LACP communication, the state is IE (default).
- **Question**: What is the forwarding state of the lag1(mc)?
- **Answer**: The port is *up* instead of lacp-blocked. This was the result of the LACP fallback command.

Site Assignment

Assign both access switches to the site *site-campus-main* and label *sw-edge*.

1. In Aruba Central, navigate to Context: **Global**> Navigation: **Organization**> Top: **Network Structure**> **Sites.**
2. Select both access switches 6300; drag them to the site **site-campus-main**.

Label Assignment

1. In Aruba Central, navigate to Context: **Global**> Navigation: **Organization**> Top: **Network Structure**> **Labels**.
2. Select both access switches 6300; drag them to the label **sw-edge**.

Task 2: Configure access switches with Central UI group

In this task you will use an Aruba Central UI group to configure the access switches. When compare your previously built configurations for each access switch, you notice many similar configuration items. These can be considered *common* configuration settings:

- Administrator password
- Global VLANs
- Every switch will have an uplink LAG 256 with LACP.
- DNS/NTP/time zone configuration
- Default gateway IP address

There are also several *unique* configuration settings for each switch:

- Hostname
- The uplink ports that are assigned to the uplink LAG. This may be the same, but it could be different if you have some 24-port and some 48-port switches, or different combinations in a VSF stack.
- Switch IP address
- Access port configurations

You can apply common configuration settings at the group level. These settings are automatically inherited by every switch assigned to the group. The unique per-switch settings can then be applied to each individual switch.

Conflicts

Typically, the set of configuration items at the group and device level will be different, so there should not be a conflict between the 2 levels. If an item is configured at both group and device level, the device level configuration is applied.

Objectives

- Use an Aruba Central UI group.
- Apply settings at group and device level.
- Verify the configuration synchronization.

Steps

Create Switch UI Group and Apply Base configuration

1. In Aruba Central, navigate to: Context: **Global**> Navigation: **Organization**> Top: **Network Structure**> **Groups.**
2. Click the + icon to create a new group.
3. For the name, enter **campus-sw-edge-ui**.
4. Select **Switches;** do not select the template option.
5. Click **Next**.
6. Select **AOS-CX only** and click **Add**.
7. Verify the new group is listed under Groups.

Apply Basic Group Configuration

1. In Aruba Central, navigate to Context: **Groups / campus-sw-edge-ui** > Navigation: **Devices**> Top: **Switches** > Right top: **config**. For a new group, a popup is displayed with an administrator password prompt.
2. Set the password to **Aruba123!** and click **Save**.

Configure System Properties

1. Click the **System** > **Properties** link.
2. Set the time zone to **Detroit (UTC -05:00)**.
3. Set DNS servers to **10.254.1.21.**
4. Set NTP servers to **10.254.1.21.**

Note

You just configured the admin password with the popup. If required, the group level admin password can be changed in this screen.

5. Click **Save**.

Configure the Uplink LAG

1. Click **Interfaces > Ports & Link Aggregations**.
2. At the right-top, click the **+** sign to add a port.

Note

You may need to scroll to the right to see the + icon.

3. On the Add LAG screen, enter these values:
   - Name: **lag256**
   - Do *not* configure port members. This will be done at the device level.
   - VLAN Mode: **trunk**
   - Native VLAN: **1**
   - Allowed VLANs: **all**
   - Admin Up: **Yes** (checked)
   - Aggregation Mode: **LACP active**
4. Click **Add** to save the LAG.
5. Click the **Back** button (left arrow) to return to the main menu (see Figure 7-48).

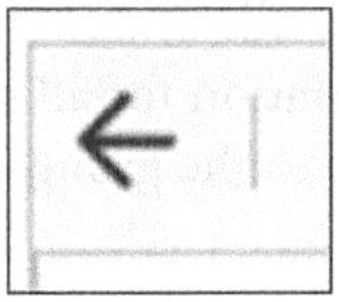

Figure 7-48: Back button

Configure the Global VLANs.

1. Click **Bridging > VLANs.**
2. At the right-top, click the + sign to add a VLAN.
3. In the Add VLAN screen enter these values:
   - ID: **3**
   - Name: **v3-mgmt**
   - Leave admin Up enabled
4. Click **Add**.
5. Add another VLAN with these values:
   - ID: **11**
   - Name: **v11-employee**
6. Click **Add**.
7. Verify the new VLANs 3 and 11 are displayed in the VLANs list.
8. Click the **Back** button (left arrow) to return to the main configuration menu.

Configure Default Gateway

1. Click **Routing > Static Routing**.
2. Click + icon to add a new entry:
   - Destination: **0.0.0.0/0**
   - Next Hop: **10.1.3.1**
3. Click **Save**.
4. Click the **Back** button (left arrow) to return to the main configuration menu.

You have now completed the basic configuration for all switches that will be assigned to this group. Now you are ready to move the first switch to the group.

Move Two Access Switches to the Group

In this section you will move the two access switches to the new group. When you move the devices, the group level settings (as defined in the previous steps) are applied to both switches.

1. In Aruba Central, navigate to Context: **Global>** Navigation: **Organization>** Top: **Network Structure> Groups.**
2. Expand **Unprovisioned Devices.**
3. Select *both* Aruba CX switches (multi-select is automatically enabled in this screen).
4. Click the **Move Devices** button (see Figure 7-49).

Figure 7-49: Move devices button

5. In the destination group field, enter or select **campus-sw-edge-ui**, then click **Move.**

Note
Do *not* check the 'Retain CX-switch configuration'. The switches are factory default at this point in the lab. This option can be used to import existing switch configurations (Brownfield deployment) or when a switch is moved from a template to a UI Group.

Configure the Per-Switch Settings for sw-edge1

In this section you will configure the **per-device** settings, such as the hostname, LAG port members, and a static IP address for the switch.

6. In Aruba Central, navigate to Context: **Global>** Navigation: **Devices>** Top: **Switches** > Right top: **List.**
7. Use the Serial column (scroll to the right) to identify your sw-edge1 (see Figure 7-50).

Device Name		Config Status	Last Seen	Usage	IP Address	Group	Serial/Stack ID
6300	: PoE ...	in sync	-	17 kbps	10.1.3.4	campus-sw-edge-ui	SG12KN105X

Figure 7-50: Verify the Serial/Stack ID column

8. Click the **6300** device name at the start of the row. This takes you to the device context.
9. In the Navigation menu click **Device** to access the device level configuration.

Configure System Properties

1. Click the **System** > **Properties** link.
2. Set the name to **sw-edge1**.
3. Verify that time zone, DNS, and NTP were received from the group level configuration.
4. Click **Save**.

Configure the Uplink LAG

1. Click **Interfaces** > **Ports & Link Aggregations**. You see the interface list, including the lag256.
2. At the right end of the lag256 line, click the **pencil** icon to edit the lag256.
3. In the Edit lag256 window, select **Port Members**. Check ports **1/1/27 and 1/1/28**.
4. Click **Save**.
5. Click the **Back** button to return to the main configuration screen.

Uplink Assignment (for Monitoring purpose only)

Aruba Central can provide dedicated statistics monitoring for the uplink ports of a device. In these steps you will set the uplink LAG member ports to 1/1/27 and 1/1/28.

1. In Aruba Central, navigate to: Context: **Global**> Navigation: **Devices**> Top: **Switches** > Right top: **List.**
2. Move your mouse over the line with sw-edge1. AT the right a link called UPLINKS will appear. Click the link **UPLINKS**.
3. In the *Assigned Uplink Ports* dropdown, select the ports **1/1/27 and 1/1/28**. These are the member ports of the uplink LAG.
4. Click **Assign**.

You have now completed the sw-edge1 onboarding configuration after ZTP was used for the initial setup. Sw-edge1 is now connected with an LACP LAG with 2 member ports; the LAG is configured as a VLAN trunk and it supports VLANs 3 and 11.

Practice with sw-edge2

You may practice the per-device configuration for sw-edge2.

1. Configure these values at the device level:
   - Hostname: **sw-edge2**
   - LAG 256 ports: **1/1/27, 1/1/28**

2. Configure the Aruba Central uplink ports with LAG256. Use the previous section if you are unsure about the steps.

Configure PC Access Ports

Now that both access switches have been onboarded successfully into a UI group, you can use the UI to make configuration changes, such as the VLAN access port settings. In this section you will configure the sw-edge1 port for PC1 (1/1/1) to become an access port in VLAN 11.

1. In Aruba Central, navigate to: Context: **Global**> Navigation: **Devices**> Top: **Switches** > Right top: **List.**
2. Click the **sw-edge1**.
3. In the Navigation menu, click **Device.**
4. Click **Interfaces** > **Ports & Link Aggregations**. You will see the interface list.
5. Click port **1/1/1;** use the **pencil** icon at the right-hand side to edit the record.

Note
You may need to scroll to the right.

---

6. Apply these settings to the port 1/1/1:
   - Description : **pc1**
   - VLAN Mode : **Access**
   - Access VLAN : **11**
7. Click **Save.**

Verify the Configuration.

1. In the Navigation menu, click **LAN.**
2. In the Ports list, port 1/1/1 should now be listed with the name pc1 and Native VLAN 11 (see Figure 7-51).

Port	Name	STATUS	LAG	MAC Address	VLAN	Native VLAN
1/1/1	pc1	Up		8c:85:c1:65:0e:e7		11

Figure 7-51: Verify the port configuration

Note

It may take several minutes for Aruba Central to discover the ports on the switch. If you don't see the ports yet, you may continue the lab and check the ports at a later time.

3. On PC1, bounce the Lab NIC (disable/enabled) and verify the PC has received an IP address in VLAN 11.
4. On PC1, open a command prompt (**cmd**) and check the IP address configuration using **ipconfig**.

```
C:> ipconfig

Windows IP Configuration

...
Ethernet adapter Lab NIC - 6300-A Port 1:

Connection-specific DNS Suffix  . : aruba-training.com
IPv4 Address. . . . . . . . . . .  : 10.1.11.50
Subnet Mask . . . . . . . . . . .  : 255.255.255.0
Default Gateway . . . . . . . . .  : 10.1.11.1
```

Task 3: Configure access switches with Central UI group MultiEdit

In this task you will use MultiEdit to make a configuration change to the access switches. While the UI configuration tiles are convenient for common features, not all configurations are available with the tiles.

Aruba Central provides a CLI-like interface to manage the AOS-CX switch: MultiEdit, which is part of a UI configuration group. In this task you use MultiEdit to make an example change to the devices.

Objectives

- Use Central MultiEdit to change the switch configuration.
- Verify the configuration synchronization status.

Steps

1. In Aruba Central, navigate to: Context: **Groups / campus-sw-edge-ui** > Navigation: **Devices**> Top: **Switches** > Right top: **Config.**
2. At the top of the screen, the MultiEdit slider is displayed (see Figure 7-52).

Figure 7-52: MultiEdit slider

3. Move the slider to the *right* to enable MultiEdit.

View CLI-Like Switch Configuration: Single Switch

1. In the list of switches, select the **sw-edge1**.
2. At the right-bottom, click **VIEW CONFIG** in the option screen (see Figure 7-53).

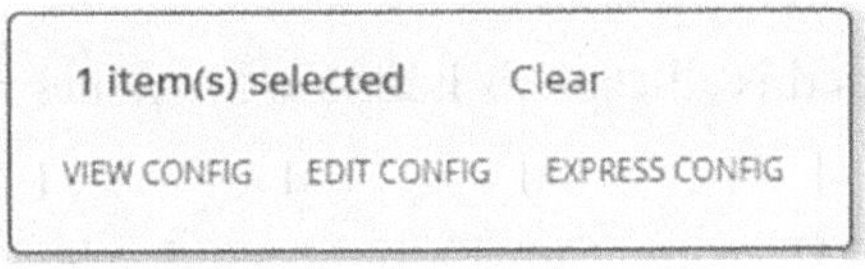

Figure 7-53: Option screen

3. After a few moments, you will see the current switch configuration. In this screen you could easily select the text-based configuration and copy the text to save or compare the configuration.
   - **Question**: What is the description and VLAN set on interface 1/1/1?
   - **Answer**: Interface 1/1/1 was previously assigned description *pc1* and is an access port in VLAN 11 (see Figure 7-54).

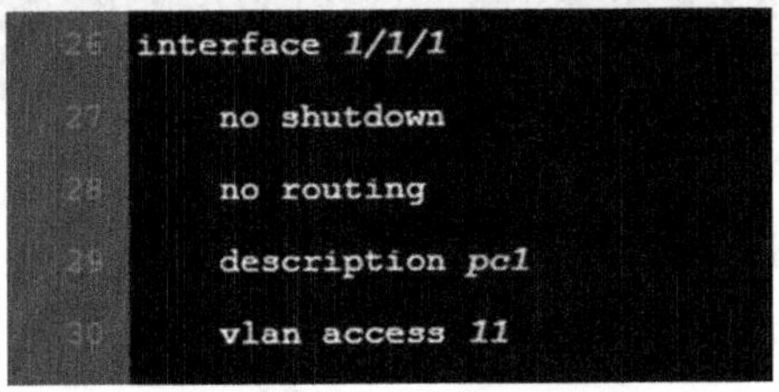

Figure 7-54: Verify port 1/1/1's configuration

4. Click the **Back** button to return to the MultiEdit switch list.

View CLI-Like Switch Configuration: Multiple Switches

When you need to compare two or more switch configurations, the MultiEdit interface can be used to quickly detect what values or settings have been applied to what switch.

1. Select *both* **sw-edge1 and sw-edge2** by clicking on the first, then the second switch.

> **Note**
> To unselect, click the device again.

2. At the right-bottom, notice the text *2 items* selected, since you have two switches selected.
3. Click **VIEW CONFIG** in the option screen.
4. In the configuration screen, notice the first line:

```
hostname HOSTNAME
```

- **Question**: Why is the HOSTNAME in all-capitals?
- **Answer**: MultiEdit uses this all-capitals notation for values that are different on the selected switches.

5. Right-click the **HOSTNAME** value. On the right-hand side, the **View Parameter** window is displayed.

- **Question**: What is the value for hostname set on sw-edge1 and sw-edge2?
- **Answer**: sw-edge1 and sw-edge2.

6. Click the **X** to close the View Parameter window.
7. Scroll down to interface 1/1/1 (see Figure 7-55).

Figure 7-55: Re-verify port 1/1/1's configuration

- **Question**: How can you see that the line **description pc1** only exists on one switch?
- **Answer**: At the end of the line, sw-edge1 is shown. This indicates that the line only exists on this switch. When there is no text at the end of the line, the line exists on all selected switches (see Figure 7-56).

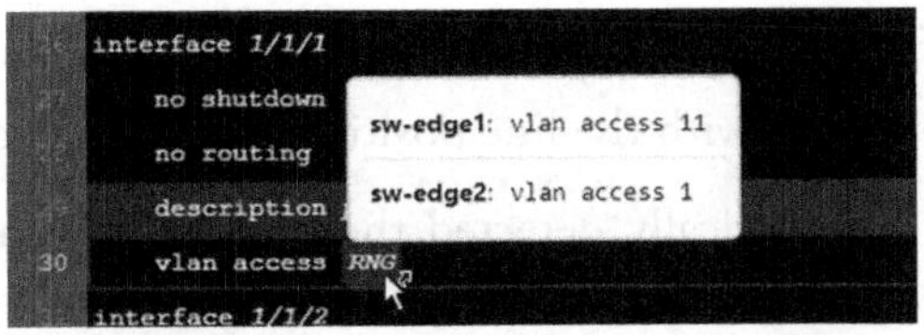

Figure 7-56: Viewing multiple switch settings

- **Question**: What do you see when you right-click **RNG** in the line **vlan access RNG**? (You may also hoover with the mouse over the value RNG.)
- **Answer**: sw-edge1 has port 1/1/1 assigned to VLAN 11, while on sw-edge2 it is still assigned to VLAN 1.

8. Click the **Back** button to return to the MultiEdit switch list.

Change Switch Configuration: Multiple Switches

You can also use the MultiEdit interface to make configuration changes. This can be convenient to push changes to multiple devices. With the parameter concept, it is still possible to push per-device settings as well in the MultiEdit interface.

In this section you will create the AP management VLAN 4 and assign port 1/1/2 on both sw-edge1 and sw-edge2 to this VLAN. This is a change that should be made on both switches. Next you will set the port description on sw-edge1 to ap1, and on sw-edge2 to ap2.

1. Select *both* **sw-edge1** and **sw-edge2** by clicking on the first, then the second switch.
2. Click **EDIT CONFIG** in the option screen.

Create VLAN 4

1. Scroll down to the line with 'vlan 1'. Put the cursor after 'vlan 1' and press **<Enter>**.

Note

This will put you on a new, blank line. Pay attention that the cursor is now left-justified—at the start of the line. This represents that you are currently at the global configuration context.

Note

MultiEdit is a CLI-like interface, but it is not a traditional CLI like the switch CLI. Since you see the entire configuration, you will need to *assist* the CLI by entering a sub-context with a preceding space if you need to enter a sub-context mode.

---

2. Type **vlan 4** and press **<Enter>**.
   - **Question**: What happened with the line position and the cursor?
   - **Answer**: MultiEdit automatically detected the new VLAN should have been placed after VLAN 3 and moved the line. The cursor is back at the global configuration level.
3. Press the <**Space Bar**>: this is the *hint* for MultiEdit that you want to *enter* the VLAN context.
4. Type **name v4-ap** and press **<Enter>** (see Figure 7-57).

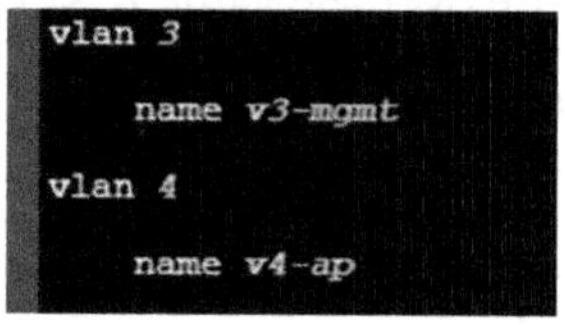

Figure 7-57: Verify the VLAN setting

   - **Question**: What happened with the cursor?
   - **Answer**: MultiEdit detected that the cursor was inside the VLAN context; it kept the cursor at this context level.

Create VLANs 12-15

VLANs 11-15 are used to support the lab's bridged WLANs. You have previously created the VLAN 11 in the UI. You will now create the VLANs 12-15 in the MultiEdit interface.

1. Move the cursor to the start of the line (global configuration) and configure VLANs 12-15, press **<Enter>**.

```
vlan 12-15
```

Assign Port 1/1/2 to VLAN 4

Now you will assign port 1/1/2 as a VLAN trunk port with native VLAN 4 and assign a unique description on each switch.

2. Scroll down to interface 1/1/2.
3. Under interface 1/1/2, change the line *vlan access 1* to **vlan trunk native 4** and press **<Enter>**. VLAN 4 is the AP management VLAN. When APs boot, they send an untagged DHCP request, which is handled by VLAN 4 on the switch. Since the cursor was under the interface context, it will now remain inside this context.

```
vlan trunk native 4
```

4. Configure the VLAN trunk allowed list. A VLAN trunk is required for bridged WLANs. The AP will support several WLANs (employee, guest, PSK), and it can tag the wireless traffic with a VLAN tag. These VLAN tags must be accepted (allowed) by the switch. In the lab, the bridged WLANs will use VLAN IDs 11-15.

```
vlan trunk allowed 4,11-15
```

Important

Pay attention for the **vlan trunk allowed** command in MultiEdit. The configured list will be the actual, final allowed VLAN list. If you entered **vlan trunk allowed 11-15**, you would not just add those VLANs, but you would also remove VLAN 4 from the list.

This is *different* from the switch CLI, where the command **vlan trunk allowed 11-15** would simply add those VLANs to the allowed list, but it would not affect the other allowed VLANs on that trunk port.

5. Enter **description ap** on the next line and press **<Enter>**.

Note

You can use command completion with the **<Tab>** key in MultiEdit.

6. Right-click the **ap** text. This reveals the Modify Parameter window at the right. Apply these values (see Figure 7-58):
   - For sw-edge1: **ap1**
   - For sw-edge2: **ap2**

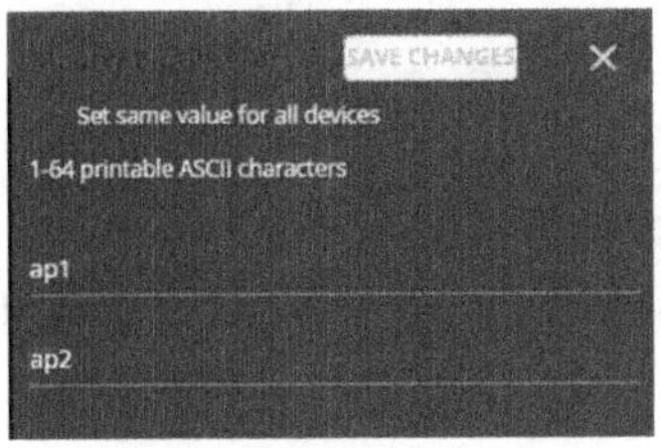

Figure 7-58: Changing the interface descriptions

7. Click **Save Changes** to close the Modify Parameters window.
8. Verify in the MultiEdit CLI that port 1/1/2 now has **description DSC** (see Figure 7-59). *DSC* indicates there are different values for the selected switches.

```
interface 1/1/2
    no shutdown
    no routing
    description DSC
    vlan trunk native 4
    vlan trunk allowed 4,11-15
```

Figure 7-59: Verifying the interface description

9. You are now ready to push the configuration to the two switches. Scroll down and click the **Save** button. Central returns to the switch device list.

10. In the list of switches, check the *Config Status* column, which dynamically updates as the configuration is pushed to the switches (see Figure 7-60).

Devices (2)

Name	Firmware Version	Config Modified	Status	Config Status
sw-edge1	10.09.1040	Oct 20, 2022, 15:04:30	• Online	Not in sync (Pushing ...
sw-edge2	10.09.1040	Oct 20, 2022, 15:02:38	• Online	Not in sync (Pushing ...

Figure 7-60: Pushing the configuration to the switches

11. Wait for the *Config Status* to be *Sync* (see Figure 7-61). This indicates the configuration on the device is synchronized.

Devices (2)

Name	Firmware Version	Config Modified	Status	Config Status
sw-edge1	10.09.1040	Oct 20, 2022, 15:18:05	• Online	Sync
sw-edge2	10.09.1040	Oct 20, 2022, 15:18:04	• Online	Sync

Figure 7-61: Central has synched the configuration to the switches

12. You have now configured two switches using MultiEdit.

Note

MultiEdit changes the *device level* configuration. These changes will not be shown at the group-level configuration, but they will appear in the device-level configuration tiles in the UI. (For features that are supported by the UI tiles).

If you want to switch back to using the UI tiles, at the top of the screen, set the slider for MultiEdit to be disabled.

Task 4: Configure access switches with Central template group

In this task you will configure a template group for the edge switches. The template has been prepared by your senior colleague; you should be able to configure the group and apply the template to the group.

Objectives

- Configure edge switches with a template group.

Steps

1. Use the lab dashboard to open a console session to sw-edge1. Login using **admin** and a blank password.
2. Take note of the *serial number*. (Example output: your output may be different.)

```
6300# show system
Hostname                        : 6300
System Description              : FL.10.09.1040
System Contact                  :
System Location                 :
Vendor                          : Aruba

Product Name                    : JL666A 6300F 24G CL4 PoE 4SFP56
Sw
Chassis Serial Nbr              : SG11KN501V
Base MAC Address                : 8c85c1-4920c0
ArubaOS-CX Version              : FL.10.09.1040

Time Zone                       : UTC

Up Time                         : 1 hour, 13 minutes
CPU Util (%)                    : 18
Memory Usage (%)                : 21
```

   a. Take note of the serial number: _______________.

   b. Take note of the Product Name part number (JL666A, JL662A or similar): _______________.

3. In Aruba Central, navigate to: Context: **Global** > Navigation: **Organization** > Top: **Network Structure** > Tile: **Groups.**
4. Create new Group with name **campus-sw-edge-tpl** template switches.
5. For Switches in group, select **AOS-CX only** and click **Add**.
6. Under **All connected devices**, select both edge 6300 switches and move them to the new group **campus-sw-edge-tpl**.

Assign Per-Device hostname Variable

1. In Aruba Central, navigate to: Context: **Groups / campus-sw-edge-tpl** > Navigation: **Devices>** Top: **Switches** > Right top: **Config.**

2. Click **Variables.** Each device that has been assigned to this group (in this example two access switches) is listed with three variables. The Device Serial Number column contains the serial number. See Figure 7-62.

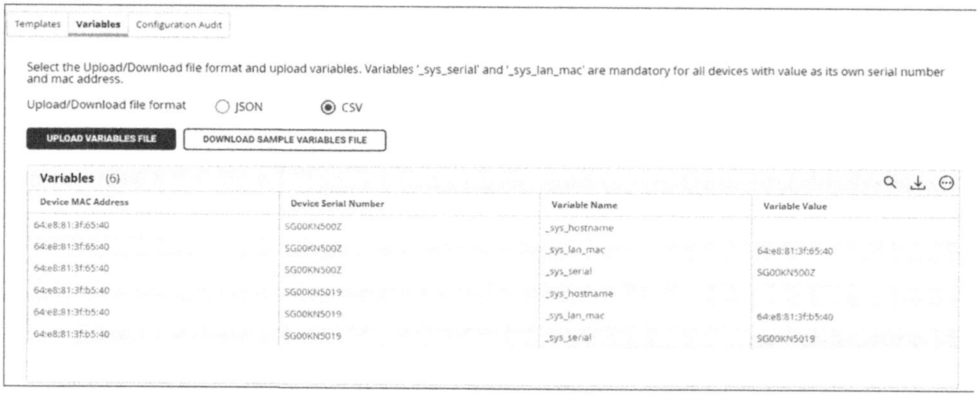

Figure 7-62: Variable screen

3. Hover with the mouse over the line that contains your sw-edge1 serial and the variable _sys_hostname. At the end of the line, a pencil icon will appear. Click the **pencil** icon to edit the value (see Figure 7-63).

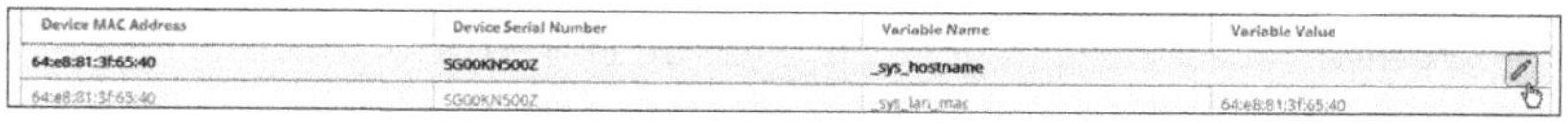

Figure 7-63: Viewing the variables

4. In the New Value field, enter **sw-edge1** (see Figure 7-64).

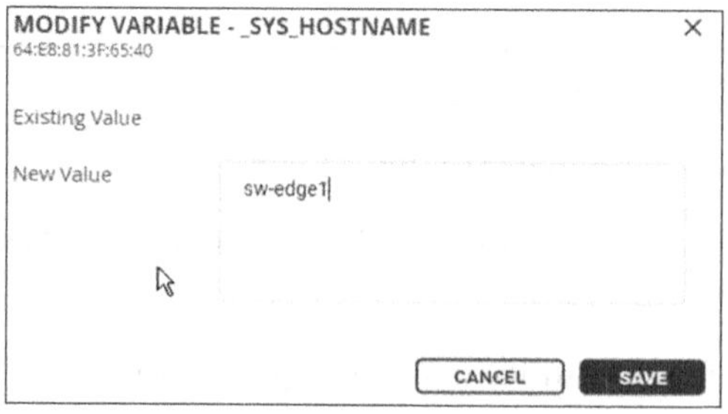

Figure 7-64: Editing a variable

Important

The access switch template contains conditions that check on the correct hostname. Please double-check the hostname is *exactly* **sw-edge1** (use dash ["-"] and not the underscore ["_"]).

5. Click **Save** to submit the value.
6. Change the other serial _sys_hostname value to **sw-edge2** (and double-check the name). It takes a few seconds to submit and process the values.
7. Refresh your browser to refresh the variable list.
8. Verify the resulting variable list shows both hostnames **sw-edge1** and **sw-edge2** (see Figure 7-65).

Variables

Device MAC Address	Device Serial Number	Variable Name	Variable Value
64:e8:81:3f:65:40	SG00KN500Z	_sys_hostname	sw-edge1
64:e8:81:3f:65:40	SG00KN500Z	_sys_lan_mac	64:e8:81:3f:65:40
64:e8:81:3f:65:40	SG00KN500Z	_sys_serial	SG00KN500Z
64:e8:81:3f:b5:40	SG00KN5019	_sys_hostname	sw-edge2
64:e8:81:3f:b5:40	SG00KN5019	_sys_lan_mac	64:e8:81:3f:b5:40
64:e8:81:3f:b5:40	SG00KN5019	_sys_serial	SG00KN5019

Figure 7-65: Verifying the variable list

Template Configuration

Now you will upload the prepared template for the edge switches. Based on the hostname, each switch will receive the correct configuration from the template.

1. Click **Templates**.
2. At the top-right, click the + icon to add a new template.
3. For Template Name use **sw-edge**.
4. The Device Type should be Aruba CX, based on the group configured switch type.
5. Click **Next**.
6. On the PC-MGMT, in the ACAF Student Files, open **lab07.02-sw-edge-tpl-JLxxxx.txt** for your switch model. For example, if your switch model was *JL666a*, you should open lab07.02-sw-edge-tpl-*jl666a*.txt.

Note

Refer to the **show system** output in step 2 of this task if you are unsure about the product number.

7. Copy all the contents. Paste the contents of the file into the template.
8. Click **Save**.

Verification

1. Click **Configuration Audit**. The Configuration status will show two devices *Not In Sync*.
2. Since the template changes the switch management IP address, the switches will need to re-establish their connection to Aruba Central and check in with the new IP address. This will take one to two minutes to complete.
3. After about two minutes (when the switches have reconnected to Aruba Central), the status should show *0* devices Not in Sync (see Figure 7-66).

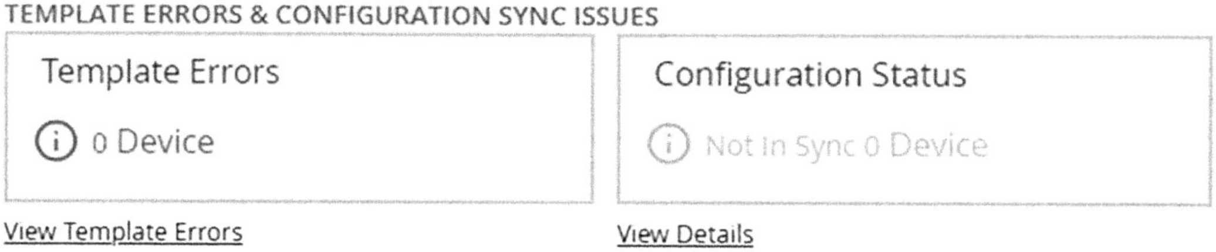

Figure 7-66: Verifying the configuration status

Learning check

Discuss these questions.

1. Choose the statements below that accurately describe Aruba Central device onboarding. (Select three.)
   a. Zero touch provisioning (ZTP) is an onboarding method that relies on the cloud-based Aruba Activate service.
   b. Manually adding devices to Aruba Central is often considered the easiest method, because it does not rely on other services.
   c. You first onboard devices to HPE GreenLake, and then assign them to the Aruba Central app.
   d. Before a customer can configure and monitor devices in Aruba Central, those devices must be onboarded.
2. Your manager asks you about Aruba Central licensing. Which of the options below would help your manager understand this aspect of your deployment? (Select five.)
   a. Foundation keys are available for gateways, APs, and switches.
   b. Licenses are available in 2-year increments.
   c. You need one license per managed device.

d. You can purchase additional Advanced licenses to add functionality and features.

e. You must add subscription keys to Aruba Central, and then assign them to network devices.

f. The auto-subscribe feature allows you to assign licenses to all new devices automatically.

3. Which of the options below offer true comparisons of Aruba Central Groups vs Aruba Central Sites? (Select three.)

   a. Both can help ease monitor, filter, and report functions.

   b. Devices can only belong to one site, but can belong to multiple groups.

   c. Groups come in two types: a UI group and a template group.

   d. Only groups include a device persona, which represents the device's role in the network.

4. Match the task involved in group management with the order in which you should perform that task (Step 1, Step 2, and Step 3).

   ____

   a. Define the group persona.

   b. Create the group.

   c. Add devices to the group.

5. Labels help you filter devices for Aruba Central monitor, manage, and report purposes. You can assign multiple labels to a single device.

   a. True

   b. False

Summary

Congratulations! You have completed Chapter 7. You should now be able to:

- Perform device onboarding.
- Create Central Groups.
- Use the UI configuration mode.
- Use template configuration mode.
- Describe Central licensing.

8 Device Profiles and AP Onboarding

OBJECTIVES

Chapter 8 introduces you to AOS-CX device profiles and AP onboarding. To avail yourself of Aruba Central management features, you must, of course, onboard your Aruba networking devices. You'll also find device profiling to be a vital part of your Aruba switch operations.

After completing this chapter, you will be able to use device profiling to help simplify and secure the deployment. This includes both MAC and LLDP-based profiling. You will also be able to connect your APs to Aruba Central and perform the initial AP setup.

After completing this chapter, you should be able to:

✓ Use device profiling.

✓ Describe LLDP and MAC profiling.

✓ Connect an AP to Aruba Central.

✓ Perform initial AP setup.

Device profiling

Let's begin this chapter with a discussion of device profiling.

Profiling introduction

Profiling and endpoint classification allows the network infrastructure to adjust settings and security policies dynamically, based on a device's type/model (see Figure 8-1). For example, when an Aruba infrastructure device detects a wireless or wired VoIP phone, it can optimize the connection with the correct quality of service (QoS) settings for voice traffic. Aruba-based networks can also automatically adapt to various IoT device types.

Assigning roles based on identity might work if the company manages all the devices, placing them in correct groups. But it can be simpler and more reliable to let the Aruba solution *automatically* categorize devices and assign roles based on those categories.

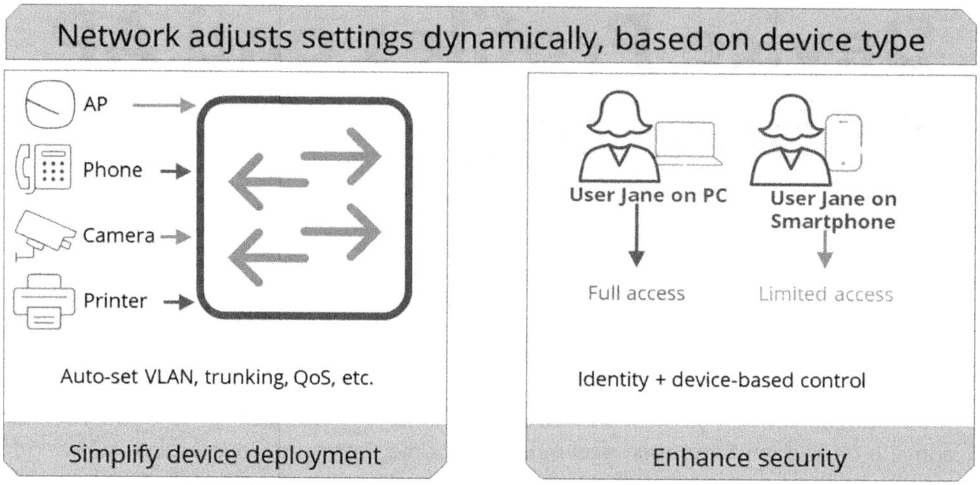

Figure 8-1: Device profiling introduction

And here's another use case for profiling: sometimes identity-based authentication does not provide the granular control that you need. This is because users often connect to the network using multiple devices. If you only use the user's *identity* to assign roles, the same user always gets the same rights, regardless of device type they use with their network connection (see the right-side of Figure 8-1).

But smartphones, for example, tend to be less secure against some exploits and employees tend to use them for less demanding work applications, like browsing the Internet or checking email. A more limited access profile makes sense for when the user connects using a smartphone versus their company laptop. Therefore, by combining identity-based and endpoint category-based criteria, a user gets normal corporate access from their laptop, but only Internet access from their smartphone.

Device provisioning

When you enable profiling for device provisioning, the network devices, like Aruba switches, APs, and gateways, can dynamically assign roles to the user devices. This section discusses the components of device provisioning:

- LLDP, CDP, and/or MAC address groups
- Roles
- Device profiles

Roles

A *role* is a collection or group of settings that can be assigned to a switch port: a VLAN identifier, PoE priority, QoS trust mode (DSCP and 802.1p prioritizations), security policy, bandwidth restrictions (rate limiting), custom captive (web) portal page, and etcetera (see Figure 8-2). Devices can now be plug-n-play with no specific port configuration required on an Aruba switch, for example: the switch, instead, learns about the device that is connecting and dynamically assigns the role to associated switch port.

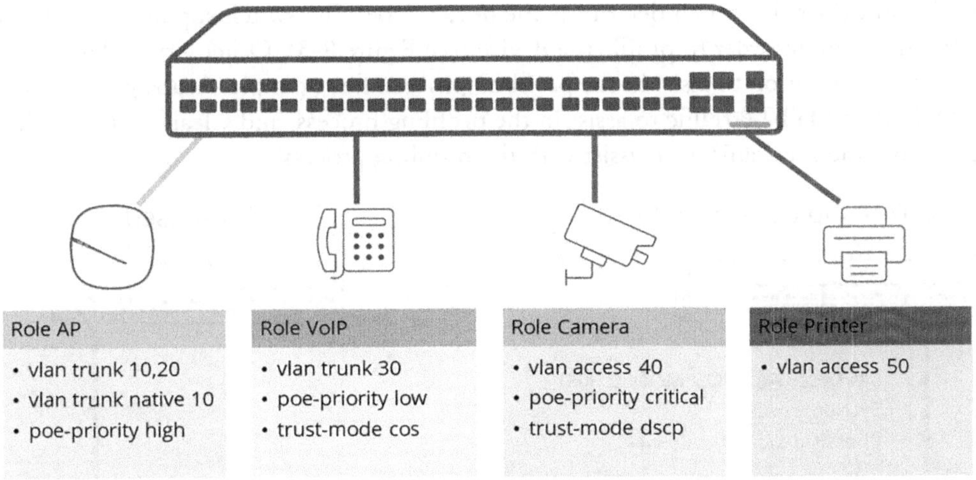

Figure 8-2: Aruba user roles

The benefits of assigning roles to users and/or devices are well known within the Aruba wireless world. Roles have been adapted from the Aruba wireless products to the Aruba switching portfolio. Just as with Aruba wireless, roles simplify your configuration. Roles group access and security policies in a single configuration that can be referenced by many devices and/or user types. Look at the role configuration example here:

```
switch(config)# port-access role AP
switch(config-pa-role)# ?
<omitted for clarity>
  description            Description for this Role.
  device-traffic-class   Configure device traffic class for this Role.
  gateway-zone           Configure gateway parameters for the Role.
  mtu                    Configure MTU for this Role.
  poe-priority           Configure POE priority for this Role.
  private-vlan           Configure Private-VLAN.
  reauth-period          Configure reauth period for this Role.
  session-timeout        Configure session timeout for this Role.
  trust-mode             Configure trust mode for this Role.
  vlan                   Configure VLAN mode for this Role.
```

Notice that you have many options for polices that will be associated to the connected user/device. The most common setting is the VLAN identifier that will be assigned to the user. You'll explore role configuration in the lab later in the chapter.

LLDP vs MAC-based profiling

Recall that roles enable predefined configurations to be applied to the user's access, based on the user's identity as well as their connected device type. But when you connect a device to an Aruba switch, for example, how does the switch determine the device type? The switch supports LLDP, CDP, and local MAC matching in order to profile the device (see Figure 8-3). Other Aruba devices might use other protocol information to assist in the profiling process. For example, Aruba gateways can examine user DHCP and HTTP traffic to assist in the profiling process, and ClearPass can look at over a dozen protocols and mechanisms to assist with the profiling process.

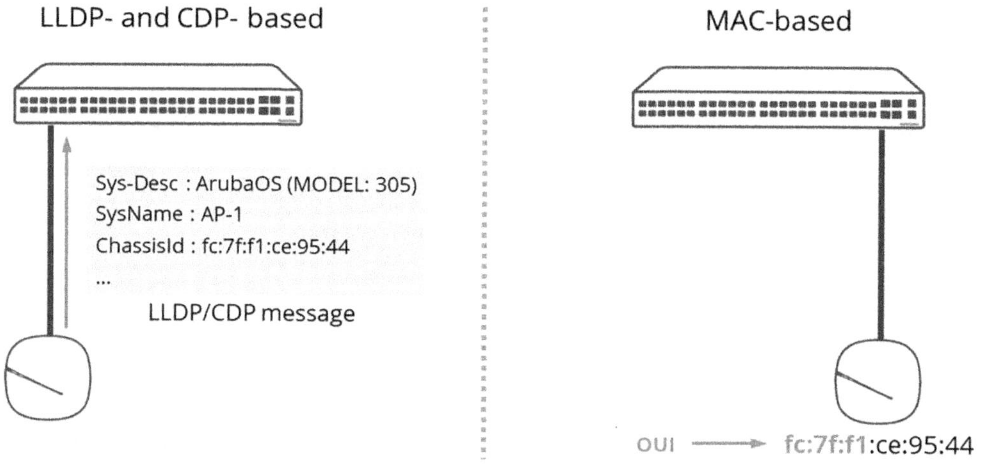

Figure 8-3: LLDP, CDP, and MAC-based profiling

LLDP or CDP messages tell the switch what type of device that it is (you were introduced to LLDP previously in Chapter 2). LLDP is an open-stand protocol while CDP (Cisco Discovery Protocol) is typically used by Cisco devices: both protocols share information, via layer 2, about the device's capabilities. LLDP and CDP messages contain information about the device product, model, software or firmware version, and many other items. The switch can also glean information by focusing on the connected device's MAC address: the first three bytes of every MAC address is the Organizationally Unique Identifier (OUI), which is the manufacturer of the device NIC.

Connected devices are identified using corresponding protocol packets. When the protocol information on the interface ages, the profile or role is revoked from the interface. When a device of a configured type is connected to an interface, the switch automatically applies the corresponding role.

If the switch enforces 802.1X authentication port security, Aruba APs support 802.1X in supplicant mode so they can function as 802.1X authentication clients just like PCs. The authentication server can also pass back the appropriate role that is to be applied. And if the authentication server is ClearPass, ClearPass has extensive profiling capabilities that can be leveraged in determining the role that it should pass to the switch, which, in turn, the switch will apply to the user device's access.

LLDP device groups

LLDP device groups are typically used when you are profiling capabilities, but are not using an authentication server, like ClearPass, to do the profiling and to dynamically pass back a respective user role. Figure 8-4 shows how to create a device group using LLDP. You give the group a name, and specify the group's matching criteria, which is used to determine group membership. LLDP device groups allow you to match on the criteria highlighted in Figure 8-4. Multiple TLVs can be matched or ignored simultaneously; use *seq* numbers to control the order of matching rules.

```
Switch(config)# port-access lldp-group ArubaAP-LLDP
Switch(config-lldp-group)# match
  sys-desc     Configure LLDP system description.
  sysname      Configure LLDP system name.
  vendor-oui   Configure LLDP vendor OUI.
Switch(config-lldp-group)# ignore sysname AP10
Switch(config-lldp-group)# match sys-desc ArubaOS
```

```
Switch(config-lldp-group)# show running-config current-context
port-access lldp-group ArubaAP-LLDP
     seq 10 ignore sysname AP10
     seq 20 match sys-desc ArubaOS
```

Figure 8-4: LLDP device groups

Note

Creating CDP device groups is very similar to creating LLDP device groups. It's configuration, though, is beyond the scope of this book.

MAC device groups

With MAC device groups, you can match or ignore specific MAC addresses as well as part of the MAC address, like the MAC OUI (leading 24 bits), and/or specify a range of MAC addresses. See Figure 8-5 as an example.

```
Switch(config)# mac-group ArubaAP-MAC
Switch(config-mac-group)# match
  mac        Configure MAC address
  mac-mask   Configure MAC mask
  mac-oui    Configure MAC OUI
Switch(config-mac-group)# match mac-oui f0:5c:19
Switch(config-mac-group)# match mac 20:4c:03:d4:3b:e3
```

```
Switch(config-mac-group)# show running-config current-context
mac-group ArubaAP-MAC
     seq 10 match mac-oui f0:5c:19
     seq 20 match mac 20:4c:03:d4:3b:e3
```

Figure 8-5: MAC device groups

Device profiles

You have enabled the switch to determine group membership. Now, what does the switch do with that information? You must configure a device profile to ensure that switch ports are properly configured for group members.

The device profile associates a device group and the user role that should be applied to devices that match. Recall that the role assigns the user's access policies, like the VLAN and/or security policy applied to the user's access (you'll learn more about user roles in Chapter 10, as it relates to wireless traffic).

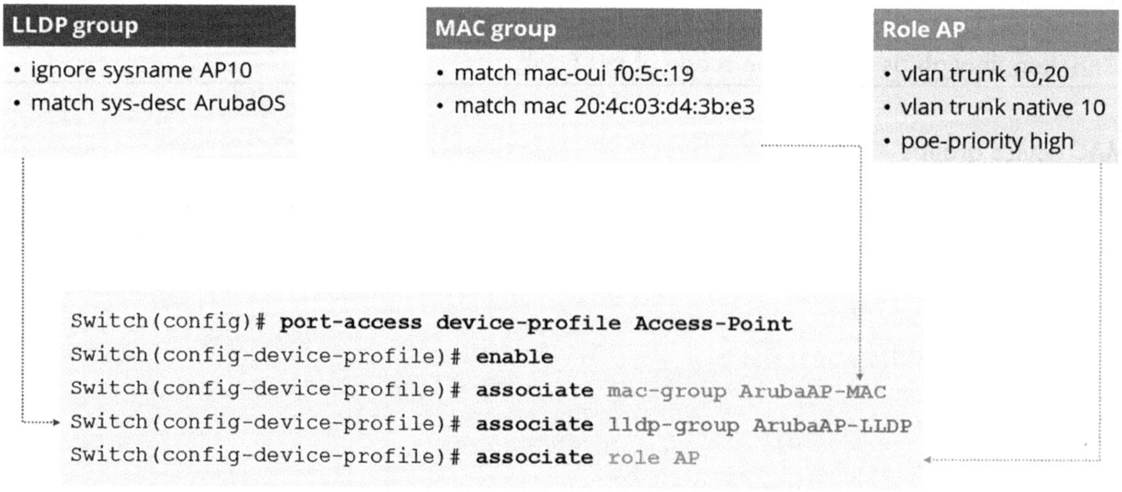

Figure 8-6: Device profiles

In Figure 8-6, a device profile named *Access-Point* has been created and is associated with the two groups you previously saw in Figures 8-4 and 8-5. The device profile is then associated with the role named *AP*. Therefore, the switch knows which role to assign to a port when a connected device matches a device group (in this case, an Aruba AP). If you configure a match-criteria that matches across multiple device groups, then the priority considered is LLDP, and then local MAC matching; that is, LLDP has precedence over local MAC matching.

Profiling verification

Use the **show port-access device-profile interface all** command to verify profiling. An example is shown in Figure 8-7.

```
Switch(config)# show port-access device-profile interface all
  Port 1/1/2, Neighbor-Mac  20:4c:03:d4:3b:e3
    Profile Name:              : Access-Point
    LLDP Group:                :
    CDP Group:                 :
    MAC Group:                 : ArubaAP-MAC
    Role:                      : AP
    State:                     : applied
    Failure Reason:            :
```

Figure 8-7: Profiling verification

Connecting APs to Aruba Central

Let's move on and discuss connecting Aruba APs to Aruba Central.

AOS 8 and AOS 10 comparison

Aruba AOS 10 simplifies access point deployment methods with just two AP options: AP and micro-branch AP (see Figure 8-8 for a comparison of AOS 8 and AOS 10 wireless operations and devices). AOS 10 does not include an IAP cluster or IAP virtual controller concept. Additionally, AOS 10 does not include a mobility conductor (formerly known as a mobility master) since all those features are now located within Aruba Central: basically, you get a simpler network architecture utilizing AOS 10.

Note

In AOS 10, Aruba has renamed controllers to gateways, and merged the firmware code for SD-Branch and mobility implementations.

AOS 8	AOS 10
IAPs, CAP, RAP	AP, microbranch AP
Mobility conductor	No mobility conductor
Mobility controllers	Gateways
SD-Branch gateways	SD-Branch and mobility
Local or cloud management	Cloud Management

Figure 8-8: AOS 8 and 10 comparison

AP and Aruba Central communications

Central uses a secure HTTPs connection and provides a strong mutual authentication mechanism, based on digital certificates, for all communication with Aruba devices in your network, including the APs (see Figure 8-9). These certificates ensure the highest level of protection. You will learn how to manage APs and groups during the lab activities.

Figure 8-9: AP and Aruba Central communications

AP system configuration

In Aruba Central, select the appropriate context in the left navigation pane. Then navigate to **Devices > Access Points**. Click the **System** tab to see essential system configurations: AP country code, time zone, preferred band, NTP server, and much more (see Figure 8-10).

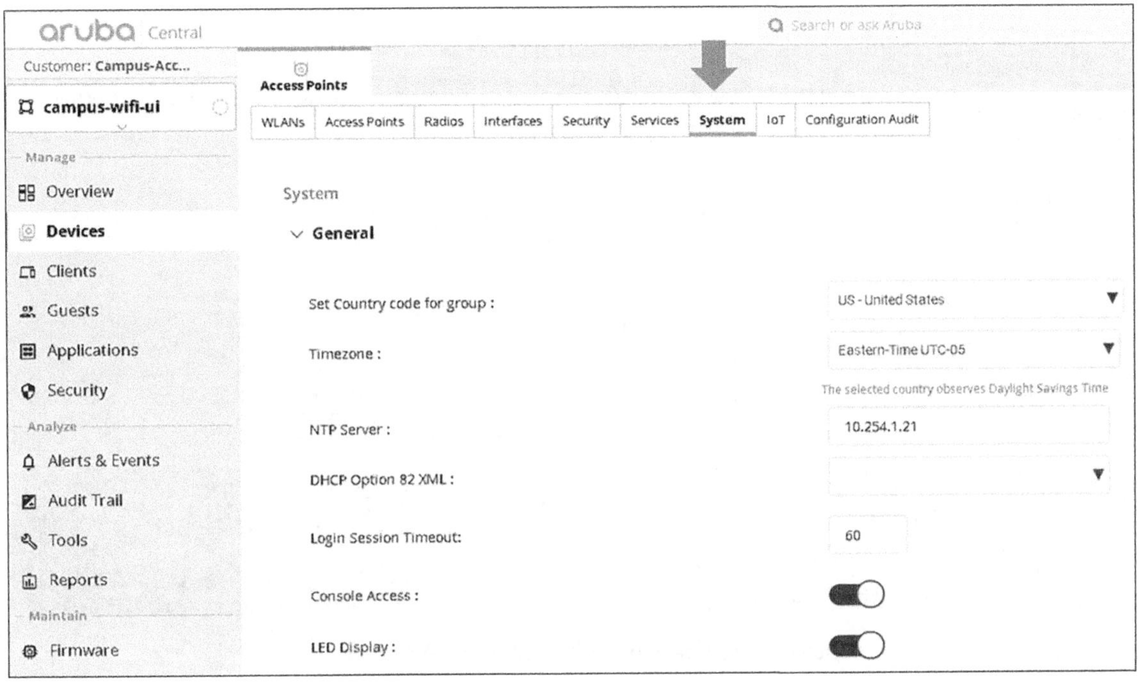

Figure 8-10: AP system configuration

Creating a floorplan

In Aruba Central, select the appropriate context in the left navigation pane. Then select **Overview** in the left navigation pane. Click the **Floor Plan** tab to plan sites, create and manage floorplans, and provision access points (see Figure 8-11). You'll need to import a picture of a floorplan, which can be something as simple as a GIF or BMP file, for example, or as extensive as an Auto-CAD file. In the lab you will see the wizard-based process that is used to create a floorplan.

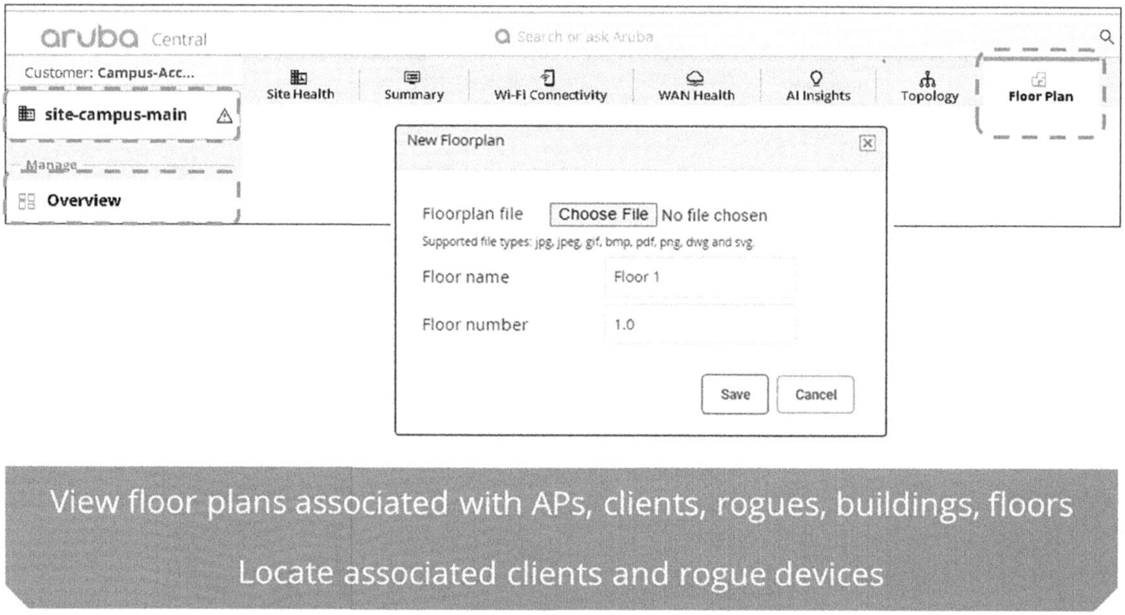

Figure 8-11: Creating a floorplan for AP placement

You can only access the floorplan dashboard from a *site* context. Here's what you can do as it relates to using floorplans:

- View floorplans associated with APs, associated clients, rogues, buildings, and floors.
- Accurately locate associated client devices using RF data from those devices.
- Accurately locate rogue devices (as classified by RAPIDS) using RF data from your devices.

Adding APs to a floorplan

The floorplan shows a real-time picture of the radio environment, as well as allowing you to plan the wireless coverage for new sites (referred to as a *greenfield* installation). To really understand your wireless network, you must know the location of wireless devices and users and the health of your RF environment. With integrated mapping and location data, there's a wealth of information at your fingertips when you use the floorplan feature.

The floorplan feature uses sophisticated RF fingerprinting to accurately display coverage patterns and calculate the location of every wireless device in range (see Figure 8-12). There is no need for dedicated RF sensors or a costly additional location appliance: Aruba Central gathers all required information from your existing APs. First, though, you must place your APs on the floorplan based on where they are physically installed. This can be done when you initially create a floorplan or after the fact.

Figure 8-12: Adding APs to a floorplan

Lab 8: Onboarding APs

In this lab, you will onboard APs in Aruba Central.

Overview

Now that you have completed the wired network rollout, you are ready to deploy the wireless network. The first phase of the rollout is to ensure:

- The APs are powered up.
- The APs are connected to the correct AP management VLAN.
- The respective switch ports are configured with a VLAN trunk port to support the VLANs for the bridged SSIDs.

Once the APs can reach Aruba Central, you need to prepare a configuration group and apply basic settings to the APs, such as the country code and the AP names. In the last section of this lab, you will prepare a floorplan in Aruba Central and position APs on the floorplan.

Objectives

- Verify AP deployment.
- Verify AP connection to Aruba Central.
- Assign the AP name and country code.
- Assign the AP site information.
- Configure a floorplan.

Task 1: Verify AP wired onboarding

The switch port that will support APs with bridged SSIDs needs to support the user VLANs for these SSIDs. This is achieved by configuring the port as a VLAN trunk port.

The AP will send out a DHCP request to obtain a management IP address. This DHCP request is sent untagged by default. This means that the DHCP request will be received on the switch port on the configured native untagged VLAN. The switch port trunk native untagged VLAN should be configured to the desired AP management VLAN. In this lab environment, VLAN 4 is the AP management VLAN.

You can configure switch ports manually or automatically. Many enterprise customers prefer to use a centralized AAA solution, such as Aruba ClearPass, to assign switch port configurations. In a smaller deployment, you can use device profiles, which you configure locally on the switch. With device profiles, the switch can recognize APs based on either LLDP, CDP, or MAC address information and assign the correct port configuration. When an AP is removed, the port configuration reverts to the port configuration as set in the running-configuration.

In a production deployment, the APs will typically be powered by the access switches. In this remote lab environment, the APs are powered in a centralized cage by an external power-over-Ethernet (PoE) device. This means you will not see any PoE activity on your access switch. It also means that a shut-down of the port on your lab switch will not be noticed by the AP, since it is connected to a transit (intermediate) switch, which provides the PoE power to the APs. For example, if the AP is rebooted, your access switch port will not lose power and move from a down to an up state (i.e., the AP is not directly connected to the switch shown in the lab topology in Lab 1). Please remember this when troubleshooting issues in the lab!

Objectives

- Verify AP connection on the switch.

Steps

PoE

Although our lab APs are not powered by your access switches, it is good idea to review the PoE commands.

1. Use PC-MGMT to open an SSH connection to the **sw-edge1**.
2. Review the overall PoE power.

```
sw-edge1# show power-over-ethernet

System Power Status for member 1

  PoE Power Status                : No redundancy
  Operational Power Status        : No redundancy
  Total Available Power           : 370.00 W
  Total Failover Pwr Avl          :   0.00 W
  Total Redundancy Power          :   0.00 W
  Total Power Drawn               :   0.00 W
  Total Power Reserved            :   0.00 W
  Total Remaining Power           : 370.00 W
  Trap Threshold                  :  80 %
  Trap Enabled                    : Yes
  Always-on PoE Enabled           : 1/1
  Quick PoE Enabled               : None


 Internal Power:
        Total Power
  PS    (Watts)        Status
  ----- ------------   ---------------------
  1/1   950            OK
```

- **Question**: What is the total watts power available on power supply PS1/1?
- **Answer**: PS 1/1 supports 370 watts for PoE.

3. Review the PoE interface status.

```
sw-edge1# show power-over-ethernet brief

   Member 1 Power Status
     Available: 370.00 W Reserved: 0.00 W Remaining: 370.00 W
     Always-on PoE Enabled:1/1
     Quick PoE Enabled:None

PoE     Pwr Power    Pre-std  Alloc    PSE Pwr PD Pwr  PoE Port
PD     Cls Type
Port    Ena Priority Detect   Act      Rsrvd   Draw    Status
Sign
-------- --- ------ -------  -------- ------- ------- ----------
1/1/1   Yes low      Off      usage    0.0 W   0.0 W   searching
N/A    N/A N/A
1/1/2   Yes high     Off      usage    0.0 W   0.0 W   searching
N/A    N/A N/A
1/1/3   Yes low      Off      usage    0.0 W   0.0 W   searching
N/A    N/A N/A
1/1/4   Yes low      Off      usage    0.0 W   0.0 W   searching
N/A    N/A N/A
...
```

Question: What does *searching* indicate?

Answer: There is currently no PoE device connected to the port: the switch is searching for a PoE-capable device on the port.

Device profiles

You can use device profiles to automatically assign the correct switch port configuration when an AP is connected to the associated port. Your senior colleague has included the device profile in the switch template configuration.

1. On sw-edge1, review the running-configuration of port 1/1/2, which is connected to AP1.

```
sw-edge1# show running-config interface 1/1/2
interface 1/1/2
     no shutdown
     description ap1
     no routing
     vlan access 1
     spanning-tree bpdu-guard
     no aaa authentication port-access allow-lldp-auth
     aaa authentication port-access client-limit 5
     aaa authentication port-access mac-auth
         enable
     exit
```

- **Question**: What is the configured VLAN mode on port 1/1/2?
- **Answer**: Port 1/1/2 is configured as access port in VLAN 1.

2. Review the operational VLAN status for port 1/1/2.

```
sw-edge1# show vlan port 1/1/2

-------------------------------------------------------
VLAN Name                   Mode             Mapping
-------------------------------------------------------
4    VLAN4                  native-untagged port-access
11   VLAN11                 trunk            port-access
12   VLAN12                 trunk            port-access
13   VLAN13                 trunk            port-access
14   VLAN14                 trunk            port-access
15   VLAN15                 trunk            port-access

Overridden VLAN list: 1
```

- **Question**: What do you notice?
- **Answer**: VLAN 4 is the native-untagged VLAN. Other VLANs are allowed as well on the trunk port.
- **Question**: Why is this different from the port configuration?
- **Answer**: A device profile applied a different configuration to the port since it discovered an AP MAC address. If the AP would disappear, the port would revert to its local port configuration.

3. Review the list of configured device profiles.

```
sw-edge1# show port-access device-profile

    Profile Name        : aruba_ap
    LLDP Groups         :
    CDP Groups          :
    MAC Groups          : aruba_ap
    Role                : dev_aruba_ap
    State               : Enabled
```

4. Review the applied list of device profiles.

```
sw-edge1# show port-access device-profile interface all
  Port 1/1/2, Neighbor-Mac 20:4c:03:c6:09:50
    Profile Name:             : aruba_ap
    LLDP Group:               :
    CDP Group:                :
    MAC Group:                : aruba_ap
    Role:                     : dev_aruba_ap
    State:                    : applied
    Failure Reason:           :
```

- **Question**: What is the configured MAC device group?
- **Answer**: *aruba_ap*
- **Question**: What is the applied role?
- **Answer**: *dev_aruba_ap*. This role contains all the settings that will be applied to the port.

5. Review the running configuration for the port-access section.

```
sw-edge1# show running-config port-access
port-access role aruba_ap
port-access role dev_aruba_ap
     auth-mode device-mode
     poe-priority high
     trust-mode dscp
     vlan trunk native 4
     vlan trunk allowed 4,11-15
port-access device-profile aruba_ap
     enable
     associate role dev_aruba_ap
     associate mac-group aruba_ap
aaa authentication port-access mac-auth
     enable
interface 1/1/2
     no aaa authentication port-access allow-lldp-auth
     aaa authentication port-access client-limit 5
     aaa authentication port-access mac-auth
         enable
```

6. Review the configured MAC device groups.

```
sw-edge1# show running-config mac-group
mac-group aruba_ap
     seq 10 match mac-oui 20:4c:03
     seq 20 match mac-oui 18:7a:3b
```

Verify the application of the device profile

In this section you will force an access port client *logoff* on interface 1/1/2. This allows you to see events related to the device profile when a matching device is detected.

1. On sw-edge1, review the list of active port access clients.

```
sw-edge1# show port-access clients

Port Access Clients

Status Codes: d device-mode, c client-mode, m multi-domain

-------------------------------------------------------------------
------------------------------------
Port       MAC-Address        Onboarding     Status    Role
Device Type
                              Method
-------------------------------------------------------------------
-------------------------------------
d 1/1/2    20:4c:03:c6:09:50 device-profile Success   dev_aruba_ap
```

2. Enable terminal-monitor to see live events in your session.

```
sw-edge1# terminal-monitor
Terminal-monitor is enabled successfully
```

Note

terminal-monitor is not supported on the console. If you received an error, make sure to use PC-MGMT and open an SSH connection to the sw-edge1 instead of a console connection.

3. Logoff the client on port 1/1/2. Since the AP almost continuously sends some traffic, it will be back online within a few seconds.

```
sw-edge1# port-access log-off client interface 1/1/2
```

Note

This command should be executed in the manager context (#), if you access the configuration mode, return to manager mode with the **end** command. If you do want to run manager mode commands from the configuration context, you can prepend the command with the keyword **do**. For example: **do port-access log-off client interface 1/1/2**.

4. Review the resulting event logs.

```
2022-07-29T10:48:24.346774+0000 port-accessd[4355] <INFO>
Event|10502|LOG_INFO|CDTR|1|Port 1/1/2 is blocked by port-access
2022-07-29T10:48:24.394887+0000 ops-switchd[686] <INFO>
Event|2107|LOG_INFO|CDTR|1|The mode for port 1/1/2 changed from
native-untagged to access on VLAN 1
```

```
2022-07-29T10:48:24.722076+0000 port-accessd[4355] <INFO>
Event|10503|LOG_INFO|CDTR|1|Port 1/1/2 is unblocked by port-access
2022-07-29T10:48:24.743632+0000 ops-switchd[686] <INFO>
Event|2107|LOG_INFO|CDTR|1|The mode for port 1/1/2 changed from
access to native-untagged on VLAN 4
2022-07-29T10:48:24.751112+0000 hpe-mstpd[3446] <INFO>
Event|2012|LOG_INFO|CDTR|1|CIST - Topology Change generated on
port 1/1/2 going in to forwarding
```

5. Stop the terminal monitor.

```
sw-edge1# no terminal-monitor
```

Verify the AP management VLAN MAC addresses

To verify that the APs are online and associated with the correct VLAN, you can check the MAC address table of VLAN 4.

1. Check the MAC address table of VLAN 4.

```
sw-edge1# show mac-address-table vlan 4
MAC age-time              : 300 seconds
Number of MAC addresses : 5

MAC Address          VLAN Type                    Port
-----------------------------------------------------------
20:4c:03:c6:09:50  4    port-access-security   1/1/2
b8:d4:e7:da:80:00  4    dynamic                lag256
b8:d4:e7:d9:e5:00  4    dynamic                lag256
12:01:00:00:01:00  4    dynamic                lag256
20:4c:03:c6:04:90  4    dynamic                lag256
```

- **Question**: Do you see a MAC address starting with 20:4c:03 on port 1/1/2?
- **Answer**: Yes.

Task 2: Prepare Aruba Central AP group

In this task you will prepare a group in Aruba Central for the AP.

Objectives

- Create an AP group in Aruba Central.

Steps

1. Open or switch to your Aruba Central session.
2. In Aruba Central, navigate to: Context: **Global**> Navigation: **Organization**> Top: **Network Structure**> **Groups.**
3. Group name **campus-wifi-ui**, select only APs. Do not enable templates.

Note

For AOS 10, UI groups are the recommended best practice.

4. Click **Next**.
5. For Architecture, leave it set to **ArubaOS 10.** For the Network Role leave it as **Campus/Branch**.

Note

You could choose AOS 8 to configure an Instant AP-based deployment, but this lab uses the cloud AP deployment with AOS 10.

6. Click **Add**.

Configure basic group settings

In the next steps, you will apply some essential configuration settings for the new group.

1. In Aruba Central, navigate to: Context: **Groups / campus-wifi-ui** > Navigation: **Devices**> Top: **Access Points**> Right top: **Config.**
2. Since this is the first time you access the AP Group configuration, and since no password has been set, the system prompts you to set the admin password. This is the password for the 'admin' user account on the APs that belong to this group.
    a. For the password enter **Aruba123!**
    b. Confirm the password and click **Set Password**.
3. By default, the basic configuration UI is displayed. In these labs you will also use some other settings, so you can enable the Advanced View. Click **Show Advanced**. You see additional configuration tabs.
4. Click **System.** The **General** section opens by default (see Figure 8-13).

Figure 8-13: System > General

5. Configure these values under *General*:
    - Country Code for Group : **US**
    - Time zone: **Eastern-Time UTC-05**
    - NTP server: **10.254.1.21**
6. Click **Save Settings**.
7. Navigate to the **System** > **Administrator** section (see Figure 8-14). This allows you to change the *admin* password that you previously configured when accessing the group for the first time.

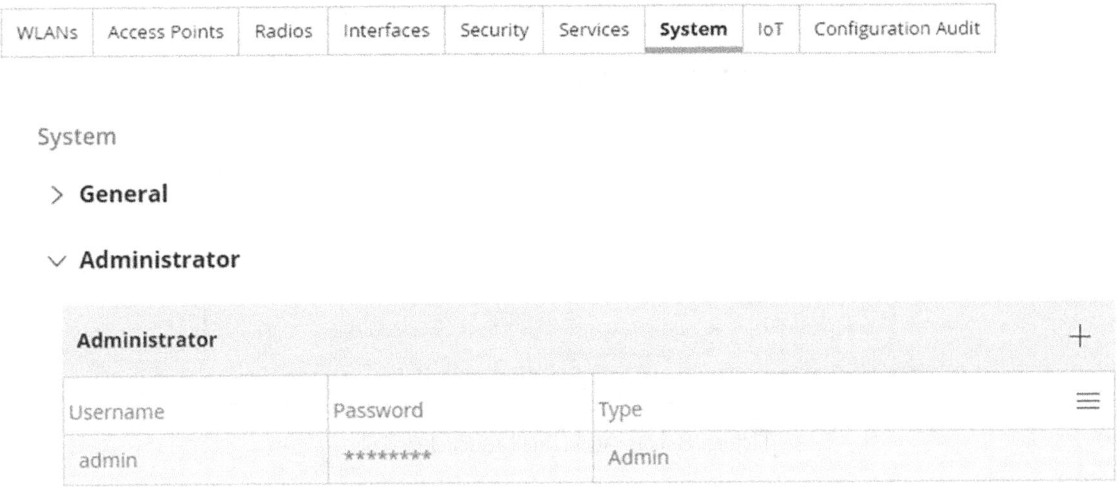

Figure 8-14: System > Administrator

Note
Make sure to leave the admin password as **Aruba123!**.

Configuration Audit Trail

Use the audit trail feature to track AP configuration changes in Aruba Central. This feature helps to verify that the UI was translated to the correct CLI commands when Aruba Central sends the commands to the APs.

1. In Aruba Central, navigate to: Context: **Groups / campus-wifi-ui** > Navigation: **Audit Trail>** Top: **Audit Trail**. (See Figure 8-15 for an example.)

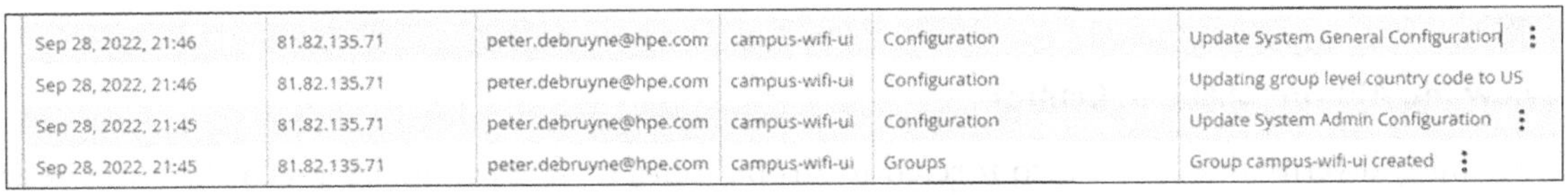

Sep 28, 2022, 21:46	81.82.135.71	peter.debruyne@hpe.com	campus-wifi-ui	Configuration	Update System General Configuration
Sep 28, 2022, 21:46	81.82.135.71	peter.debruyne@hpe.com	campus-wifi-ui	Configuration	Updating group level country code to US
Sep 28, 2022, 21:45	81.82.135.71	peter.debruyne@hpe.com	campus-wifi-ui	Configuration	Update System Admin Configuration
Sep 28, 2022, 21:45	81.82.135.71	peter.debruyne@hpe.com	campus-wifi-ui	Groups	Group campus-wifi-ui created

Figure 8-15: Audit Trail example

2. Click the **three-dots** icon at the end of the various records to see the configuration commands that you have generated. See Figure 8-16 for an example.

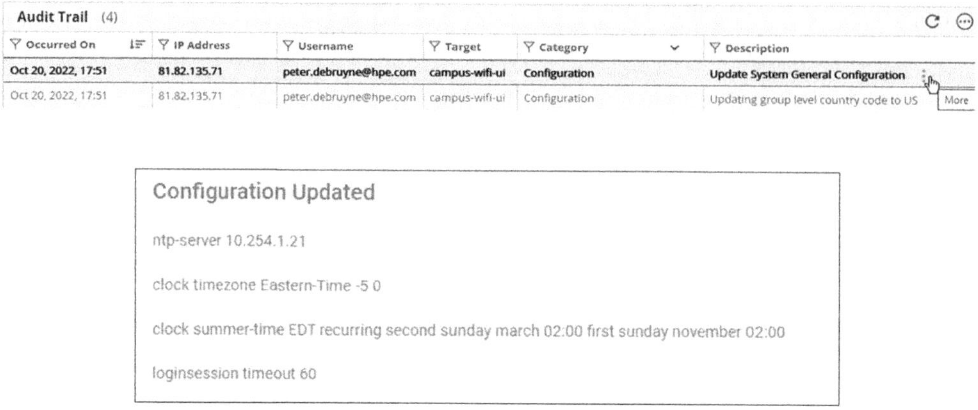

Figure 8-16: Audit Trail record details

3. **Close** the audit details window.

Task 3: Move APs to the *campus-wifi-ui* group

In this task you will move the APs to the group that you just prepared. When you move the APs, Central pushes basic settings to the APs—settings that you prepared at the group level, like the admin password and country settings.

Objectives

- Move APs to the correct AP group.

Steps

Review the APs are online in Central

First, you will verify that APs are connected to Aruba Central. You will make a record of their serial numbers and MAC addresses, which are used during the labs to identify the different APs.

1. In Aruba Central, navigate to: Context: **Global**> Navigation: **Devices**> Top: **Access Points.**
2. This is the global list of APs that have ever come online on this account. You should see 2 APs with their MAC address as the device name.
3. Compare the APs MAC and serial with the inventory information you have collected during Lab 1.
   - AP1:
     - MAC: ______________________________
     - Serial #: ____________________________

- AP2:
  - MAC: ____________________________
  - Serial #: ____________________________
- **Question**: To what group are these APs assigned?
- **Answer**: They are assigned to the group default.

Note
You may need to scroll to the right to see the group column.

Move the APs to the campus-wifi-ui group

1. In Aruba Central, navigate to Context: **Global**> Navigation: **Organization**> Top: **Network Structure**> **Groups.**
2. Under **All connected devices**, select both APs.
3. On the right-hand side, you see an overlay window with the Move Devices action button. Click the **Move Devices** button.
4. Click the **Destination Group** field.
5. Select the group **campus-wifi-ui.**
6. Click the **Move** button to continue.

Verify the group change

1. In Aruba Central, navigate to: Context: **Global**> Navigation: **Audit Trail**
   - **Question**: Do you see records with Moving Device in the description column?
   - **Answer**: Yes, there are two entries. If you don't see two entries, wait a few seconds and use the Refresh button at the right-top of the page (see Figure 8-17).

Figure 8-17: Refresh button

- **Question**: What is listed under the Target column for these records?
- **Answer**: The serial number of the AP.

Assign APs to a site

Assign both APs to the site *site-campus-main*. A label(s) is not required for the APs, but would be typically assigned in a production network.

1. In Aruba Central, navigate to: Context: **Global**> Navigation: **Organization**> Top: **Network Structure**> **Sites.**
2. Select both APs (use the <Ctrl> key to select both) and drag them to the site **site-campus-main**.
3. Confirm the site assignment with **Yes**.

Per-AP settings: AP name

To simplify monitoring and troubleshooting, it is good idea to give each AP its own, unique name. This is a *per-AP setting*. This means that the value is stored and saved on the AP: even if you would move the AP to another Aruba Central group, this setting would stay the same. You will now access the per-AP settings for the 2 new APs.

1. In Aruba Central, navigate to: Context: **Groups / campus-wifi-ui** > Navigation: **Devices**> Top: **Access Points**> Right top: **Config.**

Note

Although named "Per AP", these settings are typically still configured using the group context.

Note

You collected inventory information in Lab 1 (**mfginfo**) showing the MAC addresses and serial numbers from the AP consoles. Use this information to assign the correct names to the APs.

---

2. In the configuration menu, click **Access Points**. You will see a list with your two APs.
3. Move the mouse over the first AP. Use the **pencil** icon at the end of the line to **EDIT** the record.
4. Set the *Name* to **ap1**.
5. Click **Save Settings**.
6. Repeat this procedure for AP2 to assign the **name ap2**.
7. Check the latest entry in the audit trail (Navigation: **Audit Trail**) for the per-AP change. You may need to refresh the page to see the latest audit entries. It may take a moment for the records to appear. See Figure 8-18 for an example.

Figure 8-18: Audit Trail verification

8. **Close** the window with the audit details.

Verify the Aruba Central connection and AP settings

In these steps, you will review the AP to Aruba Central cloud connection.

1. Use the lab dashboard to open a console session to AP1. Login with a username of **admin** and a password of **Aruba123!**.
2. Review the Aruba Central cloud connection.

```
ap1# show ap debug cloud-server
IAP mgmt mode              :athena-mgmt
cloud config recved        :TRUE
autojoin mode              :disable
state diff                 :disable
Device Cert status         :SUCCESS
Cert Verify                :enable
Domain Name Verify         :enable
Device info send           :SUCCESS
Aruba Central server                    :device-uswest4.central.
arubanetworks.com
Aruba Central server path               :/ws
Aruba Central proxy server              :None
Aruba Central redirect from             :device-uswest4.central.
arubanetworks.com
Aruba Central Protocol                  :WSS
Aruba Central uptimes                   :8m:24s
Aruba Central status                    :Login_done
```

```
Cloud Debug Statistics
-----------------------
Key                        Value
---                        -----
Connect establish success 1(1)
Login done times           1(1)
Connect retry times        1(1)
Device Info send           1(1)
Domain list receive        1(1)
Domain response send       1(1)

Cloud Last connect status
-------------------------
Last connect ID      :1
Last connect time    :2022-10-20 12:04:16
Last connect trigger :athena redirect

Cloud Last login done status
-------------------------
Last connect done    :2022-10-20 12:04:19
```

3. Review the running configuration.

```
ap1# show running-config
version 10.3.1.0-10.3.1
virtual-controller-country US
virtual-controller-key d079d04f01be93b88362d5b4428638522776
f58654d2f73bec
name campus-wifi-ui
terminal-access
loginsession timeout 60
ntp-server 10.254.1.21
clock timezone Eastern-Time -05 00
clock summer-time EDT recurring second sunday march 02:00 first
sunday november 02:00
rf-band all
...
extended-ssid

hash-mgmt-password
hash-mgmt-user admin password hash
9a0b99b502a062503b4051ef3d548d660821e55b5cfdbea45034201b3acb6e2b-
c9f7627fe2
...
```

- **Question**: What is the virtual-controller country?
- **Answer**: As you have configured for the group, this is set to US.

- **Question**: What is the NTP server?
- **Answer**: 10.254.1.21
- **Question**: Do you see a line with **hash-mgmt-user**?
- **Answer**: Yes, this is the admin password, which Aruba Central applied to the configuration.

4. Review the per-AP settings. These values are stored in AP flash memory, separate from the running/startup configuration.

```
ap1# show ap-env
Antenna Type:Internal
Need USB field:Yes
name:ap1
```

- **Question**: Do you see the AP name?
- **Answer**: Yes, the name value is set to *ap1*.

5. Close the console connection.

Task 4: Configure the site floorplan with APs

In this task, you will create a floorplan for the campus site. In the floorplan, you can place the two deployed APs. This will allow Aruba Central to provide context for associated wireless clients and allows you to visualize their location.

Objectives

- Create a floorplan for a site in Aruba Central.
- Assign APs to a floorplan.

Steps

Verify AP site assignment

Before adding the floorplan, you will verify that your APs have been assigned to the correct site. This should have been completed earlier in the lab, so this is only a verification.

1. In Aruba Central, navigate to: Context: **Sites / site-campus-main** > Navigation: **Devices>** Top: **Access Points.**
2. Verify that both **ap1** and **ap2** are listed.

Note

If one of the APs is not listed, refer to the previous task in this lab "Move APs to the *campus-wifi-ui* group". This is where you assigned APs to the correct site.

Configure the site floorplan

Floorplans are bound to a physical building or location. Thus, the floorplan configuration is situated under an Aruba Central *site* object.

1. In Aruba Central, navigate to: Context: **Sites / site-campus-main** > Navigation: **Overview>** Top: **Floor Plan**. By default, there is no floorplan defined for a site.
2. Click **Add Floor** to access the floorplan Edit Mode.

Measurement system

1. On the right-hand site, examine the Ceiling Height value (see Figure 8-19).

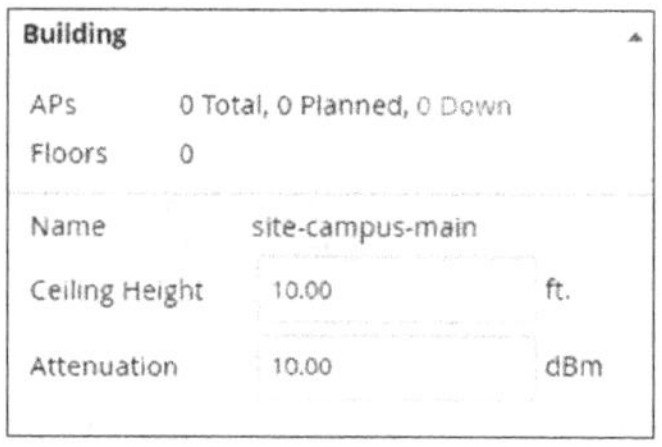

Figure 8-19: Ceiling height and attenuation

- **Question**: Is this value in feet or meters?
- **Answer**: By default, the imperial system is applied: height is reported in feet (ft).

Optional: Change to metric system

If you are based in a region that uses the metric system, you can follow the next steps. Other students can skip this optional section and move to the *Create New Floor Plan* section after the next page.

To change to the metric system, you will need to add a floorplan once, save it, and exit. You will be able to change the measurement system from the floorplan view.

Important

The next steps only apply if you want to change to the metric system! Make sure to continue the lab after this optional section!

1. Right-click the area and click **New Floorplan** (see Figure 8-20).

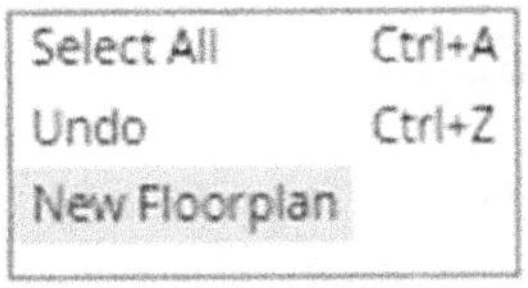

Figure 8-20: Floorplan selection

2. For Floorplan file, click **Choose File**.
3. From the student files, select the file **lab08.01-floorplan.png**.
4. You may leave the floor name and floor number at their default values.
5. Click **Save**. This uploads the file and starts the new floor wizard, which takes a few seconds to complete.
6. You don't need to complete the steps of the wizard for this lab; click **Finish**.
7. Next click **Exit Edit Mode** to return to the view mode.
8. At the top-right, click the **view** icon to open the View Settings (see Figure 8-21).

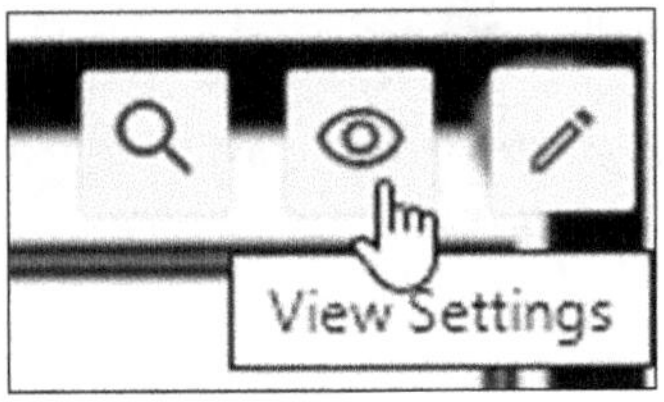

Figure 8-21: View settings

9. At the bottom, there is a toggle for Meter and Feet. Select the **Meter** radio button to set the metric system (see Figure 8-22).

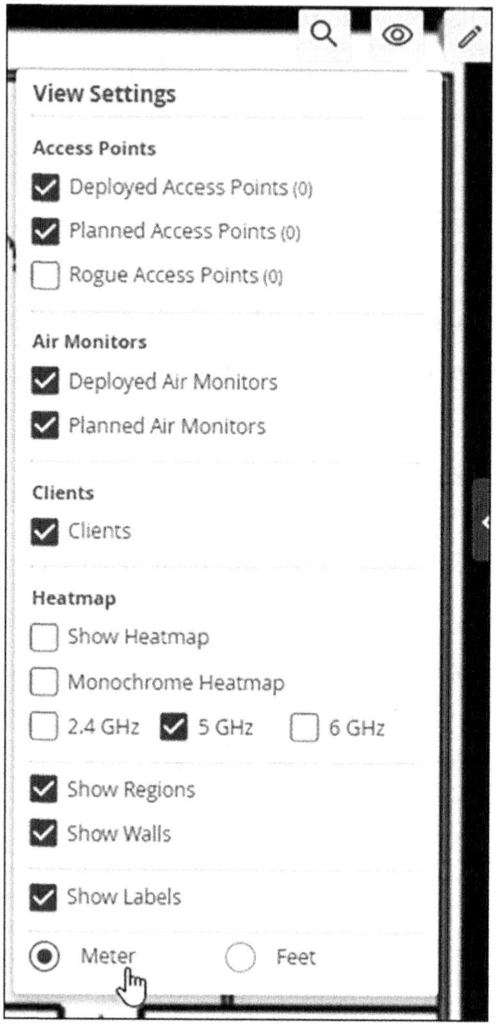

Figure 8-22: Changing feet to meters

10. At the top-right, click **List** to see a list of floors. Your temporary floor is shown.
11. At the end of the record, use the **trashcan** icon to remove the floor (see Figure 8-23).

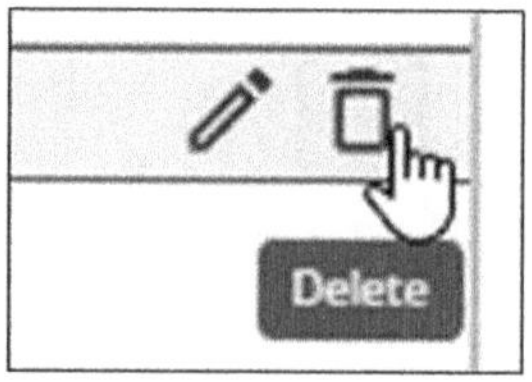

Figure 8-23: Trashcan icon

12. Confirm by clicking **Delete**.
13. At the top-right, click **Summary** to switch the view (see Figure 8-24).

Figure 8-24: Switching to the Summary View

14. Click **Add Floor** to access the floorplan *Edit Mode*.

Note

This is the end of the optional section (Change to the metric system). Make sure to continue the lab from this point!

Create a New Floorplan

These steps apply to all readers again (both metric and imperial system).

1. Right-click the gray background area and click **New Floorplan**. See Figure 8-20.
2. For Floorplan file, click **Choose File**.
3. From the ACAF Student Files folder on the PC-MGMT desktop, select the file **lab08.01-floorplan.png**.
4. You may leave the floor name and number to their default values.
5. Click **Save** to upload the file and start the new floor wizard.

Set the scale

You must set a known distance on the floorplan. This will be used by Aruba Central to perform distance calculations on the map, such as the locations of devices on the RF heatmap.

1. Click the **Measure** button. The mouse pointer changes to a *plus* sign.
2. Click and drag from the top-left corner of the building to the top-right corner.
3. When you release the mouse button, the Enter Distance window will popup. Enter **100** feet or **30** meters (depending on your measurement configuration).
4. Click **OK**.
5. Click **Next.** This takes you to the Region step in the wizard.

Note

If you need to perform AP deployment planning, this can be configured using the Planning Regions feature. In this lab you will focus on monitoring the active APs, so there is no need to create planning regions.

Complete the Floor Plan wizard

1. Click **Next** to move to CAD Layer step. If a CAD drawing is available for the floor, you can upload it here. There is no need to change this in the lab, however.
2. Click **Next** to move to the Access Points step.
3. Click **Add deployed APs**. Both your APs should be shown.

Note

If your APs are not listed, they have not been assigned to this site. Only APs that have been assigned to this site can be added to the floorplan of this site!

4. Drag and drop each AP to an appropriate place on the floorplan map.

Note

Since this is a lab environment, the actual position on the map does not matter; however, this is not the case for a production network, where the actual placement on the floor plan is important.

5. Click **Finish** to complete the new floor wizard.

Edit an existing floor

Once a floor has been added, you can still add or change the APs, their locations, and/or the floor properties.

Note

By default, the location of APs is locked. This prevents accidentally moving an AP on the map. Clicking the **lock** icon at the top of the floorplan: this unlocks edit mode, which allows you to make the appropriate changes to the floorplan. In a production network, it is best to leave it locked once you've set up the floorplan to your needs.

6. To move an AP, you must first unlock the dragging function. Click the **lock** icon shown in Figure 8-25 to unlock the edit function.

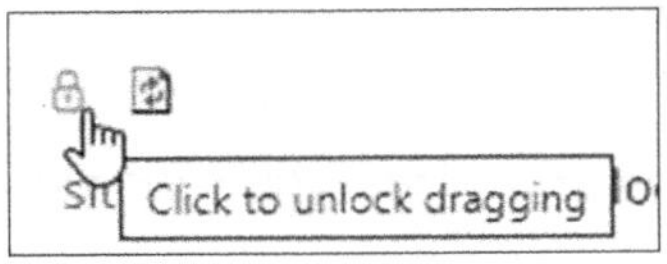

Figure 8-25: Unlocking a floorplan's edit mode

7. Now you can drag and drop an AP to a new location on the map.

Heatmap

Aruba Central can display the wireless RF heatmap based on AP locations. This is based on a calculation; therefore, it is important to correctly position APs as accurately as possible on the map.

1. Click somewhere on the map to activate the floorplan Properties view. On the right side, the Floor 1 properties should be displayed (see Figure 8-26).

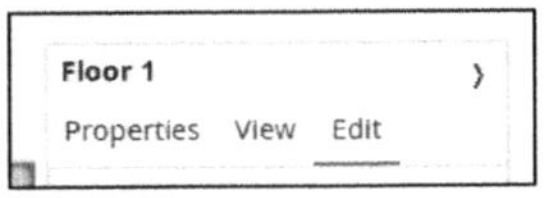

Figure 8-26: Properties view

2. Click **View**.
3. Clicking the **Heatmap** label enables or disables the heatmap view. Make sure the heatmap is *enabled*.

Note

The heatmap can only be displayed when Aruba Central has received the AP channel and transmit power from the AP. As the AP has not yet been configured with a WLAN, the radios are still disabled, and thus you will not see a heatmap...yet. You may continue with the lab, however. In Chapter 10 you will configure your first WLAN (for employee access).

4. Figure 8-27 shows an example of how the heatmap will appear once the AP radios are enabled (this will be done in the next lab activity).

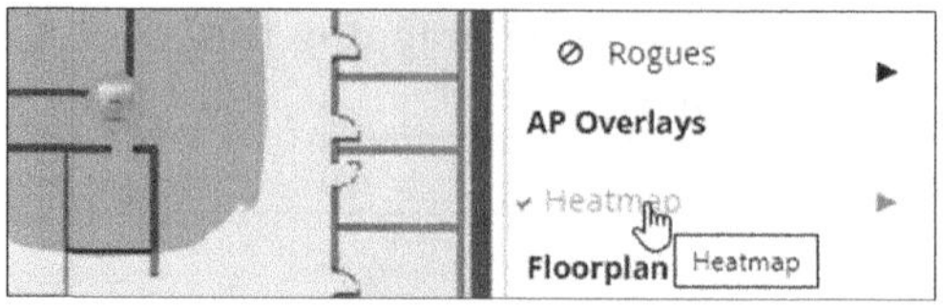

Figure 8-27: Heatmap example view

5. Expand the **Heatmap** section.
6. This section allows you to choose the desired band for the heatmap – 2.4, 5, or 6GHz (see Figure 8-28).

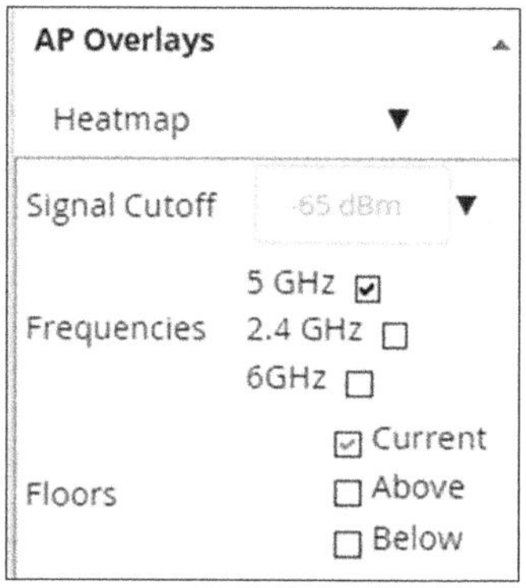

Figure 8-28: Heatmap view properties

7. After you add or move an AP, you can manually refresh the heatmap display (see Figure 8-29).

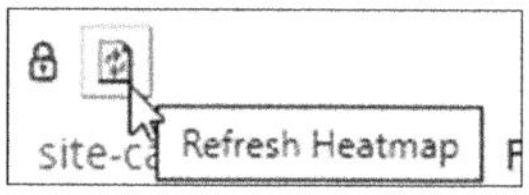

Figure 8-29: Refresh the heatmap view

8. Hover your mouse over an AP to see a popup with current channel information (see Figure 8-30).

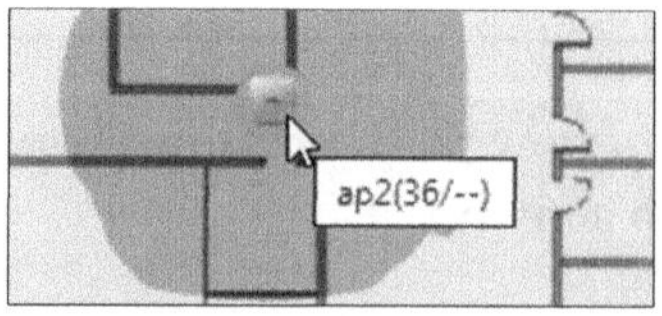

Figure 8-30: Verify the AP in the heatmap view

Note
The channel view option will not be available yet, since the AP radios are still disabled.

9. Click an AP to access the AP-level window (see Figure 8-31).

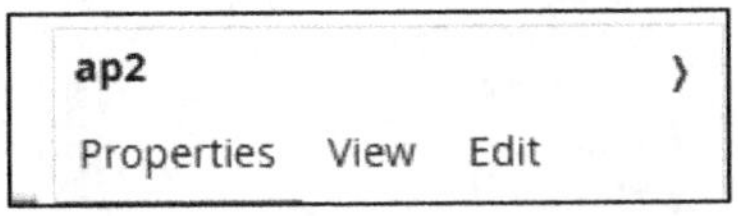

Figure 8-31: Accessing the AP-level window

10. Click the **Properties** link to view the channel and transmit power. Figure 8-32 shows an example:

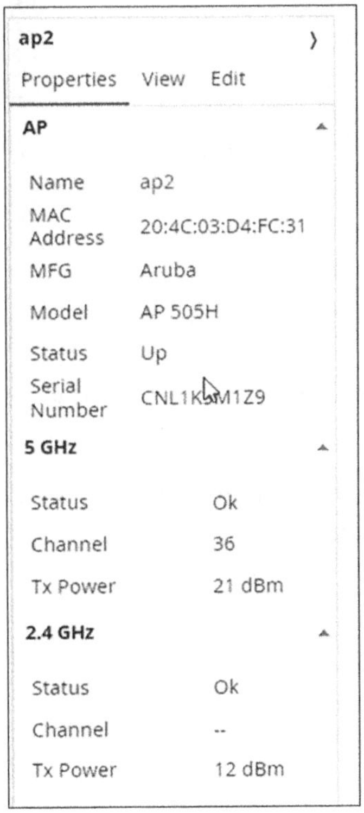

Figure 8-32: Properties view of a selected AP

11. Click **Exit Edit Mode** to switch to the View Mode.

View mode

In the default view mode of the floorplan, several View Settings can be configured.

1. At the top-right, click the **View Settings** icon (see Figure 8-33).

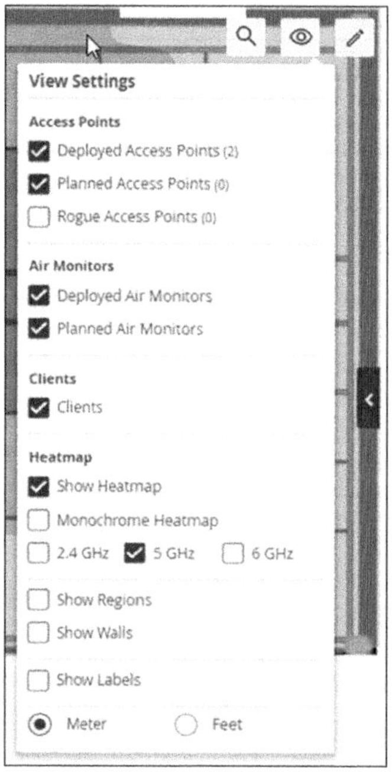

Figure 8-33: View Settings window

2. The View Settings window allows you to choose whether clients, APs, and/or Air Monitors should be displayed in the heatmap.

Learning check

Discuss these questions.

1. What statements below are true concerning device profiling? (Select two.)
   a. Profiling and endpoint classification allows the network infrastructure to adjust settings and security policies dynamically, based on device types.
   b. You can use profiling with device provisioning, so when a device connects, switches are automatically configured with the correct VLAN, PoE priority, security policies, and more.
   c. End systems can only be profiled based on LLDP information.
   d. A switch knows which role to assign to a port when a connected device matches a device group that you define.
2. AOS 8 supports Instant APs (IAPs), while AOS 10 does not. However, AOS 10 simplifies the network architecture, because the Mobility Conductor features are now in Aruba Central.
   a. True
   b. False

Summary

Congratulations! You have completed Chapter 8. You should now be able to:

- Use device profiling.
- Describe LLDP and MAC profiling.
- Connect an AP to Aruba Central.
- Perform initial AP setup.

9 WLAN Fundamentals

OBJECTIVES

Chapter 9 focuses on WLAN fundamentals. Reliable, high-performance WLANs are key to successful deployments. After completing this chapter, you will be able to explain RF bands and channels and describe radio frequency (RF) basics and WLAN mobility. You will be able to explain IEEE 801.11 frames and describe key aspects of good RF design. Finally, you will be able to configure WLANs.

After completing this chapter, you should be able to:

✓ Explain RF bands and channels.

✓ Describe RF basics and WLAN mobility.

✓ Explain 802.11frames.

✓ Describe aspects of RF design.

✓ Configure WLANs.

RF bands and channels

Let's begin the discussion with RF bands and channels.

WLAN organizations

Several organizations are responsible for wireless local area network (WLAN) standards. These regulatory bodies place restrictions on RF technology and how you use it in a WLAN environment (see Figure 9-1):

- The IETF is the international Internet standards organization, responsible for many Internet standards, as documented in Request for Comments (RFCs). There are several Wi-Fi standards related to WLAN control and provisioning.
- Each area of the world is governed by local RF government regulations. The FCC regulates RF spectrum in the United States, ETSI regulates Europe, and so on. These bodies dictate RF power limits and usable WLAN frequencies. These organizations are largely concerned with minimizing RF interference.

- Within these limitations, the IEEE created the 802.11 standards. 802.11 channels must fall within FCC unlicensed ranges. Therefore, the FCC regulates power and channels, and IEEE standardizes how to use those channels: layer 1 (physical layer) transmission methods and layer 2 frame types and protocols.
- The Wi-Fi alliance is just that—an alliance of WLAN hardware and software manufacturers, including Aruba. They provide interoperability standards to assist in ensuring different vendors' APs and clients can work together. This often center on security and quality of service (QoS) mechanisms.

For now, let's focus on the IEEE standards.

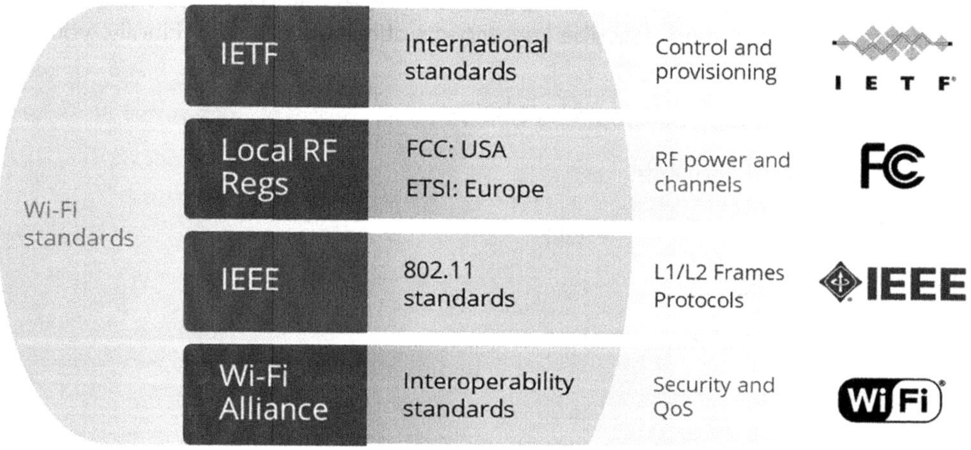

Figure 9-1: WLAN organizations

IEEE 802.11 standards

Early IEEE 802.11 standards, like the original 802.11, 802.11b, a, and g standards (see Figure 9-2), were great when they were introduced, with data rates of up to 54 Mbps, and support in both the 2.4 and 5 GHz frequency bands. However, these standards are slow and inefficient by today's standards.

The 802.11n amendment added high throughput (HT) speeds, theoretically up to 600 Mbps, with 300 Mbps being a typically speed. It accomplished these speeds by increasing the channel width, using the advanced 64-quadrature amplitude modulation (QAM) mechanism, and techniques like multiple input multiple output (MIMO).

802.11ac only operates in the 5 GHz range with theoretical data rates of up to 6.93 Gbps, using very high throughput (VHT) methods, with higher channel width and more complex modulation techniques. 802.11ac is sometimes been referred to as *Wi-Fi 5*.

802.11ax is the latest standard and operates in both the 5 GHz and 2.4 GHz frequencies, with a maximum data rate of 4.8 Gbps. The big news with 802.11ax is a technique called multi-user MIMO (MU-MIMO), enabling multiple transmitters at the same time. This early version of 802.11ax has also been referred to as *Wi-Fi 6.*

An 802.11ax extension was introduced in 2020, with up to 7.8 Gbps, 1024-QAM with modulation and coding scheme (MCS) 10/11. There were two important features introduced with this standard: it operates in the 6 GHz RF spectrum and supports a 1 Gbps channel width. This version of 802.11ax has been referred to as "*Wi-Fi 6E*".

Let's look at the Wi-Fi Alliance standards.

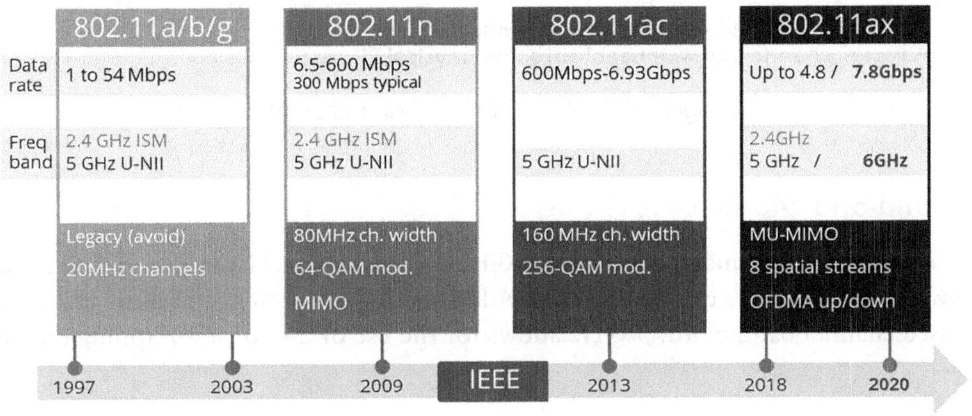

Figure 9-2: IEEE 802.11 standards

2.4 GHz band and channels

One of the things that the Wi-Fi Alliance defines is how to use RF for Wi-Fi. 802.11 channels in the 2.4 GHz spectrum are 20 MHz wide (see Figure 9-3). Clients and APs use this frequency range to transmit and receive data.

The 2.4 GHz band allows for three non-overlapping channels in some regulatory domains such as the United States. Other regulatory domains, such as Japan, allow the use of channel 14, providing four non-overlapping channels.

A WLAN signal uses spread spectrum technology and therefore adjacent channels overlap. When an AP broadcasting on channel 1 can hear an AP broadcasting on channel 2, both radios will experience adjacent channel interference or ACI. ACI is like talking over top of another person when having a conversation. Some or all the transmission may be corrupted due to the overlapping channels. Due to the nature of channel overlapping, only the use of channels 1, 6, 11 is recommended; and depending on regulatory domain, channel 14 may also be considered for use.

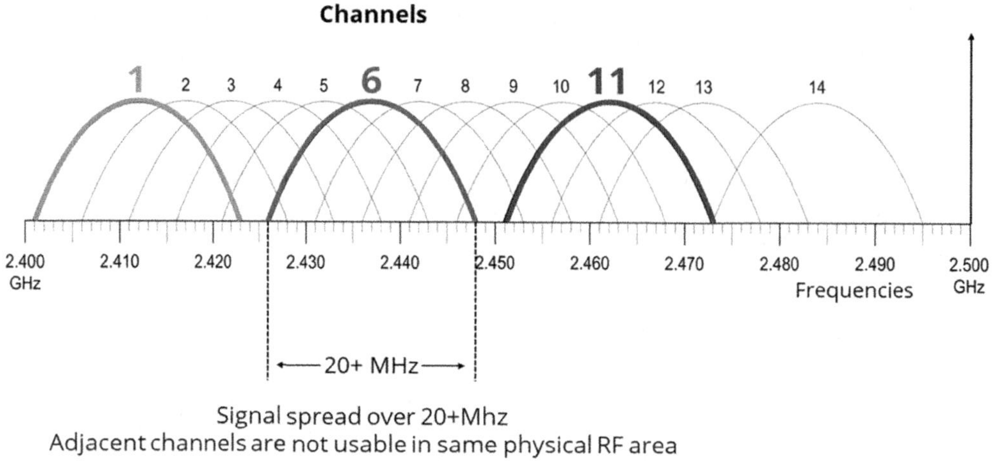

Figure 9-3: 2.4 GHz band and channels

5 GHz band and channels

802.11a, n, ac, and ax all make use of the 5 GHz RF spectrum. Given that 5 GHz is a higher frequency, higher data dates can be achieved. Additionally, the presence of more channels allows for increased channel bandwidth. 5 GHz allows for the use of 24 non-overlapping channels (see Figure 9-4).

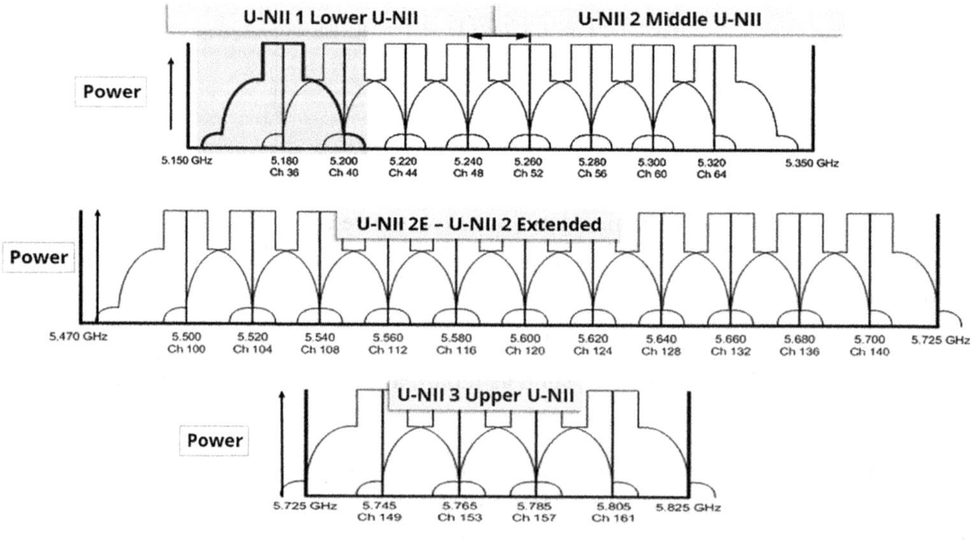

Figure 9-4: 5 GHz band and channels

Highlighted in orange, you will notice at lower power levels, 5 GHz channels do overlap. However, as power increases, so does beamwidth, eliminating ACI in the 5 GHz spectrum. Design considerations concerning low powered APs in 5 GHz are beyond the scope of this guide and will be addressed in Aruba's design certification track.

Channel bonding

As you learned previously, 802.11 channels are 20 MHz wide, and devices use this frequency range to transfer data. Channel bonding is the process of combining the bandwidth of two or more adjacent channels: for example, channel 149 bonded with channel 153 becomes channel 149+ (see Figure 9-5).

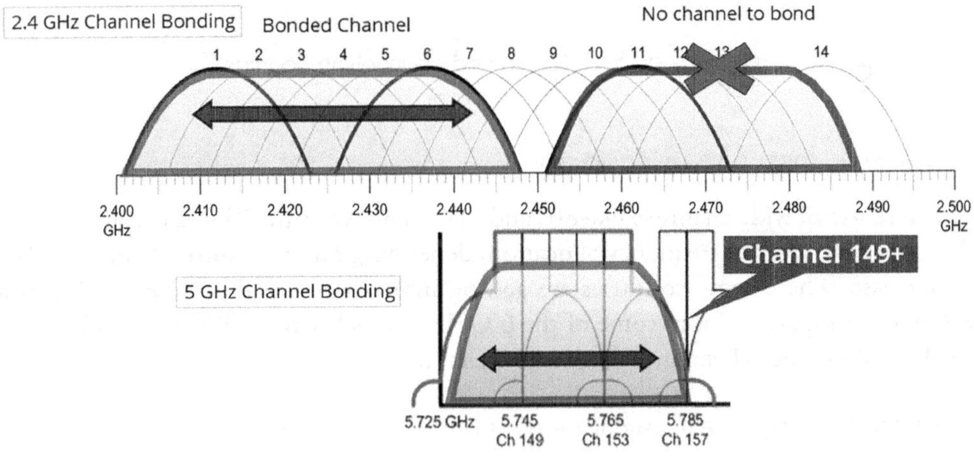

Figure 9-5: Channel bonding

When channels are bonded, devices using this now wider *channel* have twice as many frequencies (or more) to modulate their signal across, allowing for increased throughput. Channel bonding comes with a cost, however: for every bonded channel, the noise floor of the channel will increase by 3 dB (decibel). As you will soon learn, a 3 dB increase means a double in the decibel calculations.

5 GHz allowed channels in countries

Bonding channels can aid in providing more bandwidth to a channel but becomes a point of diminishing return as fewer channels are available and the noise floor increases. In most enterprise WLANs, 20 or 40 MHz channel widths are optimal. 80 and 160 MHz channels are less practical due to lack of available bandwidth and increased noise floor. Notice the difference in available channels as channel bandwidth increases (see Figure 9-6).

Also, IEEE's 802.11h dynamic frequency selection (DFS) places restrictions on which channels you can use in the 5 GHz range when co-located with satellite and/or radar signals. Plus, there might be country restrictions as to which additional channels you cannot use based at your location.

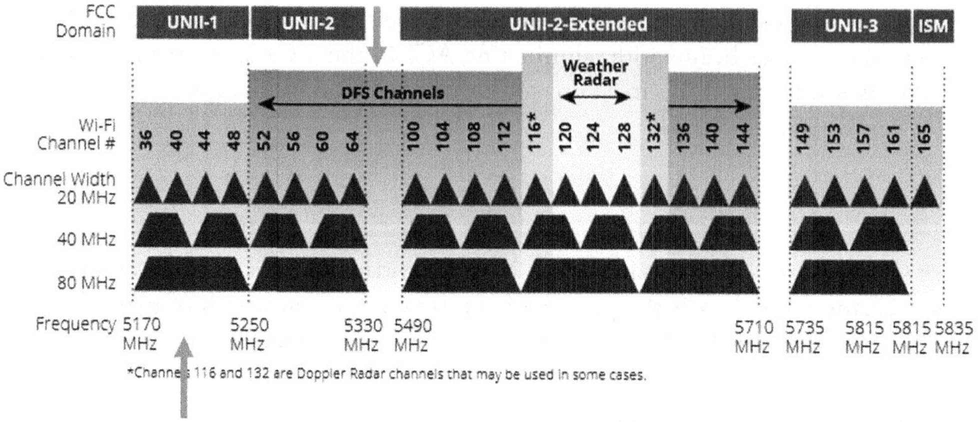

Figure 9-6: 5 GHz allowed channels in countries

6 GHz: Massive increase in capacity

Today, spatial re-use of wide channels like 80s and 160s is not realistic. This is because we are limited with only about 500 MHz of frequency allocation, depending on the country. With 6 GHz, you get a massive increase, where some countries are getting more than two times the existing amount of capacity. Notice in Figure 9-7 that some of the 6 GHz channels have a different shade of color. You will learn about these special channels later in the chapter.

- More than 2x the existing amount of unlicensed spectrum
- Up to seven 160 MHz wide channels
- 6 GHz is greenfield - no need for backwards compatibility

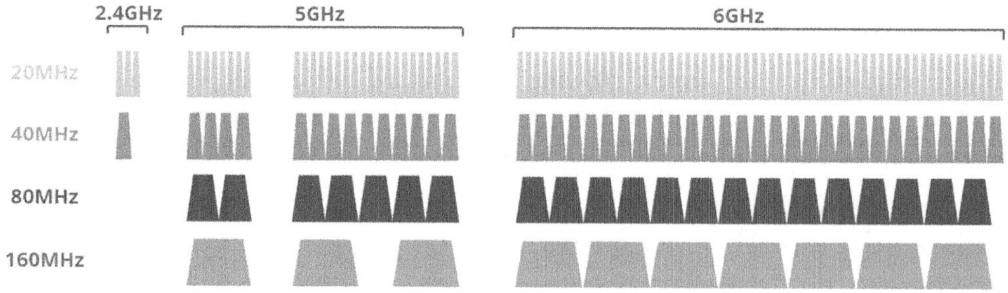

Figure 9-7: 6 GHz—massive increase in capacity

Usable 6 GHz channels in the Americas and Europe (EU)

Throughout the Americas, regulators are opening the entire 1,200 MHz frequency for unlicensed use (see Figure 9-8). To put this another way, you have more 80 MHz channels in this model than there are 40 MHz in the current 5 GHz band. This means that you will default to 80 or 160 MHz-wide channels in the 6 GHz band. For the first time, it will be practical to have wide channels and spatial re-use for enterprise deployments!

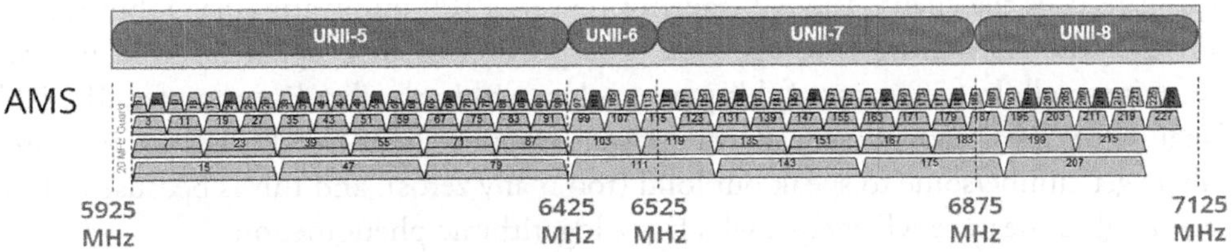

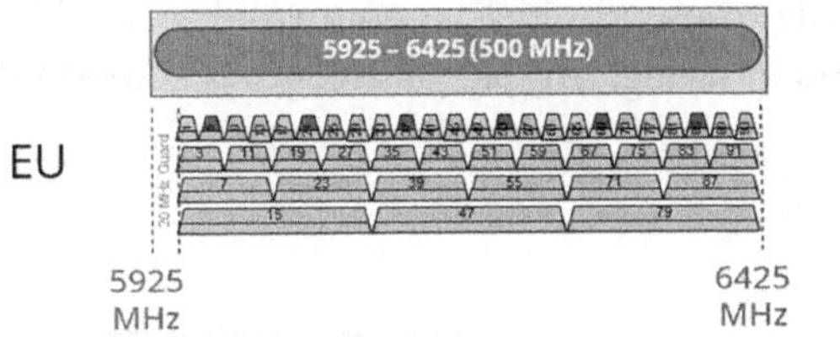

	European Model	Americas Model
20 MHz	24	58
40 MHz	12	29
80 MHz	6	14
160 MHz	3	7

Figure 9-8: Usable 6 GHz Channels in the Americas and Europe

In other regulatory domains, or the European model, only the first 500 MHz portion shall be opened for unlicensed use. With that said, the global organization, Dynamic Spectrum Alliance, published an article in 2022 urging European and other regulators to consider opening the entire band because of the importance of permitting unlicensed use of the full 1,200 MHz, since this is critical for spatial re-use.

Initial product focus is on the low power indoor (LPI) device class, where access points are limited to indoor operation only. LPI has access to all 1,200 MHz of the frequency, has no DFS requirement, has no spectrum database requirement, and has a massive amount of additional spectrum and capacity.

Note

Technically, channel 233 would be the 59th channel, and you may find references to it in other places, but power will be very low and likely realistically unusable due to band edge limitations.

RF basics

To design, deploy, and troubleshoot a wireless solution, you really should know the basics of RF energy. This section introduces you to the basics of RF.

Milliwatts and dBm

One scale used to measure RF power is milliwatts. Figure 9-9 shows a signal sent by an AP, decreasing in power as it reaches the client. The AP transmit power is 100 milliwatts (mW), but that energy is spread out over the AP's entire coverage area. The portion of that signal received by the client's antenna is quite small. In fact, most of this power is lost within the first linear meter. The table at the right of Figure 9-9 shows how drastically the power drops as client-to-AP distance increases. These tiny numbers get cumbersome to speak out loud (too many zeros), and this is because milliwatts is a linear scale, used to measure RF energy, which is a logarithmic phenomenon.

Thankfully, we have another, more appropriate scale to use: decibels as compared to 1 mW (dBM). So, 1 mw = 0 dBM, and 100 mW = 20 dBM. Instead of saying, "my signal strength is .00000001 mW", you can simply say "-80 dBM".

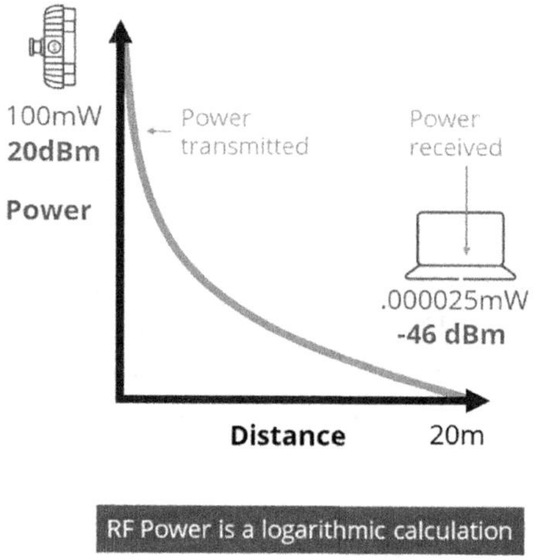

Received Signal Strength

Distance(m)	mW	dBm
	1	0
0	100	20
1	.0098911	-20
10	.0000989	-40
20	.0000247	-46
100	.0000010	-60
1000 (1km)	.0000000099	-80

dBm is much easier to work with than mW

Figure 9-9: Milliwatts and dBm

dBm and mW calculations

Let's get some perspective on these numbers: how do you use them in the real world? When deploying a WLAN, it is very important to ensure that clients will always receive a good signal, otherwise the connection may be poor, which leads to low data rates and disconnects. The amount of signal power the client receives is known as received signal strength indicator (RSSI). An RSSI value of -62 dBm or so is great for most users, while receiving a higher value is considered an excellent signal (see the left side of Figure 9-10). However, a lower power can lead to underperformance. Many wireless administrators use -75 dBm as a baseline minimum value to properly communicate. Anything lower than that is largely unusable.

Remember, dBm is decibels relative to one milliwatt: 0 dBm equals one milliwatt. When dealing decibel representations, you must know the rule of tens and threes. A 3 dBm increase is equal to double the power. A 10 dBm increase is equal to 10 times the power. Likewise, a 3dB decrease is half power, and a 10 dB decrease is one tenth of the power (see the right side of Figure 9-10).

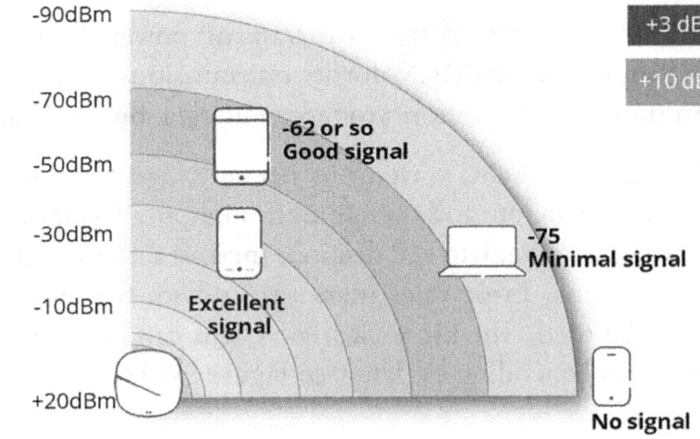

0 dBm = 1mW

+3 dBm = 2X power / -3 dBm = 1/2 power

+10 dBm = 10X power / -10 dBm = 1/10th power

dBm	mW
+20	100
+17	50
+14	25
+10	10
0	1
-10	.1
-30	.001
-50	.000001
-70	.0000001
-90	.000000001

Figure 9-10: dBm and mW calculations

Calculating effective isotropic radiated power (EIRP)

Effective or equivalent isotropic radiated power (EIRP) simply means how much power the antenna effectively radiates out into free space. So perhaps the radio generates a 100 mW (20 dBM) signal that is transmitted into the external cable, which, in turn, is connected to an external antenna (see Figure 9-11). In this example, it is a long cable, and therefore there is 3 dB of loss in the signal's power as it travels through the wire. Also, this particular antenna has a fairly *squashed* pattern and therefore has a 10 dBi (dB increase) of gain. So, 20 dBm – 3 dBm cable loss = 17 dBm, plus the 10dBi gain of the antenna resulting in a 27 dBm EIRP. As you can see in this example, the antenna provides amplification for the signal, and thus the gain in dBm. Antennas come in many shapes and sizes, and depending on the amplification and clarity you need for a signal (like using wireless to connect two buildings), the antenna itself could actually cost more than the AP itself!

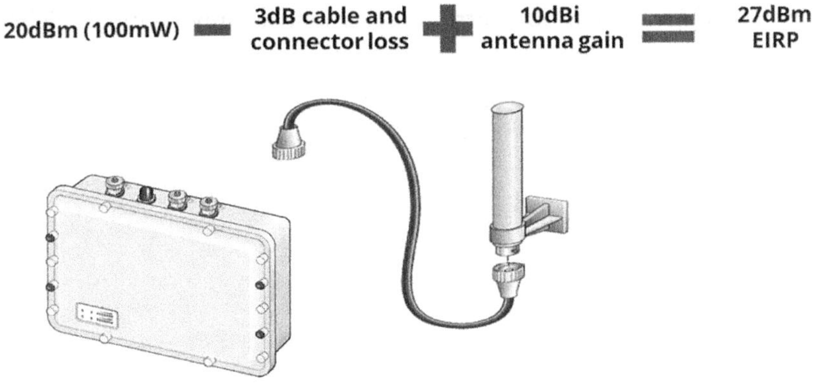

Figure 9-11: Calculating EIRP

Distance, performance, and MCS

RF transmit power, or EIRP, determines the AP's RF coverage area. The transmit power also helps the client to choose the best modulation and coding scheme (MCS) for its transmission. A modulation scheme is a way of changing (modulating) a transmitted waveform to represent data: binary 1s and 0s.

As the client-to-AP distance increases, the signal weakens, waveforms degrade, and the error rates increase. The MCS is a key factor in determining data rate and therefore throughput. The more complex the MCS, the higher the data rate. However, as client-to-AP distance increases, the signal weakens, thus distorting these very complex waveforms. Error rates increase. And so, the client must automatically drop down to a lower MCS to be used. The more clearly a client can hear an AP the more complexity it can add to its modulation and encoding of data (see Figure 9-12).

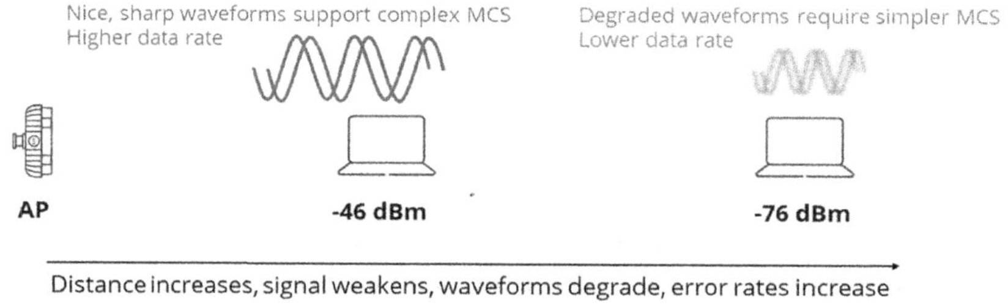

Figure 9-12: Distance, performance, and MCS

MCS versus distance

Figure 9-13 relates possible MCS encoding schemes versus the client-to-AP distance. If a client is very far away, it may have to drop down to MCS 0, which is a very simple modulation scheme called binary phase shift keying (BPSK) and only supports low data rates. As a client gets closer to the AP, it can *auto-rate-shift* up to a more complex MCS 9 encoding scheme called 256-quadrature amplitude modulation (256-QAM): a far more intricate scheme that supports much higher data rates.

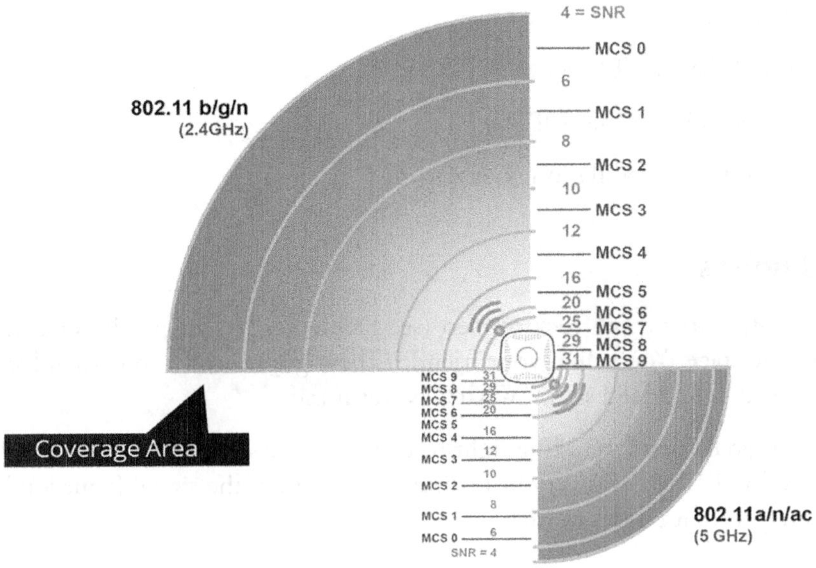

Figure 9-13: Distance, performance, and MCS

MCS versus distance

The graph in Figure 9-14 shows what signal strengths are required to achieve a certain MCS. You can see how higher MCS rates such as 256-QAM require high signal strengths and lower noise floors. As the channel bandwidth is increased, so is the noise floor. Notice how the signal strength requirements to achieve a specific MCS increase with channel bandwidth.

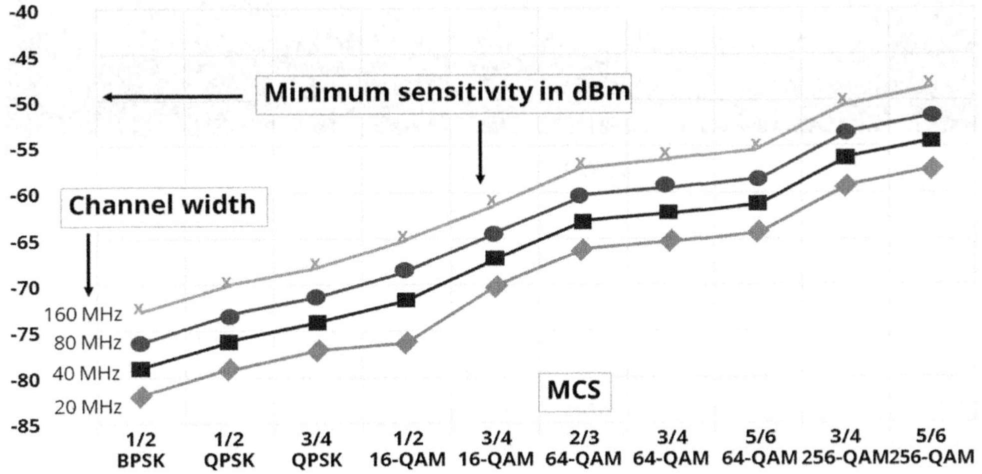

Figure 9-14: Minimum signal strength requirements by data rate

MCS solutions include:

- Binary phase shift keying (BPSK)
- Quadrature phase shift keying (QPSK)
- Quadrature amplitude modulation (QAM)

Signals and noises

Imagine you are with a friend at a rock concert (see Figure 9-15). The music is playing, and you are right in front of the stage. You yell at your friend, "These guys rock!". Your friend says, "What?" "I said, these guys rock!" The friend says, "What?" Never mind.

You were yelling to your friend, standing right next to you: the signal was very loud. But the noise in the room was also loud. There was very little difference between the signal (your loud voice) and the noise. This is an important concept in RF.

Figure 9-15: Signals and noises

Signal to noise ratio (SNR)

The ambient noise or noise floor is the sum of all undesired transmissions from devices that are either too far away for the signal to be intelligible or nearby electronic devices or materials or objects that create interference. The signal-to-noise ratio (SNR) is the difference between the ambient noise floor and the received signal strength (see Figure 9-16). Notice that you express it as a *difference*, like SNR 25 dBM, and not a ratio.

In Figure 9-16, the noise floor measures -90 dBm. The receiver's signal strength is -65 dBm. Therefore, -90 minus -65 give us an SNR of 25 dBm. The higher your SNR, the better your Modulation Coding Scheme (MCS) will be, and therefore the higher the communication data rate. Clients often require a minimum SNR of 15 dBm. However, SNRs in excess of 30 dBm provide the best data rates. Most people agree that you want at least 25 dBm or better for the SNR.

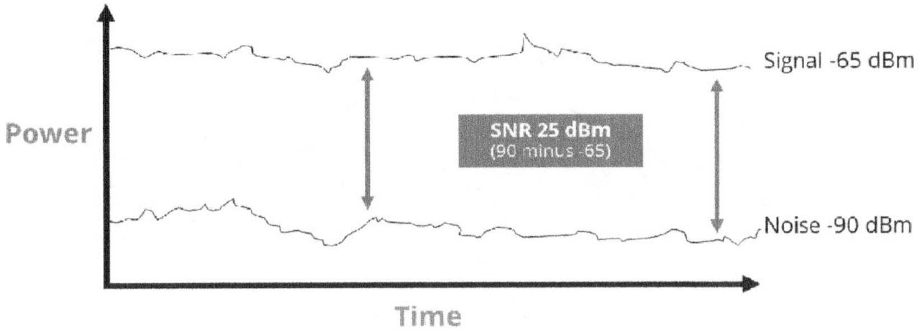

Figure 9-16: SNR

Antenna types

Aruba offers APs in two main types: integrated antennas or external antennas. External antenna APs can be customized to provide extra passive gain, increased receive sensitivity, and provide directionality for an AP's RF transmission by selecting the appropriate external antenna. Most integrated antennas are omni-directional (see Figure 9-17), meaning they will transmit approximately equal power in all directions. External antennas can be either omni-directional or directional, depending upon your need. This book will focus only on omni-directional radiation patterns.

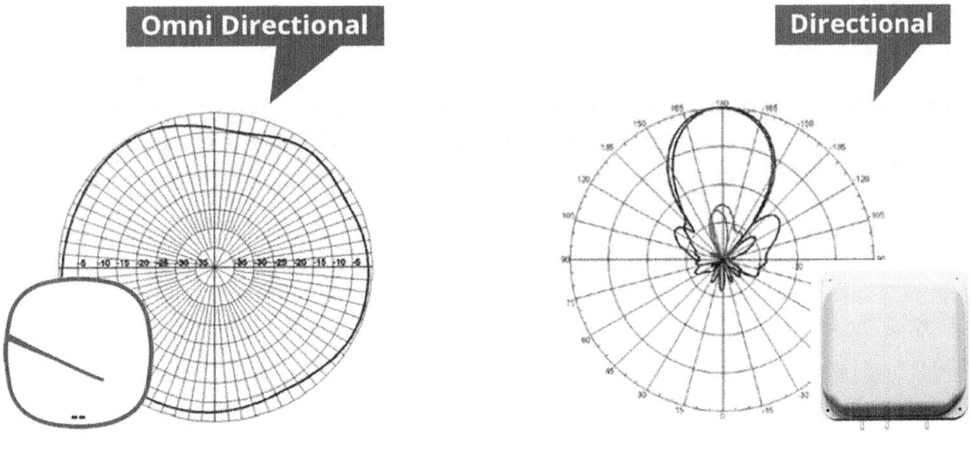

Figure 9-17: Antenna types

Horizontal plane (azimuth)

AP radiation patterns show the horizontal and vertical coverage areas. H-plane or horizontal view, also known as the azimuth, shows the radiated pattern coming from the sides of the AP (see Figure 9-18). If you have an AP, place it on the desk, white dome facing up. Draw a circle around the AP. That is your horizontal plane.

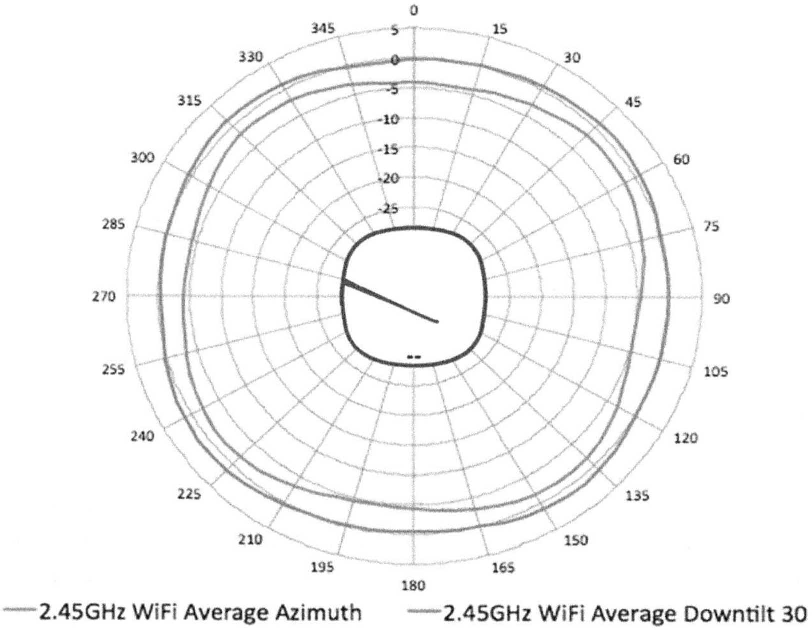

Figure 9-18: Horizontal plane (azimuth)

Vertical plane (elevation)

The elevation plane or E-plane, also known as the vertical plane, shows radiated patterns coming from the front (typically the faceplate) of the AP (see Figure 9-19). Many directional antennas sacrifice horizontal gain in favor of vertical gain to achieve connectivity at longer distances.

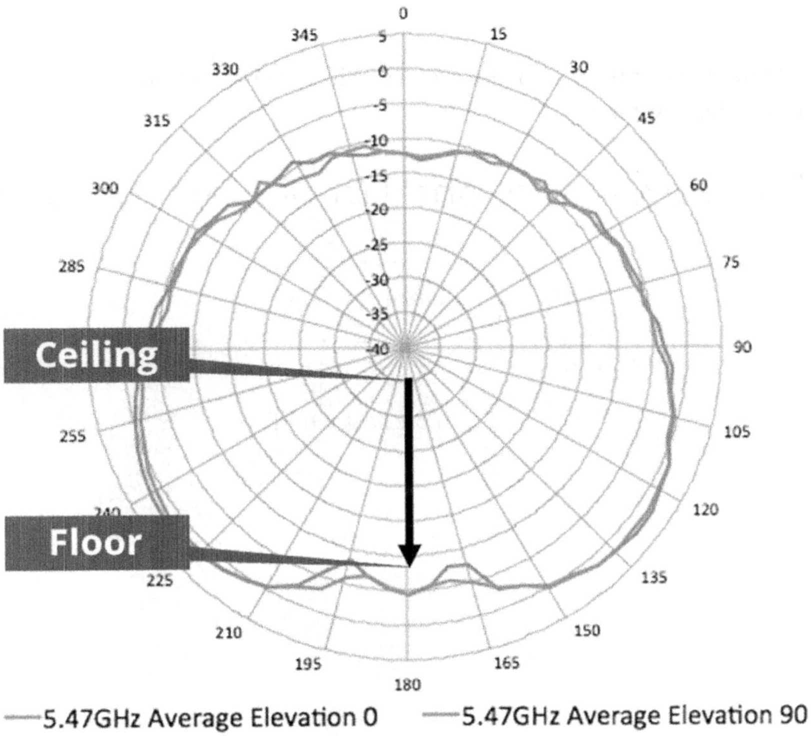

Figure 9-19: Vertical plane (elevation)

Passive E-plane antenna gain

Passive gain, omni-directional antennas, however, sacrifice vertical coverage in favor of horizontal coverage. Passive gain antennas can be used when applying coverage to a single floor instead of through a floor. In Figure 9-20, you see the side view of the affect that passive gain has on vertical coverage.

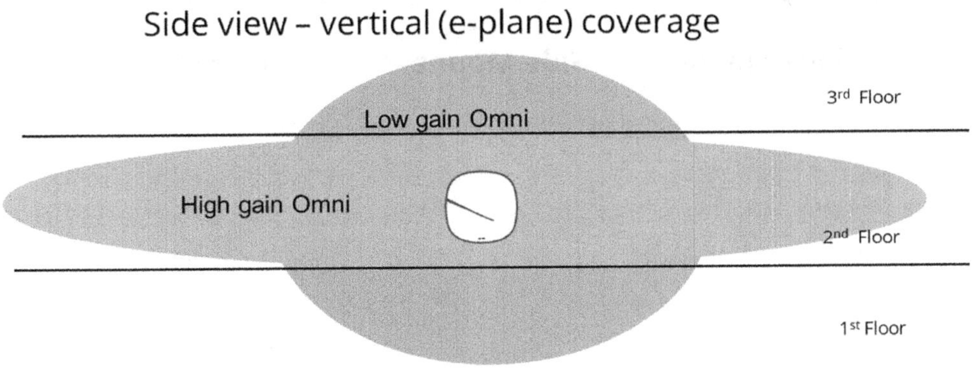

Figure 9-20: Passive E-plane antenna gain

Passive H-plane antenna gain

In Figure 9-21, look at the top view of the passive gain omni-directional antenna and how it will provide more horizontal coverage.

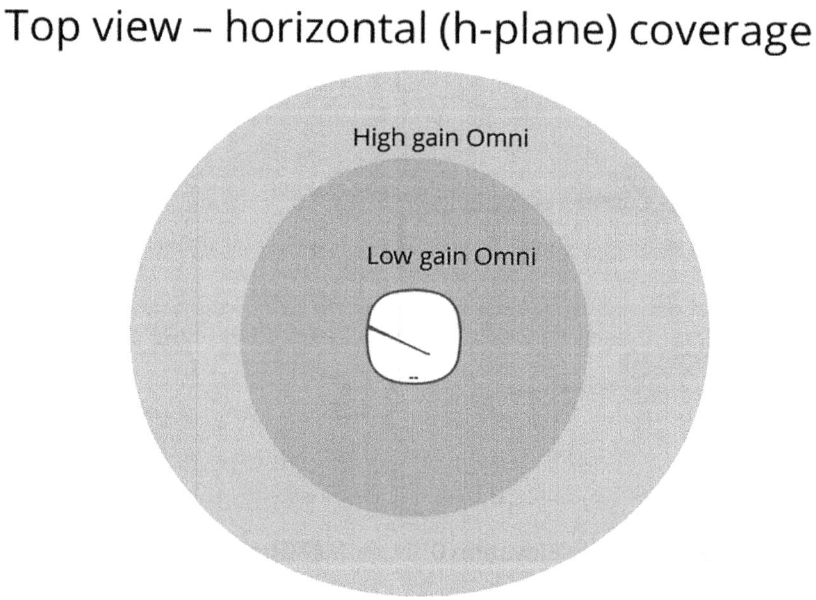

Figure 9-21: Passive H-plane antenna gain

AP antenna mounting options

APs can be mounted in one of three orientations (see Figure 9-22). The placement of your APs will vary on a situational basis. A typical drop ceiling is as close to a perfect mounting place as you can get for most AP deployments. Cables are usually easy to run above a drop ceiling and the porous nature of the material provides attenuation of the signal coming from the back of the AP.

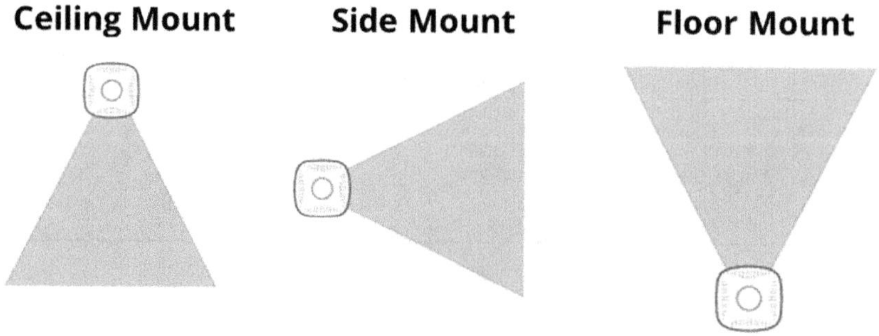

Figure 9-22: AP antenna mounting options

If a ceiling mount is not available, a side mount can be considered. Reasons to use a side mount include the ceiling being too high, forcing the AP to be too far from clients. Also, aesthetics play a key factor in AP placement as well as cable location. If the ceiling is not available due to height, cabling restrictions, or aesthetics, consider a side mount on walls or columns. Additionally, when side-mounting an AP, special mounts can be used to orientate the AP like a ceiling mount while not. Another consideration with a side mount is that the built-in antenna might not be adequate to radiate a signal an appropriate distance horizontally; therefore, you might need to use an AP that supports external antenna options.

In very high-density deployments, a floor mount may be considered. Mounting APs under the floor or seats in a stadium, for example, creates pico cells. Pico cells are very small service areas by design. You can learn more about pico cells designs in Aruba's Architect (Design) Certification track.

Note

Most Aruba APs are meant to be mounted on the ceiling, with the exceptions of the desktop and hospitality series, which are meant to be side-mounted. If you need to mount an AP, like a hospitality AP on the ceiling, you will negatively affect its RF coverage. Therefore, you need to carefully consider where you will be placing your APs to determine the correct model and whether you can use internal versus external antennas.

Antennas and transmissions

This section introduces you to how APs send and clients receive RF transmissions.

SISO: Legacy 802.11a/b/g devices

Legacy 802.11 networks use single-input single-output (SISO) radio technologies where only one antenna is used for transmitting or receiving at a time (see Figure 9-23). SISO does not make full use of a radio with multiple antennas. The transmitter sends the signal out a single antenna. While the receiver will hear the transmission on both antennas, it will only use one of those signals.

Remember, SISO is largely a legacy operating procedure. The benefits of beamforming and MIMO will be covered shortly.

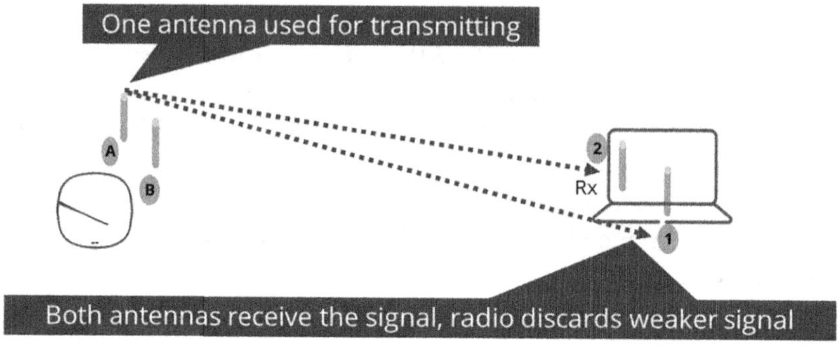

Figure 9-23: SISO transmissions

Multi-user multiple input multiple output (MU-MIMO)

Multi-user MIMO (MU-MIMO) uses beamforming technology to send traffic to multiple clients in the same transmission. The AP will use a channel sounding procedure to which the clients will respond. The AP uses this information to build a steering matrix. The AP uses the steering matrix to manipulate its transmission, coordinating in phase spatial streams for each client when the frames intended for them egress the radio (see Figure 9-24).

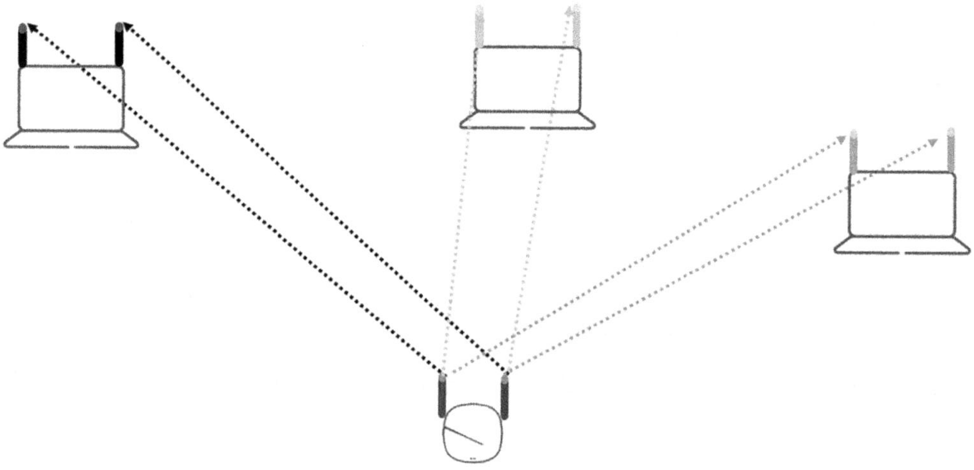

Figure 9-24: MU-MIMO transmissions

Note

Due to the nature of MU-MIMO, multi-user clients physically near each other do not receive the benefits of beamforming as their transmissions will interfere with one another.

Beamforming and MU-MIMO

Beamforming technology allows an AP to locate client's antennas and modulate its spatial streams in phase. When spatial streams are modulated in phase, two or more copies of the same signal are generated from the AP's radio. The AP times the transmission from each antenna, so the spatial streams reach the receiving antennas at the same time, in phase, increasing the received signal strength at the radio.

Remember, signal strength and noise floor are key factors in determining the data rate. Beamforming allows for increased signal strength at the client radio, without adding more power from the AP, which can increase the noise floor.

When deciding on an AP model, supported spatial streams may be a key factor: different products have different capabilities. Figure 9-25 shows an example of the Aruba 515 AP. Note the difference between MU spatial streams and MU stations: Multi-user stations is number of clients an AP can serve in one downstream with the MU-MIMO transmission. In the example shown in Figure 9-25, the AP can generate four spatial streams, but can only interact with three clients at a time. This would seem like a limitation, but in reality, the AP can send two streams to one station and a single stream to three stations. Or it could send four streams to just one station. Depending on traffic flows between the AP and client, the AP can adapt to provide the best service.

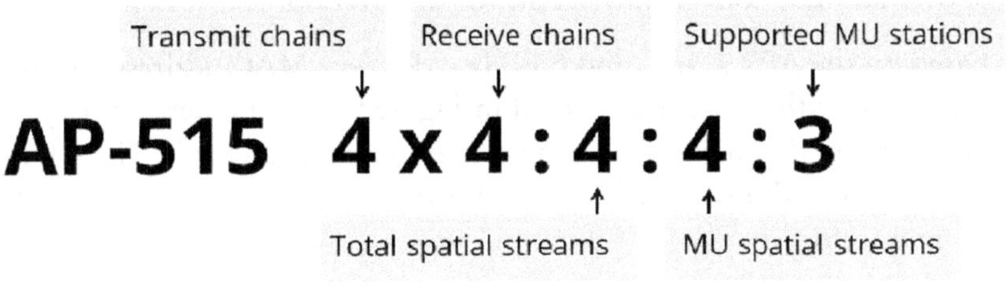

Figure 9-25: Beamforming and MU-MIMO

Obviously, your client device would need to support the number of spatial streams that your AP supports to maximize transmission speeds. Most smartphones support two spatial streams. The same is true of tablets. Most laptops will support either two to four spatial streams, depending on the Wi-Fi adapter that is installed. A good summary of client capabilities of common products can be found at https://clients.mikealbano.com/, where the information provided there is crowd-sourced.

WLAN mobility

One of the many great advantages of Wi-Fi is end-user mobility. This section briefly introduced mobility and roaming concepts.

WLAN physical and logical configuration

There are two types of configurations involved with a WLAN:

- Basic Service Set (BSS)
- Extended Service Set (ESS)

The following two sections provide an introduction to these configurations.

Basic Service Set (BSS)

A BSS includes an AP radio and all associated clients in a coverage area. An AP with two or more radios therefore has more than one BSS. When you define an SSID (service set identifier, i.e., the name of the WLAN), it is assigned a different 48-bit MAC address on each of the AP radios, which is used to identify the SSID *and* the radio it is advertised on. The radio MAC address, known as the BSSID, identifies the BSS. All WLAN clients associated to the radio use this BSSID as the destination MAC address on 802.11 frames; and the AP understands, based on the received radio and the BSSID MAC address, with which SSID the 802.11 frame is associated.

An ESS includes all APs and clients associated to the *same* logical network name, commonly referred to as an SSID. The SSID name is case-sensitive and identifies the WLAN to the client. An AP beacon sends both the BSSID and the ESSID into the air for discovery of the WLAN, by default.

All BSSIDs are based on the AP's 2.4, 5, and 6 GHz radio *base* MAC address. So the first SSID you define on AP1 might end with aa:aa:aa, as shown in Figure 9-26. The next SSID you define on that AP might end with aa:aa:ab, the third with aa:aa:ac, and so on and so forth. The 5 GHz radio's SSIDs would be based on its MAC address, like bb:bb:bb for the first SSID, bb:bb:bc for the second MAC address, and so on and so forth.

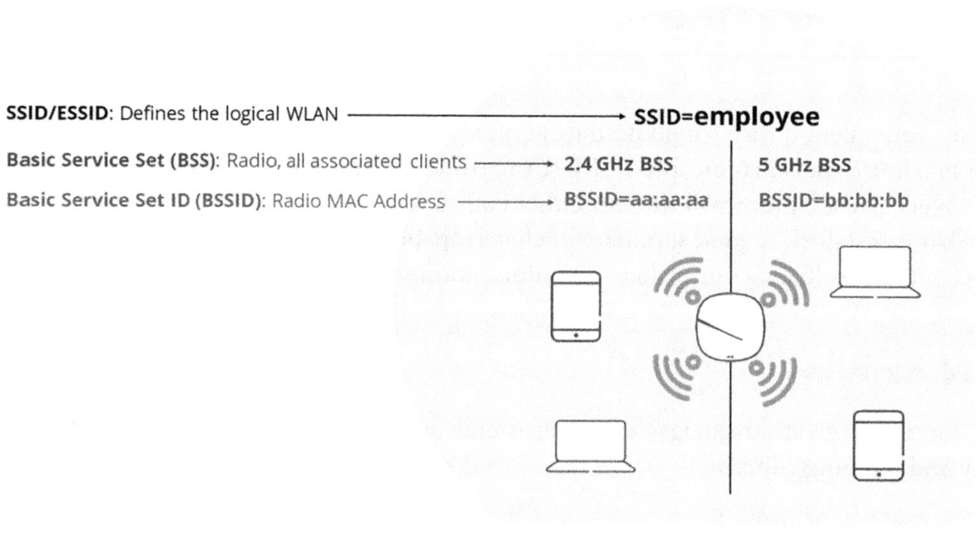

Figure 9-26: BSS

Extended Service Set (ESS)

Figure 9-27 shows an ESS architecture applied to multiple APs. Note that the WLAN advertises the same SSID across multiple APs (*employee*). However, each AP radio will have its own unique BSSID MAC address for that SSID. This allows the APs to understand when traffic is sent to one of their own radios versus a different AP's radio.

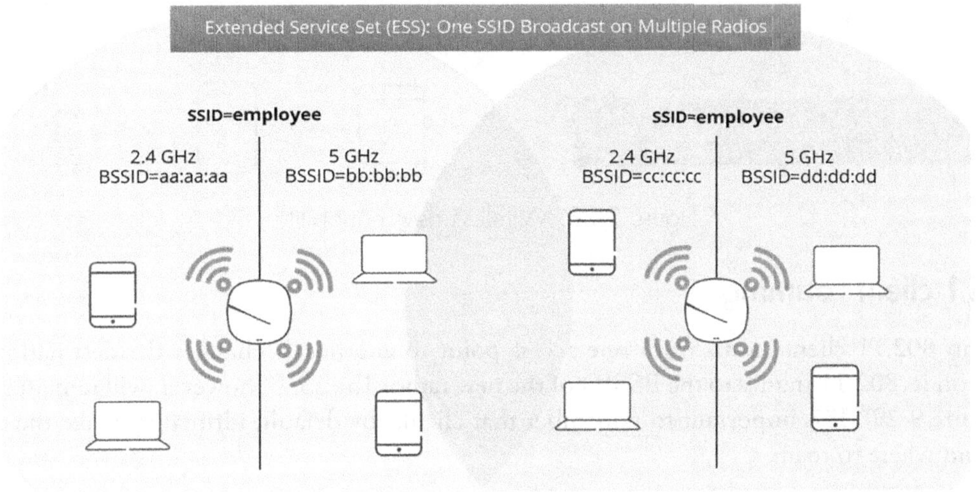

Figure 9-27: ESS

Wireless device mobility

Different devices have different functions on the network as well as different roaming capabilities (see Figure 9-28):

- Stationary devices, such as printers and smart TVs, almost never move when associated to a WLAN; therefore, they are not expected to roam from one AP to another while in use.
- Somewhat mobile devices, such as handheld scanners, laptops, and some IoT devices, may move when in use and therefore might roam from one AP to another.
- Highly mobile devices, such as tablets, phones, and some IoT devices, are more likely to require smooth roaming as they are moving around the network.

These client definitions are generic and subject to change based on the actual function of the device. For example, an iPad in a lobby that is bolted to a kiosk, which is used to register visitors, is a stationary device, even though it is a tablet. Understanding the roaming capabilities of your devices is very important in designing an optional wireless solution.

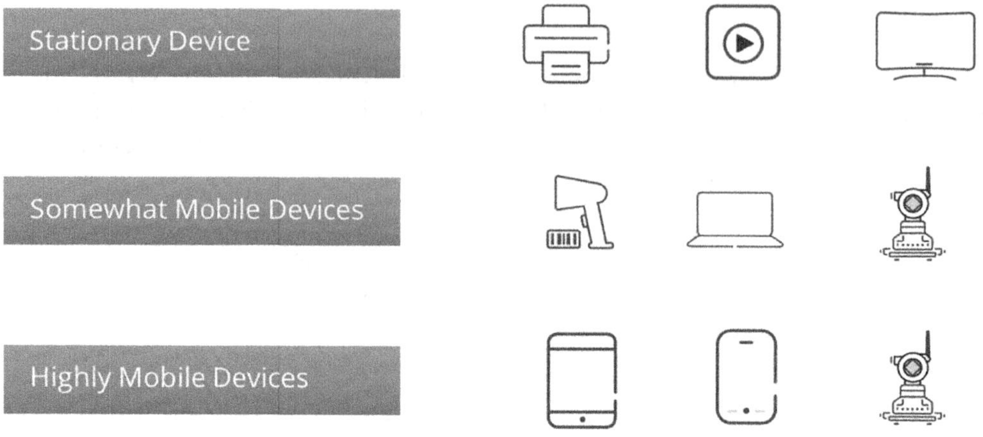

Figure 9-28: Wireless device mobility

802.11 client roaming

When an 802.11 client roams from one access point to another, it changes the destination MAC address on its 802.11 frames to the BSSID of the new radio. The *SSID*, however, will remain the same (see Figure 9-29). It is important to remember that clients, by default, ultimately make the decision when and where to roam.

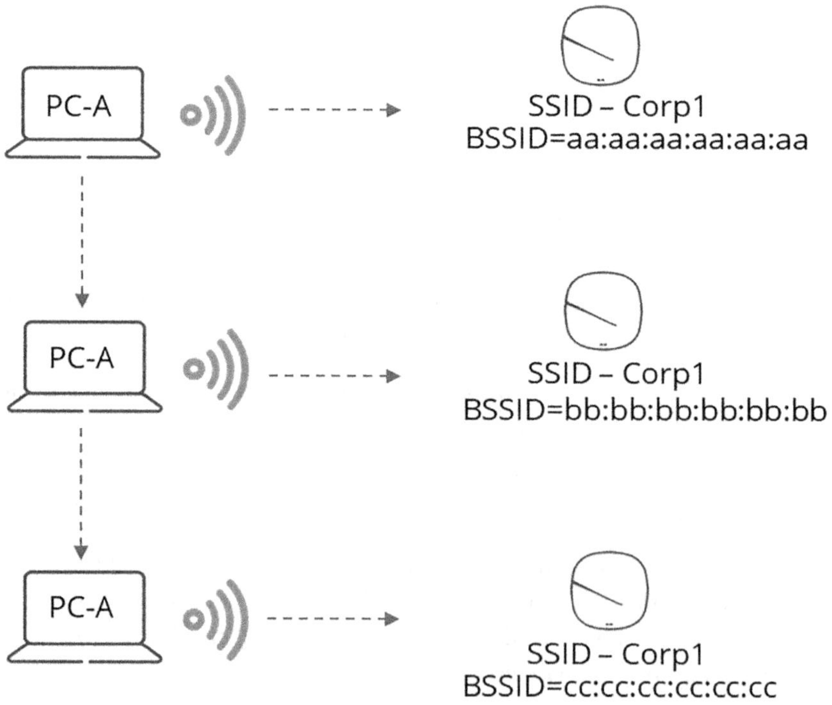

Figure 9-29: Roaming

802.11 frames

This section introduces 802.11 frames and their types, and how these are used in AP communications, like a client connecting (*associating*) to an AP.

802.11 frame types

All 802.11 frames fall under one of the three types (see Figure 9-30):

- **Management**: Management frames are used to manage the wireless cell or BSS. This includes probing, associating, roaming, and disconnecting clients from the BSS.
- **Control**: Control frames are used to control access to the medium and are used for frame acknowledgement. Control frames only contain a header and trailer (no body).
- **Data**: Data frames are used to carry user data.

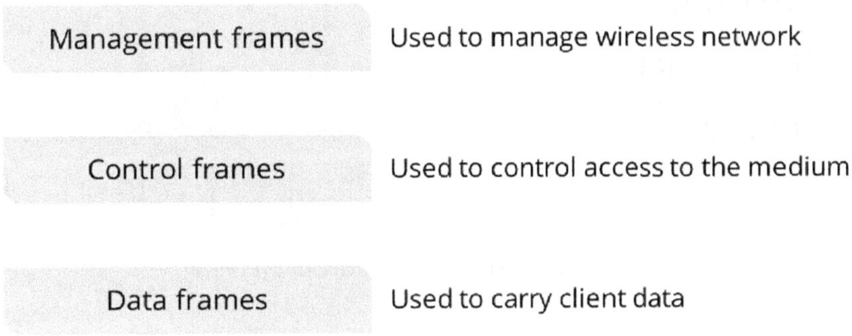

Figure 9-30: 802.11 frame types

AP communication overview

Figure 9-31 has an overview of how a client connects to an AP. Client Wi-Fi devices automatically scan the respective RF spectrum to see the available SSIDs. They do this by looking for beacons that APs broadcast. APs transmit a beacon every 100ms, by default. These beacons contain basic information about the AP and the services offered. Among other things, the following information is included in each beacon:

- **Timestamp**: Synchronize devices
- **Beacon interval**: The time interval between beacons
- **A list of available SSIDs**: Control which supported SSIDs are advertised in beacons and which are not. You will typically allow your APs to broadcast guest SSIDs and disallow them from advertising corporate and employee SSIDs.
- **Supported data rates**: Speed of the client and AP communications.

Beacons allow end stations to passively discover available SSIDs. When you view the list of available Wi-Fi networks on your smartphone or laptop, you are seeing the SSIDs that have been broadcast by APs in the area; this is perfect for guests to easily discover and access your guest network.

But what about those corporate SSIDs that are not broadcast in the beacon for all to hear? Your employee devices are manually configured with the appropriate SSID, so those stations need not passively wait to hear a beacon. They actively transmit a probe request frame to connect to the employee network.

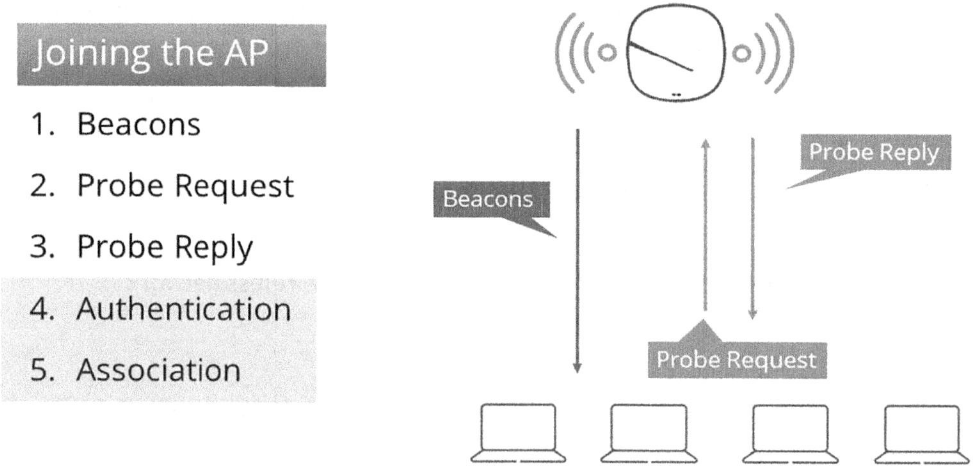

Figure 9-31: AP communication overview

When you try connecting to the WLAN, your device will send a probe request to the AP. The AP will then reply with a probe response, confirming the network information learned from the beacon. At this time, negotiation of the authentication protocol is done, like open authentication or WPA, WPA2, WPA3 personal or enterprise authentication (you'll learn more about the WPA standards in chapter 12) and the association can begin.

Note

It is best to limit the number of SSIDs advertised in beacon frames. Every SSID increases the size of the beacon frame, which then takes more *air time* from everything else. Too many APs transmitting too many SSIDs, every 100ms, can begin to impact performance. One or two SSIDs are perfect. Most wireless administrators agree that staying below four is a reasonable target in a corporate production environment.

Connecting to the AP

The Wireshark frame capture in Figure 9-32 shows a WLAN client connecting to an SSID (CorpWLAN). The client first discovers the network with beacons and probes. After discovery is complete, 802.11 authentication negotiation and association begin.

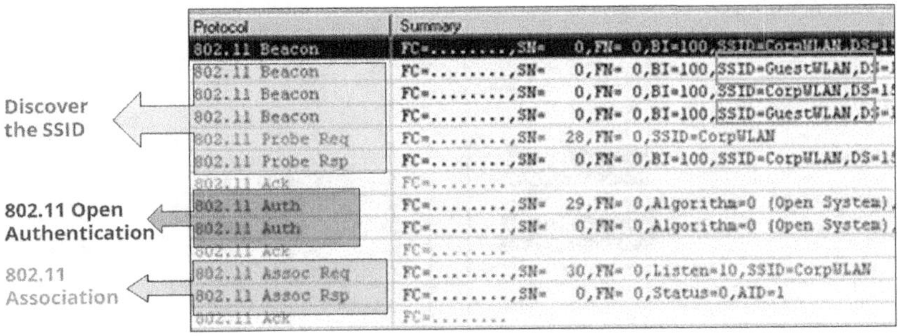

Figure 9-32: 802.11 Association

Leaving the AP

When a client leaves an AP, it notifies the AP with a de-authentication and disassociation handshake (see Figure 9-33). For example, clients might decide to leave an AP because they are too far away and cannot meet minimum data rate requirements, and thus will roam to a different AP radio.

APs can also initiate the de-authentication and disassociation frames to clients in order to move a client to a better AP or to disconnect the client from the WLAN, perhaps because of a security or other policy violation.

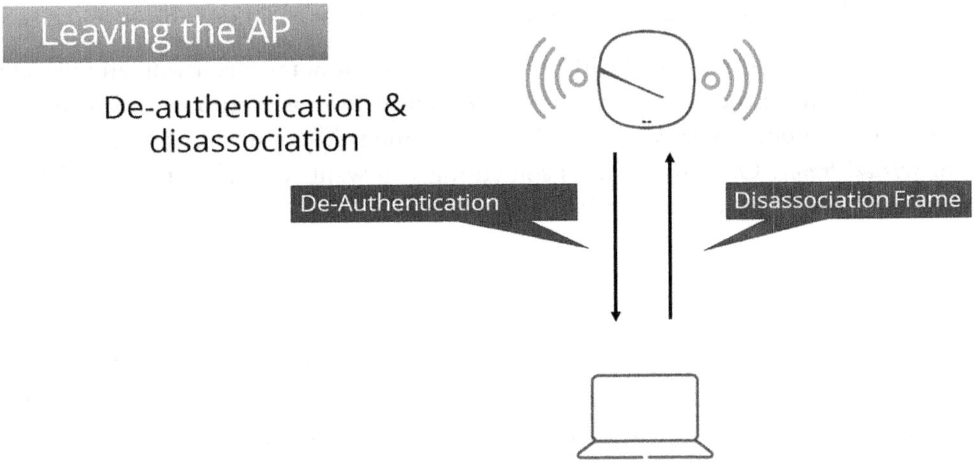

Figure 9-33: De-authentication and disassociation

802.11 CSMA-CA control frames and timers

An RF channel is a contention-based, collision avoidance medium (see Figure 9-34). This means that all clients are competing for a chance to send their traffic. Let's discuss how the collision avoidance medium works with Wi-Fi control frames.

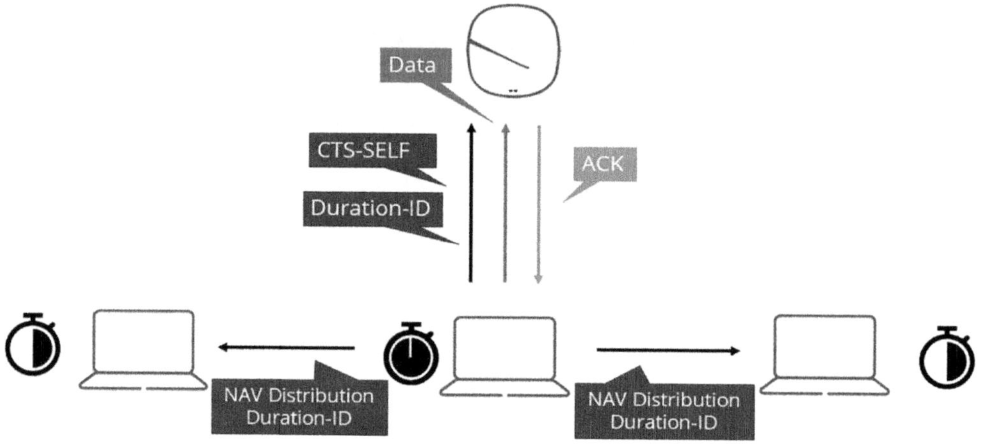

Figure 9-34: CSMA-CA control frames and timers

The first mechanism you must understand is the countdown timer on each client. When the countdown reaches 0, the client has a transmit opportunity or TXOP. When a client reaches its TXOP state, it will alert the AP and all other clients in the cell with a clear-to-send (CTS) message: The CTS-to-self announcement contains a network allocation vector or NAV. The duration-ID portion informs the other clients of how long the transmission will take. It is important that the other clients can hear this notification. If two or more clients attempt to use the medium at the same time a collision will occur and both clients must re-attempt the transmission again.

Immediately after sending CTS-to-self with the NAV distribution, the client will send its data frames and receive an ACK from the AP. If any part of the transmission is corrupted due to high noise or collisions, the entire process must start over. When a frame is lost and must be resent, it is called a L2-retry or *retried frame*. Obviously, this is something you want to track, since a high number of retried frames will negatively impact the WLAN performance.

WLAN design

This section introduces you to coverage versus capacity design and things you should consider in your design, like channel utilization, RF interferers, and noise. Lastly, you'll be introduced to some Aruba Wi-Fi features, AirMatch and ClientMatch, that will make your users' Wi-Fi experience much better.

Coverage vs. capacity design

WLAN designs are often more complex than simply asking "is it a coverage or capacity design?" For the purpose of this book, you will look at coverage and capacity in a WLAN from a high-level perspective (see Figure 9-35).

A WLAN designed for coverage can be implemented in environments where a few clients are spread across a large area, such as a warehouse, a shipping yard, or marina. These APs will have higher transmit power levels and typically only a few clients associated to them.

Note

During this discussion, when referencing the *AP*, of course, it means the radio(s) within the AP.

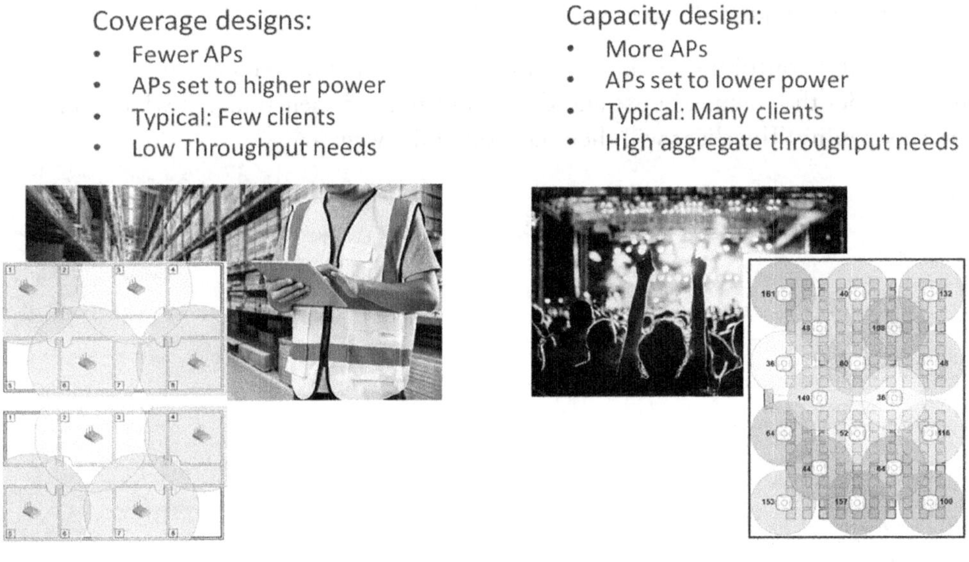

Figure 9-35: Coverage versus capacity design

A capacity design requires serving many clients that are close together. Auditoriums and stadiums are examples of areas that require a high-capacity design. High-capacity designs require that APs use lower power settings. Lower power settings mean the APs cover a smaller footprint on the floorplan, therefore servicing fewer clients. Now instead of, say 90 clients sharing the bandwidth of 1 AP, perhaps only 20 or 30 clients share that bandwidth. Yes, you spend more money on APs for this high-density deployment, but using fewer APs than needed is a bit like getting a great deal on a pair of shoes that do not fit, and the pain everyone feels would be very aggravating.

Also, when you squeeze APs closer together, you need to carefully plan the layout and power settings to avoid co-channel interference. Two APs on the same or even adjacent channels can interfere with each other, drastically reducing performance. This is all about proper placement of APs within the physical space, along with proper tuning of power and channel settings.

Many seasoned Wi-Fi design and deployment engineers have been successful by placing APs where their power settings are optimal at about 30-50% of maximum power. This, of course, can vary greatly based on several factors beyond the scope of this book. This power setting range ensures that AP power can be adjusted up or down to accommodate inevitable changes in the RF space, like an AP that needs to reboot because its firmware was updated.

Channel utilization

Channel availability is a challenge to wireless network performance. High channel utilization will lead to a lack of transmit opportunities for clients. A channel is considered utilized when a device is making a transmission on the channel (see Figure 9-36).

Interference will also contribute to channel utilization. While interference on a channel is present, clients may consider the channel in use and decide not to send their traffic at what would have been a transmit opportunity. The client will then back off and try again.

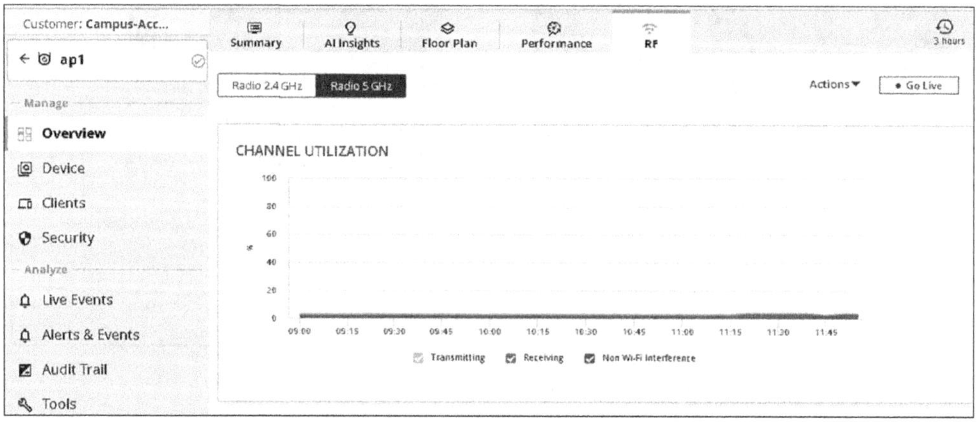

Figure 9-36: Channel utilization

RF WLAN interferers

Video cameras, cordless phones, third-party APs, microwave ovens, and Bluetooth devices are some of the most common sources of WLAN interference (see Figure 9-37). A spectrum analyzer is an effective tool that identifies and locates potential sources of interference. Spectrum analysis is covered in more depth in chapter 14.

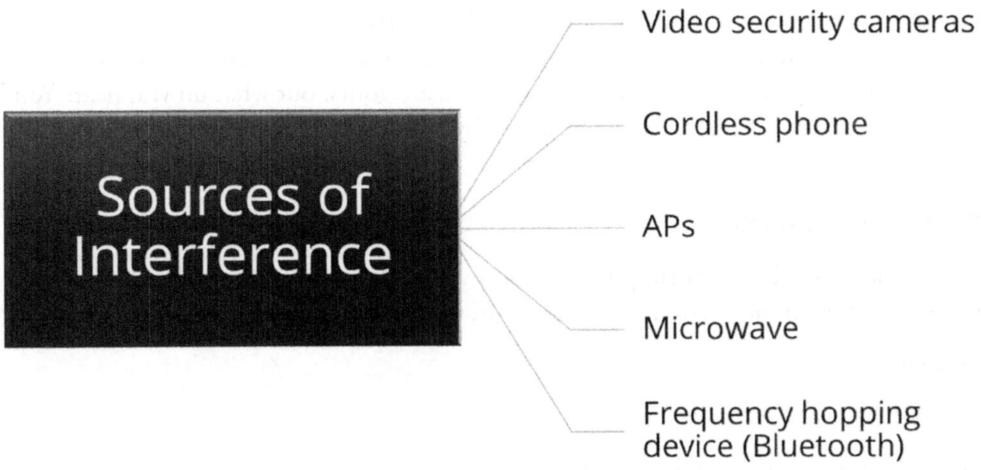

Figure 9-37: RF WLAN interferers

Things that negatively affect RF signals

One of many things that affects an RF signal as it travels is free space path loss (FSPL). Examine the left side of Figure 9-38. Note how a fresh signal from the AP weakens and broadens as it moves further from the AP. As the signal's power decreases, the signal-to-noise ratio (SNR) decreases, and the data rate drops.

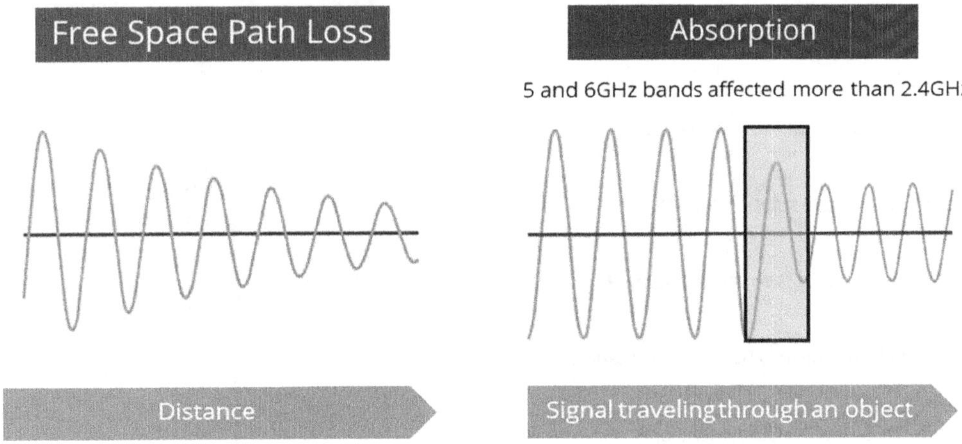

Figure 9-38: Things that negatively affect RF signals

Another factor that negatively affects an RF signal is absorption (see the right side of Figure 9-38): the signal power decreases as it moves through an object, like a wall. The denser the object (and the material from which it is made), the more the power decreases.

Note that the higher frequency bands like 5 and 6 GHz are more affected by absorption than the lower frequency 2.4 GHz band. Relate this to an audio system. Someone in a car near you might be playing their music obnoxiously loud. Their windows are up, as are yours, but what do you hear? You hear the low frequency bass sounds, which are less attenuated by obstacles than the high-frequency treble sounds.

Aruba WLAN features

Aruba has two features their APs support that deal with RF signal issues and ensuring users get the best experience when connected to an Aruba wireless solution:

- AirMatch
- ClientMatch

The next two sections introduce these features.

AirMatch

AirMatch provides automated RF optimization to improve a user's experience by providing a stable network that dynamically adapts to the changing RF conditions. AirMatch, which runs in Aruba Central, is a centralized RF planning and optimization service. As shown in Figure 9-39, Aruba Central receives telemetry data from APs for radio measurements, channel range, EIRP range, operational conditions, and local RF events like radar detection or high noise. AirMatch also *minimizes* channel coupling, where adjacent radios are assigned to the same channel.

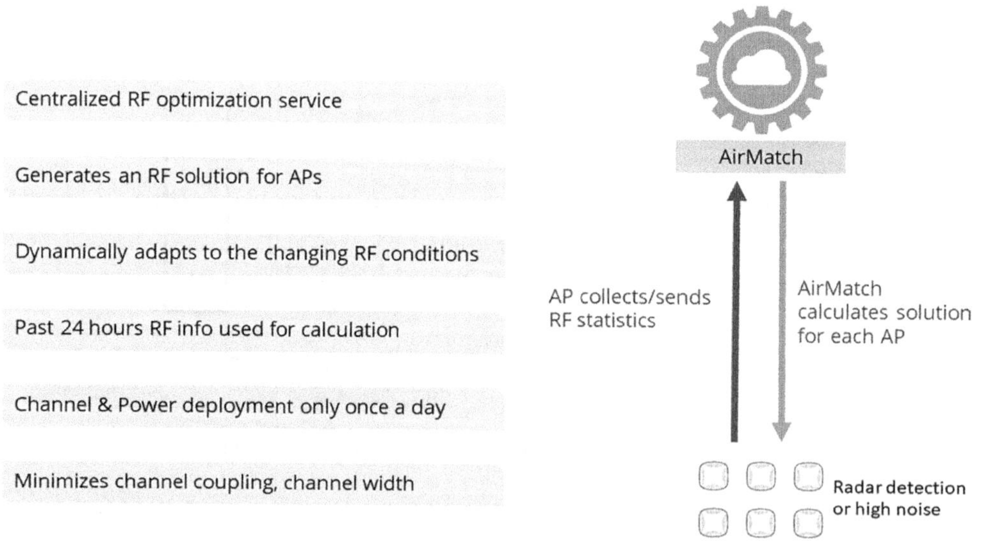

Figure 9-39: AirMatch

Aruba Central then models the RF network as a whole and devises a channel and power plan for the RF environment. AirMatch consumes the RF information and generates an RF solution that specifies new channel, bandwidth, EIRP, and mode of operation per radio. Central forwards this information to the APs and the APs apply the change in their radio configuration and operation. AirMatch calculates an RF channel and power level solution based on the past 24 hours RF information collected for each AP. The channel and power deployment process happens once a day at 5 AM, by default; however, deploying the channel plan can be changed based on your environment.

AirMatch also provides these additional features:

- Providing the neighboring AP list to key management services to implement other features.
- Provide AP partitions to implement a live upgrade service (discussed in Chapter 13).

You will learn how to configure and use AirMatch during the lab activity.

ClientMatch

ClientMatch is an Aruba patented technology that provides client load balancing across radio bands and AP radios. ClientMatch helps to influence sticky clients to connect to the best AP. When clients associate to the WLAN, their firmware used for their wireless NIC determines which radio, and on which AP, the client will connect. Sometimes the vendor's firmware is intelligent, and looks at criteria like the strength of the signal and the SNR, to name a couple, when determining which radio to associate. Other times, a vendor's firmware might use the last association the NIC had when it was in this area (this is called a *sticky* client). A lot of things could have changed since then, plus, the client doesn't understand what's actually happening on a radio when it connects: for example, it doesn't know how many other devices are currently associated to that radio.

ClientMatch is a feature that matches a client to the most appropriate AP and frequency in the area. When APs see a probe request from a user, they all report that to Aruba Central. Aruba Central then looks at a lot of criteria to determine which AP and radio would be best and this is shared with the original AP that received the probe. The AP will then redirect the user to the more appropriate radio, assuming that the most appropriate radio is not itself (see Figure 9-40). This is best done if the client supports IEEE 802.11v, which allows for this interaction between the client and the AP, which any client product sold today should support. ClientMatch thus load balances clients and if a client tries to connect to a less preferred frequency, like 2.4 GHz, the client can be redirected to a more appropriate 5 GHz or 6 GHz radio.

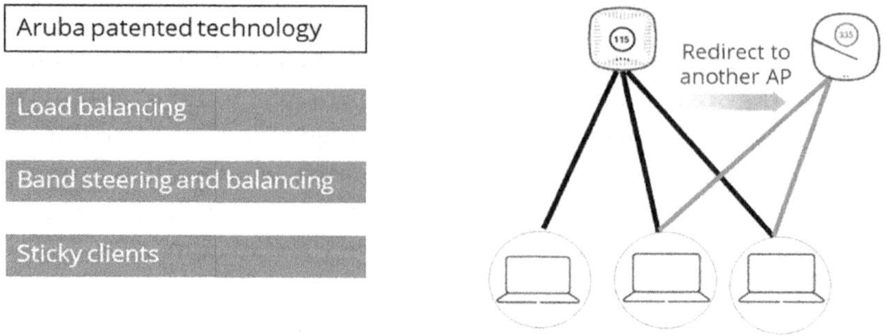

Figure 9-40: ClientMatch

WLAN setup

Now that you have an understanding of the basics of wireless LANs, let's talk about how you configure a WLAN on your Aruba APs using Central.

WLAN wizard

The WLAN wizard (see Figure 9-41) guides you through the SSID creation process. There are five tabs in this wizard:

- General
- VLANs
- Security
- Access
- Summary

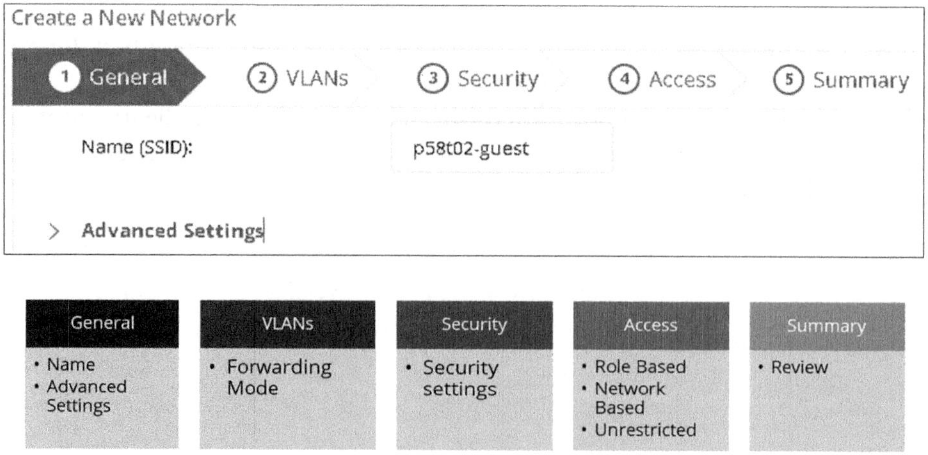

Figure 9-41: WLAN wizard

Under the General tab you can give the new SSID a name. Notice that the SSID name is the same as the one your APs will broadcast and it is case-sensitive. Under the advanced settings you may fine-tune the new SSID setting.

Under the VLANs tab, you need to choose the traffic forwarding mode and the client VLAN assignment.

Note

This book covers the bridge forwarding mode. The professional-level certification covers the other two forwarding modes.

Under the Security tab, select the SSID security level:

- Enterprise security means the network will use 802.1X for user authentication
- Personal means you'll use a pre-shared key.
- Visitors uses captive portal to authenticate users.
- Open means that no user or device authentication required.

Under the Access tab, you will configure the firewall settings which includes three options:

- Role-based, where a different firewall role could be applied to each user.
- Network-based, where all the users connected to the SSID will receive the same firewall role.
- Unrestricted, where no enforcement will be applied (no restrictions).

Note

For simplicity, the same steps used to create WLANs for AOS 10 bridge, tunnel, mixed mode SSIDs, and AOS 8 IAPs WLANs are basically the same, with just a few small changes in the wizard based on the implementation you choose.

Traffic forwarding mode

As mentioned in the last section, this chapter will focus on the *bridge* traffic forwarding method. When you select bridge for the forwarding mode, you have three client VLAN assignment options (How the client device acquires IP addressing information):

- Static
- Dynamic
- Native VLAN

The next three sections cover these options.

Bridge and native VLAN

If you select a *Native VLAN* for the client VLAN assignment, the corporate network will provide the client with an IP address on the same subnet as the AP. In Figure 9-42, the corporate DHCP server provides the AP the 10.1.4.49 address. The user associates to the AP and receives the 10.1.4.57 address, which is in the same subnet as the standalone AOS 10 AP. This option is only available for bridged mode WLANs. This mode is commonly used for small-office/home-office (SOHO) environments.

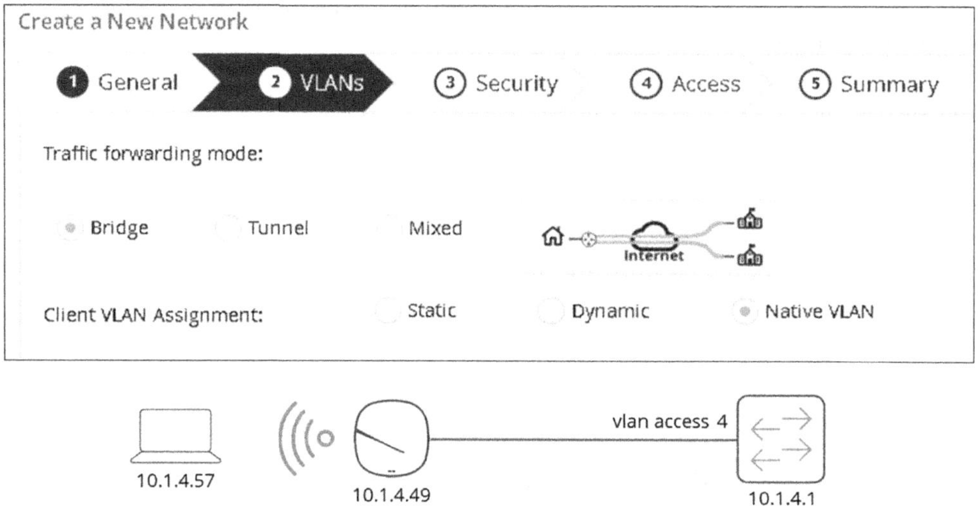

Figure 9-42: Traffic forwarding mode—bridge and native VLAN

Bridge and Static VLANs

If you select *Static* for the client VLAN assignment, you can statically set which VLANs the users are assigned, based on their SSID. In this configuration, the AP's IP address is in its own subnet. The administrator configures the SSID with a specific VLAN(s). If the WLAN is in bridge mode, then the VLANs are defined on the local switch's trunk port connected to the AP.

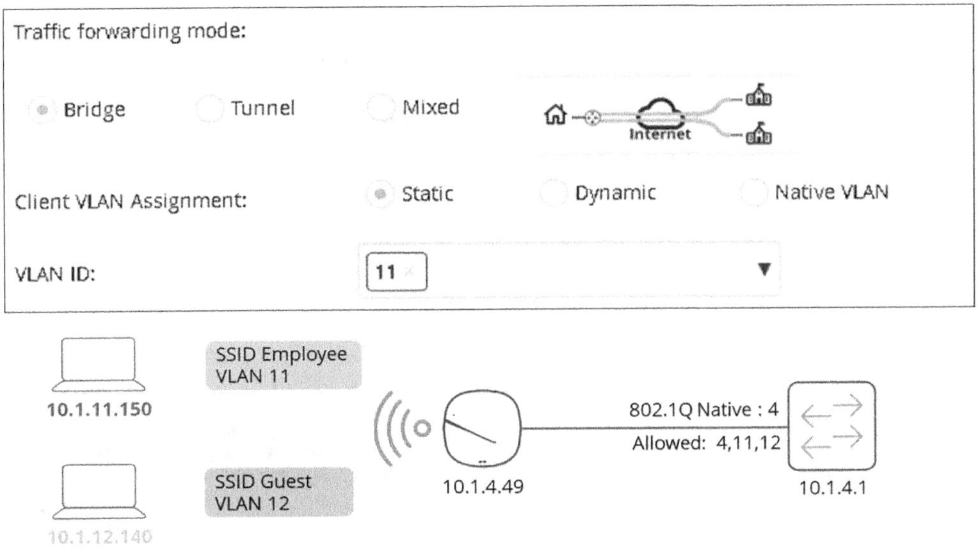

Figure 9-43: Traffic forwarding mode—Bridge and Static VLANs

In Figure 9-43, the AP is in VLAN 4. The users associated with the Employee SSID are assigned to VLAN 11 and the users associated with the Guest SSID are assigned to VLAN 12. For this to work, the switch port connected to the AP must be configured as a trunk port. The AP management VLAN is the trunk's native VLAN (untagged), and the client VLANs (11 and 12) are tagged on the trunk.

Bridge and Dynamic VLANs

If you select *Dynamic* for the client VLAN assignment, the VLAN is derived based on the client connection parameters, such as the client's user account, MAC address, device profiling, location, and other attributes. A centralized security server, like ClearPass, looks at this information and determines the VLAN assignment it will send in its RADIUS response to the AP. The RADIUS attribute the server replies with is *Aruba-User-VLAN* which is an Aruba-specific attribute The administrator needs to set up the VLAN derivation rules on the AP to ensure this is used. You can set a default VLAN in case the server does not return a VLAN value. Dynamic VLAN assignment is available for bridged and tunneled mode WLANs.

In Figure 9-44, if the user associating with the SSID receives a RADIUS attribute Aruba-User-VLAN 16, the user will be placed in VLAN 16. If the associating user has a MAC address that contains the following characters: 00:11:22 it will be assigned to VLAN 17. If the associating user does not match any of the assignment rules it will be assigned to VLAN 11 (defined as the default VLAN for the WLAN). Note that most enterprise companies will use the dynamic method to implement the VLAN assignment.

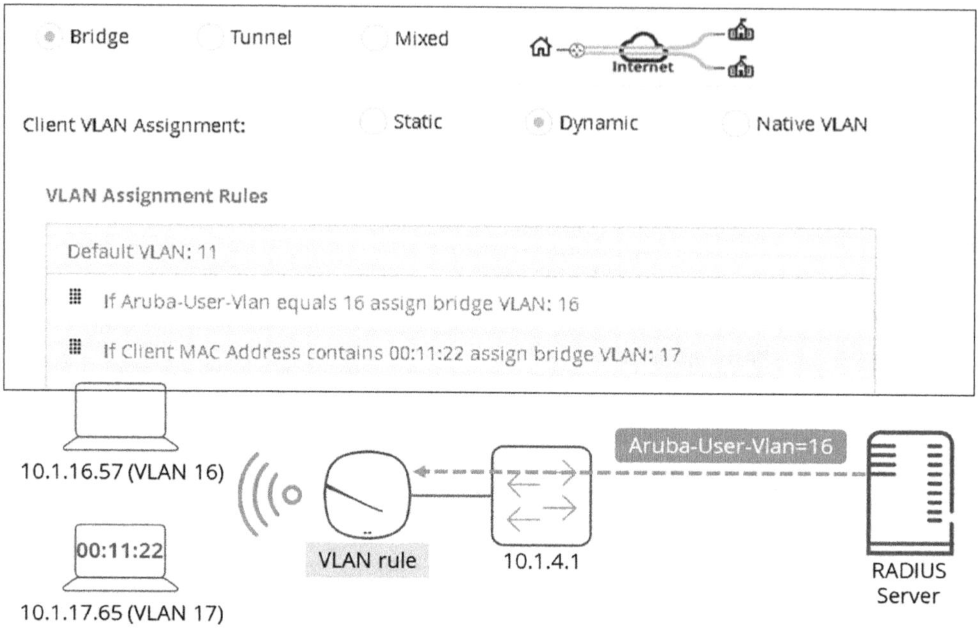

Figure 9-44: Traffic forwarding mode—Bridge and Dynamic VLANs

Security

You can configure a WLAN with the following security profile types (see Figure 9-45):

- **Enterprise:** for enterprise security, with user and/or device authentication.
- **Personal:** for personal networks, with a pre-shared key.
- **Visitors:** for guest networks with captive portal authentication.
- **Open:** for open networks, with no authentication.

Note

Keep in mind that WEP is supported for backward compatibility. It should be *avoided* on new WLAN implementations since it is very *insecure* and impacts and/or disables many of the modern functionalities such as MIMO/MU-MIMO, faster data rates, and much more. In short, WEP has been deprecated from the standards and should not be used in today's WLAN implementations.

Enterprise	• 802.1x • WPA3, WPA2, Dynamic WEP • External RADIUS or Cloud Auth
Personal	• PSK • WPA3, WPA2, Static WEP, MPSK AES, MPSK Local
Visitors	• External Captive Portal (IE: ClearPass) • Cloud Guest (Central CP) • Encryption optional
Open	• Enhanced Open • Open

Figure 9-45: Authentication options under the Security tab

These will be covered in more depth in Chapter 12.

Access

There are three main types of WLAN access rules (see Figure 9-46):

- Unrestricted
- Network-based
- Role-based

An unrestricted WLAN does not enforce any firewall policy, therefore allowing all traffic. All devices associated with this SSID can pass traffic unrestricted by the AP.

Network-based WLAN has the same rules that are imposed on *all* devices associated to this specific WLAN. For example, if there is a rule that blocks pings, then all devices on the SSID would be denied ping traffic. In other words, the firewall policies are the same for *everyone* in the SSID.

The role-based WLAN assigns roles to various devices in the SSID. Roles are derived on a user and/or device basis. This allows you to have different security policies and firewall rules for different groups of people, like sales, programmers, and accounting: each has their own firewall policies that affect what they can and cannot connect. If you have not assigned a role, then the AP will use the default role for the WLAN. For example, you may have one user receive the *Contractor* role, but the other users receive the *employee* role. In this situation both users are in the same subnet and in the same SSID, but they have different access rights that affect their traffic flows.

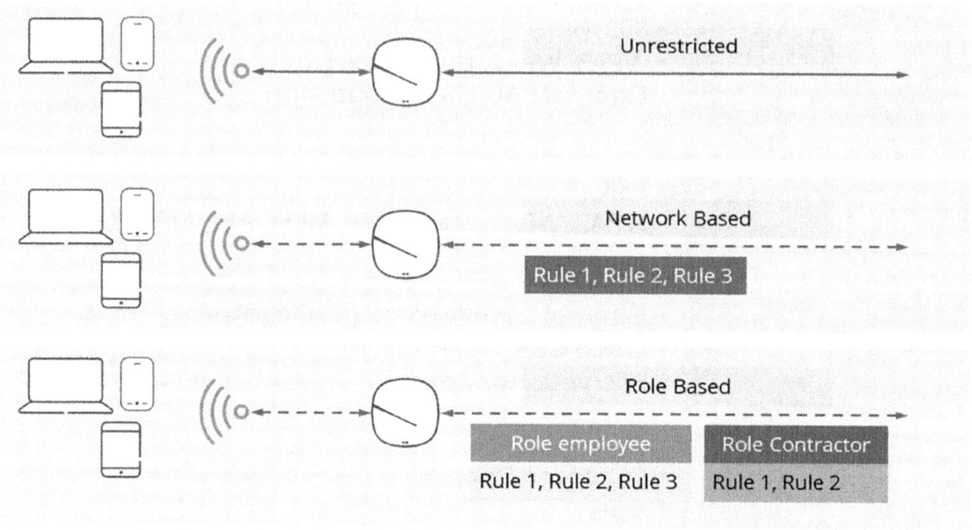

Figure 9-46: Role policy options under the Access tab

Once you have completed the configuration of this tab, you are taken to the Summary tab where you can view your configuration and then have Central apply it to the AP(s).

Monitoring clients

Figure 9-47 shows a typical List View output when monitoring clients in Aruba Central. You can also see this in a summary view. Notice that, by default, there is no way to differentiate bridged and tunneled clients. However, you may add a new column to the table by clicking the *three-dot* icon at the right side and selecting the check box for *tunneled*. You will learn how to monitor clients during the lab activity.

Client Name	Status	IP Address	VLAN	Connected To	AP Role	Gateway Role	SSID/...
employee1	Connected	10.3.91.252	91	AP515-f0	employee		employee-T
employee1	Connected	10.3.91.254	91	AP515-7a	employee	employee	employee-T
employee2	Connected	10.3.71.254	71	AP515-D2	employee		employee-B
employee1	Offline	10.3.91.253	91	AP515-D2	employee	employee	employee-T

Figure 9-47: Monitoring wireless clients in Aruba Central

Lab 9.01 WLAN fundamentals

In this lab, you will build a simple WLAN for AP network.

Overview

Now that the APs have been deployed in the network, your customer would like to understand what options Aruba Central offers with regards to RF monitoring and radio configuration. In this lab you will explore the configuration options using radio profiles to control the channels and power settings. You will review the centralized channel and power planning offered by AirMatch and how you can see the applied settings from the AP and in Aruba Central. The last section will show how to deploy a WLAN to a limited set of APs in a group.

Objectives

- Create a WLAN.
- List the current radio profile use.
- Change a radio profile.
- Review the channel and power changes.
- View the AirMatch global schedule.
- Restrict a WLAN to a subset of APs in a group.

Task 1: Review radio default channel and power

In this task you will create an example WLAN and review the default AP radio channel and power settings. You will explore AP BSSID information and see how radio channel utilization and power settings can be visualized in Aruba Central.

Objectives

- Review the default radio profile.
- Explore AP BSSID information.

Steps

Review the radio status at the group level

First review the current radio settings at the group level.

1. In Aruba Central, navigate to: Context: **Groups / campus-wifi-ui** > Navigation: **Devices** > Top: **Access Points** > Right-top: **List** > **Access Points**.
   - **Question**: Are there two access points connected to Central? How can you see that an AP is connected to Central?
   - **Answer**: Yes, both APs are connected. They have status *Online* with a green dot next of them (see Figure 9-48).

Figure 9-48: AP online status

2. Click **Radios** option.
   - **Question**: What is the status for your radios?
   - **Answer**: All radios are currently offline.

Review the radio status at the device level

You can also review the radio settings at the individual AP device level.

1. Click **ap1** in the Access Point list. Central will take you to the ap1 device-level *Overview* page. You are now at this location in Central: Context: **Device / ap1** > Navigation: **Overview** > Top: **Summary**.

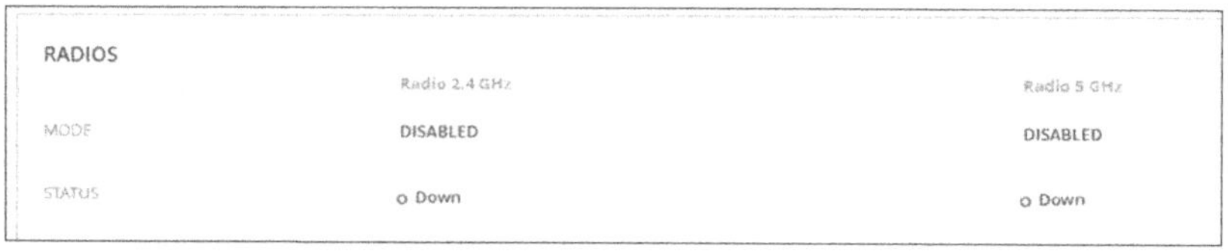

Figure 9-49: AP radio status

- **Question**: After the Device and Network applet, the Radios applet is displayed. What is the status for the radios 2.4 and 5 GHz?
- **Answer**: Both radios are *DISABLED* now.
- **Question**: Why would the radios currently be disabled?
- **Answer**: There are no WLAN SSIDs configured for the APs: the radios remain disabled (see Figure 9-49).

Configure WLAN

You will now configure a very basic, open SSID to enable the radios. Later you will apply additional configurations to this WLAN.

1. In Aruba Central, navigate to: Context: **Groups / campus-wifi-ui** > Navigation: **Devices**> Top: **Access Points**> Right top: **Config**.
2. Under the WLANs tab, click **Add SSID** to start the WLAN Wizard.
3. From the General tab, configure name **p#tX-guest**.

Note

Make sure to replace **#** with your assigned pod number and **X** with your table number (01-14, for example). For example, if you are assigned Table 07 in Pod 28, your WLAN name will be: **p28t07-guest**. This indicates Pod 28, Table 07. Please check with your instructor or lab administrator if you are not sure about the Pod and Table number. This procedure ensures that you will be able to recognize your own SSID in the WLAN list, even when multiple pods (classrooms) are active at the same time.

4. Click **Next**.

5. On the VLANs page, leave the traffic forwarding mode to bridge mode (default).

Note

The use of tunnel and mixed forwarding requires an Aruba gateway in the deployment. At the Aruba associate level, only the bridge mode option is used. In the professional-level Aruba Campus Access course and book, you will learn about Aruba gateways.

6. For the Client VLAN Assignment, select **Static**.

7. Set VLAN ID to **15**.

8. **Remove** VLAN 1.

Note

Make sure to remove the default VLAN 1 from the VLAN list. If VLAN 1 were listed together with VLAN 15, the wireless clients would be distributed over *both* VLANs.

9. Click **Next**.

10. From the Security tab, move the slider to **open**. Leave other values at their defaults.

Note

By default, the *Enhanced Open* option is enabled. This provides automatic encryption on an open WLAN for clients that support Enhanced Open. Clients that do not support this feature will still connect without wireless encryption to this WLAN. This feature uses an additional hidden SSID for the encrypted traffic. The feature is known as Opportunistic Wireless Encryption (OWE), which is used in the hidden SSID.

11. Click **Next**.

12. From the Access tab, leave the default as **unrestricted.**

Note

The Access page controls the access control to the network, such as the firewall rules. The unrestricted setting translates to a default rule 'Allow any traffic to all destinations'.

13. Click **Next**.
14. To complete the configuration, click **Finish**.
15. Wait for the deployment wizard to complete and confirm by clicking **OK**.

Verify configuration push

After making a configuration change, you can verify that the updated configuration is pushed (synchronized) to the devices in the group.

1. Click the **Configuration Audit** tab (see Figure 9-50).

Figure 9-50: Accessing the Configuration Audit status tile

2. The Configuration Status tile should show: *Not in Sync 0 Device*. This indicates that all APs have synced the configuration *successfully*.

Note

It may take a minute for the changes to be queued. You might have to refresh the page a few times to see the configuration status change.

Verify the status of the new WLAN in Central

Now that the new WLAN has been created and pushed to the APs, you can use Aruba Central to verify the operational status of the WLAN.

1. In Aruba Central, navigate to: Context: **Groups / campus-wifi-ui** > Navigation: **Overview** > Top: **Summary**.

2. Scroll down to the WLANs table.
   - **Question**: Do you see the new WLAN in the list?
   - **Answer**: Yes, the configured guest WLAN is listed.

Note

It may take a few minutes for the WLAN to show up in the list. If any *SetMeUp* WLANs are listed, they will eventually disappear.

Review the base radio MAC addresses

Each AP-advertised WLAN is advertised with a user-friendly name, the ESSID, and a unique MAC address for that specific AP radio, the BSS. The endpoint should know the name (ESSID) to which it should connect. Then the endpoint's WLAN NIC queries the APs. Each AP responds with a list of the actual MAC addresses (BSS table) to which the endpoint can connect. Each AP radio has a MAC address range. These addresses are used to advertise configured WLANs.

In these steps you review the base MAC addresses of your lab environment radios. Each radio (2.4 and 5 GHz in this lab) has its own base MAC address.

1. In Aruba Central, navigate to: Context: **Groups / campus-wifi-ui** > Navigation: **Devices**> Top: **Access Points** > Top-Right: **List.**
2. Click the **Radios** tab (see Figure 9-51).

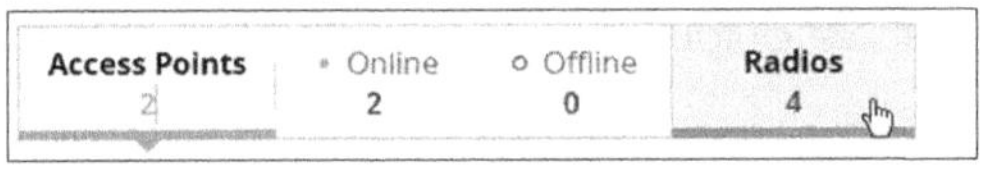

Figure 9-51: Accessing the APs' radios

3. A list of AP radios for the APs in this group is displayed (see Figure 9-52).

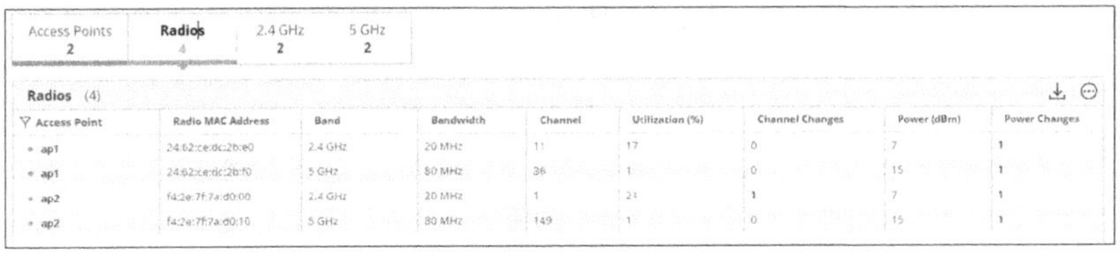

Figure 9-52: Viewing the APs' radios

- **Question**: How many radios are listed?
- **Answer**: There are 4 radios, 2 for each AP.
- Take note of your ap1 2.4 GHz radio MAC address: ______________________________
- Take note of your ap1 5 GHz radio MAC address: ______________________________

Note

You can also download a CSV file with the AP names, radio base MAC and band information using the download icon at the right-top (see Figure 9-53). An example output is shown in Figure 9-54.

Figure 9-53: Downloading the AP information

	A	B	C	D	E	F	G	H	I	J	K
1	ACCESS POINT	AP STATUS	RADIO MAC ADDRESS	BAND	BANDWIDTH	CHANNEL	UTILIZATION (%)	CHANNEL CHANGES	POWER (dBm)	POWER CHANGES	NOISE FLOOR (dBm)
2	ap1	Up	24:62:ce:dc:2e:10	5 GHz	80 MHz	52	0	0	18	0	-95
3	ap1	Up	24:62:ce:dc:2e:00	2.4 GHz	20 MHz	11	21	0	9	0	-94
4	ap2	Up	24:62:ce:dc:2e:d0	5 GHz	80 MHz	36	4	0	18	0	-92
5	ap2	Up	24:62:ce:dc:2e:c0	2.4 GHz	20 MHz	11	15	0	9	0	-93
6											

Figure 9-54: Sample output of the AP information in the file download

- **Question**: For each AP, what is the relationship between the radio base MAC address of the 2.4 and 5 GHz?
- **Answer**: The second, least-significant number is incremented with 1. Since MAC addresses are hexadecimal numbers, this is an increment of 16.

Here is an example radio MAC list (see Figure 9-55):

Access Point	Radio MAC Address	Band
ap1	cc:d0:83:ce:0a:00	2.4 GHz
ap1	cc:d0:83:ce:0a:10	5 GHz
ap2	cc:d0:83:d4:5e:40	2.4 GHz
ap2	cc:d0:83:d4:5e:50	5 GHz

Figure 9-55: Sample output of the AP information in Central

- **Question**: What is the significance of the number 16?
- **Answer**: The base radio MAC Address is the start of a range of MAC addresses. Each SSID (WLAN name) that this radio shall advertise gets the next available MAC address assigned as the BSS MAC address.

Review the BSSID MAC used by the WLAN

You recently saw the base radio MAC address. Next you see how these MAC addresses are allocated on the test WLAN that you created.

1. In the list of radios, click the AP labeled **ap1**. You will see ap1 device details screen. You are now at this location: Context: **device / ap1** > Navigation: **Overview** > Top: **Summary**.
2. Scroll down to see the WLANs table at the bottom of this screen.
   - **Question**: You have only configured one WLAN. Why do you see two ESSID names?
   - **Answer**: The *p#tx-guest* is the open WLAN. Since you have Enhanced Open enabled, an additional, hidden SSID was created to support Opportunistic Wireless Encryption (OWE). These WLANs start with *_owe*. They are effectively using the same settings as the open WLAN, but they have encryption enabled without need for a pre-shared key (PSK).
3. Expand the **p#tx-guest** WLAN to see the BSSID MAC that is used by this AP (ap1). Figure 9-56 shows an example screenshot.

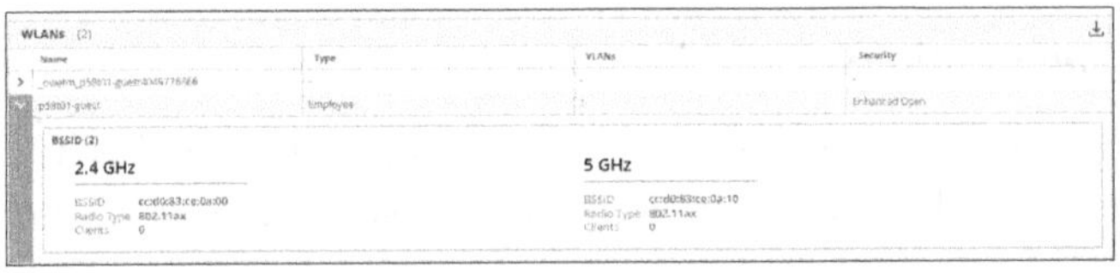

Figure 9-56: Example screenshot of the BSSID MAC addresses

- **Question**: Do you see a relation with the base radio MAC address?
- **Answer**: Yes, since this is the first SSID that is advertised by this radio, the first address from the radio MAC range is used.

4. Expand the WLAN that starts with **_owe**. See Figure 9-56 for an example.

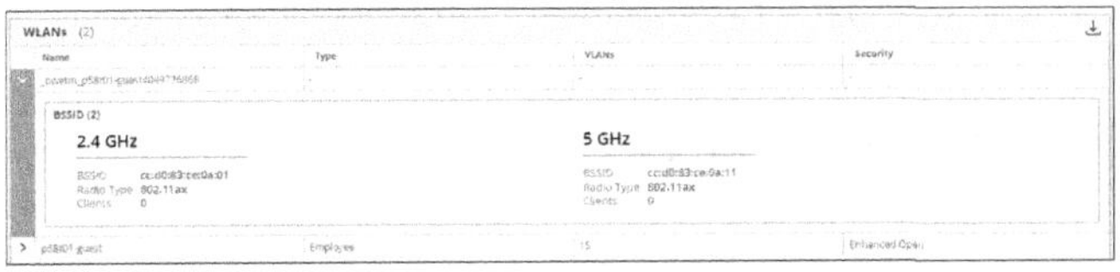

Figure 9-57: Example screenshot of the BSSID MAC addresses for the _owe SSID

- **Question**: What is the difference between this BSSID and the previous WLAN BSSID?
- **Answer**: Since this is the second SSID that is advertised by the radio, the next BSSID MAC in the range is used (see Figure 9-57).

Connect with client

Next, you will have PC1 to connect to your WLAN. Since PC1 has both a wired and wireless NIC, first check its network connections. The wired NIC should be disabled and the wireless NIC should be enabled.

1. Open a desktop session to PC1.
2. Right-click the **Network** icon in the Windows notification bar.
3. Click **Open Network & Internet Settings**, as shown in Figure 9-58.

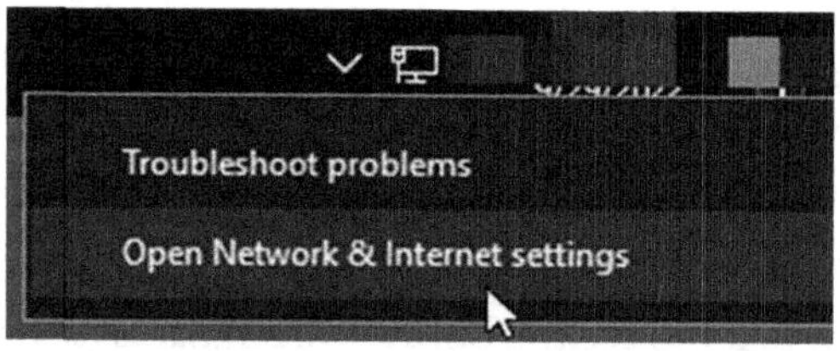

Figure 9-58: Accessing Network & Internet settings

4. Click **Change adapter options**.
5. In the Network Connections, **disable** the Lab NIC (this is the wired NIC connected to sw-edge1 port 1/1/1). (Right-click the wired NIC and select **Disable**, as shown in Figure 9-59.)

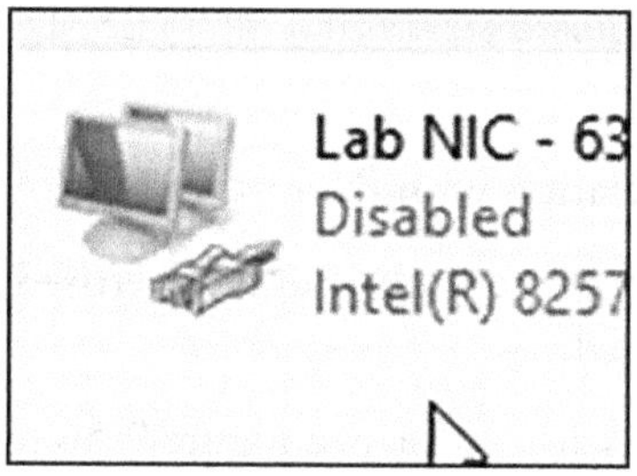

Figure 9-59: Disable the wired NIC

6. Make sure the Wireless NIC is enabled (see Figure 9-60).

Figure 9-60: Enable the wireless NIC

7. Click the **Network** icon in the notification bar to see the current list of WLANS (see Figure 9-61).

Figure 9-61: View the WLANs

8. In the WLAN list, look for your WLAN (*p#tX-guest*) and connect to it.
9. Open a Windows command prompt (cmd) and review your IP address using ipconfig.

```
C:> ipconfig
...
Wireless LAN adapter Wi-Fi Lab:

   Connection-specific DNS Suffix  .   : aruba-training.com
   Link-local IPv6 Address . . . . . : fe80::753c:42f5:cdf5:6f7a%19
   IPv4 Address. . . . . . . . . . . : 10.1.15.50
   Subnet Mask . . . . . . . . . . . : 255.255.255.0
   Default Gateway . . . . . . . . . : 10.1.15.1
...
```

- **Question**: What IP address is assigned to the wireless NIC?
- **Answer**: The wireless NIC should have received an IP address in the 10.1.15.0/24 range, for example 10.1.15.50.

10. To generate some traffic, start a continuous ping to 10.254.1.21 (your lab DHCP/DNS server). The ping should be successful.

```
ping 10.254.1.21 -t
```

Review the WLAN on the AP

You can review the BSS entries using the **show ap bss-table** command on the AP. Using Aruba Central, it is easy to send common CLI diagnostics commands to one or more devices.

1. In Aruba Central, navigate to: Context: **Groups / campus-wifi-ui** > Navigation: **Tools** > Top: **Commands**.
2. In the **Device Type** dropdown select **Access Point**.
3. Click **Available Devices** and select both **ap1** and **ap2**.
4. In the **Categories** list, select **All Category**. This will show all available commands.
5. In the **Commands** list, click the **filter** icon and type **bss**.
6. Select the **show ap bss-table** command, then click **Add** (see Figure 9-62).

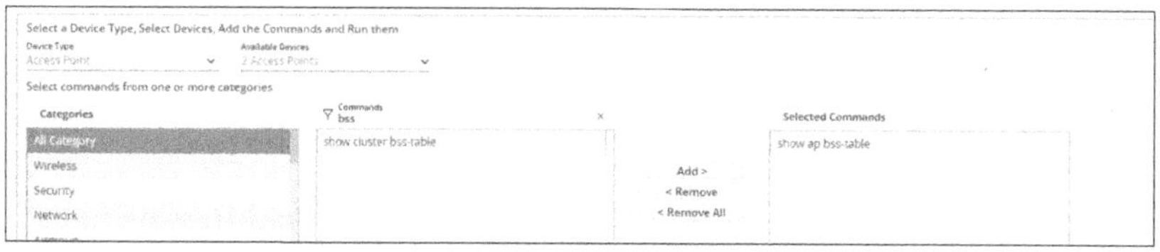

Figure 9-62: Preparing to send a command to an AP

7. Click **Run** to initiate the request. You will now see rotating dots while Central collects command output on both ap1 and ap2 (see Figure 9-63).

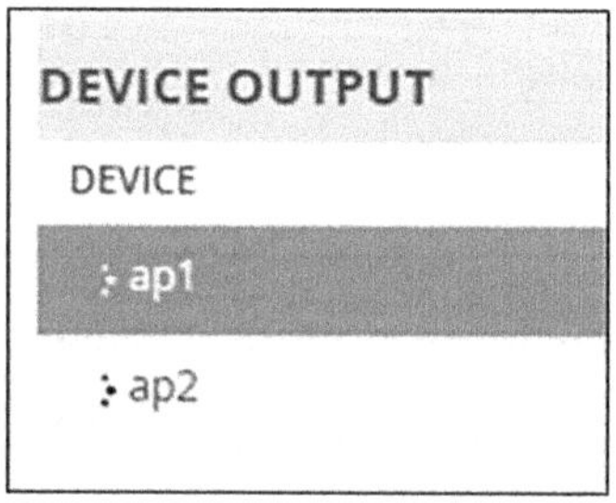

Figure 9-63: Waiting for the command output from an AP

8. After a few moments, the output will be collected. You can detect this based on the confirmation icon next to the device names (see Figure 9-64).

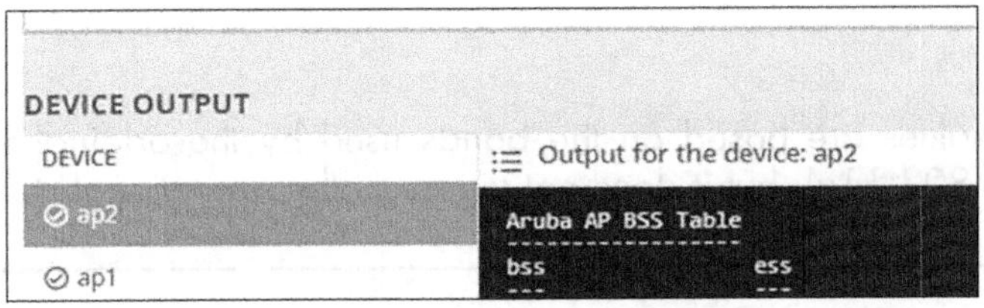

Figure 9-64: Command output from an AP

9. Under Device Output, click the **ap1** entry to see the results for ap1.

Tip

You can maximize the command output using the maximize button (see Figure 9-65). To resize, use the close button (see Figure 9-66).

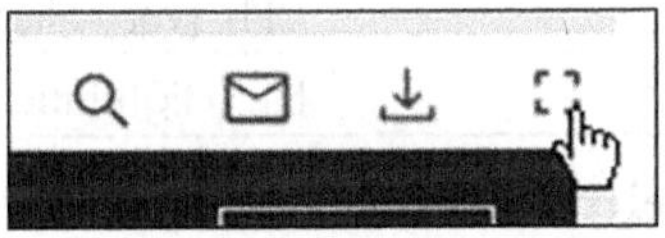

Figure 9-65: Maximize the output

Figure 9-66: Resize the output

- **Question**: How many entries do you see in the BSS list?
- **Answer**: Four.
- **Question**: What do these BSS MAC addresses represent?
- **Answer**: These are the MAC address for this AP for this ESSID, reported as ESS in the output. This is the same BSS information you have seen in the AP device details screen.
- **Question**: What "phy" is reported?
- **Answer**: This depends on the AP models used in your lab setup. For example, **a**-HE or **a**-VHT, which represents the 5 GHz band, and **g**-HE or **g**-HT, which represents the 2.4 GHz band.

Note

The *a* and g values are based on the bands used by the original wireless standards (802.11a and 802.11g), but it does *not* represent the operational standard used by the WLAN.

- **Question**: What do HE, HT, or VHT represent?
- **Answer**: This is the operational WLAN standard in use by the radio. Table 9-1 shows the WLAN standards and how they are displayed on the AP **phy**. This allows you to determine the standard used by the radio.

Table 9-1. Definitions of *Phy* in the output.

Standard	Phy
802.11n	HT (High Throughput)
802.11ac	VHT (Very High Throughput)
802.11ax	HE (High Efficiency)

- **Question**: What is the **ch/eirp** column?
- **Answer**: This represents the channel and the EIRP, which represents the AP's effective transmit power.
- Take note of the channel and power for the 2.4 GHz: ____________________
- Take note of the channel and power for the 5 GHz: ____________________

10. Check the output for ap2.
    - Take note of the channel and power for the 2.4 GHz: ____________________
    - Take note of the channel and power for the 5 GHz radio: ____________________

Aruba Central radio information

Now you will review the same radio information from the Central UI. While the CLI works fine for a single AP, Central provides an easy overview and dashboard for a group of AP radios.

1. In Aruba Central, navigate to: Context: **Groups / campus-wifi-ui** > Navigation: **Devices** > Top: **Access Points** > Right-top: **List**.
2. The top shows the Access Points and Radios. Click the **Radios** tab (see Figure 9-67).

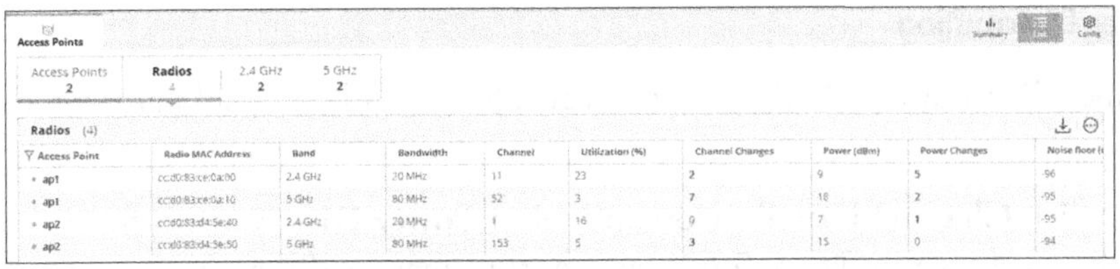

Figure 9-67: The Radios tab

- **Question**: Compare the channel and power to the command output you previously captured. Do you see the same information?
- **Answer**: Yes, the AP reports the actual channel, power, and utilization to Aruba Central, which is known as telemetry data.

Visualize Channel Distribution

Aruba Central also provides a convenient dashboard overview with the channels and bands that are in use by the selected APs.

1. At the top- right, click **Summary**.
2. At the top, click the **Radios** tab. This provides a graphical representation of the number of radios that are using a channel, as well as the configured channel width (see Figure 9-68).

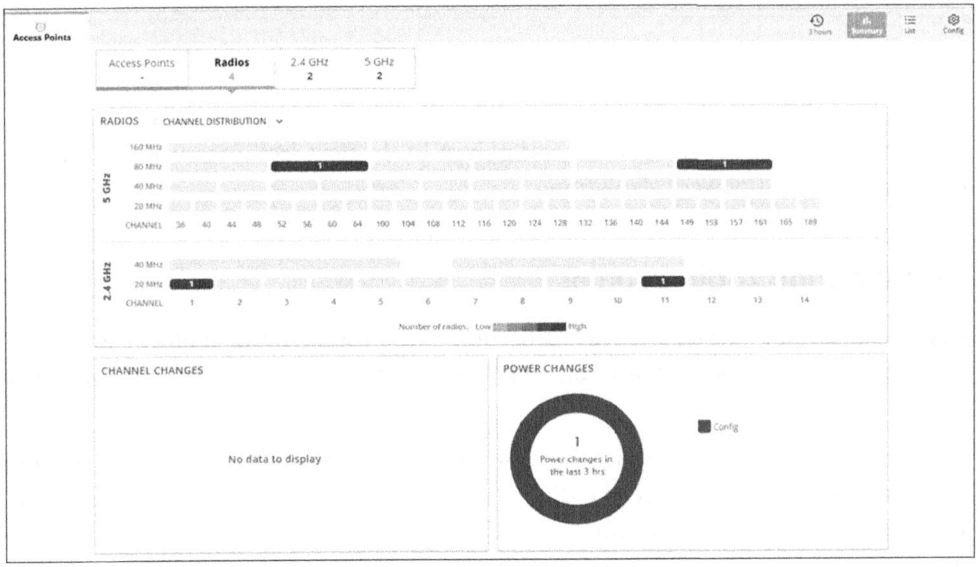

Figure 9-68: Channel Distribution

Visualize Power Distribution

Like the channel distribution, Aruba Central can also provide a dashboard for the radio power settings.

3. At the top, click **Channel Distribution** and select **Power Distribution** from the list. An example screenshot is shown in Figure 9-69 (your output may be different).

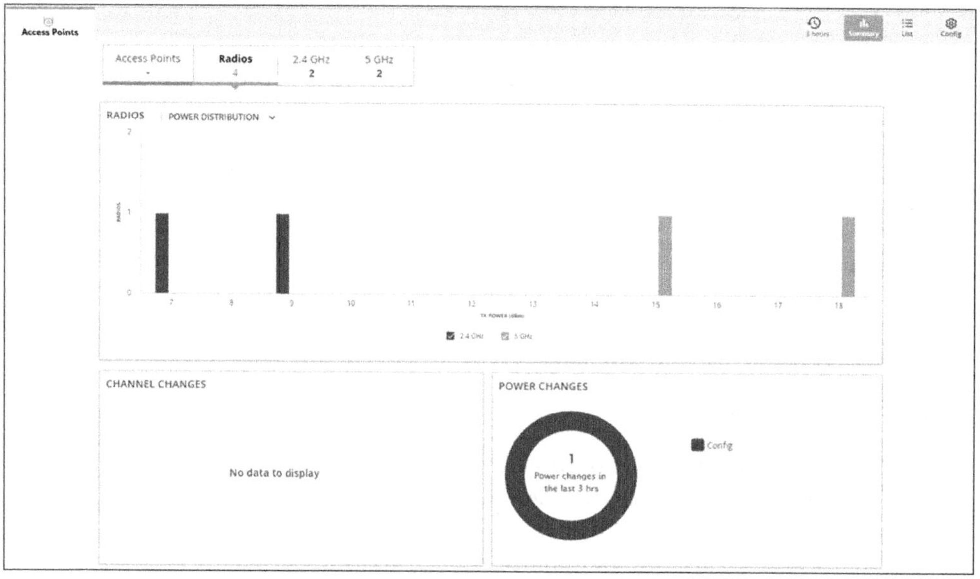

Figure 9-69: Power Distribution

Now you have observed the default channel and power. In the next section you will use a radio profile to control the channel and power calculation.

Task 2: Configuring radio profiles

Your customer would like to know if they can control the use of some channels and the transmit power of the radios. After contact with your senior colleague, you learn that you can control these settings using a radio profile.

Within an Aruba Central configuration group, multiple AP radio profiles could exist, to handle some APs in special areas that may have special radio tweaking requirements. However, at the AP level, only one radio profile can be applied. This must be one of the radio profiles you have created at the group level. In the next steps you will first create a new radio profile, and then apply it to your APs.

Objectives

- Create AP radio profiles.
- Apply AP radio profiles.
- Verify the AP radio profile.

Steps

Create radio profile

In the next steps you will create a new radio profile to control the channel use of the AP radios.

1. In Aruba Central, navigate to Context: **Groups / campus-wifi-ui** > Navigation: **Devices** > Top: **Access Points** > Right top: **Config**.
2. Click the **Radios** tab.
3. Click **Add Profile**. Define a name of **campus-table**.

Note

Do not adjust the Transmit Power settings at this point. They will be adjusted in the next task.

---

4. Click the allowed channel list for the 5 GHz radio. (Click the list with the channel numbers, as shown in Figure 9-70). This will open the Allowed Channels window (see Figure 9-71).

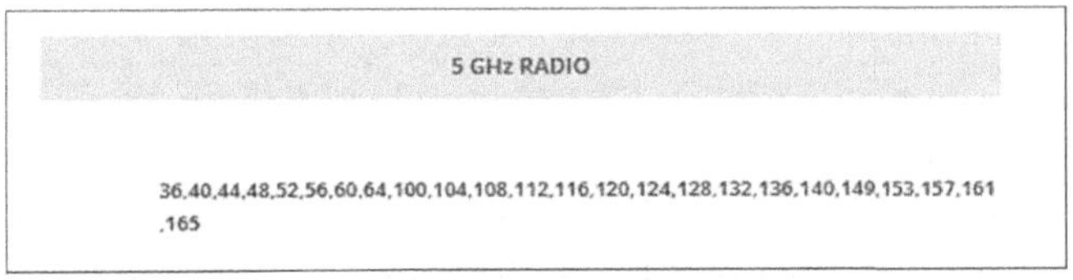

Figure 9-70: Channel listing

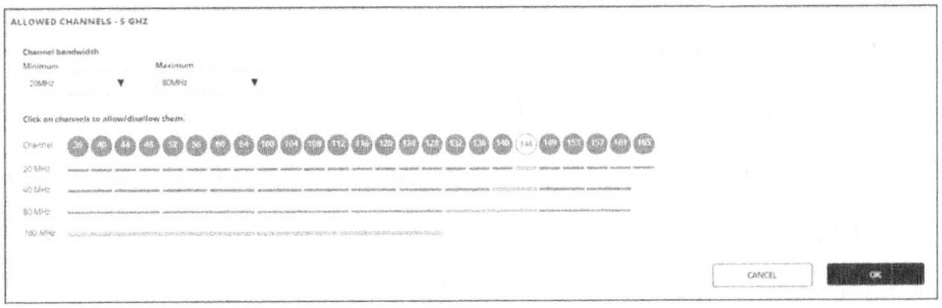

Figure 9-71: Allowed Channels window

5. Set the 5 GHz Maximum to **20 Mhz**. Leave the Minimum to the default **20 Mhz**.

Note

In the remote lab environment, there are *many* APs. Reducing the channel width helps to reduce the interference within this lab (pod) other lab environments (other pods).

---

6. For testing purposes, adjust the Allow channel list so your profile only contains two 20Mhz channels. You can click a channel number to enable or disable it. Disable all channels, except the two channels from the table below for your assigned table (see Table 9-2). Figure 9-72 has a configuration example based on Table 1.

Table 9-2. Channels your assigned table number should use.

Your assigned table #	Channels to use
1 / 5 / 9 / 13	36-40
2 / 6 / 10 / 14	44-48
3 / 7 / 11	52-56
4 / 8 / 12	60-64

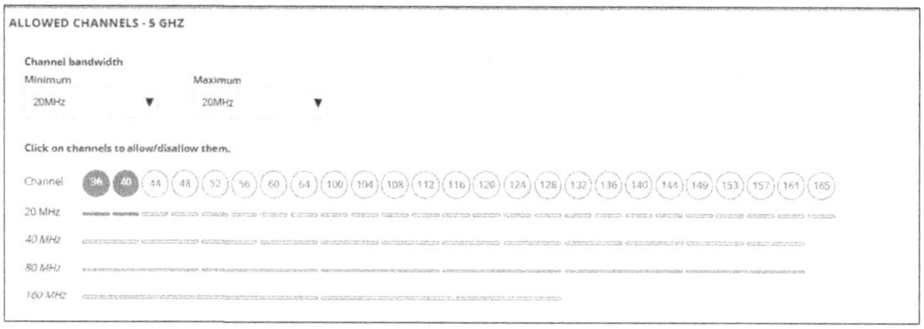

Figure 9-72: Table 1 channel configuration example

7. Click **OK** to submit the allowed channels.
8. Click **Show Advanced Settings**.

- **Question**: What does the *Very High Throughput* option represent?
- **Answer**: VHT represents the radio *phy* operation; it represents the activation of 802.1ac.
- **Question**: What does the option *Set Second Radio Differently* indicate?
- **Answer**: Some Aruba APs can have 2 5 GHz radios. With this option it is possible to apply different radio properties to the second 5 GHz radio. By default, the second 5 GHz radio will use the same settings as the primary 5 GHz radio.

Note

This setting only applies to APs with a second 5 GHz radio; it will be ignored by other APs.

9. Click **Save** to commit the new radio profile.
10. Verify that the new radio profile is shown in the list.

Apply the radio profile

Now you have created a new radio profile, but it is not yet in use. In the next steps you will activate the new radio profile on your APs. Only one radio profile can be active per radio; this defaults to the *default* radio profile.

1. Click the **Access Points** tab.
2. Click *both* **APs**. Both APs should be selected now.
3. At the bottom, click the **pencil** icon to edit the Access Points.
4. In the *Radio Profile* column, select the **campus-table** profile.
5. Enable the **Apply to all access points** checkbox, then click **Apply** (see Figure 9-73).

Figure 9-73: Changing the radio profile

6. Click the **WLANs** column.
7. Enable the **Apply to all access points** checkbox, then click **Apply** (see Figure 9-74).

Figure 9-74: Applying the radio profile

8. Click **Save Settings**.

Verify the generated configuration

After performing a configuration change in Aruba Central, you can review and analyze the generated commands. In the next steps you explore the radio profile configuration commands.

1. In Aruba Central, navigate to: Context: **Groups / campus-wifi-ui** > Navigation: **Audit Trail.**
2. In the list, look for an entry with description *Add/Update RF Radio Profile*. Click the **three dots** at the end of this record to see the details (see Figure 9-75).

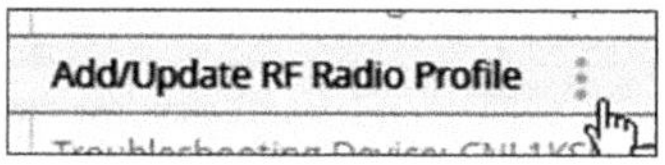

Figure 9-75: Add/update the radio profile

3. Review the commands that were generated for the radio profile. Example output is shown here (your table may use different channels):

```
rf dot11g-radio-profile campus-table
 disable-arm-wids-functions off
 exit

rf dot11a-radio-profile campus-table
 disable-arm-wids-functions off
 allowed-channels 44,48
 ch-bw-range 20MHz 20MHz
 exit

rf dot11a-secondary-radio-profile campus-table
 disable-arm-wids-functions off
 allowed-channels 44,48
 ch-bw-range 20MHz 20MHz
 exit
```

- Question: What do the dot11g and dot11a represent?
- Answer: dot11g represents the 2.4 GHz radio profile; dot11a represents the 5 GHz radio profile.
- Question: What is the dot11a-secondary radio profile?
- Answer: Some Aruba APs support two 5 GHz radios. This profile would be applied to the second 5 GHz radio. By default, it is the same as the primary 5 GHz radio. Different settings could have been applied in the radio profile with the advanced option *Set Second Radio Differently*. (Not applicable to this lab).
- Question: What is controlled by the **ch-bw-range** command?
- Answer: This command controls the channel bandwidth. You have configured the maximum channel width to 20 MHz in the radio profile.
- Question: Which channels do you see for the allowed-channels command?
- Answer: These are the channels you have selected in the radio profile.
- Question: Why is the allowed-channels command missing for the dot11g radio profile?
- Answer: You only configured the 5 GHz radio profile. No changes were made for the 2.4 GHz radio profile.

4. **Close** the audit entry details window.

Verify the configuration received by the AP

In the previous steps you have seen how Aruba Central generates the configuration for the AP based on your UI configuration steps. These commands are then sent to the AP, applied to the running configuration, and saved locally on the AP. Locally on the AP, you can also monitor received configuration commands from the cloud (Aruba Central). In the next steps you will open a console connection to the AP and review the received commands.

1. In Aruba Central, navigate to: Context: **Groups / campus-wifi-ui** > Navigation: **Tools** > Top: **Commands**.
2. In the Device Type dropdown select **Access Point**.
3. Click **Available Devices** and select ap1.
4. In the Categories list, select **All Category** to see all available commands.
5. In the Commands list, click the **filter** icon and type **cloud**.
6. Select the **show ap debug cloud-config-received** command, then click **Add**.
7. Click **Run** to start the request.

8. Wait a few moments until the Output for the device is updated and review the commands.
9. Starting from the end, look back until you see a timestamp message. This was the last time configuration changes were received. Here is an example output:

```
timestamp: 2022-09-06 05:17:12
rf dot11g-radio-profile campus-table: OK
disable-arm-wids-functions off: OK
exit: OK
rf dot11a-radio-profile campus-table: OK
disable-arm-wids-functions off: OK
allowed-channels 44,48: OK
ch-bw-range 20MHz 20MHz: OK
exit: OK
rf dot11a-secondary-radio-profile campus-table: OK
disable-arm-wids-functions off: OK
allowed-channels 44,48: OK
ch-bw-range 20MHz 20MHz: OK
exit: OK
```

- **Question**: Review the commands. Do you recognize them compared to the audit entry?
- **Answer**: The commands on the AP should be the same as the commands that were added to the audit entry.
- **Question**: Were the commands successfully processed by the AP?
- **Answer**: For each line that is successfully processed by the AP, the AP will add the *OK* keyword at the end.

You have now verified that the configuration was pushed successfully.

Verify the applied settings on the AP

In the previous steps you have verified that the commands have been pushed to the AP. In the next steps you will verify that the radio profile settings are operational on the AP.

You will do this by reviewing:

- The active radio profile in use by the AP radios
- The current radio channel in use by the AP 5 GHz radio

1. In Aruba Central, you should still be at: Context: **Groups / campus-wifi-ui** > Navigation: **Tools** > Top: **Commands** with ap1 as selected device.
2. Click **Remove All** to clear the Selected Commands window.
3. Set the Commands filter to **radio**.
4. In the Commands list, select **show radio config** and click **Add**.

5. Set the Commands filter to **bss**.

6. In the Commands list, select **show ap bss-table** and click **Add**.

Tip

You can clear the previous command output by clicking the **CLEAR** button. This ensures you will only see the output of the currently requested command.

7. Click **Run**. Figure 9-76 has an example output.

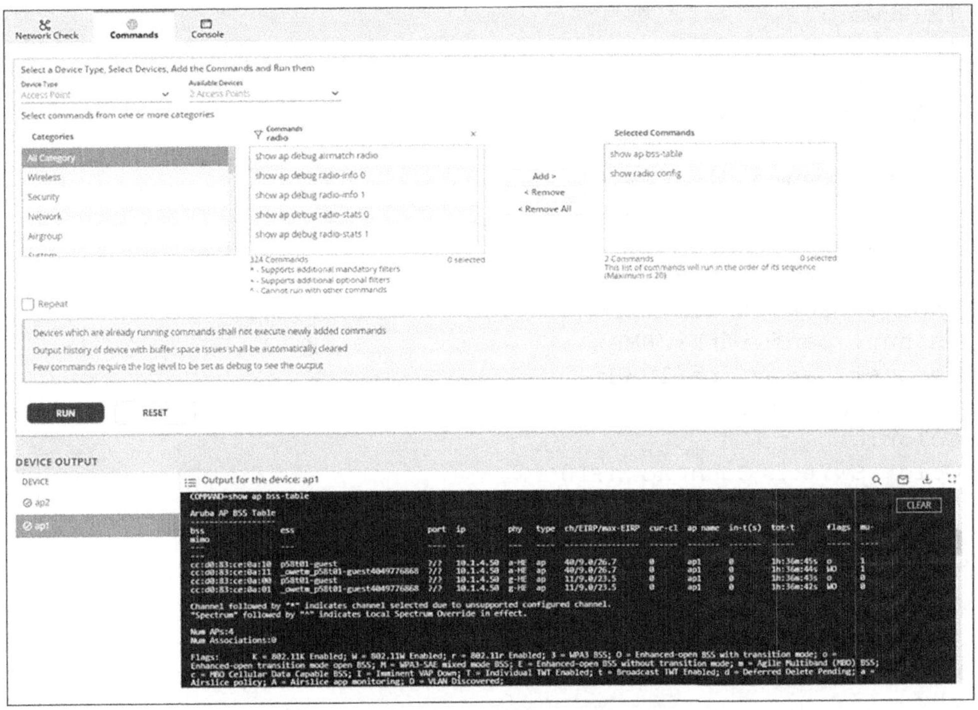

Figure 9-76: **show ap bss-table** output

1. Review the deployed radio profile. Here is an example output (your channels may be different):

```
COMMAND=show radio config

-------------------------------------------------------------
2.4 GHz:
Profile Name:campus-table
Zone Name:
```

```
Legacy Mode:disable
Single Chain Legacy:disable
Beacon Interval:100
802.11b-protection:enable
802.11d/802.11h:disable
Interference Immunity Level:2
Channel Switch Announcement Count:0
MAX Distance:0
Channel Reuse Type:disable
Channel Reuse Threshold:0
Background Spectrum Monitor:disable
Max Tx Power:12
Min Tx Power:6
Cell Size Reduction:0
Smart Antenna:disable
WIDS Override:off
40MHZ intolerance:disable
Honor 40MHZ intolerance:enable
Channel quality threshold:70
Channel quality wait time:120
High noise backoff time:720
Max channel bandwidth:20MHz
Min channel bandwidth:20MHz
Allowed channels:1,6,11
BSS Color:0
BSS Color Switch Countdown:10
Free Channel Index:25
Scanning-disable:No
Aggressive-scan:enable
Active-scan:disable
IoT Coexistence :enable
-----------------------------------------------------------
5.0 GHz:
Profile Name:campus-table
Zone Name:
Very High Throughput:enable
Legacy Mode:disable
Single Chain Legacy:disable
Beacon Interval:100
802.11d/802.11h:disable
Interference Immunity Level:2
Channel Switch Announcement Count:0
MAX Distance:0
Channel Reuse Type:disable
Channel Reuse Threshold:0
```

```
Background Spectrum Monitor:disable
Max Tx Power:21
Min Tx Power:15
Cell Size Reduction:0
Smart Antenna:disable
WIDS Override:off
Channel quality threshold:70
Channel quality wait time:120
High noise backoff time:720
Max channel bandwidth:20MHz
Min channel bandwidth:20MHz
Radar backoff time:720
Allowed channels:44,48
BSS Color:0
BSS Color Switch Countdown:10
Free Channel Index:25
Scanning-disable:No
Aggressive-scan:enable
Active-scan:disable
Zero Wait DFS:disable
-----------------------------------------------------------
5.0 GHz:
Profile Name:campus-table
Zone Name:
Very High Throughput:enable
Legacy Mode:disable
Single Chain Legacy:disable
Beacon Interval:100
802.11d/802.11h:disable
Interference Immunity Level:2
Channel Switch Announcement Count:0
MAX Distance:0
Channel Reuse Type:disable
Channel Reuse Threshold:0
Background Spectrum Monitor:disable
Max Tx Power:21
Min Tx Power:15
Cell Size Reduction:0
Smart Antenna:disable
WIDS Override:off
Channel quality threshold:70
Channel quality wait time:120
High noise backoff time:720
Max channel bandwidth:20MHz
Min channel bandwidth:20MHz
```

```
Radar backoff time:720
Allowed channels:44,48
BSS Color:0
BSS Color Switch Countdown:10
Free Channel Index:25
Scanning-disable:No
Aggressive-scan:enable
Active-scan:disable
Zero Wait DFS:disable
```

- **Question**: What is the Max channel bandwidth for the 5 GHz radio?
- **Answer**: 20 MHz.
- **Question**: What are the allowed channels?
- **Answer**: This depends on your table number, but it should reflect your two assigned channels (for this lab setup).

8. Review the **BSS table** command output.
    - **Question**: What is the current active channel of the 5 GHz radio (the 'phy' with a-HE or a-VHT)?
    - **Answer**: The active channel should be one of the two channels you have allowed in the radio profile.
9. Click **Remove All** to clear the commands.
10. In the Command filter, enter **arm**.
    - **Question**: What does ARM represent?
    - **Answer**: Adaptive Radio Management (ARM) is part of the process that analyzes channel and power use, reports information to Aruba Central, and applies the changes as requested by Aruba Central. ARM is the older method of dynamic radio management. Previously in the chapter you learned about AirMatch, which is its replacement. However, in a lot of the commands and command output, you'll still see the term *arm*.
11. Add the command **show ap arm history** and click **Run**. Here is an example output:

```
COMMAND=show ap arm history

Interface :wifi0
ARM History
-----------
Time of  Change  Old Channel  New Channel  Old Power  New Power Reason
Result
--------------  ---------  ----------  --------  ---------
------      ------
```

```
2022-09-08 04:03:31  44          40          18.0       18.0       AM RD
-
2022-09-08 03:35:01  44          44          9.0        18.0       AM
MinEIRP  -
2022-09-08 03:28:03  0           44          0.0        9.0        AM
Init     -

Interface :wifi1
ARM History
-----------
Time of Change       Old Channel New Channel Old Power New Power Reason
Result
--------------       ----------- ----------- --------- ---------
------   ------
2022-09-08 03:28:38  2+          11          0.0        9.0        AM RD
-
2022-09-08 03:28:03  0           11          9.0        9.0        AM
Init  -

I: Interference, R: Radar detection, N: Noise exceeded, Question:
Bad Channel Quality E: Error threshold exceeded, INV: Invalid
Channel, G: Rogue AP Containment, M: Empty Channel, P+: Increase
Power, P-: Decrease Power, 40INT: 40MHZ intol detected on 2.4G,
NO40INT: 40MHz intol cleared on 2.4G, OFF(R): Turn off Radio due to
Radar, OFF(CONFIG): Turn off Radio due to Wrong Config, ON: Turn on
Radio, D: Dynamic Bandwidth Switch, I*: CCA Interference, C: Radar
cleared, DM: Dynamic Mode Change, O: Opmode change, AIRMATCH:
AirMatch Event, AM Solver: AirMatch(AM) service selected channel/
power, AM Init: Initialized channel/power from flash, AM N: Noise
exceeded, AM NC: Noise Cleared, AM RD: Reg-Domain Profile Change, AM
Rogue: Rogue AP Containment, AM DRT: DRT File Change, AM MinEIRP:
Min EIRP Change, AM MaxEIRP: Max EIRP Change, AM Freeze: set static
channel/power, AM Unfreeze: unset static channel/powerNC: Noise
Cleared, Random: Random Channel, RMC: Radio Mode Change, RCP: Radio
Client Preference Change
```

- **Question**: What are the two interfaces shown in the output?
- **Answer**: wifi0 and wifi1.Wifi0 represents the 5 GHz radio. The Channel columns can be used as a hint (channels 1-11 for 2.4 GHz).
- **Question**: This list shows the latest channel and/or power changes for each radio. What code is listed in the Reason column for the last change of the *wifi0* interface (5 GHz)?
- **Answer**: The reason is AM RD.
- **Question**: What does this mean? Use the legend at the bottom to find the explanation.
- **Answer**: AM RD indicates regulatory domain profile change. This is used when the allowed channel list has changed.

Verify the applied settings in Central

Now you review the same radio information from the Central UI.

1. In Aruba Central, navigate to: Context: **Groups / campus-wifi-ui** > Navigation: **Devices** > Top: **Access Points** > Right-top: **Summary**.
2. The top shows Access Points and Radios. Click the **Radios** tab. Figure 9-77 shows an example screenshot. Your table channels may be different from the screenshot.

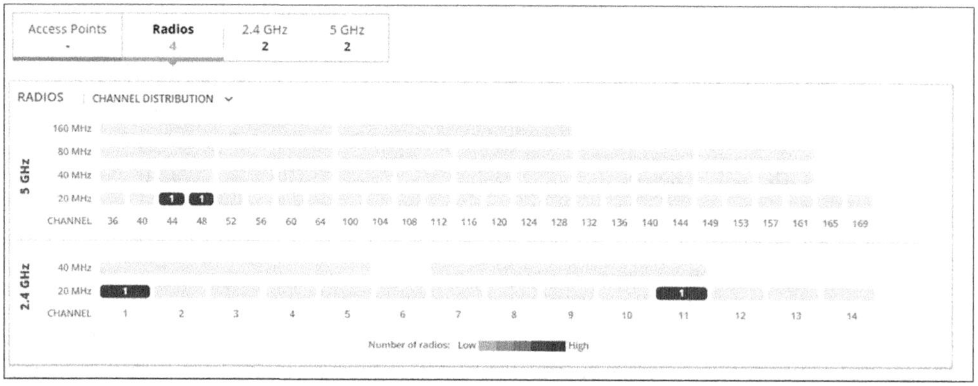

Figure 9-77: Radio channel distribution

- **Question**: Do you see radios active on the channels you configured?
- **Answer**: Yes, after the radio profile configuration, the AP radio moved to one of the configured channels.

3. Under channel distribution, the channel changes are reported (see Figure 9-78).

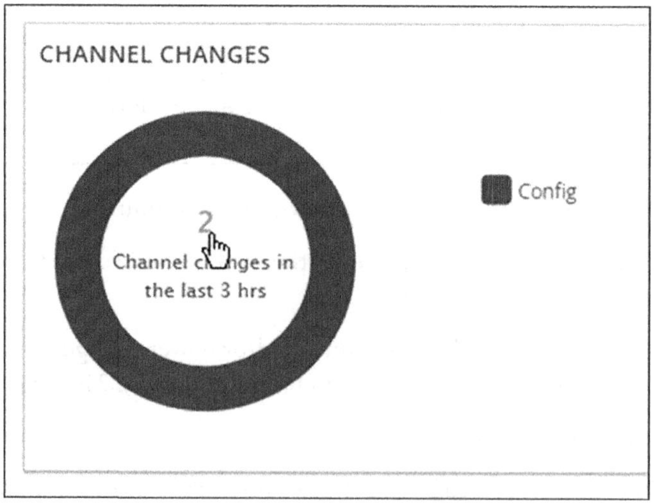

Figure 9-78: Channel changes

4. Click the number inside the circle of the Channel Changes pane to see the list of changes. Figure 9-79 shows an example screenshot.

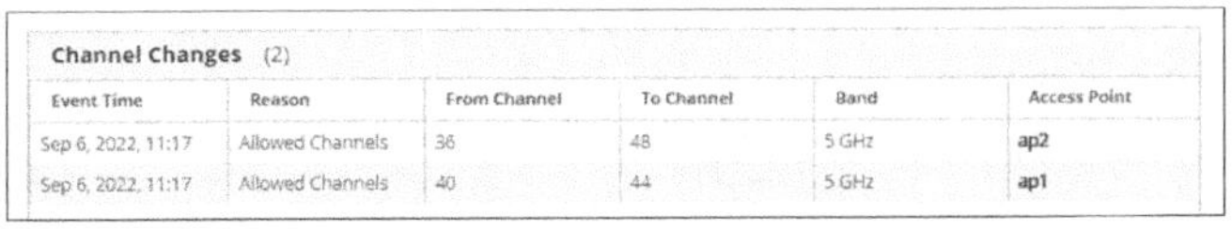

Channel Changes (2)

Event Time	Reason	From Channel	To Channel	Band	Access Point
Sep 6, 2022, 11:17	Allowed Channels	36	48	5 GHz	ap2
Sep 6, 2022, 11:17	Allowed Channels	40	44	5 GHz	ap1

Figure 9-79: Verifying your channel changes

- **Question**: For your *ap1*, what is the reason for the last channel change?
- **Answer**: The Allowed Channels was changed by the radio profile.
- **Question**: What is the new channel for your *ap1*? Does it match the output from the **show arm history** command?
- **Answer**: Yes, the AP reports the new channel and ARM history to Aruba Central.

5. Close the Channel Changes window. This concludes the radio profile channel configuration.

Changing radio power

In this section you will update the allowed transmit power in the radio profile.

1. In Aruba Central, navigate to: Context: **Groups / campus-wifi-ui** > Navigation: **Devices** > Top: **Access Points**> Right top: **Config**.
2. Click the **Radios** tab.
3. For the *campus-table* record, click the **pencil** icon to edit the profile.
4. Change the default power values to a lower value for this lab environment (see Figure 9-80 for an example).
   - 2.4 GHz: **6-9**
   - 5 GHz: **9-12**

Tip

Using the mouse may be difficult to move the slider correctly. You can also click the start or end marker and use the arrow keys (left and right) to change values.

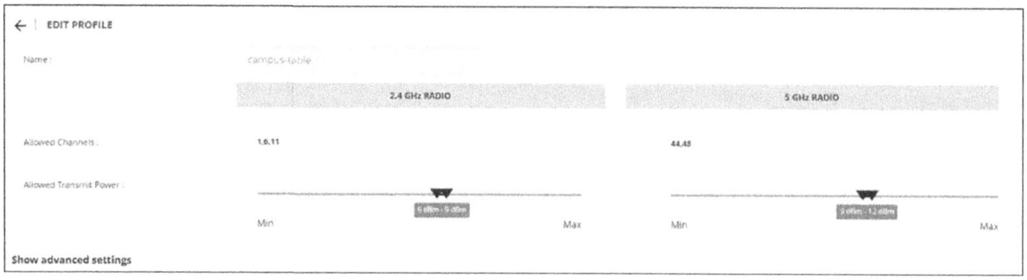

Figure 9-80: Changing the radio power settings

5. Click **Save** for the profile.
6. Use the Audit Trail to review the commands that affect the transmit power (see Figure 9-81).

```
Configuration Updated

rf dot11g-radio-profile campus-table
max-tx-power 9
exit
rf dot11a-radio-profile campus-table
max-tx-power 12
min-tx-power 9
exit
rf dot11a-secondary-radio-profile campus-table
max-tx-power 12
min-tx-power 9
exit
```

Figure 9-81: Audit Trail output

7. Navigate to **Tools > Commands** for access point ap1 and select the command **show radio config** to review the updated radio profile.

```
COMMAND=show radio config

------------------------------------------------------------
2.4 GHz:
Profile Name:campus-table
...
Max Tx Power:9
Min Tx Power:6
...
------------------------------------------------------------
5.0 GHz:
Profile Name:campus-table
...
Max Tx Power:12
Min Tx Power:9
...

=== Troubleshooting session completed ===
```

8. Use **Tools > Commands** for access point ap1 with the command **show arm history** to review the radio change history. This should include a power change now for the wifi0 interface (5 GHz radio).

```
COMMAND=show ap arm history

Interface :wifi0
ARM History
-----------
Time of Change      Old Channel New Channel Old Power New Power Reason
Result
-------------- ----------- ----------- --------- --------- ------ ------
2022-09-08 04:26:12 40          40          18.0      12.0      AM
MaxEIRP  -
2022-09-08 04:03:31 44          40          18.0      18.0      AM RD
-
2022-09-08 03:35:01 44          44          9.0       18.0      AM
MinEIRP  -
2022-09-08 03:28:03 0           44          0.0       9.0       AM
Init     -
...

I: Interference, R: Radar detection, N: Noise exceeded, Question:
Bad Channel Quality E: Error threshold exceeded, INV: Invalid
Channel, G: Rogue AP Containment, M: Empty Channel, P+: Increase
```

```
Power, P-: Decrease Power, 40INT: 40MHZ intol detected on 2.4G,
NO40INT: 40MHz intol cleared on 2.4G, OFF(R): Turn off Radio due to
Radar, OFF(CONFIG): Turn off Radio due to Wrong Config, ON: Turn on
Radio, D: Dynamic Bandwidth Switch, I*: CCA Interference, C: Radar
cleared, DM: Dynamic Mode Change, O: Opmode change, AIRMATCH:
AirMatch Event, AM Solver: AirMatch(AM) service selected channel/
power, AM Init: Initialized channel/power from flash, AM N: Noise
exceeded, AM NC: Noise Cleared, AM RD: Reg-Domain Profile Change, AM
Rogue: Rogue AP Containment, AM DRT: DRT File Change, AM MinEIRP:
Min EIRP Change, AM MaxEIRP: Max EIRP Change, AM Freeze: set static
channel/power, AM Unfreeze: unset static channel/powerNC: Noise
Cleared, Random: Random Channel, RMC: Radio Mode Change, RCP: Radio
Client Preference Change
```

- **Question**: What was the old and new power for the last change on the wifi0 interface?
- **Answer**: The old power depends on your environment, but the new power should be in the range of 9-12 dBm.
- **Question**: What is the reason of the last change?
- **Answer**: AM MaxEIRP: Max EIRP change.
- **Question**: If your wifi1 (2.4 GHz) interface does not report a new change, what could be the reason?
- **Answer**: The wifi1 interface may already have been using a transmit power in the range of 6-9. Therefore, a power change was not required.

Review the power change in Aruba Central

Now you will review the same radio power information from the Central UI.

1. In Aruba Central, navigate to: Context: **Groups / campus-wifi-ui** > Navigation: **Devices**> Top: **Access Points** > Right-top: **Summary**.
2. The top shows Access Points and Radios. Click the **Radios** tab.
3. Change the selector Channel Distribution to **Power Distribution** (see Figure 9-82). Figure 9-83 shows an example screenshot. Your table power distribution may be different from the screenshot.

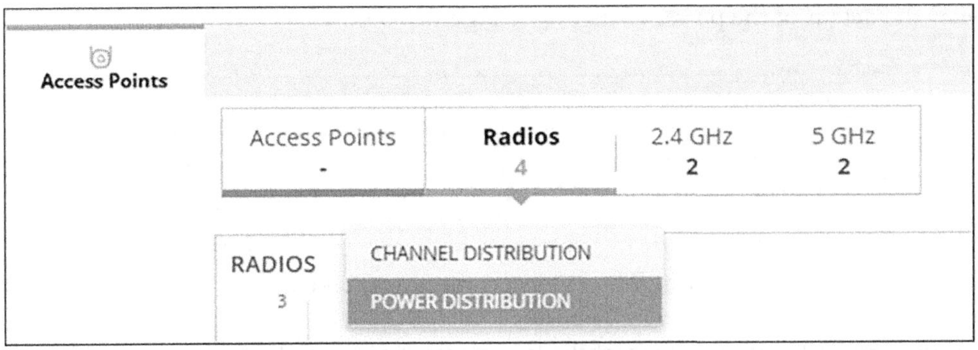

Figure 9-82: Examining the power distribution

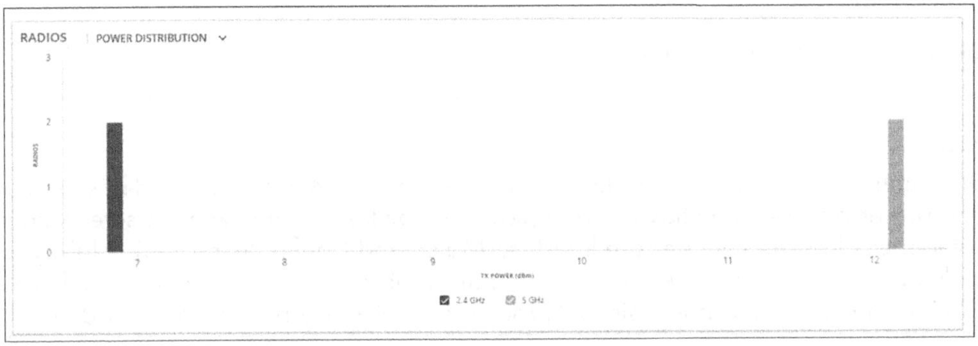

Figure 9-83: Power distribution example

4. Under the radio list, the number of power changes is shown. Click the *number* in the circle to see the details (see Figure 9-84).

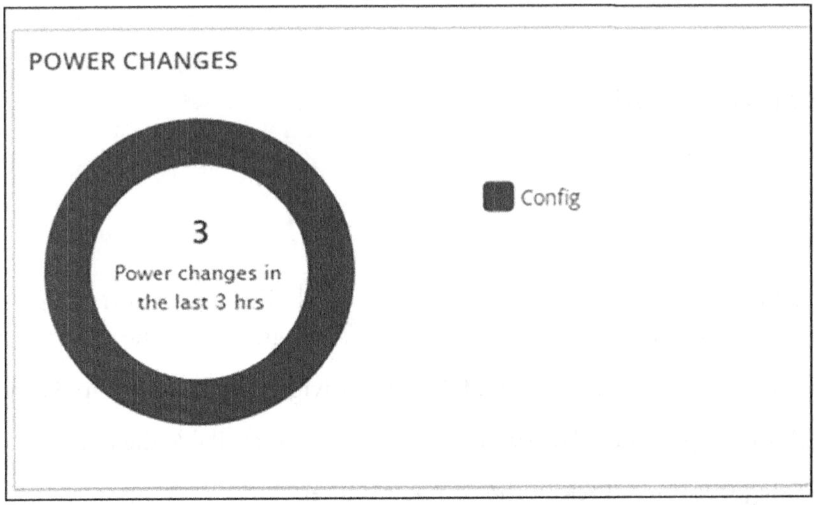

Figure 9-84: Accessing the power changes

5. Review the details of the power changes (see Figure 9-85).

Power Changes (3)

Event Time	Reason	From Power (dBm)	To Power (dBm)	Band	Access Point
Sep 8, 2022, 10:27	Power Range Change	18	12	5 GHz	ap2
Sep 8, 2022, 10:26	Power Range Change	18	12	5 GHz	ap1
Sep 8, 2022, 09:35	Power Range Change	9	18	5 GHz	ap1

Figure 9-85: Viewing the power changes

- **Question**: What is the reason of the last power change on the 5 GHz for ap1?
- **Answer**: 'Power Range Change', since the radio profile applied a maximum of 12 for the 5 GHz radio.

6. **Close** the power changes window.

Note

Depending on the time of day that you do the lab, AirMatch (the cloud-based channel and power planner) may have also deployed a solution. In the example screenshot, two categories of power changes are listed: AirMatch and Config. In the current lab, you are looking for the 'config' power changes. In most lab environments, you will only see the Config category at this point, since AirMatch typically only runs once every day.

Task 3: Configuring AirMatch

AirMatch is a service in Aruba Central that receives information from APs about activity on various channels. It processes this information into a channel plan for all the APs in your environment.

Objectives

- List the AirMatch global schedule.
- List the AP time zone.

Steps

The AirMatch deployment schedule is set at the global level. The application of the AirMatch solver settings will be applied based on the AP local time zone settings. Review the current time on your AP.

1. In Aruba Central, navigate to: Context: **Global** > Navigation: **Tools** > Top: **Commands**.
2. Select the Device Type **Access Points** and select **ap1** from available devices.
3. Select **All Category**; enter **clock** in the command filter.

4. **Add** the commands **show clock** and **show clock timezone** and **Run** the commands. Here is an example output:

```
COMMAND=show clock

Current Time     :2022-09-08 04:39:14


===================================
Output Time: 2022-09-08 08:39:16 UTC


COMMAND=show clock timezone

Current Timezone
----------------
Country        Timezone  DST Name  DST Recurring
-------        --------  --------  -------------
Eastern-Time   UTC-05    EDT       second sunday march 02:00 first
sunday november 02:00
```

5. Take note of the current time.
   - **Current Time**: ____________
   - **Question**: What is the time zone; how was this applied to the AP?
   - **Answer**: The time zone was configured at the AP group level > system settings. The APs automatically inherit the time zone from the group.
6. In Aruba Central, navigate to: Context: **Global** > Navigation: **Devices** > Top: **Access Points** > Top right: **config**.

Note

You do this at the Global context level. The AirMatch deployment schedule is set once, as a global parameter in the system. The actual deployment to the devices will consider the device local time.

---

7. As an example, in the lab, set the automatic deploy at the next top of the hour interval based on the AP time. For example, if the AP time showed 3:30, you may set the deploy time to 4:00.
8. Click **Save.**

Task 4: Client Association, Live Events, and Location information

Your customer would like to know how they can monitor, troubleshoot, and see events about clients that connect to the WLAN. You explain to the customer that Aruba Central provides client floorplan map location, historical events, and live events information. In this task you will explore the client information and client events.

Open Client Details Context

You can find the various client events and monitoring options in the client details page. Let's start there.

1. In Aruba Central, navigate to: Context: **Global** > Navigation: **Clients** > **Top: Clients**. Your wireless client should be in the list with a status of **connected**.
2. Click your *client name* in the list to see the Client Details > Overview page.

Client location on floorplan

You can see client locations on the floorplan from the client details context.

1. Use the navigation menu to select **Overview** > Top: **Location**. You see the site / floor of the client-connected AP. On the floorplan, only the current AP is displayed (see Figure 9-86).

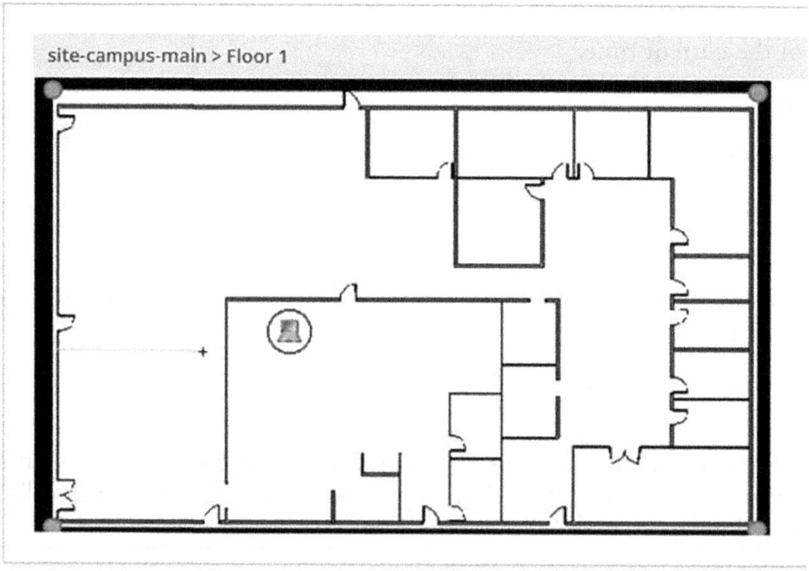

Figure 9-86: AP floorplan view

Association Events and Roaming latency

In these steps you will explore the historical association events for a client.

1. Use the navigation menu to click **Overview** > Top: **Summary**.
2. Scroll down to the Roaming Experience tile.

3. On the top-right of this tile, click the **table** icon.
4. You see a historical list of association events and roaming latency (see Figure 9-87).

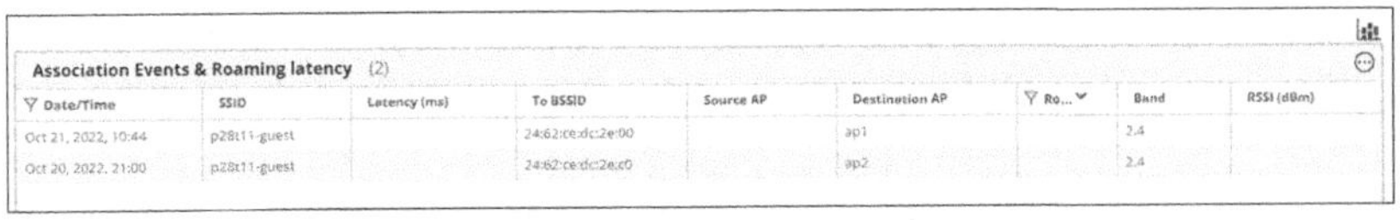

Association Events & Roaming latency (2)

Date/Time	SSID	Latency (ms)	To BSSID	Source AP	Destination AP	Ro...	Band	RSSI (dBm)
Oct 21, 2022, 10:44	p28t11-guest		24:62:ce:dc:2e:00		ap1		2.4	
Oct 20, 2022, 21:00	p28t11-guest		24:62:ce:dc:2e:c0		ap2		2.4	

Figure 9-87: Association and roaming events

- **Question**: Must the client be connected to see this events table?
- **Answer**: No, even when the client is offline, this history event table can be reviewed.

Clients Events

In the next steps you review client events recorded by Aruba Central.

1. You are still at the client device context. Use the navigation menu to click **Events** > Top: **Events**.
2. In this list you can explore various events for the client.
3. **Optional step**: You can change the displayed time period using the time selector at the top-right of the screen (see Figure 9-88).

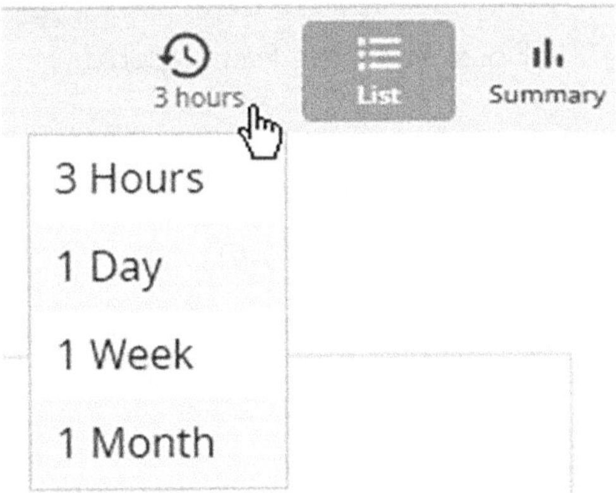

Figure 9-88: Changing the time display period

Client Live Events

Aruba Central supports client Live Events for more advanced troubleshooting. This shows live client events as they connect, disconnect, or roam.

1. Use the navigation menu and click **Live Events.**
2. The Live Events system automatically starts the troubleshooting session for the client MAC address (see Figure 9-89).

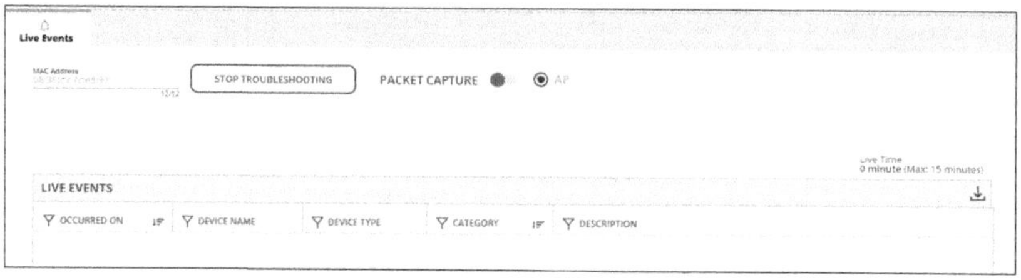

Figure 9-89: Viewing Live Events

3. On PC1, disconnect from the guest WLAN.
4. In Aruba Central, the client disassociation messages are displayed (see Figure 9-90).

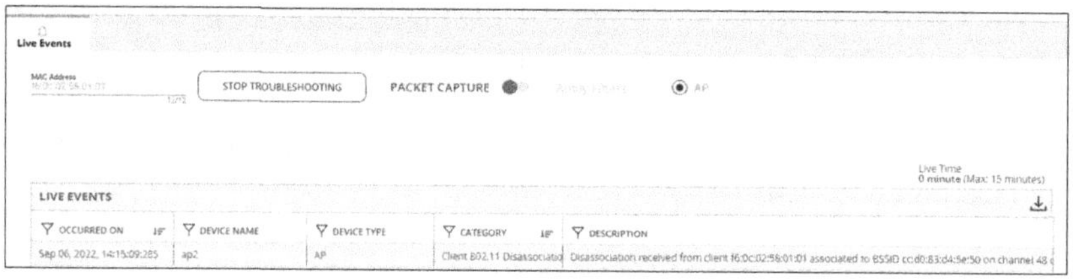

Figure 9-90: Live Events update

5. On PC1, connect to the *p#tX-guest* WLAN again.
6. In Aruba Central, review the events (see Figure 9-91).

Figure 9-91: Live Events association events

- **Question**: What is the flow of the events?
- **Answer**: Auth, Association, Role assignment, and L3 address using DHCP

Note

You can export the events using the download CSV button, which can make it easier to share the data with a colleague.

Task 5: Configuring WLAN zones

The customer wants to ensure that the test guest WLAN is not advertised on the AP2. Here are the different strategies with the pros and cons:

- **New Group**: Complete a new configuration. Complex to maintain the common settings with the other group.
- **Device Level Override**: Works for single device. Difficult to deploy to several devices.
- **AP Zones**: Easy to deploy. Only applies to the WLAN activation feature, no other configuration items can be pushed using this system.

In this task, the AP zones feature will be used to ensure that AP2 does not advertise the test guest WLAN anymore.

Objectives

- Understand AP zones.
- Configure zones.

Steps

1. In Aruba Central, navigate to: Context: **Groups / campus-wifi-ui** > Navigation: **Devices** > Top: **Access Points**> Right top: **Config**.
2. Click the **Access Points** tab.
    - **Question**: What do you see under the WLANs column for your APs?
    - **Answer**: Both APs show 'All SSIDs selected'.
3. Select **AP2** and use the **pencil** icon at the end of the record to edit the AP.
4. Click the **WLANs** tab to see the list of WLAN bindings.
    - **Question**: What WLANs are enabled?
    - **Answer**: The guest WLAN is enabled.
5. **Uncheck** the guest WLAN.

6. Click **Save Settings** to submit the configuration.
7. You will be returned to the AP list.
   - **Question**: What WLANs are listed for AP2 now?
   - **Answer**: There are no WLANs active for AP2.

Verify the BSS list on AP2

1. In Aruba Central, navigate to: Context: **Groups / campus-wifi-ui** > Navigation: **Tools** > Top: **Commands**.
2. In the **Device Type** dropdown select **Access Point**.
3. Click Available Devices and select ap2.
4. In the Categories list, select **All Category**. This will show all available commands.
5. In the Commands list, click the **filter** icon and type '**bss**'.
6. Select the **show ap bss-table** command, then click **Add**.
7. Click **Run**.
8. After a few moments, review the commands.
   - **Question**: Are there any BSS entries active in the list?
   - **Answer**: No, AP2 does not have any WLANs active now.

Task: Cleanup

2. Use the same procedure to enable all WLANs again on ap2.

Lab Cleanup: Disable the guest SSID

To reduce the number of active SSIDs broadcasting in the remote lab, you may disable the guest SSID.

1. In Aruba Central, navigate to: Context: **Groups / campus-wifi-ui** > Navigation: **Devices** > Top: **Access Points**> Top right: **config**.
2. In the WLAN list, check the Network Enabled column.
   - **Question**: What is the current value?
   - **Answer**: Enabled.
3. Hover your mouse over the SSID; icons will appear at the end of the row (see Figure 9-92).

Figure 9-92: SSID icons

4. Click the **Wi-Fi** icon to disable the WLAN.
5. Confirm the action with **Yes**. See Figure 9-93 to verify your results.
    - **Question**: What is the value now for the Network Enabled column?
    - **Answer**: The value is "No" since you disabled the WLAN.

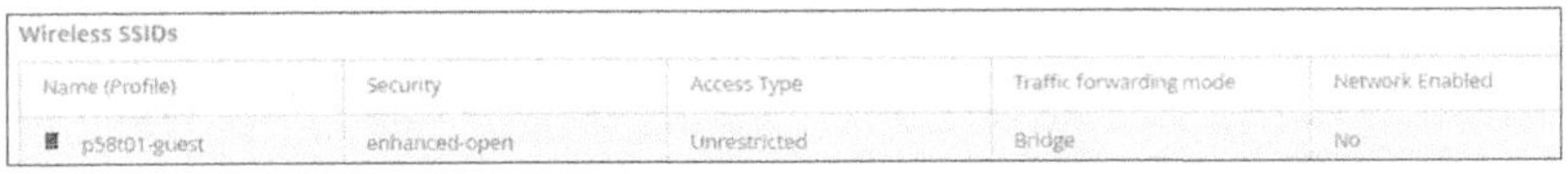

Wireless SSIDs

Name (Profile)	Security	Access Type	Traffic forwarding mode	Network Enabled
p58t01-guest	enhanced-open	Unrestricted	Bridge	No

Figure 9-93: Verifying that the SSID has been disabled

Note
The same option can be used to enable the WLAN again.

Learning check

Discuss these questions.

1. Choose the options below that accurately compare the Wi-Fi radio frequency (RF) bands. (Select two.)
    a. 2.4 GHz RF energy is less affected by absorption, as is caused when signals flow through office walls and other obstructions.
    b. 5 GHz RF energy supports more channels, which means there are less problems with co-channel interference.
    c. The 6 GHz band is supported in the 802.11ac and 802.ax standards.
    d. In a typical corporate deployment, channel bonding is used to improve performance for all RF bands.
2. Which statements below accurately compare Aruba AirMatch vs Aruba ClientMatch? (Select three.)
    a. AirMatch automatically tunes AP settings to optimize your RF environment.
    b. ClientMatch helps balance the client load across APs.

c. ClientMatch helps balance the client load across RF bands.

d. ClientMatch minimizes channel coupling, where adjacent radios are assigned to the same channel.

3. When defining a WLAN, you can statically configure the VLAN, or allow the VLAN to be derived based on the client MAC address or RADIUS response attributes.

   a. True

   b. False

4. Which of the following security mechanisms are most suitable for employees connecting to an enterprise-class deployment? (Select three.)

   a. 802.1X

   b. WPA3

   c. PSK

   d. Captive Portal

   e. External RADIUS server authentication

Summary

Congratulations! You have completed Chapter 9. You should now be able to:

- Explain RF bands & channels.
- Describe RF basics and WLAN mobility.
- Explain 802.11frames.
- Describe aspects of RF design.
- Configure WLANs.

10 Implementing Secure WLANs

OBJECTIVES

WLAN RF data emanates from each AP's radios in all directions. Security of this resource is vital. After completing this chapter, you will be able to explain authentication, authorization, and accounting (AAA) services and describe 802.1X authentication. You will also be able to configure secure WLANs and describe roles and access rules.

After completing this chapter, you should be able to:

✓ Explain AAA.

✓ Describe 802.1X authentication.

✓ Configure secure WLANs.

✓ Discuss roles and access rules.

Introduction to AAA

This chapter begins with an introduction to AAA: what it is and the basics of how it is implemented.

Why network authentication?

You have learned about Aruba's ESP platform previously in the book. Zero-trust networking is one of the ESP elements for delivering security in zero trust environments. A key component of this security strategy involves network authentication (see Figure 10-1). By identifying every user and device at the internal perimeter, a company gains control *and* visibility into their environment. Such visibility lays the foundation for control. First, the company can take action to control access to the internal network, ensuring that only authorized users and devices can connect. Next, the company can enhance this control by assigning the appropriate access controls to each client, typically by a policy management server like ClearPass, based on knowledge about who is using the device, as well as other factors such as the device type (like a laptop, desktop, tablet or smartphone, for example) and location (the main campus versus a branch office, for example). This is called network access control (NAC).

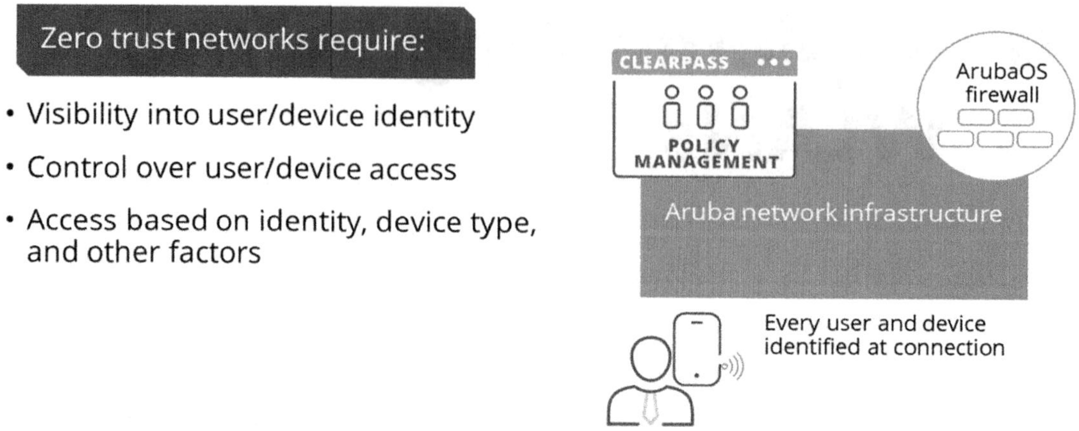

Figure 10-1: Why network authentication?

AAA services

A common NAC solution is based around AAA services running on an authentication server. So what are the three A's of AAA services (see Figure 10-2)?

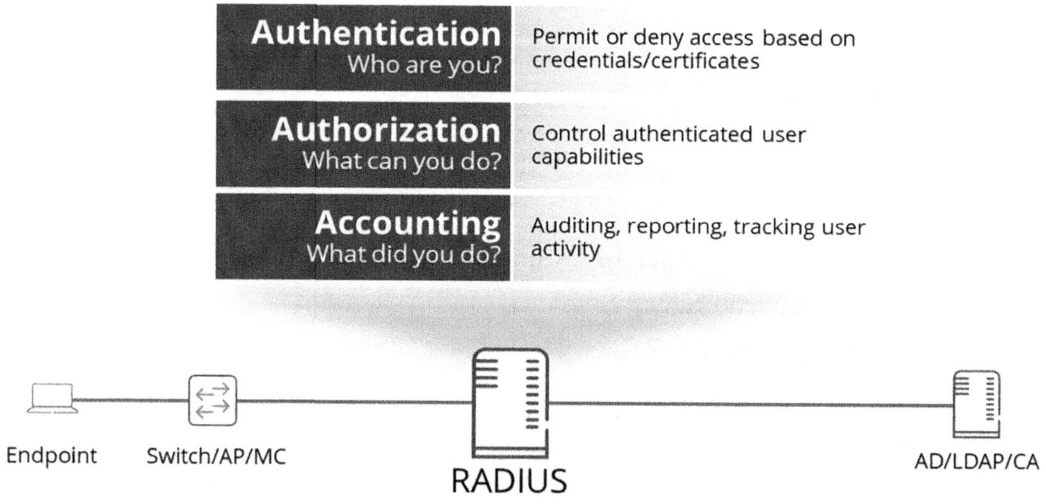

Figure 10-2: AAA services?

First up is *authentication*: who are you? The AAA server, commonly using the RADIUS or TACACS+ protocols, permits or denies access based on your credentials. Perhaps your username is *Maria*, with a password of *Secret123*: if those are valid credentials, access is granted. Authentication can be performed by the AAA server itself or more commonly handed off to a back-end directory server, like Microsoft's Active Directory (AD) or a directory service using the Lightweight Directory Access

Protocol (LDAP). The user might alternatively submit a digital certificate that includes a valid username and is signed by a trusted certificate authority (CA).

Now that you're on the network, what can you do? That's defined by *authorization*. The user Maria is a member of the marketing group, so she can access all standard business applications, plus the marketing application and database. Other resources are off-limits for Maria. You might define a policy to authorize Maria to be assigned to a certain VLAN, with certain firewall policies, and with certain quality of service (QoS) settings, which the networking device dynamically applies.

You have controlled who can access the network, and what they can do. Now you need *accounting*: what did they do? With accounting, you have auditing, reporting, and tracking of user activity: when and from where they logged in, when they logged out, and what resources they accessed during that time. One thing you're trying to eliminate here is *plausible deniability*: If the log file indicates that someone accessed a human resources (HR) database at 2:00 am, it is difficult to deny that fact.

Controlling network access control (NAC)

When a user joins an organization, they are given a user account (see Figure 10-3). This account holds details about their relationship to the organization and their authentication credentials. These credentials are often stored in a Microsoft AD or an LDAP database. Aruba's ClearPass supports many external connections either natively or through its powerful built-in API capability.

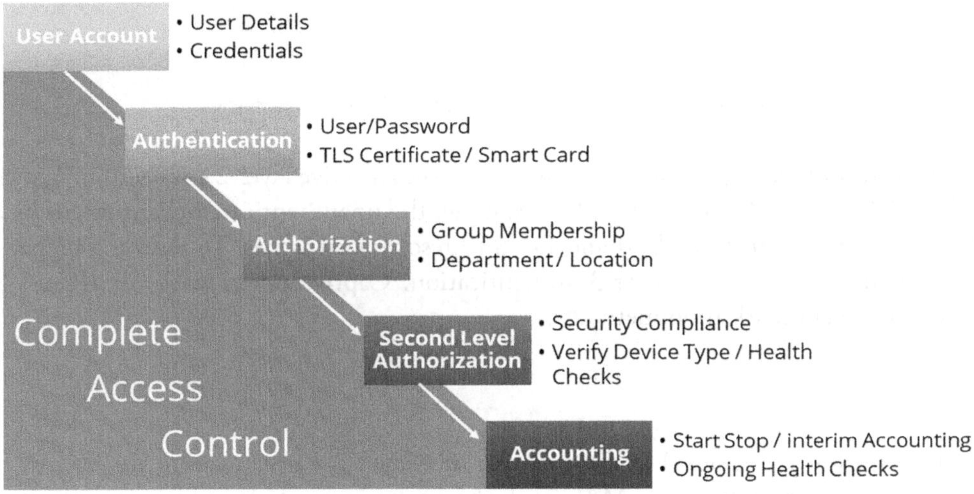

Figure 10-3: Controlling NAC?

During authentication, the user provides credentials, which the network validates against the data store. This is typically done via a AAA server like ClearPass. Then network authorizes the user's account details and client device. The network may grant or deny access based on a complete picture of the client. If access is granted, the system continues to monitor a user's activity using start, stop, and interim accounting messages, with optional ongoing compliance checks.

By leveraging all available information and implementing sound access control rules, a ClearPass AAA-based deployment provides a complete access control system, with three common authentication methods.

Layer 2 versus layer 3 authentication

You can divide network authentication methods into two broad categories: layer 2 and layer 3 (see Figure 10-4). Layer 2 authentication methods authenticate users and devices before they receive an active layer 2 connection. If the client fails authentication, the client is disconnected at layer 2, preventing the client from sending or receiving any traffic at all. MAC authentication (MAC-auth) and IEEE 802.1X are examples of layer 2 authentication methods.

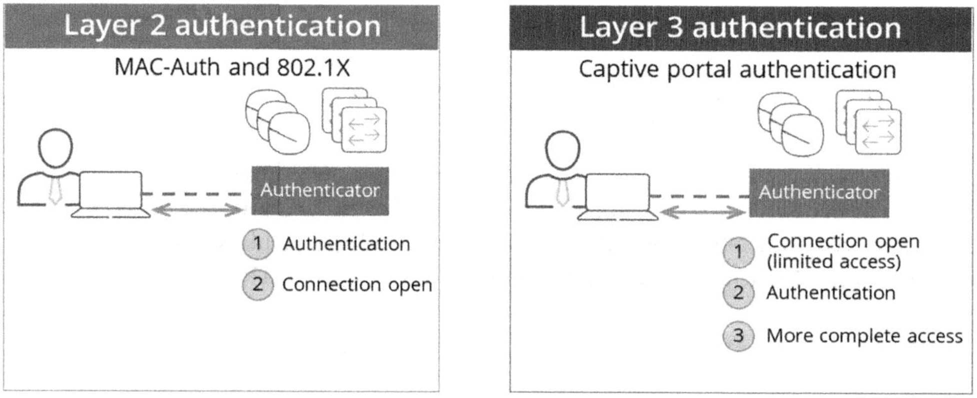

Figure 10-4: Layer 2 versus layer 3 authentication

Layer 3 authentication takes place after the client receives an active layer 2 connection. Typically, the client will initially receive limited access. For example, the unauthenticated client might be allowed to use DHCP to obtain an IP address and DNS to resolve hostnames. To receive a higher level of access, the client must complete layer 3 authentication. Captive portal (web) or IPsec VPNs are examples of layer 3 authentication methods.

MAC-auth

MAC authentication (MAC-auth) is one example of a layer 2 authentication method. A network infrastructure device that implements MAC-auth checks the source MAC address in a client's incoming traffic. Sometimes the authenticator has a local list of authorized MAC addresses. Sometimes, as shown in Figure 10-5, the authenticator submits the MAC address to an authentication server, which uses the MAC address to determine whether to grant the access to the client. For example, the server might have an allow list of MAC addresses that belong to the company: it only grants access to MAC addresses in that list.

Because wireless clients always include their MAC address in the 802.11 frame or wired clients include their source MAC address in Ethernet frames, the client does not need to be configured to submit credentials. Therefore, this authentication method provides an option for devices that do not have a user interface, sometimes called *headless devices*, or lack support for 802.1X.

However, MAC-auth does not provide rigorous security. First it identifies the device only, not the user using the device. Worse, it is often trivial for a hacker to discover MAC addresses used in your system, whether by eavesdropping or accessing a legitimate device. The hacker can then easily spoof an authorized MAC address. For example, many operating systems, like Microsoft Windows, let you change the MAC address in the NIC properties. Finally, it can be difficult to create and manage an allowed list of the MAC addresses of all the company's authorized devices. Apple and Android devices are now randomize their MAC addresses when they connect to different SSIDs, making it even harder to create a definitive allow list.

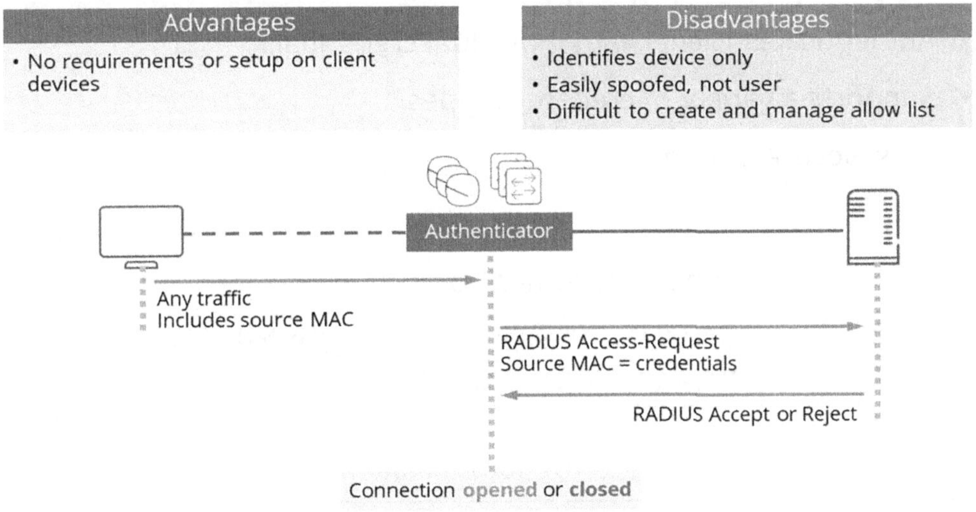

Figure 10-5: MAC-auth

Rather than a security measure, MAC-auth is perhaps most useful for simplifying the provisioning for devices such as printers, internet of things (IoT) devices, IP phones, and security cameras.

802.1X overview

Another layer 2 authentication method is 802.1X, which is recommended for secure authentication in enterprises. 802.1X, an Institute of Electrical and Electronics Engineers (IEEE) standard for NAC, enables the network infrastructure to authenticate and authorize endpoints for network access. 802.1X requires users to authenticate as soon as a layer 2 connection is established. In a wired network, 802.1X takes effect as soon as the Ethernet connection comes up. In a wireless network, it takes effect after the initial 802.11 association.

To restrict the unauthenticated client's access, 802.1X divides the port into two logical ports. The controlled port permits all types of traffic (see Figure 10-6). Its state depends on authentication status. Both sides of the connection should block the controlled port until the peer has successfully authenticated. After successful authentication, all types of traffic may flow through the controlled port from the point of view of 802.1X; however, the network infrastructure can further restrict access with firewall policies and access control lists (ACLs). The uncontrolled port is always open, but it permits only Extensible Authentication Protocol (EAP) data units, by default, which is used to perform the authentication process. At this point, the port does not allow *any* other form of traffic, even DHCP messages and DNS messages, by default. 802.1X allows an endpoint to authenticate *before* it has acquired an IP address, providing a high level of security.

- ✓ Requires clients to authenticate to a RADIUS server as soon as the endpoint connects
- ✓ Authenticates clients with individual credentials
- ✓ Supports a variety of credential types
- ✓ Most secure option

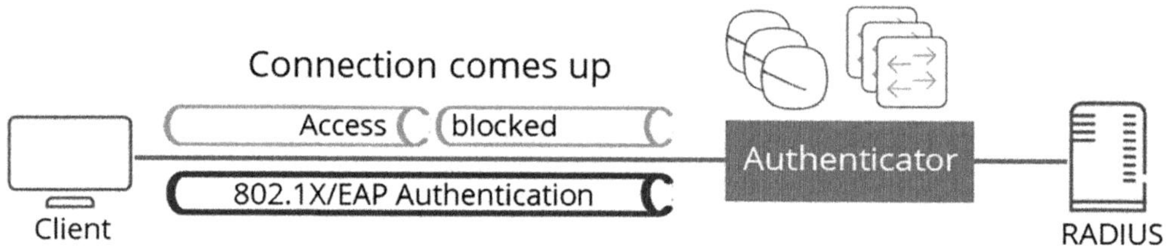

Figure 10-6: 802.1X overview

Note

Some vendors allow you to make exceptions to only EAP traffic on an uncontrolled port. For example, AOS-CX switches allow you to make exceptions for LLDP and/or CDP traffic. This might be necessary to provide the necessary power for power-over-Ethernet (PoE), for example, to an AP, IP phone, or security camera. AOS-CX switches have this feature disabled, by default, however.

802.1X authentication

If you want the best authentication in a wireless and wired campus network, you need to implement 802.1X authentication. This section covers the protocol in more depth.

802.1X roles

The 802.1X process includes three parties (see Figure 10-7):

- Supplicant
- Authenticator
- Authentication server

To understand these parties, you will consider an analogy in which a person is attempting to get into an exclusive, very important person (VIP) event in a secure building.

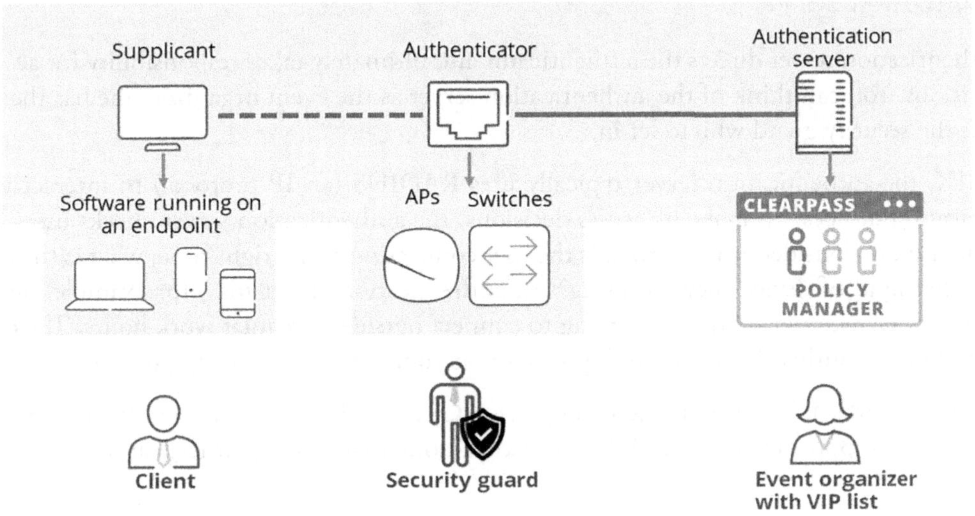

Figure 10-7: 802.1X roles

Supplicant

The supplicant is the 802.1X component that runs on the endpoint, which is attempting to connect to the network. In our analogy, Bob, the person attempting to get into the event, is the supplicant. He is the one who wants to gain access to something.

In the real world, 802.1X supplicants are software that runs on client devices. For example, Windows and MAC operating systems have native 802.1X supplicants, as do smartphone operating systems such as Apple iOS and Android. Administrators can also install third-party supplicant software on endpoints, which might provide additional security features lacking in an operating system's native 802.1X supplicant software.

Authenticator

The authenticator controls network access. It enforces decisions made by the authentication server, either permitting or denying access. In our analogy, the authenticator is the security guard at the building in which the event is being held. The security guard is responsible for guarding the building and only allowing authorized people in. Similarly, in 802.1X the authenticator stands at the internal network perimeter and either grants or denies access to the network. In Aruba networks, the authenticator could be an Aruba AP, an Aruba gateway, or an Aruba switch (older AOS switches as well as the newer AOS-CX switches).

Authentication server

The authentication server directs the authenticator and ultimately takes responsibility for access control decisions. You can think of the authentication server as the event organizer. She has the VIP list and tells the security guard who to let in.

In 802.1X, the authentication server typically uses RADIUS (an IP protocol) to interact with the authentication devices. To make its access decisions, the authentication server checks users' credentials, but also can consider other factors. Is the user connecting at the right time, what is the location, and/or what kind of device they are using? An authentication server may, for example, deny users with legitimate credentials if they are trying to connect outside of regular work hours. The decisions all depend on the authentication server's policies that your company security policy dictates.

For Aruba networks, it is recommended that you use ClearPass Policy Manager (ClearPass or CPPM), which includes support for the RADIUS protocol, among many, many other features.

802.1X high-level analogy

Let's now look at the 802.1X process. To help you understand the process, return to Bob, who is trying to get into the VIP event (see Figure 10-8). Bob comes to the building and is met by the security guard. The guard informs him that the event is closed to the public and asks for Bob's name. Bob gives him his name and says that he is on the VIP list.

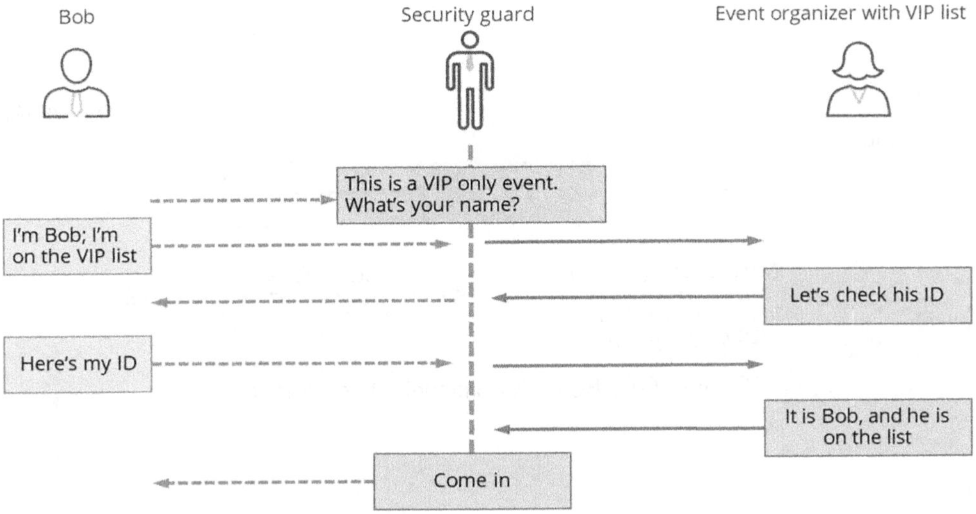

Figure 10-8: High-level analogy

The security guard has an audio/visual communication channel open with the event organizer, who is inside the building. He's going to act as a courier for everything that Bob says, passing it on to the event organizer and similarly, passing on the event organizer's requests to Bob. The guard informs the event organizer that a person claiming to be Bob is outside. The event organizer tells the guard to collect Bob's ID. The guard passes on the message and has Bob hold his ID up to the camera for the event organizer to check. Checking the ID and her list, she sees that the person at the door is indeed Bob, who is on her list. She tells the security guard to let him in, which the guard does.

Note

Please see Aruba's security courses for more detailed information about 802.1X.

802.1X communication protocols

In a moment, you will see how this process translates to 802.1X. To understand the process, you need to understand the two communication protocols used in 802.1X (see Figure 10-9):

- EAPOL (Extensible Authentication Protocol Over LAN)
- RADIUS (Remote Authentication Dial-in User)

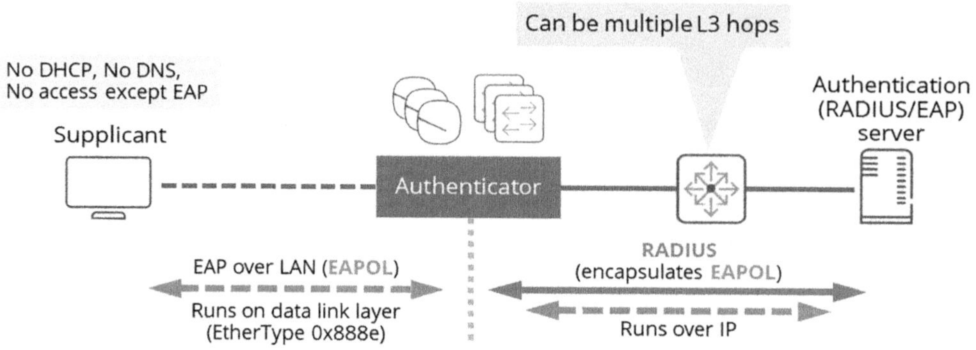

Figure 10-9: 802.1X communication protocols

EAPOL

The supplicant and authenticator communicate with EAPOL messages. EAPOL operates at the data link layer and uses EtherType 0x888e. Note that, despite the name, even wireless connections use EAPOL messages. In practice, the messages are often simply referred to as *EAP* messages.

802.1X almost entirely shuts down the layer 2 link for unauthenticated clients. The client cannot even send DHCP or DNS messages into the network. The authenticator accepts *nothing* except EAP, by default, which is encapsulated in an 802.1X frame. This means that usually the client has no IP address. It can only communicate on the local link using a data link layer protocol. The supplicant is either directly connected to the authenticator or if the authenticator is an Aruba gateway, an AP tunnels the supplicant's messages to the authenticator (gateway).

RADIUS

The authenticator communicates with the AAA server over an IP network, and multiple layer 3 hops can exist between these two devices (see Figure 10-9). The authenticator acts as a relay between the supplicant and the authentication server, much like the security guard in the previous example. The authenticator encapsulates EAP messages from the client with RADIUS access-requests, which it forwards to the AAA server. Similarly, the AAA sever encapsulates EAP messages for the client in RADIUS-challenge messages, which it forwards to the authenticator. The authenticator decapsulates the EAP messages and passes them on to the client.

In this way, the supplicant and authentication server act as the two endpoints in an EAP exchange with the authenticator transporting messages between them.

Note that, in RADIUS terms, the authenticator is the network access server (NAS), network access device (NAD), or RADIUS client. However, the rest of this discussion will use "authenticator" to refer to the device for consistency.

Examples of EAP methods

Many specific EAP authentication methods exist within the flexible framework. Some common ones are listed in Table 10-1. One important point is that each of these methods provides mutual authentication—both the server and supplicant authenticate. It is important for the supplicant to authenticate the authentication server to avoid man in the middle (MITM) attacks in which the supplicant connects to a hacker device rather than the legitimate network. EAP-TLS is the preferred method and the one that you should implement for enterprise networks.

Table 10-1. Examples of some EAP methods

EAP method	Credentials	Comments
EAP-TLS	Server = Certificate Supplicant = Certificate	• Recommended option • Widely supported - Windows native • TLS handshake authenticates server and supplicant
PEAP	Server = Certificate Supplicant = Username and password (typical)	• TLS handshake creates secure tunnel (outer method) • A less secure authentication (MSCHAPv2) handles inner method
EAP-TTLS	Server = Certificate Supplicant = Username and password (typical)	• Similar to PEAP in concept • Less secure
EAP-TEAP	Server = Certificate Supplicant = Username and password or certificate	• Emerging method (Windows 10, ClearPass 6.9 support) • TLS handshake creates a secure tunnel (inner method) • Enables "chaining" of machine and user auth with tokens

EAP-Tunneled Layer Security (TLS)

Considered one of the most secure EAP methods, EAP-TLS uses a TLS 1.2 handshake to authenticate the authentication server and supplicant with digital certificates. The Windows native client supports EAP-TLS, as do many other supplicants. ClearPass also supports EAP-TLS. Note that both the supplicant and the authentication server need a certificate from a common certificate authority (CA). While deploying certificates to clients can be challenging, methods exist for easing and automating this deployment. You will look at some of these methods later.

Protected EAP (PEAP) and EAP-Tunneled TLS (TTLS)

PEAP was developed to provide some of the security of EAP-TLS without forcing stations to have digital certificates. It is the option that the Windows native client uses when the user logs in with a username and password.

Tunneled Transport Layer Security (TTLS) is a similar, but less commonly used, method. (TTLS was developed by Funk Software and Certicom; PEAP was developed by Microsoft, Cisco Systems, and RSA Security.) ClearPass supports both PEAP and EAP-TTLS.

TTLS and PEAP function in similar ways. Both methods involve a two-phase authentication process. In the first phase, the outer method creates a secure tunnel using TLS. The authentication server then initiates a second authentication method, called the inner method, inside the secure TLS tunnel. PEAP and EAP-TTLS can use several different options for the inner method. (In fact, they can even use EAP-TLS as the inner method; however, doing so does not typically make sense because the goal of PEAP and TTLS is to protect a less secure inner method.) EAP-Microsoft CHAPv2 (MSCHAPv2) is the most common option. EAP-MSCHAPv2 is not secure enough to run over an untrusted network, but is acceptable when protected by the TLS tunnel.

EAP-Tunneled EAP (EAP-TEAP)

Like PEAP and EAP-TTLS, EAP-TEAP uses TLS as an outer method to create a secure tunnel. It then uses an inner method to authenticate the client. Depending on the inner method implemented, the client can authenticate with a username/password or a certificate. EAP-TEAP also supports *chaining* of multiple authentications, which can be useful when a company wants to ensure that users connect on authorized devices. The client can authenticate with device or *machine* credentials first, proving the device is authorized. The client obtains a token during this authentication. The client can then authenticate with user credentials, including the token from the machine authentication, proving that both the user and device are authorized. EAP-TEAP is supported in Windows 10 and ClearPass 6.9.

802.11 with 802.1X

Before you look at EAP in more detail, it will be useful for you to review the steps in establishing an 802.11 wireless connection and the authentication that occurs with 802.1X (see Figure 10-11). An AP periodically sends out beacons to advertise the WLANs that it supports. The beacons include information about the network name, or SSID, and the security options. A client typically also sends a probe request for the specific SSID to which it wants to connect. If the AP supports that SSID with compatible options, it will send a probe response.

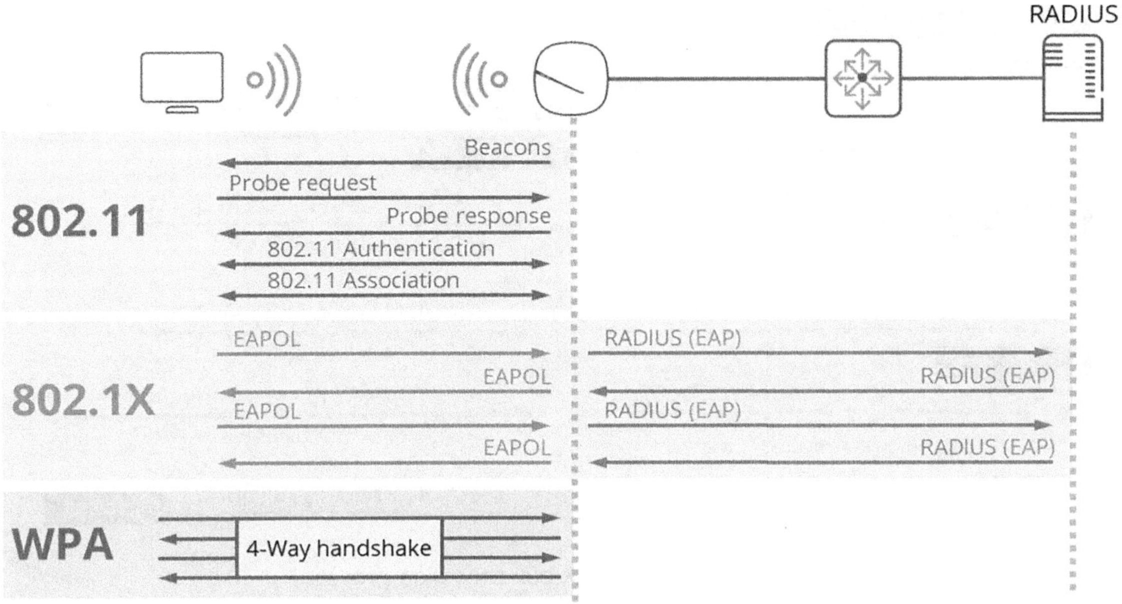

Figure 10-10: 802.11 with 802.1X

The client then sends an 802.11 authentication request. 802.11 authentication often operates in open mode, which means that the AP permits all clients to pass 802.11 authentication. The real 802.1X authentication will occur later. Next, the 802.11 association occurs. The client sends an association request, and the AP responds.

If the WLAN is using enterprise mode, 802.1X authentication follows the 802.11 association. The AP is the 802.1X authenticator. As you learned earlier, the 802.1X process can involve many packets, but in the end, the client is authenticated.

The final step for all WLANs that use encryption is an exchange called the four-way handshake (WPA). The client and AP use this handshake to derive the keys used to secure communications; this handshake might also provide implicit authentication, as you will see. After the handshake, the 802.11 connection is up, and the client can send and receive data on it.

802.1X authentication process

Let's look at the 802.1X authentication process, in detail, to help improve your theoretical knowledge and real-world diagnostic abilities. 802.1X occurs *after* a successful 802.11 association. 802.1X carries EAP messages which are used for authentication. EAP defines a flexible framework that supports several authentication methods, to meet your particular environment and policy needs. Figure 10-10 shows this general framework. As you follow the process, remember that the NAS authenticator relays all messages from the client to the authentication server, encapsulating them with RADIUS as necessary. The vertical line underneath the switch shows the point at which the authenticator encapsulates the messages.

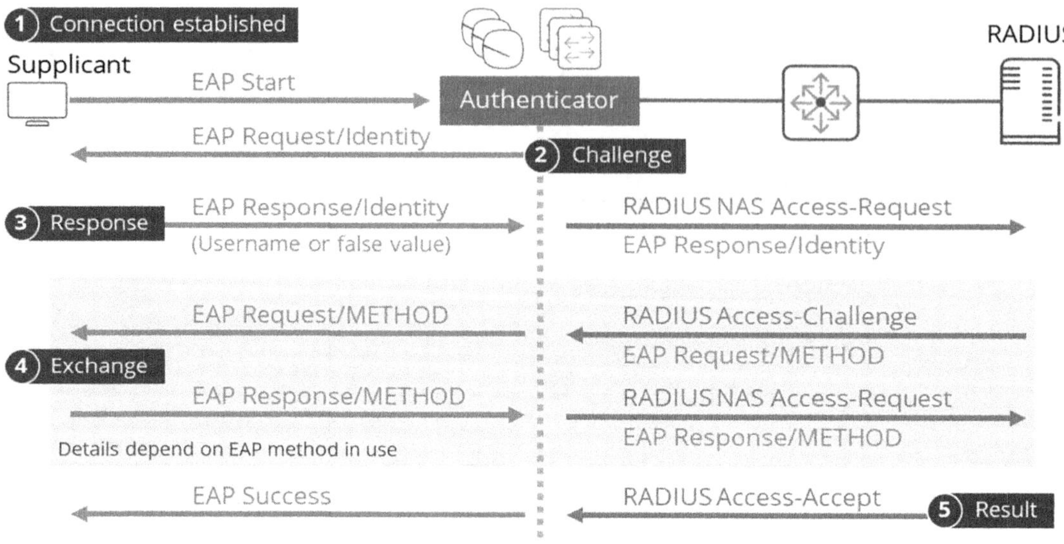

Figure 10-11: 802.1X authentication process

The client establishes a connection. Networks access is controlled, blocking all traffic except EAP, and sends an EAP request/identity challenge. This initiates authentication and asks the client only to identify itself: do not send any other information.

The client responds with an EAP response/identity response, which typically includes a username. This is encapsulated in a RADIUS message, and forwarded to the RADIUS authentication server. The authenticator includes the user's identity in all future frames so that the server can keep track of which EAP messages belong to which user. The user's identity also marks any accounting frames for the connection: this is important for companies with rigorous auditing requirements. Depending on the EAP method, though, the username submitted at this point might not be the actual username. Some methods have the supplicant submit a false name at this point and the true name later in the process after a secure, encrypted session is established. This protects the username from hackers since the connection isn't protect yet and the supplicant doesn't know if it should trust the network (authentication server).

The server then initiates an EAP method for exchanging credentials. Message exchange details, flow, and authentication credentials depend on the EAP method. The authentication server also passes back its own certificate so that the supplicant can authenticate the network and provide a secure connection for the passing the supplicant's EAP credentials. Based on information received in the exchange, the authentication server grants or denies access and informs the switch. The authenticator then either activates the connection and transmits an EAP-Success or leaves the connection deactivated and transmits an EAP-Failure. Step 4 will be covered in more depth a bit later in this section since what occurs in step 4 slightly differs based on whether PEAP or EAP-TLS is employed.

EAP and RADIUS together

Now let's look at how the authenticator relays messages between the client supplicant and authentication server (see Figure 10-12). You must configure the authenticator (sometimes referred to as a network access server, or NAS for short) and the authentication server to work together. You must configure the NAS with the authentication server's IP address, so the NAS knows where to send packets. And you must also configure the authentication server with the IP addresses of acceptable NAS devices, so it knows who to accept requests from. Both the authenticator and the authentication server must be configured with a matching shared secret: a simple alphanumeric string of your choice.

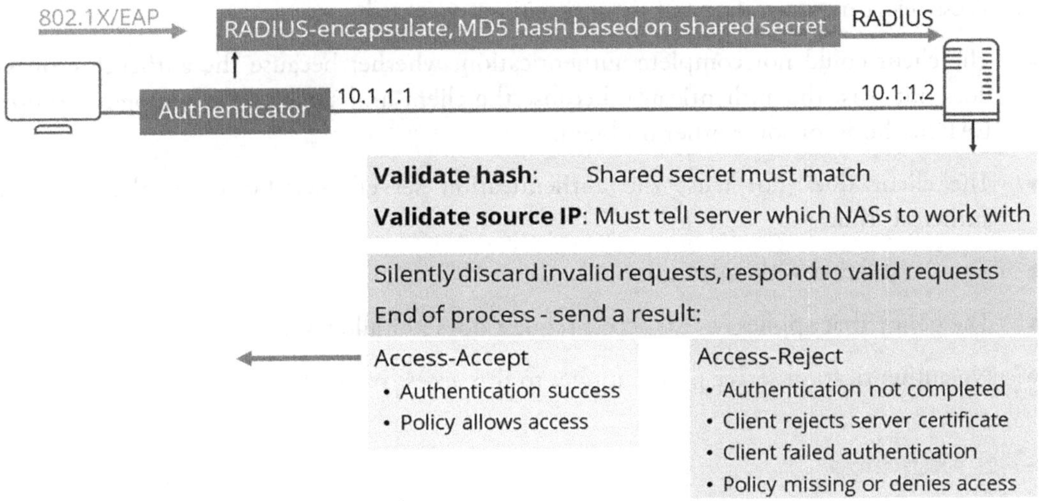

Figure 10-12: EAP and RADIUS together

The authenticator encapsulates a supplicant's EAP message in a RADIUS access-request packet. It also hashes the packet with MD5, using the RADIUS shared secret as the key. It adds the hash to a message authenticator field. The authenticator then transmits the packet to the authentication server. This communication is between the NAS IP and the authentication server IP addresses: the client does not need an IP address or any network access for this process to work, since the IP communication is between the NAS and the server.

The authentication server checks that the message has arrived from a valid network device (the NAS or sometimes called a RADIUS client); that is, the NAS IP address must match a network device address configured on the server. The authentication server also checks the message authenticator field by hashing the packet with its shared secret. If the request is not from a known network device with a correct shared secret, the authentication server does not respond to the message: it silently discards the request.

If the authenticator is known and the secret is correct, the server decapsulates the EAP messages and formulates the appropriate response, which it encapsulates in RADIUS challenge messages. These messages are also authenticated with the message authenticator field.

At the end of the process, the authentication server sends either:

- An access-accept message, if two requirements are met:
  - The client has successfully authenticated.
  - The policy that applies to this access request allows access.
- An access-reject message, if one of these situations occurred:
  - The client could not complete authentication, whether because the authentication server does not trust the authenticator, because the client and server do not support compatible EAP methods, or some other problem.
  - The client does not trust the authentication server's certificate, so the server failed authentication.
  - The client failed authentication.
  - The policy that applies to this access request does not allow access.
  - No authentication server policy applies to this access request.

EAP-TLS exchange

As mentioned earlier, step 4 in the EAP/RADIUS process is where users prove their identity (see Figure 10-13). The server's EAP request/method packet starts the process and indicates the method that the server requires. You can configure most authentication servers to select methods according to conditions such as user identity and location.

If the station supports the requested method, it continues the exchange. Otherwise, the station sends an EAP NAK packet, which can suggest a different method. The server, if it supports the alternative method, may then initiate the exchange with the new method. The supplicant and the authentication server must agree upon the method for the process to continue.

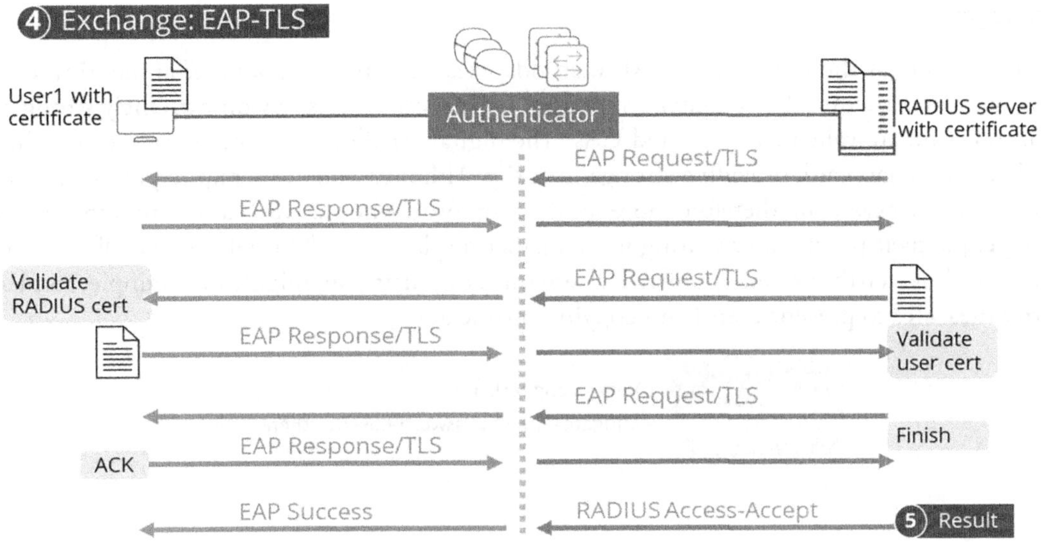

Figure 10-13: EAP-TLS exchange process

The server sends its identity certificate to the supplicant, which validates the supplicant against a local store of acceptable certificates. Typically, this certificate store contains the root certificates of the certification authorities (CAs) that have generated certificates for devices (within and/or external to the company). The EAP request/TLS and EAP response/TLS packets include information such as:

- Digital certificates and certificate verifications
- Supported encryption suites
- Values for generating encryption keys (not the keys themselves)

The digital certificate includes the supplicant's identity and it is signed by a CA. The authentication server validates the certificate submitted by the supplicant, and the supplicant validates the authentication server's certificate. Validation occurs through asymmetric encryption with public and private keys, which is beyond the scope of this book. The validation process also includes checking the CA signature, so the certificate must be signed by a CA that the device is configured to trust.

After the server validates the supplicant certificate and it sends a message to indicate a successful finish, which the client acknowledges. Recall that after the EAP exchange successfully completes, step 5 occurs: the authentication server sends a RADIUS access-accept message.

EAP-TLS considerations

Figure 10-14 summarizes the advantages and concerns related to EAP-TLS.

Advantages

EAP-TLS is one of the most secure EAP methods because it provides *mutual* authentication with public key infrastructure (PKI) digital certificates. Digital certificates rely on extremely strong asymmetric keys and are validated by trusted CAs. The digital certificate is typically stored on a hard or flash drive or a smart card, providing stronger security. Although you can set up requirements to force users to create *stronger*, and therefore more secure passwords; whereas, you cannot prevent users from telling people their password or writing it on a paper displayed in plain sight. Generally, it is harder for users to leak certificates inappropriately; systems administrators might, for example, implement security measures to prevent users from copying certificates.

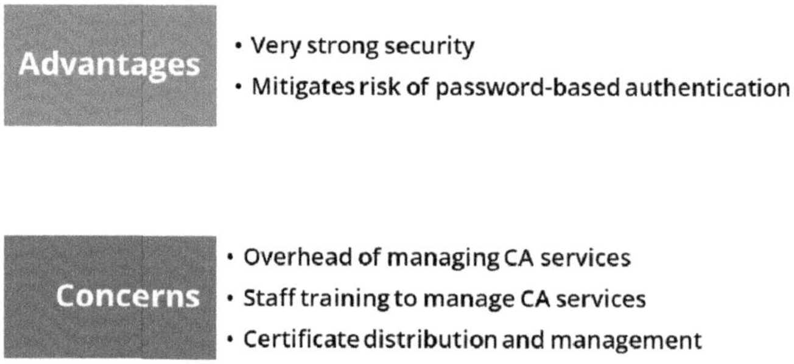

Figure 10-14: EAP-TLS considerations

Hackers often run dictionary attacks on passwords, phish for passwords, or use other social engineering techniques to get users to leak their passwords. Although no credentials can be totally secure—malware might be able to steal certificates, for example—digital certificates are more resistant to many more common attacks, though.

Someone who steals a laptop or smart card might be able to gain access to the certificates within, but they are usually protected by an additional layer of security such as a password, which the user must enter before the supplicant can use the certificate. In addition, the user can immediately report the theft, allowing you to revoke the compromised certificate.

And on top of this, only supplicants that are given certificates can connect to the network, which means administrators control which devices (company versus personal) they will allow access to corporate resources.

Concerns

EAP-TLS is not vulnerable to eavesdropping and replay attacks that affect less secure EAP methods such as EAP-MD5, but security comes at the cost of purchasing and managing the digital certificates: substantially more expensive than managing passwords. Maintaining a large number of certificates requires specialized software and trained IT staff.

In addition, with environments that have multiple operating systems, installing certificates on all clients can be difficult. Bring your own device (BYOD) environments can make managing certificates even more difficult. However, Aruba ClearPass's BYOD onboarding capabilities can help companies surmount these challenges. ClearPass Onboard can provide new user devices with limited access and helps users.

PEAP exchange

Figure 10-15, which illustrates the PEAP exchange process, starts at step 4 of the 802.1X authentication process. This example shows a PEAP MS-CHAPv2 exchange.

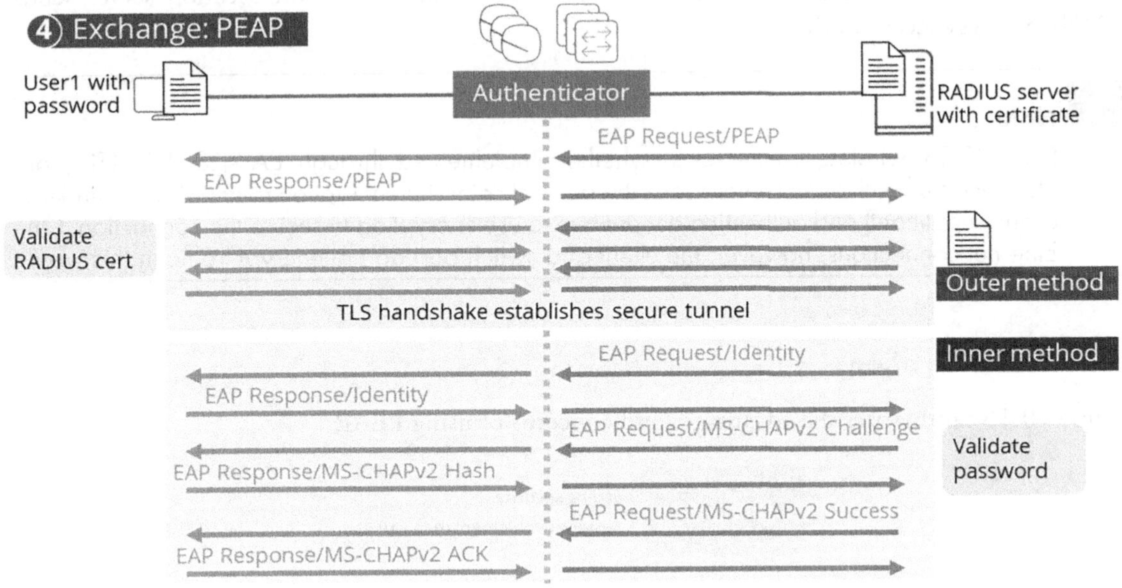

Figure 10-15: PEAP exchange process

- Step 1—outer method: Like EAP-TLS, TTLS and PEAP use a TLS handshake to generate encryption keys and negotiate a tunnel secured by those keys. Only the server authenticates itself with a digital certificate during this exchange, but you must ensure that the clients trust the CA that signed the server certificate. The encryption keys for the tunnel are derived from the public key in this certificate.
- Step 2—inner method: The client authenticates itself in the second step; it uses a weaker—and so more easily implemented—authentication method, protected by the secure tunnel established during the outer method exchange.

The authentication server sends a new EAP request/identity packet, and the supplicant responds with their identity. The server then begins the inner authentication method.

For the inner authentication method, TTLS can use a weaker EAP method, such as EAP-GTC (one-time passwords via token cards) or a legacy RADIUS method, such as CHAP, PAP, or Microsoft CHAP variants (MS-CHAP v1 or MS-CHAP v2). PEAP supports methods such as MS-CHAP v2, EAP-GTC, and TLS. Because Windows native supplicants support PEAPv0 with MS-CHAP v2, this is by far the most prevalent EAP method, which is explained next.

The authentication server sends a challenge. The supplicant responds with a hash of the password. The server validates that the hash is correct, and, if so, it sends a message indicating the success. The supplicant responds. Figure 10-15 shows only the PEAP part of the complete 802.1X authentication process. After these messages, the tunnel is taken down, and the authentication server sends a RADIUS access-accept to the switch.

Note

For 802.1X wireless connections, whether implemented through EAP-TLS, EAP-TTLS, or PEAP, the authentication server sends the authenticator the encryption material. The wireless client (supplicant) and authenticator continue to use encryption to secure the connection. On Ethernet connections, however, the client and switch port do *not* encrypt traffic in this way.

PEAP Considerations

Figure 10-16 summarizes the advantages and concerns of using PEAP.

Figure 10-16: PEAP considerations

Advantages

Like EAP-TLS, EAP-TTLS and PEAP provide strong, mutual authentication. Because TTLS and PEAP use encrypted tunnels to secure usernames and passwords (rather than require digital certificates on stations), you can implement these methods more easily than you can TLS.

TTLS has one unique benefit: it always protects the username. Depending on how PEAP is implemented, the username might be transmitted in plain text in the initial EAP/response identity packet, allowing a hacker to detect the user's identity and possibly lock the user out of his or her account.

Concerns

As compared to EAP-TLS, EAP-TTLS and PEAP have the disadvantage of a weaker authentication method: they're based on a password authentication scheme.

Options for deploying certificates on devices

As you see, ensuring that devices have the correct certificate can be challenging. Maintaining a large number of certificates requires specialized software and trained IT staff. In addition, with environments that have multiple operating systems, installing certificates on all clients can be difficult. There are many options for deploying and managing certificates. In Figure 10-17 you see just a few of those options.

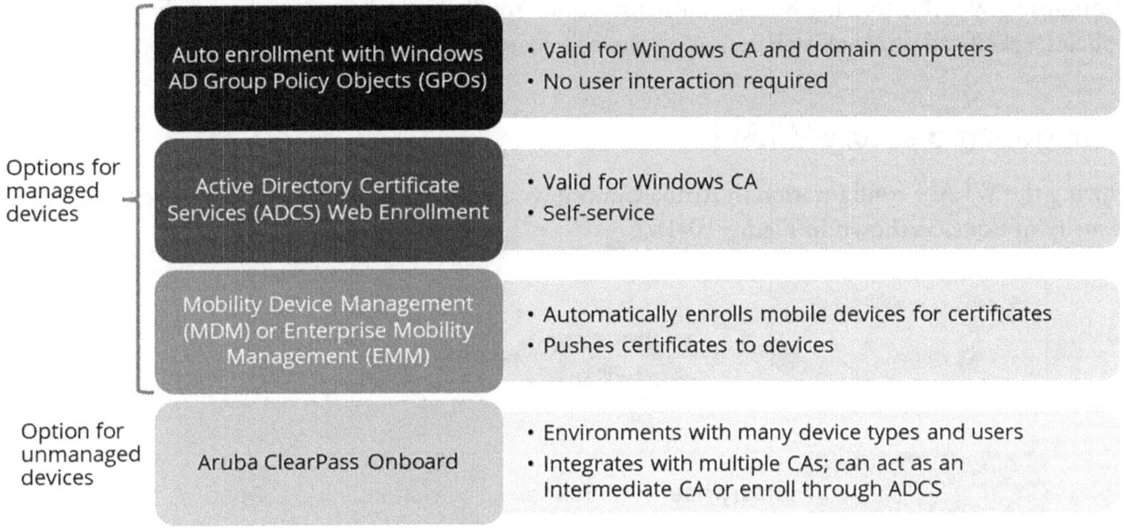

Figure 10-17: Options for deploying certificates on devices

Typically, customers set up a private CA to issue the certificates required for EAP-TLS: this includes both the authentication server and the supplicants using 802.1X. Microsoft Windows CA is a common option for Windows domains that use Active Directory (AD). AD administrators can set up group policy objects (GPOs) to automatically push certificates to devices joined to the domain. The Windows CA also supports Active Directory Certificate Services (ADCS) web enrollment, which enables users to obtain certificates through a self-service web portal.

However, the Windows CA options are geared for a world in which the only devices are computers joined to the domain. A modern network has far more types of devices, including smartphones, tablets, and internet of things (IoT) devices. BYOD environments can make managing certificates even more difficult. There are many mobility device manager (MDM) products that help to automatically enroll mobile devices for certificates and push certificates to devices.

If a company needs to deploy certificates to many unmanaged devices, Aruba ClearPass Onboard can help. ClearPass Onboard helps users to easily download the correct certificates to their devices. Onboard can integrate with multiple CAs. It can act as a root or intermediate CA, issuing certificates for devices on behalf of a root CA. When Onboard integrates with a Windows CA PKI, Onboard can alternatively act as an intermediary with ADCS (registration authority, or RA), in which case the Windows domain signing CA issues client certificates, but the Onboard client makes the process of obtaining the certificate transparent to the user. As an RA, ClearPass Onboard basically acts as a proxy between the client and the Windows root CA.

Secure WLAN setup

Now that you are familiar with the 802.1X authentication process, this section introduces you to configuring WLANs that use 802.1X authentication. You'll also be introduced to apply authorization policies, specifically firewall policies, using the role feature.

Configuring a secure WLAN

During the WLAN configuration in Aruba Central, you have the opportunity to choose from various security options, as shown in Figure 10-18.

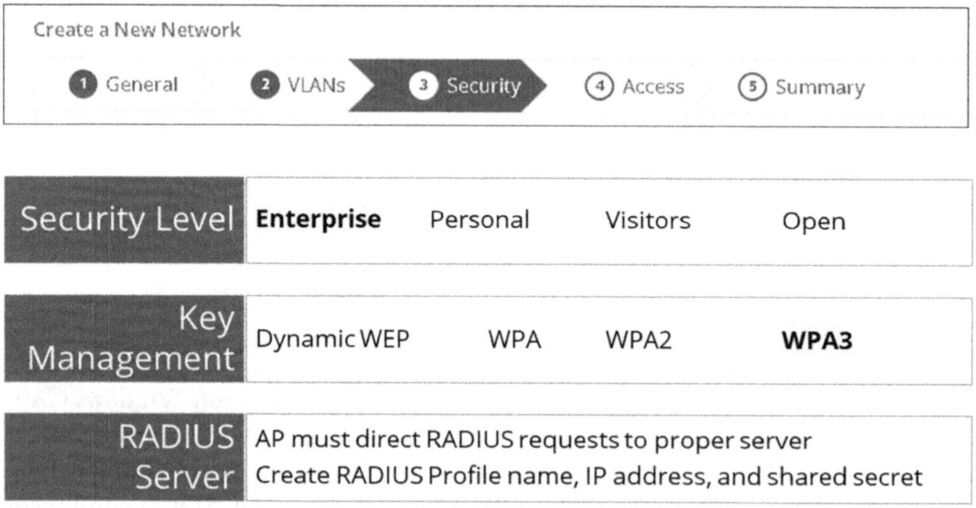

Figure 10-18: Using the wizard to configure a secure WLAN

The *Enterprise* security level is most appropriate for employee WLANs. WPA3 is the most secure key management option. You must also specify an authentication server: the AP must know where to direct RADIUS requests during client authentication. Remember, 802.1X/EAP authentication is between the end client device and the authentication server (authentication server), with the authenticator (AP) acting as an intermediary for this process.

During the lab activities later, you will perform this process, including specifying a RADIUS profile with a name, IP address, and shared secret. This shared secret must match for what the authentication server has been configured. If not, the authentication server will reject requests from the AP.

Roles and firewall rules

This section introduces Aruba roles and firewall rules: these are used to control the supplicant's access once they are connected (authenticated) to the network.

ArubaOS role-based firewall

The role-based ArubaOS firewall can enforce different rules based on user identity and assigned role, rather than simply on features in packet headers. For example, the firewall can restrict contractors from sending SSH traffic to limited ranges of IP addresses while it allows engineers' traffic to any SSH host (see Figure 10-19).

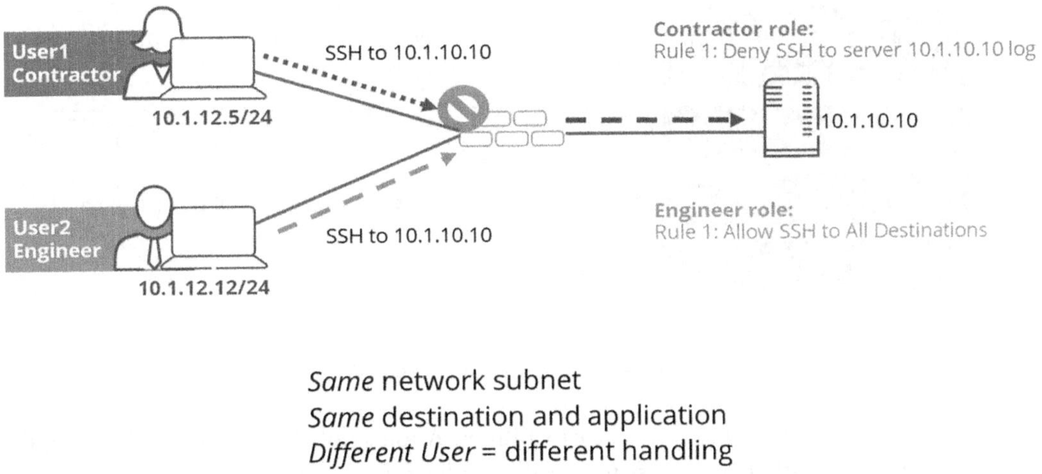

Figure 10-19: ArubaOS role-based firewall

With less sophisticated firewalls, you would create such behavior by ensuring that contractors and managers receive IP addresses in different ranges (different VLANs). You would then create different rules based on the VLAN source IP address. However, such an approach makes planning user connections more complex: you need more VLANs/subnets and a way to get users in the right VLAN every time. With the ArubaOS firewall, you can place users in the same subnets, but easily create different rules for different users.

The ArubaOS firewall determines the user role associated with a device as the device authenticates and accesses the network. It might assign this role entirely on its own, or it might have help from ClearPass, the Aruba authentication server (or any authentication server). By sending all internal user traffic through the role-based ArubaOS firewall, you can change from a model in which you protect a single perimeter to a model in which you place every user within their own unique and properly customized firewall policy.

What is a Role?

An Aruba *role* is a group of settings that controls the client access. They bundle multiple settings that enable the ArubaOS firewall to control the client's traffic (see Figure 10-20). Some of the most important settings for a role are access control rules, grouped in firewall policies. These rules filter traffic based on layers 2 through 4 packet headers: to control the destination IP addresses and TCP/UDP applications accessed by clients assigned to the role.

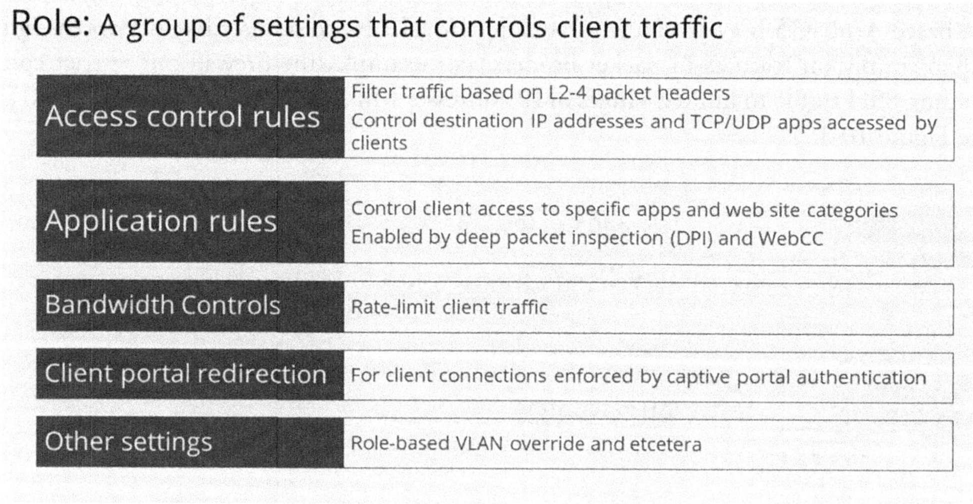

Figure 10-20: What is a role?

However, the ArubaOS firewall has far more capabilities. A role's policies can also include application rules, which let you control client access to specific applications, such as social media or gaming or specific web site categories, based on URL or reputation. These policies are enabled by deep packet inspection (DPI) and the web content classification (WebCC). The access control and application control policies are the focus of this book and provide you an introduction to their implementation.

However, you should also be aware that the role can implement bandwidth controls to rate limit clients' traffic. And, if the client connection is enforcing captive portal authentication, the role can further specify how the client is directed to the captive portal. Finally, roles define a variety of other settings, a role-based VLAN overrides the VLAN assigned in another way such as through the WLAN settings. For wireless users it is best practice to assign the VLAN through the WLAN itself and leave the role-based VLAN non-specified.

Example role definitions

Figure 10-21 shows some typical roles that you might define. Some roles come with the factory default configuration, but you can either easily customize these roles or create your own roles, like many of those shown in Figure 10-21.

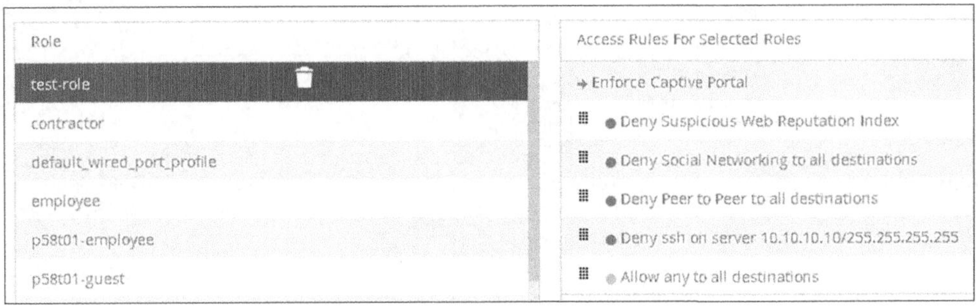

Figure 10-21: Example role definitions

Creating access control rules

When you select either a network-based or roles-based SSID you can also select which type of access rule you want to assign to users. The rule type can be one of the following:

- **Access control** is where you can choose from several services and actions.
- **VLAN assignment** is the process of assigning a configured VLAN to the users.
- **Captive portal** allows you to assign a captive portal page to the user.
- **CALEA** is for law enforcement.
- **Bandwidth contracts** are only available if you're placing a rule into a role-based SSID and not a network-based SSID.
- You can redirect to a URL page or redirect a blocked HTTPs service.

Once you've selected your rule type then you must select your Service, Action, and Destination, in the case of an access control rule (see Figure 10-22). The network service option allows you to either choose any service, or you can choose the custom option and specify TCP, UDP, or other port numbers. Many of the standard protocols have been predefined for your selection. The action can be allow, deny, source-NAT, or destination-NAT. Source-NAT and destination-NAT are forms of allowed traffic, with address translation occurring as well. The destination choices are simple and will depend on your preferences.

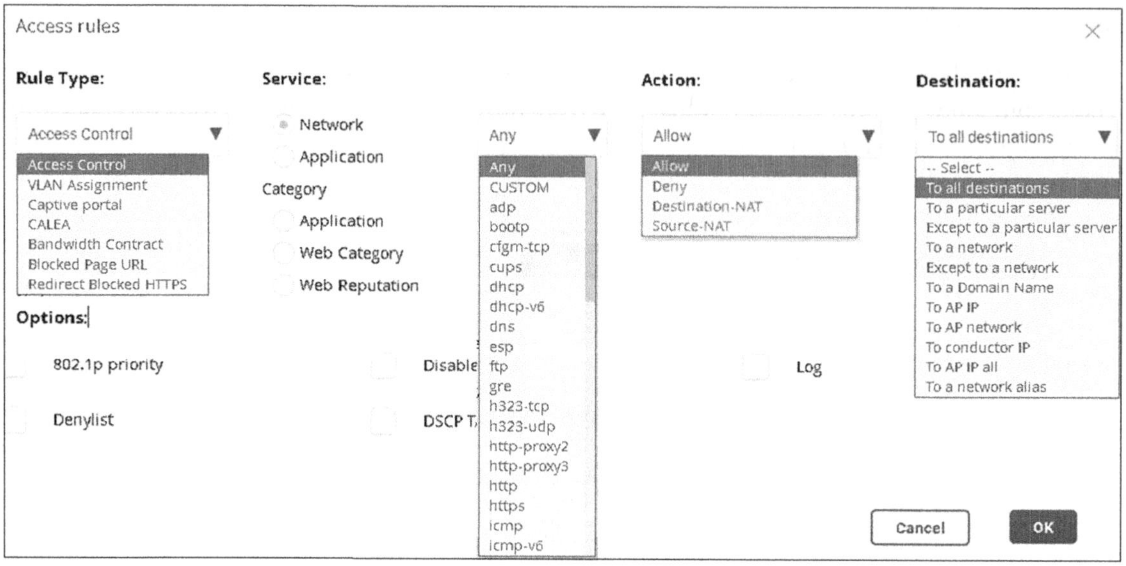

Figure 10-22: Creating access rules

Creating application rules

Aruba APs support both Aruba's built-in DPI engine (AppRF) and a cloud-based web filtering service (WebCC) as a single AppRF *visibility* feature. WebCC is part of the Aruba Central Foundation license. The DPI engine has support for over 1,500 applications.

You can use enforcement at the rule level to allow or block applications or app categories after classifying the packets (see Figure 10-23). Many application rules are included in the default configuration, but you can create your own as well. The web category and web reputation categories are defined by WebCC and are provided to Aruba from BrightCloud. These are used to control your users' access to resources on the Internet.

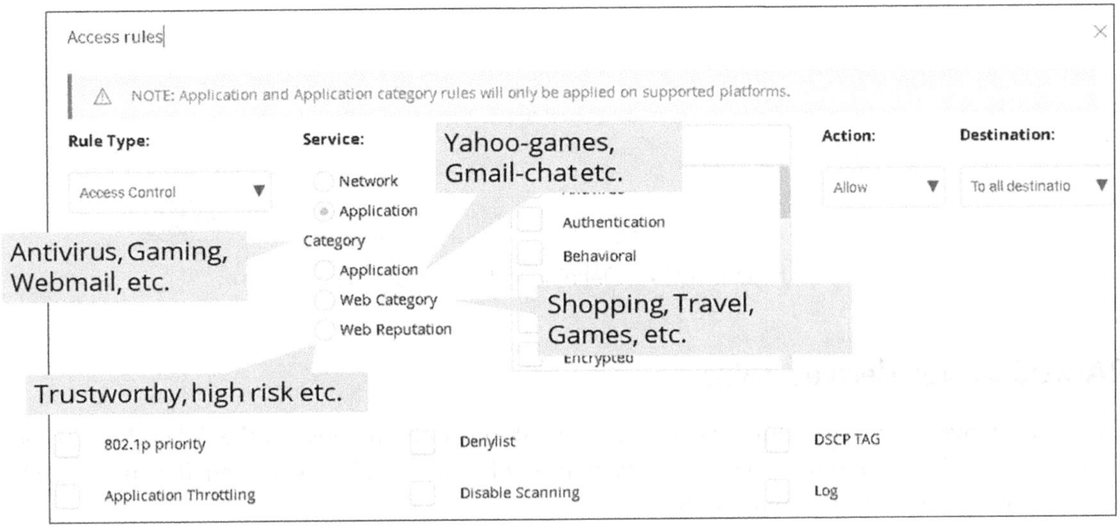

Figure 10-23: Creating application rules

Application classification requires matching packet contents, including headers and possibly payload information. AppRF needs one or more packets in both directions for application classification. You could have a maximum of 8 or 9 packets that must be inspected in order to classify the flow as a particular application or application category. Therefore, rules matching for any app rule must *leak* packets until the classification is complete. If the connection is not or cannot be classified, then it will define it as *not-classified.*

Defining extended actions

The extended actions allow you to perform actions beyond the main action, discussed earlier in the chapter (see Figure 10-24). The extended actions will vary according to the service you select. For example, if you chose the service application then there is also an option for application throttling.

Here are common actions you can typically choose:

- The *Log* action means the AP will send all logs as security logs to a server. You can configure the logging server under System > Monitoring.
- The *Blacklist* action removes this user from the network. The user (and any devices) would not be allowed back on the network unless un-blacklisted or the blacklist time has expired. The blacklist time is set in the security window as Auth failure blacklist time.
- *Disable Scanning* disables ARM scanning when the rule is invoked.
- *DCSP TAG* sets the priority DCSP priority in the IP ToS field, which can be configured from 0 to 63.
- The *802.1p priority* option tags the layer 2 frame with assigned priorities 0 through 7.

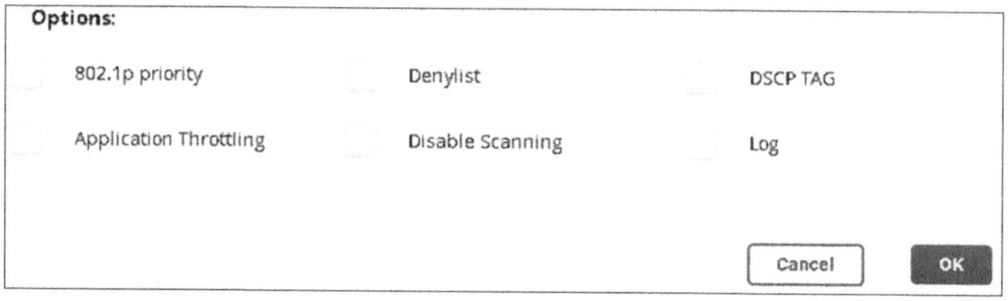

Figure 10-24: Defining extended actions

RADIUS server-derived roles

For server-derived roles, there must be an exchange of information between the AP and the authentication server. What is sent depends on what type of server you're using and if you've installed RADIUS vendor-specific attributes (VSAs).

The AP sends attributes along with the RADIUS request (see Figure 10-25). The server can use this information to derive a role and reply with Aruba VSAs or RADIUS attributes. When a server returns an Aruba VSA with the RADIUS reply, the Aruba AP or switch places the user in the specified role and/or VLAN that was sent. When the server returns a RADIUS attribute, you must configure the Aruba AP to look for this attribute. If you use an Aruba ClearPass server, the Aruba AP-to-ClearPass communication is automatic. The AP sends a RADIUS request with the Aruba VSAs.

ClearPass can aggregate the information and perform many actions, like profiling, posture assessment, and onboarding. ClearPass can also interact with the Microsoft Active Directory and many other directory services and other server types to perform authentication and/or to retrieve attributes. ClearPass will return a RADIUS message using Aruba VSAs with enforcement actions, such as a role name and VLAN identifier, enforcement of bandwidth limitations, or redirect the user to a web page, for example.

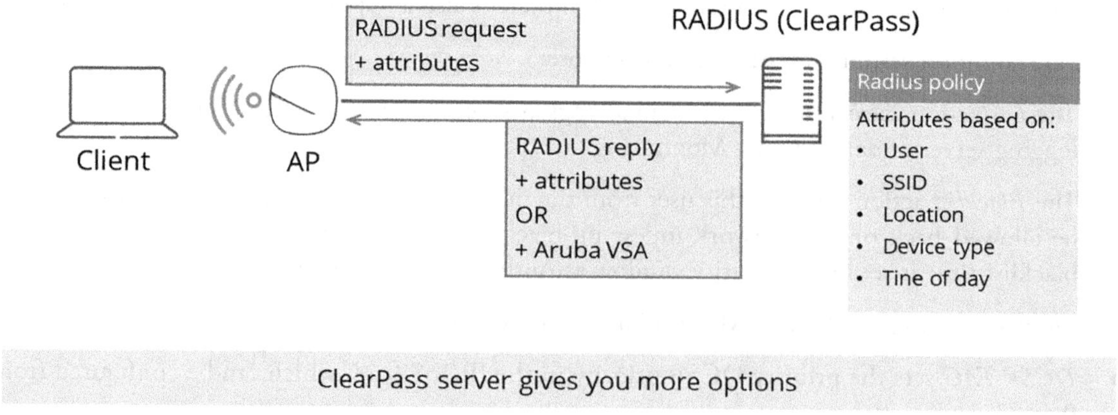

Figure 10-25: authentication server-derived roles

RADIUS role assignment

When using the wizard to create a WLAN in Aruba Central, you can use role derivation to assign a role to the users/devices associated to a WLAN network from the Access tab (see Figure 10-26). When the network receives an association request, it can assign the user/device a role based on the AP or the AP-group. If the server requires authentication, the network can assign the role from attributes it receives from the authentication server.

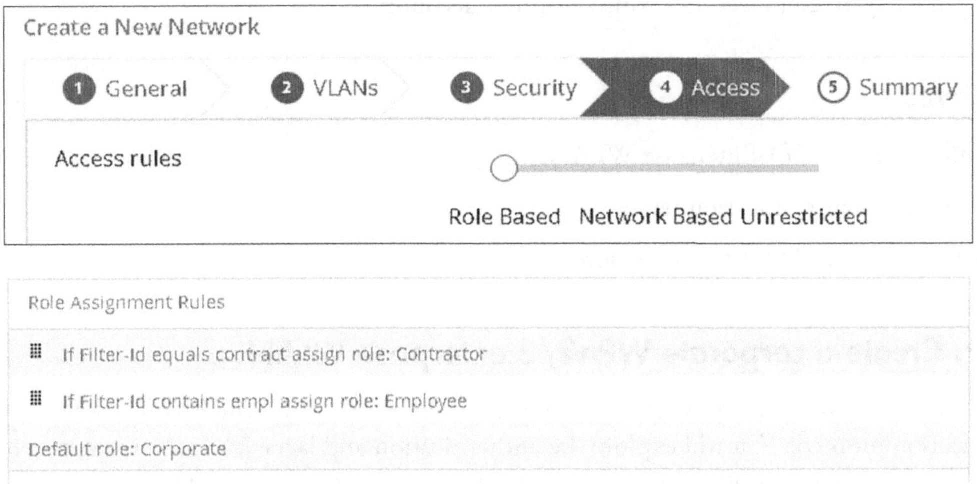

Figure 10-26: RADIUS role assignment

For example, if a user associates to an AP and the authentication server returns an Aruba-User-Role RADIUS filter-ID or VSA with value *employee*, the AP assigns the *employee* role. If RADIUS returns the value *contractor*, the AP assigns the *contractor* role. This requires coordinating RADIUS values with role assignment rules. If RADIUS does not send a filter-ID, nor an Aruba-User-Role VSA, or the value does not match with a role defined on the AP, then the AP assigns the default role defined for the WLAN. Typically, you want to use the server-derived role and not the default one. The default one should only be used as a fallback in case there is a misconfiguration on either the AP or the authentication server, and therefore, should be defined fairly strict in nature.

Once you select an attribute, the received information is compared to the strings and a role is assigned. This choice depends on operator settings: either match, equals, contain, start-with, ends-with, or not equal. When no match is found, the user device is assigned the default role.

Lab 10.01: Implementing a corporate WLAN

Your customer would like to connect the internal employees and contractors to a secured WLAN that offers authentication and encryption. You learn that this can be achieved by configuring a WPA2 or WPA3 enterprise WLAN. It uses a authentication server for centralized authentication and authorization, and it offers encryption for the WLAN traffic.

In this lab you will configure and test the employee WLAN with 802.1X authentication and test the connection using an employee and contractor user account.

Objectives

- Configure a WPA2/3-Enterprise WLAN.
- Configure an authentication server.
- Verify the secured WLAN connection.

Task 1: Create a corporate WPA2/3 enterprise WLAN

In this task, you will create a WPA2/3 Enterprise secured WLAN, which will use 802.1X as the authentication protocol. You will explore the authentication and layer 2 VLAN access for the SSID. The AP should assign the traffic for the corporate wireless clients to VLAN 11.

Objectives

- Configure the WPA2/3 Enterprise WLAN.
- Verify the deployment of the WLAN.

Steps

1. In Aruba Central, navigate to: Context: **Groups / campus-wifi-ui** > Navigation: **Devices**> Top: **Access Points**> Right top: **Config**.
2. Under the WLAN page, click **Add SSID**.

General Page

1. On the General page, configure the name **p#tX-employee**, where # is your assigned pod number and X is your assigned table number.

Note

Make sure to replace the # value with your pod number and x with your table number. For example, if you are using table 07 in pod 28, your WLAN name will be *p28t07-employee*. Check with your instructor or lab administrator if you are not sure about the pod and table number.

Optional Steps

The next steps show values as proposed in the *Aruba Validated Solution Guide*. You may optionally apply them as an example to practice the settings.

Tip

The latest version of the Aruba Validated Solution Guides can be accessed on this page: https://www.arubanetworks.com/techdocs/VSG/.

1. Expand **Advanced Settings**.
2. Open the *Broadcast/Multicast* section.
3. For Broadcast filtering, set the value to **ALL**.
4. Set Dynamic Multicast Optimization (DMO) to **enabled**.
5. Set the DMO client threshold value to **40**.
6. Open the **Transmit Rates (Legacy Only)** section.
7. For 2.4 GHz configure Minimum value as **5**.
8. For 5 GHz configure Minimum value as **18**.
9. This is the end of the optional steps. Continue with the next steps!

5 GHz-only operation

In the remote lab environment, there is a lot of interference from neighboring APs, especially on the 2.4 GHz band. You will limit this SSID to using only the 5 GHz band.

1. Expand **Advanced Settings**.
2. Open the **Miscellaneous** section.
3. For the Band option select only **5 GHz**.
4. Click **Next**.

VLANs tab

1. On the VLANs tab, for the client VLAN assignment, select **Static**.
2. In the VLAN ID field, *remove* **VLAN 1**.
3. Enter VLAN ID **11**.
4. Click **Next**.

Security tab

1. On the Security tab, move the slider to **Enterprise**.
2. Leave key management to the default (**WPA3-Enterprise CCM128**).
3. An enterprise WLAN requires a RADIUS authentication server. In the lab the ClearPass server with IP 10.254.1.23 will be used as the authentication server. It has been preconfigured to support 802.1X wireless authentication for your APs. For the Primary Server, click the + sign to add an authentication server.
4. In the New Server dialog box, enter these values:
   - Name: **cppm1**
   - IP address: **10.254.1.23**
   - Shared key: **Aruba123!**
   - Retype key: **Aruba123!**
   - Dyn authorization: **Enabled** (You may need to scroll down to see this option)
5. Click **OK** to add the server.
6. Verify that the Primary Server field now has *cppm1* selected.
7. Expand the **Advanced Settings** section.
8. RADIUS Interim Accounting ensures that the AP updates the authentication server with status updates about the authenticated clients, such as the amount of traffic generated and whether the clients are still active (online). It is recommended to enable RADIUS Accounting. Expand the **Accounting** section.
9. For the Accounting field, select **Use Authentication Servers**.
10. Set the *Accounting Interval* to **5** min.
11. Open the **Fast Roaming** section.
12. Set 802.11K to **enabled**.
13. Click **Next**.

Access tab

1. Click the Access tab. Leave the default to **unrestricted**.
2. Click **Next**.
3. To complete the configuration, click **Finish**.
4. Wait for the deployment wizard to complete and confirm by clicking **OK**. You should now see the *p#tx-employee* WLAN in the Wireless SSIDs list.

Verify configuration deployment

1. Click the **Configuration Audit** link (see Figure 10-27).

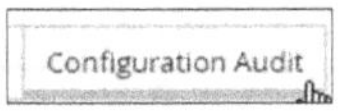

Figure 10-27: Accessing configuration audit

> **Note**
> If you don't see the Configuration Audit page, you should enable the **Advanced View**.

2. Under the LOCAL OVERRIDES & CONFIGURATION SYNC ISSUES, check the Configuration Status window. The *Not In Sync* value should be *0* (see Figure 10-28 for an example).

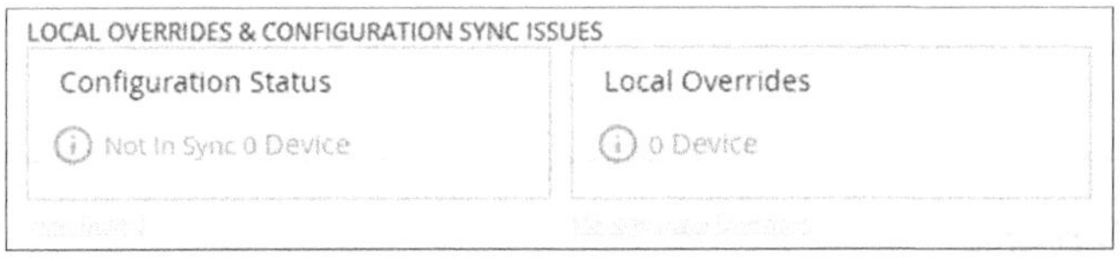

Figure 10-28: Verifying synchronization

> **Note**
> It may take a minute or so for Central to push the configuration to the devices. In the meanwhile, *Not In Sync* will report *2* devices. Refresh this page until the Not In Sync shows *0* Devices, which indicates that the synchronization was successful.

SSID configuration basics

Now you will review the commands that were generated by the UI Wizard.

1. In Aruba Central, navigate to: Context: **Groups / campus-wifi-ui** > Navigation: **Audit Trail** > Top: **Audit Trail**.

2. Look for the entry with description *Created/Updated WLAN profile*. Click the three dots at the end of this entry to see the details. There are some key sections in the configuration to note:

1.1.1.1.1 WLAN access rule

The WLAN access rule acts as an Aruba user role in the AP configuration. It is defined in the global configuration of the AP. For every WLAN that is created, there will automatically be a user role (access rule) based on the SSID name. This will be the default user role that authenticated users get assigned when they connect to this SSID.

```
wlan access-rule p58t01-employee
 utf8
 rule any any match any any any permit
 exit
```

1.1.1.1.2 WLAN authentication server

The RADIUS authentication server is defined in the global configuration of the AP. This means that the same authentication server can be used in other SSID configurations as well.

```
wlan auth-server cppm1
 ip 10.254.1.23
 key ********
 port 1812
 acctport 1813
 rfc3576
 cppm-rfc3576-port 5999
 exit
```

1.1.1.1.3 WLAN SSID profile

The SSID profile is the WLAN object that defines the wireless properties and wireless authentication settings. It also includes the default layer 2 forwarding settings (VLAN assignment) for clients that connect to this SSID.

```
wlan ssid-profile p58t01-employee
 essid p58t01-employee
 opmode wpa3-aes-ccm-128
 vlan 11
 rf-band 5.0
 type employee
 captive-portal disable
 dtim-period 1
 broadcast-filter all
```

```
   radius-accounting
   radius-interim-accounting-interval 5
   inactivity-timeout 1000
   g-min-tx-rate 5
   a-min-tx-rate 18
   max-authentication-failures 0
   blacklist
   dynamic-multicast-optimization
   dmo-channel-utilization-threshold 90
   max-clients-threshold 64
   enable
   dot11r
   utf8
   okc
   dot11k
   openflow-enable
   dmo-client-threshold 40
   auth-server cppm1
   exit
```

- **Question**: Under the WLAN **ssid-profile**, what command controls how wireless clients will see this WLAN object when they scan for wireless networks?
- **Answer**: The command **essid** controls the wireless ESSID name.

3. **Close** the audit details window.

Verify the received Aruba Cloud commands on the AP

On the AP, you can view all commands that are pushed by Aruba Central. This can be useful in case a command fails to execute.

1. In Aruba Central, navigate to: Context: **Groups / campus-wifi-ui** > Navigation: **Tools** > Top: **Commands.**
2. In the Device Type dropdown list, select **Access Point.**
3. In the Available Devices dropdown list, select **ap1**.
4. In the Categories table, scroll down in the list and select **Central**.
5. In the Commands table, select **Show Central Last Configuration Received**.
6. Click **Add** to move it to the Selected Commands column.
7. Click **Run** to run the command on the selected devices. You will need to wait a few moments for the command to be executed.

8. The output will show all the commands that the AP received and the time of execution.
   - **Question**: How can you see if the command was executed successfully?
   - **Answer**: The output includes *OK* after a line that was accepted by the AP CLI.

Note: Renaming a WLAN SSID

The WLAN profile name cannot be changed in Aruba Central; however, the ESSID name can be updated at a later point in time. When this is done, please note that in Aruba Central, the WLAN profile name will not be the same as the actual ESSID name and it will be displayed behind the ESSID name. Figure 10-29 shows an example configuration where the profile name is the *same* as SSID name. Figure 10-30 shows an example configuration with the ESSID changed to p58t01-corporate on the profile name p58t01-employee, so the profile name is *different* from the SSID name. There is no need to change the SSID name in the lab.

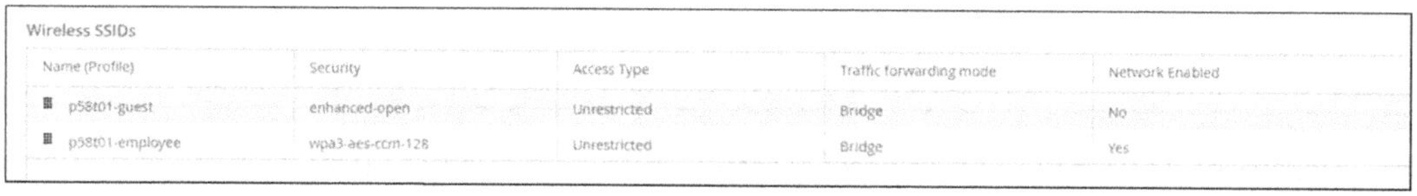

Figure 10-29: Example configuration where the profile name is the *same* as SSID name

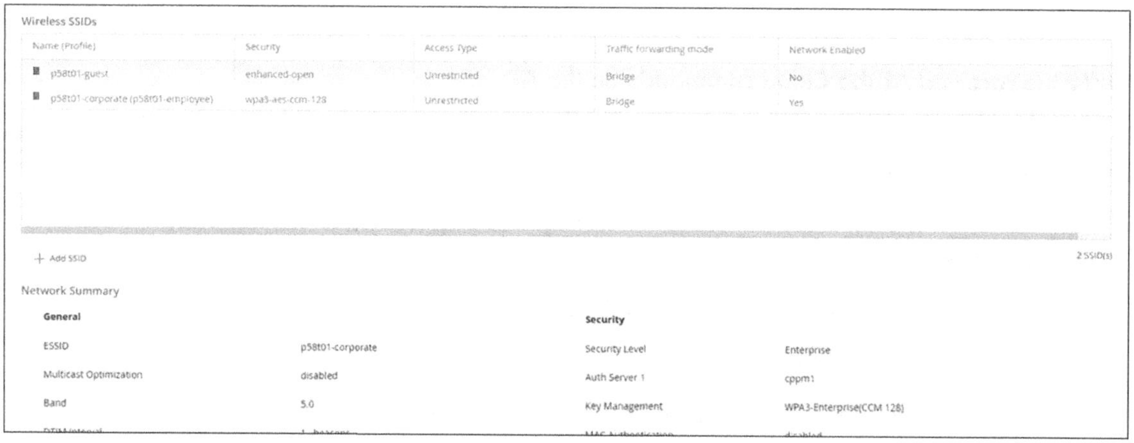

Figure 10-30: Example configuration where the profile name is different from the SSID name

Verify the SSID is being broadcasted by the AP

In this section, you will check the BSS table of AP1 to verify that the SSID is being broadcasted by the AP. You will use the command **show ap bss-table** on the AP.

1. Click **Remove** to clear the Selected Commands window.
2. In the Categories table, select **Wireless.**
3. In the Commands table, select **AP BSSID Table**.
4. Click **Add** to move it to the Selected Commands column.
5. Click **Run** to run the command on the selected devices. You will need to wait a few moments for the command to be executed. Once the output is displayed, you can maximize the output screen. Click the maximize button at the right-top of the command output.
   - **Question**: Do you see the employee ESS broadcasted on the ap1?
   - **Answer**: Yes, the ESSID is the name of the employee SSID as configured in the previous steps. Since it was only enabled on the 5 GHz radio, it appears only on the *a phy* (radio).
6. Exit the maximized view by clicking on the **X** button at the right-top.

Test the AAA server connection

On the AP, you can use the **aaa test-server** command to verify authentication server communication. This will send a RADIUS access request to the configured authentication server. This test allows you to verify that the shared secret and RADIUS client settings are matching between the AP and the authentication server. You must know an example username and password to run the test. In this lab, use username **employee** with password: **Aruba123!**

The **aaa test-server** command is not included in the standard command list, therefore you will open an SSH console session to the AP using Aruba Central.

1. In Aruba Central, navigate to: Context: **Groups / campus-wifi-ui** > Navigation: **Tools** > Top: **Console**.
2. In the Device Type dropdown list, select **Access Point.**
3. In the Available Devices dropdown list, select **ap1**.
4. For the username enter **admin**.
5. For the password enter **Aruba123!**
6. Click **Create New Session** to request the console session. After a few seconds, the console connection will appear.

Tip

If you have issues with this connection in your own production environment, make sure to check the ***Information*** icon (the "I" with a circle around it) at the right-top. The console connection is a reverse SSH connection over port 443 to an Aruba Central cloud host. Some new-generation firewalls will detect that this is not HTTPS over port 443 and they may block this communication. You must make sure that this is explicitly allowed in the firewall. Close the information box with the **X** button.

---

7. In the AP console connection, run the **aaa test-server** command. You can use ? and the <Tab> key for command completion. The server **cppm1** is the name you used during the WLAN wizard.

```
aaa test-server cppm1 username employee password Aruba123! auth-type pap
```

- **Question**: What is the result of the command?
- **Answer**: The AP reports a success authentication.

Note

If the test is not successful, you should run the WLAN wizard again and verify the CPPM1 IP address (10.254.1.23) and the shared secret (Aruba123!) from the Security tab of the wizard.

Note

The PAP method (cleartext Password Authentication Protocol) is different from the EAP-PEAP-MSCHAPv2 authentication method that is used by the wireless client. Your authentication server must be configured with *PAP* support to run this test command successfully. The lab ClearPass server has been configured with a test service for this purpose.

---

8. You will now verify on the ClearPass authentication server that this test request was processed. On PC-MGMT, open a browser to **https://10.254.1.23/tips**.
9. Login with username **admin**/password **Aruba123!**
10. In the left pane, expand the **Monitoring** section.
11. Under Monitoring, expand **Live Monitoring**.
12. Click **Access Tracker**. Access Tracker shows the latest authentication requests that were processed by the server. This should include the test request you have made in the previous steps. There will be a record with username *employee*.
    - **Question**: What is the *Login Status* for this entry?
    - **Answer**: The result for this request is ACCEPT: the server has accepted the login request. This resulted in the success message for the AP **aaa test-server** command.
13. This concludes the test procedure. You may leave the ClearPass session open.

Task 2: Verify corporate access with wireless client

In this task, you will connect to the corporate wireless network using PC1. The authentication method will be EAP-PEAP-MSCHAPv2. The wireless client in the lab supports only WPA2, therefore only WPA2 can be tested.

Objectives

- Connect with a Windows client to a WPA2 or WPA3-Enterprise network.
- Verify the Wi-Fi client connection in Aruba Central.
- Verify the Wi-Fi client connection in the wired network.
- Verify the Wi-Fi client connection on the AP.

Steps

Use PC1 to make a wireless connection

1. Open a connection to your PC1 system.
2. In the wireless LAN list, select **p#tx-employee**, connect with username **employee** password **Aruba123!** Figures 10-31 and 10-32 are example screenshots: your pod and table may be different.

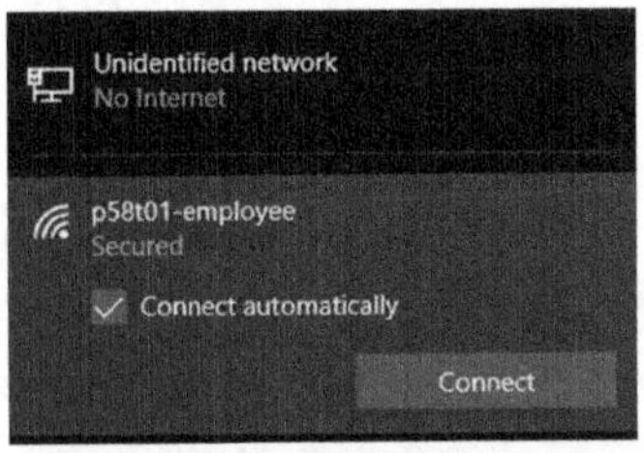

Figure 10-31: Connect to the WLAN

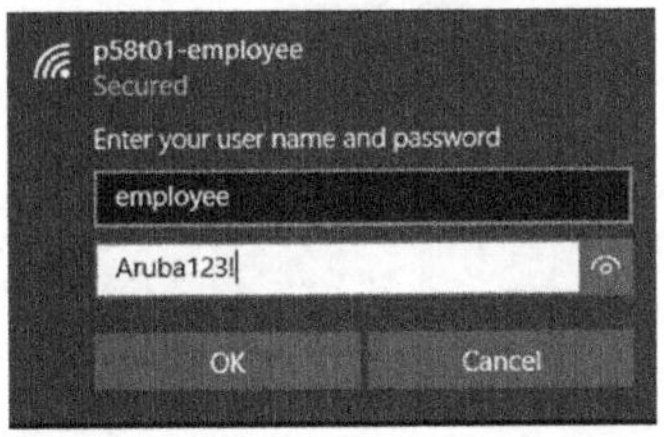

Figure 10-32: Enter the user credentials

3. Click **OK** to submit your credentials.
4. Click **Connect** to accept the certificate warning. You are seeing this because ClearPass has a self-signed certificate. In a production network, a CA should generate a certificate for your AAA server and your supplicant should have the CA's root certificate installed in order to validate the AAA server's certificate.

Note

The lab test PC is not part of a managed environment. In a managed, corporate environment the WLAN profile can be provisioned using ClearPass Onboard, Windows Group Policies, or an MDM product, for example.

Check the assigned IP address on the client

1. Right-click the **wireless** icon in the Windows status bar.
2. Click **Open Network & Internet Settings**.
3. Click **Change Adapter Options**.
4. Right-click **Wi-Fi** connection and select **Status.**
5. In the Status dialog select **Details**. Figure 10-33 displays an example.

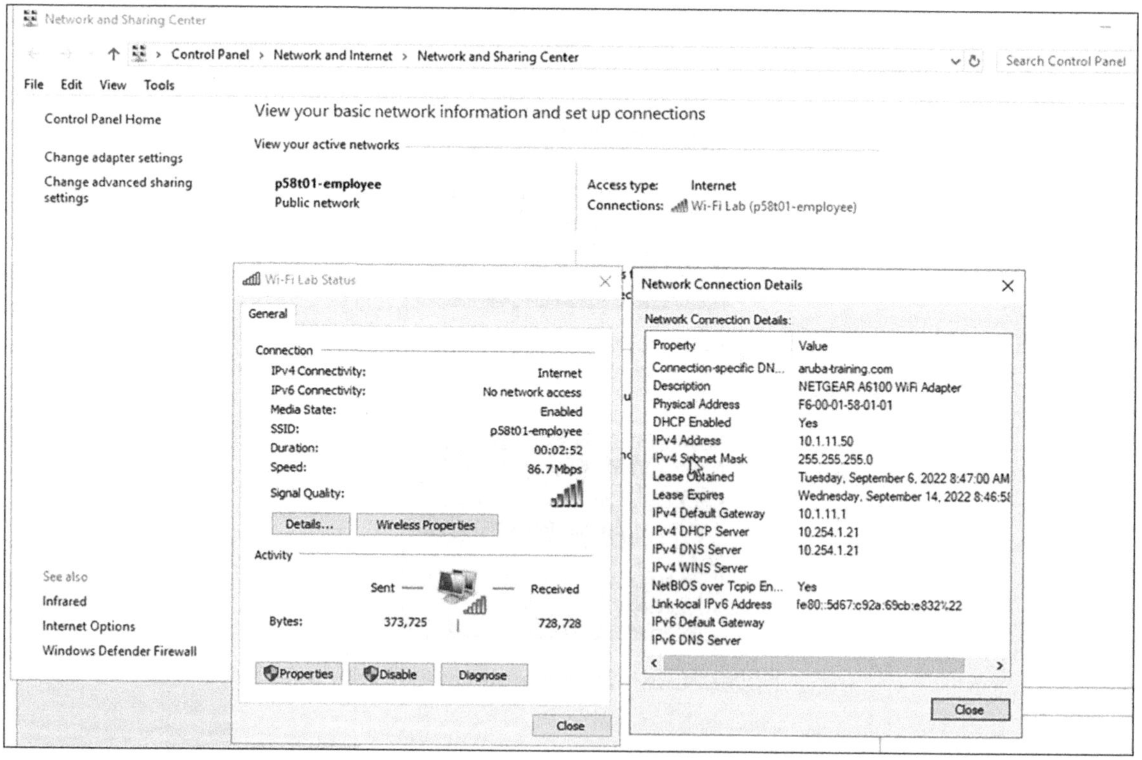

Figure 10-33: Accessing the Network Connection Details window

- **Question**: Take note of the physical address for the client?
- **Answer**: ________________________________
- **Question**: What is the IPv4 address the client received? Does this match with the configured VLAN?
- **Answer**: The IP address should be in the 10.1.11.0/24 subnet. This is indeed VLAN 11.

6. Close the **Details** and **Status** windows.

Verify the client connection in Aruba Central

1. In Aruba Central, navigate to: Context: **Groups / campus-wifi-ui** > Navigation: **Clients**> Top: **Clients**.
   - **Question**: Do you see the client employee in the list?
   - **Answer**: Yes, the client employee is listed.
   - **Question**: What is the status for the client?
   - **Answer**: The status is *Connected*. This indicates that the user is now online.

Note

Aruba Central can also show the historical clients when they are no longer connected, this is convenient for troubleshooting. You can use the top-right time period icon to control how far back the historical clients should be displayed. See Figure 10-34 for an example.

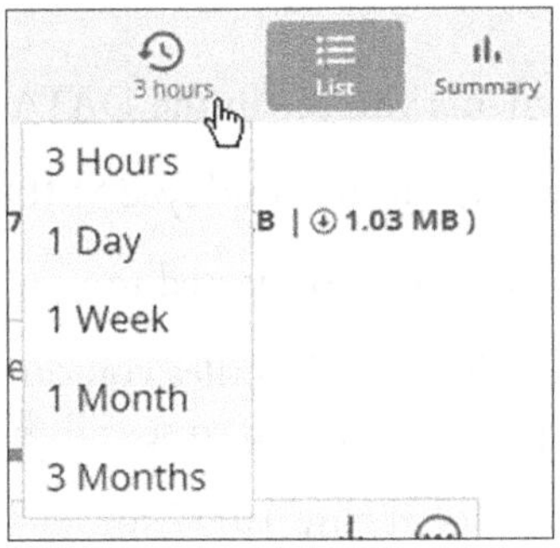

Figure 10-34: Changing the time period for client information

- **Question**: What is the VLAN and IP address that was assigned to the client?
- **Answer**: The VLAN ID is 11, the IP address should match the address you found on the PC1 client.

Client details

The previous screen shows a list of all the clients connected at the group level. You will now investigate client details for an individual client.

1. Click the client name **employee** to access the client details. The context will change to the client name (see Figure 10-35).

> **Note**
>
> You can also use the AI search at the top of the window. Simply type **employee** and press **<Enter>** and the client will be listed in the output. Click the name **employee** to access the client details page.

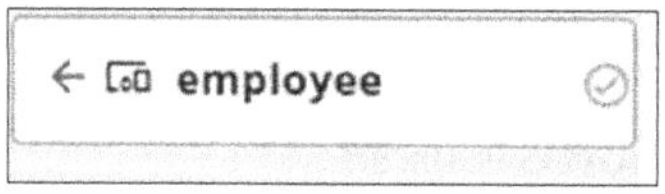

Figure 10-35: Client context

2. You are now on the client employee details > Overview > Summary page (see Figure 10-36).

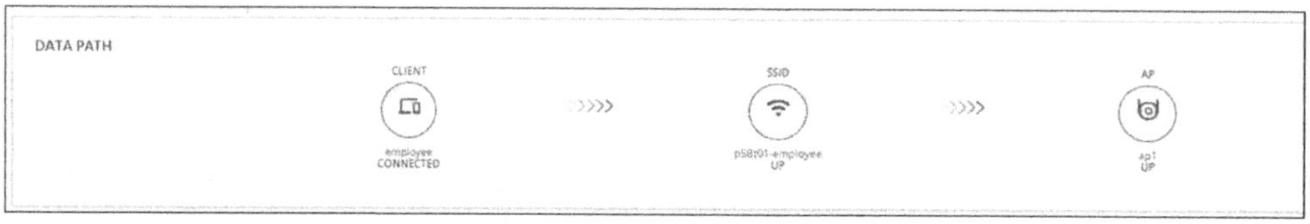

Figure 10-36: Client details Summary page

- **Question**: What are the elements you see in the DATA PATH?
- **Answer**: The client is connected via the employee SSID to either AP1 or AP2.

3. Take note of the access point the client is connected to: ______________

- **Question**: What are the other information tiles you can see in the Client details Overview > Summary page? Make sure you scroll down to see all the tiles.
- **Answer**: The other tiles show (see Figure 10-37 for an example):
  - Client information
  - Network
  - Connection
  - Throughput

- Health
- Signal Quality
- Retry Frames
- Tx/Rx Rate
- Roaming Experience

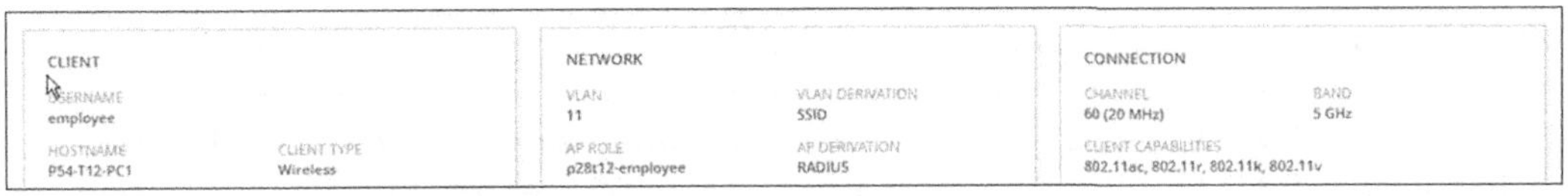

Figure 10-37: Client details Summary page (cont.)

Verify the client connection in the wired network

You will now look at the access switch details to verify you can see the client MAC address. The wireless client MAC address can be seen on the access switch since the wireless client is connected to a bridged WLAN.

1. In the AI search bar, enter the name of the access switch. Use **sw-edge1 or sw-edge2**, depending on which AP your client was connected to.
2. In the AI search result, click the *name* of your switch.
3. While on the edge switch details page, navigate to: Navigation: **Tools**> Top: **Console**
4. For the username enter **admin**, password **Aruba123!**
5. Click **Create New Session**.
6. When the console session is connected, check the VLANs on the AP port.

```
sw-edge1/2# show vlan port 1/1/2

------------------------------------------------------------------
------
VLAN  Name                                Mode             Mapping
------------------------------------------------------------------
------
4     VLAN4                               native-untagged port-access
11    VLAN11                              trunk            port-access
12    VLAN12                              trunk            port-access
13    VLAN13                              trunk            port-access
14    VLAN14                              trunk            port-access
15    VLAN15                              trunk            port-access

Overridden VLAN list: 1
```

- **Question**: Is VLAN 11 listed, what is the mode for VLAN 11?
- **Answer**: Yes, VLAN 11 is listed and it is in trunk mode: this is VLAN tagged traffic.

7. Check the MAC-address table on the AP port.

```
sw-edge1/2# show mac-address-table port 1/1/2
MAC age-time               : 300 seconds
Number of MAC addresses : 3

MAC Address          VLAN     Type                      Port
--------------------------------------------------------------
20:4c:03:c6:09:50    4        port-access-security      1/1/2
20:4c:03:c6:09:50    11       port-access-security      1/1/2
08:36:c9:7c:e5:97    11       port-access-security      1/1/2
```

- **Question**: Do you see the PC wireless MAC address in the list and in the correct VLAN?
- **Answer**: Yes, the PC wireless MAC is listed in VLAN 11.

8. Exit the console connection (**exit** command).

Verify the client connection to the AP

In this section, you will verify on the AP the connected client status.

1. Use the AI search box to navigate to the **employee** client details page.
2. In the DATAPATH tile, click the **AP**. Central will take you to the AP device details page.

Note

It takes some time for the device links and topology to be established. Therefore, it is possible that you don't see the AP connected to the edge switch, or you may even see the AP as connected to the aggregation switch. You may ignore this and continue with the next steps. This will be corrected by Aruba Central over time.

3. While on the AP details page, navigate to: Navigation: **Tools**> Top: **Console**.
4. For username enter **admin**, password **Aruba123!**
5. Click **Create New Session**.

6. When the console session is connected, check the clients.

```
ap1# show clients
Client List
-----------
Name       IP Address  MAC Address        OS       ESSID            Access
Point  Channel  Type  Role              IPv6 Address                Signal
Speed (mbps)
-------------------------------------------------------------------------------
employee  10.1.11.52  08:36:c9:7c:e5:97  Win 10  p28t11-employee  ap1
56       AC    p28t11-employee  fe80::753c:42f5:cdf5:6f7a  69(good)
173(good)
Number of Clients    :1
Info timestamp       :69628
```

- **Question**: Do you see the client's name and MAC address?
- **Answer**: Yes

7. Check the detailed output of the clients using show clients debug.

```
ap1# show clients debug
Client List
-----------
Name       Host Name    IP Address  MAC Address        OS      ESSID
Access Point  Channel  Type  Role              IPv6 Address
Signal    Speed (mbps)  *Reauth Age  *Reauth Interval  *Reauth ESSID
Auth Type    *Authenticated  DEL  Age  Vlan       *ESSID
*Private role info  Accouting Session Name  BSSID              Idle
Timeout  csum       *mcast groups  *Acct Interval  *Class Attribute
*Dhcp-Opt Vlan  *Dhcp-Opt role  Intercept  *Offline  *FB Token  *FB
RxBytes   *FB TxBytes  *SLAAC IP Address  *Link Local IP Address     *DHCP
Status  *DHCP v6 Status
----       ---------   ----------  -----------        --      -----
------------  -------  ----  ----              ------------
------    ------------  -----------  ----------------  -------------
---------    --------------  ---  ---  ----
------     ------------------ ---------------------- -----     ------------
---- -------------------------------------------          -------------
--------------  ---------  --------  ---------  -----------
-----------  -----------------  ----------------------
------------  ---------------
employee  P54-T11-PC1  10.1.11.52  08:36:c9:7c:e5:97  Win 10
p28t11-employee  ap1          56        AC     p28t11-employee
fe80::753c:42f5:cdf5:6f7a  69(good)  173(good)     432          0
p28t11-employee  802.1X/PEAP  yes             no   13   11(SSID)
p28t11-employee(EAP-PEAP)  148(RADIUS-7fff)     employee
24:62:ce:dc:2e:10  1000           2ed05adc  (0)              300
5e95b071935f47e09e15952e26b2116ab90b00000000000052303030303030312d30312
d363335323762633500000000000000000000000  0,NONE          ,,0-0
no         no          null        null          null        0.0.0.0
fe80::753c:42f5:cdf5:6f7a  Completed     Soliciting
```

- **Hint**: You may want to maximize the screen.
- **Question**: What is the VLAN this client is connected to?
- **Answer**: The VLAN shows 11.

8. Check the MAC address table on the AP using the **show datapath bridge** command. You may also apply a filter to include lines with the text 11 (only include output lines that have "11" in them) to reduce the output. Here's an example:

```
ap1# show datapath bridge | include 11
8     bond0                 3     1      0/0/0    106  1500   117049
79215  FTQBW
25    aruba000              1     11    148/0/0     0  1500      915
809   B
12:01:00:00:01:00  11    11           dev8              G        0
B8:D4:E7:D9:E5:00  11    11           dev8                       0
B8:D4:E7:DAnswer:80:00  11    11           dev8                  0
08:36:C9:7C:E5:97  11    11           dev25                      0
```

- **Question**: Do you see the client MAC address in the list?
- **Answer**: Yes, it is listed in VLAN 11.
- **Question**: Are there any other MAC addresses in VLAN 11?
- **Answer**: Yes, the aggregation switch's active gateway MAC address 12:01:00:00:01:00 is also listed. This is the default gateway for the clients in VLAN 11. (The MAC address you see in the lab might be different.)

9. This concludes the AP client review. Exit the console connection with the **exit** command.

Task 3: Connect to corporate network with contractor user

In this task, you will connect to the corporate wireless network using PC4 as a contractor user.

Objectives

- Practice the client connection.

Steps

Use PC4 to make a wireless connection

1. Open a connection to your PC4 system.
2. Verify that the *LAN* interface is disabled. Disable it if needed.
3. Verify that the *Wi-Fi* interface is enabled. Enable it if needed.

> ⚠ **Tip**
> Use the steps from the previous task if you need help to find the network adapters in Windows.

4. In the PC4 Windows taskbar wireless LAN list, select **p#tx-employee** and connect with a username of **contractor** and a password of **Aruba123!**
5. In Aruba Central, navigate to: Context: **Groups / campus-wifi-ui** > Navigation: **Clients**> Top: **Clients**
6. Verify that the contractor user is connected to the network. You can use the **Refresh** button to refresh the client list. Figure 10-38 shows an example client list:

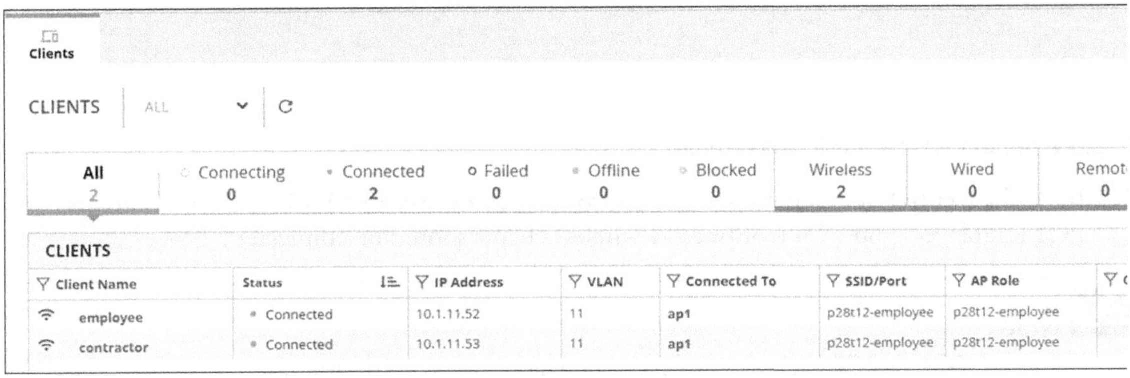

Figure 10-38: Verifying the contractor connection

Lab 10.02: Implementing access control

Your senior colleague is pleased with how you configured and validated the 802.1X secured WLAN. Now the customer would like to enforce some access control policies to the wireless clients. In this lab, you will explore different access control options for AOS 10 bridged WLANs. This will include the default role configuration, creating new user roles for the employee and contractor roles, applying access control using roles, and assigning VLANs.

Objectives

- Explore the default SSID user role access.
- Leverage authentication-based user roles.
- Use Aruba user role VSAs.
- Perform user role VLAN assignment.
- Explore AppRF and application statistics in Central's dashboard.

Task 1: Explore the default SSID user role access

In this task, you will explore the default SSID-based user role access. When you create a WLAN, a default user role is automatically created for that WLAN, based on the WLAN (SSID) name. The rules defined in this default SSID role are applied to any device that connects to this SSID by default. The default role can be overruled (overridden) by RADIUS (ClearPass) assigned roles or custom role derivation rules. You will set up a RADIUS role derivation in the upcoming tasks. In this task, you will also explore the default SSID role.

Objectives

- Understand the default SSID user role.
- Apply network access control using the default role.

Steps

Explore the default SSID role assignment.

1. In Aruba Central, navigate to: Context: **Groups / campus-wifi-ui** > Navigation: **Clients**. Both PC1 (employee) and PC4 (contractor) wireless clients should be connected.

Note
If either PC1 or PC4 lacks a wireless connection, re-establish that now.

---

2. The client list includes an AP role column. You may need to scroll to the right to see it.

Note
You can change the column display order by grabbing the top line of the column header and moving it left or right.

- **Question**: What is the AP role that is currently assigned to the employee and contractor users?
- **Answer**: Both users currently have the role that is named *p#tx-employee*.
- **Question**: What does the role name seem to match?
- **Answer**: The role name matches the WLAN object name. When you create a new WLAN, a new *default access role* is automatically created, based on the WLAN name.

Review the default SSID Role

1. In Aruba Central, navigate to: Context: **Groups: campus-wifi-ui** > Navigation: **Devices**> Top: **Access Points**> Right top: **Config**.
2. Click the **Security** tab.
3. Expand the **Roles** section.
   - **Question**: What roles currently exist in the configuration?
   - **Answer**: There are currently four roles:
     - Default_wired_port_profile
     - p#tx-employee
     - p#tx-guest
     - wired-SetMeUp
   - Two of these are default. The employee and guest roles were created when you added a new WLAN.

Review the p#tx-employee Role

1. Click the role name **p#tx-employee**.
   - **Question**: What are the access rules for this role?
   - **Answer**: There is one access rule: "Allow any to all destinations". This means the role does not have any restrictions and allows unrestricted network access.

Configure network-based access control

The customer wants to start with a simple access control measure. They want to ensure that whoever connects to the corporate WLAN cannot access IP subnet 10.1.0.0/24. This is the subnet with aggregation switch loopback IP addresses you configured in an earlier lab. You have been asked to configure this network access control on the employee WLAN.

Set up test traffic for access control

Configure both clients with a continuous ping to the 10.1.0.2 IP address (the loopback IP address of sw-agg1).

1. On PC1 (the employee PC), start a continuous ping to **10.1.0.2**. This should be successful. Leave the ping running.

```
C:> ping 10.1.0.2 -t
```

2. On PC4 (the contractor PC), start a continuous ping to **10.1.0.2**. This should be successful. Leave the ping running.

```
C:> ping 10.1.0.2 -t
```

Configure the access control

Now you will implement access control to block access to the *off-limits* subnet: 10.1.0.0/24.

1. In Aruba Central, navigate to: Context: **Groups / campus-wifi-ui** > Navigation: **Devices**> Top: **Access Points**> Right top: **Config**
2. On the **WLANs** page, select the **employee** SSID and edit it using the **pencil** icon.
3. Click the **Access** tab.
   - **Question**: What is the current access rules slider status?
   - **Answer**: **Unrestricted**. This reflects the access rule: **allow any to all destinations** in the default user role.
4. Move the slider to Network Based.
   - **Question**: What are the listed access rules?
   - **Answer**: There is one access rule: **Allow any to all destinations.**
5. Click **Add Rule** to create a new entry.
6. In the Rule Type dropdown list, select **Access Control.**
7. For Service, leave the default Network: **Any**.
8. For Action, select **Deny**.
9. For Destination, select To a network: IP **10.1.0.0** and netmask **255.255.255.0**.
10. Click **OK** to add the rule.
11. Verify in the Access Rules table that the Deny rule is listed *before* the Allow any rule (see Figure 10-40).

Note

You can use the **select** icon (see Figure 10-39) at the start of the line to change the order of entries by dragging and dropping the rule to a different location in the list.

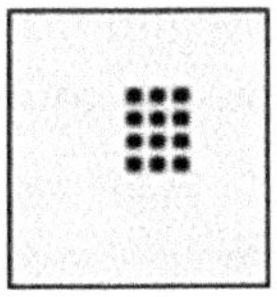

Figure 10-39: Select icon

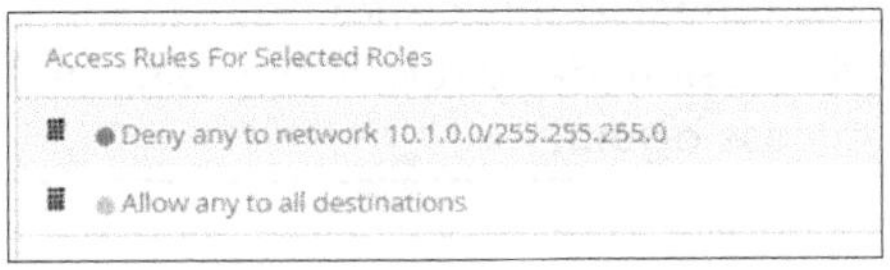

Figure 10-40: Verifying the access rule update

- **Question**: Why is it important that the *deny* rule is listed first?
- **Answer**: Rule sets are processed from top to bottom. When there is a match on the condition, the action is applied. If the *Allow any to all destinations* were the first entry, all traffic would match this rule and be allowed. No traffic would ever hit the next rule, and so no traffic would be denied.

12. Click **Save Settings** to save the configuration.
13. Wait for the configuration wizard to complete and confirm with **OK**.

Verify the access control

1. On PC1, verify that the ping no longer responds.
2. On PC4, verify that the ping no longer responds.
3. In Aruba Central, navigate to: Context: **Groups / campus-wifi-ui** > Navigation: **Clients**.
   - **Question**: Did the user role for the employee and contractor change?
   - **Answer**: No, they still have the *p#tx-employee* role assigned to them.
4. In Aruba Central, navigate to: Context: **Groups / campus-wifi-ui** > Navigation: **Devices** > Top: **Access Points** > Right top: **Config**.
5. Click the **Security** tab, open the **Roles** section.
6. Select the role **p#tx-employee**.
   - **Question**: What do you notice for the access rules for this role?
   - **Answer**: The deny access rule was added to the rule set of this role.
   - **Question**: What do you know now about the WLAN wizard **Network Based** access and **Unrestricted** options?

- **Answer**: The unrestricted option configures the *allow any* to all destinations in the default SSID role, while the network-based access configures user-defined rules in the default SSID role.

Note

You could also have configured the custom rule directly under Security > Roles. The WLAN Wizard would have reflected this custom rule in the UI as network-based access. Similarly, if you removed the custom deny rule in the role, the WLAN Wizard UI automatically shows unrestricted access again.

Task 2: Authentication-based user roles

The customer is happy that you managed to implement this access control, but they realized that this access control is not specific enough for their requirements. They want the employee users to have full access to the network, while the contractors should be denied access to subnets 10.1.0.0/24 and 10.1.3.0/24. The customer wants an easy way to block additional contractor destinations, should the need arise.

Your senior colleague explains that each client can be assigned a different user role, based on RADIUS authentication. To do this, you leverage an Aruba vendor-specific RADIUS attribute (VSA), named *Aruba-User-Role*. In fact, your colleague has already configured the ClearPass Policy Manager to return the user role called *employee* for the user employee and the Aruba-User-Role *contractor* for the user contractor when these people authenticate using 802.1X.

Steps

Review the assigned RADIUS attributes

In these steps, you will log in to the ClearPass authentication server and review the RADIUS attribute that is returned when employee and contractor users authenticate successfully when connecting to the network.

1. Using the PC-MGMT, open a browser and connect to ClearPass Policy Manager: **https://10.254.1.23/tips**.
2. Login using **admin / Aruba123!**
3. In the left pane, navigate to: **Monitoring > Live Monitoring > Access Tracker**.
4. Click the latest **employee** record in the list (the list is sorted from newest to oldest entries) to open its **Summary** page (see Figure 10-41).
5. Click the **Output** tab and expand the **RADIUS Response**.

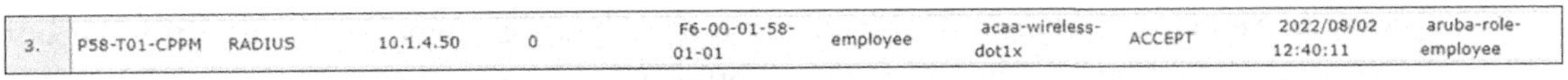

Figure 10-41: Access Tracker entries

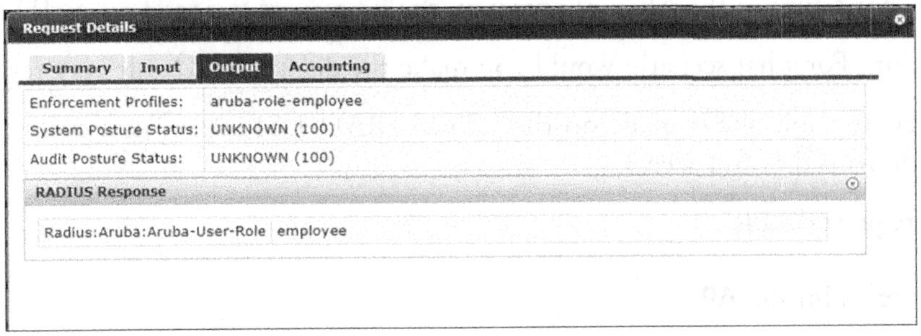

Figure 10-42: Access Tracker entry details

- **Question**: What is the Aruba-User-Role RADIUS attribute that ClearPass returns (output) to the AP when the employee connects?
- **Answer**: The value is *employee.*

6. Click **Close** for the Request details.
7. Repeat this for the **contractor** authentication. Notice the role assignment in Figure 10-43.

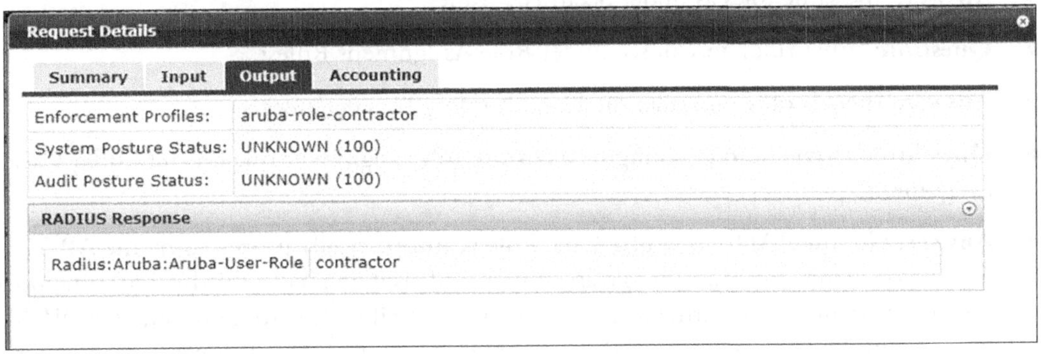

Figure 10-43: Access Tracker entry details for contractors

- **Question**: What is the Aruba-User-Role RADIUS attribute that ClearPass returns (output) to the AP when the contractor connects?
- **Answer**: The value is *contractor.*

- **Question**: Since the authentication server was already assigning a specific user role for both employee and contractor, why didn't these connections receive the contractor and employee user roles assigned to them?
- **Answer**: The AP tried to process the requested role, but since the roles employee and contractor don't exist in the AP configuration, the users were assigned the SSID default role.
- **Question**: For what scenario would you make the default SSID role very restrictive?
- **Answer**: If a mistake is made on the CPPM (RADIUS) server, user access would be very limited, instead of full access.

8. **Close** the request details.

Configure user roles for the AP

In this section, you will create additional user roles for employee and contractor.

1. In Aruba Central, navigate to: Context: **Groups / campus-wifi-ui** > Navigation: **Devices**> Top: **Access Points** > Right top: **Config**.
2. On the **WLANs** tab, select the **employee** SSID and **edit** it using the pencil.
3. Click the **Access** tab.
    1. Move the slider to **Role Based**.
    - **Question**: What roles do you see in the Role list?
    - **Answer**: The same roles as under Security > Roles.
    - **Question**: What rules do you see under Role Assignment Rules?
    - **Answer**: There is only one rule: the Default role: *p#tx-employee*.
    - **Question**: Do you need to configure a custom role assignment rule for the Aruba-User-Role assignment?
    - **Answer**: No, the RADIUS Aruba-User-Role is automatically processed by the AP. The only reason you would create a new role assignment rule is if you had a third-party authentication server that returns some other attribute, such as the Filter-ID. To map some RADIUS attributes to an Aruba user role, use the role assignment rules. This is not required in a typical deployment: typically, the authentication server would assign the Aruba-User-Role attribute during authentication.
4. In the role list, click the + sign to add a Role.
5. For the role name, enter **employee** and click **OK**.
6. Add another role, with name **contractor** and click **OK**.
7. Verify in the role list that the two new roles are listed.

Important

Although you can also create the roles under Security > Roles, it is recommended to create new roles under the WLAN wizard. In deployments where a gateway is used for tunneled WLANs, the wizard ensures that roles are created in both the AP and the gateway configuration.

8. Click the **employee** role.
   - **Question**: What is the default access rule in a new role?
   - **Answer**: There is one default rule: *Allow any to all destinations.*
9. Click **Save Settings**.
10. Wait a few seconds for the wizard to complete, then confirm with **OK**.

Review the WLAN configuration access UI slider

Important

The Access type slider in the UI for the WLAN configuration will only change when a custom rule is added for the role mapping. This does *not* mean that Aruba user roles are not processed by the WLAN!

1. In the WLAN list, click the **p#tx-employee** WLAN.
2. Use the **pencil** icon to edit the WLAN.
3. Click the **Access** tab.
   - **Question**: What does the access rule slider show?
   - **Answer**: The slider is set to *Network Based.*
   - **Question**: Does this mean that RADIUS-assigned Aruba user roles are not applied?
   - **Answer**: No, the RADIUS VSA Aruba-User-Role is automatically accepted by the AP. There is no need to make any custom role assignment rule to use the Aruba-User-Role VSA.

Re-Authenticate the clients and verify the new role assignments

Simply adding the roles to the AP configuration will not reauthenticate the clients that are already connected. You will now disconnect the two clients and then verify that they have received the new roles.

1. In Aruba Central, navigate to: Context: **Groups / campus-wifi-ui** > Navigation: **Clients** > Top: **Clients** > Right top: **List.**
2. Move the mouse over the **employee** record. At the right end of the line the option Disconnect from AP should be shown.

3. Click the link **Disconnect from AP** link and confirm the disconnect with **Yes**.
4. Repeat the disconnect for the **contractor** client.
5. Wait a few moments, then refresh the clients list using the **Refresh** button next to the CLIENTS title (see Figure 10-44).

> **Note**
> You may need to repeat this refresh a few times until you see that the AP Role for both employee and contractor user has changed.

> **Note**
> Some of the remote lab wireless NICs do not automatically reconnect to the WLAN after a disconnect (even when the Connect Automatically checkbox is enabled). Please check the wireless connection status on your PC1 and PC4; you may need to manually reconnect to the employee or contractor WLAN!

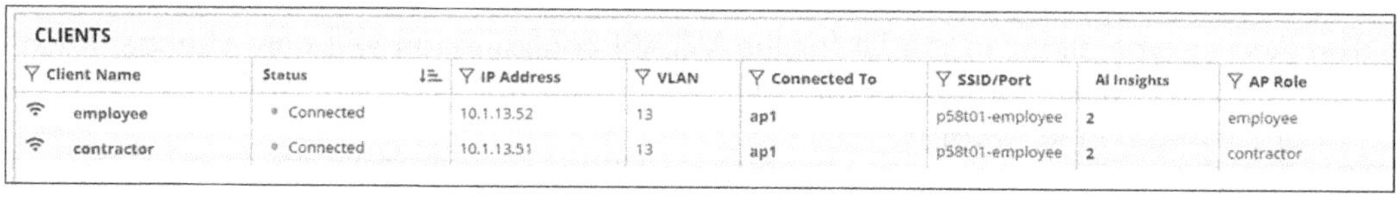

CLIENTS

Client Name	Status	IP Address	VLAN	Connected To	SSID/Port	AI Insights	AP Role
employee	Connected	10.1.13.52	13	ap1	p58t01-employee	2	employee
contractor	Connected	10.1.13.51	13	ap1	p58t01-employee	2	contractor

Figure 10-44: Client list in Aruba Central

6. On PC1, verify the ping to 10.1.0.2 works.
7. On PC4 (contractor), verify that the ping also works, for now. This was not the desired outcome, but you will adjust this configuration in the next section.

Configuring an alias for access control use

The customer requested that you block several IP subnets for contractors, and they may want to block additional subnets in the future. They want a simple administration method to maintain this list of blocked IP ranges. After discussing this request with your senior colleague, you learn that this is possible using an *alias*.

In the AP global configuration, you will create an alias for the *contractor-blocked* subnets and use this alias for the contractor's deny rule. If the customer must block additional subnets later, they simply add them to the alias.

Define a network alias

In this section, you will define an alias with blocked subnets.

1. In Aruba Central, navigate to: Context: **Groups / campus-wifi-ui** > Navigation: **Devices** > Top: **Access Points** > Right top: **Config**.
2. Click the **Security** tab to open the Aliases section.
3. Click the + sign to add a new alias.
4. For the Name, use **contractor-blocked**.
5. Click the + sign to add an item.
6. In the Type dropdown list, select **Network**.
7. For Network Prefix, enter **10.1.0.0**, and a network mask of **255.255.255.0**.
8. Click **OK** to add the item.
9. Click the + sign to add another item.
10. In the Type dropdown list, select **Network**.
11. For Network Prefix, enter **10.1.3.0**, with a network mask of **255.255.255.0**.
12. Click **OK** to add the item.
13. Click **OK** to add the alias.
14. Click **Save Settings** to submit the configuration.

Modify the contractor role

You can change rules for a role in the WLAN Wizard, but also directly in the role under Security.

1. On the Security tab, expand the **Roles** section.
2. Select the role **contractor**.
3. On the Access Rules table, click the + sign to add a new rule.
4. In the Rule Type dropdown list, leave the default **Access Control.**
5. For Service, leave the default **Network – Any**.
6. For Action, select **Deny**.
7. For Destination, select **To a network alias**.
8. For Network Alias, select **contractor-blocked** from the list.
9. Click **Save** to add the rule. Verify that your new rule is displayed in the Access Rules list.

- **Question**: What is the current position for the rule?
- **Answer**: When adding a rule directly in the role, the rule is added to the end of the ruleset.
- **Question**: What would be the result of this order?
- **Answer**: Rules are inspected top to bottom, and the action of the first matching rule is applied. Traffic is not blocked since all traffic matches the *any to all destinations* and is allowed.

10. At the end of the **Deny** rule, use the arrow up button to move the line upwards.
11. Now verify in the Access Rules table that the Deny rule is now listed *before* the Allow any rule. See Figure 10-45.

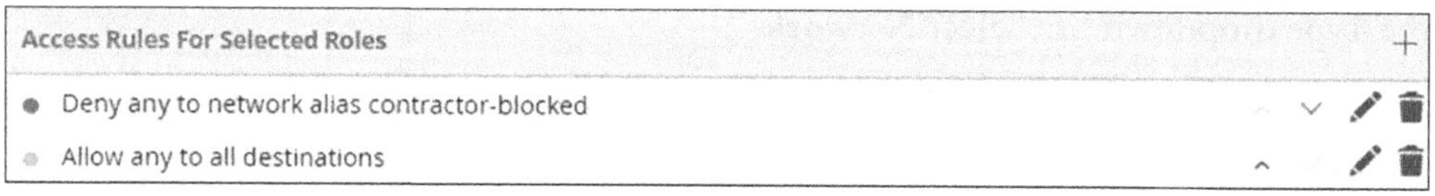

Figure 10-45: Verify the new access rule configuration

12. Click **Save Settings** to save the role.

Verify the updated contractor role access control

1. In Aruba Central, in the **AI search** bar, enter **contractor** and click the name contractor. This takes you to the contractor user detail page.
2. At the top, click **Sessions** to see the client's current firewall sessions.
   - **Question**: Do you see any ICMP entries?
   - **Answer**: Yes, one or more ICMP sessions can be seen.
   - **Question**: What is the Action, based on the Action column?
   - **Answer**: The Deny action was applied.
   - **Question**: What is the Last refreshed time on this page?
   - **Answer**: That will be close to your current time. You should note that Central does not store firewall session information. The information on this page is retrieved on demand from the device. Click the **refresh** icon next to the Last Refreshed time to get an updated list of firewall sessions.
3. On PC4, verify that the ping to 10.1.0.2 no longer works.
4. From the contractor access, attempt to ping to 10.1.3.2. This should also fail.

Optional: Alias update

Note
Make sure to continue the lab after these optional steps.

As you are completing and testing the configuration, the customer already has a new subnet to block for the contractors: 10.1.4.0/24. Since your configuration is now ready for this request, you can simply change the alias and add the new subnet.

This section does not include the detailed steps. Use the steps from the previous section if you need assistance. The remainder of the labs do not depend on the change in this section.

1. Using PC4, start a continuous ping to the IP address 10.1.4.2.

```
C:> ping 10.1.4.2 -t
```

2. Add the **10.1.4.0/24** subnet to the contractors-blocked alias.
3. Using PC4, verify that you can no longer access the IP address 10.1.4.2. This confirms the change was successfully applied using the alias.
4. Stop the ping on PC4.
5. This is the end of the optional steps.

Task 3: User role-based VLAN assignment

Your customer is pleased with the contractor access control that you configured. Now they want to ensure that contractors are separated using VLAN 12. The default VLAN for the corporate WLAN should still be VLAN 11.

Objectives

- Learn about VLAN WLAN assignment.
- Configure a VLAN in a user role.
- Verify the VLAN assignment in Aruba Central.

Steps

Check the default VLAN derivation method

In these steps, you review how the default client VLAN derivation is reported in Central.

1. In Aruba Central, enter **contractor** in the AI search bar.
2. Click the **contractor** name in the search output.
3. On the contractor user details page, review the NETWORK tile (see Figure 10-46).

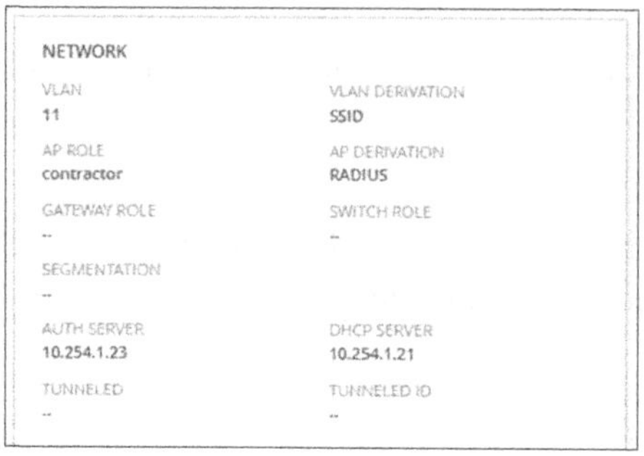

Figure 10-46: Verify the contractor VLAN in Central

- **Question**: What is the VLAN ID and VLAN derivation method for the contractor?
- **Answer**: VLAN 11, the method is SSID. This means that the client is assigned to VLAN 11 based on the WLAN's default VLAN configuration.

Configure the contractor user role to use VLAN 12

Now you will configure the contractor user role with VLAN 12. Thus, any client assigned to the contractor user role is placed in VLAN 12.

1. In Aruba Central, navigate to: Context: **Groups / campus-wifi-ui** > Navigation: **Devices**> Top: **Access Points**> Right top: **Config**.
2. Click the Security tab and open the **Roles** section.
3. Select the **contractor** role.
4. Click the **+** sign to add a new Access Rule.
5. In the Rule Type dropdown list, select **VLAN Assignment**.
6. In the **VLAN ID** field, enter **12**.
7. Click **Save** to add the Rule.

8. Verify the new rule is in the list (see Figure 10-47).

Figure 10-47: Verify VLAN policy configuration change

9. Click **Save Settings** to save the role.

Reauthenticate the client

Since the AP caches the client VLAN information, a regular disconnect will not assign the client to the new VLAN. You must do a hard disconnect from the connected AP's console using the **disconnect-user** command:

1. In Aruba, in the AI Search bar enter **contractor** and press **<Enter>**.
2. In the result list, click the name **contractor**.
3. In the DATAPATH tile, click the AP. This will take you to the AP device details page.
4. Click navigation: **Tools** > Top: **Console**
5. Enter these credentials: username **admin** / password **Aruba123!**
6. Click **Create New Session**.
7. Once the session is active, review the client list.

```
apX# show client
```

8. Take note of the IP address for the contractor user? ________________
9. Use the disconnect-user command with the IP Address of the contractor user.

```
apX# disconnect-user 10.1.11.x
```

10. Check the client list again until you see the contractor with an IP address in VLAN 12 (10.1.12.0/24).

```
apX# show client
```

Note

If you don't see the client reconnect, switch to the PC4 Windows system. The PC4 wireless client should automatically reconnect but may need a manual connect in the lab environment. If the PC4 is not connected to the wireless network anymore, make the connection manually.

11. In Central, enter in the AI Search bar **contractor** and press ENTER.
12. In the result list, click the name **contractor** to see the client details page.
13. On the CLIENT DETAILS page, check the NETWORK tile (see Figure 10-48).

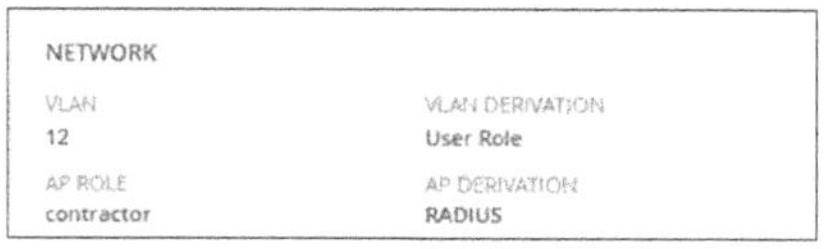

Figure 10-48: Verify the new VLAN assignment for the contractor

- **Question**: What is the VLAN ID and the VLAN derivation method for the client?
- **Answer**: The contractor is now in VLAN 12, the VLAN derivation is now *User Role*. Previously, the derivation was SSID based, since the user role did not have a VLAN assigned to it. The user role VLAN assignment overrules the default SSID-based VLAN assignment.

Note

NOTE: It may take a few minutes before the information in Central reflects the changed access. Check the status again after a few minutes.

Task 4: AppRF and application statistics

Your customer has heard that the Aruba WLAN solution includes DPI, an application recognition system that can filter and report on application use. In this task, you will review DPI activation and the options to review DPI statistics.

Objectives

- Enable AppRF on the APs.
- Use Aruba Central to monitor application usage.
- Implement application-based access control.

Steps

Enable Deep Packet Inspection on the AP

1. In Aruba Central, navigate to: Context: **Groups / campus-wifi-ui** > Navigation: **Devices**> Top: **Access Points**> Right top: **Config.**
2. Click the **Services** page, open the **AppRF** section.
3. Set the Deep Packet Inspection value to **All**.
   - **Question**: What are the options under Deep Packet Inspection?
   - **Answer**: None, App, WebCC and All. App indicates Application inspection and classification. WebCC indicates Web Content Classification. This can be used for URL filtering and URL reputation filters.
4. Click **Save Settings**.

Reboot the APs

1. Reboot AP1 and AP2. This is required for DPI to go into effect. Examine the lab access guide to determine how this is done (it will either be an option for the web topology portal page or you will have to disable power on the switch's associated PoE ports to which the APs are connected.

Optional: Test Access Control

In these optional steps, you will configure the contractor role to block access to the Streaming Application category. You will then use the PC4 (contractor) to attempt to open a streaming site.

1. In Aruba Central, navigate to: Context: **Groups / campus-wifi-ui** > Navigation: **Devices**> Top: **Access Points**> Right top: **Config**
2. On the Security tab, expand **Roles**.
3. Select the **contractor** role.
4. Add a new Access Rule
   - Type: **Access Control**
   - Service Application Category: enable checkbox for **Streaming**
   - Action: **Deny**
   - Destination: **To all destinations**
5. Click **Save**.
6. Make sure the new Access rule is at the top of the access control list (see Figure 10-49).

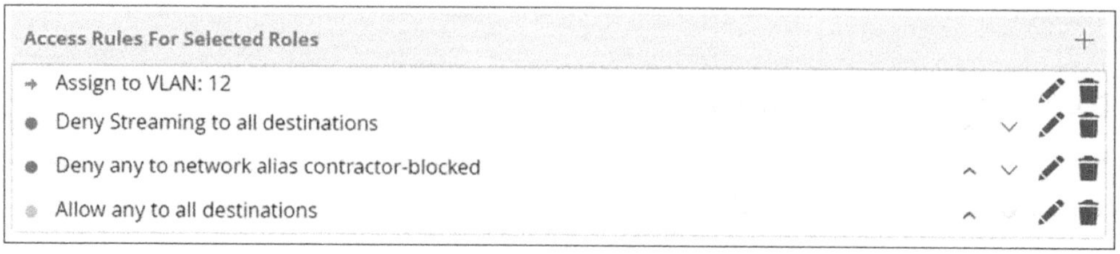

Figure 10-49: Verify order of the access control rules

Test the new access rules using a client

1. On PC4, attempt to navigate to a streaming site (for example to **www.netflix.com**).
2. This connection should be blocked by the AP.
3. This is the end of the optional steps. Continue with the next steps.

Review application information

The application monitoring report pages in Central need some time in order to aggregate the application layer information.

1. In Aruba Central, navigate to: Context: **Groups / campus-wifi-ui** > Navigation: **Applications**> Top: **Visibility**
2. Click the **Applications** tab under Visibility.
3. To see historical statistics, you can change how far back in time the data should be presented. At the right-top, change the history period to, say 3 months, for example (see Figure 10-50).

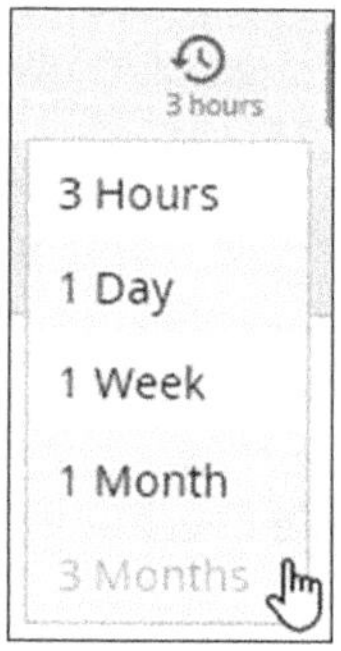

Figure 10-50: Changing the time range

4. This reveals the applications that have been used in the requested period. Figure 10-51 shows an example.

Note

Please note that in the lab environment, the collection has only recently been started. You will probably not see the same number of applications as the example below, or the output may even be completely blank. You may continue the lab and check these screens again at a later time.

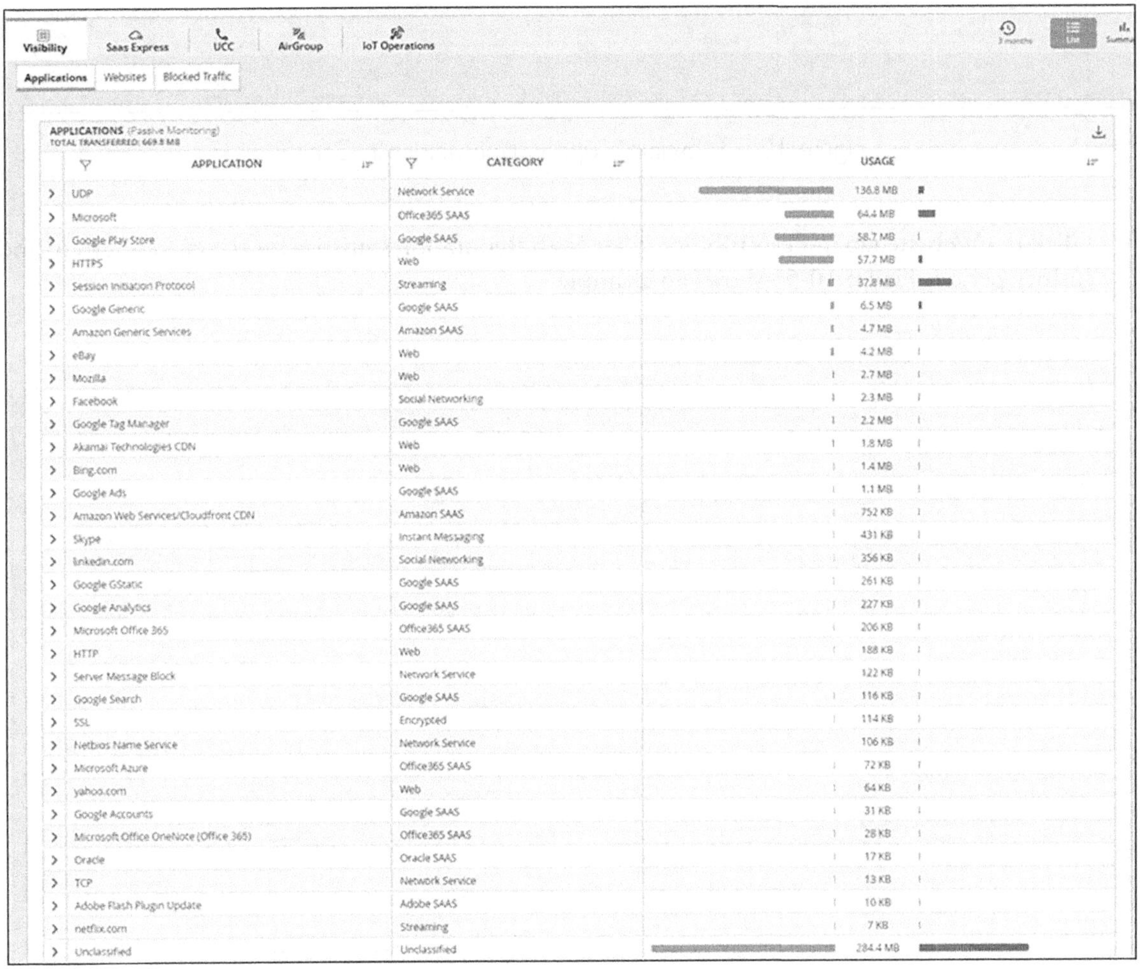

APPLICATION	CATEGORY	USAGE
UDP	Network Service	136.8 MB
Microsoft	Office365 SAAS	64.4 MB
Google Play Store	Google SAAS	58.7 MB
HTTPS	Web	57.7 MB
Session Initiation Protocol	Streaming	37.8 MB
Google Generic	Google SAAS	6.5 MB
Amazon Generic Services	Amazon SAAS	4.7 MB
eBay	Web	4.2 MB
Mozilla	Web	2.7 MB
Facebook	Social Networking	2.3 MB
Google Tag Manager	Google SAAS	2.2 MB
Akamai Technologies CDN	Web	1.8 MB
Bing.com	Web	1.4 MB
Google Ads	Google SAAS	1.1 MB
Amazon Web Services/Cloudfront CDN	Amazon SAAS	752 KB
Skype	Instant Messaging	431 KB
linkedin.com	Social Networking	356 KB
Google GStatic	Google SAAS	261 KB
Google Analytics	Google SAAS	227 KB
Microsoft Office 365	Office365 SAAS	206 KB
HTTP	Web	188 KB
Server Message Block	Network Service	122 KB
Google Search	Google SAAS	116 KB
SSL	Encrypted	114 KB
Netbios Name Service	Network Service	106 KB
Microsoft Azure	Office365 SAAS	72 KB
yahoo.com	Web	64 KB
Google Accounts	Google SAAS	31 KB
Microsoft Office OneNote (Office 365)	Office365 SAAS	28 KB
Oracle	Oracle SAAS	17 KB
TCP	Network Service	13 KB
Adobe Flash Plugin Update	Adobe SAAS	10 KB
netflix.com	Streaming	7 KB
Unclassified	Unclassified	284.4 MB

Figure 10-51: Example application list in Central

5. You can expand any application in the list to see application details. Figure 10-52 illustrates an example:

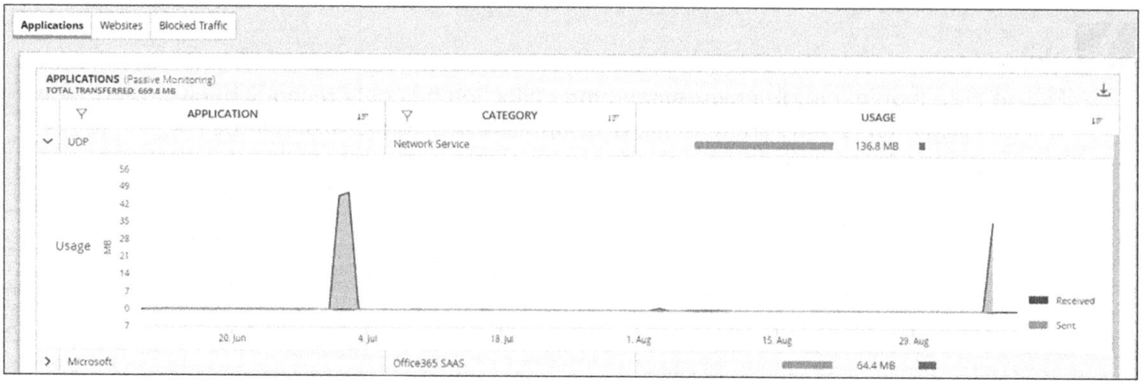

Figure 10-52: Example application details in Central

6. Click the **Website** tab under Visibility to see Web Reputation statistics for the current group and history period. Figure 10-53 shows an example.

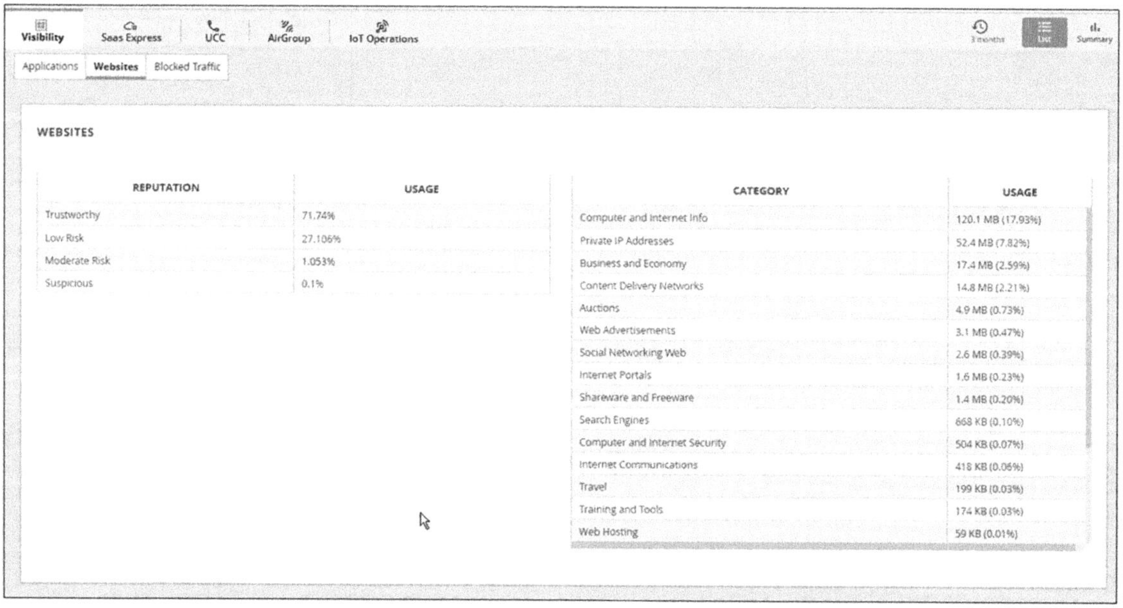

Figure 10-53: Example web application details in Central

Review blocked sessions

You can collect a report about the sessions that were blocked by the firewall.

1. In Aruba Central, navigate to: Context: **Groups / campus-wifi-ui** > Navigation: **Applications**> Top: **Visibility**.
2. Click **Blocked Traffic.**
3. If blocked traffic has been registered, you can download the list.
4. When available, click **Download CSV**, you can open the downloaded CSV file with Excel or a text editor (see Figure 10-54).

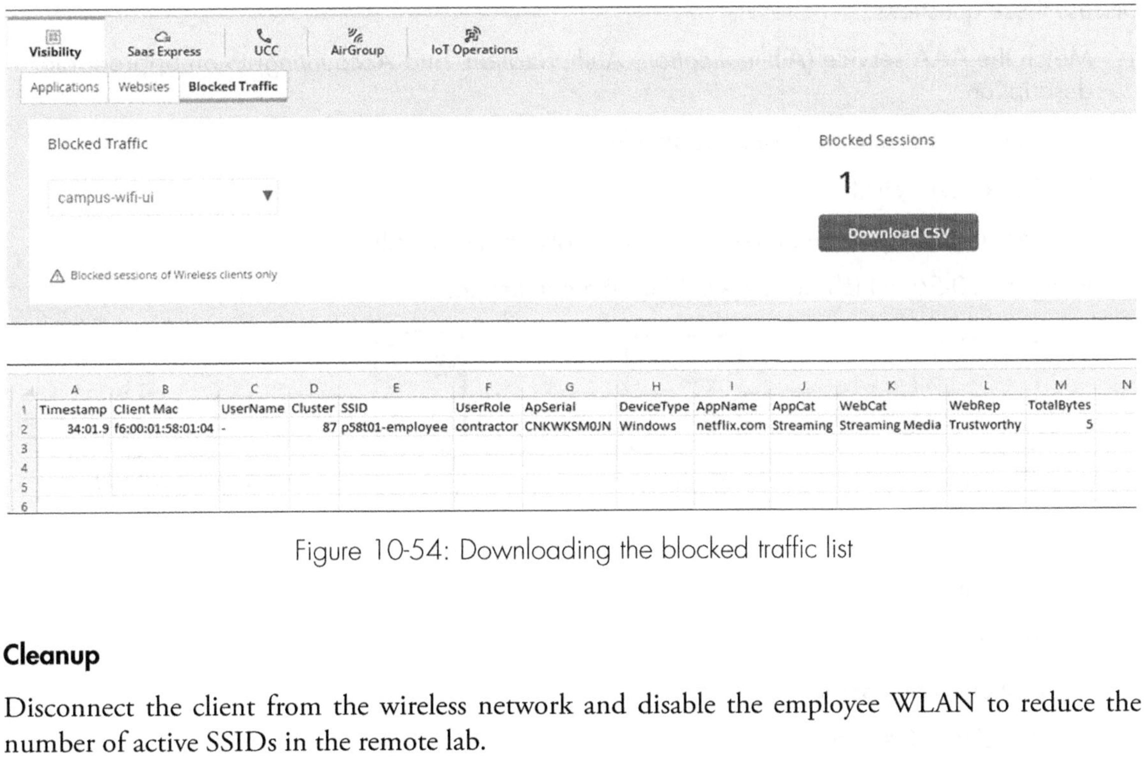

Timestamp	Client Mac	UserName	Cluster	SSID	UserRole	ApSerial	DeviceType	AppName	AppCat	WebCat	WebRep	TotalBytes
34:01.9	f6:00:01:58:01:04	-	87	p58t01-employee	contractor	CNKWKSM0JN	Windows	netflix.com	Streaming	Streaming Media	Trustworthy	5

Figure 10-54: Downloading the blocked traffic list

Cleanup

Disconnect the client from the wireless network and disable the employee WLAN to reduce the number of active SSIDs in the remote lab.

1. On PC1 and PC4, disconnect from the employee WLAN.
2. In Aruba Central, navigate to: Context: **Groups / campus-wifi-ui** > Navigation: **Devices**> Top: **Access Points**> Right top: **Config.**
3. In the WLAN list, click the **employee** WLAN.
4. On the right end of the record, click the **WLANs** icon to disable the WLAN (see Figure 10-55).

Figure 10-55: Disable the WLAN in Central

5. Confirm the Action with **Yes**.
6. Verify in the WLAN list that the *Network Enabled* column reports **No** for the employee WLAN.

Learning check

Discuss these questions.

1. Match the AAA service (*Authentication*, *Authorization*, and *Accounting*) to an appropriate description.
   a. Control authenticated user capabilities
   b. Track user activity
   c. Permit or deny access based on credentials and/or certificates
2. In Figure 10-56, which device would be the acuthenticator?

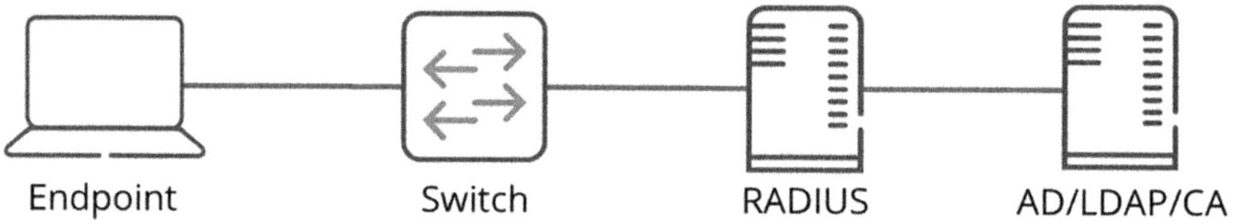

Figure 10-56: Question 2

   a. Endpoint
   b. Switch
   c. Authentication server
   d. AD/LDAP/CA server
3. Which of the options below accurately compares 802.1X authentication vs. captive portal authentication? (Select three.)
   a. They both occur at layer 3 of the OSI model.
   b. 802.1X authentication is typically used for corporate employee access.
   c. Captive portal authentication is often used for guest access.
   d. 802.1X authentication requires clients to authenticate to a authentication server as soon as the endpoint connects.

4. During 802.1X client authentication, which devices can perform the authenticator role? (Select two.)
    a. Supplicant
    b. AP
    c. Switch
    d. Authentication server
    e. RADIUS
5. There are a few popular Extensible Authentication Protocols (EAP). Match the advantage or disadvantage with the appropriate EAP method (*EAP-TLS*, *PEAP*, and *EAP-TEAP*).
    a. Very strong security, mitigates risk of password authentication.
    b. Easier to implement than some other methods, but slightly weaker security.
    c. May have relatively limited support as compared to other methods.
    d. Enables chaining of machine and user authentication with tokens.
    e. Requires the overhead of managing CA services, certificate distribution and management
6. Your colleague is asking about the role-based ArubaOS Firewall. Which statements below can you leverage to improve their understanding? (Select three.)
    a. It can enforce different rules based on user identity and assigned role but increases the reliance on VLANs and subnets for security.
    b. A role is a group of settings that controls client traffic.
    c. A role can include rules that leverage Deep Packet Inspection (DPI) and Web Content Classification (WebCC).
    d. It can leverage roles derived from a authentication server.

Summary

Congratulations! You have completed Chapter 10. You should now be able to:

- Explain AAA.
- Describe 802.1X authentication.
- Configure secure WLANs.
- Discuss roles and access rules.

11 Implementing Guest Access

OBJECTIVES

The last chapter introduced you to securing your WLAN using 802.1X. This chapter will focus on guest access, where a web browser is used for the authentication process. This is commonly referred to as *captive portal*. Proper guest access deployment helps to accommodate less trusted users that require limited access: typically internet only. It also improves security by helping to maintain strict separation between employee and guest users. Captive portal is commonly used to secure guest access as well as onboard employee devices.

After completing this chapter, you should be able to:

✓ Describe guest access.

✓ Set up captive portal authentication.

✓ Configure guest WLANs.

Guest access introduction

This section introduces guest access with Aruba's guest access solutions and captive portal options.

Aruba's guest access solution

Aruba allows for a simple collapsed architecture: you get differentiated access based on user and device characteristics. This is the basis for several guest access features. Guest access is often configured as a software option. No new hardware is required for basic guest access beyond the Aruba gateway and/or APs used for the guest WLAN. Guest access is supported for Aruba AOS and CX switches as well, but is beyond the scope of this book.

In contrast to some vendors, where you must configure the LAN to add a VLAN for guest access at every LAN switch where an AP connects, you can add Aruba's user-centric networks as an overlay on the existing wired LAN (see Figure 11-1). Access point traffic is directed via secure tunnels straight to the mobility gateway, where an integral stateful firewall maintains strict segregation between different traffic classes (employee and guest, for example).

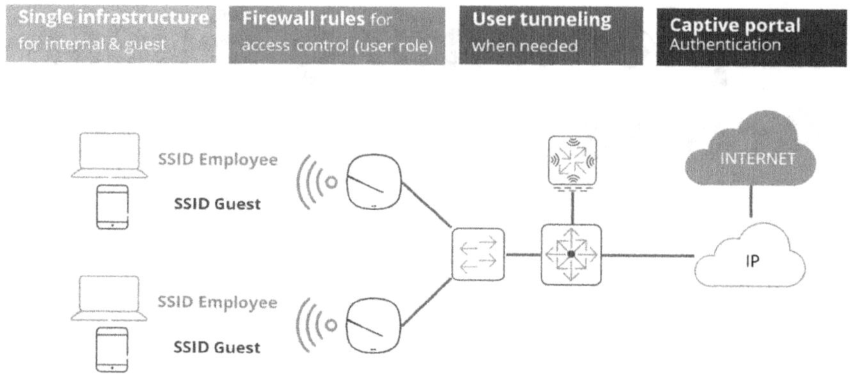

Figure 11-1: Aruba's guest access solution

Internal (employee) traffic can reach the core LAN and corporate resources, while guest traffic travels through a secure tunnel to a mobility gateway in the DMZ, and then on to the Internet. Or, you can let the AP route traffic to the gateway or directly to the Internet: this is referred to as *split tunnel mode* for an AP group.

Captive portal login screens and web forms for administration are served directly from the mobility gateway or the AP. Or use the Aruba ClearPass server for more sophisticated guest access solutions. ClearPass provides advanced guest features like credit card processing, access code authorization, property management systems, advertising, detailed captive portal page construction, and much, much more.

Captive portal options

There are two options to deploy a guest network with captive portal authentication (see Figure 11-2):

- **Cloud guest**: with the cloud guest service from Aruba Central, a network admin can easily implement captive portal authentication in a multisite environment with consistency and central management. The splash page is customizable and includes great features like guest self-registration and social login.
- **External captive portal**: with the external captive portal option, you may use an authentication solution that's not part of Aruba Central or that does not run on the gateways and APs. An example is the Aruba ClearPass Policy Manager and Guest components, which can provide guest captive portal authentication with more customization options and scalability.

You will configure a guest WLAN to use both options in the upcoming labs.

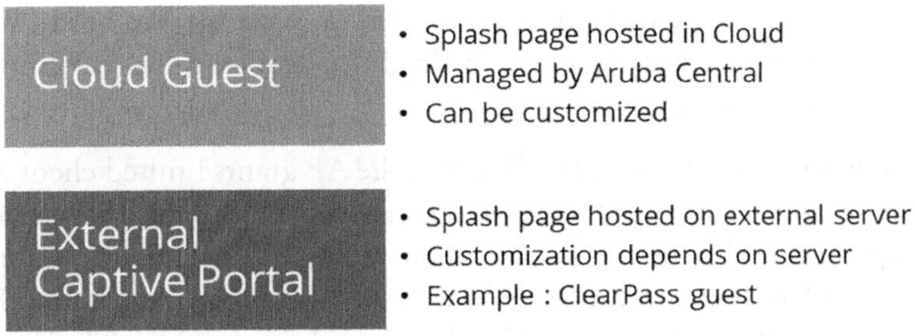

Figure 11-2: Aruba's captive portal options

Captive portal authentication

The section begins by providing an introduction to how captive portal authentication works. The remainder of the section then covers how to create the captive portal page, commonly called a *splash page*.

Captive portal process

Captive portal authentication is the primary example of layer 3 authentication and is often implemented for guest wireless networks: airports, hotels, and coffee shops, for example. It's an easy way for users to log into a network without special client device requirements or user expertise: they simply connect to the guest network and open a web browser, where the browser is redirected to a login page (commonly called a splash page). The user enters credentials or maybe just accepts an agreement to log into the network. The user's device then receives more complete network access, like access to the Internet.

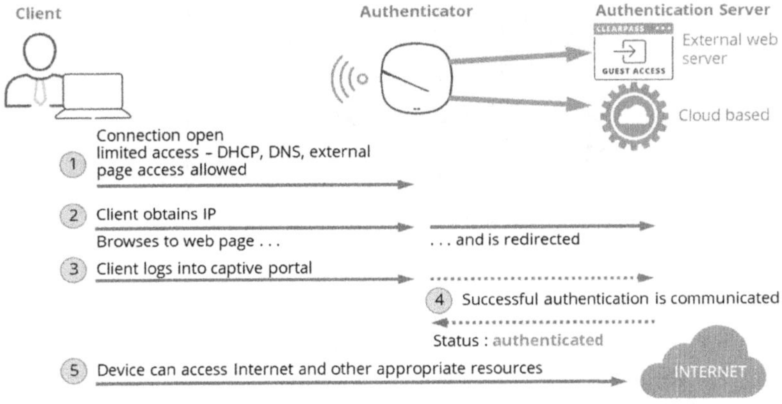

Figure 11-3: Captive portal process

Captive portal deployments can be quite nuanced, based on your specific needs, but let's examine a high-level process here, illustrated in Figure 11-3. In this example, the client connects to an Aruba AP, acting as the AAA authenticator.

Upon initial connection to the captive portal WLAN, the AP grants limited client access: enough for the client to get an IP address via DHCP, resolve names via DNS, and just enough web access to access the guest captive portal page. Once the client receives IP addressing information, the user opens a web browser and attempts to browse to a web page. This communication flows through the ArubaOS firewall on the AP, which redirects the client to a login page on ClearPass Guest or Aruba Cloud in Central. The user logs in. Then a RADIUS message is communicated from the external server or Aruba Cloud back to the AP (the authenticator). The RADIUS attribute(s) will typically change the firewall policy, minimally so that the guest can browse the Internet or use other network resources, as dictated by their updated user role.

Captive portal splash pages

You can use the Aruba Central Guest app to configure splash page profiles. The splash page originates from the Central account. You can configure several splash pages to allow different groups or guest SSIDs to send different splash pages as needed.

The guest management feature allows guest users to connect to the network and allows you to control guest user access. You create a splash page profile for guest users in Aruba Central. Then guest users access the Internet by providing either credentials, which can be configured by a *guest operator*, or the user's respective social networking login credentials.

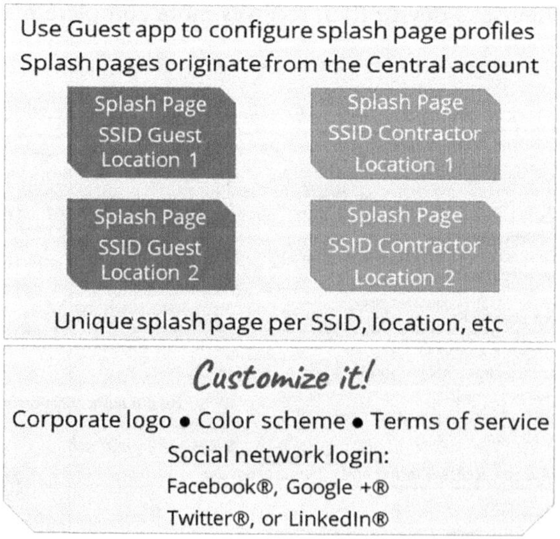

Figure 11-4: Captive portal splash page

You can easily customize your splash page with a corporate logo, color scheme, and terms of service (see Figure 11-4). Enable user login authentication with social networking services, like Facebook®, Google +®, Twitter®, or LinkedIn®. Facebook's Wi-Fi feature also allows you to pair your network with a Facebook business page.

Authentication type

To create a new splash page, select a group in the left navigation pane and then select **Guests**. At the top, navigate to **Guest Access > Splash Pages**. First assign the splash page a name. Then select the type of splash page: Anonymous, Authenticated or Facebook Wi-Fi (see Figure 11-5).

- The *Anonymous* login method allows guest users to log in to the splash page *without* providing any credentials. Minimally, though, you want the guests to accept the terms of use for guest access. You can also enable a pre-shared key for anonymous user authentication to allow access. To enable a pre-shared key-based authentication, set the Guest Key to **On** and specify a password.
- *Authenticated* requires you to configure authentication and authorization attributes and login credentials that enable users to access the Internet as guests. You can configure an authentication method based on sponsored access and social networking login profiles.
- *Facebook Wi-Fi* gives you the option to enable network access through the free Wi-Fi service Facebook offers. This feature allows you to pair your network with a Facebook business page, thereby allowing the guest users to log in from Wi-Fi hotspots using their Facebook credentials. If you have a Facebook Wi-Fi business page set up, when the users try to access the Internet, the browser redirects the user to the Facebook page. The user can log in with their Facebook account credentials and can either check in to access free Internet or skip checking in and continue.

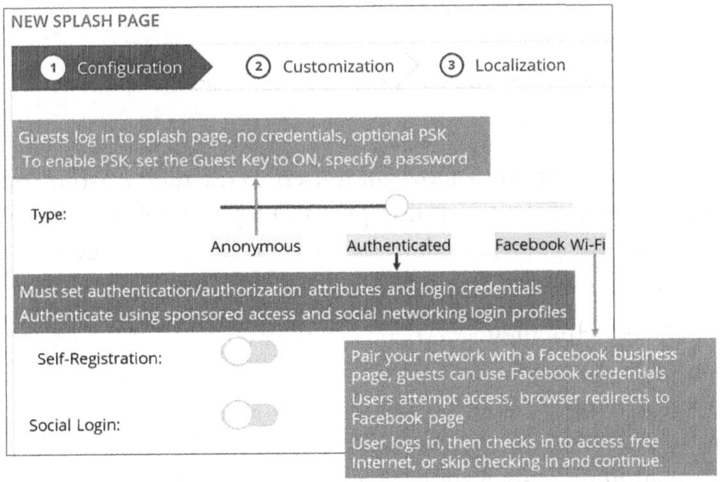

Figure 11-5: Splash page authentication type

Authentication method

When you set the splash page type to *Authenticated*, you should configure an authentication method (see Figure 11-6):

- By default, when you select username/password, an administrator or guest administrator will need to manually configure the guest user accounts.
- An option to avoid an overhead of creating and maintaining guest accounts is to enable *self-registration*, which allows guest users to register their own accounts. In addition, as a security measure, email and phone verification are available to validate the guests' identities.
- Another option is to enable social login, which uses single sign-on with a social media page to authenticate the guests.

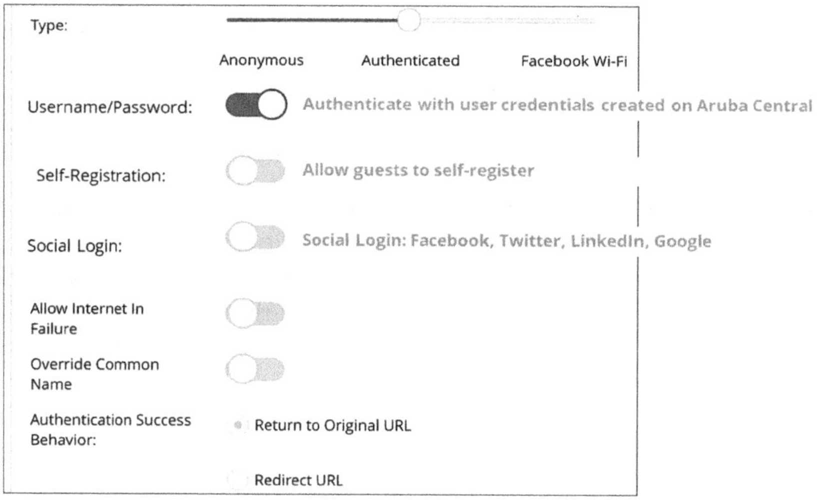

Figure 11-6: Splash page authentication method

You can enable the social login option to allow guest users to use their existing login credentials from their social networking profiles and sign in through a third-party website. When Central configures a social login-based profile, the guest user does not have to have a new guest account to access the guest network or third-party websites.

Facebook, Twitter, and LinkedIn allow guests to use their social media credentials to log in to the Aruba Central splash page. To enable site integration, you must create an app for each service you wish to use and obtain the app ID and individual secret key as well for the app. To complete the integration, you will need to enter the app ID and secret key for the client ID and client secret respectively (for more information on app creation, search "creating a Facebook app", "creating a Twitter app", or "creating a LinkedIn app").

Similarly, Google+ allows guests to use their Google+ credentials to log in to the splash page. To enable Google+ integration, you must create a Google app and obtain the app ID and secret key (for more information, search "creating a Google app").

To restrict authentication attempts to only Google host domain members, enter the domain name in the *Gmail for Work Domain* text box. Ensure that you have a valid Google Domain or Google Apps licensed domain account. For more information see:

- https://apps.google.com/intx/en_in/
- https://domains.google.com/about/

Then specify the text for the login button.

Guest splash page customization

After creating a guest splash page, you must customize it to your specifications. For example, you can set your own colors and enter your logo and background image (see Figure 11-7). You can also set up the terms and conditions and add an image with a URL to an external site.

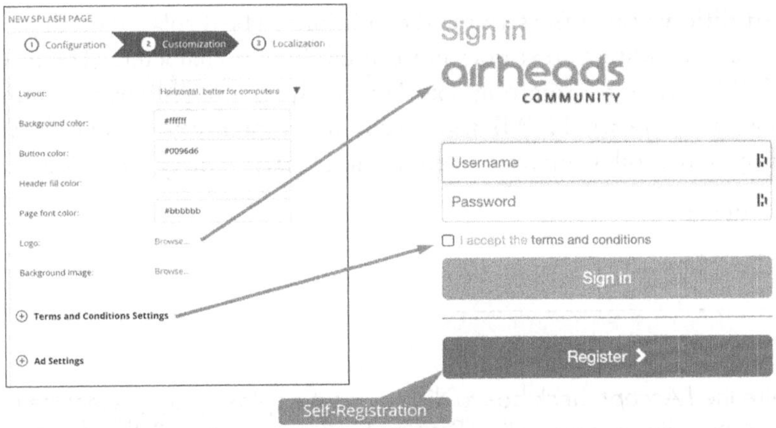

Figure 11-7: Splash page authentication customization

Here are the items you can customize:

- **Layout**: To customize thepagelayout based on the device type, specify a layout by selecting one of the following options. The horizontal layout is selected by default. To change the layout, click the drop-down list and select the required layout type.
  - *Horizontal*, which is better for computers
  - *Vertical*, which is better for phones

- **Background color**: To change the splash page color, select a color from the Background Color palette.
- **Button color**: To change the sign-in button color, select a color from the Button Color palette.
- **Header fill color**: Select the splash page header's fill color from the Header fill color palette.
- **Page font color**: To change the text font color on the splash page, select a color from the Page font color palette.
- **Page font color**: Select the splash page's font color from the palette.
- **Logo**: To upload a logo, click **Browse**, and browse the image file. Ensure that the image file size does not exceed 256 KB.
- **Background image**: Click **Browse** to upload a background image. Ensure that the background image file size does not exceed 512 KB.
- **Pagetitle**: Add a suitable title for the splash page.
- **Welcome text**: Enter the welcome text to be displayed on the splash page. Ensure that the welcome text does not exceed 20,000 characters.
- **Terms and conditions**: Enter the terms and conditions to be displayed on thesplashpage. Ensure that the terms and conditions text does not exceed 20,000 characters. The text box also allows you to use HTML tags for formatting text. For example, to highlight text with italics, you can wrap the text with the <i> </i> HTML tag. Specify an acceptance criterion for terms and condition by selecting any of the following options from the *Display "I Accept"* check box:
- No, accept by default
- Yes, display check box

Note

If you require the *I Accept* check box to display on the splash page, select the display format for terms and conditions. Ensure that **Display Option For Terms & Conditions** has the *Inline Text* option auto-selected and displayed as an un-editable text.

- **Ad settings**: If you want to display advertisements on the splash page, enter the URL in the *Advertisement URL*.
- **Advertisement image**: click **Browse** and upload the image.

Localization

The localization page (see Figure 11-8) allows you to localize (or translate) the cloud guest portal content to the guest's local language: configure the parameters and enter titles for the various buttons.

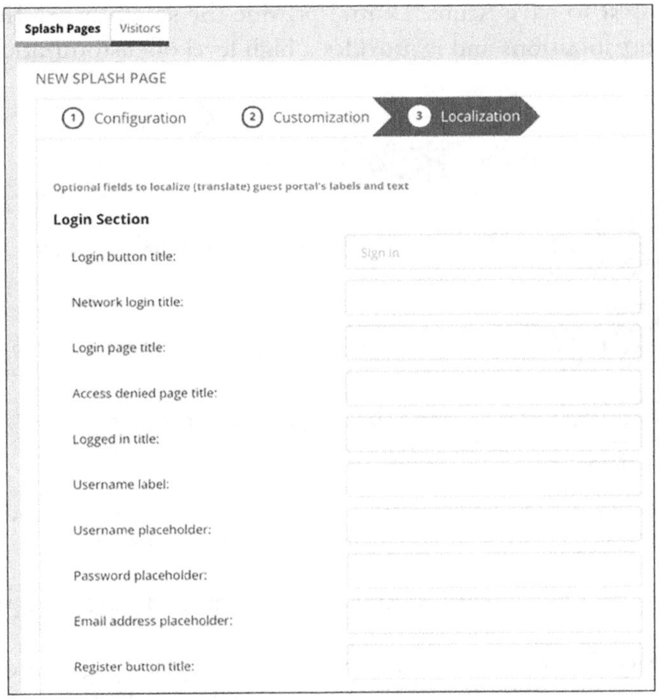

Figure 11-8: Splash page authentication localization

Guest WLAN setup

Configuring a guest WLAN is very similar to creating an employee WLAN: the main difference is that the captive portal process is used (and thus the need for a splash) page, instead of 802.1X. This section introduces you to the creation of the guest WLAN.

Guest WLAN: Cloud guest

The main difference, from a configuration perspective, for creating a guest WLAN versus an employee WLAN is the need for a splash page and the information you define under the Security tab in the WLAN wizard. When creating a new WLAN for guests, the visitors' security level is chosen under the *Security* tab: it enables the SSID's captive portal options (see Figure 11-9).

There are then two access type options:

- Choose **External Captive Portal** if you use some external server to host the splash page for authentication. You might prefer this option if your organization has very complex requirements, or requires heavy customization. You must configure a *captive portal profile* with an authentication type, server address, and URL of the splash page (typically on a server, like ClearPass or a web proxy). You must also configure a RADIUS server, allowing the APs and gateways communication with the RADIUS server to authenticate the guest users.

- Choose **Cloud Guest** to have Aruba Central provide the splash page. This is usually powerful enough for most organizations and is provides a high level of customization.

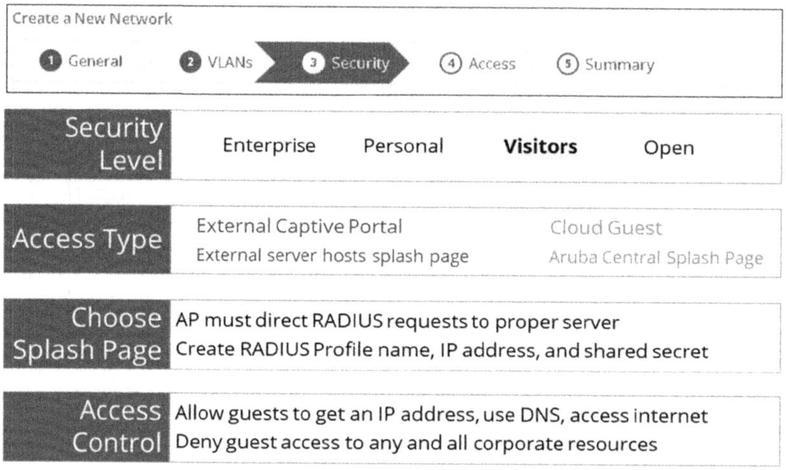

Figure 11-9: WLAN wizard—Security tab and visitor access

Then choose a splash page. Some customers may have different splash pages for different locations. You will experience how to create splash pages in Aruba Central during the lab activities, along with other required steps to create fully functional guest WLANs using Aruba ClearPass and Aruba Central.

As a best practice, you may also navigate to the WLAN wizard's *Access* tab to configure firewall rules: to restrict guests from accessing the corporate network. For example, you might want to allow guests to obtain an IP address using DHCP and have access to DNS servers. However, you want to deny any and all access to corporate resources. The lab explores this in much more depth.

Monitoring guest visitors

Figure 11-10 shows typical information on guest session and account details of the visitors who access the splash page. During the lab activities, you will learn to create and manage visitors. You will also learn to monitor their sessions and get precisely the information you need, using filters and the search tool.

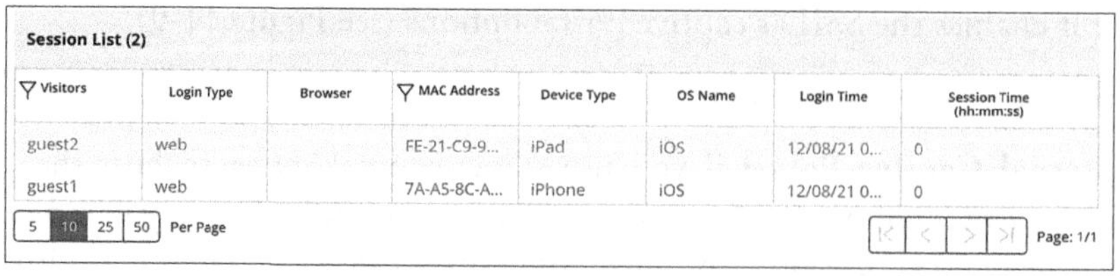

Figure 11-10: Monitoring guest visitors in Central

Lab 11.01: Implementing guest access

Your customer has heard that Aruba Central includes a Cloud Guest captive portal solution. They have asked if you can configure an example Guest WLAN with Cloud Guest. They want to have a basic captive portal page where they can show some legal information about the use of the guest network and guests should agree with those terms. They don't require formal username and password registration for this example.

Objectives

- Configure a splash page in Aruba Cloud Guest.
- Configure a guest WLAN using Aruba Cloud Guest.
- Verify the operation of the guest solution.

Task 1: Configure a cloud guest splash page

In this task you will prepare the webpage that guest users see when they connect to the guest WLAN. This is also known as a splash page. Aruba Central provides integrated, customizable splash pages. The Aruba Central solution also includes guest user account management and a cloud-based RADIUS system to support guest logins.

A Cloud Guest splash page could be used by AP WLANs of multiple AP groups (configurations), but, by default, it is only available to the AP group where it is created.

Objectives

- Configure a Cloud Guest splash page.

Steps

Configure the cloud guest splash page

1. In Aruba Central, navigate to: Context: **Groups / campus-wifi-ui** > Navigation: **Guests** > Top: **Guest Access** > **Splash Pages**. The list of configured splash pages is displayed.
    - **Question**: How many splash pages do you see?
    - **Answer**: Only one, the default page.
    - **Question**: What is the type for this page?
    - **Answer**: The default page is configured as type *Anonymous*. This means that guest users need not enter credentials to access the guest network. The system automatically creates a user account and they automatically login with these generated credentials.
    - **Question**: What is the state of the page?

- **Answer**: The default page is *Shared*. This means the page can be use by all groups in Central: it is not exclusive to this group. This allows you to configure and customize a splash page and use that same splash page on multiple AP groups.
- If you don't want the splash page to be available on other Aruba Central configuration groups, it should not be marked as *Shared*.

2. Click the **+** sign to create a new Splash Page. A wizard with 3 main steps will appear: Configuration – Customization – Localization (see Figure 11-11).

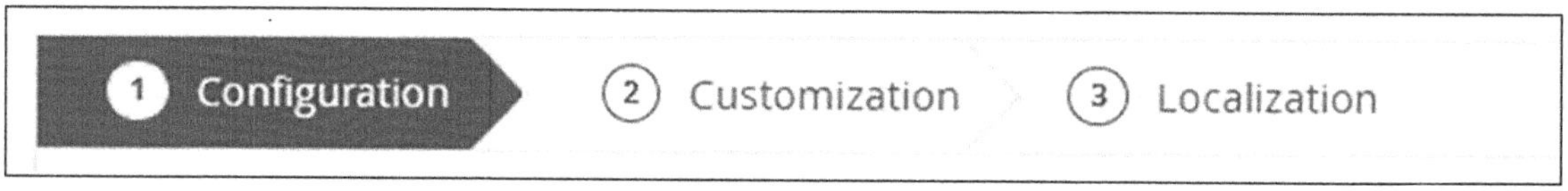

Figure 11-11: Creating a new splash page

Configuration page

1. From the Configuration tab, configure these settings:
   - *Name*: **p#tX-guest-page**

Note
Make sure to replace the # value with your pod number and X with your table number. For example, if you are using table 07 in pod 28, your page name will be p**28**t**07**-guest-page. Check with your instructor or lab administrator if you are not sure about the pod and table number.

   - Type: Leave the default (**Anonymous**)
   - Authentication Success: **Redirect URL**
   - Redirect URL: **https://www.arubanetworks.com**

Note
This configuration forces all authenticated guest to this website. By default, the guest is redirected to the original requested web page.

   - Session timeout: for lab purposes, set it to **5** minutes.

Note
The session timeout means that five minutes after guest users have successfully connected and authenticated to the network, they will be logged out and presented with the login page again. This is only set to a low value for lab testing purposes.

- Share This Profile: leave it at the default (**enabled**). This ensures the same page and page customization can be used over multiple AP Groups.

2. Leave the other settings default. Click **Next**.

Customization page

1. On the **Customization** tab, configure these settings:
    - *Optional step*: Feel free to change the colors and logo on this page.
2. Expand the **Terms and Conditions Settings** section.
3. Enter some terms. For example, enter: **Use at your own risk**.

Note

For a company or organization, you should involve your corporate legal department in drafting this text.

---

4. Set Display 'I Accept' to **Yes, display checkbox**.
5. Click **Next**.

Localization page

1. On the Localization page, the local language translation can be configured.
2. For the **Login page title** enter **Aruba Guest Access P#Tx**.

Note

Feel free to complete some of the other fields as well.

---

3. At the bottom of the page, click **Preview**.

Note

Make sure that popups are enabled in your browser to see the page preview.

---

4. Pay attention to the URL. This is the same base URL the guest users are shown when they connect to the guest WLAN.

Note

If you need to make any changes, you can still go back and enter different values in the splash page configuration.

---

5. When you have completed the review, close the preview browser tab.
6. In the splash page wizard, click **Finish** to complete the creation of the page.
7. Verify that the new splash page is listed.

Task: Configure WLAN profile with cloud guest splash page

In this task you will configure your guest WLAN profile with the splash page that was just created.

Objectives

- Enable the guest captive portal on a WLAN.

Steps

1. In Aruba Central, navigate to: Context **Groups: campus-wifi-ui** > Navigation **Devices**> Top **Access Points**> Right top: **Config**.
2. In the WLAN list, move the mouse over the **p#tX-guest** and click the **WLAN** icon to enable it (see Figure 11-12). (You created this open WLAN in Lab 9.01: WLAN Fundamentals.)

Figure 11-12: Enable the WLAN

Note
You have previously disabled this WLAN to reduce the number of broadcasting WLANs in the remote lab environment.

3. Confirm with **Yes**.
4. In the WLAN list, use the **pencil** icon to edit the *p#tX-guest* WLAN profile (see Figure 11-13).

Figure 11-13: Edit the WLAN profile

5. You are now on the **General** page of the WLAN configuration.
6. Expand **Advanced Settings**.

Note

In a production deployment you should apply the recommended best practice settings based on the latest *Aruba Validated Solution Guide*, such as the broadcast and multicast optimization. In this lab task you will skip this best practice configuration.

---

7. In the remote lab environment, there is much interference from neighboring APs. Therefore, you will limit this SSID to the 5 Ghz band.
8. Expand the **Miscellaneous** section.
9. For the Band option select only **5 GHz**.
10. Click **Save Settings**.

Guest WLAN security and access configuration

1. In the WLAN list, use the **pencil** icon to edit the *p#tX-guest* WLAN profile.
2. From the **Security** tab, move the *Security Level* slider to **Visitors**.
3. Set the Type to **Cloud Guest**.
4. Set the Guest Captive Portal Profile to **p#tX-guest-page**. This is the name of the splash page you created in the previous task. Figure 11-14 shows an example.

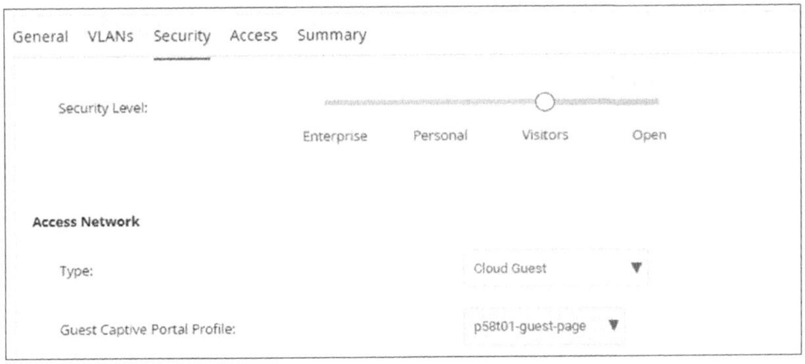

Figure 11-14: Guest WLAN security configuration

5. Click the **Access** tab.
6. Move the slider to **Network Based**.
   - **Question**: What would be the effect of leaving the access rule to Unrestricted?
   - **Answer**: The guest users would have full access to internal, corporate network resources.
   - **Question**: How can you easily block access to internal resources and allow access to the rest of the internet?
   - **Answer**: Most businesses use an internal IP range from some of the subnets of RFC1918:
     - 10.0.0.0/8
     - 172.16.0.0/12
     - 192.168.0.0/16
7. For guests, you will block access to these subnets and allow access to everything else. Click **Add Rule**, deny access to the network **10.0.0.0** mask **255.0.0.0**

Note

10.0.0.0/8 is the only range you need for the lab environment. If you are unsure in your production environment, you can add the other RFC 1918 ranges as necessary.

8. While this rule works fine, it also blocks access to the lab DNS server. In this lab environment, you should add an exception for traffic to 10.254.1.21. Also, you use this host for connectivity testing, so you should allow any traffic to this host (do not limit the rule to DNS).
9. Click **Add Rule**; allow Network service **any** to a particular server **10.254.1.21**.
10. Review the order of the rules (see Figure 11-15).

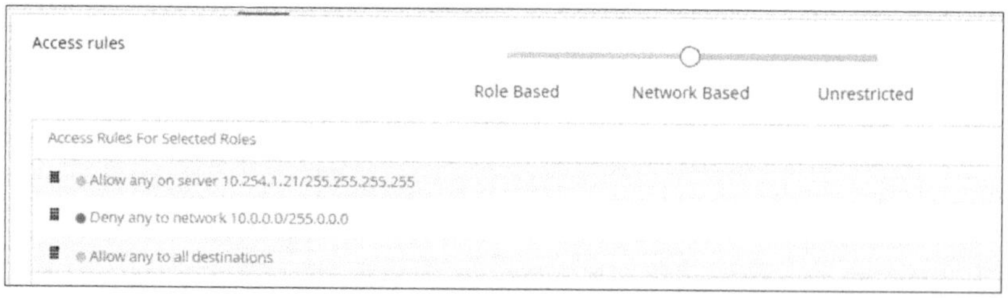

Figure 11-15: Creating the necessary access rules for guest access

Note

Remember rules are processed top to bottom. If the rule is at the top, the guest users will have full access to the internal resources. Similarly, if the deny 10.0.0.0/8 is on top of the allow any to server 10.254.1.21, all DNS traffic will be blocked.

11. Click **Save Settings** to save the WLAN security and access changes.
12. After a few seconds, click **OK** to confirm the wizard has completed.

Task: Test Cloud Guest access

In this task you will use PC1 to make a connection to the guest WLAN that has now been configured with the Cloud Guest captive portal.

Objectives

- Test the Cloud Guest connection.

Steps

1. Open a connection to PC1.

> **Note**
> After performing the next step, a browser page will popup. Do *not* log in yet. You will first explore the current status of the client known as the pre-auth (before the client is authenticated) stage.

2. Establish a wireless connection to **p#tX-guest**. *Do not login yet.*
3. In Aruba Central, navigate to: Context: **Groups / campus-wifi-ui** > Navigation: **Clients** (see Figure 11-16).

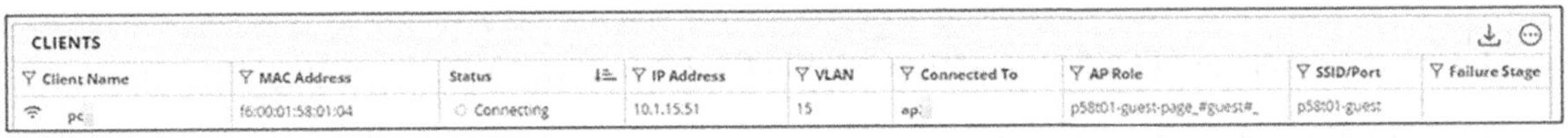

Figure 11-16: Viewing the connected guest

- **Question**: What is the status of the PC1 client?
- **Answer**: The client state is connecting. Note that the client WLAN state is already connected, but since the client has not completed the captive portal login, the status will be reported as connecting.
- **Question**: What is currently the assigned AP Role?
- **Answer**: The role is *p#tX-guest-page_#guest#_*.
- **Question**: Did you create this role?
- Answer: No, this role was automatically created when you enabled the captive portal on the WLAN profile. This role contains the captive portal redirect instructions to the splash page named *p#tX-guest-page*.

4. On the PC1 client, open a new tab on the browser and navigate to a secure (HTTPS) website: https://www.arubanetworks.com.
    - **Question**: What do you notice?
    - **Answer**: A certificate warning is displayed (see Figure 11-17).

Figure 11-17: Certificate warning

    - **Question**: Why would there be something wrong with the certificate of www.arubanetworks.com?
    - **Answer**: There is nothing wrong with the actual webserver. However, at this point, the guest user does not have full access, so the HTTPS session is captured by the AP. Since the AP does not have the www.arubanetworks.com certificate, the guest browser will show the certificate error.
    - **Question**: Can you avoid or solve the certificate warning when guest users make an initial HTTPS connection?
    - **Answer**: No, the AP can never have the original certificate of the initial HTTPS connection destination.
5. Attempt to open an HTTP connection (**not** HTTPS) to http://1.1.1.1 or http://neverssl.com.
    - **Question**: What do you notice?
    - **Answer**: The session is now redirected to the Cloud Guest page without certificate warning. This is because the HTTP connection can be hijacked by the AP and redirected to the correct HTTPS website that Cloud Guest is using.
    - **Question**: Does this mean that every guest user needs to open a browser to an HTTP site manually?
    - **Answer**: Theoretically yes, however that would be challenging to explain to every guest user. This is why most operating system and browser vendors have solved this by making a test HTTP connection automatically for you when you connect to the WLAN. In the lab example, a Windows 10 client automatically makes a test connection for you be connecting to http://www.msftconnecttest.com/connecttest.txt. When the test fails to download the correct test file, your windows client will report that it is not connected to the internet (see Figure 11-18):

Figure 11-18: Test guest connection prior to authentication

After successfully connecting (don't log in yet!) to the guest portal and authenticating, the test connection is successful, and the client reports a normal connection (see Figure 11-19):

Figure 11-19: Successfully authenticated guest

Note

The remote lab blocks access to microsoft.com domains, so the connectiontest.txt download file is hosted on the 10.254.1.21 host instead.

Review the captive portal user role

In these steps you will review the configuration that was applied to this captive portal user role. You will first try to use the UI:

1. In Aruba Central, navigate to: Context: **Groups / campus-wifi-ui** > Navigation: **Devices** > Top: **Access Points** > Right top: **Config**.
2. Click **Security**, expand **Roles**.
    - **Question**: Do you see the *p#tX-guest-page_#guest#_* user role in the roles list?
    - **Answer**: No, this role is automatically created for you and it is hidden in the UI.

Explore the _#guest#_ Role

In the next steps you will explore this system *_#guest#_* role in the AP configuration.

1. In Aruba Central, navigate to: Context: **Groups / campus-wifi-ui** > Navigation: **Tools** > Top: **Commands**
2. Select Device Type: **Access Point**, Available devices: **ap1**.
3. Select Category: **Security**
4. Select Commands: **AP Access Rule Table**.
5. Add the command and click **Run**.
6. Use the **maximize** button to increase the output window size.

7. In the AP1 output, check the access rule for the role p#tX-guest-page_#guest#_, this access rule will be at the end of the output.

```
...
Access Rule Name :p58t01-guest-page_#guest#_
In Use           :Yes
ACL Vlan     :0
ACL Captive Portal:external
ACL ECP Profile    :p58t01-guest-page_#guest#_
CALEA              :disable
Redirect Blocked HTTPS Traffic  :disable
DPI error page URL:
Bandwidth Limit    :downstream disable upstream disable
airslice-application-list        :
monitoring-application-list      :
Access Rules
------------
Dest IP  Dest Mask              Eth Type  Dest Match  Protocol
(id:sport:eport)  Application  Action  Log  TOS  802.1P  Blacklist
App Throttle (Up:Down)  Mirror  DisScan  ClassifyMedia  TimeRange
-------------- ---------------------------------------------------
alias    licdn.com              IPv4/6    match       https
permit
ClassifyMedia
alias    twimg.com              IPv4/6    match       https
permit
ClassifyMedia
alias    bam.nr-data.net        IPv4/6    match       https
permit
ClassifyMedia
alias    js-agent.newrelic.com  IPv4/6    match       https
permit
ClassifyMedia
alias    symcb.com              IPv4/6    match       http
permit
ClassifyMedia
alias    symcd.com              IPv4/6    match       http
permit
ClassifyMedia
alias    digicert.com           IPv4/6    match       http
permit
ClassifyMedia
alias    any                    IPv4/6    match       http
permit
ClassifyMedia
```

Captive portal

Now you will login with the guest user and verify the updated access to the network.

1. On PC1, click the **I accept** checkbox and **Log in** on the captive portal webpage. After the successful authentication, the browser will redirect you to the configured welcome page (https://www.arubanetworks.com).
2. In Aruba Central, navigate to: Context: **Groups / campus-wifi-ui** > Navigation: **Clients.**
3. Click the Aruba Central **Refresh** button to display the latest client information.
   - **Question**: What is the status for the client now?
   - **Answer**: After the successful captive portal authentication, the status changes to *Connected.*
   - **Question**: What is the guest client name now?
   - **Answer**: The client's name changed from the original hostname to the username. Cloud Guest with anonymous access is using username *anonymous,* which is now shown as the Client name.
   - **Question**: What is the AP role for this user?
   - **Answer**: The AP user role is *p#tX-guest.* After Cloud Guest returns a RADIUS accept to the AP. The AP assigns the user the SSID default role. The pre-authentication roles were based on the splash page name. The post-authentication default role is the default SSID user role. See Table 11-1 for the phase and role names.

Table 11-1. User rules during guest access

Phase	Role name
Pre-authentication	p#tX-guest-page_#guest#_ Splash Page name + _#guest#_
Post-authentication	p#tX-guest (SSID default role)

Verify guest access control

During configuration of the guest WLAN, you configured the default SSID role (p#tX-guest) with some restrictions. The goal was to ensure that the authenticated guest users cannot access the internal network resources. You also blocked access to the entire 10.0.0.0/8 network, except host 10.254.1.21, a DNS/DHCP server that also hosts the msfconnecttest.txt.

1. On PC1, attempt to ping to **10.254.1.21**. This should be successful.
2. Attempt to ping to **10.254.1.23** (the ClearPass server); this connection is blocked, along with all else in the 10.0.0.0/8 range.
3. Based on your configuration, 5 minutes after the guest user logged in on the network, the session will expire.

Optional: Review the guest session details—Remaining time

Make sure to continue the lab after these optional steps!

1. In Aruba Central, navigate to: Context: **Groups / campus-wifi-ui** > Navigation: **Clients.**
2. Click your guest client to open the client details page.
3. On the client details page, navigate to **Tools > Commands**. Central automatically selects Device Type: Access Point and Available Devices: ap1 or ap2 for you, since the client is connected to this AP.
4. In Categories, select **All Categories.**
5. In the Commands window, use the filter **clients**. Thus, you only see commands that contain the string "clients" (see Figure 11-20).

Figure 11-20: Filtering clients

6. Add **show clients debug** to the Selected Commands window, then click **Run** to execute the command. After a few moments, the command output is shown.

Tip

The output may be easier to analyze if you use the maximize button.

- **Question**: What is the *Reauth Interval* for your client?
- **Answer**: This should be 300 (seconds). The Reauth Interval is based on the Cloud Guest RADIUS session timeout value. You have set the Cloud Guest session timeout to 5 minutes which equals 300 seconds.
- **Question**: What is the *Reauth Age* for your client?
- **Answer**: This value is different for each client. This is the number of seconds since the client's last authentication, but it should be lower than the Reauth interval.

Note

If 5 minutes has passed since you logged in as a guest, the client is logged out and it reverts to the pre-authentication role. You see this as the authenticated value = No, and the Role is p#tX-guest-page_#guest#_ (the pre-authentication role). In that case, make a new connection on PC1 to your guest SSID and login again. Then repeat the **show clients debug** command and check the output based on the instructions above.

7. End of the optional steps: continue with the next steps!

Cleanup

Disconnect the client from the wireless network and disable the Guest WLAN.

1. On PC1, disconnect from the guest WLAN.
2. In Aruba Central, navigate to: Context: **Groups / campus-wifi-ui** > Navigation: **Devices**> Top: **Access Points**> Right top: **Config**.
3. In the WLAN list, click the **guest** WLAN.
4. At the right end of the record, click the **WLAN** icon to disable the WLAN (see Figures 11-12 and 11-13).
5. Confirm the Action with **Yes**.
6. Verify in the WLAN list that the Network Enabled column reports **No** for the guest WLAN.

Lab 11.02: Guest access with ClearPass guest

The customer is pleased with your demonstration of Cloud Guest. They are planning to complete the migration to AOS 10 over the next year, at which point they will also migrate to Cloud Guest. During the migration, however, they would like to have the same Guest experience as they have today on their sites with AOS 8 deployments, using a ClearPass guest solution. You have been asked to setup an AOS 10 guest WLAN that integrates with the existing ClearPass guest solution, using ClearPass as the external captive portal server.

Objectives

- Review the ClearPass Guest page configuration.
- Configure a Guest WLAN using an external captive portal server.
- Configure ClearPass as an external captive portal.
- Verify access to the guest WLAN.

Task 1: Verify the existing ClearPass guest page

In this task you will review the existing guest page configuration on the ClearPass guest server.

Objectives

- Understand the URL of a ClearPass guest page.
- Understand the NAS vendor settings on the ClearPass guest page for a default AOS 10 AP.

Steps

Review the ClearPass guest page

7. On the PC-MGMT, open a command prompt and ping to cppm.aruba-training.com

```
C:> ping cppm.aruba-training.com
```

- **Question**: What is the IP address for this host?
- **Answer**: The name resolves to IP address 10.254.1.23. This is the IP address of the ClearPass server.

8. On the PC-MGMT, open a browser to ClearPass to: **https://cppm.aruba-training.com/guest**.

Note
The FQDN cppm.aruba-training.com is only available inside the remote lab environment.

9. Login with a username of **admin** and a password of **Aruba123!**
10. On the left side, navigate to **Configuration** > **Pages** > **Self-Registrations** (see Figure 11-21).

Figure 11-21: ClearPass Guest module

- **Question**: What are the page names you see in the list?
- **Answer**: There is a **Guest Self-Registration** and **acaa-wifi** page. The **acaa-wifi** was created for this training lab.

11. Click the **acaa-wifi** entry. Some action buttons will appear under the line.
12. Click **Launch** to see an example of the self-registration page. Pay attention to click **Launch**, *not* on "Launch Self-Service" or "Launch Login" (see Figure 11-22).

Figure 11-22: Viewing the portal page

Note

You should not enter any credentials in the form at this point, since you are connected using the PC-MGMT; you are not using a guest client PC at this point, just previewing the web page!

Question: What is the full URL of the page?

Answer: The full URL is https://cppm.aruba-training.com/guest/acaa-wifi.php. Take note of this URL, since you will need this URL for the guest WLAN redirect page configuration.

Note

ClearPass is using a server certificate that was signed by a private Lab CA in this lab environment.

- **Question**: What is the FQDN used to reach the ClearPass guest URL?
- **Answer**: The FQDN is *cppm.aruba-training.com*.

13. After the preview, you may **Close** the preview web page.

NAS vendor settings

On the external captive portal, the administrator needs to provide the hostname to which the guest browser will submit the guest credentials. The AOS 10 APs are by default configured with a public certificate with subject name of *securelogin.hpe.com*. This is automatically installed and configured by Aruba Central on the AOS 10 APs. The captive portal administrator only needs to refer to this name in the captive portal configuration. In the next steps you will review that this information was set on the guest captive portal page.

1. In the Self-Registrations list, click **Edit** for the **acaa-wifi** page (see Figure 11-23).

Figure 11-23: Edit the self-registration configuration

2. On the customize self-registration page, click the **NAS Vendor Settings** (see Figure 11-24).

Figure 11-24: Edit the *NAS Vendor Settings*

3. Review the value that has been set in the Address field (see Figure 11-25).

Figure 11-25: Examine the *Address* settings

4. This value must match the captive portal certificate name that has been installed on the APs. In this lab setup, it matches the AOS10 default: the public certificate with subject name of securelogin.hpe.com. In your own environment, you would need to match ClearPass Guest's Address field to what you actually have on your APs' certificates. This concludes the ClearPass page review. You may close the web browser on PC-MGMT.

Task 2: Configure WLAN profile with a ClearPass guest splash page

In this task you will configure a new guest WLAN profile with the ClearPass splash page that you have just reviewed.

Objectives

- Enable guest captive portal on a WLAN.
- Define the external splash page with a ClearPass guest page.

Steps

Create the guest CPPM WLAN

1. In Aruba Central, navigate to: Context: **Groups / campus-wifi-ui** > Navigation: **Devices**> Top: **Access Points** > Right top: **Config**.

2. From the WLAN tab, click **Add SSID**.
3. From the **General** tab, in the **Name** field, enter **p#tX-guest-cppm**.

Note

Make sure to replace the # value with your pod number and X with your table number. For example, if you are using table 07 in pod 28, your WLAN name will be **p28t07-guest-cppm**. Check with your instructor or lab administrator if you are not sure about the pod and table number.

---

4. Expand the **Advanced Settings** section. Under Miscellaneous set band to **5 GHz**.

Note

In a real deployment you should also apply the best practices. These steps are omitted in this task. Refer to the previous labs to see the WLAN general best practices.

---

5. Click **Next.**
6. From the **VLANs** tab, set **Static** and VLAN ID **15**.
7. Click **Next.**

Configure the external captive portal

In the next steps you will define the ClearPass Guest page as an external captive portal.

1. From the **Security** tab, move the **Security Level** slider to **Visitors**.
2. Set the Type to **External Captive Portal**.
3. For the Captive Portal Profile, use the **+** button to add a new profile.
4. In the External Captive Portal New window enter these settings:
   - Name: **cppm-guest**
   - IP or hostname: **cppm.aruba-training.com**
   - URL: **/guest/acaa-wifi.php**
   - Port: leave default to **443**
5. Click **OK** to save the profile.
   - **Question**: Why did you configure cppm.aruba-training.com as the hostname?
   - **Answer**: This is the DNS name that was registered in the lab environment for the ClearPass guest server.

- **Question**: Why are you using /guest/acaa-wifi.php as the URL name?
- **Answer**: You have reviewed on ClearPass guest the splash page. Based on the launch example, you have seen that the page name was /guest/acaa-wifi.php. You need to update this field to match the page name of the ClearPass system.

6. In the Captive Portal Profile field, verify that **cppm-guest** is now selected.
7. For the primary server, select **cppm1**.
   - **Question**: When was this RADIUS server created?
   - **Answer**: You created the cppm1 RADIUS server during the employee WLAN lab activity in the last chapter.
8. Navigate to **Advanced Settings > Accounting**.
9. Set the Accounting to **Use Authentication Servers**.
10. Set the Accounting Interval to **5** min.
11. Click **Next.**

Authenticated guest Access Control

Using the Access Control, you can control the level of network access that the guest users will have on the network. In this example, some basic restrictions will be applied.

1. From the Access tab, move the slider to **network-based**. Make sure authenticated guests can only reach the host 10.254.1.21 (for any service) and block traffic to all other 10.0.0.0/8 IP addresses.
2. Click **Add rule** and create a rule with these settings:
   - Type: **Access Control**
   - Services: **Any (default)**
   - Action: **Deny**
   - Destination network IP: **10.0.0.0**
   - Destination subnet mask: **255.0.0.0**
3. Click **OK** to save the rule.
4. Click **Add rule** and create a rule with these settings:
   - Type: **Access Control**
   - Services: **Any** (default)
   - Action: **Allow** (default)
   - Destination (particular server): **10.254.1.21**

5. Click **OK** to save the rule.
6. Click **Next** to move to the Summary page.
7. From the summary tab, click **Finish**.
8. After a few seconds, click **OK** to confirm success configuration message.

Optional: Check the Audit Trail

Make sure to continue the steps after this optional section!

Check the Audit Trail configuration changes that have been pushed to the AP.

1. In Aruba Central, navigate to: Context: **Groups / campus-wifi-ui** > Navigation: **Audit-Trail**
2. After a few moments, the new configuration will be shown. This may take a minute to show up, refresh the page until you see the new audit entry.
3. Click the **three dots** to see the details for the configuration. The WLAN access-rule is the network access role that controls the IP access for the guest users.

```
wlan access-rule p58t01-guest-cppm
 utf8
 rule 10.254.1.21 255.255.255.255 match any any any permit
 rule 10.0.0.0 255.0.0.0 match any any any deny
 rule any any match any any any permit
 exit
```

The SSID-profile controls the WLAN object, the SSID name and the VLAN assignment. It also references the external captive portal profile to ClearPass.

```
wlan ssid-profile p58t01-guest-cppm
 essid p58t01-guest-cppm
 opmode enhanced-open
 vlan 15
 rf-band 5.0
 type guest
 captive-portal external profile cppm-guest
 dtim-period 1
 broadcast-filter none
 radius-accounting
 radius-interim-accounting-interval 5
 inactivity-timeout 1000

 ax-authentication-failures 0
 blacklist
```

```
  dmo-channel-utilization-threshold 90
  max-clients-threshold 64
  enable
  utf8
  openflow-enable
  auth-server cppm1
  exit
```

The external captive portal refers to the ClearPass FQDN and the URL for the guest page.

```
wlan external-captive-portal cppm-guest
  server cppm.aruba-training.com
  url "/guest/acaa-wifi.php"
  port 443
  auto-whitelist-disable
  https
  exit
```

4. End of the optional steps: continue with the next steps!

Task 3: Test ClearPass guest access

In this task you will use a wireless client to test the guest access.

Objectives

- Test the access to a Guest WLAN with external captive portal.

Steps

Make the WLAN connection

In these steps you will connect to the *guest-cppm* WLAN, but you will *not login* yet.

1. Open a connection to PC1.
2. Make a wireless connection to the guest CPPM WLAN: **p#tX-guest-cppm.**

> **Note**
> A browser page will pop up at this point, do not log in yet. You will first explore the status of the client during the pre-authentication phase.

3. In Aruba Central, navigate to: Context: **Groups / campus-wifi-ui** > Navigation: **Clients**
   - **Question**: What is the status of the PC1 client?
   - **Answer**: The client state is connecting. Note that the client WLAN state is already connected, but since the client has not completed the captive portal login, the status in Aruba Central will be reported as *connecting*. You may need to wait a few moments for the client status to be updated in Aruba Central.
   - **Question**: What is currently the assigned AP role?
   - **Answer**: The role is *External CP*.
   - **Question**: Did you create this role?
   - **Answer**: No, this is an internal role that is used when an external captive portal (external CP) is enabled for a WLAN profile.

Log in to the ClearPass captive portal

In these steps you will login to the captive portal page.

1. On the PC1, complete the guest self-registration form
   - Full name: **guest**
   - Email: **guest@aruba.local**
2. Click the **I accept** checkbox and **Register** button to create a new account. A new guest account is now created, and the details will be displayed.
3. Click **Log In** to connect to the network.
4. You should now have access to the internet. After the successful authentication, the browser will redirect you to the configured welcome page.

Verify the connectivity

In these steps you will check the updated client status in Aruba Central after the guest login has completed.

1. In Aruba Central, navigate to: Context: **Groups / campus-wifi-ui** > Navigation: **Clients**.

Note

It may take a minute before you see the updated client information and username in Aruba Central. Use the refresh button to get the latest status.

- **Question**: What is the guest client name now?
- **Answer**: The client's name changed from the original hostname to the username that was entered during the self-registration. If you used guest@aruba.local, this will be shown as the client's name.
- **Question**: What is the AP role for this user?
- **Answer**: The AP user role is *p#tX-guest-cppm*. After ClearPass returns a RADIUS accept to the AP, the AP assigns the user the SSID default role (see Figure 11-26).

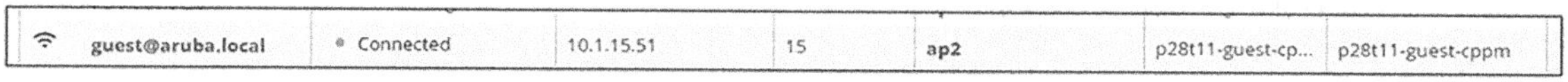

Figure 11-26: Verifying the guest role assignment

2. This completes the guest WLAN with ClearPass Guest configuration.

Cleanup

Disconnect the client from the wireless network and disable the Guest CPPM WLAN to reduce the number of active SSIDs in the remote lab.

1. On PC1, disconnect from the guest WLAN.
2. In Aruba Central, navigate to: Context: **Groups / campus-wifi-ui** > Navigation: **Devices** > Top: **Access Points** > Right top: **Config**
3. In the WLAN list, click the **p#tX-guest-cppm** WLAN.
4. On the right end of the record, click the **WLAN** icon to disable the WLAN (see Figure 11-12).
5. Confirm the Action with **Yes.**
6. Verify in the WLAN list that the Network Enabled column reports **No** for the guest-cppm WLAN.

Learning check

Discuss these questions.

1. Which options below can help you explain Aruba guest access captive portal options to your colleagues? (Select four.)
   a. Cloud Guest is relatively simple to implement, even for multisite environments.
   b. Cloud Guest includes self-registration and social login features.
   c. The Cloud Guest splash page is not customizable.

   d. External captive portal can be implemented with ClearPass or a third-party solution.
   e. The external captive portal performs client authentication, while the splash page remains hosted on Aruba Central.
   f. You might use the external captive portal if you have very high customization and/or scalability requirements.
2. You can use the Guest app to configure profiles for splash pages that originate from your Aruba Central account.
   a. True
   b. False
3. For guest users to access the internet, they must enter a username and password.
   a. True
   b. False
4. You can set a guest splash page to Type = Authenticated, so guests can login using sponsored access or social networking credentials.
   a. True
   b. False
5. Instead of creating guest user credentials yourself, you can allow guests to self-register.
   a. True
   b. False

Summary

Congratulations! You have completed Chapter 11. You should now be able to:

- Describe guest access.
- Set up captive portal authentication.
- Configure guest WLANs.

12 WLAN Security

OBJECTIVES

In this chapter you will learn how to secure WLAN systems. After completing this chapter, you will be able to describe WLAN security concepts, explain certificates as relates to WLAN security, and describe cloud authentication.

After completing this chapter, you should be able to:

✓ Describe WLAN security.

✓ Explain certificates.

✓ Describe cloud authentication.

WLAN security

This chapter begins by discussing the need for wireless security and exploring the different, common options that are used in today's WLANs.

Need for secure wireless communications

Because it is trivial for anyone in the near area to collect wireless communications, you must take extra measures to secure all communications with encryption and authentication. Otherwise, you are susceptible to hacks and attacks (see Figure 12-1).

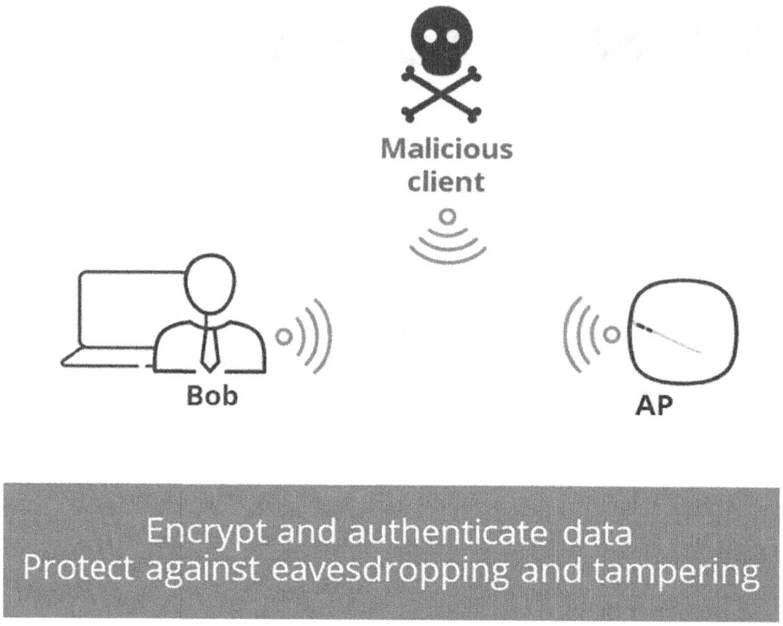

Figure 12-1: Need for secure wireless communications

WLAN security options

Every WLAN operates in one of three broad categories (see Figure 12-2), regardless of the additional authentication types you layer on it.

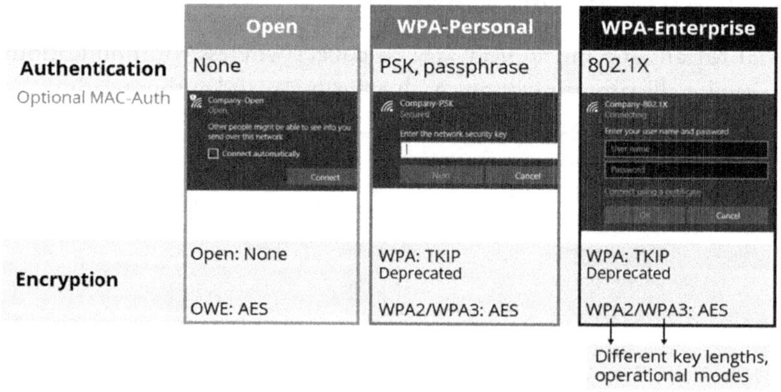

Figure 12-2: WLAN security options

802.11 WLANs can be *open*: users need not submit authentication credentials to connect. The client requests a connection, and the AP permits that connection, which is very simple to implement. Traditionally, open networks offer no encryption. But today's Opportunistic Wireless Encryption

(OWE) standard now provides Advanced Encryption Standard (AES) encryption for open networks, as you will soon learn.

Some WLANs are secured with Wi-Fi Protected Access (WPA), in one of two flavors.

- WPA Personal requires users to enter a passphrase to access the WLAN: a shared secret for all authorized users to connect. The passphrase is sometimes called a pre-shared key (PSK) :it is used as the foundation to encrypt and authenticate traffic.
- WPA Enterprise is recommended, due to its superior, 802.1X-based authentication. With 802.1X, a client authenticates to a backend authentication server using individual credentials: a username and password and/or a digital certificate.

Both the Personal and Enterprise flavors do authentication and encryption. WPA comes in three versions, and the version determines the supported encryption algorithm. WPA originally used Temporal Key Integrity Protocol (TKIP), based on RC4 encryption. WPA and TKIP are now deprecated.

WPA2 and WPA3 both support AES encryption, but WPA3 can impose more stringent requirements for the AES key length and mode of operation. Note that later versions of WPA also have backward compatibility with earlier versions. For example, a WPA2 WLAN could permit clients to use TKIP or AES for encryption; however, for security, it *should* only allow AES.

The WPA versions have additional differences, as you will soon learn. For now, just understand that the user experience for all three versions is similar. The user enters a passphrase to gain access to a WPA Personal WLAN, but submits individual credentials to gain access to a WPA Enterprise WLAN. Finally, note that you can implement MAC-auth together with any of these options. However, the main focus of this chapter is the base security options.

Important

Wired Equivalent Privacy (WEP) is a pre-WPA security standard. Static WEP used a passphrase to authenticate users and derive a key used for encrypting data. Dynamic WEP used 802.1X for authentication and a key established during the 802.1X exchange to encrypt data. However, *both* forms of WEP are proven to be *very* insecure and should **not** be used in today's networks.

WPA key hierarchy

With this brief background in wireless encryption, you are ready to look at the three categories of security in more detail. Let's begin an introduction to WPA. To understand the four-way handshake and data encryption and authentication keys used for all encrypted WLANs, you must have an understanding of the WPA key hierarchy.

At the top (see Figure 12-3) is the Pairwise Master Key (PMK). The PMK itself does not encrypt or decrypt data. Wireless devices derive a Pairwise Temporal Key (PTK) from the PMK. Beside the PMK, other inputs for the PTK creation are a random number used once (nonce) from the AP, a random nonce from the client, the AP's MAC address, and the client's MAC address. The ability to derive different PTKs from the PMK at different times keeps encryption keys from being reused too much.

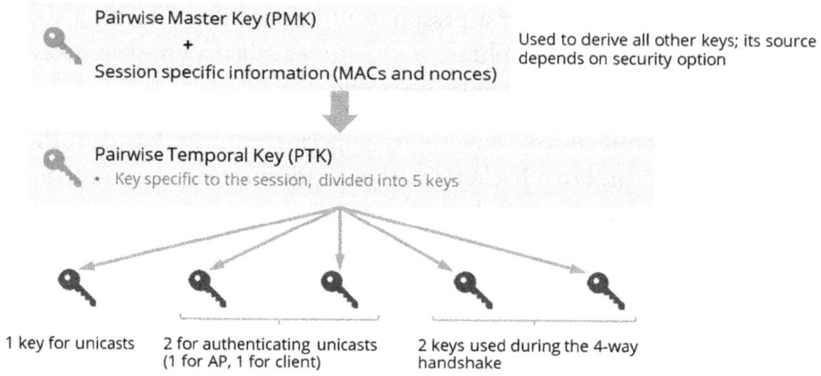

Figure 12-3: WPA key hierarchy

Wireless devices then derive five more keys from the PTK. You do not need to memorize the names of these keys, but you should understand their basic purpose:

- Temporal Key (TK), which the client and AP will use to encrypt unicasts in this session
- MIC transmit key, which the AP will use to provide integrity for data that it transmits in this session (client will use this key to check the integrity)
- MIC receive key, which the client will use to authenticate data that it transmits in this session (AP will use this key to check the integrity)
- Key Confirmation Key (KCK), used during the four-way handshake
- Key Encryption Key (KEK), used during the four-way handshake

WPA four-way handshake

The four-way handshake provides several functions (see Figure 12-4). First, it lets the client and AP exchange the nonces, which are needed to derive the PTK from the PMK.

Second, it enables the station and AP to prove that they know the PMK without transmitting the PMK. They do this by using the KCK to create a message integrity code (MIC), which authenticates the third and fourth packets in the handshake. The authentication will only check out if the device knows the correct KCK, which demonstrates that it knows the PMK.

Third, during the handshake, the AP also distributes a groupwise transient key (GTK) to the client, encrypting this key with the KEK. The GTK is shared by the AP and all clients connected to the WLAN. The AP uses it to encrypt broadcasts and multicasts.

Finally, the handshake enables the AP and client to agree on the first initialization vector (IV) for their encrypted messages.

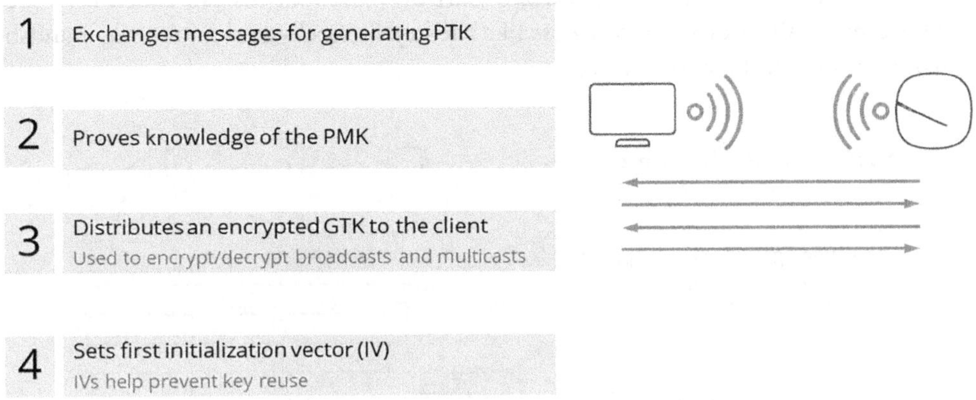

Figure 12-4: WPA four-way handshake

Open WLAN

An open WLAN simply allows clients to connect (see Figure 12-5). The client passes through 802.11 open-system authentication, which permits all clients, and then 802.11 association. No further steps occur; the 802.11 connection is up, and the client can receive IP addressing information and start communicating. As you can see, this is a fairly simple process.

After that, though, layer 3 authentication often takes place. For example, companies often implement captive portal authentication on open networks. Companies also sometimes combine open WLANs with MAC-auth to provide minimal authentication for devices with no user interface or 802.1X support.

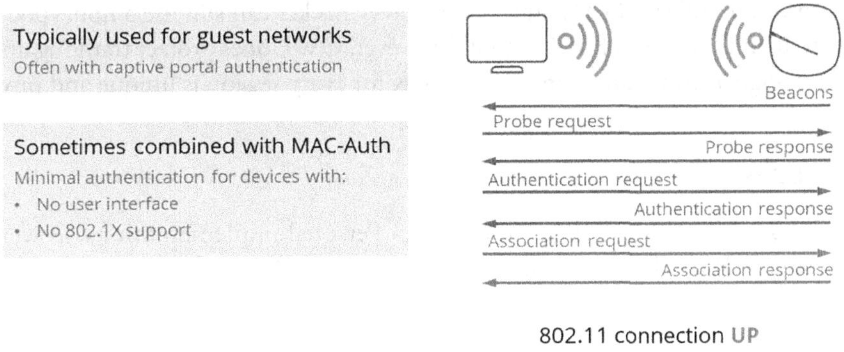

Figure 12-5: Open WLAN

Opportunistic Wireless Encryption (OWE)

Let's now examine how Opportunistic Wireless Encryption (OWE), defined in RFC 8110, which enhances open WLANs. OWE is not technically part of WPA, but instead rolled into the 802.11 standard. However, OWE uses WPA concepts like the key hierarchy and a four-way handshake (see Figure 12-6), discussed in the next two sections.

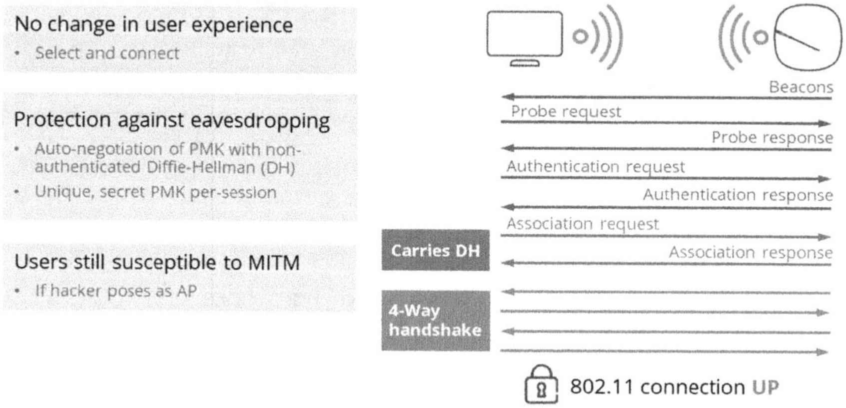

Figure 12-6: OWE

An AP's beacons for an OWE WLAN indicate that it supports OWE. When a station connects to the WLAN, it passes through open-system authentication as usual. But the 802.11 association frames include extra TLVs, which carry an unauthenticated Diffie-Hellman (DH) exchange to derive a PMK. The station and AP then conduct the four-way handshake to derive all the keys for encrypting and authenticating future data. From the user's viewpoint, an OWE SSID works like earlier forms of open SSIDs. The user clicks *connect*, and the connection happens without the user having to submit any credentials. However, the traffic is encrypted and authenticated (digitally signed).

OWE provides best effort security—not true security. Unauthenticated Diffie-Hellman exchanges are subject to man-in-the-middle (MITM) attacks, so a savvy hacker can still use a honeypot AP to lure clients to connect to it and generate a PMK with it. However, OWE does protect traffic against more casual snooping. Even though anyone can connect, the PMK for every session is unique and protected.

Personal security

This section will cover the WPA, WPA2, and WPA3 Personal implementations, as well as the issues involved with using WPA/WPA2 Personal and the benefits of WPA3 Personal.

WPA/WPA2 Personal

Let's first look at WPA/WPA2 Personal WLANs. A client that wants to connect to a WPA/WPA2 Personal WLAN passes through open-system 802.11 authentication and then 802.11 association. It then conducts the four-way handshake (see Figure 12-7). Remember that part of the four-way handshake is proving knowledge of the PMK. For WPA/WPA2 Personal, the PMK is expanded out in a predictable way from the passphrase configured for the WLAN. Thus the four-way handshake proves that the client and AP are using the same passphrase, providing implicit authentication. If the four-way handshake is successful, the connection comes up.

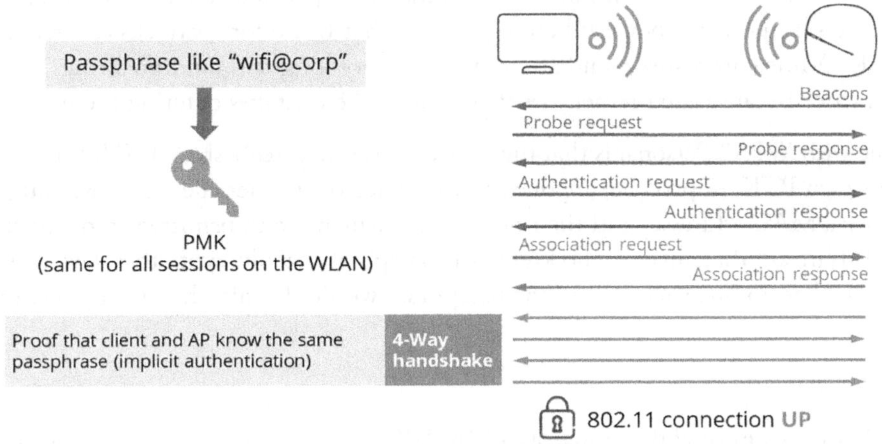

Figure 12-7: WPA/WPA2 Personal

WPA/WPA2 Personal issues

WPA2-PSK has two main, related flaws (see Figure 12-8). First, the PMK and all other keys ultimately derive from a human-selected passphrase. Humans rarely select good passphrases because they cannot remember good passphrases. Therefore, it is important that any key that derives from a passphrase that might be in a dictionary that would be more easily guessed. However, WPA2-PSK *is* subject to brute force dictionary attacks!

PMK derived directly from admin-selected passphrase

- Subject to offline dictionary attacks from captured handshakes

PMK shared by all users

- Passphrase often shared publicly
- Easy to derive another client's PTK

Figure 12-8: WPA/WPA2 Personal issues

An attacker can capture a valid handshake between a client that knows the password and the AP. Two of the handshake messages include a MIC, which is a hash of the plaintext message with the KCK (one of the keys derived from the PMK). An attacker takes this data offline and uses the massive parallel processing power now available with GPUs and hacking farms to simply try every possible passphrase until one produces a key(s) that creates the right MIC.

Even if the passphrase is not a common word and includes special characters, if it is eight characters or under, processing power is powerful enough today that brute force attacks can relatively quickly crack the PSK. While companies could attempt to use very long and random strings, doing so is not feasible in practice because can't remember them and make mistakes entering them.

Another issue with WPA2 Personal is that the password is often freely shared. While every client technically has its own PTK, in practice, anyone can easily derive another client's PTK. You just need the client's and AP's MAC addresses and the nonce values which are transmitted in plaintext during the handshake. This means that anyone who knows the passphrase can decrypt other devices' traffic! While you could hope that no one trusted with the passphrase would do this, that is often *not* the case.

WPA3 Personal

Instead of using a passphrase as the source of a PMK, WPA3 Personal uses Simultaneous Authentication of Equals (SAE) to generate a different PMK each time that a station connects (see Figure 12-9).

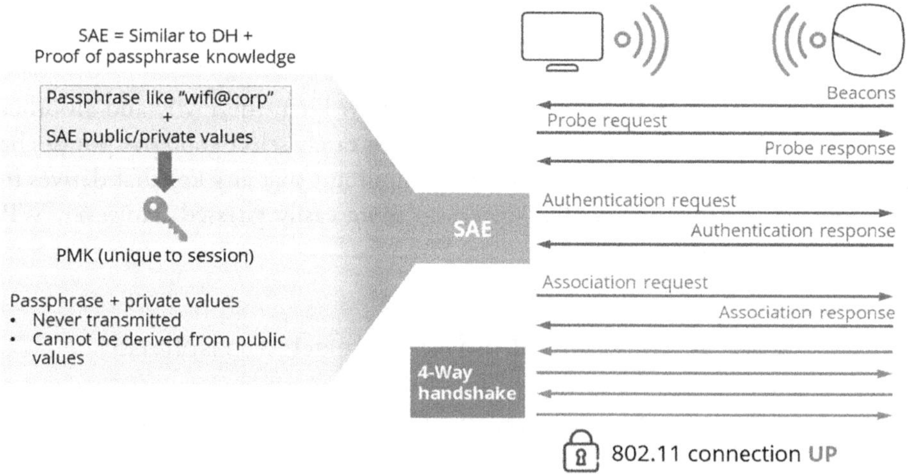

Figure 12-9: WPA3 Personal

The AP advertises SAE as the authentication method for a WPA3 Personal SSID in its beacon and probe response frames. This is a new 802.11 authentication method (value 3); WPA3 Personal is the only security option that uses an 802.11 authentication method other than open-system. (Shared-keys is a deprecated option that is insecure and should never be used.)

The 802.11 SAE process involves four frames. You do not need to know the precise exchange but should understand the high-level concepts. SAE is based on the Dragonfly Key Exchange defined in IETF RFC 7664. This key exchange process uses some of the same cryptographic concepts as Diffie-Hellman and can use either Finite Field Cryptography (FFC) or Elliptic Curve Cryptography (ECC). ArubaOS uses group 19, a 256-bit elliptic curve group. The added twist, distinguishing SAE from Diffie-Hellman, is that the peers can only derive the same shared key (the PMK) if they also know the same passphrase.

After going through SAE, the client and AP have authenticated each other and derived a PMK that is unique to the session. The client associates. The familiar four-way handshake then occurs.

WPA3 Personal benefits

WPA3 Personal offers many improvements over WPA/WPA2 Personal (see Figure 12-10).

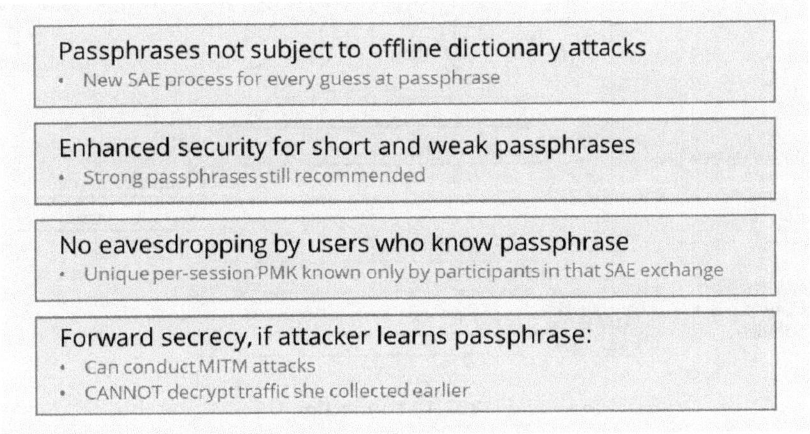

Figure 12-10: WPA3 Personal benefits

The SAE exchange is not subject to offline dictionary attacks because it is computationally infeasible to derive the passphrase from the values transmitted in the SAE exchange. While attacker can still take the four-way handshake offline, the PMK is no longer directly derived from the passphrase, so it is more resistant to a dictionary attack. And cracking the PMK will not be useful because it is *different* for each session. If the attacker wants to launch a dictionary attack, they must actually go through the SAE process and handshake for each guess of the passphrase, which makes the attack infeasible. Therefore, administrators can choose weaker passphrases while still maintaining good security. In fact, SAE is designed to work with a large pool of passphrases known to hackers that they used for hacking WPA/WPA2 Personal networks. However, it is still best practice to choose good passphrases.

WPA3 Personal also protects users' communications from eavesdropping by other authorized users. Every session has a unique PMK, known only to the client and AP that established the session.

WPA3 Personal further provides forward secrecy, which means that a future compromise of the passphrase does *not* compromise data transmitted earlier. If the attacker learns the passphrase somehow—perhaps through a social engineering attack—they can implement a honeypot AP and MITM attacks. However, they cannot go back to encrypted data that was collected earlier and decrypt it. This is because the attacker was not part of the SAE exchange that derived the PMK for those exchanges.

Aruba mPSK

Some companies have *headless* devices that do not support 802.1X authentication, so these devices must connect to WLANs using personal mode. However, the devices might also lack support for WPA3, so WPA2 Personal, with all its flaws, is the only option for them. Aruba provides multi-pre-shared keys (mPSK) within a WLAN to enhance security for these devices (see Figure 12-11).

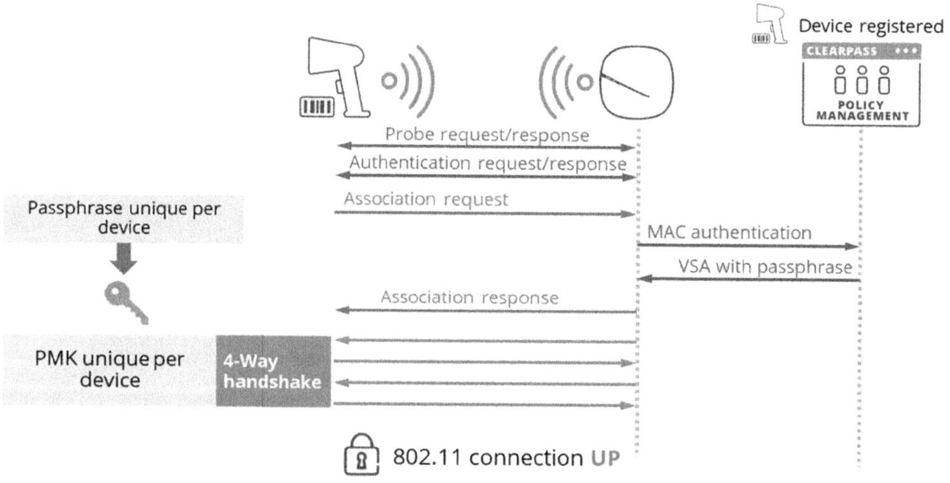

Figure 12-11: Aruba mPSK

A detailed explanation of the mPSK configuration is beyond the scope of this book; however, Figure 12-11 shows, at a high level, how the solution typically works.

- **Step 1:** Before attempting to connect the headless device, a user connects to the ClearPass Self Service Registration page on a different connection. On that page, the user authenticates and specifies the device's MAC address. ClearPass assigns the device a passphrase, or PSK, and displays that passphrase to the user.
- **Step 2**: The user then configures the client device to use the passphrase.
- **Step 3**: The client can now connect to the WPA2 Personal WLAN. From the client's point of view the connection proceeds like a typical WPA2 Personal connection.
- **Step 4**: However, at the point at which the client sends its 802.11 association request, the Aruba AP or gateway performs MAC-auth to ClearPass. ClearPass responds with the passphrase assigned

to the client in an Aruba-MPSK-Passphrase VSA (based on the registered MAC address). The ArubaOS AP or gateway then sends the 802.11 association response and performs the four-way handshake using the PMK derived from the passphrase individual to the client device.

This is a really nice feature since if one key is hacked, it only affects one device and not the entire WLAN in the case of WPA/WPA2 Personal.

WPA/WPA2/WPA3 Enterprise

This section introduces WPA, WPA2, and WPA3 Enterprise WLANs. You were introduced to Enterprise in Chapter 10.

WPA/WPA Enterprise

Let's start first with WPA/WPA2 Enterprise. A connection to a WLAN that enforces this form of security begins with open-system 802.11 authentication and an 802.11 association (see Figure 12-12). 802.1X then shuts down the connection to all but EAP traffic. EAP authentication proceeds as you examined in Chapter 10, enabling the client and AAA RADIUS server to authenticate each other. In addition to authentication, the EAP process also generates a TLS master secret (whether EAP-TLS, PEAP, or another EAP method is used). The client and RADIUS/EAP server each use this secret to derive the PMK.

Because the AP or gateway is the authenticator that needs the PMK for encrypting and decrypting wireless traffic, the RADIUS server sends the PMK to the authenticator in the RADIUS Access-Accept message. The server encrypts the PMK with the RADIUS shared secret. Now both the client and the authenticator know the PMK for that session. The client and AP then complete the four-way handshake, and the connection is up.

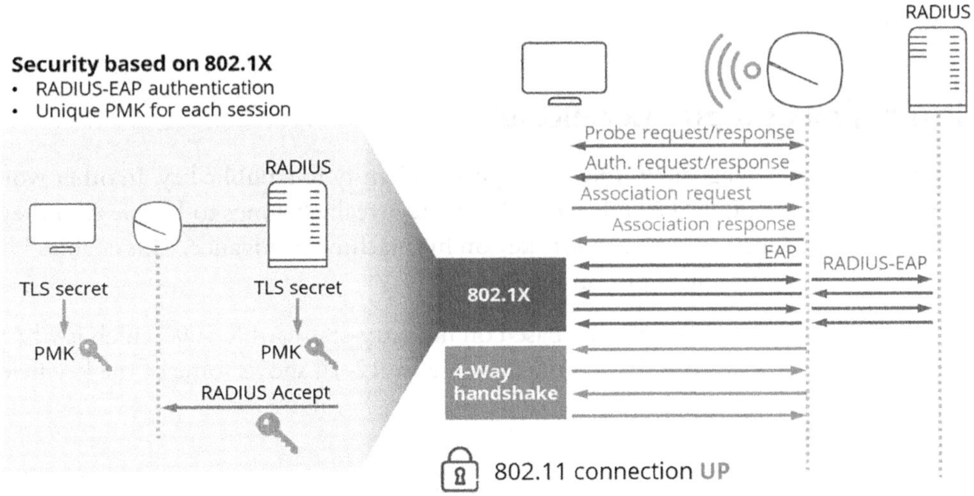

Figure 12-12: WPA/WPA2/WPA3 Enterprise

WPA3 Enterprise overview

WPA3 Enterprise continues to work much like WPA2 Enterprise. Clients use 802.1X with EAP to authenticate, and the PMK is derived during the exchange. However, WPA3 Enterprise has added support for new data encryption and authentication algorithms. WPA2 Enterprise used AES CCMP with a 128-bit key for encryption and a 64-bit key for ensuring integrity. WPA3 Enterprise still supports this mode, but also supports AES-GCM-256, which simultaneously encrypts and ensures integrity for data with 256-bit keys. The key new feature is Commercial National Security Algorithm (CNSA) mode (see Figure 12-13).

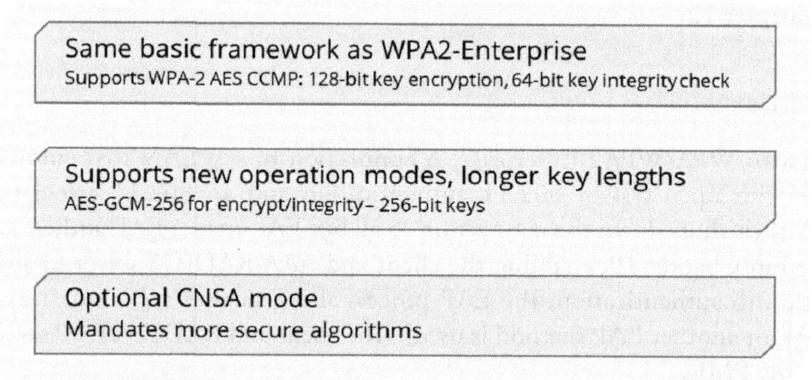

Figure 12-13: WPA3 Enterprise overview

Certificates

Security is greatly enhanced through the use of digital certificates. This section introduces what a certificate is used and the infrastructure required to support the use of certificates during authentication, like 802.1X.

What is a digital or X.509 Certificate?

Certificates provide key management: they tie a specific identity to a public key. In other words, they verify to everyone that the public key posing as "User1 key" really belongs to User1. And they do this without User2 having to install User1's public key on his machine in advance. This enables User1 and User2 to set up secure communications on the fly.

Digital certificates follow a standard format based on industry-standard X.509, which is why you will sometimes hear certificates called X.509 certificates. Figure 12-14 shows some of the key fields on an X.509 certificate.

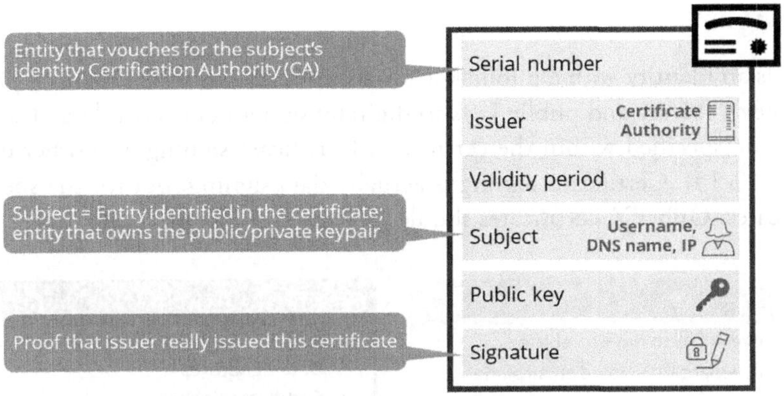

Figure 12-14: What is a digital certificate?

The *serial number* uniquely identifies the certificate among all certificates issued by the same CA. The validity period gives the certificate a specified period in which the certificate is valid. The longer that a device can retain a certificate, the less secure the certificate becomes. Perhaps the private key is compromised or the certificate owner is no longer approved for the rights that the certificate provides. Just as you must renew your driver's license or passport periodically, devices must renew their certificates periodically.

The *issuer* identifies who issued this certificate to the subject. This is a crucial component of certificate trustworthiness. Hackers could create a certificate with his own public key and any identity he wants to claim. But unless the certificate has been issued by a source that you trust, you would know to reject the certificate.

The *subject* identifies the entity that owns the certificate and the keypair associated with the certificate.

The *public key* is the subject's public key.

A certificate also includes the issuer's *signature*. Much like watermarks help people to validate that the proper governmental body truly created an ID like a driver's license, the certificate signature validates that the issuer claimed by the certificate truly issued the certificate.

Just like specific governmental bodies are responsible for issuing documents such as driver's licenses and passports, certification authorities (CAs) are responsible for issuing digital certificates. We trust that the government has taken the proper steps to validate the passport holder's identity, so we trust the passport. Similarly, devices and applications trust that the CA has validated the subject's identity, so they trust the certificate issued by the CA. Certificates can be obtained by purchasing them from public CAs, but there are many products you can use in the marketplace that allow you to set up your own CA, like ClearPass, for example.

Certificate usage

A certificate binds an identity with a public key (and implicitly with the identity's private key paired with the public key). Private and public keys to digitally sign and encrypt data. These are called key usages, and they are specified within the certificate. Certificate signing is another example of a key usage (see Figure 12-15). Certificate signing is actually data signing, but for the specialized purpose of issuing certificates. Only CA certificates should have the certificate signing usage.

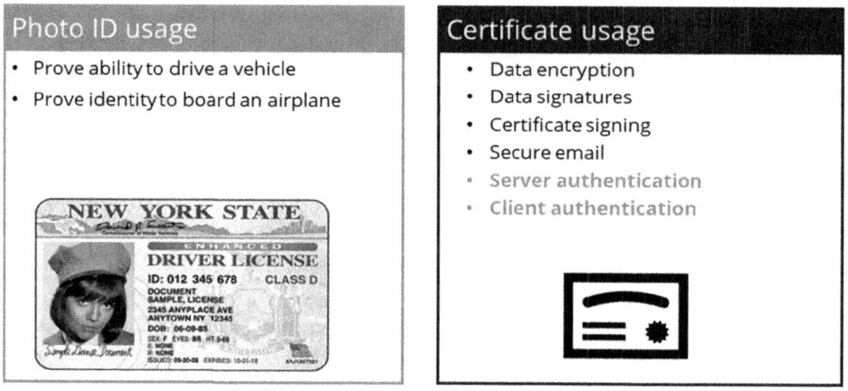

Figure 12-15: Certificate usage

Also, consider a broader viewpoint: *why* the device must encrypt or sign data and *how* the device will use the certificate. People might use a driver's license to prove they can drive a car, or to prove their identity when boarding an airplane. Likewise, users and devices can use their certificates for multiple purposes, which are defined in extended key usages (EKUs), which can also be included in the certificate. For example, the public key in a certificate with the secure email extended key usage can be used to encrypt an email destined to the owner of certificate.

The most relevant EKUs for your purposes are server authentication and client authentication. To authenticate itself with a certificate, either a server or a client signs data with the private key. By signing the data, the device proves that it owns the private key associated with the certificate and is the source of the signed communications.

Note

It is optional to define key usage: certificates can function without a usage defined. But this means the certificate can be used for anything! This widens the possibility of abuse. When defining how certificates will be created, it is a best practice to lock down how they can be used by specifying the usage for the certificate.

Certificate example

In Figure 12-16 you see an actual certificate, as shown on a Windows device. Some of the important fields include:

- Subject name, which identifies the owner of the certificate
- Issuing entity (the name of the CA that signed the certificate)
- Certificate serial number, which is unique to this certificate and issuing entity
- Validity dates (just like passwords should expire, secure certificates should expire too, forcing a renewal in which the entity generates a new keypair and once again proves its identity)
- Public key (remember that the certificate includes the public key itself)

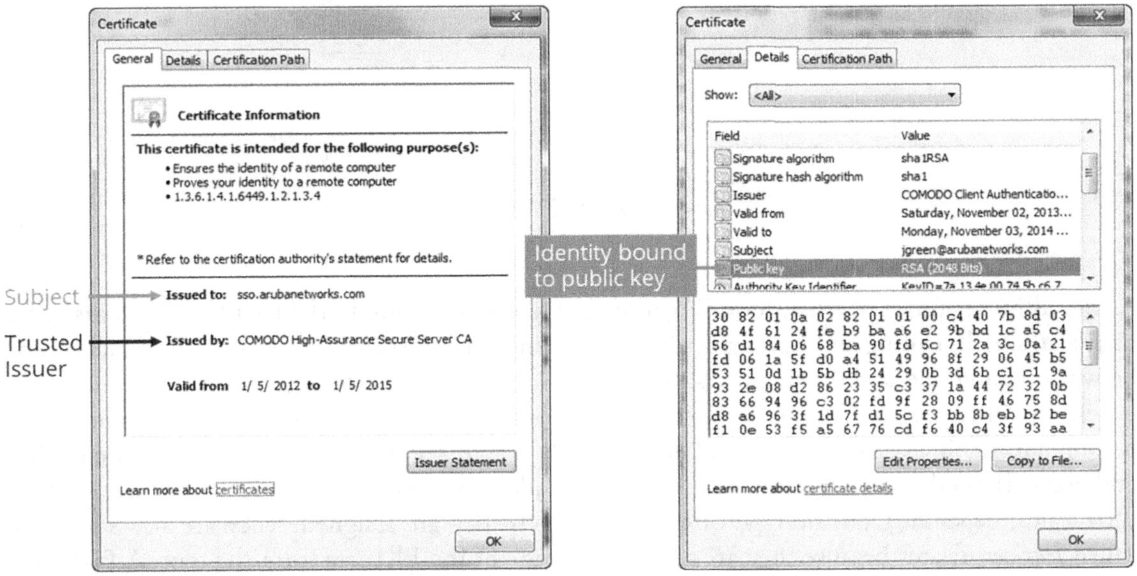

Figure 12-16: Certificate example

You can also look through these fields to see the key usages and EKUs.

Certificate authorities (CAs)

Now that you have looked at certificates, you examine the mechanisms for issuing those certificates.

CAs and the public key infrastructure (PKI)

Certificates exist within a public key infrastructure (PKI) hierarchy that enables devices to determine which certificates they should trust. Every certificate within the PKI is signed by an entity that vouches for the authenticity of the certificate holder's identity.

Root CAs are at the top of the hierarchy (see Figure 12-17). They sign their own certificates, called root CA certificates. Root CAs are well-known entities that are widely believed to be trustworthy. For example, DigiCert and Comodo are examples of root CAs. A company can also set up its own internal or private root CA, which is believed to be trustworthy within the company.

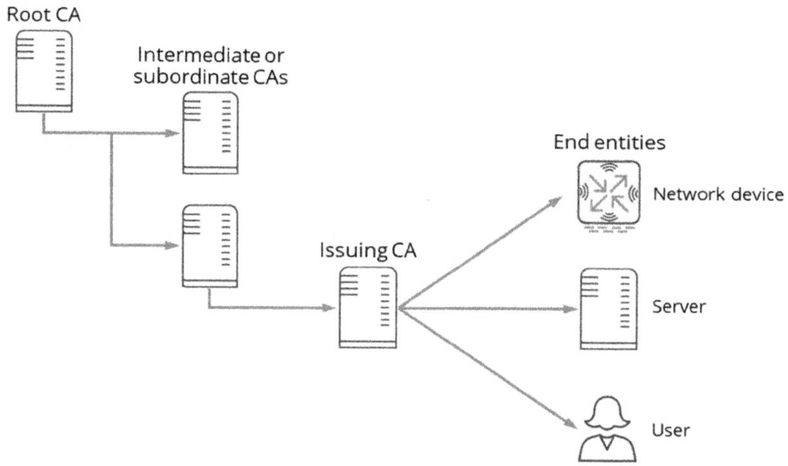

Figure 12-17: CA hierarchy

As the ultimate source of trust, root CAs must be rigorously secured. The root CA issues just a few certificates itself, issuing and signing certificates for intermediate CAs. This ensures minimal contact between the root CA and other entities. In fact, the root CA is typically kept offline for security purposes. The intermediate CAs, sometimes also called subordinate or signing CAs, are responsible for issuing and signing certificates for end entities such as web servers, network infrastructure devices, and users. The end entity certificates are sometimes called server certificates, user certificates, or client certificates, depending on the type of entity to which they are assigned. They are also sometimes called *leaf* certificates because they are at the other side of the PKI *tree* from the root. A PKI could also have multiple tiers of intermediate CAs, as shown in Figure 12-17.

Chain of trust

You can also look at the PKI as a chain of trust, which flows back up through the intermediate CAs to the root CA (see Figure 12-18). Devices and applications can trust the server's certificate because they trust the intermediate CA's certificate. In other words, to validate the server's certificate, applications check the signature in the certificate using the issuing intermediate CA's certificate.

In turn, devices and application trust the intermediate CA's certificate because they trust the root CA's certificate. Because the root CA certificate is self-signed, applications and devices must accept this certificate as trusted through another mechanism.

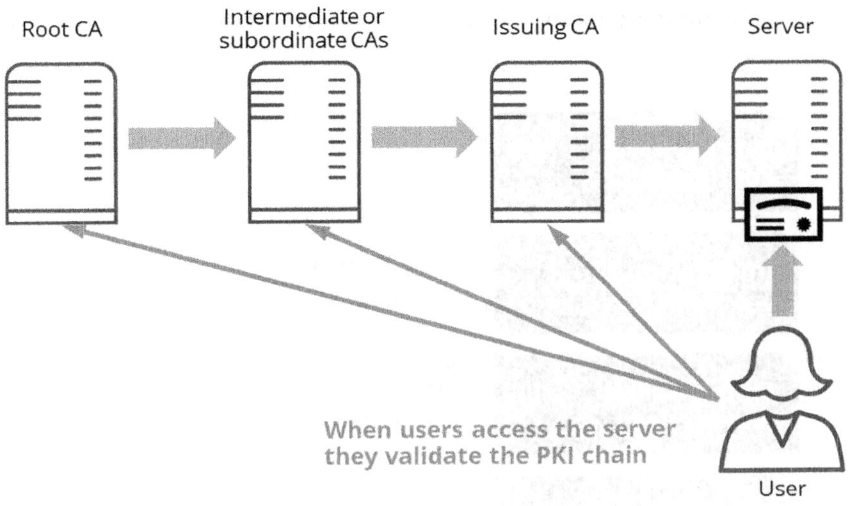

Figure 12-18: Chain of trust

Typically, they are pre-configured to do so. This structure solves the practical problem of using asymmetric encryption. Rather than having to learn the public key for every entity with which you might want to communicate over a secure channel in advance, you need only learn the certificate, and included public key, for the *root* CA.

Cloud authentication

Cloud authentication is convenient and scalable. This section introduces the benefits and features of cloud authentication and how it is implemented in Aruba Central.

Cloud authentication benefits and features

It is critical for a company to apply consistent security controls and ensure users have seamless access to applications and data in the office, at home, and on the go. Aruba Central simplifies this IT process with cloud-based authentication, extending its ability to deliver a single point of visibility and control over all network infrastructure and related security services. Featuring an easy-to-use interface and dashboard, Aruba Central makes it easy to onboard new clients, and to monitor and troubleshoot user connection issues.

You configure fine-grained policies in Aruba Central. Thus, when end users authenticate, they are authorized for appropriate network access. To help enhance security, you get MAC-based authentications and AI-based client insights, which captures and profiles all connected devices. This gives you continuous, in-depth visibility into the connected client behavior, enabling a stronger security posture than by using MAC addresses alone.

See Figure 12-19 for a summary of the benefits and features of cloud authentication with Aruba Central:

Figure 12-19: Cloud authentication benefits and features

Cloud authentication and policies for Aruba Central

Cloud authentication and policy creation in Aruba Central involves four main components (see Figure 12-20):

- Aruba onboard for user devices
- Aruba onboard for Aruba devices (previously covered in the book when discussing zero-touch provisioning or ZTP)
- Network management (covered throughout the book)
- Cloud identity

This section will focus on the items that were not previously covered in this book.

Cloud identity

Aruba Central enables end users to connect to Wi-Fi networks securely and automatically. The application integrates with a company's existing cloud identity store such as Google Workspace or Azure Active Directory to authenticate user information and assign appropriate access. You get frictionless Wi-Fi connectivity for known and new users, visibility into traffic patterns, and fine-grained policies to control access.

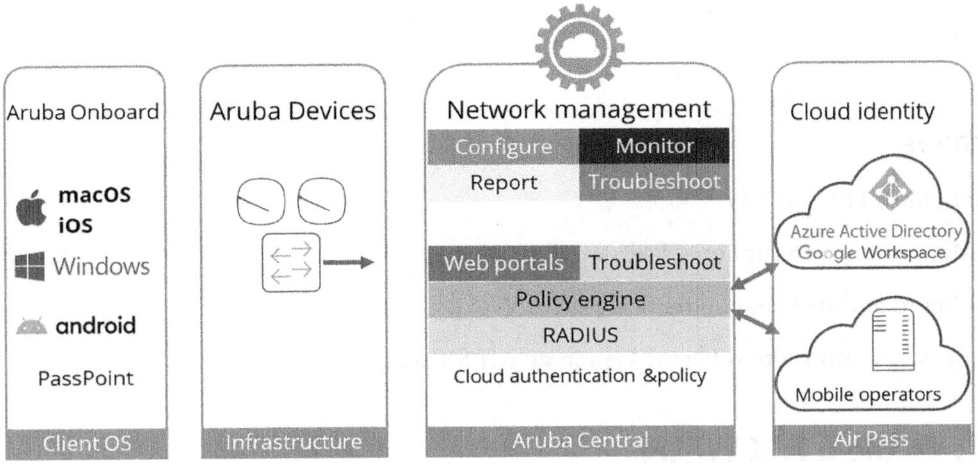

Figure 12-20: Cloud authentication and policies for Aruba Central

End-user device onboarding

Easily configure employee devices for seamless wired and wireless connections. Managed through Aruba Central, end users are authenticated through the company's cloud identity store with an enrollment link provided on Aruba Central.

Using company credentials to log in, the user is redirected to the identity store for authentication. Configure client devices using Aruba Onboard, a client app that is supported on macOS, Windows, iOS, and Android operating systems. With the enterprise PassPoint profile installed on the device, when user walks into network range, the device automatically connects with the appropriate access, based on your Aruba Central configuration. The client app provides automatic renewals: no additional onboarding steps and upkeep from the end user are necessary. And you can change and update policies at any time.

Automated onboarding for guests and visitors

Aruba Air Pass is the industry's first seamless cellular roaming solution designed to unify enterprise and mobile network experiences. You manage Air Pass in Aruba Central, extending the 5G experience to the enterprise using Aruba Wi-Fi technology. Air Pass uses pre-negotiated agreements with mobile network operators (MNOs) powered by PassPoint. Air Pass gives guests and visitors an uninterrupted Wi-Fi experience in indoor venues with the added benefit of enterprise-grade security

Lab 12.01: WLAN security features

Your customer has purchased several IoT devices that must be connected to the wireless network. The network configuration of the IoT devices is limited: the devices do not support 802.1X or WPA2 or WPA3 enterprise authentication. They only support an open or pre-shared key (PSK) configuration.

The customer would like to connect IoT devices to a PSK network and they have asked you to build this WLAN.

Objectives

- Configure a PSK WLAN.
- Configure access control on a PSK WLAN.
- Configure multiple keys on the PSK WLAN.
- Apply access control to different keys in an MPSK setup.

Task 1: Create a PSK WLAN

In this task you will create a PSK WLAN that can be used by the IoT devices. The initial configuration in this task will use a single key for the PSK WLAN.

Objectives

- Create a PSK WLAN.
- Verify access control on the PSK WLAN.

Steps

Create the PSK WLAN

In the next steps you will create the PSK WLAN.

1. In Aruba Central, navigate to: Context: **Groups / campus-wifi-ui** > Navigation: **Devices**> Top: **Access Points**> Right top: **Config**.
2. On the WLAN page, click **Add SSID** to add a new WLAN profile.
3. From the **General** tab, complete the Name(SSID): **p#tX-psk**.

Note

Make sure to replace the **#** value with your pod number and **X** with your table number. For example, if you are using table 07 in pod 28, your WLAN name will be p**28**t**07**-employee. Check with your lab administrator or instructor if you are not sure about the pod and table number.

---

4. Expand the **Advanced Settings** section. In the remote lab environment, there is much interference from neighboring APs. Therefore, you will limit this SSID to the 5 Ghz band.

Note

Many IoT devices only support the 2.4 GHz band; you should not disable this in a production environment. Since you will use the lab PCs to simulate IoT devices, your lab devices can work on 5 GHz only.

---

5. Expand the **Miscellaneous** section.
6. For the Band option select only **5 GHz**.
7. Click **Next**.
8. From the VLANs tab, configure **Static** VLAN **14.**
9. Make sure to *remove* **VLAN 1**.
10. Click **Next**.

PSK WLAN security configuration

In the next steps you will configure the PSK Security and apply the PSK key.

1. From the **Security** tab, verify the Security Level slider is set to **Personal**.
2. Leave the Key Management to the default: **WPA3 Personal**.
3. For the Passphrase, enter **Aruba123!**
4. Expand the **Advanced Settings** section.
5. Verify that the WPA3 Transition is **enabled**. This is the default setting.

Note

Many current and future IoT devices may not support WPA3 yet. To support WPA2 clients on a WPA3 WLAN, the WPA3 *Transition* setting must be enabled.

---

6. Click **Next**.

Access control

In the next steps you will set the network access control.

1. From the **Access** tab you can select the IoT devices level of access.
2. The customer has indicated that IoT devices can only access the internet and the server subnet 10.254.1.0/24. Access to the rest of the internal network must be blocked.
3. Move the slider to **Network based**.

4. Click **Add rule** and create a rule with these settings:
    - Type: **Access Control**
    - Services Network: **Any** (default)
    - Action: **Deny**
    - Destination network IP: **10.0.0.0** netmask: **255.0.0.0**

Note

Pay attention to configure the correct subnet mask!

---

5. Click **OK** to save the rule.
6. Click **Add rule** and create a rule with these settings:
    - Type: **Access Control**
    - Services: **Any** (default)
    - Action: **Allow** (default)
    - Destination network IP: **10.254.1.0** netmask **255.255.255.0**
7. Click **OK** to save the rule.
8. Verify the rules have been added in the correct order (see Figure 12-21).

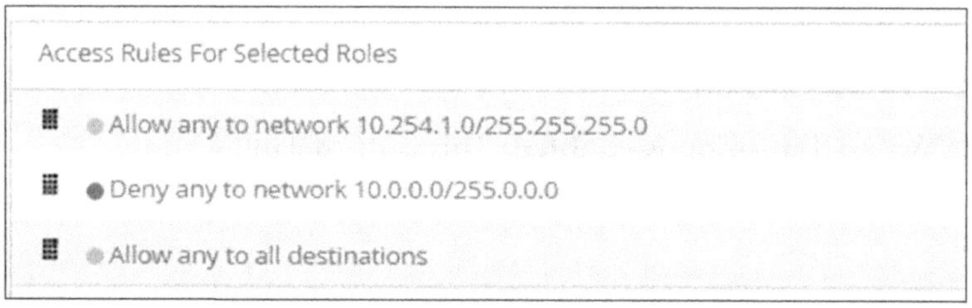

Figure 12-21: Verify access rule order

9. Click **Next** to move to the Summary page.
10. From the Summary tab, click **Finish**.
11. After a few seconds, click **OK** to confirm the configuration message.

Task 2: Test PSK access with an IoT air sensor

The customer is planning to deploy some IoT air sensors, and they want you to test access to the PSK network. To simulate the air sensor, you will apply a different WLAN MAC to the PC1.

Objectives

- Test access to a PSK WLAN.
- Verify PSK WLAN access control.

Steps

Verify the WLAN access

In the next steps, you will change the WLAN MAC.

12. On PC1's desktop, open the ACAF student folder in Windows Explorer.
13. Double-click the file **host-wifi-pc1-iot-sensor-air.cmd**. This will set your WLAN MAC to F60C02 *P#Tx*01. (C02 is the MAC hint for the air sensor.)

Note

Setting a different client MAC is not strictly required for this lab to function. It only ensures that Aruba Central sees the client as a new device (new MAC address). This ensures there will be no client history or IP address information from previous lab activities. The pod # will be replaced with your pod number, for example 28; table "x" reference will be replaced with your table number, for example 01 or 12.

---

14. Connect to the WLAN **p#tX-psk** using the key **Aruba123!**
15. In Aruba Central, navigate to: Context: **Groups / campus-wifi-ui** > Navigation: **Clients** > Top: **Clients** > Right top: **List**. See Figure 12-22 for an example.

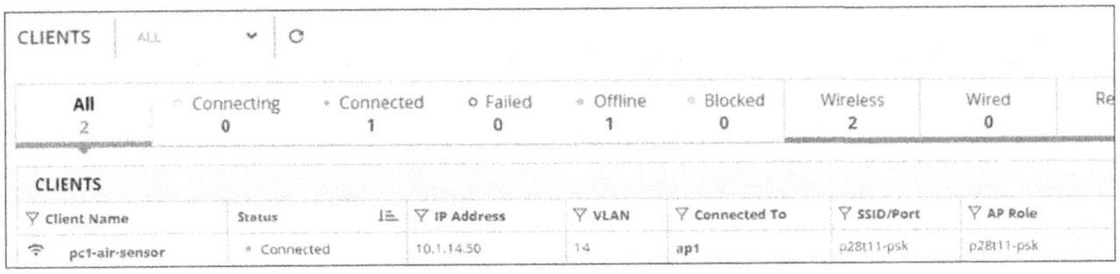

Figure 12-22: View the connected user

- **Question**: Do you have a connected client on the PSK SSID? (use the SSID/Port column to see the SSID)
- **Answer**: Yes, the client (simulated air sensor) is now connected.

- **Question**: To which VLAN is the client connected?
- **Answer**: The client is on VLAN 14, based on the WLAN default VLAN configuration.

Verify the access control

The customer requested access control for the IoT devices, so you validate access control in the following steps. IoT devices should be able to access the internet (8.8.8.8 for example), the server subnet (10.254.1.21), but no other devices on the internal network (10.1.14.1 for example).

1. On PC1, test the configured access control.
2. Open a command prompt and **ping** the following:
    - **8.8.8.8**: should be successful.
    - **10.254.1.21**: should be successful.
    - **10.1.14.1**: should fail.

Task 3: MPSK with differentiated access control

The customer was happy with the easy access control configuration for IoT air sensors. Now they want a new set of HVAC air conditioning (AC) devices that must connect to the WLAN. They want the AC units to connect with a PSK (**Airco123!**) that is different from the air sensors. The AC units also have different access control needs. They should only be able to access the AC unit management software on server 10.254.1.21. They also want to make sure that anyone using the default PSK key Aruba123! can only access the internet.

After consulting with your senior colleague, you learn about the MPSK local feature on the APs. This feature allows you to configure multiple Pre-Shared Keys (MPSK) for the same WLAN. The feature also enables different user roles (access control) for the different keys. The MPSK configuration will consist of the following steps:

- Configure user roles for *iot-air-sensor* and *iot-ac* with unique access control.
- Configure an MPSK rule set to link unique PSK keys to the user roles.
- Change the default SSID role access control to allow internet access only.

Objectives

- Configure MPSK local on a PSK WLAN.
- Configure unique access control with MPSK.

Steps

Update the default access

In these steps you will limit access for the default user role on the PSK WLAN.

1. In Aruba Central, navigate to: Context: **Groups / campus-wifi-ui** > Navigation: **Devices**> Top: **Access Points**> Right top: **Config**.
2. Select the **p#tX-psk** WLAN; use the pencil icon to **edit**.
3. Click the **Access** tab to see the network-based access.
4. Move the slider to **Role Based**.
   - **Question**: What role is automatically selected when you moved the slider?
   - **Answer**: This is the default SSID role, based on the WLAN profile name: **p#tx-psk**.
   - **Question**: What is the current rule set for this role?
   - **Answer**: Based on the requirements in the previous task, the default role allows:
     - Allow access to the 10.254.1.0/24 network
     - Blocks all other access to the 10.0.0.0/8 network.
5. Update the rule set so only internet access is allowed. Remember that access to the host 10.254.1.21 is always required in the lab for several network services such as DNS and DHCP.
6. Select the rule **Allow any to network 10.254.1.0/255.255.255.0**; use the **pencil** edit the rule.
7. Set the Destination to particular server: **10.254.1.21**.
8. Click **OK** to save the rule.

Create new user roles for iot-sensor-air

The air sensor will now get its own user role. You will apply the same access control as that previously applied to the default SSID:

- Allow Access to the server subnet 10.254.1.0/24
- Deny Access to internal network 10.0.0.0/8
- Allow Access to Internet

1. Click the **Add Role** button to create a new role.
2. For the name, enter **iot-sensor-air** and click **OK**.
3. Verify that the new user role iot-sensor-air is now selected.

4. Click **Add rule** and create a rule with these settings:
   - Type: **Access Control**
   - Services: **Any** (default)
   - Action: **Deny**
   - Destination network IP: **10.0.0.0** netmask: **255.0.0.0**
5. Click **OK** to save the rule.
6. Click **Add rule** and create a rule with these settings:
   - Type: **Access Control**
   - Services: **Any** (default)
   - Action: **Allow** (default)
   - Destination network IP: **10.254.1.0** netmask **255.255.255.0**
7. Click **OK** to save the rule.
8. Verify the rules for iot-sensor-air are listed in the correct order (see Figure 12-23). *If necessary*, you can adjust the order by dragging the rules up or down in the list using the **dots** at the start of the line (see Figure 12-24).

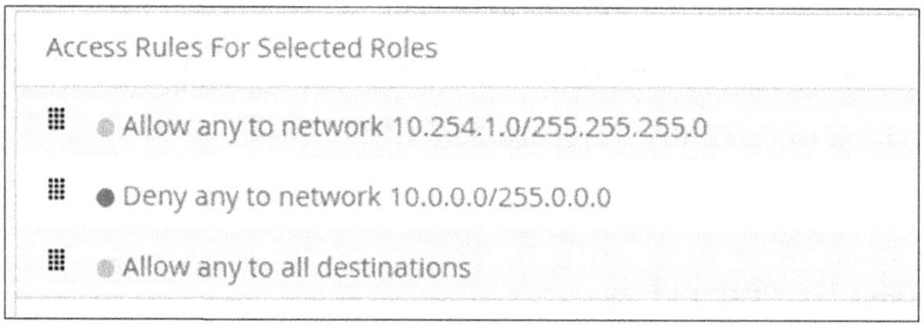

Figure 12-23: Verify the rule order

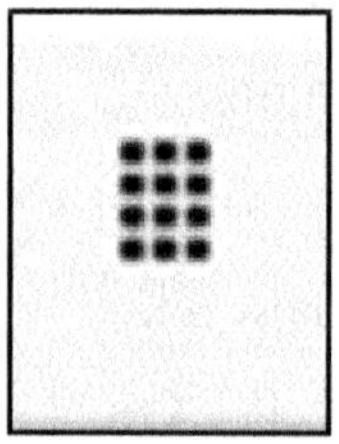

Figure 12-24: Button to re-order the rules

Create new user role for iot-ac units

The HVAC AC units will get their own user role. You will apply these access controls based on the customer requirements:

- Allow Access to server 10.254.1.21.
- Deny all other access.

9. Click the **Add Role** button to create a new role.
10. For the name, enter **iot-ac** and click **OK**.
11. Verify that the new user role *iot-sensor-air* is now selected.
12. Click **Add rule** and create a rule with these settings:
    - Type: **Access Control**
    - Services: **Any** (default)
    - Action: **Allow** (default)
    - Destination (particular server): **10.254.1.21**
13. Click **OK** to save the rule.
14. Click the rule Allow any to all destinations.
15. Use the **bin** (**trashcan**) icon to *delete* the entry.
16. Confirm the action with **yes**.
17. Verify the rule of the role (see Figure 12-25).

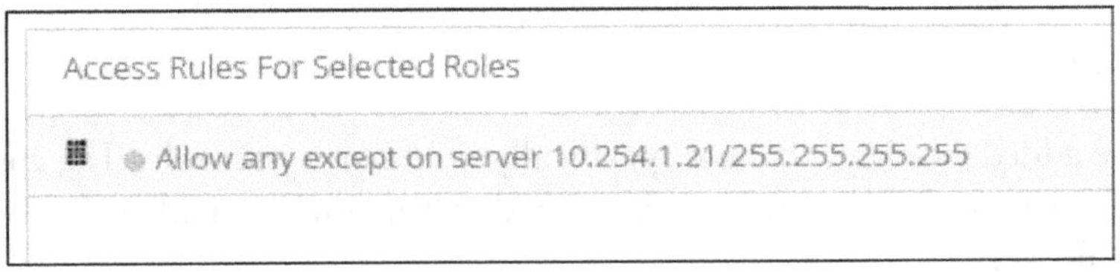

Figure 12-25: Verify the role

18. Click **Save Settings** to save the Access Control changes.

Note

You have created the new roles after moving the slider to *Role Based*, but you did not create any custom role-derivation rules. Therefore, if you would edit the WLAN again, the access control slider would still show *Network Based* instead of Role Based; this is expected.

Create MPSK Local rule set

In the previous steps you have prepared dedicated roles for the AC and the air-sensors. Now you need to prepare an MPSK Local rule set where unique PSK keys can be added and linked to unique user roles.

1. In Aruba Central, navigate to: Context: **Groups / campus-wifi-ui** > Navigation: **Devices**> Top: **Access Points**> Right top: **Config**.
2. Access the **Security** tab and expand the **MPSK Local** section.
3. Click the + button to add a new rule set.
4. For Name enter **psk**.
5. In the key list, use the + button to add a key with these settings:
    - Name: **iot-ac**
    - Passphrase: **Airco123!** (then retype the Passphrase). **IMPORTANT**: Make sure you *don't configure Aruba123!*
    - Role: **iot-ac**
6. Click **OK** to add the key.
7. In the key list, use the + button to add another key with these settings:
    - Name: **iot-sensor-air**
    - Passphrase: **Sensorair123!**
    - Retype the passphrase
    - Role: **iot-sensor-air**
8. Click **OK** to add the key.
9. Now you have added two keys that are specifically linked to a user role. The customer also wants to support the existing PSK, which can be linked to the SSID default role. Thus, you need not specify a specific role.
10. In the key list, use the + button to add a key that is not linked to a user role:
    - Name: **default**
    - Passphrase: **Aruba123!**
    - Role: <do not select a role>
11. Click **OK** to add the key.

Note
You could also have selected the **p#tX-psk** user role since that is the default user role for the *psk* WLAN.

12. Click **OK** to add the MPSK local rule set.
13. Click **Save Settings** to submit the changes.
14. You have now created an MPSK local rule set in which different keys will be linked to different user roles.

Update PSK WLAN profile with the MPSK Local rule set

In the previous steps you have created a rule set. The rule set doesn't do anything by itself; it must be enabled on a WLAN profile. You will now enable it on your PSK WLAN profile.

1. In Aruba Central, navigate to: Context: **Groups / campus-wifi-ui** > Navigation: **Devices**> Top: **Access Points**> Right top: **Config**.
2. On the WLANs page, select the **p#tx-psk** WLAN and use the **pencil** to edit the WLAN.
3. Click the **Security** tab.
4. For Key Management select **MPSK Local** from the list.
   - **Question**: The list also contains MPSK AES, what is the difference with MPSK Local?
   - **Answer**: MPSK-AES requires the use of ClearPass RADIUS server to store and manage the PSKs. It provides per user or per device keys to be generated. In this lab you will only configure MPSK Local setup.
5. In the **MPSK Local** field, select **psk**. This is the list you have prepared in the previous steps.
6. Click **Save Settings** to submit the configuration.
7. After a few seconds, click **OK** to confirm the success message.

Task 4: Test MPSK access

In this task you test the PSKs defined in the MPSK Local rule set. You verify that the correct role was assigned for each key.

Objectives

- Test access to an MPSK WLAN.
- Verify MPSK WLAN role assignment.

Steps

Verify WLAN access for the AC units

1. On PC1, forget the existing WLAN connection.
2. Click the **Wireless** icon in the Windows status bar.
3. Then right-click the active connection to see the popup menu. In the menu, select **Forget** (see Figure 12-26).

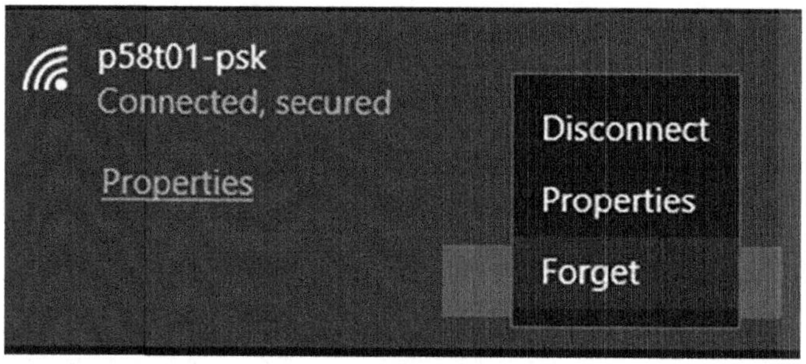

Figure 12-26: Forget the WLAN

4. Connect again to your *psk* WLAN; this time use the key **Airco123!**
5. In Aruba Central, navigate to: Context: **Groups / campus-wifi-ui** > Navigation: **Clients** > Top: **Clients** > Right top: **List**. See Figure 12-27 for an example.
   - **Question**: What AP role is assigned to the client?
   - **Answer**: Based on the PSK, the client has received the AP Role *iot-ac*.

CLIENTS

Client Name	Status	IP Address	VLAN	Connected To	SSID/Port	AP Role
pc1-air-sensor	Connected	10.1.14.50	14	ap1	p28t11-psk	iot-ac

Figure 12-27: View the client list—role iot-ac

Note

It may take a moment for Central to reflect the role update.

Verify access control for the AC units

The AC units are only allowed to access the host 10.254.1.21.

6. On the PC1, attempt to ping to these destinations:
   - **10.254.1.21**: should be successful.
   - **10.254.1.23**: should fail.
   - **8.8.8.8**: should fail.

Verify WLAN access for IoT air sensor

In these steps you will test the MPSK access with the IoT air sensor PSK.

1. On PC1, *forget* the existing PSK WLAN connection.
2. Connect to your PSK WLAN with the key **Sensorair123!**
3. In Aruba Central, navigate to: Context: **Groups / campus-wifi-ui** > Navigation: **Clients** > Top: **Clients** > Right top: **List**. See Figure 12-28 for an example.
   - **Question**: What AP role is assigned to the client?
   - **Answer**: Based on the PSK, the client has received the AP role *iot-sensor-air*.

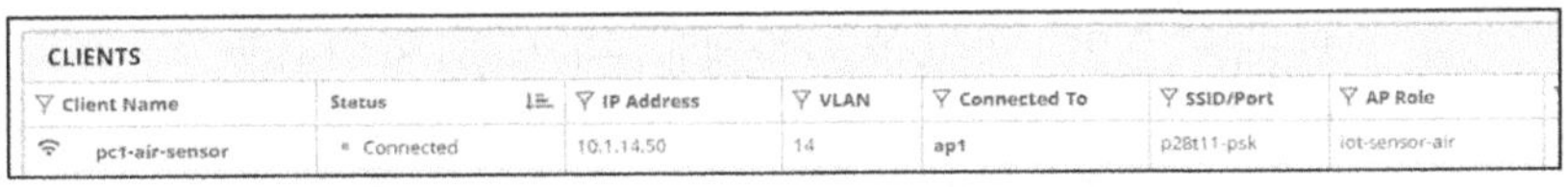

Figure 12-28: View the client list—role iot-sensor-air

Note
It may take a minute before the status of the client is updated in the Central list.

Verify access control for the IoT air sensor

The air sensors are allowed to access the server subnet 10.254.1.0/24 and the internet. Other internal networks should be blocked.

1. On the PC1, attempt to ping to these destinations:
   - **10.254.1.21**: should be successful.
   - **10.254.1.23**: should be successful.
   - **8.8.8.8**: should be successful.
   - **10.1.14.1**: should fail.

Verify WLAN access with the default key

You have configured two additional keys for the MPSK WLAN, but clients that have been configured with the original Aruba123! Should still be able to connect to the PSK. In the next steps you will verify that clients can still use the original PSK to connect to the WLAN.

1. On PC1, forget the existing WLAN connection.
2. Connect to your PSK WLAN with the key **Aruba123!**
3. In Aruba Central, navigate to: Context groups: **campus-wifi-ui** > Navigation Clients > **Top Clients** > Right top: **List**. See Figure 12-29 for an example.
    - **Question**: What is the AP role that is assigned to the client?
    - **Answer**: Based on the PSK, no role was configured for this key. Therefore, the client has received the SSID default Role *p#tX-psk*.

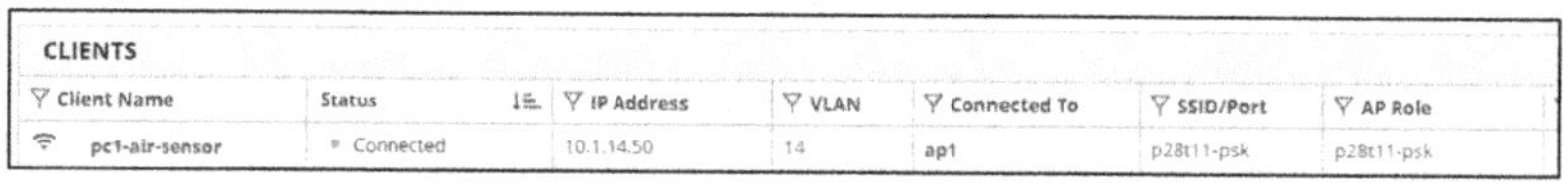

Figure 12-29: View the client list—role p28t11-psk

Optional: Verify access control for the default key

Anyone connecting with the default key Aruba123! Should only be allowed internet access.

1. On the PC1, attempt to ping to these destinations:
    - **10.254.1.21**: should be successful based on the allowed host entry.
    - **10.254.1.23**: should fail.
    - **10.1.14.1**: should fail.
    - **8.8.8.8**: should be successful.

Task 5: Troubleshoot clients connecting to PSK

In this task you will enter an invalid PSK and see how to troubleshoot this using Aruba Central.

Objectives

- Troubleshoot a PSK key mismatch.

Steps

Troubleshoot a PSK key mismatch

In these steps you will attempt to make a WLAN connection with an invalid PSK. Using the client details and Live Events in Aruba Central, you will see the issues.

1. On PC1, forget the existing WLAN connection.
2. Attempt to connect to your PSK WLAN with an invalid key, for example **test123!**
3. In Aruba Central, navigate to: Context: **Groups / campus-wifi-ui** > Navigation: **Clients** > Top: **Clients** > Right top: **List**. See Figure 12-30 for an example.

Figure 12-30: Mismatched PSK example

4. Click PC1 to access the client details page (see Figure 12-31).
   - **Question**: The DATA PATH window is shown at the top of the page. What is the *Failure Stage* for the client?
   - **Answer**: The Failure Stage is KEYEXCHANGE.

Note

If the client is not shown as failed, you can repeat entering the invalid key on the client a few times.

Figure 12-31: View the client details

5. In the Navigation pane, click **Live Events**. The troubleshooting will automatically start for a maximum duration of 15 minutes.
6. On PC1, attempt to connect again with an invalid key, for example **test123!**
7. In Aruba Central, check the event descriptions.

Tip

If the events are difficult to read, you can download the events as a CSV file and open them on your local system with a text viewer or spreadsheet application such as Excel. The *download* button at the top of the event list.

- **Question**: What is the result message for the Client exchanged key events?
- **Answer**: The reported result is MIC failure. This indicates an invalid PSK.

Cleanup

Disconnect the client from the wireless network and disable the *p#tX-psk* WLAN.

1. On PC1, disconnect from the PSK WLAN.
2. In Aruba Central, navigate to: Context: **Groups / campus-wifi-ui** > Navigation: **Devices**> Top: **Access Points**> Right top: **Config**.
3. In the **WLAN** list, click the **psk** WLAN.
4. At the right end of the record, use the WLAN icon to **disable** the WLAN (see Figure 12-32).

Figure 12-32: WLAN button

5. Confirm the Action with **Yes**.
6. Verify in the WLAN list that the *Network Enabled* column lists **No** for the PSK WLAN.

Learning check

Discuss these questions.

1. Match the authentication and encryption options to the appropriate WLAN security option (*Open*, *WPA Personal*, and *WPA Enterprise*). Options can be used more than once.
   a. Pre-shared key (PSK)
   b. WPA3: Advanced Encryption Standard (AES)
   c. Opportunistic Wireless Encryption: AES
   d. WPA2: Advanced Encryption Standard (AES)
   e. 802.1X

2. During the WPA four-way handshake, the Pairwise Master Key (PMK) is used to encrypt data, while the Message Integrity Check (MIC) keys are used for hashing, a frame integrity check.
   a. True
   b. False
3. Opportunistic Wireless Encryption (OWE) adds encryption to an open SSID that eliminates Man-in-the-Middle (MITM) attacks.
   a. True
   b. False
4. When using WPA2 Personal as your WLAN security option, your security could be subject to offline dictionary attacks. WPA3 Personal mitigates such attacks.
   a. True
   b. False
5. Aruba provides a multi-PSK (mPSK) feature to improve security for devices that lack 802.1X authentication support.
   a. True
   b. False
6. Which of the statement below help you accurately describe digital certificates to your colleagues? (Select three.)
   a. Certificates can serve to prove user and device identity, and to provide key management.
   b. Only certificates that are properly signed by a Certificate Authority (CA) should be trusted.
   c. When creating certificates, you are not required to specify how it should be used, but it is a good idea to do so.
   d. Entities must validate all certificates by communicating directly with the root Certificate Authority (CA).
7. Which options below describe significant features of Aruba Central's Cloud authentication? (Select all.)
   a. It offers a single point of visibility and control over all network infrastructure and related security services.
   b. It makes it easy to onboard, monitor, and troubleshoot client connections.
   c. It integrates in cloud identity stores like Google Workspace and Azure Active Directory.
   d. Artificial intelligence (AI) Insights give you continuous, in-depth visibility into connected client behavior.

Summary

Congratulations! You have completed Chapter 12. You should now be able to:

- Describe WLAN security.
- Explain certificates.
- Describe cloud authentication.

13 Monitoring and Maintenance

OBJECTIVES

Maximizing situational awareness means understanding your network health, who's connected it, the health of those connections, and more. This awareness prepares you to more quickly diagnose issues, often proactively, *before* they impact users.

After completing this chapter, you will be able to use Aruba Central monitoring capabilities, identify device LED status, perform firmware upgrades, and using SNMP on your devices. You will also be able to describe AI Insights, and explain Aruba's User eXperience Insights (UXI) deployments, which can really improve your awareness of end user performance issues, and quickly nail down the cause.

After completing this chapter, you should be able to:

✓ Use Aruba Central monitoring capabilities.

✓ Identify device LED status.

✓ Perform firmware upgrades.

✓ Using SNMP on devices.

✓ Describe AI Insights.

✓ Describe alerts and reports.

✓ Explain UXI.

Monitoring with Aruba

This section focuses on monitoring with Aruba Central.

Overview

The next three sections cover some of the more common pages you'll use to monitor your Aruba network.

Summary

Figure 13-1 shows the typical overview page info:

- Network health
- WAN health
- Summary
- Wi-Fi connectivity
- AI Insights

The network health information includes a site map with locations displayed in various colors. The summary information shows bandwidth usage, client count, top APs by usage, top five charts, top IAP clusters by usage and clients, application usage, as well as WLAN network detailed from a selected group. By default, Central plots the graphs for a time range of three hours, but this time interval is customizable. You will learn more about accessing this information during the lab activity.

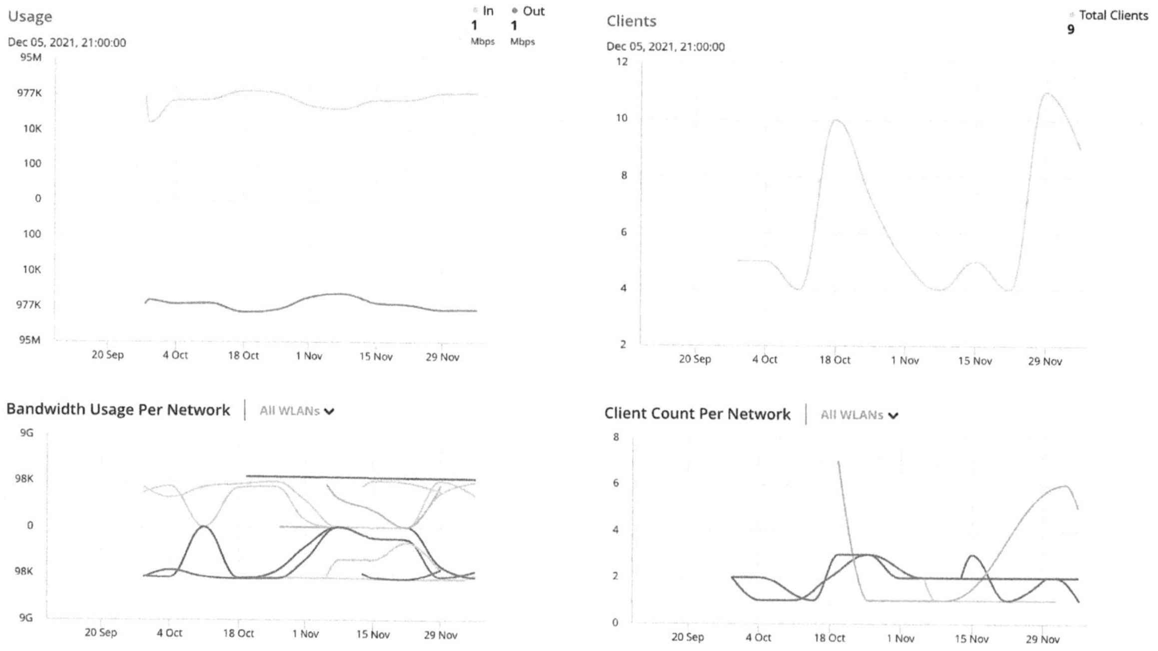

Figure 13-1: Overview—Summary

Network health

Figure 13-2 shows typical network health dashboard information. This *Summary* view shows vital network information of individual sites on cards mapped by geographical location. Sites are marked with location pins: a green pin for a site with no issues and a red pin for a site with potential issues.

Hover your mouse over a pin to view more details about a site in an overlay card view. When there are multiple sites in a region that cannot be displayed as individual pins on the map view, they appear as numbers on a round pin in the condensed view. The pin color is red if it has one site in the down status and green if all the sites are up. Click a pin to zoom in and view the sites on the map.

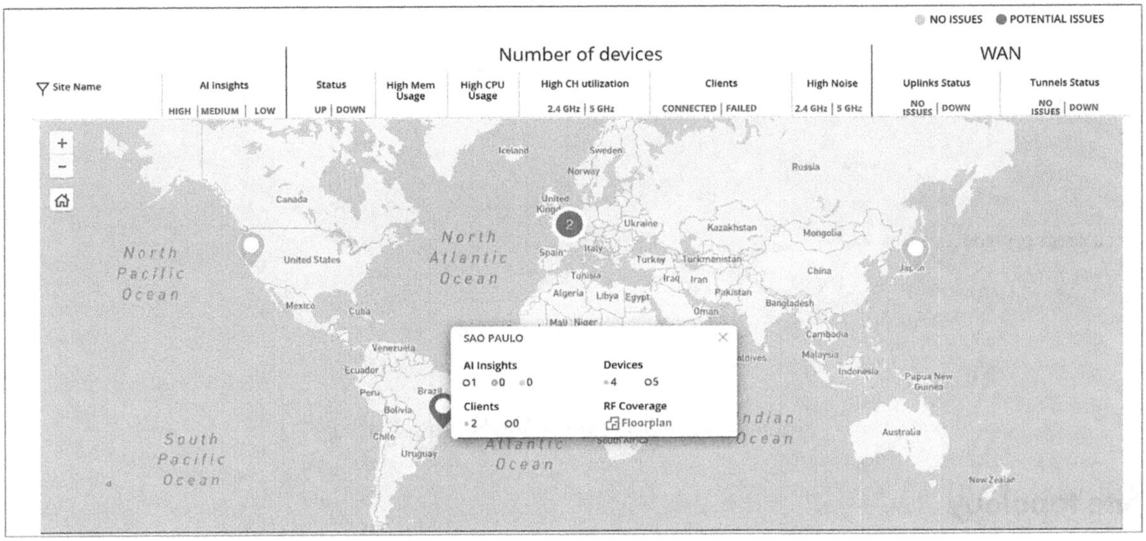

Figure 13-2: Overview—Network health

There is also a *List* view, which is not shown in Figure 13-2. With a list view, you get a global network report in a list that is sorted according to individual sites.

Wi-Fi connectivity

Aruba Central has a *Wi-Fi Connectivity* page (see Figure 13-3), formerly known as Clarity, where you can see an overall view of client connection details: those that are connected to or tried to connect to each connection phase. The connection phases include information concerning wireless *association*, *authentication*, *DHCP*, and *DNS*. You will explore this and other information during the lab activity.

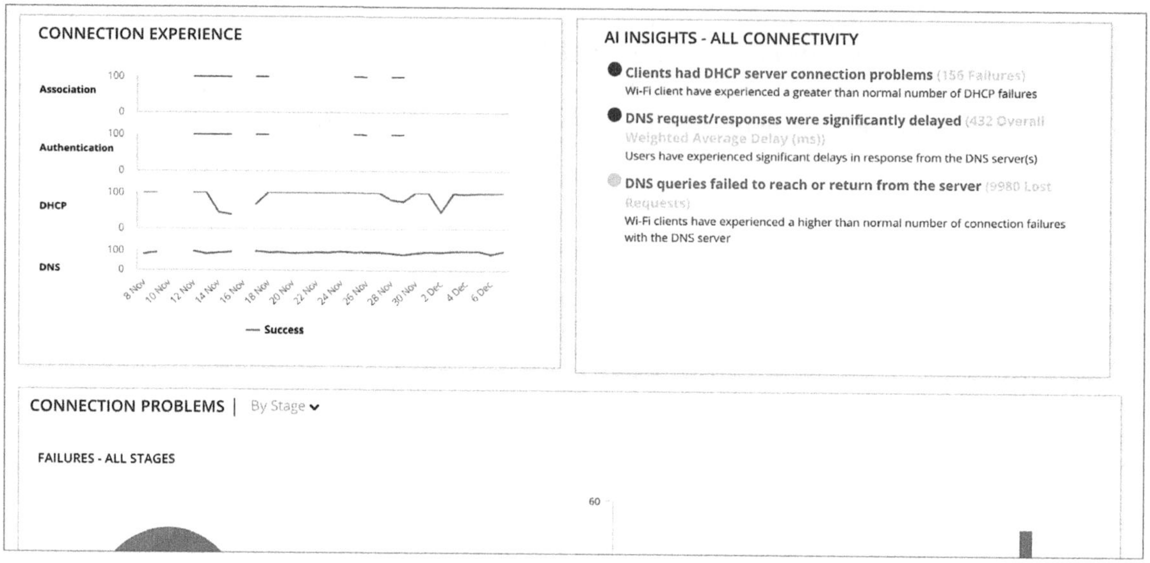

Figure 13-3: Overview—Wi-Fi connectivity

Site topology

The *Topology* pane gives you a topographical view of your network. The topology is created based on the LLDP neighbor list of the Aruba managed devices. You need to select a filter, based on labels and sites, to see the topology.

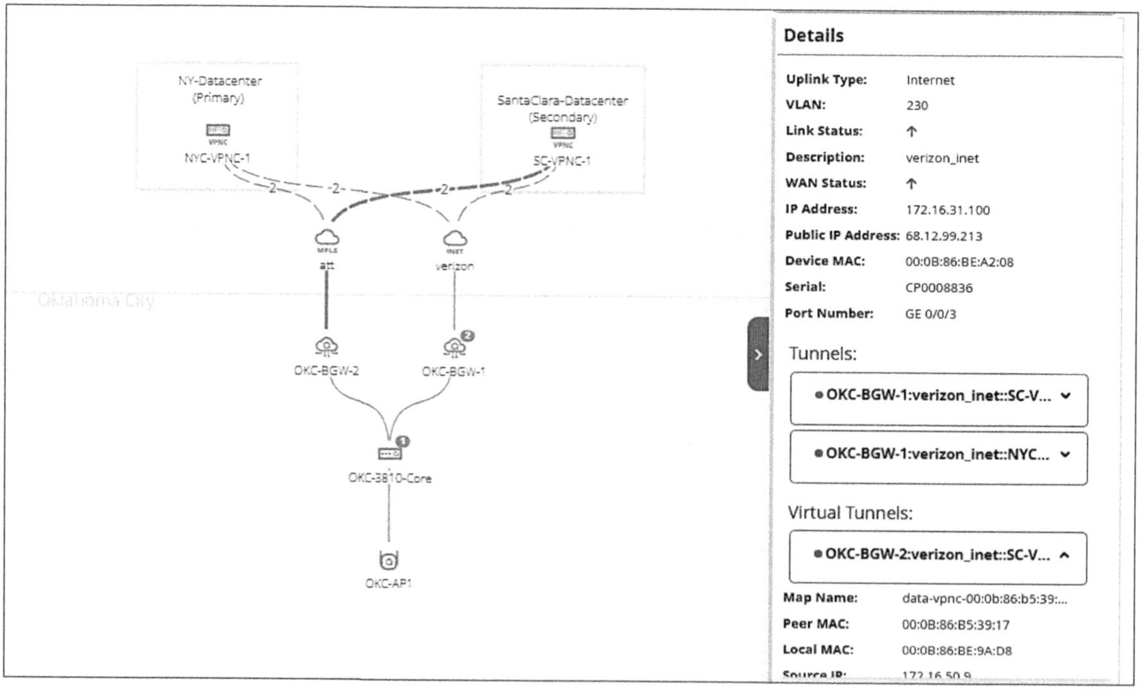

Figure 13-4: Site topology

In the example shown in Figure 13-4, you can see three IAPs and one switch. You can see the connections because the IAPs and switch are both at the Oklahoma City site. If the switch or IAP were at different sites or labels, you would only see the devices, but no connections.

If you select a device, then the device information is displayed on the right-hand side of the screen. You can use the filter option to display only the IAP, switches, gateways, or a combination of them.

Device list monitoring

The *Devices* page allows you to view managed devices. The devices are divided into their types: APs, switches, and gateways. By selecting the device type, a list of devices will be displayed, including online and offline devices. Pick a device to drill down and get more detail. The example shown in Figure 13-5 displays information for an AP. Of course, you can easily get information for any managed device. You will explore this in more depth during lab activity.

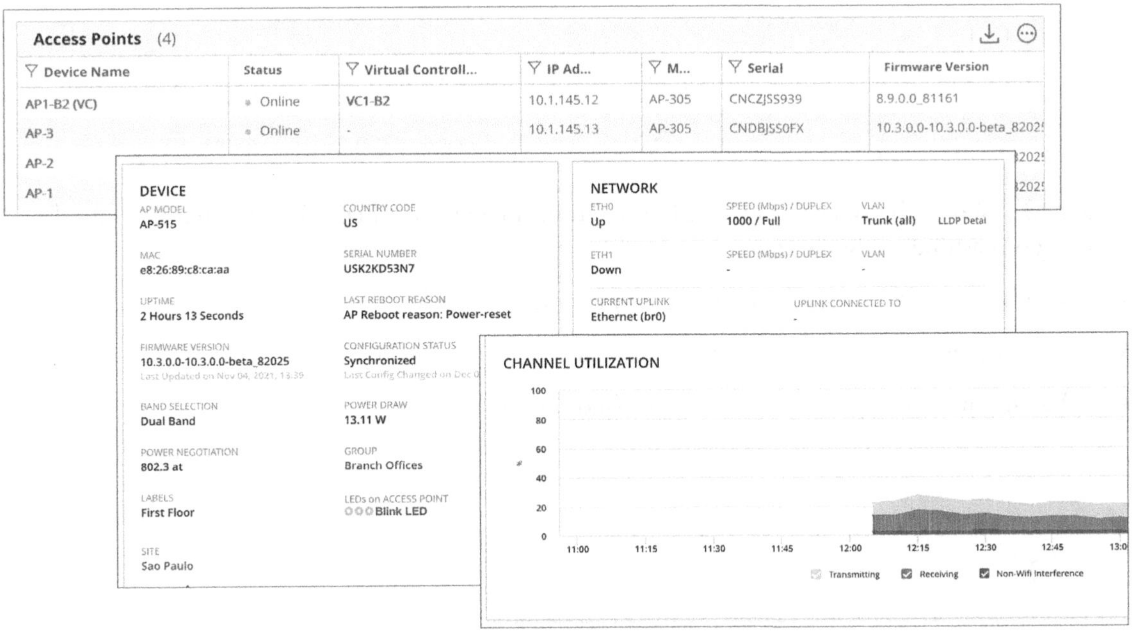

Figure 13-5: Device list monitoring

LED status

This section introduces you to the LEDs and their meanings for the Aruba APs and AOS-CX switches.

LEDs on APs

The LED displays located on the front panel of the access point indicate the system status of the AP. See Figure 13-6 for an example.

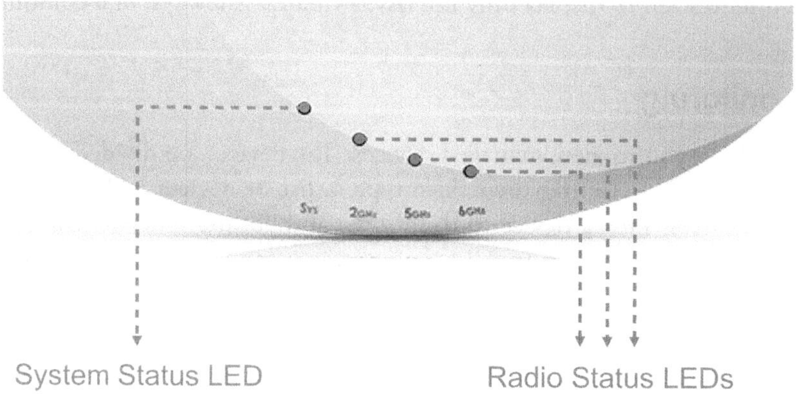

Figure 13-6: LEDs on APs

System status LED

Figure 13-7 shows the system status LED operation. Here's an explanation of what is referred to as blinking or flashing in the figure:

- **Blinking**: one second on, one second off, two-second cycle
- **Flashing off**: mostly on, fraction of a second off, two-second cycle
- **Flashing on**: mostly off, fraction of a second on, two-second cycle

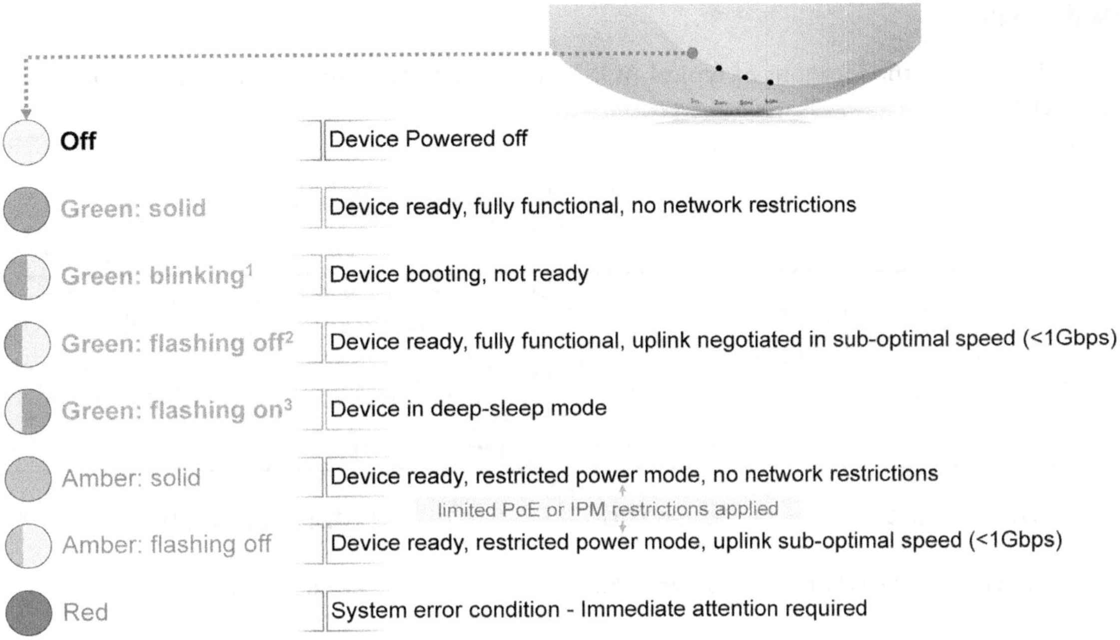

Figure 13-7: System status LED

Radio status LED

The radio status LED table shown in Figure 13-8 is applicable to 2 GHz, 5 GHz, and 6 GHz labels.

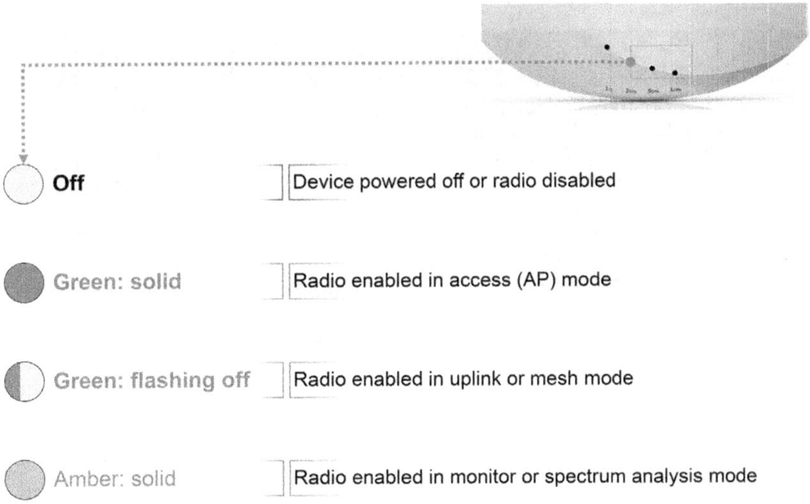

Figure 13-8: Radio status LED

Switch LEDs

Figure 13-9 shows the locations of typical AOS-CX switch LEDs. The meanings are covered in the next section.

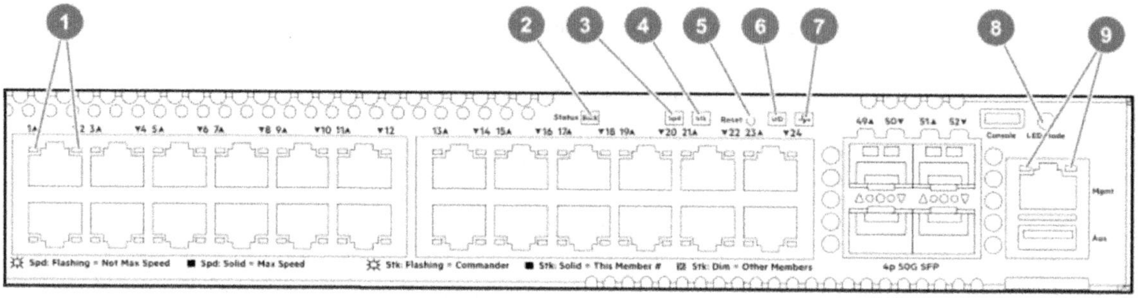

1 - Switch port LEDs
2 - Back Module status LED
3 - Speed mode selected LED
4 - Stk LED
5 - Reset button
6 - Unit Identification LED
7 - Global Status LED
8 - Mode select button
9 - OOBM port LED

Figure 13-9: Switch LEDs

Switch LED meanings

In Figure 13-10, you see operation meanings of the back and PoE LEDs

Switch LEDs	Function	State	Meaning
Back LED	Status of modular components installed in the back of the chassis (not applicable for 6300F switches	On - Green	Normal
		Slow Flash - Amber	Fault in one of the modules in the back of the chassis
PoE LED	Indicates Port LEDs are showing PoE information (not applicable for non PoE switches)	Off	PoE mode not selected
		On - Green	PoE mode selected
		Slow Flash - Amber	One or more ports experiencing PoE failure. PoE mode not selected.
		On - Amber	One or more ports experiencing PoE failure. PoE mode selected.

Figure 13-10: Back and PoE LEDs

The state of the switch port LEDs is controlled by the LED *Mode* select button. The current view mode is indicated by the mode LEDs next to the button (see Figure 13-11). To step from one view mode to the next, press the button to cycle through the different modes (speed or *Spd* and stacking or *Stk*).

Switch LEDs	Function	State	Meaning
Spd LED	Indicates Port LEDs are showing speed information	Off	Speed mode not selected
		On - Green	Speed mode selected
		Not Implemented	No fault defined
Stk LED	Indicates Port LEDs are showing stacking mode information	Off	Stacking mode not selected
		On - Green	Stacking mode selected
		On - Amber	One of the switch is experiencing stacking failure. Port LEDs will be used to indicate the number of the stacking member.
		Slow flash Amber	A port has a stacking failure. Stacking mode not selected

Figure 13-11: Back and PoE LEDs

Figure 13-12 shows UID and Global Status Indicator LEDs.

Switch LEDs	Function	State	Meaning
UID LED	Locator LED	Off	User configured the located LED : OFF
		On/Flash Blue (for 30 min)	User configured the locator LED: On/Flash
Global Status Indicator LED	Overall status of the product	Flash - Green	Self-test in progress during UBOOT and SVOS
		On - Green	Successfully booted AOS-CX
		Flash - Amber	Recoverable faults (e.g. fans, PSU fault)
		On - Amber	Critical faults (e.g. exceed temperature limit)
		Snoring (Dim - Bright periodically)	System is in hibernation

Figure 13-12: UID and Global Status LEDs

Switch reset buttons

The *Reset* button is recessed from the front panel; this design protects it from being pushed accidentally. The button is accessible through a small hole at the top of the front panel. Use a thin pointed object, such as an unbent paper clip, to press it. There are two types of resets: soft and hard. See Figure 13-13 for an explanation of these.

To Accomplish this:	Do this:	This will happen:
Soft reset	Press and release the Reset button.	The switch operating system is cleared gracefully. The switch then reboots and runs self-tests.
Hard reset	Press and hold the Reset button for more than 5 seconds, then release.	The switch reboots, similar to a power cycle. A hard reset is used, for example, when the switch CPU is in an unknown state or not responding.

Figure 13-13: Switch reset buttons

Firmware upgrades

You really should keep up with firmware upgrades and Aruba Central makes it easy. This section covers the topic of firmware upgrades.

Firmware management and upgrade overview

Aruba Central eases firmware management and upgrades. You can easily see the latest supported firmware version for your devices (see Figure 13-14). You can see the upgrade status for all the devices and if there is a new version available, with an option to upgrade all devices, or choose a specific device(s) to upgrade. Of course, you can choose from the available versions used for the upgrade.

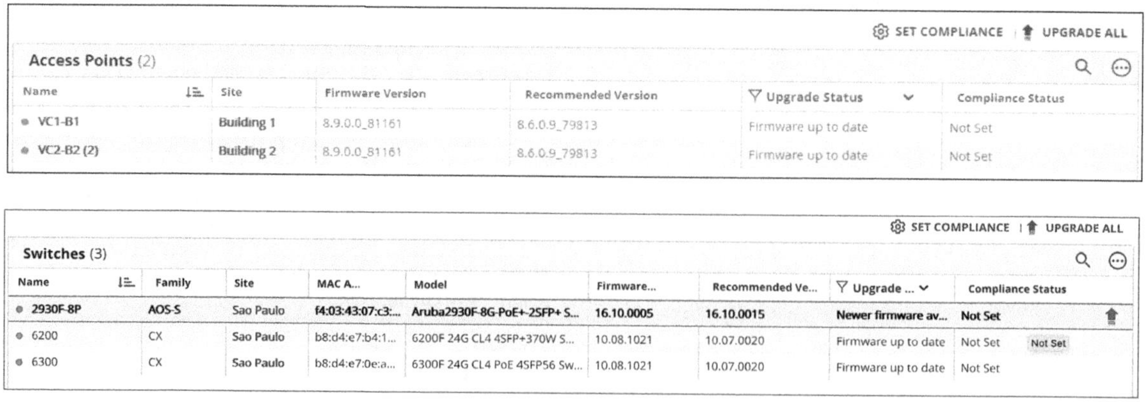

Figure 13-14: Firmware management and upgrade overview

You can also update the firmware now or schedule a time for the update. You can also update per site: updating one site without affecting the other sites.

AP firmware upgrade and compliance

When upgrading APs, there are two upgrade types (see Figure 13-15):

- A *standard* upgrade will upgrade all the selected APs simultaneously, which may disrupt connectivity. Therefore, Aruba does not recommend a standard upgrade in a production network during working or production hours. However, a good use case for standard upgrade is to upgrade APs in a maintenance window, since it will complete the upgrades faster than live upgrades.
- A *live* upgrade will upgrade APs based on AirMatch defined partitions, minimizing the impact in a production environment. The upcoming sections will explain the live upgrade process in detail. Notice in Figure 13-15 that the live upgrade feature requires advanced licenses applied on the APs.

Firmware compliance ensures that devices in a group all run a specific OS version. Choose either the APs (virtual controllers for IAPs), switches, or gateways. Firmware compliance also allows you to view a list of supported firmware versions for each device in a group.

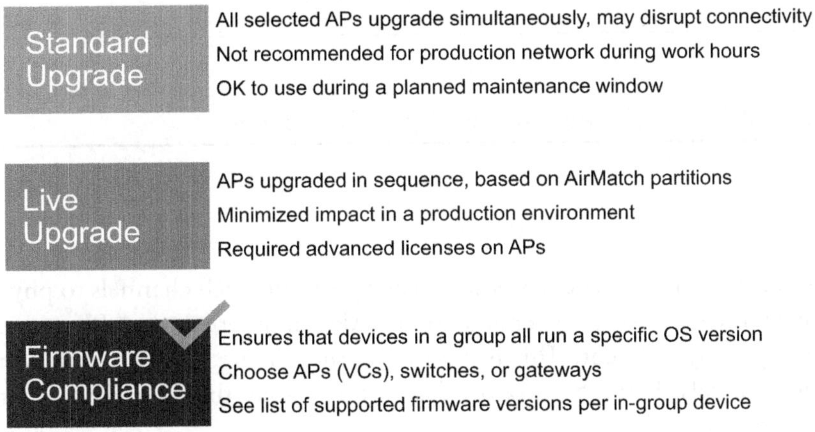

Figure 13-15: AP firmware and upgrade compliance

AP live upgrade

A live upgrade should not disrupt clients that are associated with APs. The live upgrade interfaces with the AirMatch service to partition the APs into smaller groups that will be upgraded in parallel (within the group). AirMatch partitions APs based on RF neighborhood data. APs partitions are used for clients to roam when the APs are undergoing an upgrade. A set of APs, known as *seed* APs, is selected (see Figure 13-16). Only these APs download images from the cloud; the remaining APs download images from their designated seed AP. This saves time and bandwidth since the image is only downloaded to the company once, and then distributed by the seed AP(s).

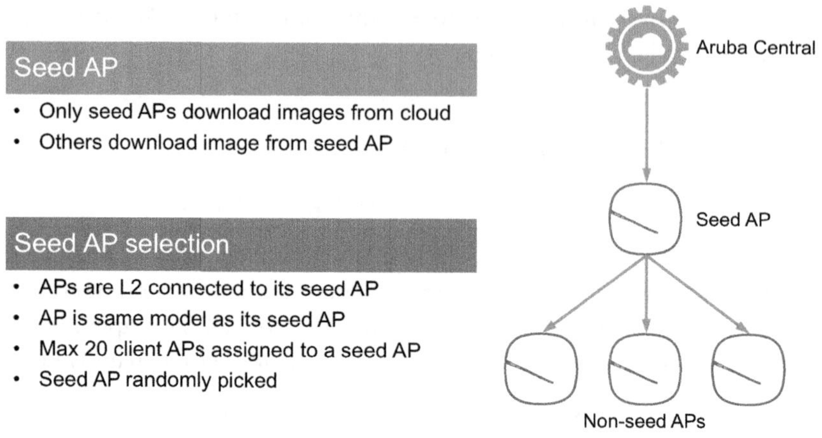

Figure 13-16: AP live upgrade

With seed AP group selection, the client AP is layer 2 connected to its seed AP. The client AP must be the same model as its seed AP and no more than 20 client APs are assigned to a seed AP. Non-seed APs use HTTPs to download the image from the seed APs.

Note

The seed APs are selected randomly by Aruba Central.

AirMatch partitions and live upgrade

AirMatch optimizes the RF coverage patterns and assigns unique RF channels to physically adjacent APs. This optimizes the RF coverage overlap between APs. In other words, APs set to the same channel will not be physically adjacent. This is especially true for the 5 GHz band, which has several non-overlapping channels. If the 5 GHz band is not available, then AirMatch uses the 2.4 GHz band (see Figure 13-17).

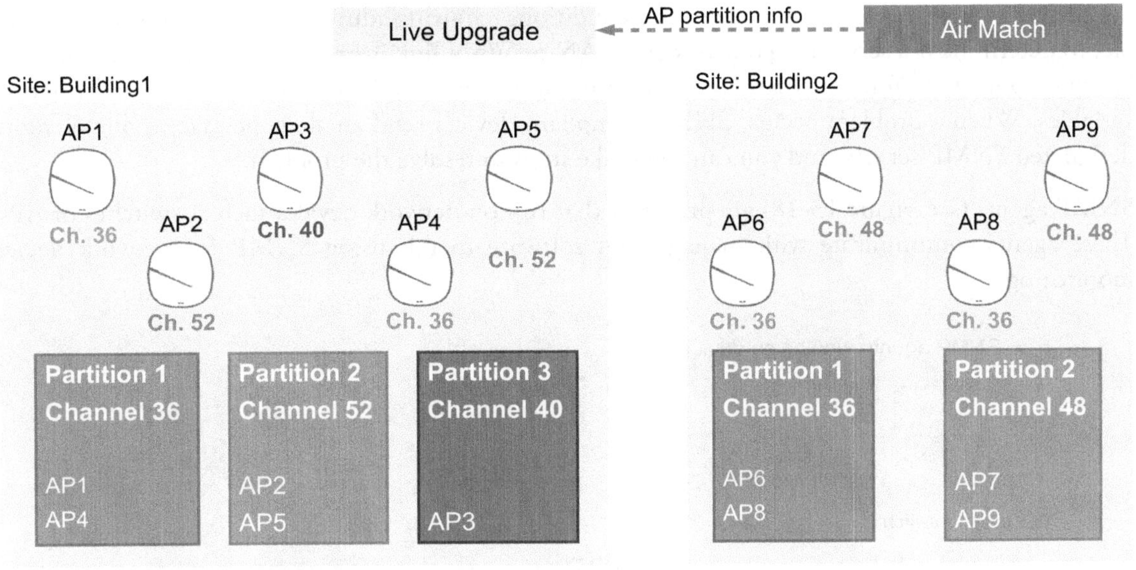

Figure 13-17: AirMatch partitions and live upgrade

AirMatch partitions APs into groups, based on their AirMatch-assigned 5 GHz channels. In Figure 13-17, AP1 and AP4 are both on channel 36, and assigned to Partition 1. AP2 and AP5 are on channel 52 and assigned to Partition 2. AP 3 is on channel 40 and is assigned to Partition 3.

If all APs in a partition reboot simultaneously, the surrounding APs (in other partitions) will provide the necessary RF coverage (they'll automatically increase their RF power output for their radio). Of course, this assumes a very healthy RF coverage pattern with ample overlap; in other words, you need to implement a capacity, and not a coverage, design.

Sites can also make partitions. Therefore, APs in the same RF channel, but in different sites, would belong to separate partitions. For this reason, it is important to properly select your sites. If you have a campus with six buildings, all in one site, then all APs in one RF channel would belong to one partition. If each building is its own site, then the AP partitions on the same RF channel would be limited to the building.

SNMP

This section examines the open-standard Simple Network Management Protocol (SNMP).

SNMP overview

SNMP is an industry-standard protocol that allows you to manage and monitor a variety of network devices from a central location. Using an SNMP application, you can configure network devices and apply consistent security and management policies to all the devices on your network. You can also use SNMP to monitor compliant devices for conditions requiring administrative attention: for example, to determine if a server has gone down, a LAN or WAN link is no longer available, or a link has become congested. You can track device uptime, link states, and many other device information variables. When a problem occurs, SNMP compliant devices send an alert message to one or more designated SNMP servers, and you can then take steps to resolve the problem.

SNMP agents (see Figure 13-18) are processes that run on network devices such as switches or APs. These agents communicate with management software that is using SNMP for network device monitoring.

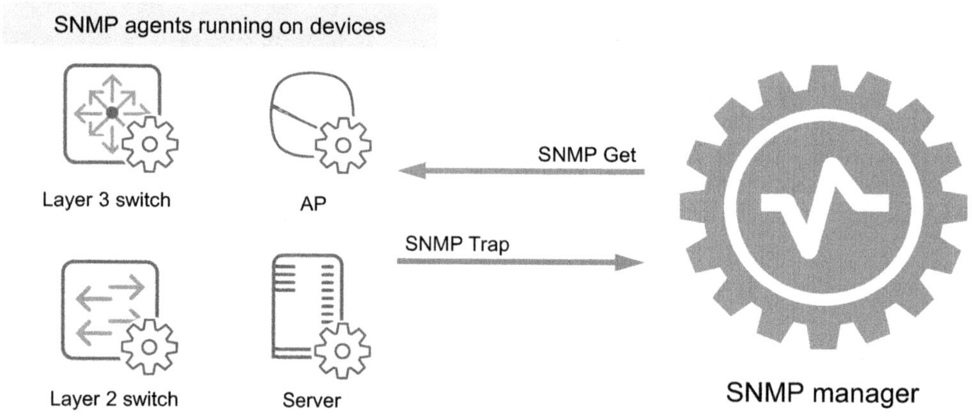

Figure 13-18: SNMP overview

SNMP MIBs

SNMP management information bases (MIBs) are the heart of SNMP and allow the storing and sharing of configuration and event-related information. This section introduces MIBs and how SNMP managers and devices interact with each other using the MIB information. Use Figure 13-19 for reference.

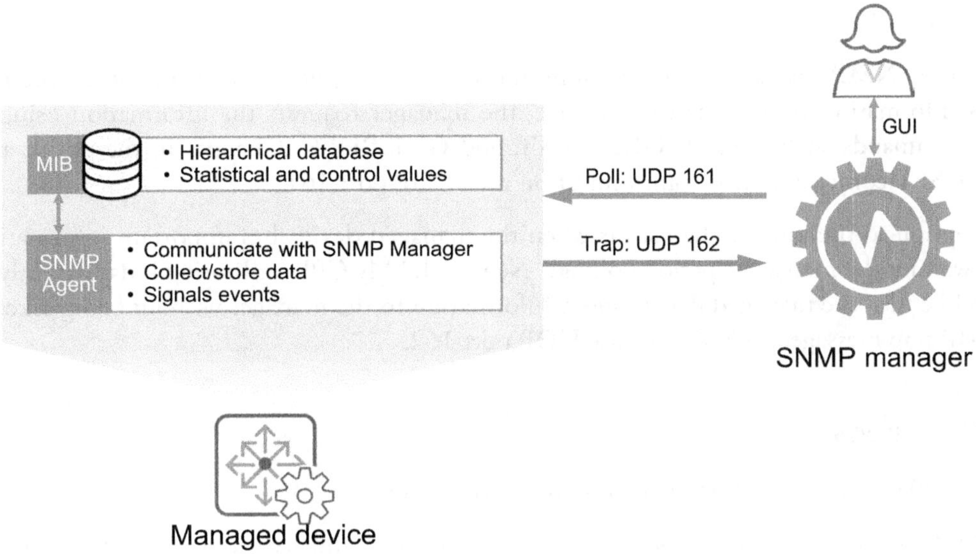

Figure 13-19: SNMP manager, managed devices, and MIBs

SNMP manager

The SNMP manager (commonly referred to as a *management station*) is typically a dedicated server that communicates with SNMP the SNMP managed device (commonly called an *agent*). The SNMP manager often has a GUI interface, so administrators can view and change parameters for many network devices from a single interface. The manager uses several message types to gather and modify network device parameters. Some of its key functions include:

- Queries agents
- Gets responses from agents
- Sets variables in agents
- Acknowledges asynchronous events from agents

Managed devices

The routers, switches, firewalls, wireless gateways, access points, and any other network devices that need to be centrally monitored, managed, and configured are referred to as managed devices or agents. Internally, these managed devices have two main SNMP components.

- The *SNMP agent* collects information and communicates with the SNMP manager about device configuration and status. It stores and retrieves this information in a MIB or MIBs.
- The *MIB* is a hierarchical database that stores device status, statistics, and configuration information. This is the information that is exchanged between the SNMP manager and the managed device.

Communication and ports

Usually, the SNMP manager-to-agent communication uses a *pull* mechanism. When the manager requires information from a managed device, the manager requests the information using simple SNMP commands such as GET, GET NEXT, and GET BULK. Changes in the device are done using the SET command. This communication uses UDP port 161.

The exception to the pull mechanism is when the managed device has important information that cannot wait for the manager's period polling: "Newsflash!! My CPU utilization is dangerously high!". In cases like this, the managed device *pushes* information to the manager without being asked, using an SNMP trap message. SNMP traps use UDP port 162.

SNMP versions

Figure 13-20 summarizes SNMP versions, as described here:

- **SNMP version 1**: First version of the SNMP protocol and uses a community-based security mechanism. The *community string* is like a simple pre-shared passcode. If the managed device has a community string of *public*, then any SNMP manager that can reach the device can access the agent's MIB, as long as it sets its community string to *public*.
- **SNMP version 2c**. This is the revised protocol, which includes enhancements to SNMPv1 in the areas of protocol packet types, transport mapping, and MIB structure elements. This version also uses the community-based security mechanism.
- **SNMP version 3**. Provides a higher level of security, including message integrity, authentication, and encryption. SNMPv3 is recommended.

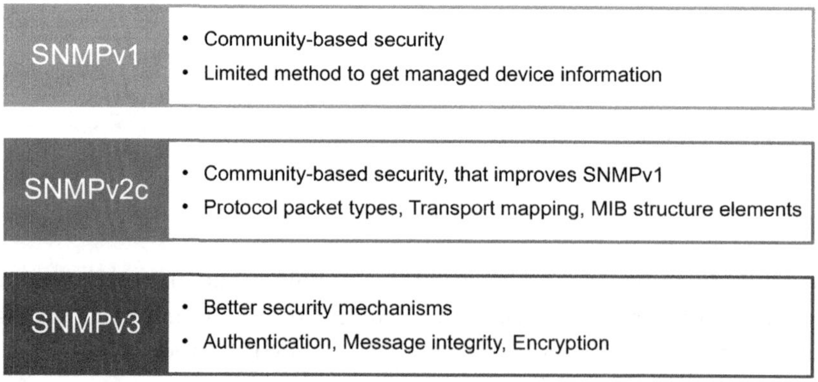

Figure 13-20: SNMP versions

Note
Because SNMPv1 and v2c send the community strings in clear text, it is *highly* recommended not to use them; instead, SNMPv3 should be used.

AI Insights

This section introduces the AI Insights feature available with Aruba Central: you will be impressed with what AI Insights can do for you.

Aruba Central AIOps portfolio

Aruba Central includes a full-service AI operations (AIOps) solution that automates common troubleshooting activities by reducing IT support tickets and their associated costs. Aruba's AIOps solution is based on AI machine learning models that are consistently trained with network performance data collected from tens of thousands of Aruba customers across every vertical, market segment, and network type.

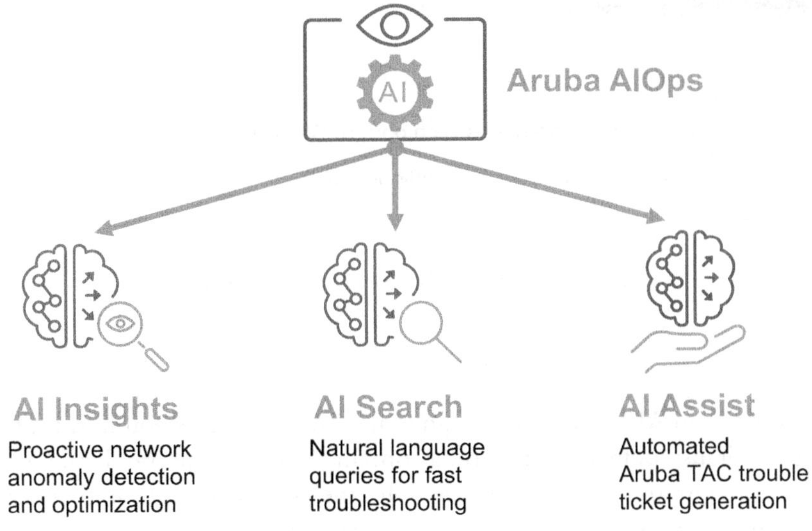

Figure 13-21: Aruba Central AIOps portfolio

The AIOps solution's core components include three important features (see Figure 13-21):

- *AI Insights* which automatically diagnose an array of common network-impacting issues by using dynamic, per-site baselines that are continuously tuned as conditions change: no manual setup or service level threshold adjustments are required by the customer. Built-in anomaly detection highlights the severity and impact of issues as they occur, helping IT pinpoint root cause and proper remediation steps with 95% accuracy!

- *AI Search* is a natural language processing (NLP) engine of the Aruba Central that points to solution guides, troubleshooting tips, and much more. Whether operators are looking for best practices on device configurations or need to isolate a problem impacting a specific user, AI Search provides fast, interactive responses that simplify day zero to day two operations.
- *AI Assist* uses event-driven automation to collect diagnostics, posts them to a shared location, and even notifies Aruba TAC for proactive customer support.

What AIOps is and isn't

AIOps doesn't replace your management solution: it augments it (see Figure 13-22). Aruba Central was designed with AIOps being part of it. Traditional configuration, management features, troubleshooting, and reporting tools are still there. The difference is that AIOps uses a wealth of telemetry information and data to find ongoing patterns that point to potential problems.

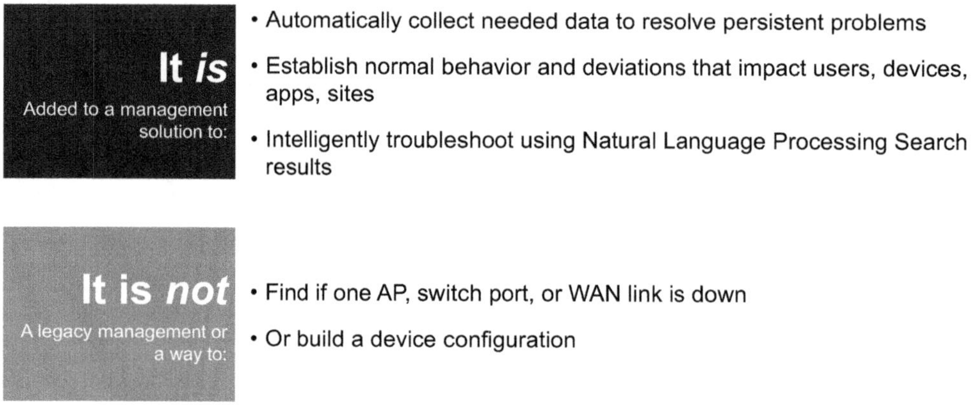

Figure 13-22: What AIOps is and isn't

AI Search

The AI Search feature is an AI-based search engine that not only helps you find problems that users are experiencing, but it also points you to Aruba content related to the problem, and where you can make changes in a configuration to fix the problem. It's the only NLP solution that responds with day zero, day one, and post deployment tips in an interactive way.

You can use keyword searches using full or partial names and addresses. Figure 13-23 illustrates an example where a search was made for IP addresses containing "192.168". AI Search then lists, in sections the devices. In the example below you can see that there are both clients and APs that have that IP address.

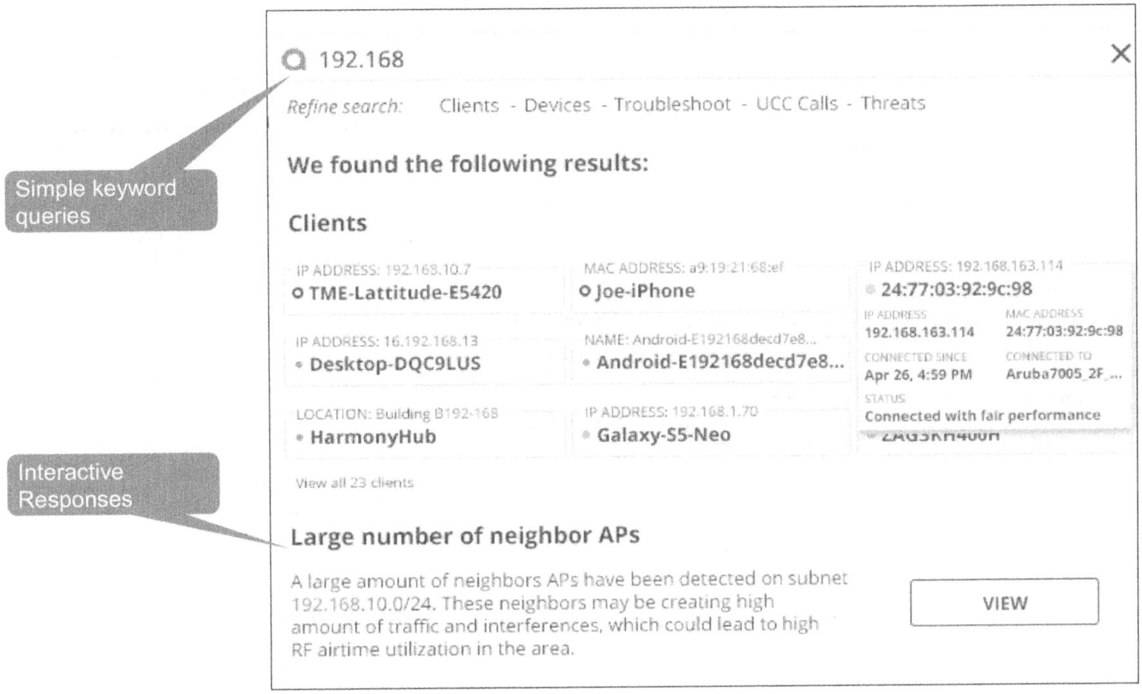

Figure 13-23: AI Search

Creating a WLAN example

AI Search provides fast, interactive responses that simplify operations from day-zero (provisioning the network) to day-two (maintaining the network). The example in Figure 13-24 shows a WLAN creation search. Notice that the NLP engine suggests many topics related to the search, including a link to the WLAN configuration page, a guided configuration tutorial, and related articles.

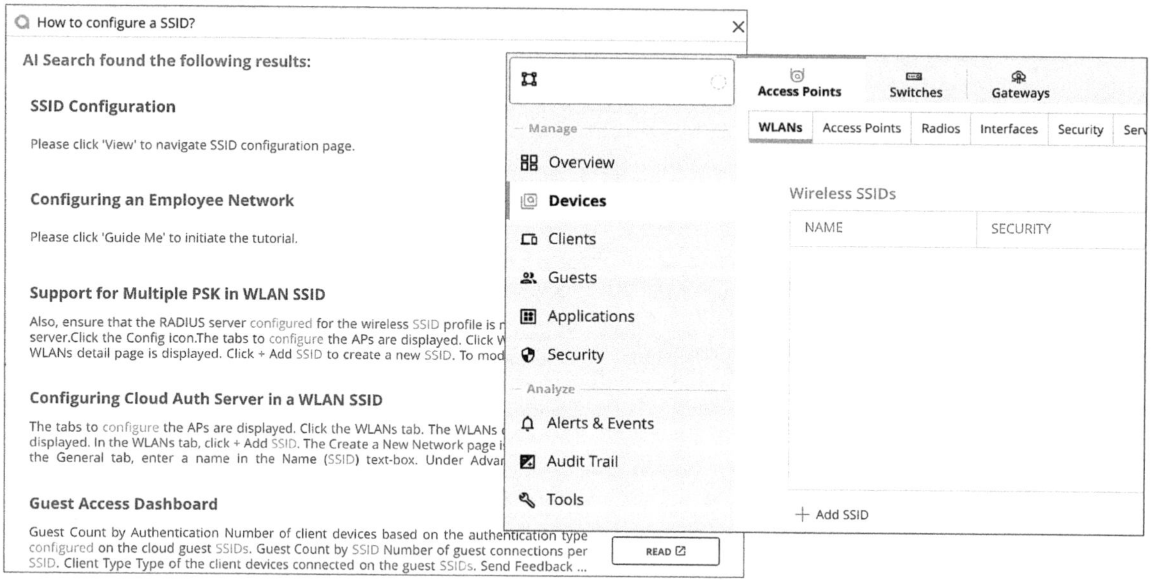

Figure 13-24: AI Search: Creating a WLAN example

AP example

One example of the AI Search feature is to search managed devices. In Figure 13-25, a search for APs returned all the APs the Aruba Central customer manages. The AP box expands by hovering the mouse over an AP, showing the AP status, name, model, and other device information. At the bottom, you will find additional actions like device configuration, accessing other Central pages to perform additional actions, like the *Network Check* action button to perform troubleshooting or the Clients button to access client page of clients connected to the AP, and many other actions.

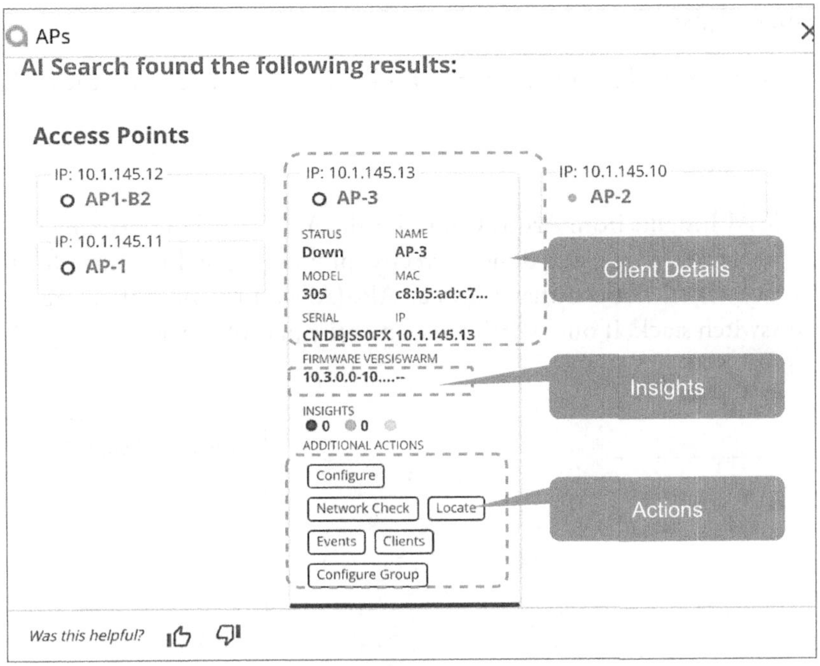

Figure 13-25: AI Search: AP example

Client search example

Figure 13-26 shows the results for an AI search for a client or clients. You can search for clients based on partial or full names, IP addresses, and MAC addresses.

Figure 13-26: AI Search: Client search

AI Insights overview

This section introduces a few of the many different ways that you can leverage the AI Insights feature.

AP to stack mapping example

One of the possible AI Insights from Aruba Central is the AP-to-stacking map. Figure 13-27 displays a site where the wireless coverage was done correctly, providing good coverage for the entire floor. However, Aruba Central AI noticed that *neighbor* APs (APs on the floorplan) are connected to the *same* switch in the switch stack. If one switch fails, an entire part of the floor may get low or no Wi-Fi coverage!

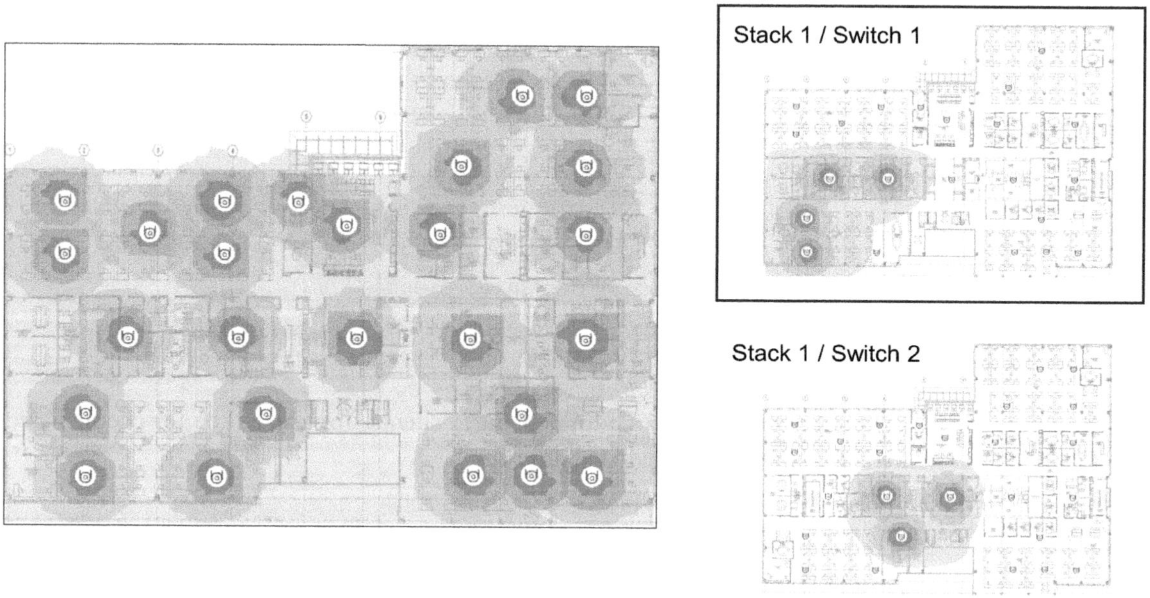

Figure 13-27: AI Insight—AP-to-stack mapping example

Per-stack RF resiliency example

Following the AI insight and physically moving APs connections from one switch to another will provide better coverage in case of failure or a reboot of one switch in the stack. See Figure 13-28 as an example.

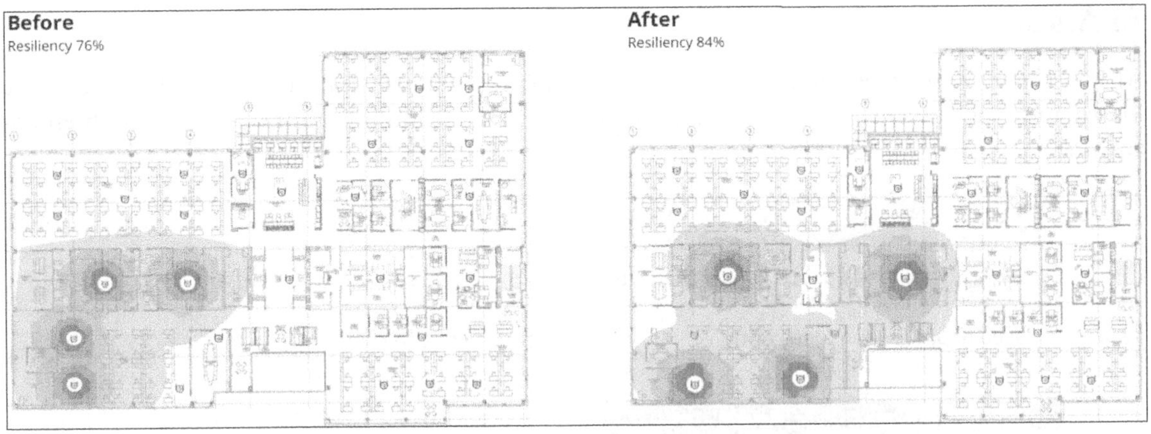

Figure 13-28: AI Insight—Per-stack RF resiliency example

DNS loss pattern clustering

With DNS loss pattern, Insight's goal is to attribute DNS losses to specific *root causes* by looking for patterns. *Patterns* are defined as a combination that caused most of the losses for the user(s) or customer. Figure 13-29 shows that 100% of the DNS losses are affecting four sites, where two servers are responsible for all the losses, allowing network administrators to quickly drill down and find the root cause of the issue.

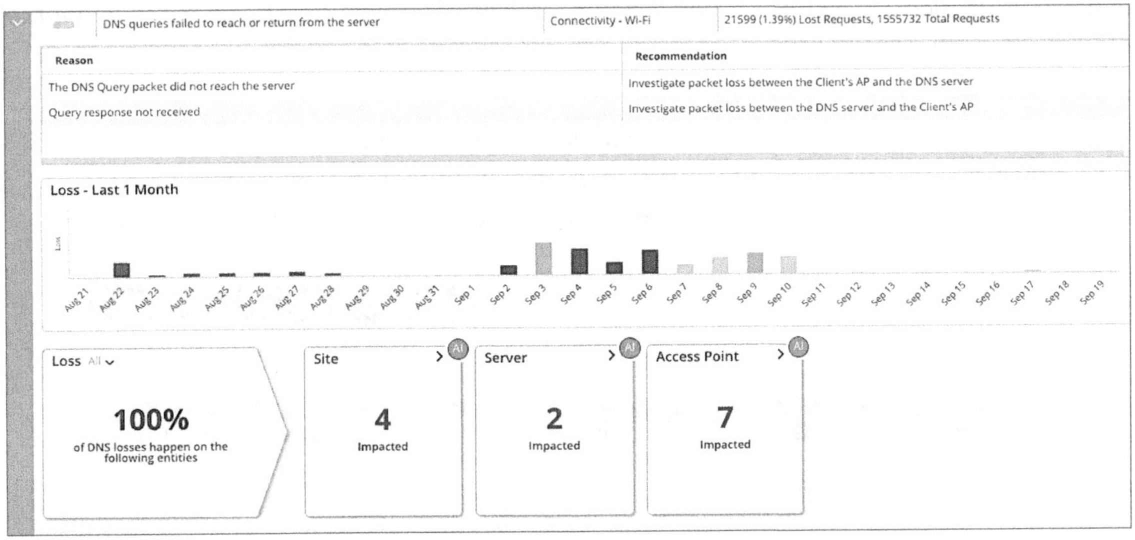

Figure 13-29: AI Insight—DNS loss pattern clustering

AI Assist

AI Assist uses event-driven processes to initiate automation that will automatically collect diagnostic information such as logs, packet captures, and etcetera, that an administrator can use to troubleshoot an issue.

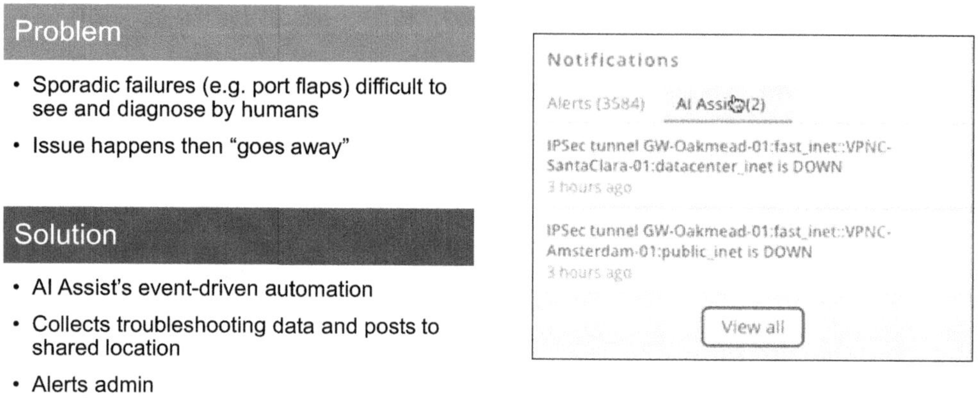

Figure 13-30: AI Assist

Here's how it works (see Figure 13-30). When an issue, such as a port flap (port constantly goes up and down) occurs, AI Assist collects diagnostic information and alerts the administrator via the *alerts* icon in the upper right of the browser window. When the administrator clicks this icon, they can see all the AI Assist alerts, including the port-flapping problem. In Figure 13-31, you see an example of an IPsec tunnel that has an issue.

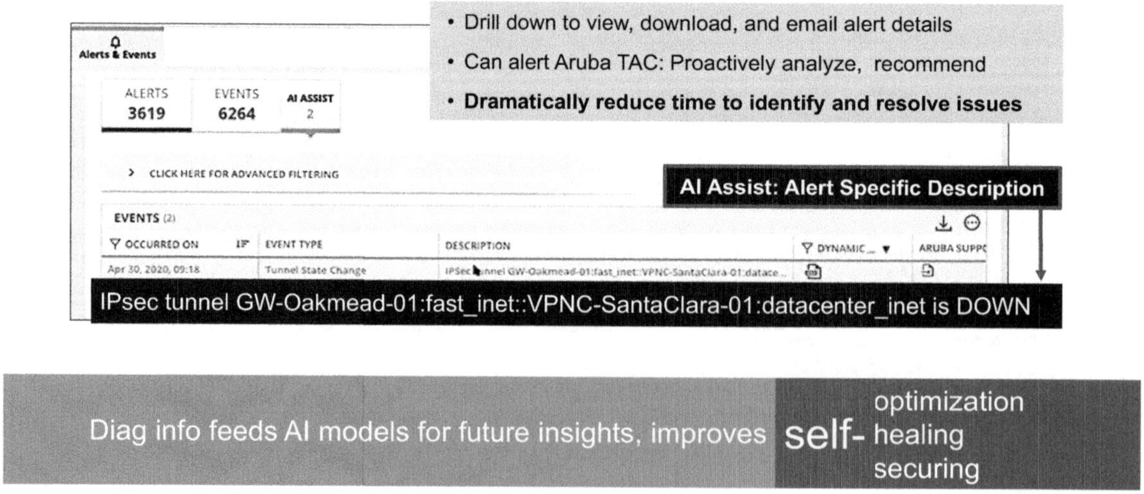

Figure 13-31: AI Assist example

Aruba Central licenses for AIOps

Aruba Central advanced licenses include all AIOps features plus the features in the foundation license, and much more (see Figure 13-32). Keep in mind that new features are added constantly; to check the latest license and feature support, navigate to the Aruba Central Documentation Center and search for licenses.

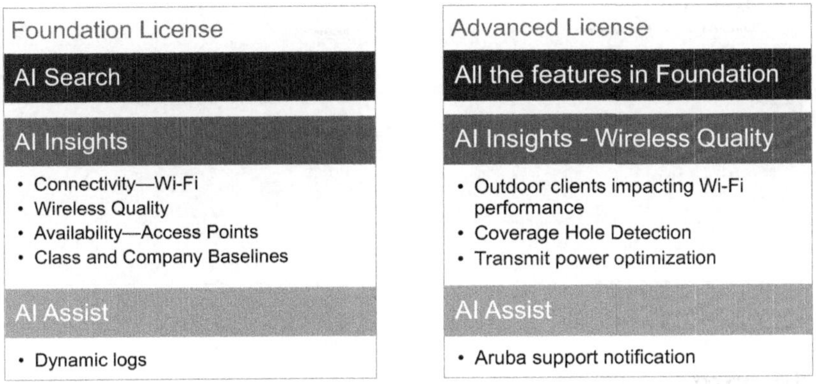

Figure 13-32: Aruba Central licenses for AIOps

Alerts and reports

This section covers the alerting and reporting features available to customers in Aruba Central.

Alerts & Events dashboard

A good starting point for any troubleshooting is the *Alerts & Events* dashboard. This dashboard displays a list of alerts and events generated concerning device provisioning, configuration, and user management (see Figure 13-33). You can view the alerts and events in the *List* view and *Summary* view. You can use the *Configuration* view to see configuration alerts, but it is only available in the Global context. The *List* view components (the columns displayed) are different for the Alerts and Events tab, whereas the Summary view displays similar components.

You can view alert details as well as acknowledge alerts. Alerts are acknowledged automatically when the event count drops below the lowest severity threshold configured for the alert. Users with administrator access can acknowledge alerts irrespective of the severity configuration. As manually acknowledging an alert does not reset the count data, the alert service continues to aggregate events. When the number of new events meets the configured threshold, an alert is triggered again.

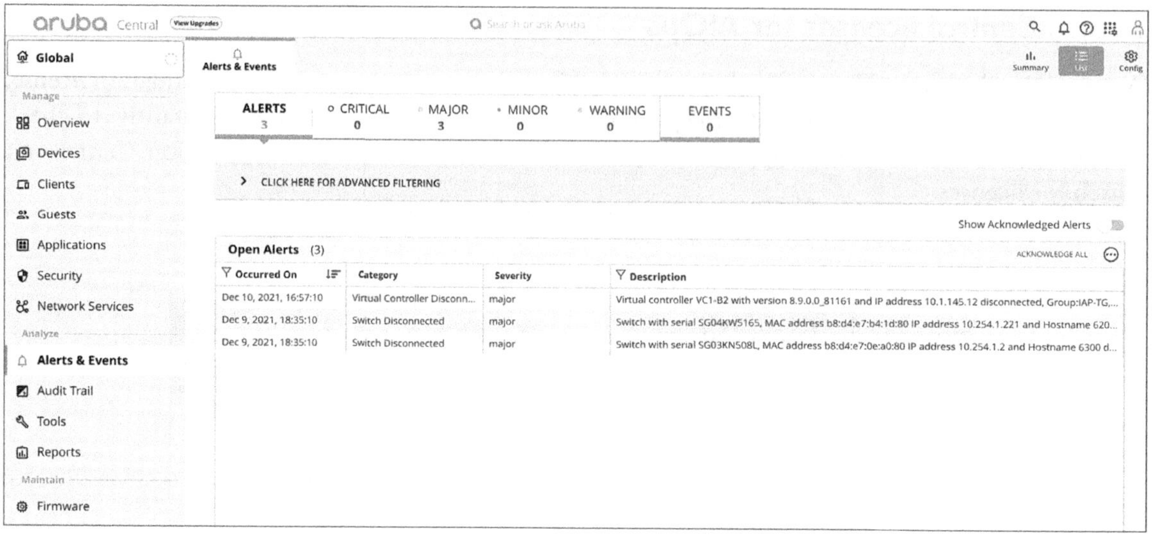

Figure 13-33: Alerts & Events dashboard

Reports dashboard

The *Reports* page allows you to create various reports which you can configure to run on demand or periodically (scheduled). Figure 13-34 shows an example of this dashboard. You must have read/write privileges or be an administrator user to create reports. The reports page includes the *browse* reports option, which displays the configured reports. The *generated* reports option displays the generated reports.

There are 24 reports in five categories from which you can choose, along with the time and date range for the report. Note that new releases of Central could introduce new reports. You can generate the report immediately or schedule it to repeat daily, weekly, or monthly. You can view the report from the Central GUI, download the report via a PDF file, or email the report.

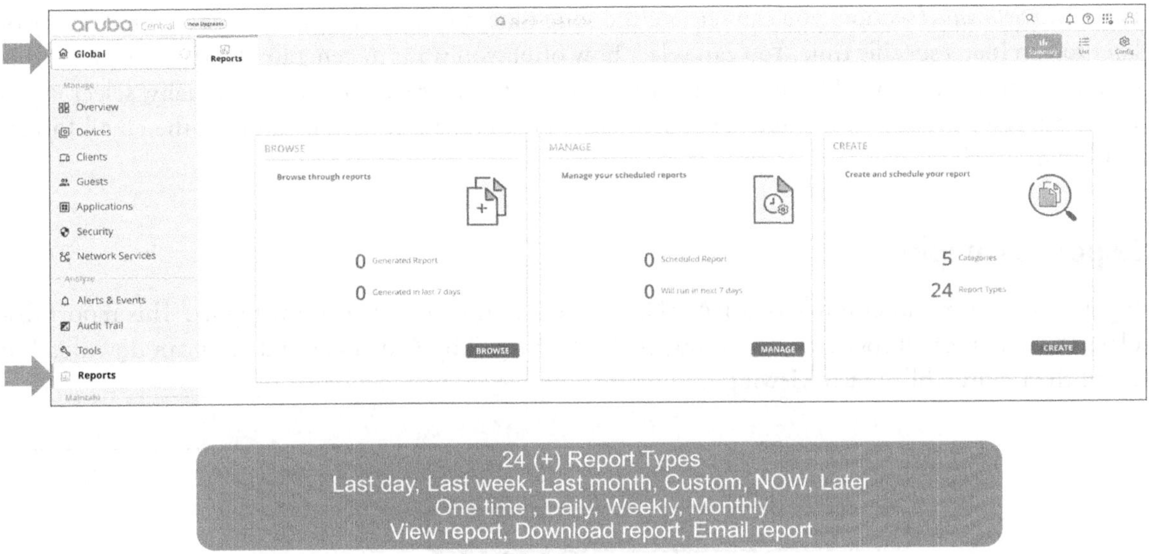

Figure 13-34: Reports dashboard

Report creation

To create a report, select from one of the five categories (see Figure 13-35), then click **Next** to navigate to the contexts. Here you can select groups, labels, or a specific site. You can then select the report to be the last day, the last 7 days, the last 30 days, or customize the date range.

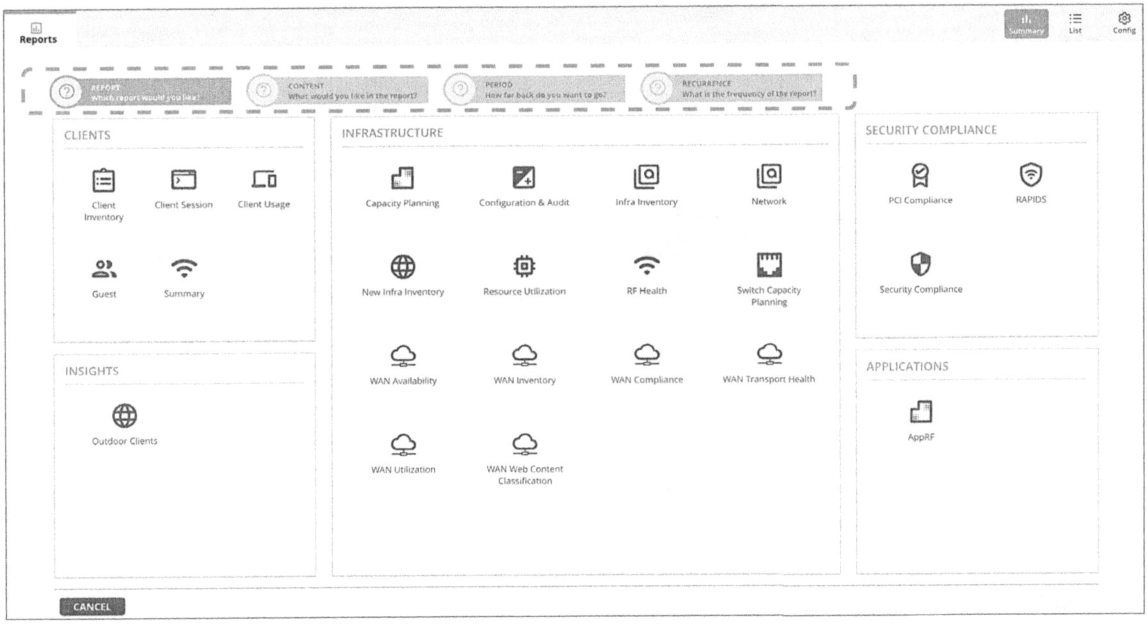

Figure 13-35: Report creation

From the *Scheduled* section, you can select either one time now to generate the report immediately or later to run it at a specific time. You can select how often you want to generate the report by choosing either one-time, daily, weekly, or monthly intervals. Enter the report name. Optionally, select if you want to receive the report via email and then enter an email address. You can pick the email format as PDF or CSV.

Report example

Figure 13-36 show an example of a RAPIDs (rogue AP intrusion detection) report. This report displays the details of all rogue or interfering devices managed by Aruba Central. You can download or email the report while you're viewing it.

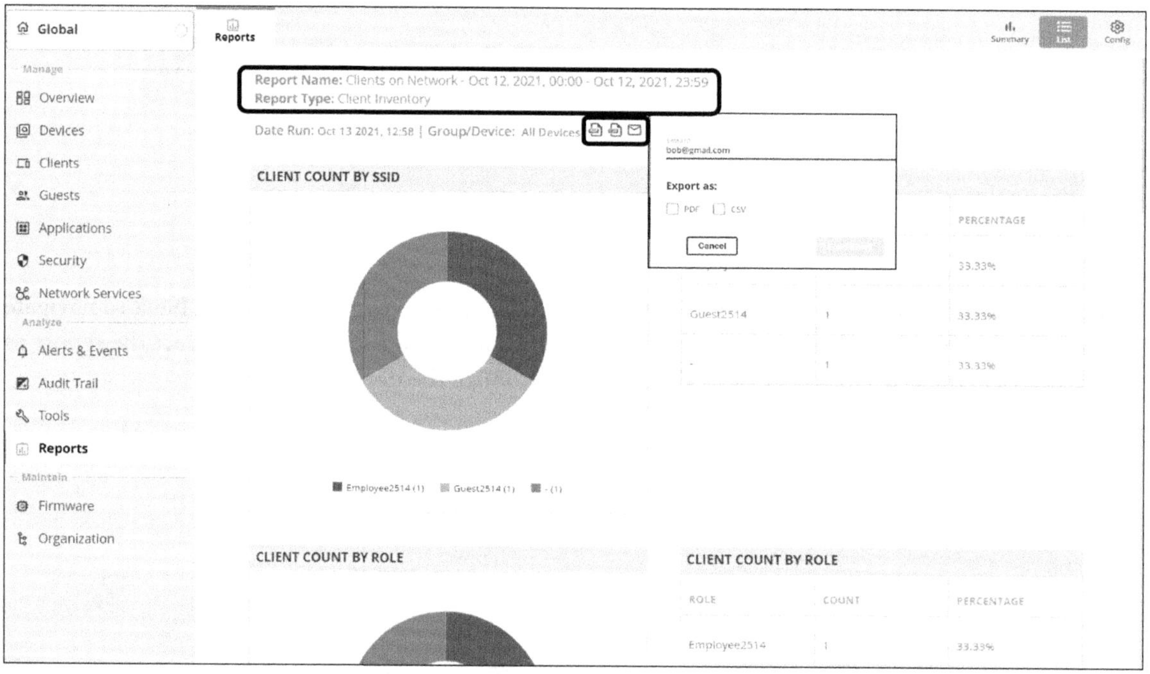

Figure 13-36: Report creation

User eXperience Insight (UXI)

This User eXperience Insights (UXI) feature set can really improve your situational awareness when managing and monitoring devices with Aruba Central. This section introduces the UXI feature.

What is UXI?

Where do IT teams start in a world of generic user statements like "the Wi-Fi is broken again" or "I've been struggling to get my app to work for the past week"? (See Figure 13-37.) There could be a myriad of things that could create these kinds of problems. What makes it more difficult is that people do not complain as soon as they have a bad experience using a device, or application on the network. Instead, they tend to wait until multiple instances of the issue have happened, or they happen to bump into someone else having a similar issue. Or sometimes, as a result of a missed goal, they will blame poor performance of the network or device they are using.

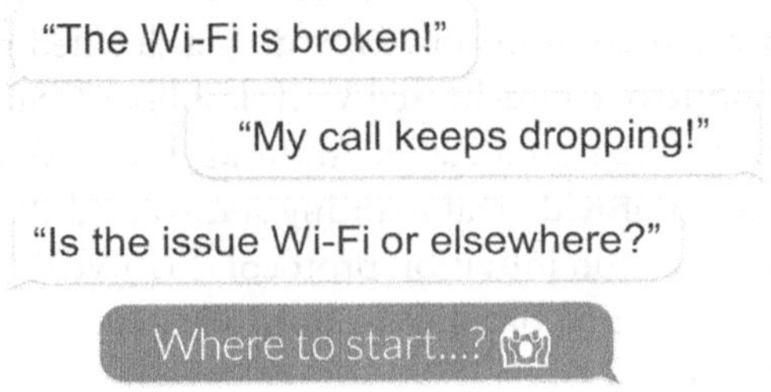

Figure 13-37: Where to start

All is not lost: UXI was designed to help assist you with this process!

UXI components

UXI consists of front-end sensors and agents and a back-end cloud-based UI (see Figure 13-38).

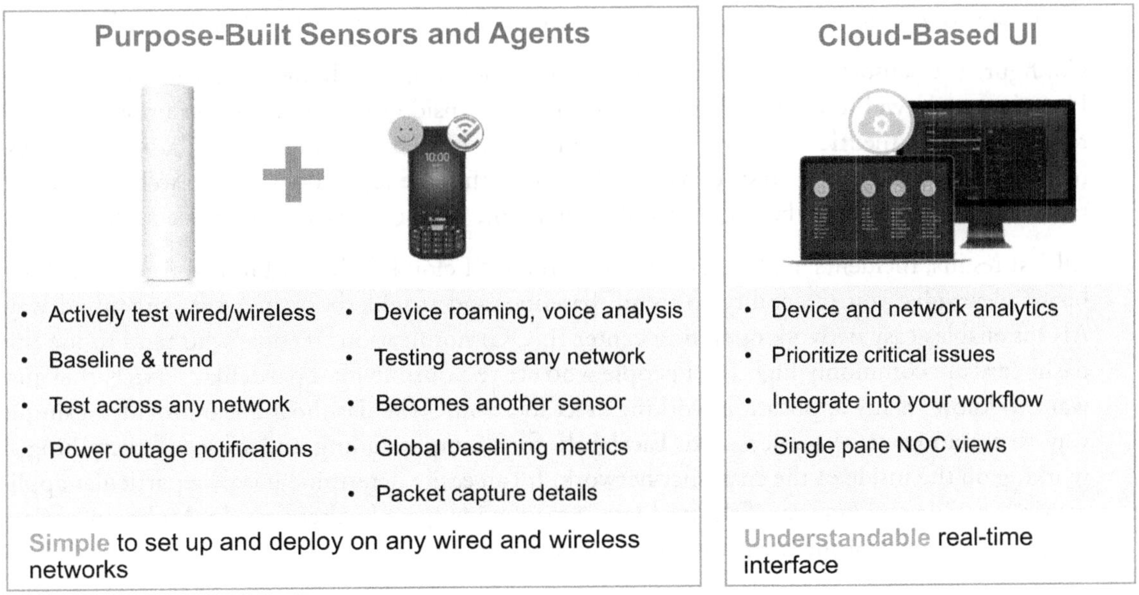

Figure 13-38: UXI components

The sensors act like an end user, creating traffic typical of an end user. Figure 13-38 summarizes the sensor and agent features and advantages.

The cloud-based system provides real-time monitoring and analytic intelligence, as summarized in the figure.

UXI process

Aruba UXI leverages a cloud-hosted dashboard and physical sensors deployed onsite. Together, they function as your remote technician. The UXI *sensors* onboard onto your network as an end user and actively monitor wireless and wired connections, providing a detailed performance analysis from an edge perspective via a highly intuitive cloud-hosted UXI dashboard. Similar to the sensor, Aruba has sensor software which can be installed on a Zebra Android device. Aruba has integrated UXI with Zebra's Wireless Insights engine to provide real roaming statistics (did it roam?, why did it roam?, or why did it fail to roam?) and true session initiation protocol (SIP) VoIP call measurement can provide phone call metrics, like if a call dropped or who dropped the call. In other words, Aruba's UXI can measure the end-to-end VoIP call.

The steps involved with UXI are explained here (see Figure 13-39):

1. Sensors onboard via cellular or standard PoE. You just plug in the sensor and it looks for the network it's supposed to test. As soon as it finds that network, the cellular function (if being used) turns off.

2. Just like a normal user, the sensor associates to a network, goes through all of the same association, DHCP, DNS, gateway (DORA) process. If there is a captive portal involved, the sensor will log in and log out of that captive portal. UXI measures load, and redirect timings (and success) of the captive portal. (Note that the agent just onboards a wireless network, not wired.)

3. Configure the sensor/agent to test services and applications which are critical to that location. First the sensor will test internal applications that are inside the firewall; for example, it might test printer or authentication server access. In the case of Zebra, it might test a basic ping access to a web server or voice analysis of VoIP calls. (Note that the agent does not do web application tests, or throughput tests, because it would impact the performance of the device itself.)

4. All test results, incidents, and alerts are sent to the UXI cloud dashboard in real-time. The dashboard views give you the ability to trend, baseline, and troubleshoot in a very organized way. AIOps enables easy network operation center (NOC) notifications. People who tend to use this dashboard are commonly high-level people who are responsible for a particular service: they just want to know "is my application working at xx location?" The dashboard also provides a simple way to see an external services, like Facebook, for example, failing and to see if everything is working on the inside of the customer network. It can easily determine if only a particular application is having an outage or if the problem is more wide spread. Alternatively, Aruba can stream all data, alerts, and/or incidents to a container of your choice.

Along the way the device assesses performance of wired and/or wireless networks, the steps to get onto a network, execute transactions, exchange data, and etcetera. Every step of the way, every metric that's possible to capture is captured.

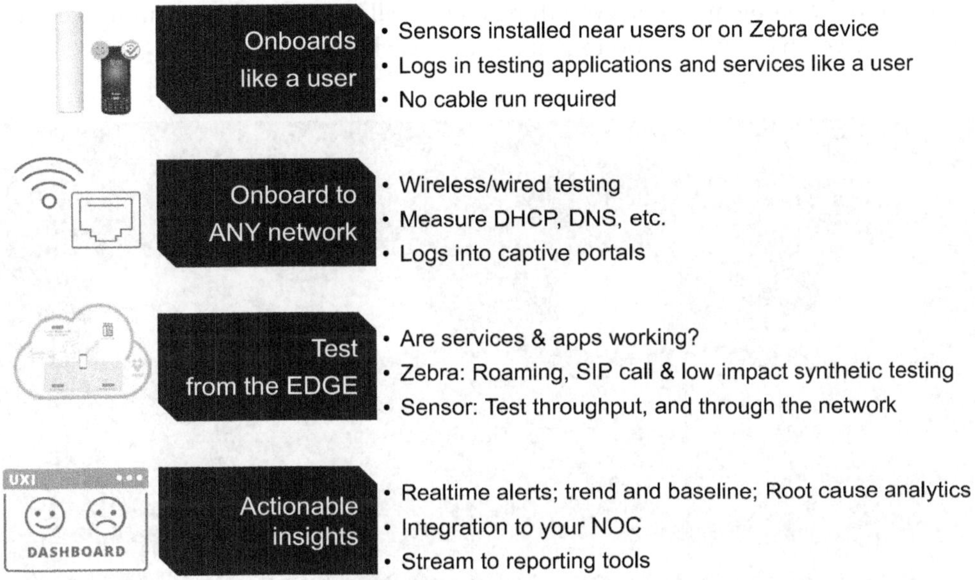

Figure 13-39: UXI process

Because the sensor sits outside the network onboarding just as a client would:

- It can measure end-to-end app performance from client's perspective
- It is not subject to any of the same points of failure as the network
- It is vendor neutral

The UXI dashboard provides a single pane-of-glass: UXI and AI Insight provide *assurance* of connectivity to and through the network with optimal performance and the appropriate Wi-Fi coverage.

As you would expect from the names and descriptions, individually UXI and AI Insight provide *insights* into how the network is performing and should perform. Collectively they provide a holistic *intelligence* from the perspective of the customers and end users and the network infrastructure. Note that AI Insight requires Aruba infrastructure, there are other requirements too (for example, ideally over 2,000 APs).

Examining UXI information

This section introduces you to the UXI dashboard, some of the tests performed by UXI, and how UXI is integrated into Aruba Central.

At-a-glance status

The UXI status dashboard (see Figure 13-40) is a simple way to see, at-a-glance, whether everything is performing well or if something needs attention. You can immediately see if there is an issue and then drill down to the problem. When everything is *green*, all is good and performing well. When you see *red*, you know you have work to do to make your users happy.

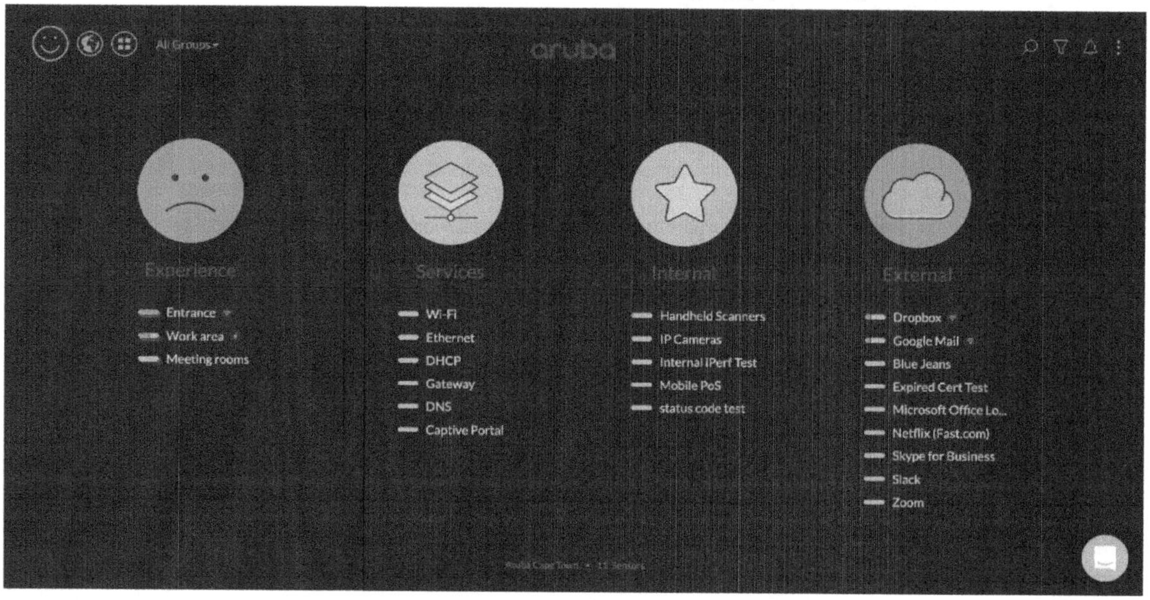

Figure 13-40: Examining UXI information

UXI tests

Figure 13-41 shows a sample list of what is tested by UXI. Note that the application list on the right is only a sample; there's a lot more applications that are supported and Aruba is constantly adding applications to the list. If there's something specific that you need to test, it's also possible to create your own test.

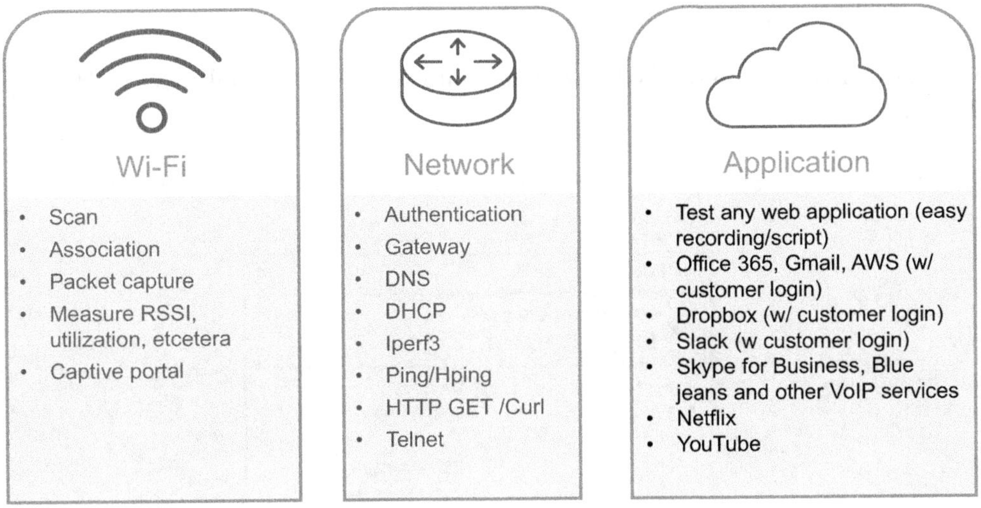

Figure 13-41: UXI tests

Not only can you test to a device or application, you can test through it. Here is a very small list of things UXI can test:

- Log into websites and applications
- Office 365, Gmail, Amazon Web Services (AWS)
- Pings, web server access
- Teams and many other VoIP services
- Netflix, YouTube, Dropbox, and etcetera
- Libraspeed and iPerf
- Voice analysis, roaming

Aruba Central integration

The UXI user experience status can be incorporated into the Aruba Central *Network Health* page (see Figure 13-42).

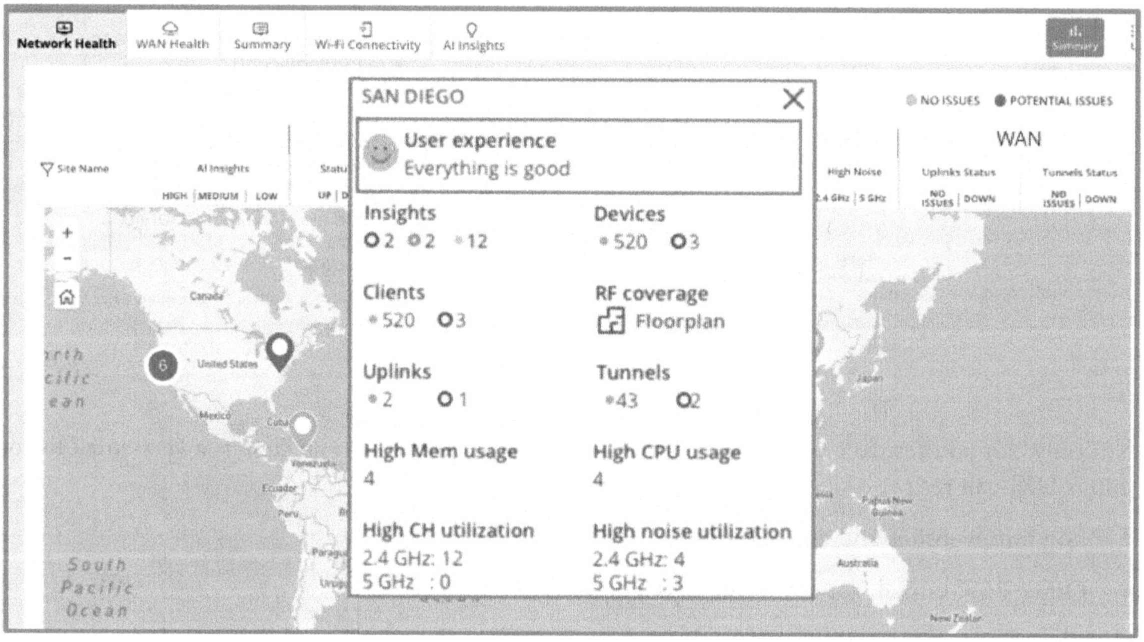

Figure 13-42: Aruba Central integration

Lab 13.01: Monitoring and maintenance

Your customer has asked you how they can monitor the network and perform maintenance on the APs and switches of their Aruba solution. In this lab you will see how you can use Aruba Central to configure alerts and generate reports. You will also explore how firmware compliance can be configured for the different device types in your setup (APs and switches).

Objectives

- Configuring alerts
- Configuring reports
- Configuring firmware compliance

Task 1: Configuring alerts

Your customer wants to use Aruba Central alerts to help with network monitoring, such as when an AP goes offline. After contacting your senior colleague, you know that you can use Aruba Central to alert on several common network and wireless deployment issues.

In this task you will configure several alerts:

- When AP is offline for more than five minutes
- When a configuration change is made in Aruba Central

The customer would like to see the alert in Aruba Central, as well as receive an email about the alert.

Objectives

- Configure alerts in Aruba Central.
- Configure default recipients for alerts.
- Review the generated alerts.

Steps

Configure the alerts

In these steps you will be configuring the following alert definitions in Aruba Central:

- **AP**: AP disconnected, major, over 5 min
- **Switch**: switch disconnected, major, over 5 min
- **Audit**: config change detected, minor

1. In Aruba Central, navigate to: Context: **Global** > Navigation: **Alerts & Events** > Top: **Alerts** > Right top: **Config**.

Set default recipients

You can set a default email address. This makes it easy to set the email destination when creating alert definitions.

1. At the top-right, click **Default Recipients**.
2. Click the **+** button to add a new recipient.
3. *Optional*: You may enter your own email address here to test the alerts. If you don't want to use your own email address, you can leave the field blank (see Figure 13-43).

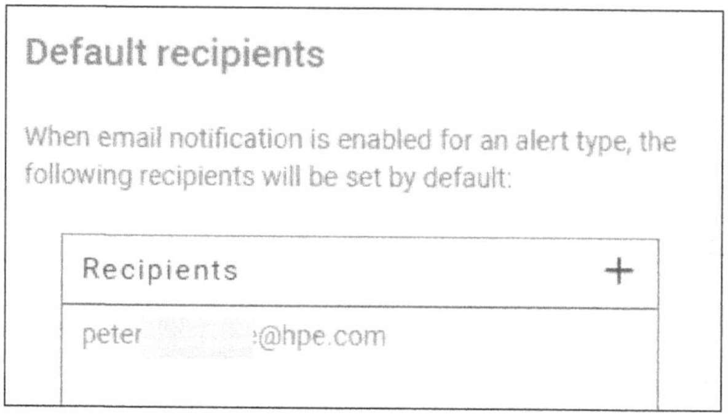

Figure 13-43: Defining the default email recipients

4. Click **Save**.

Set AP alerts

1. Several categories are used to group the alert definitions (see Figure 13-44).

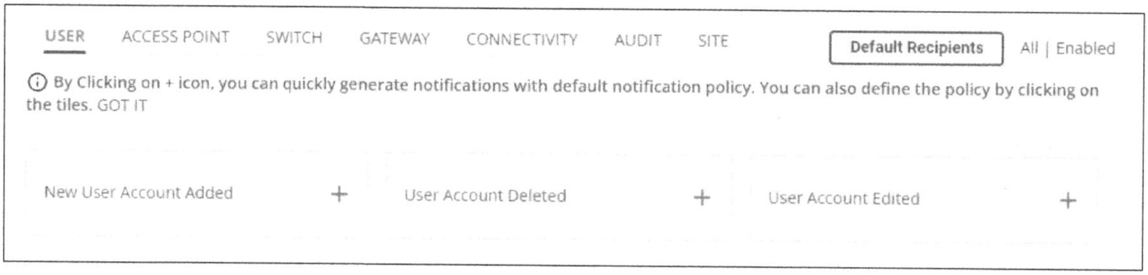

Figure 13-44: Alert categories

- **Question**: What are the categories for the Alerts?
- **Answer**: User, Access Point, Switch, Gateway, Connectivity, Audit, and Site

2. Click the **Access Point** category to see the AP Alert options.

AP disconnected alert

1. Click the **AP Disconnected** text (see Figure 13-45).

Figure 13-45: AP Disconnected selection

Note

Do not click the + sign next to the text; this will only *enable* the alert with the default settings.

2. The severity level controls how alert severity is set for this alert. Set Severity to **Major**.
3. Set Duration to **5** mins. This is the minimum value.
4. There is no need to set the device filter options.
5. *Optional*: If you have set your email, **Enable** the Email checkbox. The Default Recipients are automatically enabled. You could optionally add another email address for this specific alert.
6. Click **Save**, wait a few seconds, and confirm the message.
7. Verify that the alert is enabled; there should be a check mark next to the alert (see Figure 13-46).

Figure 13-46: Verify the AP Disconnected alert is enabled

8. Repeat the previous steps to enable the **Virtual Controller Disconnected** alert with the same settings.

Review configured alerts

You can adjust the view so you only see alerts that are enabled.

1. At the top-right, click the **Enabled** link to filter the alerts (see Figure 13-47).

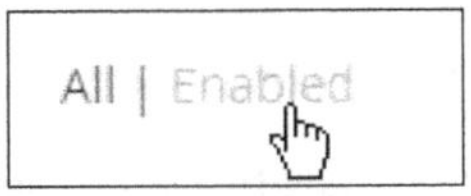

Figure 13-47: Filter the alerts

2. Now only the alerts that you have enabled are displayed.

Set a switch alert

1. Click the **Switch** category to see switch alert options.

2. **Enable** the Switch Disconnected alert.
   - Severity: **Major**
   - Duration: **5 min**
   - Email: **enabled** (optional, if you have configured your email)

Set an audit alert

To receive an alert when a device configuration changes, you can enable the audit alert.

1. Click the **Audit** category to see the switch alert options.
2. Enable the **Config Change Detected** alert.
3. Set the Severity to **Critical** and optionally set email to **enabled**.
4. **Save** and confirm the message.

Generate an action that triggers an alert

You will now make a very basic configuration change to verify the Audit Alert operation.

1. In Aruba Central, navigate to: Context: **Groups / campus-wifi-ui**> Navigation: **Devices** > Top: **Access Points** > Right top: **Config**.
2. In the WLAN list, enable or disable the guest WLAN by clicking on the **Wi-Fi** icon on the right-end side of the record (see Figure 13-48).

Figure 13-48: Wi-Fi icon

Review the alerts

In the next steps you review where you can find alerts.

1. In Aruba Central, navigate to: Context: **Global** > Navigation: **Alerts & Events** > Top: **Alerts**.
2. Review the list of **Open Alerts**.
   - **Question**: What is the category for the last alert?
   - **Answer**: Config change detected
3. Hover your mouse over the *Description* field to see alert details (see Figure 13-49 for an example).

Figure 13-49: Alert description example

4. The *Open Alerts* list reflects a list of *active* alerts. With an alert acknowledge action, the alert will be removed from the Open list.
5. Select the alert (if there are any other alerts, you may select them as well).
6. At the bottom of the page, click **Acknowledge** (see Figure 13-50).

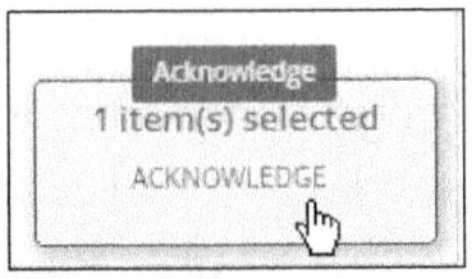

Figure 13-50: Select Acknowledge

7. Verify that the Open Alert list is now empty (see Figure 13-51).

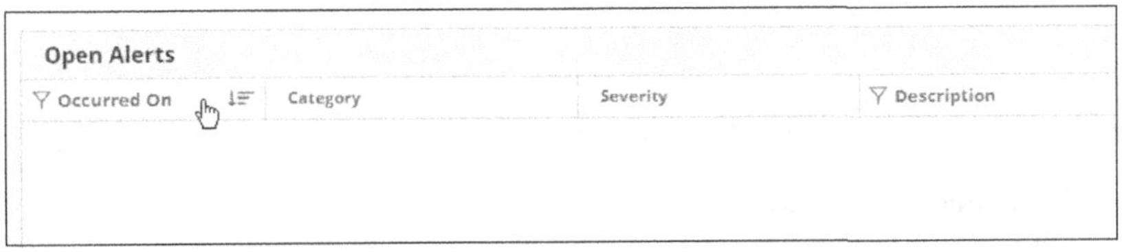

Figure 13-51: Verify the alert list

Reviewing historical alerts

If you need to review an Acknowledged Alert, you can still show all alerts.

1. Enable **Show Acknowledged Alerts** (see Figure 13-52).

Figure 13-52: Show Acknowledge Alerts slider

2. You can enable some extra columns to see who acknowledged the alert and the alert's elapsed time. The elapsed time is the time between the alert start time and the acknowledge time.
3. Click the **3 dots** to see columns display options.
4. Enable the **Acknowledged By** and **Elapsed time** columns (see Figure 13-53).

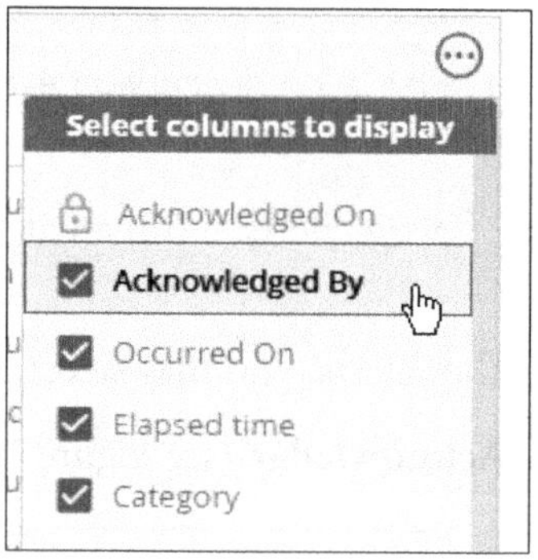

Figure 13-53: Change the columns that are displayed

5. Review the last alert about the config change. Figure 13-54 shows an example screenshot:

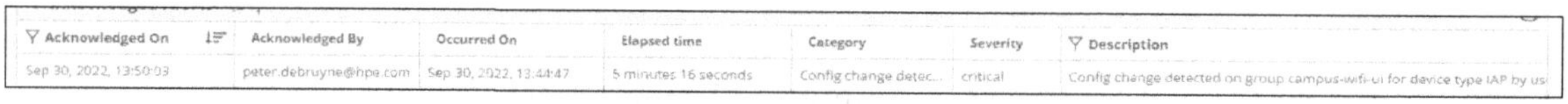

Figure 13-54: Verify the columns that are displayed

- **Question**: Who acknowledged the alert?
- **Answer**: Your login name should be listed.
- **Question**: What is the elapsed time?
- **Answer**: Depending on your lab progress, this will be time between the alert detection and the Acknowledge action.

6. *<u>Optional</u>*: If you have enabled your email in the default recipients, check your email for the new alert.

Task 2: Generating reports in Aruba Central

In this task you will use Aruba Central to generate a report about the network's RF health.

Objectives

- Use Aruba Central to generate a report.

Steps

Generate a report

1. In Aruba Central, navigate to: Context: **Global** > Navigation: **Reports** > Right-top: **Summary**.
2. Click **Create**.
3. Review report categories.
4. Under Infrastructure, click **RF Health**.
5. Click **Next**.
   - Context: **Site**
   - Site: **site-campus-main**
6. Click **Next**.
7. Set the Report Period: **Last 7 days**.
8. Select **One Time (Now)**.
9. Set the report title: **Campus-RF-Usage**.
10. *Optional*: Enter your email address, email format: enable PDF.
11. Click **Generate** and confirm the message.

Review the report

1. In Aruba Central, navigate to: Context: **Global** > Navigation: **Reports** > Right-top: **List**.

Note
It may take a few minutes to complete the report generation; repeat checking the list until you see your report.

---

2. This will show the list of generated reports. Click the entry with title **Campus-RF-Usage**.
3. Review the report. It contains details for the different bands, channel changes, channel use, errors, etc.
4. At the top of the page, you can download the report as PDF (see Figure 13-55).

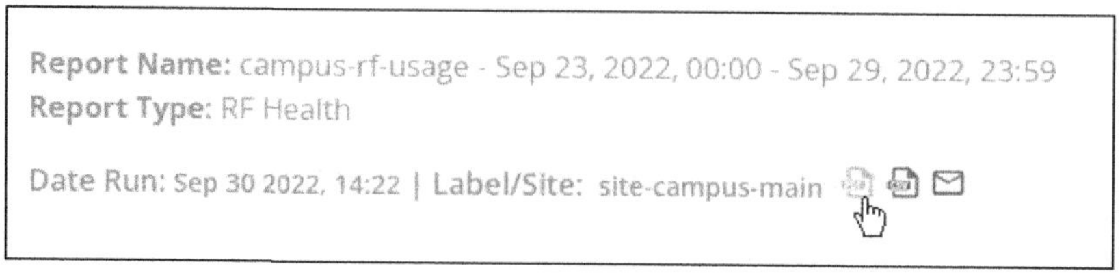

Figure 13-55: Download the report

5. Review the RF Health in the downloaded PDF document.

Task 3: Firmware compliance

Your customer is thinking about the required firmware on the devices during their deployment. They will be deploying factory default devices with Aruba Central, but when the devices are unboxed, they may be running a different firmware version compared to the validated production version.

The customer knows that they can manually update the firmware to the validated version, but they have asked you to simplify this operation. After some research, you learn that Aruba Central has a firmware compliance feature. This allows you to set a desired version on a global or group level. When a new device is added to that context, Aruba Central automatically updates the device to the desired version. This may be an upgrade or downgrade of the firmware version.

The compliance can be set for all groups or for specific groups; it is not possible to set the compliance at a site level. The logic for this is that one group represents one configuration, and a configuration may have features that depend on a specific firmware version. Within a site, you may have multiple device configurations, represented by multiple groups.

Objectives

- Set firmware compliance.
- Verify firmware compliance configuration.

Steps

Configure compliance on a new group

Important

You will ***not*** change the actual firmware version on the remote lab devices in the lab. In this task you will create a new group and set the compliance on this new group as an example of the steps. However, you will ***not*** move the devices to this new group.

The customer would like to test a new configuration and firmware version on a subset of their devices. Since they don't want to change the existing production configuration, you should not make changes to the existing group. Therefore, you will first create a copy of the existing group. On this new group you can then set the new firmware compliance version.

Note

Copying a group is fine for switches, gateways, and APs operating only in bridge mode, but must be handled with care for APs with tunneled WLANs.

1. In Aruba Central, navigate to: Context: **Global**> Navigation: **Organization**> Top: **Network Structure**> Tile: **Groups**.
2. In the list of groups, look for the group **campus-sw-edge-tpl**.
3. Move your mouse over the group to see the action icons (see Figure 13-56).

Figure 13-56: Viewing the action icons

4. Click the **Clone Group** icon to make a copy of the group (see Figure 13-57).

Figure 13-57: Click the Clone Group action icons

5. In the Clone Group dialog, for the name field, enter **campus-sw-edge-tpl-v2**.
6. Click **Clone** to copy the group.
7. Verify the new group is now displayed in the group list.

Verify the cloned configuration

1. In Aruba Central, navigate to: Context: **Groups / campus-sw-edge-tpl-v2** > Navigation: **Devices**> Top: **Switches**> Right top: **Config**.
   - **Question**: Do you see a template in the Templates list?
   - **Answer**: Yes, the sw-edge template is available on this new group. It was copied from the original group.

2. At the top, click the **Variables** tab.
   - **Question**: Are there any variables listed here?
   - **Answer**: No, since there are no devices assigned to the group yet, no variables will be displayed.
   - **Question**: Will this cause any issues if you move a device from another group? (*Hypothetical question since you will not move your devices!*).
   - **Answer**: No, Aruba Central stores the device variables at the device level. The variables will not change when the device is moved to another group.

Set the firmware compliance at group level

With the new group defined, you are now ready to set the desired firmware version. Compliance can be set at the global and group level. You will now configure the compliance at the group level.

1. In Aruba Central, navigate to: Context: **Groups / campus-sw-edge-tpl-v2** > Navigation: **Firmware**> Top: **Switches**.

> **Note**
> Make sure to access the correct device type (Access Points/Switches/Gateways) at the top of the screen!

2. At the top-right, click **Set Compliance**. The Manage Firmware compliance window will be displayed.

> **Note**
> The *Compliance* window reflects the context in which you are currently working. Since you navigated to this screen via the Group context, this screen shows that no compliance is currently set.

3. Move the slider for **Set firmware compliance** to enabled.

> **Note**
> Compliance may have been enabled already at the MSP level in Aruba Central; you can continue the steps even when it was enabled already.

4. Leave the field AOS-S version at its default.

> **Note**
> There are no AOS-S devices in your lab, so this setting has no effect in this lab.

5. Set the CX Firmware Version to **10.09.1050**.

> **Note**
> This is just an example version: any version is fine since you will ***not*** actually update the Aruba devices.

6. Leave the field When at its default (**Now**).

> **Note**
> This can be useful to control when the new compliance policy shall take effect. This should typically be set to a time in during a maintenance window.

7. Leave Install On set to **primary** partition.
8. Enable the **Automatically reboo**t checkbox.

> **Note**
> Make sure to enable this checkbox, otherwise Aruba Central will only update the firmware in the flash, but you must still decide when to reboot the devices. This may be convenient if you want to control the order of rebooting devices.

9. Click **Save** to submit the configuration.
10. You have now created a new group and configured firmware compliance for that group. Any AOS-CX switch you might add to this group is automatically updated to the configured compliance version. There is no need to do this in this lab environment.

AP compliance

You configure AP firmware compliance like the previous steps.

1. For example, in Aruba Central, navigate to: Context: **Groups / campus-wifi-ui** > Navigation: **Firmware**> Top: **Access Points**.
2. At the top-right, click **Set Compliance**. You see the Manage Firmware compliance window with the available AP firmware version.
3. Click **Cancel**.

> **Important**
> You should ***not*** set the compliance on the APs in the lab.

Learning check

Discuss these questions.

1. You can leverage Aruba Central's *Topology* pane to ease troubleshooting tasks, by quickly seeing how devices interconnect.
   a. True
   b. False
2. When analyzing system status LEDs on an Aruba AP, you see a green light repeatedly blinking on for one second, then off for one second. What does this indicate?
   a. The device is fully functional.
   b. The device is operating in restricted power mode.
   c. There is a system error.
   d. The device is booting.
3. Match the Aruba AP firmware upgrade feature (*standard upgrade*, *live upgrade*, and *firmware compliance*) with its appropriate descriptions. Options can be used more than once.
   a. See a list of supported firmware versions per in-group device.
   b. Minimize impact in a production environment.
   c. Good to use during a planned maintenance window.
   d. AP upgrade sequence is based on AirMatch partitions.
   e. Ensures that devices in a group all run a specific OS version.
   f. May disrupt connectivity.
4. A colleague asks you for advice about the SNMP. Which statements below should you use to help your colleague? (Select three.)
   a. An SNMP manager queries SNMP agents for information.
   b. The SNMP manager typically provides a GUI for network managers to review all gathered SNMP information.
   c. The SNMP manager stores a hierarchical database – a Management Information Base (MIB) for each device.
   d. For the best security, your colleague should deploy SNMP version 3.
5. Match each Aruba AIOps core component (*AI Insights*, *AI Search*, and *AI Assist*) with an appropriate description.
   a. Natural language queries for fast troubleshooting
   b. Proactive anomaly detection and optimization
   c. Automated Aruba TAC trouble ticket generation

6. Typically, you use Aruba Central's Alerts & Events page at the end of a troubleshooting session to review your activities.
   a. True
   b. False
7. Which statements below are true about Aruba's User eXperience Insight (UXI) solution? (Select three.)
   a. The solution includes front-end sensors that request and generate traffic, much like an actual end user.
   b. The solution includes a cloud-based UI and intelligent analytics.
   c. One advantage of this solution is that it provides intelligent views from a network device perspective.
   d. The solution can test the entire client experience, from the Wi-Fi connection, through the network, and up to and including end-user applications.

Summary

Congratulations! You have completed Chapter 13. You should now be able to:

- Use Aruba Central monitoring capabilities.
- Identify device LED status.
- Perform firmware upgrades.
- Using SNMP on devices.
- Describe AI Insights.
- Describe alerts and reports.
- Explain UXI.

14 Troubleshooting

OBJECTIVES

When things go astray, your ability to troubleshoot is vital. After completing this chapter, you will be able to perform password recovery and factory reset procedures, troubleshoot Aruba Central connectivity issues, and enable spectrum analysis.

After completing this chapter, you should be able to:

✓ Perform password recovery and factory reset procedures.

✓ Troubleshoot Aruba Central connectivity.

✓ Enable spectrum analysis.

✓ Explore Aruba Central troubleshooting tools.

Password recovery and factory reset

This chapter begins with a discussion on password recovery and factory reset procedures.

AOS-CX switches

Let's start first with the AOS-CX switches. The next two sections show you how to perform the password recovery process as well as how to set the switches to their factory default settings.

AOS-CX password recovery process

In case that the credentials do not work when logging into the switch, a password recovery process can be performed. Here are the basic steps you would follow to perform the password recovery process (see Figure 14-1):

1. Connect to the switch using the console port.
2. Power cycle the switch.
3. When the switch system displays the boot profile prompt, select the *Service OS console* option by typing **0**.
4. Login with the user **admin**; no password is set for this account.

5. Enter **password** command and type the *new* password for the admin account.
6. Enter the **boot** command.
7. Once the switch boots completely, login using the admin username and the password that was set in step 5.

1. Connect to the switch via console.
2. Power cycle the switch.
3. When the system prompts, select the **Service OS console** option by typing **0**.

```
Boot Profiles:

0. Service OS Console
1. Primary Software Image [FL.10.09.1031]
2. Secondary Software Image [FL.10.09.1031]

Select profile(primary): 0
```

4. Login with user **admin**. No password is set for this account.
5. Enter **password** keyword and type the new password for the admin account.

```
ServiceOS login: admin
SVOS> password
Enter password: ********
Confirm password: ********
```

6. Enter **boot**.
7. Login using the admin username and the password that was set on step 5.

Figure 14-1: AOS-CX switch password recovery process

AOS-CX factory default

The AOS-CX switch command **erase all zeroize** restores the switch to its factory default configuration (see Figure 14-2). You will be prompted before the procedure starts to ensure that you want to perform this process. Once complete, the switch will restart from the primary image with factory default settings.

Zeroize	Erase-startup
• Deletes all checkpoints in the system • Returns the system to factory default	• Deletes only the startup checkpoint in the system • No factory default

```
switch(config)# erase all zeroize
This will securely erase all customer data and reset the switch to factory
defaults. This will initiate a reboot and render the switch unavailable until
the zeroization is complete.
This should take several minutes to one hour to complete. Continue (y/n)? y
The system is going down for zeroization.
```

Figure 14-2: Setting an AOS-CX switch to factory default settings

The command **erase-startup** erases the startup checkpoint only: all other checkpoints will still remain on the switch. Using the **erase all zeroize** command, however, will erase *all* checkpoints.

Aruba APs

Now that you know how to perform the password recovery procedure and factory reset the AOS-CX switches, let's move on to the Aruba APs. The next two sections show you how to perform the password recovery process as well as how to set the APs to their factory default settings.

AP CLI password reset via Central

To reset AP password, navigate to **AP Group > Devices > System**. Scroll down to the *Administrator* section, click the **pencil** icon for the *admin* account, and change the admin password. See Figure 14-3.

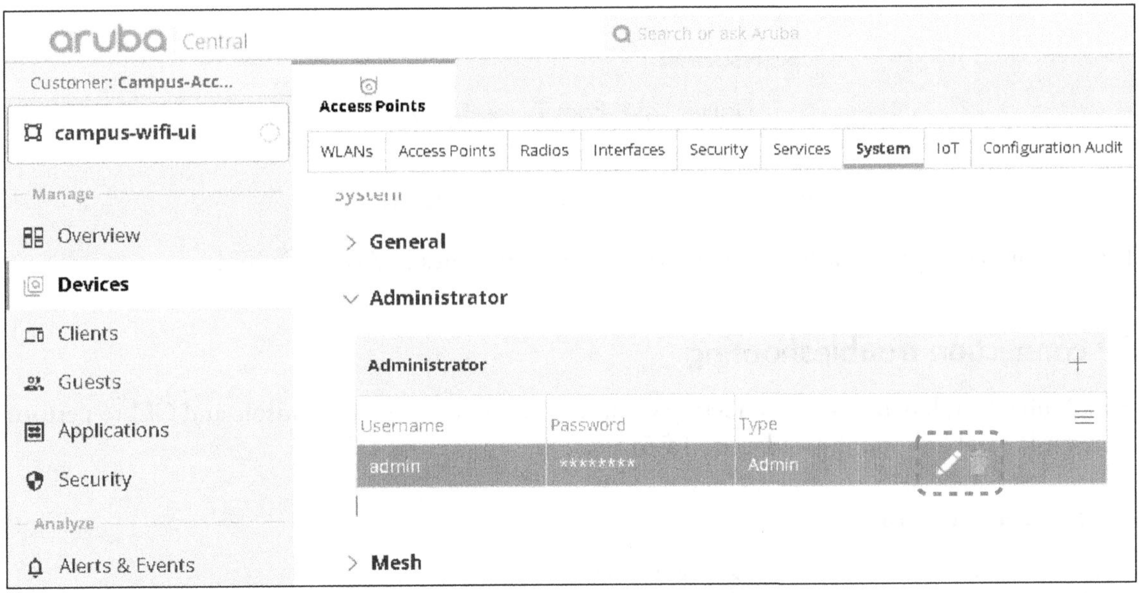

Figure 14-3: AP CLI password reset via Central

AP factory reset

To factory reset an Aruba AP, perform the following (see Figure 14-4):

1. Connect to the console port of an AP.
2. Power cycle the AP.
3. When prompted to stop the autoboot process, press <**Enter**>.

4. Enter the **factory** command to factory reset the AP.

5. Enter the **reset** command to boot the AP.

1. Connect to AP via console.
2. Power cycle the AP.
3. Interrupt boot process by pressing **<Enter>**.

```
Net:    eth0
Radio: bcm47622#0,bcm47622#1
Reset: warm
FIPS:  passed
Hit <Enter> to stop autoboot:  0
```

4. Type **factory** and press **<Enter>**.

```
apboot> factory
Clearing state...
Checking OS image and flags
...
done
```

5. Type **reset**.

Figure 14-4: Factory reset an AP

Troubleshooting Aruba Central connectivity

This section shows you how to troubleshoot connectivity issues to Aruba Central.

AP connection troubleshooting

If an Aruba AP is having issues contacting Central, you can use the AP's console and CLI to perform some basic troubleshooting (see Figure 14-5).

```
AP# show activate

IAP MAC Address              :20:4c:03:32:aa:a8
IAP Serial Number            :CNG9K2R64T
Cloud Activation Key         :GHGMVKXI
Activate Server              :device.arubanetworks.com
Activate Status              :fail-prov-no-shipped
Provision interval           :5 minutes20:4c:03:32:aa:a8#


AP-50-a8# show ap debug cloud-server
AP mgmt mode                 :local-mgmt
cloud config recved                 :FALSE
autojoin mode                       :disable
state diff                          :disable
Device Cert status                  :SUCCESS
```

No entry for AP on Aruba Activate or no internet access

Check Aruba Central AP License Assignment

Figure 14-5: AP connection troubleshooting

On the AP console, use the **show activate** command to see the Aruba Activate status. In Figure 14-5 there is a problem. The status is *fail-prov-no-shipped*, which is an indication that Activate does not know about the AP or the AP does not have internet access.

Another useful command is **show ap debug cloud-server**. In Figure 14-5, *cloud config recvd* is *FALSE*. If the AP has internet access, you should verify in Central if the AP was assigned a license.

AOS-CX switch connection troubleshooting

The following three sections cover troubleshooting AOS-CX switch and Aruba Central connection issues. Figure 14-6 shows an example of how to start troubleshooting an AOS-CX switch and its connection to Central. The **show aruba-central** command output shows that you don't have a Central location (URL) and the switch does not have a connection to Central (something is wrong either in the internal network or the Internet when trying to reach Central). Another problem could be that Activate has no provisioning rule for this switch. Alternatively, the problem could be that the Central account was not given the switch a subscription key.

```
switch# show aruba-central
Central admin state                                : enabled

Central location                                   : N/A
VRF for connection                                 : N/A
Central connection status                          : N/A

Central source                                     : activate
Central source connection status                   : connection_failure
Central source last connected on                   : N/A
System time synchronized from Activate   : True

Activate Server URL                                : devices-v2.arubanetworks.com
CLI location                                       : N/A
CLI VRF                                            : N/A

Source IP                                          : N/A
Source IP Overridden                               : False
Central support mode                               : disabled
```

No entry for the switch on Aruba Activate or no internet access

Figure 14-6: AOS-CX switch connection troubleshooting

AOS-CX switch logs

Debugging ZTP offers more information on why a switch does not connect to Aruba Central. Enable the **debug aruba-central** and **debug ztp** commands to dig deeper into the connection issue.

In Figure 14-7, the logs confirm that the switch has access to Activate but Activate has no rules for this switch. This could be an issue with Activate or Central not having a subscription key to this device. Once you correct the problem, issue the **show aruba-central** command to verify if the problem has been fixed. This command shows you whether or not the switch is connected to Central.

Note

The command to debug ZTP events on AOS-CX switches is **debug ztp**. Also, the "**-r**" parameter in the **show logging** command displays the logging events from newest to oldest (reverse chronological order).

```
switch# debug ztp all
switch# show logging -r
--------------------------------------------------
Event logs from current boot
--------------------------------------------------
2021-10-01T13:18:38.386195+00:00 6200 hpe-restd[843]:
Event|4650|LOG_INFO|AMM|-|Unable to fetch Aruba Central location  from
activate via VRF .
2021-10-01T13:18:38.123971+00:00 6200 hpe-restd[843]:
Event|7708|LOG_INFO|UKWN|1|Certificate devices-v2.arubanetworks.com verified
and accepted
2021-10-01T13:18:34.011232+00:00 6200 hpe-restd[843]:
Event|4648|LOG_INFO|AMM|-|Switch time is synced with Aruba Activate Server
https://devices-v2.arubanetworks.com.
```

Figure 14-7: AOS-CX switch logs

AOS-CX Aruba Central connection: Disable and enable

By default, the AOS-CX switch cannot be locally configured when it is managed by Aruba Central. The command **aruba-central disable** disables the connection to the Aruba Central server (see Figure 14-8). When the connection is disabled, the switch does not attempt to connect to the Aruba Central server or fetch central location from any of the three sources (CLI/Aruba Activate/DHCP). It also disconnects any active connection to the Aruba Central server. Disabling and enabling Aruba Central will trigger a configuration refresh, where the current configuration will be pushed from Central. Using this command is useful to *kickstart* the switch to reconnect to Central when troubleshooting connection issues. **aruba-central enable** re-enables the Central access.

Local config not possible when managed by Central

```
switch(config)#
  aruba-central    Configure Aruba-Central
  debug            Configure debug logging
  end              End current mode and change to enable mode
  exit             Exit current mode and change to previous mode
  list             Print command list
  no               Negate a command or set its defaults
```

Disabling and enabling Aruba Central

```
switch(config)# aruba-central
switch(config-aruba-central)# disable
switch(config-aruba-central)# enable
```

Figure 14-8: Disabling and re-enabling the Central connection

Aruba Central configuration lockout and support modes

The AOS-CX switch connects to Aruba Central in either monitor or managed mode. When the device connects in monitor mode, Aruba Central monitors the switch configurations. When it connects in managed mode, the **configuration-lockout central managed** command does not allow configuration changes from other switch configuration methods, like the switch CLI or the switch's web UI (see Figure 14-9).

The **aruba-central support-mode** command allows the device to be writable for all operations in Aruba Central lockout mode for troubleshooting. The **no** form of this command disables this activity. This is useful if a configuration was pushed down from Central and activated on the switch, but has created a connectivity issue where the switch can no longer communicate with Central. The checkpoint auto mode, discussed in Chapter 2, was specifically developed for this situation. However, if this doesn't work, you can use the above command from the CLI of the switch to fix the configuration issue.

Set device to only be writable from Aruba Central1

```
switch(config)# configuration-lockout central managed
switch(config)# show configuration-lockout
Configuration Lockout
---------------------
central: managed
```

Enable Central Support Mode

```
switch(config)# aruba-central support-mode
switch(config)# show aruba-central
<-output omitted->
Central support mode                        : enabled
```

Figure 14-9: Aruba Central configuration lockout and support modes

RF spectrum analysis

Did you ever wish you could see outside of the human range, like x-ray vision or perhaps into the microwave range? If that is the case, then you should check out spectrum analysis, which is introduced here.

Wireless networks operate in environments with electrical and RF devices that can interfere with network communications (see Figure 14-10). Microwave ovens, cordless phones, and even adjacent Wi-Fi networks are all potential sources of continuous or intermittent interference. The spectrum monitor software modules on IAPs and AOS 10 APs can examine the RF environment in which the Wi-Fi network is operating, identify interference, and classify its sources. An analysis of the results can then be used to quickly isolate issues associated with packet transmission, channel quality, and traffic congestion caused by contention with other devices operating in the same band or channel. Spectrum monitors are AP radios that gather spectrum data but do not service clients. Each SM scans and analyzes the spectrum band used by the AP's radio (2.4 GHz, 5 GHz, and/or 6 GHz). By default, an APs' radio accepts device associations. When you switch a radio to spectrum analysis mode, you can use it for troubleshooting RF issues. However, when it is in this mode, it will disconnect any associated users and deny user's access to the radio. When done troubleshooting, remember to disable spectrum analysis for the radio so that it can, again, allow users to associate to it. Alternatively, you can dedicate an AP(s) in an area that perform spectrum analysis full time. When doing this full time, these APs are commonly called spectrum monitors or SMs.

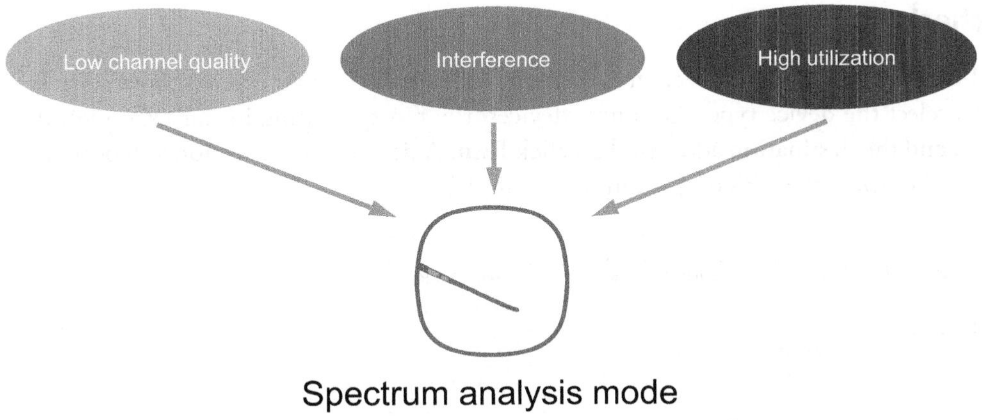

Figure 14-10: RF spectrum analysis

Aruba Central troubleshooting tools

This section introduces you to some tools that Aruba Central can use to troubleshoot connection and device issues.

Network check

Network Check tab provides information on network issues faced with devices Aruba Central manages (see Figure 14-11). You can diagnose network performance from the *Tools* option in the left navigation pane. There are various test types that you can perform, such as, ping test, speed test, TCP test, and traceroute.

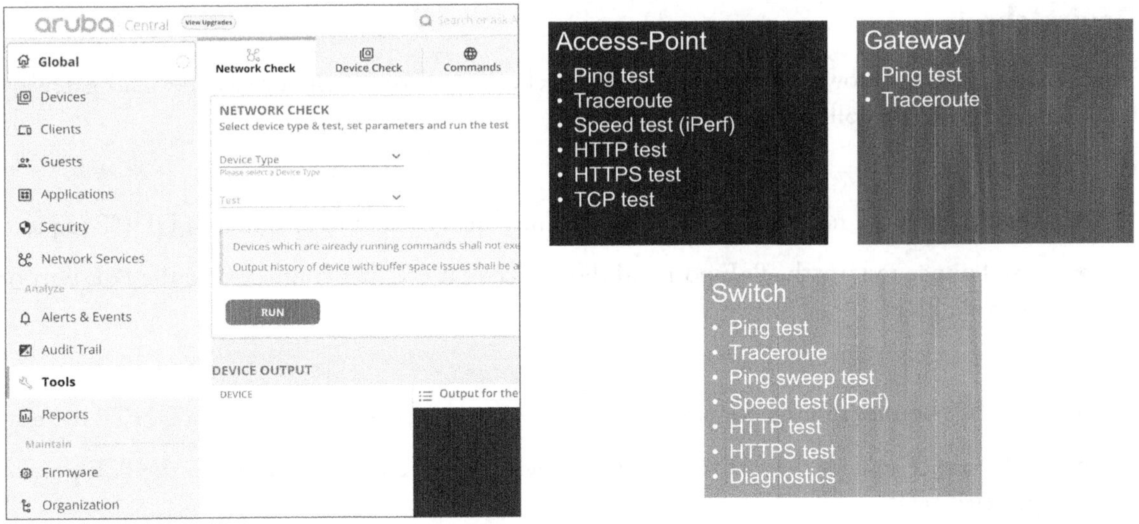

Figure 14-11: Central's network check tools

Ping check

Figure 14-12 shows a ping test executed from one managed AP to an external host/address. To run a ping test, select the device type, the source devices, the test type (ping in this example), the destination type, and the destination address, then click **Run**. A device output section will populate with the device's CLI output (the AP's output, in this example).

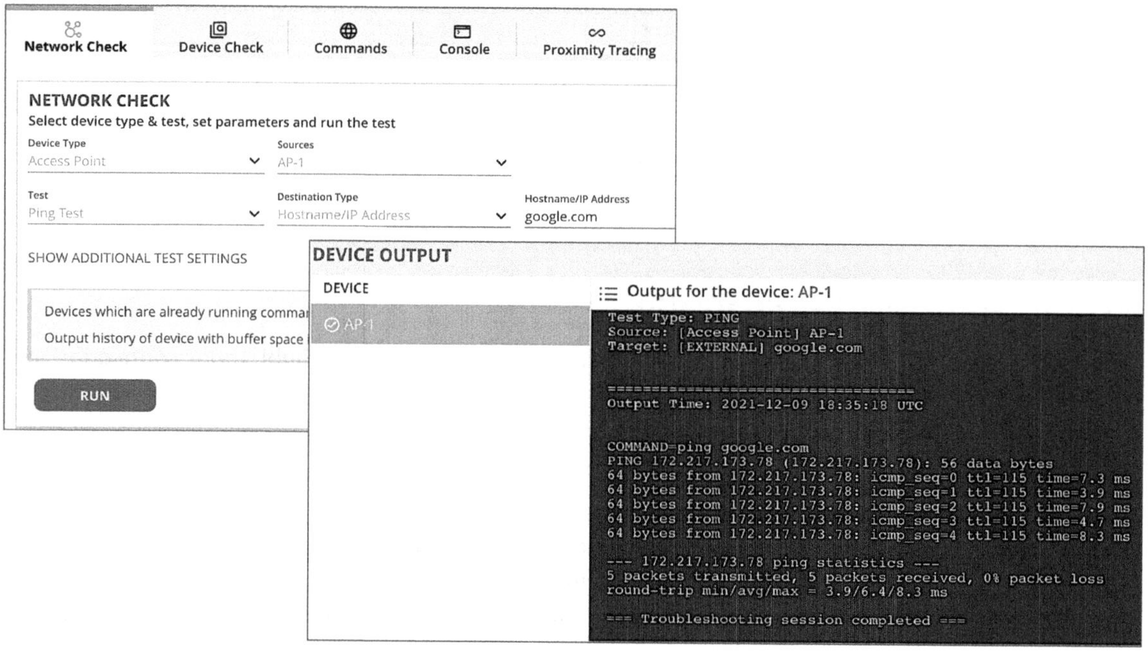

Figure 14-12: Central's network check tools: Ping

Device check

The *Device Check* tab allows you to run diagnostic checks and troubleshoot switches and gateways (see Figure 14-13). The following tests are available:

- Gateways:
  - Interface bounce restarts the port interface and forces a client to re-initiate a DHCP request.
  - PoE bounce restarts the PoE port and the device that is either connected to the PoE port or powered by it.
- Switches:
  - Interface bounce restarts the port interface and forces a client to re-initiate a DHCP request.
  - PoE bounce restarts the PoE port and the device that is either connected to the PoE port or powered by it.
  - Chassis locate activates the switch locator LED. The locator LED indicates the physical location where an Aruba switch is currently installed.

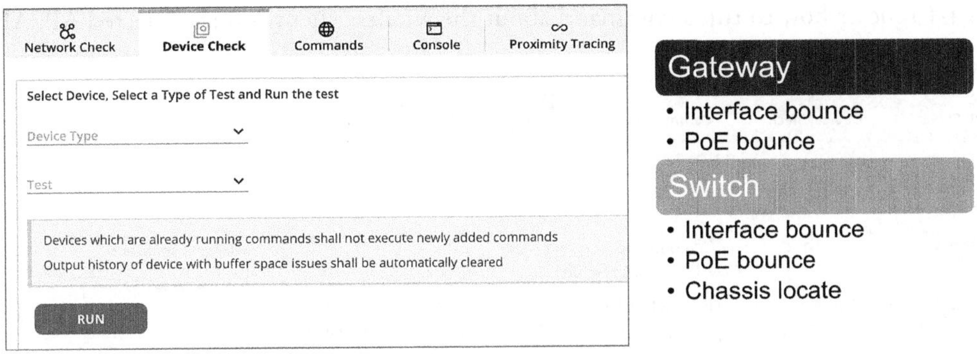

Figure 14-13: Central's device check tools

Run commands

The *Commands* tab (see Figure 14-14) allows you to perform a network health check on devices at an advanced level using command categories. Read-only users can also perform advance checks. To run a command, select the device type you want to run the command; then select the device(s). The category box will show the command category applicable to the selected device(s). Once you chose a category, the available commands will populate to the right of it. Choose the command you want to execute and click **Add**. When all the commands you want to run are in the selected commands box, click **Run**. A device output section will display the CLI output from each selected device.

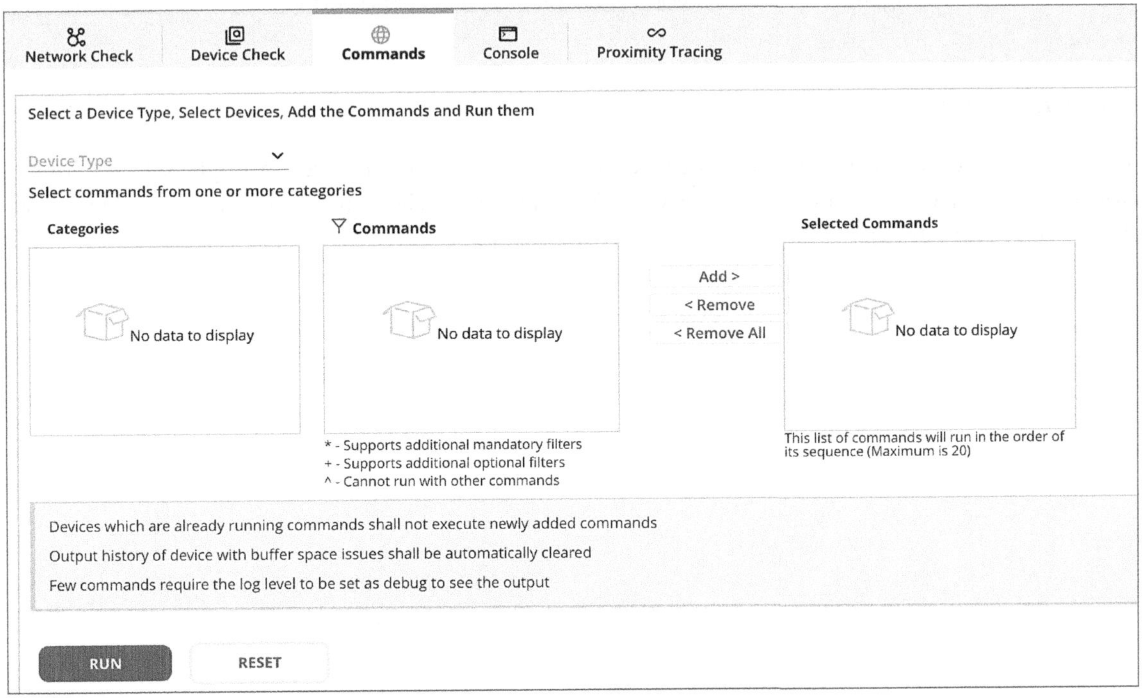

Figure 14-14: Run device commands from Central

Figure 14-15 shows how to run a command about the wireless clients on the selected AP: AP-2.

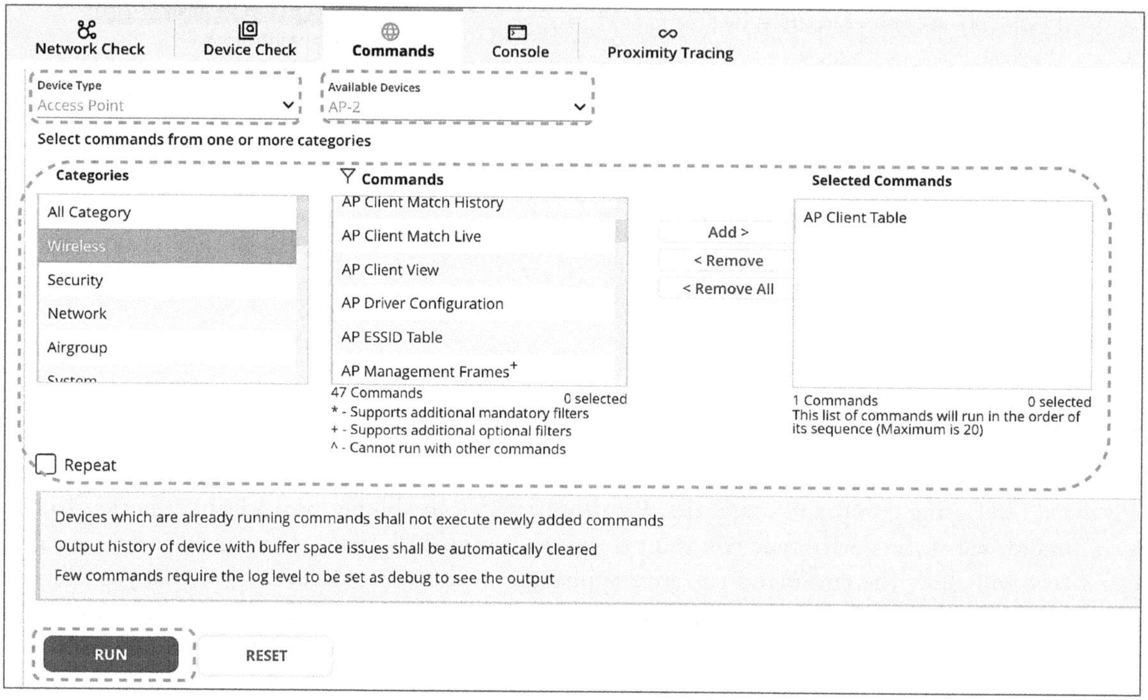

Figure 14-15: Command example: Clients connected to AP-2

Remote console

The remote console feature enables you to connect to the device directly through SSH sessions to troubleshoot any device issues in the network. You can either view saved sessions or create new sessions (see Figure 14-16).

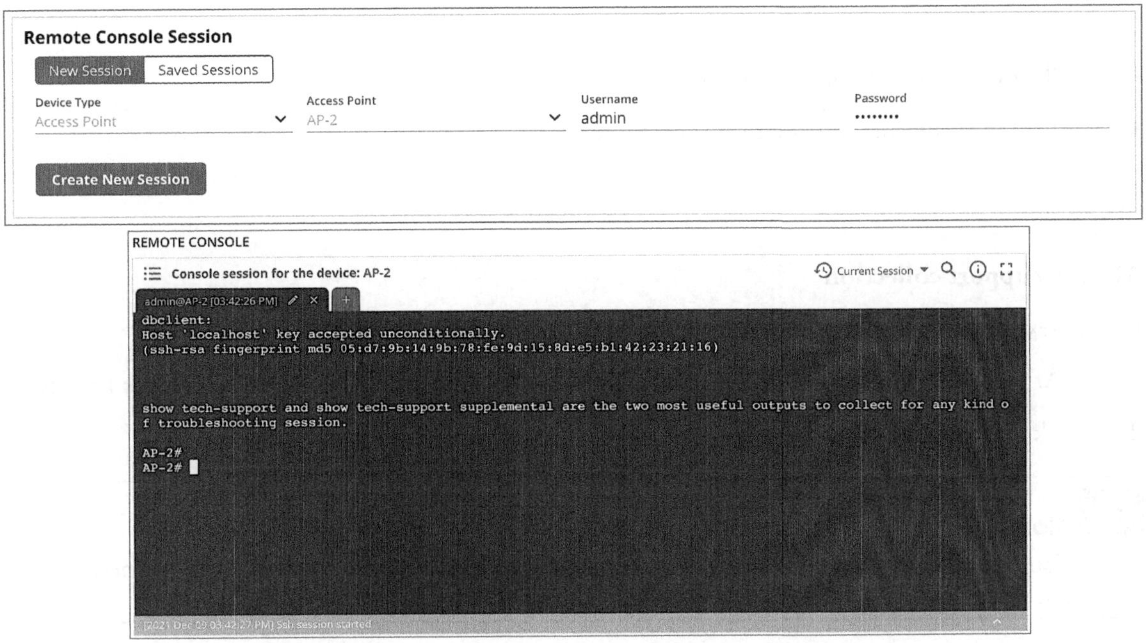

Figure 14-16: Remote console to a device from Central

Lab 14.01: Troubleshooting

Your senior colleague has asked you to become more familiar with some troubleshooting tasks and scenarios. In this lab you will collect tech support log file information from the Aruba APs and switches. These files can be used by your senior colleague or Aruba TAC to analyze the system.

You will also troubleshoot the connection between a switch and Aruba Central, and explore Aruba Central's site-level status windows.

Objectives

- Collect log information for senior support or TAC.
- Troubleshoot the connection between a switch and Aruba Central.
- Review site-level troubleshooting information.

Task 1: Collection of log information for senior support or TAC

In this task you will learn to collect tech support information from the Aruba APs and switches.

Objectives

- Collect tech support information for an AP.
- Collect tech support information for a switch.

Steps

AP tech support collection

In the next steps you will collect the support information for an AP.

1. In Aruba Central, navigate to: Context: **Global** > Navigation: **Devices** > Top: **Access Points**.
2. Click **ap1** to view its device page.

Note

You can also use the AI Search bar: enter **ap1**, press **<Enter>**, and click the AP *name* in the search result window.

---

3. You are now at this location in Aruba Central: Context: **Device / ap1** > Navigation: **Overview** > Top: **Summary**. At the top-right, click the **Actions** dropdown list and select **Tech Support** (see Figure 14-17). This action will take you to the Tools > Commands page (see Figure 14-18).

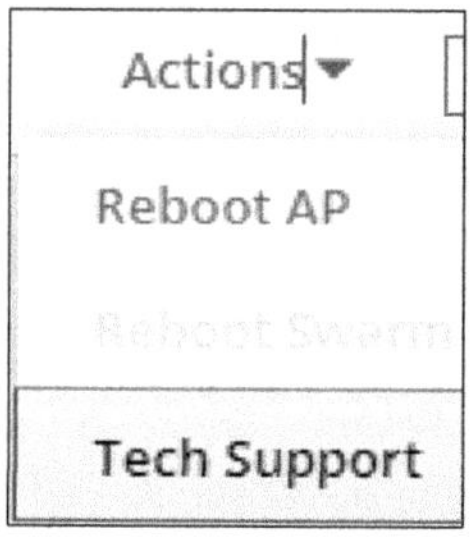

Figure 14-17: Accessing Tech Support

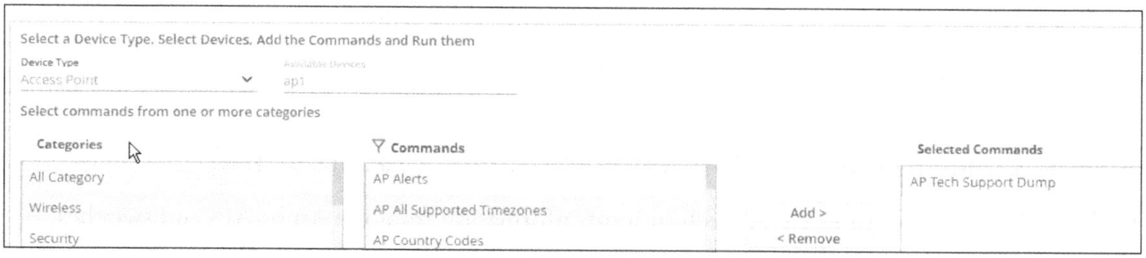

Figure 14-18: Accessing Tech Support (cont.)

- **Question**: What is the currently selected *Command*?
- **Answer**: The command for **AP Tech Support Dump** was automatically selected.

4. Click **Run**.
5. It may take some time to complete this command. Upon completion, download the command output using the **download** icon (see Figure 14-19). This downloads a zip file to your PC that contains the tech support files. See Figure 14-20 for an example. You can share this file with Aruba TAC or your senior engineer for further analysis.

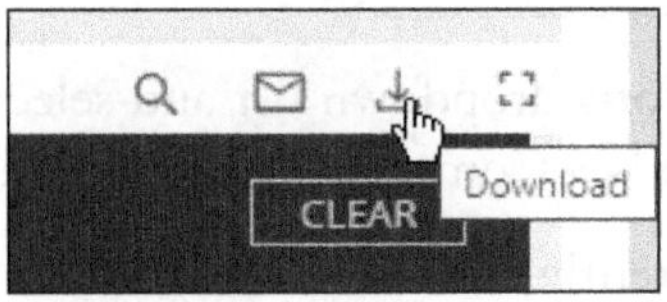

Figure 14-19: Downloading the Tech Support information

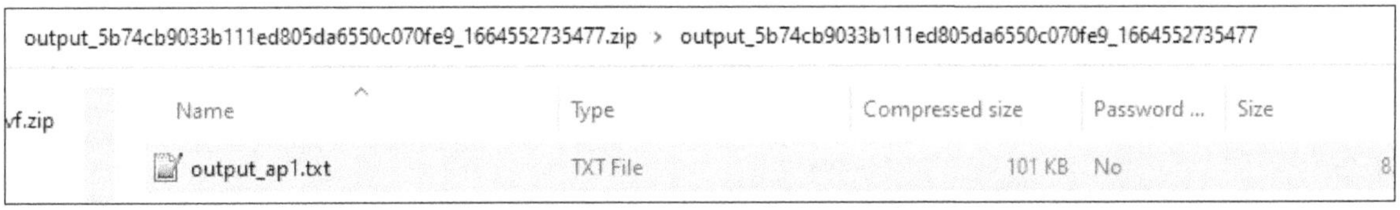

Figure 14-20: Accessing the downloaded file in Windows

Switch tech support collection

In the next steps you will collect switch support information.

1. In Aruba Central, navigate to: Context: **Global** > Navigation: **Devices** > Top: **Switches**.
2. Click **sw-edge1** to see its device page.

Note
Just like with the AP, you can also use the AI Search bar: Enter **sw-edge1**, press **<Enter>**, and click the switch name in the search result window.

3. You are now at this location in Aruba Central: Context: **Device / sw-edge1** > Navigation: **Overview** > Top: **Summary**. Take note of the product code of the switch; this typically starts with *Jxxx*. You will need this information later in this lab. Figure 14-21 shows an example output with JL666 (your output might by different).

Figure 14-21: Switch product code

- **Product code**: ____________________

4. At the top-right, click the **Actions** dropdown list and select **Tech Support** (see Figure 14-17). This action takes you to the Tools > Commands page (see Figure 14-22).
   - **Question**: What is the currently selected *Command*?
   - **Answer**: The command **show tech** was automatically selected.

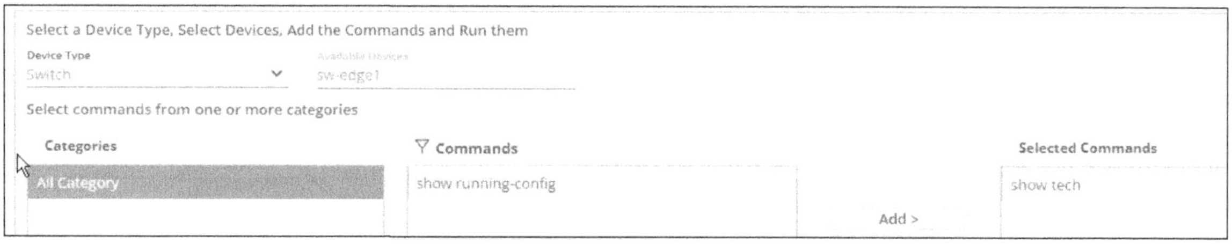

Figure 14-22: Tools > Command page

5. Click **Run**.

Note

Aruba Central reserves a limited buffer for the command output. If the buffer is full, the system will not display the output, but provide you with a download or email option.

6. It may take some time to complete this command. Then download the command output using the **download** icon (see Figure 14-23). This downloads a zip file to your PC that contains the tech support files. Again, you can share this file with Aruba TAC or your senior engineer for further analysis.

Figure 14-23: Download the command output

Task 2: Troubleshoot switch-to-Aruba Central connections

In this task you will learn how to troubleshoot the connection between an AOS-CX switch and Aruba Central. In the scenario, the customer has uploaded a template to Aruba Central that contains a configuration error. As a result of the configuration error, the switch can no longer connect to Aruba Central. The customer wants you to resolve this issue: to ensure that the switch connects to Aruba Central again.

Objectives

- Verify the Aruba Central connection.
- Understand the function of the Aruba support command.

Steps

Initially, the lab environment is still functional. You will explore the admin access level you have on the switches.

Explore default configuration status

When a device is managed by Aruba Central, access to configuration mode is limited. This prevents changing the configuration locally on the switches; all configuration changes must be done via Aruba Central, by default.

1. Use PC-MGMT to open an SSH connection to *sw-edge1*.
2. Login using **admin** and **Aruba123!**
3. Review the current Aruba Central output.

```
sw-edge1# show aruba-central
Central admin state                          : enabled

Central location                             : de-
vice-uswest4-d2.central.arubanetworks.com
```

```
VRF for connection                          : default
Shared Token                                : N/A
Central connection status                   : connected

Central source                              : activate
Central source connection status            : connected
Central source last connected on            : Thu Oct 20 14:44:59 UTC
2022
System time synchronized from Activate      : True

Activate Server URL                         : devices-v2.arubanetworks.
com
CLI location                                : N/A
CLI VRF                                     : N/A

Source IP                                   : 10.1.3.4
Source IP Overridden                        : False

Central support mode                        : disabled
```

- **Question**: What is the Aruba Central connection status?
- **Answer**: Connected.

Aruba Central configuration lockout

Check the status of the configuration lockout.

4. Review the running configuration; look for the configuration lockout command.

```
sw-edge1# show running-config | include config
Current configuration:
configuration-lockout central managed
```

- **Question**: What does it mean?
- **Answer**: When the device is connected to Aruba Central, it is not possible to make configuration changes.

5. Attempt to access the configuration mode and review the available options.

```
sw-edge1# configure
sw-edge1(config)# ?
  aruba-central  Configure Aruba-Central
  debug          Configure debug logging
  end            End current mode and change to enable mode
  exit           Exit current mode and change to previous mode
  list           Print command list
  no             Negate a command or set its defaults
```

6. Attempt to create a new VLAN.

```
sw-edge1(config)# vlan 1000
Invalid input: vlan
```

- **Question**: What do you observe?
- **Answer**: The command is not available. The device is currently managed by Aruba Central, to which it has an active connection. You are now locked out of making configuration changes.

Note
Please note that it is still possible to disable the Aruba Central feature on the switch in this mode. If you do this, the switch reverts to local configuration mode.

Aruba support mode

In the next steps you will review the default Aruba Support mode status.

7. Review the previous **show aruba-central** output.

```
sw-edge1(config)# show aruba-central
Central admin state                          : enabled

Central location                             : de-
vice-uswest4-d2.central.arubanetworks.com
VRF for connection                           : default
Shared Token                                 : N/A
Central connection status                    : connected
Central source                               : activate
Central source connection status             : connected
Central source last connected on             : Thu Oct 20 14:44:59 UTC
2022
System time synchronized from Activate       : True

Activate Server URL                          : devices-v2.arubanet-
works.com
CLI location                                 : N/A
CLI VRF                                      : N/A

Source IP                                    : 10.1.3.4
Source IP Overridden                         : False

Central support mode                         : disabled
```

- **Question**: What is the current support mode?
- **Answer**: Disabled.

- **Question**: What does it mean?
- **Answer**: With Aruba support mode enabled, Aruba Central will not perform configuration sync and it will be possible to make local configuration changes. After disabling the support mode, Aruba Central will push the latest version of the configuration (known to Aruba Central) to the device again. You will use this command later in this task.

Apply a template with a Configuration Error

In the next steps you will load a template that contains a configuration error.

1. In Aruba Central, navigate to: Context: **Groups / campus-sw-edge-tpl** > Navigation: **Devices** > Top: **Switches** > Right-top: **Config**.
2. Click **Templates**.
3. From the ACAF Student Files folder, open **lab14.01-sw-edge-tpl-troubleshoot-Jxxxx.txt**. Replace *xxxx* with your switch product code. You have registered this product code earlier in this lab (see Figure 14-24 for an example).

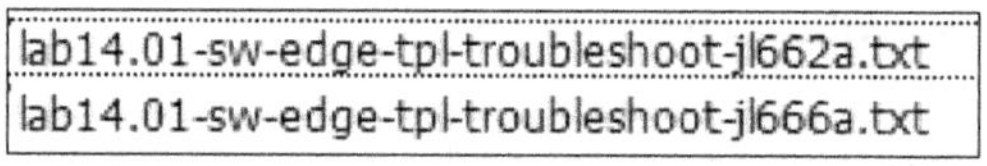

Figure 14-24: Locating template with the configuration error

4. Copy the contents of the file.
5. In Aruba Central, edit the template **sw-edge** using the **pencil** icon.
6. **Replace** the contents of the template with the contents of the troubleshoot text file.
7. Click **Save**.

Verify the problem on the switch

In the next steps you will verify that the switch has a problem connecting to Aruba Central.

1. Switch to the PC-MGMT SSH connection to sw-edge1.
2. Wait about one minute for the template to get pushed to the device. Attempt to reach an internet IP. This should fail now, because of the template error.

```
sw-edge1(config)# ping 8.8.8.8
PING 8.8.8.8 (8.8.8.8) 100(128) bytes of data.
From 10.1.3.4 icmp_seq=1 Destination Host Unreachable
From 10.1.3.4 icmp_seq=2 Destination Host Unreachable
From 10.1.3.4 icmp_seq=3 Destination Host Unreachable
```

```
--- 8.8.8.8 ping statistics ---
5 packets transmitted, 0 received, +3 errors, 100% packet loss,
time 4024ms
```

3. Review the Aruba Central status.

```
sw-edge1(config)# show aruba-central
Central admin state                          : enabled

Central location                             : de-
vice-uswest4-d2.central.arubanetworks.com
VRF for connection                           : default
Shared Token                                 : N/A
Central connection status                    : connected
...
```

4. Repeat the command every 30 seconds. The connection should be disconnected after about a minute.

```
sw-edge1(config)# show aruba-central
Central admin state                          : enabled

Central location                             : de-
vice-uswest4-d2.central.arubanetworks.com
VRF for connection                           : default
Shared Token                                 : N/A
Central connection status                    : connection_failure
...
```

Verify the configuration access

Now verify your access to the configuration mode and the available commands.

5. While in configuration mode, check the available options.

```
sw-edge1(config)# ?
  aaa                          Configure Authentication, Authorization
and
                               Accounting feature
   access-list                 Access control list (ACL)
   alias                       Create a short name for the specified
                               command(s).
   allow-unsafe-updates        Allow non-failsafe updates of programmable
                               devices
  allow-unsupported-transceiver Allow unsupported transceivers
```

```
   apply                       Apply a configuration record
   aruba-central               Configure Aruba-Central
   banner                      Customize login banner
   bfd                         Enable Bidirectional Forwarding Detection
                               (BFD)
   bluetooth                   Configure Bluetooth wireless management
...
```

6. Attempt to create a VLAN.

```
sw-edge1(config)# vlan 1000
sw-edge1(config-vlan-1000)# exit
```

- **Question**: Did this work?
- **Answer**: Yes, when the Aruba Central connection is disconnected, local configuration changes can be made. These local changes will be overwritten by the Aruba Central version again when the connection is restored.

7. *Optional step*: You may attempt to troubleshoot the configuration issue yourself. If you get stuck, you can move to the next page to implement the solution.

First attempt to correct the issue

You will now discover the configuration issue that was introduced by the template.

1. Review the IP routing table.

```
sw-edge1(config)# show ip route

Displaying ipv4 routes selected for forwarding

Origin Codes: C - connected, S - static, L - local
              R - RIP, B - BGP, O - OSPF
Type Codes:   E - External BGP, I - Internal BGP, V - VPN, EV - EVPN
              IA - OSPF internal area, E1 - OSPF external type 1
              E2 - OSPF external type 2

VRF: default

Prefix               Nexthop              Interface     VRF(egress)
Origin/   Distance/   Age
                                                                   Type
Metric
------------------------------------------------------------------
-------------------------------
```

```
0.0.0.0/0            10.1.3.254        vlan3          -           S
[1/0]        00h:03m:15s
10.1.3.0/24          -                 vlan3          -           C
[0/0]        -
10.1.3.4/32          -                 vlan3          -           L
[0/0]        -
Total Route Count : 3
```

- **Question**: What is the default route next hop?
- **Answer**: The default route 0.0.0.0/0 points to 10.1.3.254.

2. Review the route commands in the running configuration.

```
sw-edge1(config)# show run | include route
ip route 0.0.0.0/0 10.1.3.254
```

3. Since you have access to the configuration, you can apply the correct command to the configuration.

```
sw-edge1(config)# ip route 0.0.0.0/0 10.1.3.1
sw-edge1(config)# exit
```

4. Verify you can reach the internet again. Ping 8.8.8.8; use 180 repetitions.

Note

When you expect longer command output, such as a ping with many repetitions, your command output will be stopped due to the paging system. Example:

```
-- MORE --, next page: Space, next line: Enter, quit: q
```

You can disable the paging before running your command to prevent your command from being stopped:

```
no page
```

```
sw-edge1# no page
sw-edge1# ping 8.8.8.8 repetitions 180
PING 8.8.8.8 (8.8.8.8) 100(128) bytes of data.
76 bytes from 8.8.8.8: icmp_seq=1 ttl=114 (truncated)
76 bytes from 8.8.8.8: icmp_seq=2 ttl=114 (truncated)
76 bytes from 8.8.8.8: icmp_seq=3 ttl=114 (truncated)
76 bytes from 8.8.8.8: icmp_seq=4 ttl=114 (truncated)
...
```

- **Question**: What do you observe?
- **Answer**: The ping works again.

5. Leave the ping running; wait about 1-3 minutes. The switch will re-establish the connection to Aruba Central in this period.

```
76 bytes from 8.8.8.8: icmp_seq=23 ttl=114 (truncated)
76 bytes from 8.8.8.8: icmp_seq=24 ttl=114 (truncated)
76 bytes from 8.8.8.8: icmp_seq=25 ttl=114 (truncated)
76 bytes from 8.8.8.8: icmp_seq=26 ttl=114 (truncated)
76 bytes from 8.8.8.8: icmp_seq=27 ttl=114 (truncated)
From 10.1.3.4 icmp_seq=28 Destination Host Unreachable
From 10.1.3.4 icmp_seq=29 Destination Host Unreachable
From 10.1.3.4 icmp_seq=31 Destination Host Unreachable
From 10.1.3.4 icmp_seq=32 Destination Host Unreachable
From 10.1.3.4 icmp_seq=34 Destination Host Unreachable
From 10.1.3.4 icmp_seq=35 Destination Host Unreachable
```

- **Question**: What happens with the ping after about 1-2 minutes? (This may occur before earlier as well; it depends on when the connection to Aruba Central was restored.)
- **Answer**: It stopped again with a *destination host unreachable* message.
- **Question**: What happened?
- **Answer**: When internet access was restored, the switch connected to Aruba Central and received the configuration version from Aruba Central again. This re-introduced the issue in the switch configuration.

Second attempt to correct the issue

You will now try this again, but now you will use the correct procedure. To give you time to troubleshoot, you will enable Aruba Support mode. This *prevents* the configuration push, even when the Aruba Central connection is active.

1. Enable aruba support mode.

```
sw-edge1# aruba-central support-mode
```

2. Change the default route again.

```
sw-edge1# config
sw-edge1(config)# ip route 0.0.0.0/0 10.1.3.1
```

3. Verify Aruba Central is connected again (this may take a few minutes).

```
show aruba-central
```

Note
You can force the connection attempt by disabling and enabling the Aruba Central feature on the switch.

```
sw-edge1(config)# aruba-central
sw-edge1(config-aruba-central)# disable
sw-edge1(config-aruba-central)# enable
sw-edge1(config-aruba-central)# exit
```

4. As you have now 'found and corrected' the issue on the switch (for now), the next step is to apply this change to the template in Aruba Central.
5. In Aruba Central, correct the template. Navigate to: Context: **Groups / campus-sw-edge-tpl** > Navigation: **Devices** > Top: **Switches** > Right-top: **Config.**
6. Click **Templates.**
7. You can either change the configuration yourself in the template or upload the known-good text from the lab drive.
8. From the ACAF Student Files, open **lab07.02-sw-edge-tpl-jxxxx.txt**. (Replace Jxxxx with your product code.)
9. Copy the contents of the file.
10. In Aruba Central, edit the template **sw-edge** using the **pencil** icon.
11. Replace the contents of the template.
12. Click **Save.**

Verify the configuration push and the resulting status: In sync

1. On the PC-MGMT, switch to the SSH connection to **sw-edge1**.
2. Disable Aruba support mode. This will trigger the template configuration push.

```
no aruba-central support-mode
```

3. In Aruba Central, navigate to: Context: **Groups / campus-sw-edge-tpl** > Navigation: **Devices** > Top: **Switches** > Right-top: **List.**

4. In the Config Status column, the value should be *In Sync*.
5. On the PC-MGMT, switch to the SSH connection to sw-edge1.

```
sw-edge1(config)# show aruba-central
Central admin state                        : enabled

Central location                           : de-
vice-uswest4-d2.central.arubanetworks.com
VRF for connection                         : default
Shared Token                               : N/A
Central connection status                  : connected
...
```

6. Verify the connection to Aruba Central is still active.

Task 3: Site-level troubleshooting

When an issue occurs at a site, it may be useful to see what configuration changes were made to site devices in recent history. In this task you will learn how to see when configuration changes were made, on what devices they were made, and what those changes were.

Objectives

- Explore site level change logs in Aruba Central.

Steps

1. In Aruba Central, navigate to: Context: **Site / site-campus-main** > Navigation: **Overview** > Top: **Site Health**.
2. Under the Summary Statistics, click the *Change Log* tile.
3. At the top-right, change the period history to **1 day** (see Figure 14-25).

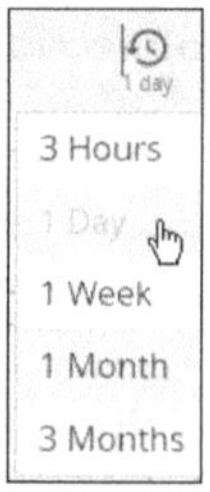

Figure 14-25: Change the period history

4. The change log shows configuration, firmware, and reboot events that were initiated from Aruba Central (see Figure 14-26). In the previous lab activity (Monitoring) you disabled the guest WLAN. This was registered by Aruba Central as a configuration change.

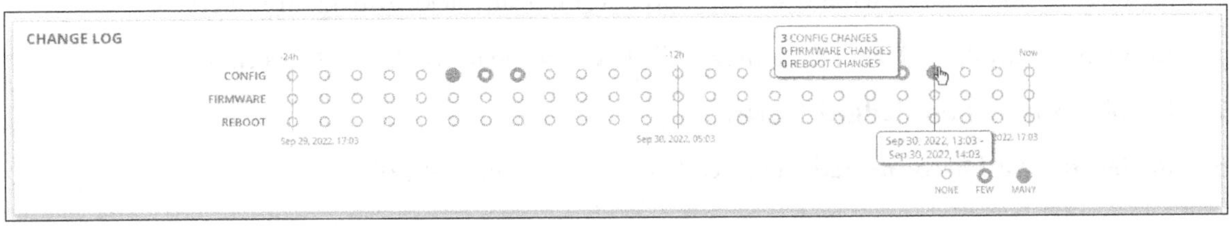

Figure 14-26: The change log

5. Hover your mouse over the config row over specific dots (days and times of the day). An overlay will appear and will show how many configuration changes were made, if any. Click the **dot** that contains a config change (see Figure 14-27).

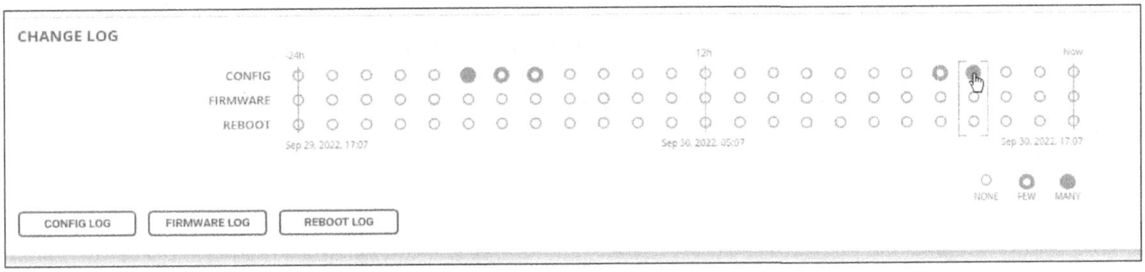

Figure 14-27: Accessing configuration changes

6. At the bottom-left you see the config log. Click **Config Log**.
7. Expand the entry (or entries) in the log to see the details (see Figure 14-28).

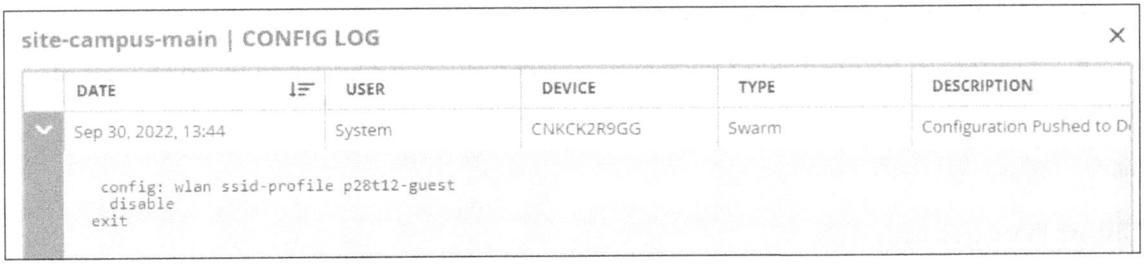

Figure 14-28: Configuration changes details

8. This may help you to collect data about the problem at the site.

Review the alert system acknowledgement

In the previous lab you enabled alerts when an AP is disconnected. With the switch template issues that were introduced in one of the previous tasks, your AP was disconnected for a while, and then it reconnected. Aruba Central will notice this and automatically acknowledge this alert.

1. In Aruba Central, navigate to: Context: **Global** > Navigation: **Alerts & Events**.
2. Enable **Show Acknowledged Alerts**.
3. Review the **AP Disconnected** alert (see Figures 14-29 and 14-30).

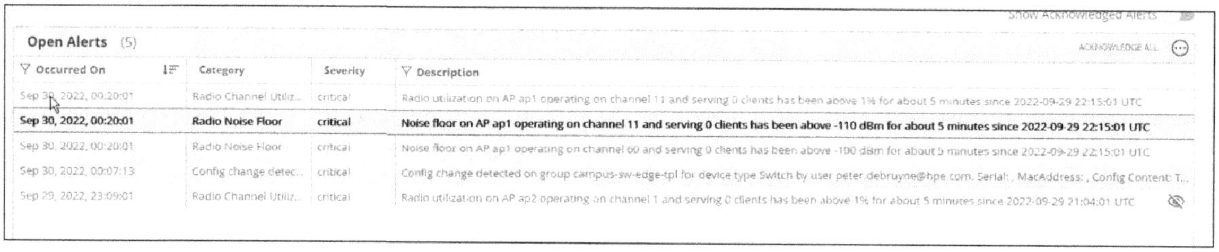

Figure 14-29: Viewing open alerts

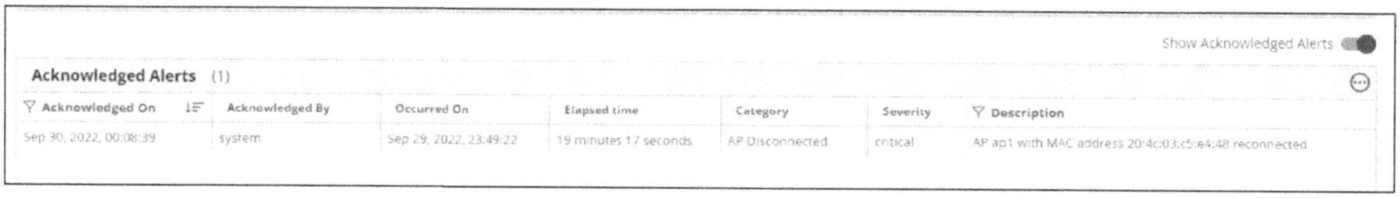

Figure 14-30: Viewing the AP Disconnected alert

- **Question**: What is the user that acknowledged the AP Disconnected alert?
- **Answer**: System. When Aruba Central detects that the AP is online again, it automatically acknowledges the alert.
- **Question**: What is the elapsed time?
- **Answer**: This is the time it took you to correct the switch connection to Aruba Central.

Learning check

Discuss these questions.

1. To perform the password recovery process on an AOS-CX device, you must connect via the console port and power cycle the switch.
   a. True
   b. False

2. Both the **erase all zeroize** and **erase-startup** commands delete the startup checkpoint.

   a. True

   b. False

3. You can reset an AP's CLI password from Aruba Central.

   a. True

   b. False

4. When using the command **show logging -r**, the "-r" means to list log entries in reverse chronological order.

   a. True

   b. False

5. Examine CLI output. What does the highlighted output indicate? (Select two.)

```
AP-50-a8# show activate
IAP MAC Address          :20:4c:03:32:aa:a8
IAP Serial Number              :CNG9K2R64T
Cloud Activation Key           :GHGMVKXI
Activate Server                :device.arubanetworks.com
Activate Status                :fail-prov-no-shipped
Provision interval             :5 minutes20:4c:03:32:aa:a8#
```

   a. There is no entry for this AP in Aruba Activate.

   b. The AP has no internet access.

   c. Aruba's Activate service is offline.

   d. The AP is currently being provisioned by Aruba Central.

6. Aruba Instant AP (IAP) radios can act as spectrum monitors, to gather spectrum data while servicing clients.

   a. True

   b. False

7. You are using Aruba Central's troubleshooting tools to analyze some network issues. Which actions can you perform using these tools? (Select all that apply.)

   a. Ping test

   b. Bounce an interface

   c. Blink the switch locator LED

   d. Start a remote console session to a device

Summary

Congratulations! You have completed Chapter 14. You should now be able to:

- Perform password recovery and factory reset procedures.
- Troubleshoot Aruba Central connectivity.
- Enable spectrum analysis.
- Explore Aruba Central troubleshooting tools.

15 HPE6-A85 Sample Exam

OBJECTIVES

The purpose of this sample exam is to:

✓ Reinforce the material you have learned.

✓ Become familiar with the type and style of questions found on the Aruba Campus Access Associate (ACAA) exam.

Overview

This chapter provides sample test questions based on the style and difficulty you would find on the real HPE6-A85 exam for the Aruba Campus Access Associate certification exam.

Exam overview

This certification validates the candidate's basic wired and wireless network knowledge, including routing and switching, basic RF concepts, basic security, connectivity, monitoring, and reporting.

The ideal candidate has 1+ years of practical experience with networking and a vendor agnostic understanding of basic network protocols. Under the direction of a Professional or Expert, the candidate should apply a recommended configuration and verify the status of a campus network.

Ideal candidate for this exam

The typical candidate for this certification is a network field technician, deployment technician, network operation center (NOC) level 1, or junior network engineer with *hands-on* experience with the Aruba products covered in this study guide. It is suggested all candidates take the related course for this exam.

Exam contents

This exam has 60 questions. Here are types of questions to expect:

- Multiple choice (multiple responses)
- Multiple choice (single response)

Advice to help you take this exam

Complete the training and review all course materials and documents before you take the exam. Exam items are based on expected knowledge acquired from job experience, an expected level of industry-standard knowledge, or other prerequisites (events, supplemental materials, etc.).

Successful completion of the course (Aruba Campus Access Fundamentals, Rev. 22.41) or study materials alone does not ensure you will pass the exam. Aruba and HPE have recently changed their philosophy about how they will be testing candidates. Historically, all questions were based on the material found in the related course and HPE Press study guide. This is still true for the bulk of the testing material. However, you should expect questions on the exam objectives that are from *other* sources, including the following, than this study guide or the related official courseware, to name a few:

- Practical experience of working with the Aruba AOS-CX, AP, Central, and other related Aruba products
- Information found in the Aruba ESP Campus Design Validation Guide: https://www.arubanetworks.com/techdocs/VSG/docs/010-campus-design/esp-campus-design-000/
- Information found in the Aruba AOS-CX and AP user and configuration guides found on Aruba's support site

Please note that this list is by no means comprehensive, since anything listed in the exam objectives (found in the next section) is fair game for the exam. Since hands-on experience is emphasized for the exam preparation, the more time you have configuring Aruba solutions, the better prepared you will be for the exam. If you take the related course, lab equipment is provided during the week on which you can practice. Likewise, you can purchase lab time directly from Aruba, typically in four-hour time blocks, for additional lab time. The remote labs Aruba provides are based on the actual labs listed in this study guide and which is used for the actual physical/virtual course or eLearning alternative. As of the writing of this book, the related course and lab access was not posted on Aruba's web site, but should be listed sometime in January 2023, when the course is officially released to the public. Information about the course and lab access, when added, will be found here: https://www.arubanetworks.com/support-services/training-services/.

Exam objectives and exam information

This section covers the objectives for the HPE6-A85 exam, as well as information about the exam.

HPE6-A85 exam objectives

Table 15-1 breaks down the objectives covered on the exam.

Table 15-1. HPE6-A85 exam objectives.

Percentage of exam	Sections/Objectives
7%	Network stack 1.1 Describe and differentiate various OSI Model layers and architectural elements. 1.2 Describe and differentiate TCP/IP stack.
10%	Connectivity 2.1 Identify foundational networking architectures and technologies. 2.2 Describe network infrastructure deployment.
8%	Network resiliency and virtualization 3.1 Describe mechanisms for resiliency, redundancy, and fault tolerance. 3.2 Manage an infrastructure device (AP, gateway, switch).
19%	Switching 4.1 Describe the basics of layer 2/3 technologies. 4.2 Review and apply a campus architecture and elements based on a given template.
18%	WLAN 5.1 Describe fundamental RF attributes and wireless functions. 5.2 Create an AP GUI configuration group.
9%	Routing 6.1 Differentiate between basic routing topologies and functions. 6.2 Evaluate and deploy basic routing topologies.
9%	Security 7.1 Define fundamental security standards and concepts. 7.2 Integrate wireless SSID into an existing ClearPass deployment.
6%	Authentication and authorization 8.1 Describe and differentiate between AAA concepts.
7%	Management and monitoring 9.1 Define and differentiate between common tools used for network monitoring. 9.2 Evaluate and interpret UXI test results. 9.3 Use Central to monitor network health.
4%	Troubleshooting 10.1 Define basic troubleshooting methodologies.
3%	Performance optimization 11.1 Describe QoS.

Exam information

Table 15-2 summarizes the information to register for the exam. You need an HPE Learner ID and a Pearson VUE login and password. No reference material is allowed at the testing site. This exam may also contain beta test items for experimental purposes. During the exam, you can make comments about the exam items. HPE/Aruba welcomes these comments as part of our continuous improvement process. To register for the exam, visit http://www.pearsonue.com/hpe.

Table 15-2. HPE6-A85 exam information.

Exam ID	HPE6-A85
Exam type	Proctored
Exam duration	1 hour 30 minutes
Exam length	60 questions
Passing score	67%
Delivery languages	English
Supporting resources	Aruba Campus Access Fundamentals course, Rev. 22.41, available in physical, virtual, and eLearning formats

Note

More and current information concerning the HPE6-A85 exam can be found here: https://certification-learning.hpe.com/TR/datacard/Exam/HPE6-A85.

Test preparation questions

The questions listed below are similar to the types of questions you would find on the HPE6-A85 exam. For the sample example provided here, there are two to three questions each from Chapters 1 through 14, totaling 30 questions. Answers can be found in Appendix A. Good luck!

1. Which OSI Reference Model layer is responsible for media access and error detection?
   a. Transport layer
   b. Network layer
   c. Data link layer
   d. Physical layer

2. What components of TCP allow it to implement effective flow control? (Select two.)
   a. Sequencing and acknowledgements
   b. Time-to-live (TTL)
   c. Type of service (TOS)
   d. Windows size
   e. 802.1Q tag
3. Which of the following is true concerning the interfaces of an 6400 AOS-CX switch?
   a. The slot number is always 1.
   b. Interface 1/1/1 is the management interface, by default.
   c. All interfaces are routed ports, by default.
   d. All ports belong to VLAN 0, by default.
4. An administrator is connecting port 1/1/1 on an AOS-CX 6300 switch to an Aruba 6400 switch. The 6400 switch's port is in its default mode and has an IP address assigned to it. The administrator is attempting to configure an IP address on interface 1/1/1 of the 6300, but the switch's CLI displays an error message when executing the **ip address** command utilizing the correct syntax. What must the administrator do to assign an IP address to interface 1/1/1?
   a. Execute the **routing** command on the interface.
   b. Convert the interface into an SVI.
   c. Include the interface in an OSPF process.
   d. Use the **ip route** command to define the IP address.
5. An edge device is in a VLAN where an AOS-CX switch-pair is configured for VSX. The edge device is configured for DHCP, but is not receiving IP addressing information from a DHCP server located in a different VLAN. What command should the administrator configure on the edge VLAN interface of the AOS-CX VSX switches to allow the edge device to dynamically acquire IP addressing information?
   a. ip helper-address
   b. router vrrp enable
   c. active-gateway ip
   d. vsx-sync active-gateways

6. What is true concerning OSPF routers in the same area?
   a. They have the same router ID (RID).
   b. The must be in the same subnet.
   c. There is only one DR/BDR elected for the area.
   d. They have the same LSDB entries.
7. An administrator is setting up OSPF on AOS-CX switches in a single-area design. The administrator will be connecting both user and server devices to the switches in the area. The administrator is concerned about the AOS-CX switches forming an adjacency to the edge devices. What command can the administrator configure for the edge VLANs to ensure that the AOS-CX switches do not form adjacencies with the edge user and server devices, but this network is still advertised to the rest of the area?
   a. ip ospf network point-to-point
   b. ip ospf passive
   c. no router ospf
   d. no area
8. An AOS-CX switch has STP globally enabled. It has two paths to reach the root, and both paths have the same accumulated path cost. What is the first tie breaker the switch will use to pick a root port?
   a. Lowest sender bridge ID
   b. Lowest MAC address
   c. Lowest port priority
   d. Lowest port ID
9. An administrator has a pair of AOS-CX 6400 switches used for a core, where all access layer switches will connect directly to this core. The administrator wants to ensure that there is default gateway redundancy where both switches can simultaneously forward traffic for the access layer VLANs. What feature should the administrator configure on the 6400 switches?
   a. VRRP
   b. IP helper
   c. VSX active-active gateway
   d. VSF

10. An administrator has four AOS-CX 6200 switches that are configured in a ring topology for VSF. The switches are interconnected using port 24. These switches are 24-port switches with two uplinks each. The administrator is trying to use the auto-stacking feature for forming a VSF stack, but the stack is not forming. What should the administrator do to fix this issue?

    a. Define the member number on each switch.

    b. Connect the switches using port 25 or 26.

    c. Define the VSF link with the **link** command.

    d. Enable MAD.

11. An administrator has configured a stack of eight AOS-CX switches using VSF. The administrator is concerned about a split stack and wants to implement MAD. Which interface should the administrator use to detect split-brain issues using MAD?

    a. mgmt

    b. Any uplink port

    c. SVI

    d. Loopback

12. Which AOS-CX feature can a company leverage to be proactive with the monitoring of their Aruba switches using a RESTful API?

    a. VSX

    b. Dynamic segmentation

    c. SNMP

    d. Network Analytics Engine

13. What is one difference between the 600-series Aruba APs compared to the 300 and 500 series?

    a. It supports 6 GHz radios.

    b. It supports more spatial streams.

    c. It supports more antennas.

    d. It supports 1, 2.5, and 5 gigabit Ethernet ports.

14. Which Aruba component helps a device, when booting, find Central in order to download its configuration?

    a. ClearPass

    b. Central on Premise (CoP)

    c. DHCP

    d. Activate

15. A company has purchased some new Aruba switches and APs and wants to manage these in Aruba Central. Where would the company's administrator navigate in order to manually assign subscription licenses for these new devices?

    a. Aruba Activate
    b. Aruba Central dashboard
    c. Aruba Central device configuration page
    d. HPE GreenLake dashboard

16. A small company wants to dynamically provision ports when devices are connected to them, like VoIP phones, security cameras, and APs. The company is not using Central for management and does not have an AAA server. What profiling configuration must the administrator create to define settings like VLAN, QoS trust mode, and security polices to these ports dynamically?

    a. Device profiles
    b. Aruba user roles
    c. LLDP group
    d. SNMP group

17. An administrator is using Central to manage APs that were newly added. The APs were at their factory default configuration. The administrator wants to place these APs on floorplans in Central; however, the administrator only sees the AP names as MAC addresses, and thus doesn't know which AP is which when assigning them to the floorplan location. What command on the AP console can the administrator use to see the AP's MAC address and serial number from its boot mode?

    a. show version
    b. mfginfo
    c. show system
    d. show serial-info

18. What technology is used to implement multi-user MIMO, such that multiple signals can be generated from an antenna simultaneously?

    a. Beamforming
    b. Airmatch
    c. SISO
    d. Azimuth

19. Which of the following information can be seen in a Wi-Fi beacon that is generated by an AP? (Select two.)
    a. Beacon interval
    b. WPA methodology supported
    c. Radio type
    d. Number of currently connected devices on the AP
    e. Supported data rates
20. An administrator is managing a network of Aruba APs using Central. An employee WLAN was previously created using Central's WLAN wizard. Under which WLAN wizard tab would the administrator navigate in order to create a new role for a new department that was recently added within the company?
    a. Access
    b. General
    c. Security
    d. VLANs
21. What protocol is used between a supplicant and an authentication server?
    a. 802.1X
    b. EAP
    c. LLDP
    d. RADIUS
22. An administrator is creating a new WLAN using Central's WLAN Wizard. Under which tab can the administrator configure the broadcast/multicast and transmission rates for the new WLAN?
    a. Access
    b. General
    c. Security
    d. VLANs

23. A company wants to implement guest Wi-Fi access, where the company wants to authenticate guests, but not maintain guest accounts. The company administrator notices that social login authentication is supported by Aruba guest Wi-Fi access. What social login services can the company use to implement this authentication? (Select two.)

    a. Microsoft 365

    b. Google

    c. Twitter

    d. DropBox

    e. SAP

24. An administrator is creating a new guest WLAN for their company. The company mandates that guests can only connect using the company's APs on their 2.4 GHz radios. When running the WLAN wizard, where would the administrator implement this restriction?

    a. Access

    b. General

    c. Security

    d. VLANs

25. A company is concerned about using an open guest network. Which 802.11 standard should the company implement to encrypt guest traffic, where guests would still authenticate using a captive portal using a social login service?

    a. mPSK

    b. OWE

    c. WPA3 Personal

    d. WPA3 Enterprise

26. A company will be implementing 802.1X using EAP-TLS. What information on the root certificate can a supplicant use to verify the authentication server's identity?

    a. Issuing entity

    b. Public key

    c. Serial number

    d. Subject name

27. An administrator purchased new non-Aruba low-end switches for their network. These switches support PoE. The administrator plugs in the APs to these switches. After a couple of hours, the administrator notices that all the Aruba APs connected to these switches have the system status LED set to a solid amber. What does this indicate?

    a. The APs are running in restricted power mode.

    b. The APs are stuck in a reboot loop.

    c. The APs uplinks are running at a suboptimal speed.

    d. The APs are experiencing a system error condition.

28. Which component of Aruba Central's AIOps portfolio uses event-driven processes to initiate automation that will automatically collect diagnostic information that an administrator can use to troubleshoot an issue?

    a. AI Assist

    b. AI Insight

    c. AI Search

    d. Live Upgrade

29. An administrator has inherited a handful of AOS-CX switches at a newly acquired branch office. The administrator doesn't know the username and password to log into the switches. How can the administrator gain access to the switches' CLI in order to change the password information?

    a. From the username and password prompt, enter a username of "password" and a password of "forgetme!".

    b. From the username and password prompt, press <Cntrl>C and select password recovery mode.

    c. From the username and password prompt, press <Cntrl>P and select the admin account.

    d. Reboot the switches and change the password from the Service OS console.

30. An administrator was using Aruba Central to configure AOS-CX switches. The administrator made a VLAN configuration mistake for a group of switches, pushing the incorrect native VLAN to these switches. This misconfiguration caused the switches to lose access to Central. The administrator fixed the template configuration in Central with the correct native VLAN, but the AOS-CX switches still cannot connect to Central. The switches' rollback feature was not implemented. What should the administrator do to fix this issue?

    a. Boot to the Service OS Console and enable the Central recovery option.

    b. Enter lockout mode on the switches and manually fix the native VLAN on the trunk port.

    c. Enable disaster recovery mode on the switches and manually fix the native VLAN on the trunk port.

    d. Execute the **show aruba-central** command and wait for a connection failure. Then manually fix the native VLAN on the trunk port.

Appendix
Appendix A

Overview

This appendix contains the answers to all the Learning Check sections found in the previous chapters and the answers to the sample questions in Chapter 15.

Chapter 1: Networking Fundamentals

This section contains the Learning Check answers for Chapter 1.

Learning Check Answers

1. ☑ **C and D**
2. ☑

Layer	Description
Physical	**Deals with bits, wires, and frames**
Network	**Involves routers that send packets**
Application	**End users interact with this layer**
Transport	**Involves protocols like TCP and UDP**
Data link	**Involves switches that send frames**

3. ☑ **True**
4. ☑ **A and D**

5. ☑

Layer	Description
Network interface	**You might have an Ethernet or wireless network Interface card at this layer**
Application	**Maps to three specific layers of the OSI model**
Transport	**Passes data to the application layer**
Internet	**Deals with IP addresses**

6. ☑ **B and D**

Chapter 2: AOS-CX Switching Fundamentals

This section contains the Learning Check answers for Chapter 2.

Learning Check Answers

1. ☑ **A and C**
2. ☑ **B**
3. ☑ **D**
4. ☑ **A, B, and D**
5. ☑ **B and D**
6. ☑ **B**
7. ☑ **A and B**
8. ☑ **A, B, and D**

Chapter 3: Basic IP Setup

This section contains the Learning Check answers for Chapter 3.

Learning Check Answers

1. ☑ **The host portion of the address is "42".**

2. ☑

Description	Interface type
Involves an IP address	**Routed port**
Used by multi-layer switches for internal IP routing between VLANS	**Switch Virtual Interface (SVI)**
Assigned as a member of a VLAN	**Switching port**
A physical interface with an assigned IP address	**Routing port**
Involves a MAC address	**Switching port**

3. ☑ **B and D**

4. ☑ **C**

5. ☑ **C and D**

6. ☑

Description	Table
A list of directly connected OSPF routers	**Neighbor table**
A list of the best paths for each destination	**Routing table**
A list of every possible path in the OSPF area	**Topology database**
Is the same for every OSPF router in the area	**Topology database**

Chapter 4: Network Redundancy

This section contains the Learning Check answers for Chapter 4.

Learning Check Answers

1. ☑ **A, B, and D**
2. ☑

Description	Standard
You map certain VLANs to a particular STP instance	**802.1s MSTP**
Default operation between AOS-CX switches	**802.1w RSTP**
Default mode for AOS-CX switches	**802.1s MSTP**
Improved convergence for older versions	**802.1w RSTP**
Slow convergence	**802.1d STP**
Not recommended	**802.1d STP**

3. ☑ **D**
4. ☑ **B and D**
5. ☑ **A and C**
6. ☑ **A, B, and C**

Chapter 5: VSF

This section contains the Learning Check answers for Chapter 5.

Learning Check Answers

1. ☑ **B, C, D, and F**
2. ☑ **D**
3. ☑ **A**

Chapter 6: Introduction to Aruba Solutions

This section contains the Learning Check answers for Chapter 6.

Learning Check Answers

1. ☑ **A, B, C, and D**
2. ☑ **C**
3. ☑ **B**
4. ☑ **D**
5. ☑ **B and D**
6. ☑ **B**
7. ☑ **A**

Chapter 7: Central for Device Management

This section contains the Learning Check answers for Chapter 7.

Learning Check Answers

1. ☑ **A, C, and D**
2. ☑ **A, C, D, E, and F**
3. ☑ **A, C, and D**
4. ☑

Step	Description
Step 1	**Create the group**
Step 2	**Define the group persona**
Step 3	**Add devices to the group**

5. ☑ **A**

Chapter 8: AP Onboarding and Device Profiles

This section contains the Learning Check answers for Chapter 8.

Learning Check Answers

1. ☑ **A, B, and D**
2. ☑ **A**

Chapter 9: WLAN Fundamentals

This section contains the Learning Check answers for Chapter 9.

Learning Check Answers

1. ☑ **A and B**
2. ☑ **A, B, and C**
3. ☑ **A**
4. ☑ **A, B, and E**

Chapter 10: Implementing Secure WLANs

This section contains the Learning Check answers for Chapter 10.

Learning Check Answers

1. ☑

Term	Definition
Authentication	**Permit or deny access based on credentials and/or digital certificates**
Authorization	**Control authenticated user capabilities**
Accounting	**Track user activity**

2. ☑ **C**

3. ☑ **B, C, and D**

4. ☑ **B and C**

5. ☑

Description	EAP method
Very strong security, mitigates risk of password authentication	**EAP-TLS**
Easier to implement than some other methods, but slightly weaker security	**PEAP**
May have relatively limited support as compared to other methods	**EAP-TEAP**
Enables chaining of machine and user authentication with tokens	**EAP-TEAP**
Requires the overhead of managing CA services, certificate distribution, and management	**EAP-TLS**

6. ☑ **B, C, and D**

Chapter 11: Guest Access

This section contains the Learning Check answers for Chapter 11.

Learning Check Answers

1. ☑ **A, B, D, and F**

2. ☑ **A**

3. ☑ **B**

4. ☑ **A**

5. ☑ **A**

Chapter 12: WLAN Security

This section contains the Learning Check answers for Chapter 12.

Learning Check Answers

1. ☑

Description	WPA method
Pre-shared key (PSK)	**WPA-Personal**
WPA3 Advanced Encryption Standard (AES)	**WPA-Enterprise** **WPA-Personal**
Opportunistic Wireless Encryption: AES	**Open**
WPA2: Advanced Encryption Standard (AES)	**WPA-Enterprise** **WPA-Personal**
802.1X	**WPA-Enterprise**

2. ☑ **B**
3. ☑ **B**
4. ☑ **A**
5. ☑ **A**
6. ☑ **A, B, and C**
7. ☑ **A, B, C, and D**

Chapter 13: Monitoring and Maintenance

This section contains the Learning Check answers for Chapter 13.

Learning Check Answers

1. ☑ **A**
2. ☑ **D**

3. ☑

Description	Feature
See a list of supported firmware versions per in-group device	**Firmware compliance**
Minimize impact in a production network	**Live upgrade**
Good to use during a planned maintenance window	**Standard upgrade**
AP upgrade sequence is based on AirMatch partitions	**Live upgrade**
Ensures that devices in a group all run a specific OS version	**Firmware compliance**
May disrupt connectivity	**Standard upgrade**

4. ☑ **A, B, and D**

5. ☑

Definition	Term
Natural language queries for fast troubleshooting	**AI Search**
Proactive anomaly detection and optimization	**AI Insights**
Automated Aruba TAC trouble ticket generation	**AI Assist**

6. ☑ **B**

7. ☑ **A, B, and D**

Chapter 14: Troubleshooting

This section contains the Learning Check answers for Chapter 14.

Learning Check Answers

1. ☑ **A**

2. ☑ **A**

3. ☑ **A**

4. ☑ **A**

5. ☑ **A and B**

6. ☑ **B**

7. ☑ **A, B, C, and D**

Chapter 15: HPE6-A85 Test Preparation Answers

This section contains the answers to the questions found in Chapter 15. Besides listing the correct answer, the wrong answers have explanations as to why they are wrong.

1. ☑ **C**. The data link layer provides three main functions: media access control, link layer addressing, and error detection. **Reference: Chapter 1**.

   ☒ **A** is incorrect because the transport layer controls the reliability of a given link through segmentation, de-segmentation, and error control. **B** is incorrect because the network layer establishes device communications across multiple LANs or WANs, using the best available path based on logical addressing. **D** is incorrect because the physical layer dictates the physical aspects of how signals are transmitted and received across some med.

2. ☑ **A and D**. TCP uses sequencing and acknowledgements to identify sent traffic is received by the destination. Windowing (window size) allows the source and destination control how much information can be sent before waiting for an acknowledgement. The window size can be dynamically reduced in times of congestion and increased when there is no congestion. **Reference: Chapter 1**.

   ☒ **B and C** are incorrect because this information is found in the IP, and not the TCP, header. Answer **E** is incorrect because the 802.1Q tag is found in the Ethernet header.

3. ☑ **C**. All ports on the 6400 are in routed mode, by default. **Reference: Chapters 2 and 6**.

   ☒ **A** is incorrect because the 6400 is a modular switch. **B** is incorrect because the 6400 has a physical management (mgmt) interface. **D** is incorrect because at factory default, all ports belong to VLAN 1 on AOS-CX switches.

4. ☑ **A**. All ports on the 6300 are, by default, in layer 2 mode and must be changed to routed mode with the routing command on the respective interface in order to assign an IP address to the physical interface. **Reference: Chapter 2**.

   ☒ **B** is incorrect. An SVI is a logical interface. Layer 2 physical interfaces can be associated to an SVI, but cannot be converted into an SVI. **C** is incorrect because including a layer 3 interface in OSPF routing will not allow you to assign an IP address to it. **D** is incorrect because the **ip route** command defines a static route.

5. ☑ **A**. The **ip helper-address** command is configured on an AOS-CX layer 3 interface to relay DHCP requests from edge devices to a DHCP server (s). **Reference: Chapter 3**.

   ☒ **B** is incorrect because this is a global command and enables VRRP for default gateway redundancy. **C** is incorrect because this command enables default gateway redundancy for an AOS-CX VSX switch pairing. **D** is incorrect because this command synchronizes the active gateway configuration from a VSX primary to a VSX secondary switch.

6. ☑ **D**. The LSDB entries will be the same for OSPF routers in the same area. **Reference: Chapter 3**.

   ☒ **A** is incorrect because every OSPF router in the same routing process needs a unique RID. **B** is incorrect because they can be in different subnets in an area. **C** is incorrect because a DR/BDR is elected for each broadcast domain (VLAN).

7. ☑ **B**. The **ip ospf passive** interface command defines a layer 3 interface as passive, which means that the switch will not form adjacencies to any OSPF routers connected to that interface, but will still advertise that interface to other OSPF routers connected to other layer 3 interfaces. **Reference: Chapter 3**.

   ☒ **A** is incorrect because **ip ospf network point-to-point** is used between two OSPF directly connected routers to speed up the adjacency process by denoting that no DR/BDR is needed because of the connection type. **C** is incorrect because the **no router ospf** command will delete an entire OSPF routing process. **D** is incorrect because the no area command will completely remove an interface from an area: it will not form adjacencies and will not advertise the network.

8. ☑ **A**. If the accumulated path cost is the same, the neighboring bridge with the lowest BID is selected. **Reference: Chapter 4**.

   ☒ **B** is incorrect because the BID is used, where the BID is comprised of a priority and system MAC address of a switch. **C** is incorrect because this is the second tie breaker. **D** is incorrect because this is the third tie breaker.

9. ☑ **C**. VSX is supported on the 6400 and higher switches and the active-active gateway feature provides both redundancy and forwarding by both gateway switches. **Reference: Chapter 4**.

   ☒ **A** is incorrect because VRRP only allows one active gateway per instance. **B** is incorrect because IP helper allows an AOS-CX switch to relay DHCP requests. **D** is incorrect because the 6400 doesn't support VSF.

10. ☑ **B**. Auto-stacking for VSF requires that the uplink ports (ports 25 and/or 26 on a 24-port switch) be used for the VSF links in the VSF topology (ring or chain). **Reference: Chapter 5**.

    ☒ **A** is incorrect because the member number only needs to be defined if you are manually setting up a VSF stack. **C** is incorrect because the link command is only necessary if you are manually setting up a VSF stack. **D** is only used for dealing with split stack (split-brain) issues in a VSF topology and has nothing to do with the formation of a stack.

11. ☑ **A**. The OOBM *mgmt* interface must be used to detect split-brain conditions when implementing MAD when using VSF on supported AOS-CX switches. **Reference: Chapter 5**.

    ☒ **B, C, and D** are incorrect because these interfaces are not supported for MAD.

12. ☑ **D**. NAE uses Python-based agents to automate network monitoring and troubleshooting operations. You can automate the NAE through the REST API using tools such as Postman, Ansible, or a suitable tool of your choice. **Reference: Chapter 6**.

 ☒ **A** is incorrect because VSX provides a high level of redundancy and resiliency for aggregation and core switches. **B** is incorrect because dynamic segmentation is used to implement user access control, where the AOS-CX switches authenticate users, but tunnel their traffic to Aruba gateways, which apply authorization policies, like firewall filtering. **C** is incorrect because SNMP uses UDP, not HTTPS, for communications.

13. ☑ **A**. The 600 series APs support 6 GHz radios (Wi-Fi 6E). **Reference: Chapter 6**.

 ☒ **B and C** are incorrect because, depending on the AP model, APs in all three series have the same numbers. **D** is incorrect because the 600 series only has 2.5 for 5 gigabit Ethernet ports, not 1 GbE.

14. ☑ **D**. Rules defined in Aruba's free Activate service allow a device to either connect to Central or Airwave to download its configuration when the device is in its factory-default state. **Reference: Chapter 7**.

 ☒ **A** is incorrect because ClearPass performs AAA with access control and doesn't help Aruba devices find Central when booting from a factory-default state. **B** is incorrect because CoP is basically Central running on servers within a company's network. **C** is incorrect because DHCP cannot pass Central information to an Aruba device.

15. ☑ **D**. You view and mange HPE GreenLake subscription keys at the Subscription Management page (**HPE GreenLake > Activities & Tasks > Subscriptions**). **Reference: Chapter 7**.

 ☒ **A, B, and C** are incorrect because you cannot manage subscriptions/licenses from either Activate or Central.

16. ☑ **B**. Roles group access and security policies in a single configuration that can be referenced by many devices and/or user types. Roles allow you to assign many access properties, like QoS, dynamic segmentation tunneling, POE priority, VLAN information, and much more. **Reference: Chapter 8**.

 ☒ **A** is incorrect because a device profile binds together the LLDP/CDP/MAC group with the role that is to be applied. **C** is incorrect because an LLDP group identifies the devices that will be in the group. **D** is incorrect because SNMP is not supported by Aruba devices for device profiling.

17. ☑ **B**. The **mfginfo** boot mode command can be used from an AP's console to see the serial number and MAC address of the AP. **Reference: Chapters 8 and 1**.

 ☒ **A, C, and D** are incorrect because the **show** command isn't supported in boot mode.

18. ☑ **A**. Beamforming technology allows an AP to locate client's antennas and modulate its spatial streams in phase. **Reference: Chapter 9**.

    ☒ **B** is incorrect because Airmatch is an Aruba feature to dynamically adapt the power and channels usage for radios based on RF conditions. **C** is incorrect because SISO doesn't allow multiple simultaneous signals, but a single signal (used in legacy Wi-Fi technologies like 802.11a, b, and g). **D** is incorrect because the azimuth measures the horizontal distance a Wi-Fi signal travels.

19. ☑ **A and E**. In an AP's beacons, you can see the timestamp, beacon interval, a list of available SSIDs, and supported data rates. **Reference: Chapter 9**.

    ☒ **B, C, and D** are incorrect because this information is not found in beacon messages.

20. ☑ **A**. The Access tab is where the role assignment can be defined for a WLAN: unrestricted, network-based, or role-based. **Reference: Chapter 9**.

    ☒ **B** is incorrect because under the General tab is where you can give the SSID a name. **C** is incorrect because the security tab is where you define the WPA method (enterprise or personal) or captivate portal method. **D** is incorrect because the VLAN tab is where you define the traffic forwarding mode and client VLAN assignment policies.

21. ☑ **B**. EAP is used to transport credential information between the supplicant and the authentication server. **Reference: Chapter 10**.

    ☒ **A** is incorrect because 802.1X is used between the supplicant and authenticator. **C** is incorrect because LLDP is not used in 802.1X/EAP communications. **D** is incorrect because RADIUS is typically used between the authenticator and authentication server.

22. ☑ **B**. The General tab is where you can give the SSID a name and, optionally, under the Advanced Settings section, you can define additional properties like the broadcast/multicast characteristics and the transmit rates. **Reference: Chapter 10**.

    ☒ **A** is incorrect because the Access tab is where the role assignment can be defined for a WLAN. **C** is incorrect because the security tab is where you define the WPA method (enterprise or personal) or captivate portal method. **D** is incorrect because the VLAN tab is where you define the traffic forwarding mode and client VLAN assignment policies.

23. ☑ **B and C**. Central's guest Social Login feature supports Facebook, Twitter, LinkedIn, and Google. **Reference: Chapter 11**.

    ☒ **A, D, and E** are incorrect because these services are not supported for Social Login with guest services in Central.

24. ☑ **B**. The General tab is where you can give the SSID a name and, optionally, under the Advanced Settings section, you can define additional properties like the broadcast/multicast characteristics, the transmit rates, and the radio type allowed (radio frequency). **Reference: Chapter 11**.

   ☒ **A** is incorrect because the Access tab is where the role assignment can be defined for a WLAN. **C** is incorrect because the security tab is where you define the WPA method (enterprise or personal) or captivate portal method. **D** is incorrect because the VLAN tab is where you define the traffic forwarding mode and client VLAN assignment policies.

25. ☑ **B**. OWE is not technically part of WPA, but instead rolled into the 802.11 standard. OWE. When a station connects to the WLAN, it passes through open-system authentication as usual. But the 802.11 association frames include extra TLVs, which carry an unauthenticated Diffie-Hellman (DH) exchange to derive a PMK. The station and AP then conduct the four-way handshake to derive all the keys for encrypting and authenticating future data. **Reference: Chapter 12**.

   ☒ **A** is incorrect because mPSK is an Aruba feature that allows multiple pre-shared keys to be used for WLAN using WPA2/3 Personal. **C** is incorrect because WPA3 Personal uses a single PSK to authenticate and secure access to the WLAN. **D** is incorrect because WPA3 Enterprise uses 802.1X/EAP to restrict access to the WLAN.

26. ☑ **B**. The public key on the root certificate can be used to validate the signature on any certificate that the root CA creates. **Reference: Chapter 12**.

   ☒ **A, C, and D** are incorrect because these fields cannot be used to validate the identity of any certificate that the CA generates.

27. ☑ **A**. If the switch port does not provide enough power to the AP, the system status LED will be a solid orange color. **Reference: Chapter 13**.

   ☒ **B** is incorrect because when the APs boot, the system status LED would be blinking green. **C is** incorrect because the system status LED would be blinking amber. **D** is incorrect because the system status LED would be red.

28. ☑ **A**. AI Assist uses event-driven processes to initiate automation that will automatically collect diagnostic information such as logs, packet captures, and etcetera, that an administrator can use to troubleshoot an issue. **Reference: Chapter 13**.

   ☒ **B** is incorrect. AI Insights automatically diagnoses an array of common network-impacting issues by using dynamic, per-site baselines that are continuously tuned as conditions change. **C** is incorrect because the AI Search feature is an AI-based search engine that not only helps you find problems that users are experiencing, but it also points you to Aruba content related to the problem, and where you can make changes in a configuration to fix the problem. **D** is incorrect because Live Upgrade is used to upgrade Aruba APs and gateways.

29. ☑ **D**. When the switch system displays the boot profile prompt, select the *Service OS console* option by typing **0**. Login with the user **admin**; no password is set for this account. Enter the **password** command and type the *new* password for the admin account. **Reference: Chapter 14**.

    ☒ **A, B, and C** are incorrect. These are not supported methods of performing the password recovery procedure on AOS-CX switches.

30. ☑ **D**. When the Aruba Central connection is disconnected, local configuration changes can be made. **Reference: Chapter 14**.

    ☒ **A and C** are incorrect because these are not options for AOS-CX switches. **B** is incorrect because you first have to change the support mode for Central on the switch using the **aruba-central** command. Then you can change the lockout mode. But this process is unnecessary since this will automatically happen once the AOS-CX switch loses connectivity to Central.

Appendix B
Glossary

Term	Definition	Chapter
802.11ac	802.11ac only operates in the 5 GHz range with theoretical data rates of up to 6.93 Gbps, using very high throughput (VHT) methods, with higher channel width and more complex modulation techniques. 802.11ac has sometimes been referred to as *Wi-Fi 5*.	9
802.11ax	802.11ax is the latest wireless standard and operates in both the 5 GHz and 2.4 GHz frequencies, with a maximum data rate of 4.8 Gbps. The big news with 802.11ax is a technique called multi-user MIMO (MU-MIMO), enabling multiple transmitters at the same time. This early version of 802.11ax has also been referred to as *Wi-Fi 6*.	9
802.11n	The 802.11n amendment added high throughput (HT) speeds, theoretically up to 600 Mbps, with 300 Mbps being a typical speed. It accomplished these speeds by increasing the channel width, using the advanced 64-quadrature amplitude modulation (QAM) mechanism, and techniques like multiple input multiple output (MIMO).	9
802.1D	The IEEE 802.1D standard version of the STP was developed to build and maintain redundant, yet loop-free networks. With STP, you eliminate single points of failure, while avoiding loops and MAC table instability.	4
802.1Q tagging	IEEE 802.1Q adds an additional field between the length and payload fields in the Ethernet header: the 802.1Q tag field. The most important part of this field is the VLAN tag or VLAN ID (or simply VID).	2
802.1X	802.1X, an IEEE standard for network access control (NAC), enables the network infrastructure to authenticate and authorize endpoints for network access. 802.1X requires users to authenticate as soon as a layer 2 connection is established. In a wired network, 802.1X takes effect as soon as the Ethernet connection comes up. In a wireless network, it takes effect after the initial 802.11 association. Authenticating the network side involves validating the network's certificate. The user side can be authenticated with usernames and passwords or digital certificates.	10
Activate	Aruba Activate is a key component for provisioning Aruba devices. Aruba Activate is a cloud-based service designed to enable customers to efficiently manage their Aruba device inventory and facilitate the provisioning of their Aruba devices.	7
Address resolution protocol (ARP)	ARP maps layer 3 IP addresses to layer 2 MAC addresses.	2

Continued

Term	Definition	Chapter
Administrative distance	The administrative distance weights networks based on where it was learned or derived. The protocol or method with the lowest administrative distance is used (a lower number is considered better or more reliable).	3
AI Assist	AI Assist uses event-driven automation to collect diagnostics, posts them to a shared location, and even notifies Aruba TAC for proactive customer support.	13
AI Insights	AI Insights automatically diagnoses an array of common network-impacting issues by using dynamic, per-site baselines that are continuously tuned as conditions change: no manual setup or service level threshold adjustments are required by the customer. Built-in anomaly detection highlights the severity and impact of issues as they occur, helping IT pinpoint root cause and proper remediation steps with 95% accuracy!	13
AI Operations (AIOps)	Aruba Central includes a full-service AI operations (AIOps) solution that automates common troubleshooting activities by reducing IT support tickets and their associated costs. Aruba's AIOps solution is based on AI machine learning models that are consistently trained with network performance data collected from tens of thousands of Aruba customers across every vertical, market segment, and network type.	13
AI Search	AI Search is a natural language processing (NLP) engine of the Aruba Central that points to solution guides, troubleshooting tips, and much more. Whether operators are looking for best practices on device configurations or need to isolate a problem impacting a specific user, AI Search provides fast, interactive responses that simplify day zero to day two operations.	13
AirMatch	AirMatch provides automated RF optimization to improve a user's experience by providing a stable network that dynamically adapts to the changing RF conditions. AirMatch, which runs in Aruba Central, is a centralized RF planning and optimization service. Aruba Central receives telemetry data from APs for radio measurements, channel range, EIRP range, operational conditions, and local RF events like radar detection or high noise. AirMatch also minimizes channel coupling, where adjacent radios are assigned to the same channel.	9
Alternative port	Any port, other than the root port, that provides an alternative path to reach the root	4
Application layer	The application layer is the closest to the end user, which means that both the OSI application layer and the user interact directly with the software application.	1
Area	An area is a group of OSPF routers that share the same LSDB. All routers must be part of an area. When you split a large network into separate areas, you reduce the size of the LSDB in each router, lower CPU utilization, and increase overall network stability and scalability. This is because each router in an area must only maintain the topology for *that* area.	3
Authentication, authorization, and accounting (AAA)	A common network access control (NAC) solution is based around AAA services running on an authentication server. It provides authentication (validating a device or person's identify), authorization (assigning policies for the user/device access), and accounting (a record of what occurred) services	10

Continued

Term	Definition	Chapter
Authentication server	The authentication server directs the 802.1X authenticator and ultimately takes responsibility for access control decisions. ClearPass is an example of an authentication server.	10
Authenticator	In 802.1X, the authenticator controls network access. It enforces decisions made by the authentication server, either permitting or denying access. In Aruba networks, the authenticator could be an Aruba AP, an Aruba gateway, or an Aruba switch (older AOS switches as well as the newer AOS-CX switches).	10
Autonomous system (AS)	An AS is a collection of routing devices under a common administrative domain. Your Internet service provider (ISP) owns their own internal network: they have autonomy over this system and it is their AS. Each corporation has autonomy over their network: it is their AS.	3
Azimuth	H-plane or horizontal view, also known as the azimuth, shows the radiated RF pattern coming from the sides of the AP.	9
Backup designated router (BDR)	See Designated router.	3
Backup port	A backup port is where two or more ports on the switch connect to the same segment, and the backup port has a higher cost (less preferred).	4
Basic Service Set (BSS)	A BSS includes an AP radio and all associated clients in a coverage area. An AP with two or more radios therefore has more than one BSS. When you define an SSID (service set identifier, i.e., the name of the WLAN), it is assigned a different 48-bit MAC address on each of the AP radios, which is used to identify the SSID *and* the radio it is advertised on. The radio MAC address, known as the BSSID, identifies the BSS. All WLAN clients associated to the radio use this BSSID as the destination MAC address on 802.11 frames; and the AP understands, based on the received radio and the BSSID MAC address, with which SSID the 802.11 frame is associated.	9
Beacon	A beacon is a management frame that an AP generates concerning the SSIDs that it supports.	9
Beamforming	Beamforming technology allows an AP to locate client's antennas and modulate its spatial streams in phase. When spatial streams are modulated in phase, two or more copies of the same signal are generated from the AP's radio. The AP times the transmission from each antenna, so the spatial streams reach the receiving antennas at the same time, in phase, increasing the received signal strength at the radio.	9
Bridge identifier (BID)	STP assigns each switch a unique identifier called a bridge ID (BID). This identifier is composed of a 2-byte priority and 6-byte MAC address. The priority defaults to 32,768, so by default, all switches have the same priority. However, each switch has a unique MAC address, and so each BID will be unique. The switch with the lowest *BID* wins the election to be the root switch.	4
Bridge protocol data unit (BPDU)	A BPDU is a message from a bridge/switch running STP. This includes hellos and topology notifications, to name a couple.	4

Continued

Term	Definition	Chapter
BPDU Guard	See BPDU protection.	4
BPDU protection	BPDU protection (sometimes referred to as BPDU Guard) is used to detect switches that are not supposed to be part of the network. It does this by looking for BPDUs on ports where these should not be seen (typically edge ports).	4
Broadcast	Broadcast traffic communication is one-to-all.	1
Captive portal	Captive portal is Aruba's term for describing the authentication method based on using a web browser, commonly used for guest and onboarding access.	11
Central	Aruba Central is a cloud-based platform for managing Aruba-based wireless, WAN, and wired networks. Aruba Central supports group configuration to provision and manage similar devices with less work. Continuous monitoring and AI-based analytics provide real-time visibility and insight. Machine learning leverages a growing pool of network data and deep experience. It is very secure, with HTTPS connections, certificate-based, and cloud authentication and policies.	6
Central On Premises (COP)	Aruba COP delivers most of the same services and capabilities of Aruba Central. It has the same user interface, same code, and same user experience and workflows: it's just powered by purpose-built server appliances that are deployed at a customer's data center environment.	6
Certificates	Certificates provide key management: they tie a specific identity to a public key. Digital certificates follow a standard format based on industry-standard X.509. The serial number uniquely identifies the certificate among certificates issued by the same CA. The validity period gives the certificate a specified period in which the certificate is valid. The issuer identifies who issued this certificate to the subject.	12
Certificate authority (CA)	*Root* CAs are at the top of the PKI hierarchy. They sign their own certificates, called root CA certificates. Root CAs are well-known entities that are widely believed to be trustworthy. A company can also set up its own internal or private root CA, which is believed to be trustworthy within the company. The root CA issues just a few certificates itself, issuing and signing certificates for intermediate CAs. This ensures minimal contact between the root CA and other entities.	12
Channel bonding	Channel bonding is the process of combining the bandwidth of two or more adjacent RF channels.	9
Checkpoint	A checkpoint is a snapshot of the AOS-CX switch running configuration and its relevant metadata during the time of creation of the configuration. You can use checkpoints to apply the switch configuration stored within a checkpoint whenever needed, such as revert to a previous working configuration. A checkpoint is flexible since it can be applied to other switches of the same model.	2
ClearPass	Aruba ClearPass is an authentication server that can easily handle all network access control (NAC) functions. It is a decision-making point when clients connect to the network and go through an authentication and authorization process. It supports advanced features like guest access, onboarding, and posture assessment with OnGuard.	6

Continued

Term	Definition	Chapter
ClientMatch	ClientMatch is an Aruba patented technology that provides client load balancing across radio bands and Aruba AP radios. ClientMatch helps to influence sticky clients to connect to the best radio and AP. When clients associate to the WLAN, their firmware used for their wireless NIC determines which radio, and on which AP, the client will connect. ClientMatch is a feature that matches a client to the most appropriate AP and frequency in the area.	9
Common spanning tree (CST)	Common spanning tree (CST) defines one instance of spanning tree for the entire network, regardless of the number of VLANs. All BPDUs use untagged in the native VLAN (VLAN 1, by default), and so VLANs are not considered.	4
Connected route	A subnet physically connected to a device	3
Cost	OSPF routers must run the SPF (Dijkstra's) algorithm to find the *best OSPF paths* to each destination network. The best path is the one with the lowest cost, and cost is based on an inverse of the bandwidth of the interfaces.	3
Data link layer	The data link layer provides media access control, link layer addressing, and error detection.	1
dBm	dBm is decibels relative to one milliwatt: 0 dBm equals one milliwatt. It is a more human-friendly way of determining the strength of an RF signal.	9
Default gateway	A default gateway (DG) is the device that routes to all network destinations for the endpoint devices in a broadcast domain or network.	3
Designated port (DP)	The DP is designated to accept traffic from downstream root ports. Basically, from a segment's perspective, it's the switch port that has the lowest cost to reach the root.	4
Designated router (DR) and backup designated router (BDR)	A client/server design is implemented in OSPF on each broadcast segment. For each multi-access broadcast segment, such as Ethernet, there is a designated router (DR) and a backup designated router (BDR) as well as other OSPF routers, called DROTHERs (designated router other). In a broadcast environment, all routing updates are sent to the DR, who, in turn, disseminates this information to the other routes in the segment.	3
Device onboarding	First, you must add any device that you want Aruba Central to manage to the device inventory list. Once you've added the devices, you must assign each device a subscription. This enables Aruba Central to manage the device. And third, you configure your devices.	7
Device profiling	Profiling and endpoint classification allows the network infrastructure to adjust settings and security policies dynamically, based on a device's type/model and its operating system. This information can then be used to assign policies to the device when it is connected to the network, like an Aruba user role.	8
Dynamic Host Configuration Protocol (DHCP)	Typically, most endpoints automatically get IP addressing information using DHCP. This information includes an IP address, subnet mask, default gateway and DNS addresses, as well as other IP information.	3
DHCP relay	Using DHCP relay, a DHCP helper address is configured on each router interface where DHCP request broadcasts would be received. That router interface typically serves as the default gateway for endpoints as well. The DHCP relay device relays endpoint DHCP requests in one subnet to the location where the DHCP server resides (a different VLAN).	3

Continued

Term	Definition	Chapter
Direct attach copper (DAC) cable	Direct attached copper (DAC) cables are twinax cables that are factory-terminated to a transceiver. You use them for relatively short connections, often from a switch in one equipment rack to another switch, server, or other device in the same rack.	1
Distance vector routing protocol	The earliest methods devised to exchange IP information between routing devices are based on distance vector algorithms. Each router advertises its distance from each network, and the vector or direction packets must travel to get to that network. These algorithms were very simple in their construction and operation.	3
Duplex	A duplex communication system has two or more connected parties that communicate in both directions. With half duplex, both parties do not communicate simultaneously. With full duplex, both parties communicate simultaneously.	1
Dynamic segmentation	With dynamic segmentation, the switch authenticates the user by interacting with an authentication, authorization, and accounting (AAA) server, like ClearPass. ClearPass profiles each device, deems appropriate endpoints as acceptable, and sends accept messages to the switch. The messages include tunnel information, the assigned VLAN, and secondary role. The switch uses this information to establish the tunnel to the Aruba gateway (mobility controller, or MC for short). The user's traffic is tunneled to the gateway where the gateway applies the appropriate policies to it, like firewall policies.	6
Edge Services Platform (ESP)	ESP is based on four core attributes. **Unified Infrastructure**: manages different types of devices in different locations using a single interface. It eliminates silos and streamlines operations. **Zero trust security**: means that every single device that plugs into the network gets authenticated and authorized. You are not inherently trusted and zero trust security includes role-based security to minimize the need for manual configuration. **Artificial intelligence operations**: gives the network a *sixth sense*, resolving issues *before* they impact the business or at least resolve them faster. **Aruba Central**: powers ESP, which is a *software as a service* (SaaS) management solution, ready to monitor and manage any size company. Aruba Central's flexibility gives you greater choice and control over how you consume and pay for technology to accomplish their goals.	6
Effective or equivalent isotropic radiated power (EIRP)	EIRP simply means how much power the antenna effectively radiates out into free space. This is affected by the power output from the radio, any cable loss, and the attenuation of the antenna.	9
Elevation	The elevation plane or E-plane, also known as the vertical plane, shows radiated patterns coming from the front (typically the faceplate) of the AP.	9

Continued

Term	Definition	Chapter
Equal Cost Multi-Path (ECMP)	ECMP is a mechanism that enables load sharing layer 3 packets of two or more equal-cost paths to the same destination network.	3
Ethernet	Ethernet is a family of computer networking technologies that are used at layer 1 and layer 2 of the OSI model.	1
Ethertype	The EtherType field in the Ethernet frame defines the layer 3 protocol inside the frame, for example IPv4 or IPv6.	1
Extended Service Set (ESS)	A WLAN (a group of related APs) advertises the same SSID across multiple APs (like *employee*), which is referred to as an ESSID. However, each AP radio will have its own unique BSSID MAC address for that SSID. This allows the APs to understand when traffic is sent to one of their own radios versus a different AP's radio.	
Extensible Authentication Protocol over LAN (EAPOL)	In 802.1X, the supplicant and authenticator communicate with EAPOL messages. EAPOL operates at the data link layer and uses EtherType 0x888e.	10
EAP-Tunneled EAP (EAP-TEAP)	Like PEAP and EAP-TTLS, EAP-TEAP uses TLS as an outer method to create a secure tunnel. It then uses an inner method to authenticate the client. Depending on the inner method implemented, the client can authenticate with a username/password or a certificate. EAP-TEAP also supports *chaining* of multiple authentications, which can be useful when a company wants to ensure that users connect on authorized devices. The client can authenticate with device or *machine* credentials first, proving the device is authorized. The client obtains a token during this authentication. The client can then authenticate with user credentials, including the token from the machine authentication, proving that both the user and device are authorized.	10
EAP-Tunneled Layer Security (TLS)	Considered one of the most secure EAP methods, EAP-TLS uses a TLS 1.2 handshake to authenticate the server and supplicant with digital certificates. The Windows native client supports EAP-TLS, as do many other supplicants. ClearPass also supports EAP-TLS. Note that both the supplicant and the authentication server need a certificate from a common certificate authority (CA).	10
EAP-Tunneled TLS (TTLS)	TTLS and PEAP function in similar ways. Both methods involve a two-phase authentication process. In the first phase, the outer method creates a secure tunnel using TLS. The authentication server then initiates a second authentication method, called the inner method, inside the secure TLS tunnel. PEAP and EAP-TTLS can use several different options for the inner method.	10
Exterior gateway protocol (EGP)	To move traffic between two different autonomous systems, an EGP is used. The only EGP currently in use today is the Border Gateway Protocol (BGP).	3
First hop routing protocol (FHRP)	FHRPs add resiliency for endpoints by using a coordinated gateway solution: no change to endpoint IP configuration, no DHCP modifications, and no end user disruption. VRRP is an example of an FHRP.	4

Continued

Term	Definition	Chapter
Forwarding information base (FIB)	See Routing table.	3
Frame	A frame refers to the encapsulation that is done in layer 2 of the OSI model.	1
Frame checksum (FCS)	The frame checksum (FCS) provides error correct to detect a frame that became corrupted during transmission from the source to the destination.	1
Gateway	Starting in AOS 10, the products formerly known as mobility controllers (MCs) are now referred to as gateways.	6
Global configuration context	The *global configuration* context ends with "(config)#" in the prompt. This is where you run commands that change the switch's configuration. These commands have a *global* effect on the device itself. But to configure specific components or features, you must access a sub-config mode.	2
GreenLake	HPE GreenLake is a cloud management and monitoring platform from HPE. It hosts and provides not only Aruba Central services but many other HPE services as well. HPE GreenLake's device management feature lets you easily onboard, view, and manage all the devices in your account.	7
Group	A Central group's main purpose is to simplify your device configuration. All devices in the same group will receive the same configuration. However, you can also use groups to monitor devices. When you select a group to monitor, you'll only see the devices in that group.	7
In-band management	In-band management is when you remotely connect to and manage switches via a normal Ethernet port.	2
Interior gateway protocol (IGP)	To route packets inside an AS, each company chooses an interior gateway protocol (IGP). An IGP is simply a routing protocol or process that runs inside an AS. Examples of IGPs include the Routing Information Protocol (RIP) and Open Shortest Path First (OSPF), and even static routing.	3
Internet Control Message Protocol (ICMP)	ICMP is used by network devices to send error and operational information related to IP communications: it's basically a support protocol for IP.	2
Internet Engineering Task Force (IETF)	The IETF is the international Internet standards organization, responsible for many Internet standards, as documented in Request for Comments (RFCs).	9
Internet Protocol (IP)	The Internet Protocol (IP) suite is a conceptual model and set of communication protocols used in Internet and computing networks.	1
IP helper	See DHCP relay.	3
IPv4 addressing	An IPv4 address is comprised of 32-bits and is expressed in a dotted decimal notation. The dots divide the address in four 8-bit (1-byte) octets. This makes it easy for humans to work with IPv4 addresses. Eight bits enables values from 0 to 255.	3
Label	Aruba Central labels are tags that are used to filter devices for monitoring, managing, and reporting purposes. A device can have multiple labels, which can help you determine device ownership, departments, and functions.	7

Continued

Term	Definition	Chapter
Link aggregation control protocol (LACP)	With IEEE 802.3ad, peer devices that implement a dynamic LAG exchange control messages to establish and maintain the LAG link. The 802.3ad implementation is commonly known as the LACP. The two peers periodically exchange LACP messages called LACP data units.	2
Link aggregation group (LAG)	Ethernet link aggregation bundles multiple physical Ethernet links into one logical link, called a link aggregation group, or LAG for short.	2
Link Layer Discovery Protocol (LLDP)	LLDP is a vendor-neutral, open standard link layer protocol (layer 2 protocol). It is used by network devices to advertise their identity and capabilities over a wired Ethernet connection. This protocol enables you to discover and document network device interconnections.	2
Link state advertisements (LSAs)	OSPF routers use link state advertisements (LSAs) to communicate with each other. One type of LSA is a hello, which is used to form neighbor relationships and as a keep-alive function. Each LSA includes information that describes the portion of the network that a router announces. LSAs are collected to make up the LSDB, which represents the topology of the network.	3
Link state database (LSDB)	In OSPF, the LSDB contains a list of every path in the known topology (area), which is also referred to as a *topological* database or table.	3
Link-state routing protocol	Link-state protocols resolve the limitations and disadvantages of distance vector protocols. Routers that run link-state protocols possess information about the network topology (not just what they are connected to but how their neighboring routing devices connect to devices beyond them). Thus, they can independently calculate their routing table using a sophisticated Dijkstra algorithm in order to select the best path for every network destination.	3
Live upgrade	A live upgrade should not disrupt clients that are associated with APs. The live upgrade interfaces with the AirMatch service to partition the APs into smaller groups that will be upgraded in parallel (within the group). AirMatch partitions APs based on RF neighborhood data. APs partitions are used for clients to roam when the APs are undergoing an upgrade. A set of APs, known as *seed* APs, is selected. Only these APs download images from the cloud; the remaining APs download images from their designated seed AP. This saves time and bandwidth since the image is only downloaded to the company once, and then distributed by the seed AP(s).	13
Local area network (LAN)	A LAN is a group of computer devices that are geographically in the same place.	1
Local route	A local route is a local IP address on an AOS-CX switch.	3
Loop protection	The loop protection feature protects edge switches from downstream loops by transmitting loop protocol packets out of the ports. This is used to detect loops that exist locally on the switch, like a cable connected between two ports of the same switch.	4
Loopback interface	A loopback interface is a logical interface that is always in an up state (unless you manually disable it). This interface enhances stability for processes and protocols that depend on the interface status to work, like OSPF.	3

Continued

Term	Definition	Chapter
Management information base (MIB)	SNMP management information bases (MIBs) are the heart of SNMP and allow the storing and sharing of configuration and event-related information. SNMP managers and devices interact with each other using the MIB information.	13
Manager context	The manager context has the hashtag symbol, where you can run nearly all **show** commands to view status and configuration, but you cannot modify that configuration unless you access the global context.	2
Maximum transmission unit (MTU)	The MTU is the maximum transmission size an interface can handle, not including the Ethernet header of 18 or 20 bytes.	1
Media access control (MAC) address	At layer 2, Ethernet provides for the assignment of 6-byte (48-bit) media access control (MAC) addresses to each device (see Figure 1-24). Each manufacturer assigns a globally unique hexadecimal MAC address to each network interface card (NIC).	
MAC authentication (MAC-auth)	MAC-auth is one example of a layer 2 authentication method. A network infrastructure device that implements MAC-auth checks the source MAC address in a client's incoming traffic to validate the device's access.	10
Media endpoint discovery (MED)	MED is an enhancement of LLDP known as LLDP-MED. LLDP-MED provides auto-discovery of LAN policies such as the voice VLAN ID, and the layer 2 priority and differentiated services for quality of service (QoS), as well as many other types of information.	2
Modulation and coding scheme (MCS)	A modulation scheme is a way of changing (modulating) a transmitted waveform to represent data: binary 1s and 0s.	9
Multi-chassis LAG (MC-LAG)	See VSX LAG.	4
Multi-pre-shared keys (mPSK)	mPSK is basically an Aruba feature that allows multiple PSKs for a WLAN, where different keys are typically used for different device types.	12
Multi-user MIMO (MU-MIMO)	MU-MIMO uses beamforming technology to send traffic to multiple clients in the same transmission. The AP will use a channel sounding procedure to which the clients will respond. The AP uses this information to build a steering matrix. The AP uses the steering matrix to manipulate its transmission, coordinating in phase spatial streams for each client when the frames intended for them egress the radio.	9
Multicast	Multicast traffic refers to one-to-many communication: a transmitter and one or more receivers.	1
Multimode fiber (MMF)	MMF uses an LED to send a signal over fiber for short distances.	1
Multiple STP (MSTP)	MSTP (802.1s) enhances RSTP by building a new topology or instance of the spanning tree protocol. Instances then can carry different VLAN traffic: you get load sharing and optimal paths for different groups (instances) of VLANs.	4

Continued

Term	Definition	Chapter
MSTP region	MSTP introduces the concept of a region, which is a group of switches that have been configured with the same MSTP configuration. The items that must match between the switches in the region include the following: region name, region revision number, VLAN-to-instance mappings.	4
Native VLAN	The 802.1Q standard allows you to select a single VLAN on which no tag or VID is included in the frame. This means that a standard, untagged Ethernet frame is sent out the port. This VLAN is known as the native VLAN or untagged VLAN.	2
Network access control (NAC)	See Authentication, authorization, and accounting (AAA).	10
Network Analytics Engine (NAE)	NAE is an Aruba feature to help you with monitoring and troubleshooting of AOS-CX switches. NAE provides network operators with distributed, real-time insights into network health for a faster, more proactive way of monitoring, detecting, and troubleshooting IT- and user-impacting issues. NAE is based on the use of Python scripts that interact with the AOS-CX switches and the use of REST API calls to pull/push information from/to the AOS-CX switches.	6
Network layer	The main goal of the OSI model network layer is to establish device communications across multiple LANs or WANs, using the best available path.	1
Open Shortest Path First (OSPF)	RFC 2328 defines OSPFv2, which describes how to route IPv4 packets. OSPF has its own IP protocol number 89 and is based on a link-state protocol implementation.	3
Open Virtual Switch Database (OVSDB)	The OVSDB database is created on the conductor switch and contains state and configuration data for the VSF stack itself. The conductor switch *synchronizes* the OVSDB content with the secondary switch to ensure that it can quickly take over the active role without interruption.	5
Open Systems Interconnection (OSI) model	The OSI model organizes computing communication in seven layers to help understand how various networking protocols work.	1
Opportunistic Wireless Encryption (OWE)	OWE, defined in RFC 8110, enhances open WLANs. OWE is not technically part of WPA, but instead rolled into the 802.11 standard. However, OWE uses WPA concepts like the key hierarchy and a four-way handshake. The station and AP then derive all the keys for encrypting and authenticating future data. This basically allows per-session keying dynamically within the WLAN.	12
Operator context	The operator context level is indicated by a greater-than (>) symbol. When you see this, know that you can run a sub-set of **show** or view commands, but cannot change the switch's configuration.	2
Organizationally Unique Identifier (OUI)	The most significant first three bytes of each MAC address are referred to as the manufacturer's Organizationally Unique Identifier (OUI).	1

Continued

Term	Definition	Chapter
Out-of-band management (OOBM)	Out-of-band management (OOBM) uses alternate, dedicated connections to network devices, separate from the end user data network.	1
Packet	A packet refers to the encapsulation that is done in layer 3.	1
Passive interface	When an OSPF interface is set to passive, it stops sending and receiving OSPF packets on that interface. However, it will continue to advertise the network for that interface out other OSPF-enabled non-passive interfaces.	3
Physical layer	The physical layer dictates the physical aspects of how signals are transmitted and received across some media.	1
Ping	The **ping** command sends ICMP echo request messages to a destination device. The remote device responds to each received echo *request* with an echo *reply* message. When the source device receives the messages, it understands that the network was properly configured since the communication succeeded between the two devices.	2
Prefix length	This number appears after an IP address, followed by a forward slash (/) and indicates the number of networking bits in the network portion of the address.	3
Presentation layer	The presentation layer of the OSI model transforms data into the formats that the application accepts. Typical processes include compression/decompression, encryption/decryption, and code translation (EBCDIC to ASCII).	1
Profiling	See Device profiling.	8
Protected EAP (PEAP)	PEAP was developed to provide some of the security of EAP-TLS without forcing end stations to have digital certificates. The authentication server is still authenticated with a certificate, but the supplicant is authenticated using a username and password.	10
Protocol	A protocol is a group of rules or instructions that computer devices follow in order to properly establish and maintain communication.	1
Protocol data unit (PDU)	A PDU is simply a structure that considers the header and payload or data for each OSI model layer.	1
Public key infrastructure (PKI)	Certificates exist within a public key infrastructure (PKI) hierarchy that enables devices to determine which certificates they should trust. Every certificate within the PKI is signed by an entity that vouches for the authenticity of the certificate holder's identity.	12
RADIUS	RADIUS is an IETF protocol that defines how an authenticator and an authentication server interact with each other. It uses UDP as a transport. RADIUS is commonly used to authenticate and control access to the network.	10
Rapid STP (RSTP)	The rapid spanning tree protocol (RSTP) was developed in 1998 to speed up convergence (IEE 802.1w). Instead of only the root switch originating hello packets, *all* switches originate them, helping with faster convergence.	4

Continued

Term	Definition	Chapter
Role	An Aruba user role is a collection or group of settings that can be assigned to a switch port: a VLAN identifier, PoE priority, QoS trust mode (DSCP and 802.1p prioritizations), security policy, bandwidth restrictions (rate limiting), custom captive (web) portal page, and etcetera. These settings are typically dynamically assigned to the user's connection based on their authentication, device profiling information, type of connection, location, and many other things.	8,10
Root port (RP)	The RP is simply the port that the local switch uses to reach the root switch: the best path toward the root switch.	4
Router ID (RID)	All OSPF routers require a unique router identifier (RID). This is a 32-bit number which is typically associated with an actual IP address on the routing device.	3
Routing table	Routing devices (routers and multilayer switches) build and maintain a routing table that informs them of the best path to any given layer 3 destination, like IP networks.	3
Segment	A segment refers to the encapsulation that is done in layer 4 of the OSI model.	1
Server-derived roles	When a server returns an Aruba VSA with the RADIUS reply, the Aruba AP or switch places the user in the specified role and/or VLAN that was sent. When the server returns a RADIUS attribute, you must configure the Aruba AP to look for this attribute. If you use an Aruba ClearPass server, the Aruba AP-to-ClearPass communication is automatic. The AP sends a RADIUS request with the Aruba VSAs.	10
Session layer	The session layer is responsible for setup, maintenance, and tear down of sessions between two computing devices. A session is a particular conversation between two computer devices.	1
Shielded twisted pair (STP)	STP cables are like UTP cables, where the wires are twisted together and then wrapped with a shielding or screening material which consists of foil wrapping or a copper braid jacket. Shielding the copper cables can have a tremendous effect on the stability, performance, speed, and longevity of the cable and ultimately your network. Because STP cables have shielding they are more resistant to electromagnetic interference, like fluorescent lighting, nearby motors, or any other electro-mechanical equipment.	1
Signal-to-noise ratio (SNR)	The SNR is the difference between the ambient noise floor and the received signal strength. You express SNR as a *difference*, like SNR 25 dBM, and not a ratio. The higher your SNR, the better your Modulation Coding Scheme (MCS) will be, and therefore the higher the communication data rate.	9
Simple Network Management Protocol (SNMP)	SNMP is an industry-standard IP protocol that allows you to manage and monitor a variety of network devices from a central location. Using an SNMP application, you can configure network devices and apply consistent security and management policies to all the devices on your network. You can also use SNMP to monitor compliant devices for conditions requiring administrative attention: for example, to determine if a server has gone down, a LAN or WAN link is no longer available, or a link has become congested. You can track device uptime, link states, and many other device information variables. When a problem occurs, SNMP compliant devices send an alert message to one or more designated SNMP servers, and you can then take steps to resolve the problem.	13

Continued

Term	Definition	Chapter
Single-input single-output (SISO)	Legacy 802.11 networks use SISO radio technologies where only one antenna is used for transmitting or receiving at a time. SISO does not make full use of a radio with multiple antennas. The transmitter sends the signal out a single antenna. While the receiver will hear the transmission on both antennas, it will only use one of those signals.	9
Single mode fiber (SMF)	SMF uses a laser to send a signal over fiber for long distances.	1
Site	An Aruba Central site refers to a physical location where a set of devices are installed; for example, campus, branch, or venue (see Figure 7-23). Sites are more than just a tool to filter views, place devices on a map, and generate reports in Central. You can use sites for several Aruba Central services such as live upgrades, AirMatch, install manager, network health dashboard, User Experience Insight (UXI), and etcetera.	7
Spectrum analysis	Wireless networks operate in environments with electrical and RF devices that can interfere with network communications. The spectrum monitor software modules on IAPs and AOS 10 APs can examine the RF environment in which the Wi-Fi network is operating, identify interference, and classify its sources. An analysis of the results can then be used to quickly isolate issues associated with packet transmission, channel quality, and traffic congestion caused by contention with other devices operating in the same band or channel. Spectrum monitors are AP radios that gather spectrum data but do not service clients. Each SM scans and analyzes the spectrum band used by the AP's radio (2.4 GHz, 5 GHz, and/or 6 GHz). By default, an APs' radio accepts device associations.	14
Splash page	A splash page is a web page used during the captive portal process.	11
Split brain	See VSF MAD.	5
Static route	A static route is a manually added route to reach a remote destination.	3
Subnet mask	Every IP address has an associated subnet mask. This mask defines the line between the network portion and the host portion of the IP address. The subnet mask is a consecutive sequence of ones followed by a consecutive sequence of zeros. Where the ones end and the zeros begin is the line that divides the network and host portions of the address.	3
Supplicant	The supplicant is the 802.1X component that runs on the endpoint, which is attempting to connect to the network.	10
Switch virtual interface (SVI)	An SVI is a logical layer 3 interface for an associated VLAN broadcast domain on AOS-CX switches.	2
TCP three-way handshake	Before TCP processes any application data, it must first contact the target device and establish a reliable, flow-controlled connection: SYN, SYN/ACK, and ACK.	1
Time-to-live (TTL) IP field	The TTL field in the IP header mitigates layer 3 routing loops. If routers are misconfigured, IP packets could be bounced around between two or more routers forever. To prevent this, the originator places a number in this field from 0 to 255. Each router that routes this packet decrements this field by 1. The router that decrements the TTL field to zero discards the packet.	1

Continued

Term	Definition	Chapter
Topology change notification (TCN)	In STP, a topology change occurs when non-edge ports move to a different state, or when BPDUs are no longer received. A switch detects this as topology change. The switch actively informs other switches of the topology change (TC). It sets the TC bit in BPDUs and transmits them. It flushes the MAC address table entries associated with all non-edge ports.	4
Traceroute	Traceroute is a variant of the ICMP protocol and it is a useful troubleshooting tool when you deploy a layer 3 feature, such as an IP routing protocol. This command tracks the path that a packet takes to reach its destination, listing the layer 3 devices that are in the path.	2
Transport layer	The transport layer controls the reliability of a given link through segmentation, de-segmentation, and error control.	1
Trunk	Many vendors use the term *trunk* when discussing a port that carries multiple VLANs. AOS-CX switches use this term for switch links that carry multiple VLANs.	2
Unicast	Unicast traffic refers to one-to-one communication: one transmitter and one receiver.	1
Unshielded twisted pair (UTP)	UTP cable has no shielding. This is the most used and most basic type of cable. The cable contains pairs of wires twisted together to help reduce and prevent electromagnetic interference.	1
User eXperience Insight (UXI)	UXI consists of front-end sensors and agents and a back-end cloud-based UI. The sensors act like an end user, creating traffic typical of an end user. The cloud-based system provides real-time monitoring and analytic intelligence.	13
Virtual IP (VIP) address	A VIP address is the result of the DG coordination in VRRP. You assign a unique, *real* IP address to each individual physical gateway, as normal. The VIP address must also be unique, but an unused IP address in the subnet. The VIP is used as the default gateway address, where one of the VRRP routers, the active forwarder, will process traffic sent to it.	4
Virtual LAN (VLAN)	A VLAN is just like a physical LAN: a group of devices in the same broadcast domain. However, unlike a physical LAN, devices can be connected to different layer 2 switch ports on one or more switches.	2
Virtual Router Redundancy Protocol (VRRP)	RFC 5798 defines the VRRP protocol: a standards-based FHRP that enables two or more routing devices to provide gateway redundancy for endpoints in a VLAN/subnet.	4
Virtual Switching eXtension (VSX)	You use Aruba VSX at the campus network *aggregation* and *core* layers. VSX maintains separate control plane and management planes, and much of the data plane. This mitigates the *shared fate* nature of common planes, for improved uptimes: you get high availability *and* redundancy. It is available on the AOS-CX 6400 series switches and higher.	4
VSX LAG	A VSX LAG (MC-LAG) is a LAG connected from an external device to the two VSX switches.	4
Virtual Switching Framework (VSF)	You use VSF primarily at the campus access layer. VSF provides one control and management plane to manage switches, while each switch utilizes its own (distributed) data plane to forward traffic. It is available on the AOS-CX 6200 and 6300 series switches and supports up to 10 switches, depending on the series.	4/5

Continued

Term	Definition	Chapter
VSF link	The VSF link is a logical connection comprised one or more Ethernet ports that connect VSF members in a chain or ring topology.	5
VSF MAD (Multiple Active Detection)	You can use VSF MAD to avoid split-brain situations. Thus, if a VSF link failure occurs, the segmented switches that do not include the primary switch verify primary member status. If the original primary is up, then all members in the fragment that does *not* include the primary member will disable all their ports, with the exception of the VSL link ports.	5
Wi-Fi 5	See 802.11ac	9
Wi-Fi 6	See 802.11ax	9
Wi-Fi 6E	An 802.11ax extension was introduced in 2020, with up to 7.8 Gbps, 1024-QAM with modulation and coding scheme (MCS) 10/11. There were two important features introduced with this standard: it operates in the 6 GHz RF spectrum and supports a 1 Gbps channel width. This version of 802.11ax has been referred to as *Wi-Fi 6E*.	9
Wi-Fi Alliance	The Wi-Fi Alliance is an alliance of WLAN hardware and software manufacturers, like Aruba that provide interoperability standards to assist in ensuring different vendors' APs and clients can work together.	9
Wide area network (WAN)	A WAN is a group of computer resources that can communicate over large geographical distances, typically a few kilometers or miles, and perhaps thousands of miles, such as the Internet.	1
Windows size	In the TCP header, there's a window size field used for flow control, which indicates how much data can be sent before waiting for an acknowledgement.	1
WPA Enterprise	WPA Enterprise uses 802.1X, where a client authenticates to a backend authentication server using individual credentials: a username and password and/or a digital certificate.	12
WPA Personal	WPA Personal requires users to enter a passphrase to access the WLAN.	12
Zero-touch provisioning (ZTP)	ZTP is a method where an Aruba AP, switch, and/or gateway can contact Aruba Activate to get provisioning information to be onboarded to the network. This information includes pointing the device to either Aruba Central or Airwave so that the device can obtain its configuration.	7

Index

B

I

N

O

S

Z